NON-CIRCULATING

Handbook of
Nanostructured Materials
and Nanotechnology

Handbook of Nanostructured Materials and Nanotechnology

Volume 1
Synthesis and Processing

Edited by

Hari Singh Nalwa, M.Sc., Ph.D.
Hitachi Research Laboratory
Hitachi Ltd., Ibaraki, Japan

ACADEMIC PRESS

A Harcourt Science and Technology Company

San Diego   San Francisco   New York   Boston
London   Sydney   Tokyo

The images for the cover of this book were reprinted with generous permission from:

(Top left) R.P. Andres, J.D. Bielefeld, J.I. Henderson, D.B. Janes, V.R. Kolagunta, C.P. Kubiak, W. Mahoney, and R.G. Osifchin, *Science* 273, 1690 (1996). Copyright 1996 American Association for the Advancement of Science.

(Top right) Bruce Godfrey, Volume 5, Chapter 12 in this series.

(Middle left) M.R. Sorensen, K.W. Jacobsen, and P. Stoltze, *Phys. Rev. B* 53, 2101–2113, © 1996 American Physical Society.

(Middle right) T.W. Ebbesen et al., *Nature* 382, 54 (1996) copyright 1996 Macmillan Magazines Ltd.

(Bottom left) R.H. Jin, T. Aida, and S. Inoue, *J. Chem. Soc., Chem. Commun.*, 1260 (1993). Copyright by The Royal Society of Chemistry.

(Bottom right) NANO*SENSORS*.

This book is printed on acid-free paper. ♾

Copyright © 2000 by Academic Press

ACADEMIC PRESS
A Harcourt Science and Technology Company
525 B Street, Suite 1900, San Diego, CA 92101-4495, USA
http://www.apnet.com

Academic Press
24–28 Oval Road, London NW1 7DX, UK
http://www.hbuk.co.uk/ap/

Library of Congress Cataloging-in-Publication Data
Nalwa, Hari Singh, 1954–
  Handbook of nanostructured materials and nanotechnology / Hari Singh Nalwa.
  p. cm.
  Includes indexes.

  ISBN 0-12-513760-5
  1. Nanostructured materials. 2. Nanotechnology. I. Title.
  TA418.9.N35 N32
  620′.5–dc21
                                                                98-43220
                                                                CIP

International Standard Book Number: 0-12-513761-3

Printed in the United States of America
  00 01 02 03 MB 9 8 7 6 5 4 3 2

To my children,
Surya, Ravina and Eric

Foreword

Nanostructured materials are becoming of major significance and the technology of their production and use is rapidly growing into a powerful industry. These fascinating materials whose dimension range for 1–100 nanometer (1 nm $= 10^{-9}$ m, i.e., one billionth of a meter) include quantum dots, wires, nanotubes, nanorods, nanofilms, nanoprecision self assemblies and thin films, nanosize metals, semiconductors, biomaterials, oligomers, polymers, functional devices, etc. etc. It is clear that the number and significance of new nanomaterials and application will grow explosively in the coming twenty-first century.

This dynamical fascinating new field of science and its derived technology clearly warranted a comprehensive treatment. Dr. Hari Singh Nalwa must be congratulated to have undertaken the task to organize and edit such a massive endeavor. His effort resulted in a truly impressive and monumental work of fine volumes on nanostructured materials covering synthesis and processing, spectroscopy and theory, electrical properties, and optical properties, as well as organics, polymers, and biological materials. One hundred forty-two authors from 16 different countries contributed 62 chapters encompassing the fundamental compendium. It is the merit of these authors, their contributions coordinated most knowledgeably and skillfully by the editor, that the emerging science and technology of nanostructured materials is enriched by such an excellent and comprehensive core-work, which will be used for many years to come by all practitioners of the field, but also will inspire many others to join in expanding its vistas and application.

Professor George A. Olah
University of Southern California
Los Angeles, USA
Nobel Laureate Chemistry, 1994

Preface

Nanotechnology is the science and engineering of making materials, functional structures and devices on the order of a nanometer scale. In scientific terms, "Nano" means 10^{-9} where 1 nanometer is equivalent to one thousandth of a micrometer, one millionth of a millimeter, and one billionth of a meter. In Greek, "nanotechnology" derives from the *nanos* which means dwarf and *technologia* means systematic treatment of an art or craft. Nanostructured inorganic, organic, and biological materials may have existed in nature since the evolution of life started on Earth. Some evident examples are micro-organisms, fine-grained minerals in rocks, and nanosize particles in bacterias and smoke. From a biological viewpoint, the DNA double-helix has a diameter of about 2 nm (20 angstrom) while ribosomes have a diameter of 25 nm. Atoms have a size of 1–4 angstrom, therefore nanostructured materials could hold tens of thousands of atoms all together. Moving to a micrometer scale, the diameter of a human hair is 50–100 μm. Advancements in microscopy technology have made it possible to visualize images of nanostructures and have largely dictated the development of nanotechnology. Manmade nanostructured materials are of recent origin whose domain sizes have been precision engineered at an atomic level simply by controlling the size of constituent grains or building blocks. About 40 years ago, the concept of atomic precision was first suggested by Physics Nobel Laureate Richard P. Feynman in a 1959 speech at the California Institute of Technology where he stated, *"The principles of physics, as far as I can see, do not speak against the possibility of maneuvering things atom by atom ..."*. Research on nanostructured materials began about two decades ago but did not gain much impetus until the late 1990s. Nanotechnology has become a very active and vital area of research which is rapidly developing in industrial sectors and spreading to almost every field of science and engineering. There are several major research and development government programs on nanostructured materials and nanotechnology in the United States, Europe, and Japan. This field of research has become of great scientific and commercial interest because of its rapid expansion to academic institutes, governmental laboratories, and industries. By the turn of this century, nanotechnology is expected to grow to a multibillion-dollar industry and will become the most dominant technology of the twenty-first century.

In this handbook, nanostructures loosely define particles, grains, functional structures, and devices with dimensions in the 1–100 nanometer range. Nanostructures include quantum dots, quantum wires, grains, particles, nanotubes, nanorods, nanofibers, nanofoams, nanocrystals, nanoprecision self-assemblies and thin films, metals, intermetallics, semiconductors, minerals, ferroelectrics, dielectrics, composites, alloys, blends, organics, organominerals, biomaterials, biomolecules, oligomers, polymers, functional structures, and devices. The fundamental physical and biological properties of materials are remarkably altered as the size of their constituent grains decreases to a nanometer scale. These novel materials made of nanosized grains or building blocks offer unique and entirely different electrical, optical, mechanical, and magnetic properties compared with conventional micro or millimeter-size materials owing to their distinctive size, shape, surface chemistry, and topology. On the other hand, organics offer tremendous possibilities of chemical modification by tethering with functional groups to enhance their responses. Nanometer-sized organic materials such as molecular wires, nanofoams, nanocrystals, and dendritic molecules have been synthesized which display unique properties compared with their counterpart conventionally sized materials. An abundance of scientific data is now available to make useful comparisons between nanosize materials and their counterpart microscale or bulk materials. For example, the hardness of nanocrystalline copper increases with decreasing grain size and 6 nm copper grains show five times hardness than the conventional copper. Cadmium selenide (CdSe) can yield any color in the spectrum simply by controlling the size of its constituent grains. There are many such examples in the literature where physical properties have been remarkably improved through nanostrucure maneuvering. Nano-structured materials and their base technologies have opened up exciting new possibilities for future applications in aerospace, automotive, cutting tools, coatings, X-ray technology, catalysts, batteries, nonvolatile memories, sensors, insulators, color imaging, printing, flat-panel displays, waveguides, modulators, computer chips, magneto-optic disks, transducers, photodetectors, optoelectronics, solar cells, lithography, holography, photoemitters, molecular-sized transistors and switches, drug delivery, medicine, medical implants, pharmacy, cosmetics, etc. Apparently, a new vision of molecular nanotechnology will develop in coming years and the twenty-first century could see technological breakthroughs in creating materials atom by atom where new inventions will have intense and widespread impact in many fields of science and engineering.

Over the past decade, extraordinary progress has been made on nanostructured materials and a dramatic increase in research activities in many different fields has created a need for a reference work on this subject. When I first thought of editing this handbook, I envisaged a reference work covering all aspects of nanometer scale science and technology dealing with synthesis, nanofabrication, processing, supramolecular chemistry, protein engineering, biotechnology, spectroscopy, theory, electronics, photonics, and other physical properties as well as devices. To achieve this interface, researchers from different disciplines of science and engineering were brought together to share their knowledge and expertise. This handbook, written by leading international experts from academia, industries, and governmental laboratories, consists of 62 chapters written by 142 authors coming from 16 different countries. It will provide the most comprehensive coverage of the whole field of nanostructured materials and nanotechnology by compiling up-to-date data and information.

Each chapter in this handbook is self-contained with cross references. Some overlap may inevitably exist in a few chapters, but it was kept to a minimum. It was rather difficult to scale the overlap that is usual for state-of-the-art reviews written by different authors. This handbook illustrates in a very clear and concise fashion the structure-property relationship to understand a broader range of nanostructured materials with exciting potential for future electronic, photonic, and biotechnology industries. It is aimed to bring together in a single reference all inorganic, organic, and biological nanostructured materials currently studied in academic and industrial research by covering all aspects from their chemistry, physics, materials science, engineering, biology, processing, spectroscopy, and technology to applications that draw on the past decade of pioneering research on nanostructured materials for the first time to offer a complete perspective on the topic. This handbook should serve as a reference source to nanostructured materials and nanotechnology. With over 10,300 bibliographic citations, the cutting edge state-of-the art review chapters containing the latest research in this field is presented in five volumes:

Volume 1: Synthesis and Processing
Volume 2: Spectroscopy and Theory
Volume 3: Electrical Properties
Volume 4: Optical Properties
Volume 5: Organics, Polymers, and Biological Materials

Volume 1 contains 13 chapters on the recent developments in synthesis, processing and fabrication of nanostructured materials. The topics include: chemical synthesis of nanostructured metals, metals alloys and semiconductors, synthesis of nanostructured coatings by high velocity oxygen fuel thermal spraying, nanoparticles from low-pressure and low-temperature plasma, low temperature compaction of nanosize powders, kinetic control of inorganic solid state reactions resulting from mechanistic studies using elementally modulated reactants, strained-layer heteroepitaxy to fabricate self-assembled semiconductor islands, nanofabrication via atom optics, preparation of nanocomposites by sol-gel methods: processing of semiconductors quantum dots, chemical preparation and characterization of nanocrystalline materials, rapid solidification processing of nanocrystalline metallic alloys, vapor processing of nanostructured materials and applications of micromachining to nanotechnology. The contents of this volume will be useful for researchers particularly involved in synthesis and processing of nanostructured materials.

Volume 2 contains 15 chapters dealing with spectroscopy and theoretical aspects of nanostructured materials. The topics covered include: nanodiffraction, FT-IR surface spectrometry of nanosized particles, specification of microstructure and characterization by scattering techniques, vibrational spectroscopy of mesoscopic systems, advanced interfaces to scanning-probe microscopes, microwave spectroscopy on quantum dots, tribological experiments with friction force microscopy, electron microscopy techniques applied to study of nanostructured ancient materials, mesoscopic magnetism in metals, tools of nanotechnology, and nanometrology. The last five chapters in this volume describe computational technology associated with the stimulation and modeling of nanostructures. The topics covered are tunneling times in nanostructures, theory of atomic-scale friction, theoretical aspects of strained-layer quantum-well lasers, carbon nanotube-based nanotechnology in an integrated modeling and stimulation environment, and wavefunction engineering: a new paradigm in quantum nanostructure modeling.

Volume 3 has 11 chapters which exclusively focus on the electrical properties of nanostructured materials. The topics covered are: electron transport and confining potentials in semiconductor nanostructures, electronic transport properties of quantum dots, electrical properties of chemically tailored nanoparticles and their applications in microelectronics, design, fabrication and electronic properties of self-assembled molecular nanostructures, silicon-based nanostructures, semiconductor nanoparticles, hybrid magnetic-semiconductor nanostructures, colloidal quantum dots of III-V semiconductors, quantization and confinement phenomena in nanostructured superconductors, properties and applications of nanocrystalline electronic junctions, and nanostructured fabrication using electron beam and its applications to nanometer devices.

Volume 4 contains 10 chapters dealing with different optical properties of nanostructured materials. The topics include: photorefractive semiconductor nanostructures, metal nanocluster composite glasses, porous silicon, 3-dimension lattices of nanostructures, fluorescence, thermoluminescence and photostimulated luminescence of nanoparticles, surface-enhanced optical phenomena in nanostructured fractal materials, linear and nonlinear optical spectroscopy of semiconductor nanocrystals, nonlinear optical properties of nanostructures, quantum-well infrared photodetectors and nanoscopic optical sensors and probes. The electronic and photonic applications of nanostructured materials are also discussed in several chapters in Volumes 3 and 4.

All nanostructured organic molecules, polymers, and biological materials are summarized in Volume 5. This volume has 13 chapters that include: Intercalation compounds in layered host lattices-supramolecular chemistry in nanodimensions, transition-metal-mediated self-assembly of discrete nanoscopic species with well-defined shapes and geometries, molecular and supramolecular nanomachines, functional nanostructures incorporating responsive modules, dendritic molecules: historical developments and future applications, carbon nanotubes, encapsulation and crystallization behavior of materials inside carbon nanotubes, fabrication and spectroscopic characterization of organic nanocrystals, polymeric nanostructures, conducting polymers as organic nanometals, biopolymers and polymers nanoparticles and their biomedical applications, and structure, behavior and manipulation of nanoscale biological assemblies and biomimetic thin films.

It is my hope that *Handbook of Nanostructured Materials and Nanotechnology* will become an invaluable source of essential information for academic, industrial, and governmental researchers working in chemistry, semiconductor physics, materials science, electrical engineering, polymer science, surface science, surface microscopy, aerosol science, spectroscopy, crystallography, microelectronics, electrochemistry, biology, microbiology, bioengineering, pharmacy, medicine, biotechnology, geology, xerography, superconductivity, electronics, photonics, device engineering and computational engineering.

I take this opportunity to thank all publishers and authors for granting us copyright permissions to use their illustrations for the handbook. The following publishers kindly provided us permissions to reproduce originally published materials: Academic Press, American Association for the Advancement of Science, American Ceramic Society, American Chemical Society, American Institute of Physics, CRC Press-LLC, Chapman & Hall, Electrochemical Society, Elsevier Science Ltd., Huthig-fachverlag, IBM, Institute of Physics (IOP) Publishing Ltd., IEEE Industry Applications Association, Japan Society of Applied Physics, Jai Press, John Wiley & Sons, Kluwer Academic Publishers, Materials Research Society, Macmillan Magazines Ltd., North-Holland, Pergamon Press, Plenum, Physical Society of Japan, Optical Society of America, Springer Verlag, Steinkopff Publishers, Technomic Publishing Co. Inc., The American Physical Society, The Mineral, Metal, and Materials Society, The Materials Information Society, The Royal Society of Chemistry, Vacuum Society of America, VSP, Wiley-Liss Inc., Wiley-VCH Verlag, World Scientific.

This handbook could not have reached fruition without the marvelous cooperation of many distinguished individuals who contributed to these volumes. I am fortunate to have leading experts devote their valuable time and effort to write excellent state-of-the-art reviews which led foundation of this handbook. I deeply express my thanks to all contributors. I am very grateful to Dr. Akio Mukoh and Dr. Shuuichi Oohara at Hitachi Research Laboratory, Hitachi Ltd., for their kind support and encouragement. I would like to give my special thanks to Professor Seizo Miyata of the Tokyo University of Agriculture and Technology (Japan), Professor J. Schoonman of the Delft University of Technology (The Netherlands), Professor Hachiro Nakanishi of the Tohoku University (Japan), Professor G. K. Surya Prakash of the University of Southern California (USA), Professor Padma Vasudevan of Indian Institute of Technology at New Delhi, Professor Toskiyuki Watanabe, Professor Richard T. Keys, Dr. Christine Peterson, and Dr. Judy Hill of Foresight Institute in California, Rakesh Misra, Krishi Pal Reghuvanshi, Rajendra Bhargava, Jagmer Singh, Ranvir Singh Chaudhary, Dr. Hans Thomann, Dr. Ho Kim, Dr. Thomas Pang, Ajit Kelkar, K. Srinivas, and other colleagues who supported my efforts in compiling this handbook. Finally, I owe my deepest appreciation to my wife, Dr. Beena Singh Nalwa, for her cooperation and patience in enduring this work at home; I thank my parents, Sri Kadam Singh and Srimati Sukh Devi, for their moral support; and I thank my children, Surya, Ravina, and Eric, for their love.

I express my sincere gratitude to Professor George A. Olah for his insightful Foreword.

Hari Singh Nalwa

Contents

Chapter 6. STRAINED-LAYER HETEROEPITAXY TO FABRICATE SELF-ASSEMBLED SEMICONDUCTOR ISLANDS

W. H. Weinberg, C. M. Reaves, B. Z. Nosho, R. I. Pelzel, S. P. DenBaars

Chapter 7. NANOFABRICATION VIA ATOM OPTICS

Jabez J. McClelland

Chapter 8. NANOCOMPOSITES PREPARED BY SOL–GEL METHODS: SYNTHESIS AND CHARACTERIZATION

Krzysztof C. Kwiatkowski, Charles M. Lukehart

Chapter 9. CHEMICAL PREPARATION AND CHARACTERIZATION OF NANOCRYSTALLINE MATERIALS

Qian Yitai

Chapter 10. SEMICONDUCTOR QUANTUM DOTS: PROGRESS IN PROCESSING

David J. Duval, Subhash H. Risbud

Chapter 11. RAPID SOLIDIFICATION PROCESSING OF NANOCRYSTALLINE METALLIC ALLOYS

I. T. H. Chang

Chapter 12. VAPOR PROCESSING OF NANOSTRUCTURED MATERIALS

K. L. Choy

Chapter 13. APPLICATIONS OF MICROMACHINING TO NANOTECHNOLOGY

Amit Lal

CONTENTS

About the Editor

Dr. Hari Singh Nalwa has been working at the Hitachi Research Laboratory, Hitachi Ltd., Japan, since 1990. He has authored over 150 scientific articles in refereed journals, books, and conference proceedings. He has 18 patents either issued or applied for on electronic and photonic materials and their based devices. Dr. Nalwa has published 18 books, including *Ferroelectric Polymers* (Marcel Dekker, 1995), *Handbook of Organic Conductive Molecules and Polymers, Volumes 1–4* (John Wiley & Sons, 1997), *Nonlinear Optics of Organic Molecules and Polymers* (CRC Press, 1997), *Organic Electroluminescent Materials and Devices* (Gordon & Breach, 1997), *Handbook of Low and High Dielectric Constant Materials and Their Applications, Volumes 1–2* (Academic Press, 1999), and *Advanced Functional Molecules and Polymers, Volumes 1–4* (Gordon & Breach, 1999).

Dr. Nalwa is the founder and Editor-in-Chief of the *Journal of Porphyrins and Phthalocyanines* published by John Wiley & Sons and serves on the editorial board of *Applied Organometallic Chemistry, Journal of Macromolecular Science-Physics, International Journal of Photoenergy*, and *Photonics Science News*. He is a referee for the *Journal of American Chemical Society, Journal of Physical Chemistry, Applied Physics Letters, Journal of Applied Physics, Chemistry of Materials, Journal of Materials Science, Coordination Chemistry Reviews, Applied Organometallic Chemistry, Journal of Porphyrins and Phthalocyanines, Journal of Macromolecular Science-Physics, Optical Communications*, and *Applied Physics*.

He is a member of the American Chemical Society (ACS), the American Association for the Advancement of Science (AAAS), and the Electrochemical Society. He has been awarded a number of prestigious fellowships in India and abroad that include National Merit Scholarship, Indian Space Research Organization (ISRO) Fellowship, Council of Scientific and Industrial Research (CSIR) Senior fellowship, NEC fellowship, and Japanese Government Science & Technology Agency (STA) fellowship. Dr. Nalwa has been cited in the *Who's Who in Science and Engineering, Who's Who in the World*, and *Dictionary of International Biography*. He was also an honorary visiting professor at the Indian Institute of Technology in New Delhi.

He was a guest scientist at Hahn-Meitner Institute in Berlin, Germany (1983), research associate at University of Southern California in Los Angeles (1984–1987) and State University of New York at Buffalo (1987–1988). He worked as a lecturer from 1988–1990 in the Tokyo University of Agriculture and Technology in the Department of Materials and Systems Engineering. Dr. Nalwa received a B.Sc. (1974) in biosciences from Meerut University, a M.Sc. (1977) in organic chemistry from University of Roorkee, and a Ph.D. (1983) in polymer science from Indian Institute of Technology in New Delhi, India. His research work encompasses ferroelectric polymers, electrically conducting polymers, electrets, organic nonlinear optical materials for integrated optics, electroluminescent materials, low and high dielectric constant materials for microelectronics packaging, nanostructured materials, organometallics, Langmuir-Blodgett films, high temperature-resistant polymer composites, stereolithography, and rapid modeling.

List of Contributors

Numbers in parenthesis indicate the pages on which the author's contribution begins.

I. T. H. CHANG (501)
School of Metallurgy and Materials, University of Birmingham, Edgbaston,
Birmingham, United Kingdom

K. L. CHOY (533)
Department of Materials, Imperial College, London, United Kingdom

JOSEP COSTA (57)
Grup de Recerca en Materials, Departament de Física, Universitat de Girona,
Girona, Spain

S. P. DENBAARS (295)
Departments of Chemical Engineering and Materials, University of California,
Santa Barbara, California, USA

DAVID J. DUVAL (481)
Department of Chemical Engineering and Materials Science, University of California,
Davis, California, USA

K. E. GONSALVES (1)
Department of Chemistry and Polymer Program, Institute of Materials Science U-136,
University of Connecticut, Storrs, Connecticut, USA

E. J. GONZALEZ (215)
Ceramics Division, National Institute of Standards and Technology, Gaithersburg,
Maryland, USA

HONGGANG JIANG (159)
Materials Science & Technology Division, Los Alamos National Laboratory,
Los Alamos, New Mexico, USA

CHRISTOPHER D. JOHNSON (251)
Department of Chemistry and Materials Science Institute, University of Oregon,
Eugene, Oregon, USA

DAVID C. JOHNSON (251)
Department of Chemistry and Materials Science Institute, University of Oregon,
Eugene, Oregon, USA

KRZYSZTOF C. KWIATKOWSKI (387)
Department of Chemistry, Vanderbilt University, Nashville, Tennessee, USA

AMIT LAL (579)
Department of Electrical and Computer Engineering, University of Wisconsin,
Madison, Wisconsin, USA

MAGGIE LAU (159)
Department of Chemical and Biochemical Engineering and Materials Science,
University of California, Irvine, California, USA

ENRIQUE J. LAVERNIA (159)
Department of Chemical and Biochemical Engineering and Materials Science,
University of California, Irvine, California, USA

CHARLES M. LUKEHART (387)
Department of Chemistry, Vanderbilt University, Nashville, Tennessee, USA

JABEZ J. MCCLELLAND (335)
Electron Physics Group, National Institute of Standards and Technology, Gaithersburg, Maryland, USA

MYUNGKEUN NOH (251)
Department of Chemistry and Materials Science Institute, University of Oregon, Eugene, Oregon, USA

B. Z. NOSHO (295)
Departments of Chemical Engineering and Materials, University of California, Santa Barbara, California, USA

R. I. PELZEL (295)
Departments of Chemical Engineering and Materials, University of California, Santa Barbara, California, USA

G. J. PIERMARINI (215)
Ceramics Division, National Institute of Standards and Technology, Gaithersburg, Maryland, USA

S. P. RANGARAJAN (1)
Department of Chemistry and Polymer Program, Institute of Materials Science U-136, University of Connecticut, Storrs, Connecticut, USA

C. M. REAVES (295)
Departments of Chemical Engineering and Materials, University of California, Santa Barbara, California, USA

SUBHASH H. RISBUD (481)
Department of Chemical Engineering and Materials Science, University of California, Davis, California, USA

ROBERT SCHNEIDMILLER (251)
Department of Chemistry and Materials Science Institute, University of Oregon, Eugene, Oregon, USA

HEIKE SELLINSCHEGG (251)
Department of Chemistry and Materials Science Institute, University of Oregon, Eugene, Oregon, USA

VICTORIA L. TELLKAMP (159)
Department of Chemical and Biochemical Engineering and Materials Science, University of California, Irvine, California, USA

J. WANG (1)
Department of Chemistry and Polymer Program, Institute of Materials Science U-136, University of Connecticut, Storrs, Connecticut, USA

W. H. WEINBERG (295)
Departments of Chemical Engineering and Materials, University of California, Santa Barbara, California, USA

QIAN YITAI (423)
Department of Chemistry, University of Science and Technology of China, Hefei, Anhui, People's Republic of China

Chapter 1

CHEMICAL SYNTHESIS OF NANOSTRUCTURED METALS, METAL ALLOYS, AND SEMICONDUCTORS

K. E. Gonsalves, S. P. Rangarajan, J. Wang

Department of Chemistry and Polymer Program, Institute of Materials Science U-136, University of Connecticut, Storrs, Connecticut, USA

Contents

1. INTRODUCTION

Ultrafine microstructures having an average phase or grain size on the order of a nanometer (10^{-9} m) are classified as nanostructured materials (NSMs) [1]. Currently, in a wider meaning of the term, any material that contains grains or clusters below 100 nm, or layers or filaments of that dimension, can be considered to be nanostructured [2]. The interest in these materials has been stimulated by the fact that, owing to the small size of the building blocks (particle, grain, or phase) and the high surface-to-volume ratio, these materials are expected to demonstrate unique mechanical, optical, electronic, and magnetic properties [3]. The properties of NSMs depend on the following four common microstructural features: (1) fine grain size and size distribution (<100 nm); (2) the chemical composition of the constituent phases; (3) the presence of interfaces, more specifically, grain boundaries, heterophase interfaces, or the free surface; and (4) interactions between the constituent domains. The presence and interplay of these four features largely determine the unique properties of NSMs.

Handbook of Nanostructured Materials and Nanotechnology, edited by H.S. Nalwa
Volume 1: Synthesis and Processing
Copyright © 2000 by Academic Press
All rights of reproduction in any form reserved.

ISBN 0-12-513761-3/$30.00

In nanophase materials, a variety of size-related effects can be incorporated by controlling the sizes of the constituent components [4]. For example, nanostructured metals and ceramics can have improved mechanical properties compared to conventional materials as a result of the ultrafine microstructure. In addition, NSMs have the capability to be sintered at much lower temperatures than conventional powders, enabling the full densification of these materials at relatively lower temperatures. Semiconductor NSMs are currently also considered to have technological applications in optoelectronic devices such as semiconductor quantum dots and photodiodes, owing to the phenomenon of "quantum size effects" caused by the spatial confinement of delocalized electrons in confined grain sizes [5]. Magnetic applications of NSMs include fabrication of devices with giant magnetoresistance (GMR) effects, the property used by magnetic heads to read data on computer hard drives, as well as the development of magnetic refrigerators that use solid magnets as refrigerants rather than compressed ozone-destroying chlorofluorocarbons [6]. In addition, nanostructured metals and ceramics seem to be candidates for new catalytic applications [7].

The development of semiconductor nanoclusters is an area of intense research efforts. These nanoclusters are often referred to as quantum dots, nanocrystals, and Q-particles [8]. In the nanometer size regime, electron–hole confinement in nanosized spherical semiconductor particles results in three-dimensional size quantization. Band gap engineering by size and dimension quantization is important because it leads to electrical, optical, magnetic, optoelectronic, and magnetooptical properties substantially different from those observed for the bulk material [5a, 9]. As an example, quantum dots can be developed to emit and absorb a desired wavelength of light by changing the particle diameters. This feature allows the construction of a finely tunable and efficient semiconductor laser.

2. SYNTHESIS OF NANOSTRUCTURED MATERIALS

The synthesis of NSMs from atomic or molecular sources depends on the control of a variety of "nanoscale" attributes desired in the final product. In general, the following four methods have been used to make nanophase materials:

1. The first technique involves the production of isolated, ultrafine crystallites having uncontaminated free surfaces followed by a consolidation process either at room or at elevated temperatures. The specific processes used to isolate the NSMs are, for example, inert-gas condensation [10], decomposition of the starting chemicals or the precursors, and precipitation from solutions.
2. Chemical vapor deposition (CVD), physical vapor deposition (PVD) [11], and some electrochemical methods [12] have been used to deposit atoms or molecules of materials on suitable substrates. Nanocomposites can be produced by depositing chemically different molecules simultaneously or consecutively.
3. By introducing defects in a formerly perfect crystal such as dislocations or grain boundaries, new classes of NSMs can be synthesized. Such deformations may be brought about by subjecting the materials to high energy by either ball milling, extrusion, shear, or high-energy irradiation [13].
4. The final approach used to make NSMs is based on crystallization or precipitation from unstable states of condensed matter such as crystallization from glasses or precipitation from supersaturated solid or liquid solutions [14].

Although these are the general methodologies employed in the synthesis of nanostructured materials, several variants of these processes have been developed to generate compounds or alloys with specific compositions and properties and also for optimized production.

There are basically two broad areas of synthetic techniques for NSMs, namely, (1) physical methods [15] and (2) chemical methods [16].

2.1. Physical Methods

Several different physical methods are currently in use for the synthesis and commercial production of NSMs. The first and the most widely used technique involves the synthesis of single-phase metals and ceramic oxides by the inert-gas evaporation technique [17]. The generation of atom clusters by gas phase condensation proceeds by evaporating a precursor material, either a single metal or a compound, in a gas maintained at a low pressure, usually below 1 atm, in an apparatus similar to that shown in Figure 1. The evaporated atoms or molecules undergo a homogeneous condensation to form atom clusters via collisions with gas atoms or molecules in the vicinity of a cold-powder collection surface. The clusters once formed must be removed from the region of deposition to prevent further aggregation and coalescence of the clusters. These clusters are readily removed from the gas condensation chamber either by natural convection of the gas or by forced gas flow. Sputtering is another technique used to produce NSMs clusters as well as a variety of thin films. This method involves the ejection of atoms or clusters of designated materials by subjecting them to an accelerated and highly focused beam of inert gas such as argon or helium. The third physical method involves generation of NSMs via severe mechanical deformation [18]. In this method, NSMs are produced not by cluster assembly but rather by structural degradation of coarser-grained structures induced by the application of high

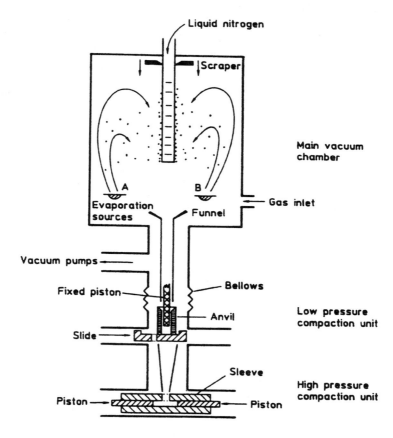

Fig. 1. Schematic drawing of a gas condensation chamber for the synthesis of nanophase materials. Precursor material evaporated from sources A and/or B condenses in the gas and is transported via convection to the liquid-nitrogen-filled cold finger. The clusters are then scraped from the cold finger, collected via the funnel, and consolidated first in the low-pressure compaction unit and then in the high-pressure compaction unit, all in vacuum. Reprinted from *Mater. Sci. Eng. A*, Siegel et al., 168, 189 (© 1993), with kind permission from Elsevier Science Ltd., The Boulevard, Langford Lane, Kidlington 0X5 1GB, UK.

mechanical energy. The nanometer-sized grains nucleate within the shear bands of the deformed materials converting a coarse-grained structure to an ultrafine powder. The heavy deformation of the coarser materials is effected by means of a high-energy ball mill or a high-energy shear process. Although this method is very useful in generating commercial quantities of the material, it suffers from the disadvantage of contamination problems resulting from the sources of the grinding media. Further details regarding the synthesis of NSMs by physical methods have been elaborated in other relevant chapters.

2.2. Chemical Methods

Chemistry has played a major role in developing new materials with novel and technologically important properties [19]. The advantage of chemical synthesis is its versatility in designing and synthesizing new materials that can be refined into the final product. The primary advantage that chemical processes offer over other methods is good chemical homogeneity, as chemical synthesis offers mixing at the molecular level. Molecular chemistry can be designed to prepare new materials by understanding how matter is assembled on an atomic and molecular level and the consequent effects on the desired material macroscopic properties. A basic understanding of the principles of crystal chemistry, thermodynamics, phase equilibrium, and reaction kinetics is important to take advantage of the many benefits that chemical processing has to offer [20].

However, there are certain difficulties in chemical processing. In some preparations, the chemistry is complex and hazardous. Contamination can also result from the byproducts being generated or side reactions in the chemical process. This should be minimized or avoided to obtain desirable properties in the final product. Agglomeration can also be a major cause of concern at any stage in a synthetic process and it can dramatically alter the properties of the materials. As an example, agglomeration frequently makes it more difficult to consolidate nanoparticles to a fully dense compact. Finally, although many chemical processes are scalable for economical production, it is not always straightforward for all systems.

Solution chemistry is used sometimes to prepare the precursor, which is subsequently converted to the nanophase particles by nonliquid phase chemical reactions. Precipitation of a solid from a solution is a common technique for the synthesis of fine particles. The general procedure involves reactions in aqueous or nonaqueous solutions containing the soluble or suspended salts. Once the solution becomes supersaturated with the product, the precipitate is formed by either homogeneous or heterogeneous nucleation. The formation of a stable material with or without the presence of a foreign species is referred to as heterogeneous or homogeneous nucleation [21]. The growth of the nuclei after formation usually proceeds by diffusion, in which case concentration gradients and reaction temperatures are very important in determining the growth rate of the particles, for example, to form monodispersed particles. For instance, to prepare unagglomerated particles with a very narrow size distribution, all the nuclei must form at nearly the same time and subsequent growth must occur without further nucleation or agglomeration of the particles.

In general, the particle size and particle size distribution, the physical properties such as crystallinity and crystal structure, and the degree of dispersion can be affected by reaction kinetics. In addition, the concentration of reactants, the reaction temperature, the pH, and the order of addition of reactants to the solution are also important. Even though a multielement material is often made by coprecipitation of batched ions, it is not always easy to coprecipitate all the desired ions simultaneously because different species may only precipitate at different pH. Thus, control of chemical homogeneity and stoichiometry requires a very careful control of reaction conditions. The problem of agglomeration may be avoided in liquid phase reactions by common methods such as spray drying and freeze drying [22, 23].

NSMs are also prepared by chemical vapor deposition (CVD) or chemical vapor condensation (CVC) [24]. In these processes, a chemical precursor is converted to the gas phase and it then undergoes decomposition at either low or atmospheric pressure to generate the nanostructured particles. These products are then subjected to transport in a carrier gas and collected on a cold substrate, from where they are scraped and collected. The CVC method may be used to produce a variety of powders and fibers of metals, compounds, or composites. The CVD method has been employed to synthesize several ceramic metals, intermetallics, and composite materials. For example, nanophase Si–N–C-containing ceramic particles were obtained by the thermal decomposition of liquid silazane precursors having the general formula $[CH_3SiHNH]_x$, $x = 3$ or 4, with 80% of the cyclic being $x = 4$. It is believed that in the pyrolysis reaction the –SiH–NH– groups were responsible for the extensive crosslinking and the nucleophilic displacements on the neighboring Si atoms, resulting in a three-dimensional network [25].

Semiconductor clusters have traditionally been prepared by use of colloids, micelles, polymers, crystalline hosts, and glasses [26]. The clusters prepared by these methods have poorly defined surfaces and a broad size distribution, which is detrimental to the properties of the semiconductor material. The synthesis of monodisperse clusters with very well defined surfaces is still a challenge to synthetic chemists. However, some of the recent approaches used to overcome these problems are: (1) the synthesis of the clusters within a porous host lattice (such as a zeolite) acting as a template and (2) the controlled fusion of clusters. In this chapter, an overview of the various synthetic methods used to make semiconductor nanoclusters is presented.

3. SYNTHESIS OF METALS, INTERMETALLICS, AND SEMICONDUCTORS

Metals and intermetallics are made by employing either aqueous or nonaqueous methods. Fine metal powders have applications in electronic and magnetic materials, explosives, catalysts, pharmaceuticals, and powder metallurgy [27]. The two main routes of chemical synthesis for the metals and intermetallics are: (1) thermal or ultrasonic decomposition of organometallic precursors to yield the respective elements or alloys and (2) reduction of inorganic or organometallic precursors by reducing agents. Semiconductor nanoclusters are made by incorporating the particles in micelles/colloids, polymers, glasses, or zeolites or by controlled cluster fusion.

Because this chapter deals with the chemical synthesis of nanophase metals and metal alloys and semiconductors, we will be describing here the various methods to prepare metals, for example, palladium, gold, chromium, molybdenum, and copper, as well as the synthesis of certain alloys such as Fe–Co, Fe–Ni, Ti–Al, and a four-component alloy of Fe, commercially known as M50 steel. The development of semiconductor clusters such as ZnO, CdS, ZnTe, and GaN by different synthetic methodologies is also described.

3.1. Chemical Synthesis of Metals

3.1.1. Thermal and Ultrasound Decomposition Methods

The most common example of thermal decomposition is the synthesis of colloidal iron dispersions by the decomposition of iron pentacarbonyl, $Fe(CO)_5$, in a high boiling solvent such as decalin. In a variation of the aforementioned procedure, dispersions of colloidal iron are produced by the decomposition of $Fe(CO)_5$ in two kinds of polymer solutions: (1) "active" polymers such as polymers having a nitrogen nucleophile and (2) "passive" polymers such as polymers having an alkenyl or benzylic functionality. Examples of the polymers used include: polybutadiene, poly(styrene-*co*-butadiene), poly[styrene-*co*-4-vinylpyridine], and poly[styrene-*co*-N-vinylpyrrolidone] [28].

The role of the polymeric catalysts in particle nucleation is their action as "dispersants," which stabilize metal particles by adsorption of a thick layer of polymer on the surface of the particles. Thus, the thermal decomposition of $Fe(CO)_5$ has been examined whereby $Fe(CO)_5$ undergoes facile valence disproportionation reactions with nitrogen nucleophiles. If analogous reactions were to occur between $Fe(CO)_5$ and nucleophilic residues [29] on a macromolecule, intermediate ligand–metal cluster compounds would be generated in the polymeric domain. It is postulated that such ligand clusters are more thermally labile than $Fe(CO)_5$ molecules in bulk solutions and thus, are precursors to colloidal iron particles. In the absence of a polymer, the mechanism and kinetics of the decomposition of $Fe(CO)_5$ in an "inert" hydrocarbon media such as decalin are quite complex [30]. Although it is not very well understood, it is certain, however, that the conversion is a sequential step-wise process in which increasingly larger clusters are formed as molecules of CO and $Fe(CO)_5$ are split out (Scheme I). Nitrogen copolymer systems have been termed "active" because the initial and overall rate of decomposition of $Fe(CO)_5$, as determined by the rate of CO production, was much faster in the presence of active polymers than in the solvent alone. Liganded polymer–metal carbonyl compounds can also be generated in passive functional substrates, that is, in molecules having alkenyl or benzylic (allylic) functionality [31]. These types of materials have been termed "passive" because they react only after the loss of a CO ligand to yield the reactive intermediate, $[Fe(CO)_4]$. The rate of decomposition of $Fe(CO)_5$ in the presence of the "passive" polymers—polystyrene, polybutadiene, and poly(Styrene-*co*-butadiene)—is initially similar to that in solvent alone. However, with the formation of the intermediate polymer–bound metal carbonyl compounds, the rate of evolution of CO increases over the rate in solvent alone.

Smith and Wychick [28] carried out the synthesis of Fe nanoparticles in the presence of butadiene-containing polymers. Uniform colloidal dispersions of approximately 70–80-Å Fe^0 particles, which are physically very stable, were obtained. In the initial phase of the thermolysis, it was felt that the major CO-evolving reaction is that resulting in the formation of $Fe_2(CO)_9$, just as is the case in the presence of decalin alone. The "catalyzed" reaction differs in that intermediate $[Fe(CO)_4]$ molecules can react with the isolated alkenyl residues on the polymer backbone. Subsequently, isomerization of the double bonds along the chain occurs, generating butadienyl–iron tricarbonyl residues in the polymer (Scheme II). Particles in the dispersion that were less than 100 Å were superparamagnetic (125 emu/g of Fe at 10,000 Oe), and particles in the 100–200-Å range

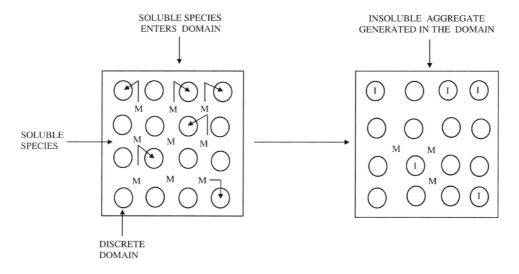

Scheme I.   (Adapted from J. Phys. Chem. 84, 1621–29 (1980), Smith et al.)

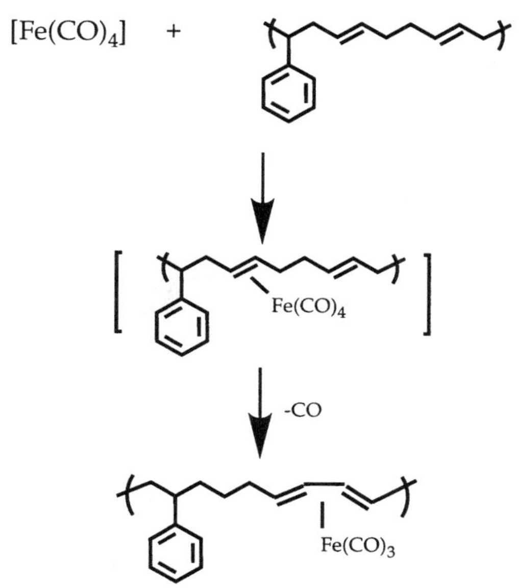

Scheme II. Reaction of $Fe(CO)_4$ with Alkenyl Residues. (Adapted from J. Phys. Chem. 84, 1621–29 (1980), Smith et al.)

had a time-dependent hysteresis. On exposure to the atmosphere, an approximately 30-Å-thick γ-Fe_2O_3 oxide film was produced on the surface of the particles. This is the "passive oxide film" detected previously by a number of techniques [32]. As water was adsorbed from the atmosphere, the chlorinated solvent-based dispersions reacted further to give β-FeOOH. This reaction was, in fact, promoted when the decompositions were carried out in a chlorinated solvent such as chlorobenzene. The magnetic moment decayed with oxidation roughly in proportion to the quantities of Fe, γ-Fe_2O_3, and β-FeOOH present.

In a slight variation of this process, Nakatani et al. [33] have developed a new metallic magnetic fluid with ε-Fe_3N fine particles dispersed in kerosene by the vapor-liquid chemical reaction between iron carbonyl and NH_3. A surfactant amine (polybutenylsuccin-polyamine) was added into the reaction to produce fine colloidal dispersions. The particles showed electron diffraction patterns indexed by the ε-Fe_3N structure. The ratio of amine to $Fe(CO)_5$ has a remarkable influence on the dispersion of the iron-nitride fine particles. Varying concentrations of the surfactant yielded magnetic fluids with different agglomerations (Fig. 2). The particles, in general, were highly uniform in size and well dispersed without agglomeration.

Iron magnetic fluids have high saturation magnetic flux densities up to 2330 G and with high relative initial permeabilities up to 160 or 180, depending on the particle diameter and particle number density as measured by the B–H loop tracer.

In another experiment, Wonterghem et al. [34] prepared a magnetic glass, or in other terms a ferrofluid [35], by the thermal decomposition of iron pentacarbonyl in decalin along with a surfactant (Sarkosyl-O). Usually, metallic glasses are prepared by the liquid quench or the vapor deposition techniques in which hot atoms are condensed onto a substrate kept at temperatures well below the glass transition temperature (T_g). In the case of glass formation by the carbonyl decomposition, the condensation process is presumably a vapor–solid transition. The condition for glass formation is that condensed atoms are prevented from diffusing more than one atomic distance at the surface before they are fixed in position by the arrival of additional atoms. It is noteworthy that the molecules forming the metallic glass particles are at a temperature that does not exceed the boiling point of the liquid

(a)   (b)

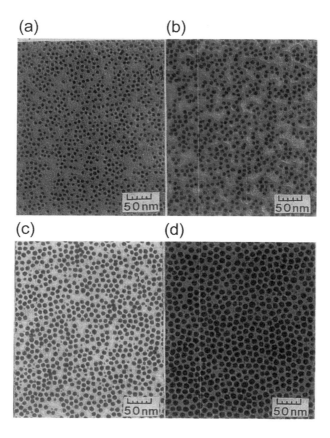

(c)   (d)

Fig. 2. Electron micrographs of iron-nitride magnetic fluids synthesized from solutions with various amounts of Fe(CO)$_5$ in 50.1 g kerosene: (a) synthesized from solution with 80 g Fe(CO)$_5$, (b) 120 g Fe(CO)$_5$, (c) 150 g Fe(CO)$_5$, and (d) 200 g Fe(CO)$_5$. Reprinted from *J. Magn. Magn. Mater.*, I. Nakatani et al., 122, 10 (© 1993), with kind permission from Elsevier Science Ltd., The Boulevard, Langford Lane, Kidlington 0X5 1GB, UK.

($\sim$460 K). Therefore, this carbonyl decomposition method is not based on rapid cooling from a high temperature, but, rather, the mechanism is based on the growth of alloy particles in a system that is kept at a temperature below T_g. Mössbauer spectra taken at room temperature, 80 K, and 166 K suggest that the particles in the colloid are not pure α-Fe but an amorphous iron–carbon alloy with 5–10 at% carbon. This type of metallic glass crystallizes into a mixture of α-Fe and iron carbides at 523 K. The Mössbauer studies [36] show that the decomposition of Fe(CO)$_5$ in the fluid results in the formation of metallic glass particles that crystallize into α-Fe and iron carbide upon heating. Figure 2 shows the room temperature Mössbauer spectrum of the particles after heating in hydrogen at 523 K. The predominant component was unambiguously identified as the spectrum of α-Fe. The remaining weak absorption lines indicate the presence of other magnetic phases with smaller magnetic hyperfine fields. The best computer fit (Fig. 3) was obtained by including three additional six-line components of low intensity. The parameters of these components are in accordance with the published values of χ-Fe$_5$C$_2$ [37]. The χ-carbide component constitutes about 8% of the spectral area corresponding to about 11 at% carbon in the particles.

Apart from thermal decomposition methods, iron nanoparticles have also been made by ultrasonic irradiation or by laser pyrolysis of iron pentacarbonyl. A dull powder was obtained in experiments performed by Cao et al. [38] on the sonochemical decomposition of Fe(CO)$_5$ in decane with varying solution concentrations. They obtained particles of varying sizes (59–243 nm), depending on the solution concentrations, with smaller particles being

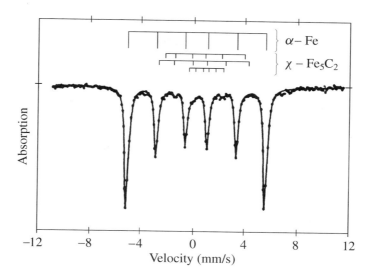

Fig. 3. Room temperature Mössbauer spectrum of the particles after heating in hydrogen at 523 K. The full line indicates a computer fit of the spectrum with a six-line component due to α-Fe and three six-line components due to χ-Fe$_5$C$_2$, as shown by the bar diagrams. Reprinted with permission from J. Wonterghem et al., *Phys. Rev. Lett.*, 55, 410 (© 1985 American Physical Society).

obtained for more dilute solutions. The nanophase iron powders do not exhibit saturation at magnetic fields up to 50 kG. The magnetic moment at 50 kG varies between 42.5 emu/g for a pure Fe(CO)$_5$ solution to 12 emu/g for the most dilute solution. This result is explained on the basis that in bulk ferromagnetic materials the local magnetic moments are organized in domains with a certain characteristic size. This yields a reduction in the magnetic energy of the demagnetizing field surrounding the sample. The total energy of the domain boundaries grows, thoroughly as a square of the effective radius of a sample, whereas the magnetic energy of a demagnetizing field is proportional to third degree. At some particle size, the domain formation is not profitable anymore and particles become single domain (all spins are oriented in a certain direction). In this case, the material becomes superparamagnetic and does not exhibit saturation. From Figure 4, it can be seen that the pure sample is the transition state from ferromagnetism to superparamagnetism because it still shows a quite flat region at high fields. Electron spin resonance (ESR) measurements on the sample showed strong signals, and it is certain that these come from the internal magnetic nature of the sample and not from the impurity contained in the amorphous iron. Interestingly, the ESR signals become stronger and sharper for more dilute solutions, implying that the magnetic exchange is becoming weaker and weaker.

By differential scanning calorimetry (DSC) measurements, it was confirmed that the crystallization temperature of iron nanoparticles obtained from pure iron pentacarbonyl was much lower than those of the other samples; that is, particles with a finer grain size had a lower crystallization temperature. This occurs because the powders produced by a decomposition of pure Fe(CO)$_5$ consist of denser particles and it is easier for denser particles to crystallize.

In our laboratory, we have synthesized nanostructured α-Fe by sonochemical as well as thermal decomposition of Fe(CO)$_5$ in decalin [39]. In a typical reaction, for the sonochemical procedure, a dispersion of 15 g (0.076 mol) of Fe(CO)$_5$ in dry decalin (200 mL) was sonicated at 50% amplitude using a Sonic and Materials VC-600 ultrasonic probe (20 kHz, 100 W cm^{-2}) for approximately 6 h at room temperature. In the thermal method, the same amounts of precursors were refluxed in decalin until the completion of the reaction. In either case, on completion of the reaction the formation of shiny metallic particles

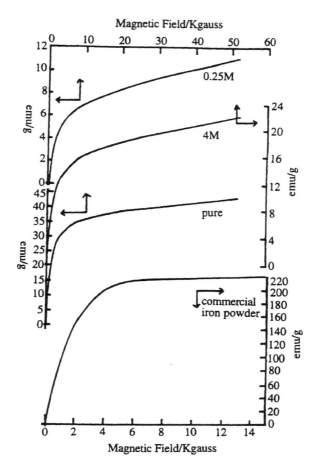

Fig. 4.   Magnetization curves of commercial iron powder and amorphous iron at 100 K. Reprinted with permission from X. Cao et al., *J. Mater. Res.*, 10, 2952 (© 1995 Materials Research Society).

was observed on the walls of the reaction vessel. Decalin was removed by decantation and the resulting black powders were isolated and dried with heating under vacuum. These iron powders were then consolidated by vacuum hot pressing at 275 MPa at 700 °C for 1 h.

The as-synthesized iron powders were amorphous by X-ray diffraction (XRD). The major peaks in the X-ray spectrum were assigned to the α-Fe phase and the average crystallite size was calculated to be approximately 40 nm as determined by line broadening analysis [40]. The morphology under scanning electron microscopy (SEM) was found to be porous coral like (Fig. 5). The consolidated iron pellet had a smooth and homogeneous microstructure as confirmed by SEM (Fig. 6). This sample had a high Rockwell C (RC) hardness of 37 as compared to the hardness of conventional iron (4–5 RC). Carbon and oxygen concentrations were 0.05% and 1.1%, respectively. Details of the sonochemical synthesis of nanostructured metals and alloys are also provided by Suslick.

A pulsed laser pyrolysis technique [41] has been used to study the gas phase thermal decomposition of iron pentacarbonyl and chromium, tungsten, and molybdenum hexacarbonyl. In essence, a pulsed infrared CO_2 laser is used to heat an absorbing gas (SF_6), which then, by collision, transfers its energy to the reactive substrate and bath gas (N_2). Laser pyrolysis offers several advantages over other techniques. Gas phase measurements avoid the possible problems created by solvent or matrix effects on the molecules. Laser heating provides a wide temperature range and well-controlled reaction times. Lewis et al. [41] have

Fig. 5.   SEM micrograph of the iron powder. Reprinted with permission from K. E. Gonsalves et al., *J. Mater. Sci. Lett.*, 15, 1261 (© 1996 Kluwer Academic Publishers).

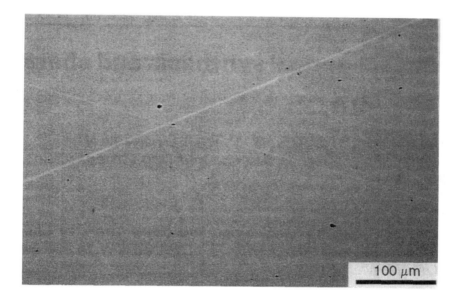

Fig. 6.   SEM micrograph of the consolidated iron sample. Reprinted with permission from K. E. Gonsalves et al., *J. Mater. Sci. Lett.*, 15, 1261 (© 1996 Kluwer Academic Publishers).

measured the bond dissociation energies of various transition metal carbonyls by use of the laser pyrolysis method.

Stabilized colloidal cobalt nanoparticles were prepared by Thomas [42] by the thermal decomposition of $Co_2(CO)_8$ in a hydrocarbon solvent containing a suitable polymeric material (methyl methacrylate–ethyl acrylate–vinylpyrrolidone terpolymer of mole proportions 33:66:1) and about 3% methyl isobutyl ketone. It should be noted that the average particle size could be easily varied from about 20 Å to about 300 Å by variations in the reagent concentrations, the temperature, and the composition of the polymeric material.

Polymers with a relatively large percentage of highly polar groups promoted the growth of smaller particles. It was also found that copolymers of reasonably high molecular weight (on the order of 10^4 and greater) were unique in furnishing a high degree of stability to such colloids. The magnetic properties of the Co particles were measured on dried films formed by evaporation of the solvent with the polymeric material acting as a binder. In general, the magnetic properties were those to be expected from single-domain particles [43]. Two unique features were observed in these nanoparticles: (1) The particles were arranged more or less in the form of a chain, and (2) they exhibited a continuous range of coercive force between the limits of 200 and 900 Oe, while retaining constant remanence-to-saturation ratios of 0.5–0.6 for randomly oriented samples and 0.8–0.9 for oriented samples.

3.1.2. Reduction Methods

Several types of reducing agents have been employed to produce fine metal particles from inorganic salts. In the following discussion, examples of various metals produced by different reducing agents are listed and brief descriptions of some property characterizations are also discussed.

Reducing agents such as $NaBEt_3H$, $LiBEt_3H$, and $NaBH_4$ have been commonly used to yield metal nanopowders. The group 6 metal chlorides $CrCl_3$, $MoCl_3$, $MoCl_4$, and WCl_4 were reduced in toluene solution with $NaBEt_3H$ at room temperature to form the corresponding metal colloids in high yield [44]. When the same metal chlorides were reduced in tetrahydrofuran (THF) solution with $LiBEt_3H$ and $NaBEt_3H$, metal carbides (M_2C) were formed in approximately 95% yield (Scheme III). The metal and metal carbide colloids were shown to comprise 1–5-nm-sized particles by transmission electron microscopy (TEM). In general, the powders were isolated as agglomerates of these primary crystallites with dimensions of approximately 400–500 nm as determined by SEM. These 400–500-nm agglomerates were, in turn, composed of the primary 1–5-nm-sized crystallites. X-ray powder diffraction studies of the black powders exhibited broad peaks for the as-synthesized powders at room temperature where the crystallite size estimated from the broadening analysis generally corresponded to the primary particle size as determined by TEM. Initially, the powders were characterized with the NaCl byproduct intact. However, when the as-precipitated black powders were washed with the deoxygenated water, the NaCl could be removed without observable oxidation.

The proposed equation for the reduction is as follows, generalized for M = Cr and Mo, $x = 3$, and M = W, $x = 4$:

$$MCl_x + xNaBEt_3H \rightarrow M + xNaCl + xBEt_3 + (x/2)H_2 \qquad (1)$$

This reaction may result from direct hydride transfer to form a metal hydride intermediate, which subsequently reductively eliminates H_2, or may be reduced directly by an

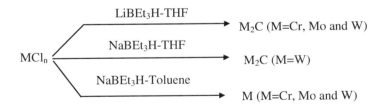

^{a} M=Cr or Mo, n=3; M=W, n=4

Scheme III.   (Adapted from Chem. Mater. 5, 689 (1993), Zeng et al.)

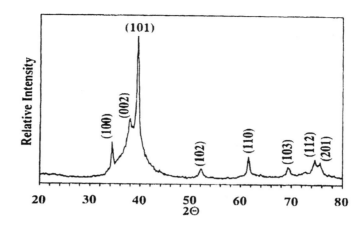

Fig. 7. X-ray powder diffraction data obtained from powder isolated from reduction of MoCl$_3$(THF)$_3$ in THF after heating to 500 °C for 4 h *in vacuo*. Reprinted with permission from D. Zeng and M. J. Hampden-Smith, *Chem. Mater.*, 5, 681 (© 1993 American Chemical Society).

electron transfer mechanism [45]. Reduction of MoCl$_4$(THF)$_2$ or MoCl$_3$(THF)$_3$ in THF at room temperature with a slight excess of stoichiometric LiBEt$_3$H resulted in the formation of a black colloid accompanied by H$_2$ gas evolution. The reaction mixture turned black immediately on addition of LiBEt$_3$H. After stirring overnight at room temperature, a black colloid was formed that has been shown to comprise 2–4-nm-sized Mo$_2$C crystallites. X-ray powder diffraction of the room temperature product showed a broad diffraction peak centered at a d spacing of approximately 2.4 Å. On heating this powder to 500 °C, the XRD pattern sharpened and corresponded to that of Mo$_2$C (Fig. 7). No evidence of Mo was observed. Elemental analysis confirmed the carbon content to be 6.3%, close to the calculated carbon percentage, 5.9%, for Mo$_2$C.

W$_2$C was isolated when WCl$_4$ was reduced in THF at room temperature with a slight excess of LiBEt$_3$H. Both SEM and TEM data revealed that the grain size of the black powder was approximately 2–4 μm and was composed of agglomerated 1-nm-sized primary particles by TEM.

On the basis of the X-ray and electron diffraction data, it appears that the peaks present are attributable to the diffraction maxima of the metal carbide phase. The origin of the carbide is still unclear. It was proposed that the formation of the carbide may have been due to transfer of an ethyl anion rather than hydride from the triethylborohydride reducing reagent to the metal center. Ethyl groups normally rearrange to liberate ethylene and result in the formation of a metal hydride [46]. Therefore, it seems likely that the common factor in the formation of the metal carbide rather than the metal is the solvent and not the reagents. Hydrogen abstraction from a metal-bonded carbon would lead to retention and strengthening of the metal carbon bond.

Buhro et al. [47] have similarly prepared copper nanopowders by the use of NaBH$_4$ or Li/NaBEt$_3$H reducing agents in THF solvents. The byproducts were removed by solvent and water washing and the product powders had average sizes ranging from 10–30 nm, depending on the reaction conditions. Copper powders stored in air for extended periods did not exhibit CuO or Cu$_2$O reflections in the XRD patterns. These powders were then hot pressed at 750 °C and 120 MPa for 2 h. Density measurements of 92% (Archimedes method) of the theoretical value were obtained. XRD and SEM analyses established that grain growth to approximately 38 nm had occurred. The compact had a yield strength of 401 MPa by three-point bending and compression tests. The same group also synthesized air-sensitive 40-nm-sized aluminum powders by the reduction of AlCl$_3$ with lithium aluminum hydride.

A series of fine metal powders has also been synthesized by Bönnemann et al. [48] using metal tetrahydroborates, $M[BH_4]$ ($M = Li, Na$). However, because all of the four hydrogen atoms in the BH_4^- anion can take part in the reducing process, the resulting metal powders are contaminated with borides [49]. Alkaline hydrotriorganoborates, $M[BR_3H]$, are also used as reducing agents [50]. Here, in contrast, the organoboron part has no reducing properties, but exclusively functions as a complexing agent to generate very soluble metal hydrides in organic media. This allows for the preparation of boride-free metal nanopowders by reducing metal salts with alkali or alkaline earth metal hydrides in organic phases. The hydrides are solubilized in organic media using BR_3 or $BR_n(OR')_{3-n}$ ($R, R' = $ alkyl or aryl, $n = 0, 1, 2$) as complexing agents to form hydrotriorganoborates of the general formula $M'H_u \cdot (BR_3)_u$ or $M'H_u[BR_n(OR')_{3-n}]_u$ ($M' = $ alkali or alkaline earth metal, $u = 1, 2$). Analogously, organogallium compounds, $GaR_n(OR')_{3-n}$ ($n = 0, 1, 2, 3$) may be used as complexing agents to give the corresponding hydrotriorganogallates in organic solvents. Tables I and II summarize the different preparation conditions for the synthesis of nanocrystalline metal powders in THF by different reducing agents. Table III summarizes the grain sizes of selected metal powders prepared by the hydrotriorganoborate reduction method, observed by TEM.

The reduction of metal halides in organic phases can also be achieved by adding only a catalytic amount of BR_3 to suspended $M'H_u$ ($M' = $ alkali or alkaline earth metal, $u = 1, 2$). The organoboron complexing agents liberated according to Eq. (2) react with further metal hydrides in suspension to regenerate the hydrotriorganoborate complexes in $situ$ [Eq. (3)]:

$$uMX_v + vM'(BR_3H)_u \rightarrow uM\downarrow + vM'Xu + uvBR_3 + uv/2H_2\uparrow \qquad (2)$$

$$uMXv + vM'Hu \rightarrow uM\downarrow + vM'Xu + uv/2H_2\uparrow \qquad (3)$$

Tetrabutylammonium hydrotriorganoborates [45], which are readily accessible from ammonium halides and alkali metal hydrotriorganoborates, may advantageously be used

Table I. Preparation of Nanocrystalline Metal Powders in THF with Formation of Soluble Borates

Number	Metal salt	Reducing agent	Condition		Product	
			t (h)	T (°C)	Metal content (%)	Boron content (%)
1	$Fe(OEt)_2$	$NaBEt_3H$	16	65	96.8	0.16
2	$Co(OH)_2$	$NaBEt_3H$	2	23	94.5	0.40
3	$Co(CN)_2$	$NaBEt_3H$	16	65	96.5	0.20
4a	$Ni(OH)_2$	$NaBEt_3H$	2	23	94.7	0.13
5	$Ni(OEt)_2$	$NaBEt_3H$	16	65	91.4	0.58
6	$CuCN$	$LiBEt_3H$	2	23	97.3	0.0
7	$CuSCN$	$NaBEt_3H$	16	65	95.0	0.23
8	$Pd(CN)_2$	$NaBEt_3H$	16	65	95.5	1.38
9	$AgCN$	$Ca(BEt_3H)_2$[a]	2	23	89.6	0.20
10	$Cd(OH)_2$	$NaBEt_3H$	2	23	97.9	0.22
11	$Pt(CN)_2$	$NaBEt_3H$	16	65	87.5	0.93
12	$AuCN$	$NaBEt_3H$	2	23	97.5	0.0

Source: H. Bönnemann et al., *J. Mol. Catal.* 86, 129 (1994).

[a] Solvent: diglyme.

Table II. Preparation of Nanocrystalline Metal Powders in THF

Number	Metal salt	Reducing agent	Condition		Product	
			t (h)	T (°C)	Metal content (%)	Boron content (%)
1	$CrCl_3$	$NaBEt_3H$	2	23	93.3	0.3
2	$MnCl_2$	$LiBEt_3H$	1	23	94.0	0.42
3	$FeCl_3$	$LiBEt_3H$	2	23	97.1	0.36
4	CoF_2	$NaBEt_3H$	16	65	96.9	0.0
5	$CoCl_2$	$NaBEt_3H$	16	65	95.1	0.0
6	$CoCl_2$	$LiH+10\% BEt_3$	16	65	95.8	0.0
7	$CoBr_2$	$LiBEt_3H$	2	23	86.6	0.0
8	$NiCl_2$	$NaBEt_3H$	16	65	96.9	0.0
9	$CuBr_2$	$LiBEt_3H$	2	23	94.9	0.0
10	$CuCl_2$	$Na(Et_2BOMe)H$	2	23	94.7	0.1
11	$ZnCl_2$	$LiBEt_3H$	12	65	97.8	0.0
12	$RuCl_3$	$NaBEt_3H$	16	65	95.2	0.52
13	$RhCl_3 \cdot 3H_2O$	$NaBEt_3H$	2	23	98.1	0.1
14	$RhCl_3$	$LiBEt_3H$	2	23	96.1	0.66
15	$PdCl_2$	$NaBEt_3H$	16	65	98.0	0.29
16	AgF	$NaB(OMe)_3H$	2	23	94.1	0.05
17	AgI	$NaBEt_3H$	2	23	95.3	0.02
18	$CdCl_2$	$LiBEt_3H$	2	23	99.4	0.0
19	$ReCl_3$	$LiBEt_3H$	2	23	95.4	0.0
20	$OsCl_3$	$NaBEt_3H$	2	23	95.8	0.0
21	$IrCl_3 \cdot 4H_2O$	$NaBEt_3H$	216	23	77.1	0.16
22	$IrCl_3$	$KBPr_3H$	2	65	94.7	0.08
23	$PtCl_2$	$NaBEt_3H$	5	23	98.2	0.21
24	$PtCl_2$	$LiH+10\% BEt_3$	12	65	98.8	0.0
25	$PtCl_2$	$LiBEt_3H$	4	65	99.0	0.0
26	$PtCl_2$	$LiBEt_3H$	2	0	99.0	0.0
27	$SnCl_2$	$LiBEt_3H$	2	23	96.7	0.0
28	$SnBr_2$	$LiBEt_3H$	2	23	87.1	0.0
29	$PdCl_2$	$Na(GaEt_2OEt)H$	2	40	92.7	Ga:0.25
30	$Pt(NH_3)_2Cl_2$	$NaBEt_3H$	2	23	97.1	0.32
31	$Pt(Py)_2Cl_2$	$LiBEt_3H$	2	23	97.1	0.02
32	$Pt(Py)_4Cl_2$	$LiBEt_3H$	2	23	97.5	0.01
33	$CODPtCl_2$	$NaBEt_3H$	2	60	97.9	0.58

Source: H. Bönnemann et al., *J. Mol. Catal.* 86, 129 (1994).

Py, pyridine; COD, 1,5-cyclooctadiene.

Table III. Selected Grain Sizes of Metal Powders Measured by TEM

Number	Metal	Starting materials	Preparation	Grain size (nm)
1	Platinum	$PtCl_2$ $LiBEt_3H$	Table II, No. 25	2–5
2	Palladium	$PdCl_2$ $LiBEt_3H$	Similar to Table II, No. 15	12–28
3	Rhodium	$RhCl_3$ $LiBEt_3H$	Table II, No. 14	1–4
4	Copper	$CuCl_2$ $LiBEt_3H$	Similar to Table II, No. 9	25–90
5	Nickel	$Ni(OH)_2$ $NaBEt_3H$	Table I, No. 4	5–15
6	Cobalt	$CoCl_2$ $NaBEt_3H$	Table II, No. 5	3–5

Source: Bönnemann et al., *J. Mol. Catal.* 86, 129 (1994).

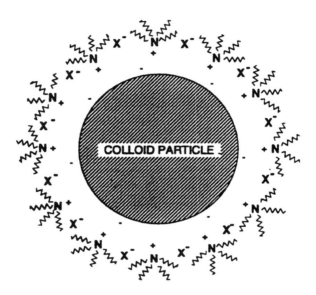

Fig. 8. Stabilization of the metal core with NR_4X. Reprinted from *J. Mol. Catal.*, H. Bönnemann et al., 86, 129 (© 1994), with kind permission of Elsevier Science-NL, Sara Burgerhartstraat 25, 105b KV Amsterdam, The Netherlands.

for the reduction of metal halides in THF. Because ammonium halides formed as the byproduct of the reaction [Eq. (4)] remain completely dissolved after the reaction, the precipitated metal powders can be easily isolated in pure form:

$$MX\nu + \nu NR_4(BEt_3H) \rightarrow M\downarrow + \nu NR_4X + \nu BEt_3 + \nu/2H_2\uparrow \qquad (4)$$

where M is a metal of groups 6–11, X is a halogen; R is *n*-butyl, and $\nu = 1, 2, 3$.

Bönnemann et al. have prepared metal colloids of elements of groups 6–11 in the organic phase. The metal salts are suspended in an organic solvent such as THF and treated with tetraalkylammonium-hydrotriorganoborates, which are readily accessible from ammonium halides and alkali–metal hydrorganoborates. Brown–red metal colloid solutions are generated with hydrogen evolution, from which only a small portion of the reduced metal precipitates. The ammonium halide salt that forms in the process functions as a protective atmosphere for the metal particles (see Fig. 8) obviating the need for external stabilizers. The elemental analysis of the isolated metal colloid and results from mass spectrometry indicate that NR_4X is present. The results hitherto suggested that tetraalkylammonium ions surround the presumably negatively charged metal nucleus. The screening of the metal nucleus by large lipophilic alkyl groups explains the remarkable solubility of the metal colloids in organic solvents and their extraordinary stability. Table IV summarizes the reaction conditions and particle properties of various metal colloids obtained by using the tetraalkylammonium salts as reducing agents.

The metal colloids can be adsorbed on the surface of supports. The metal aggregates do not coalesce during the process. TEM investigations showed that the particle size remained the same and a very uniform distribution of the metal particles on the support was achieved. The supported metal colloids are effective catalysts for the hydrogenation of unsaturated compounds such as CO, C–C, C–O, and C–N multiple-bond systems, as well as the hydrogenation of naturally occurring products and mixtures such as soya bean oil [51].

Duteil et al. [52] have prepared Ni colloids by the reaction of $Ni(acac)_2$ (acac = acetylacetonate) and PPh_3 (Ph = phenyl) in diethyl ether with the reducing agent Et_2AlH at $-40\,°C$. These colloids can be isolated in the solid state and redispersed in any concentration in polar solvents such as pyridine, owing to their ligand shell mainly consisting not of PPh_3 molecules but, instead, of PPh moieties that are generated from PPh_3 during the reaction with Et_2AlH with the formation of free benzene:

$$Et_2AlH + Ni(acac)_2 \rightarrow Et_2Al(acac) + Ni + 2H \tag{5}$$

$$PPh_3 + 2H \rightarrow C_6H_6 + PPh \tag{6}$$

High-resolution transmission electron microscopy (HRTEM) investigations of the nickel colloid, deposited on a grid from a dark-brown pyridine solution, indicate very narrow particle size distribution with an average diameter of 4 nm. The exact size of some of these particles correspond to diameters between 3.9 and 4.3 nm. Preliminary studies of the catalytic properties of the nickel colloids show very low activities in hydrogenation reactions. For instance, the turnover frequency for the hydrogenation of hex-2-yne to cis-hex-2-ene in a heterogeneous reaction is only 2 $mol_{prod}\,mol_{Ni}^{-1}\,h^{-1}$, compared with values of some hundreds using common hydrogenation catalysts of nickel or palladium. The lack of catalytic activity is due to the perfect protecting ligand shell covering the surface nickel atoms.

Another commonly used reduction procedure is called the polyol process [53]. In this method, liquid polyols such as ethylene glycol or diethylene glycol are used both as a solvent and as a reducing agent for the chemical preparation of metallic powders from various inorganic precursors. The basic reaction scheme for the synthesis of these metal powders by the polyol process involves the dissolution of the solid precursor, the reduction of the dissolved metallic species by the polyol itself, nucleation of the metallic phase, and growth of the nuclei. To obtain metal powders with a narrow size distribution, two conditions must be fulfilled: (1) A complete separation of the nucleation and growth steps is required and (2) the aggregation of metal particles must be avoided during the nucleation and growth steps.

Fievet et al. [54] have successfully used the polyol process for the synthesis of fine, highly pure, monodisperse, nonagglomerated particles of Cu. The precursor copper(II) oxide is dispersed in a given volume of ethylene glycol and the suspension is stirred at 300 rpm and heated at a rate of $6\,°C\,min^{-1}$ up to the reaction temperature ranging from 150 to 195 °C. Typical reaction times were 30 min at 195 °C or 2 h at 175 °C. To prevent particle sintering, different organic protecting agents have been tested. The best result was obtained with a solution of D-sorbitol in ethylene glycol as a reaction medium. The size of the copper particles could be strictly controlled by dissolving various amounts of sodium hydroxide in the D-sorbitol/ethylene glycol solution and then suspending CuO in this mixture. Addition of this strong base enhances the solubility of the precursor CuO and intermediate Cu_2O [55].

The XRD pattern of the solid phases present during the reduction of Cu(II) oxide in ethylene glycol is shown in Figure 9. Highly crystallized Cu powders are produced as evidenced by the sharp XRD line of Figure 9c. It is inferred from the XRD analysis of the

Table IV. Preparation of Metal Colloids in THF Solution or as an Isolated Powder

Number	Metal salt	Reducing agent	Condition t (h)	Condition T (°C)	Product colloid solution color	Workup solvent	Solvent added for precipitation	Metal content in isolated colloid (%)	Mean particle size (nm)
1	MnI_2	$N(octyl)_4BEt_3H$	1	23	Dark brown Mn completely dissolved				
2[a]	$FeBr_2$	$N(octyl)_4BEt_3H$	18	90	Dark brown to black Fe almost completely dissolved	Ethanol	Ether	11.34	3.0
3	$RuCl_3$	$N(octyl)_4BEt_3H$	2	50	Dark reddishbrown to black Ru almost completely dissolved	Ethanol	Pentane	68.72	1.3
4	$OsCl_3$	NBu_4BEt_3H	1	23	Deep red to black				
5	$CoBr_2$	$N(octyl)_4BEt_3H$	16	23	Dark brown to black Co completely dissolved	Ethanol	Ether	37.45	2.8
6	$RhCl_3$	$N(octyl)_4BEt_3H$	3	40	Deep red to black Rh completely dissolved	Ether	Ethanol	73.40	2.1
7	$IrCl_3$	$N(octyl)_4BEt_3H$	1	50	Dark red to black Ir almost completely dissolved	Ethanol	Ether	65.55	1.5
8	$NiBr_2$	$N(octyl)_4BEt_3H$	16	23	Dark red to black Ni completely dissolved	Ethanol	Ether	66.13	2.8
9	$NiBr_2$	$N(octyl)_3MeBEt_3H$	16	23	Dark red to Ni completely dissolved	Ethanol	Ether	68.29	2.8
10	$PdCl_2$	$N(octyl)_4BEt_3H$	1	23	Dark brown to black Pd completely dissolved	Ether	Ethanol	83.62	2.5
11	$PtCl_2$	$N(hexyl)_4BEt_3H$	2	23	Dark brown to black Pt up to 80% dissolved				
12	$PtCl_2$	$N(octyl)_4BEt_3H$	18	23	Dark brown to black Pt completely dissolved	Ether	Ethanol	85.13	2.8
13	$PtCl_2$	$N(hexyl)_4BEt_3H$	2	23	Dark brown to black Pt up to 80% dissolved				
14	$CuCl_2$	$N(octyl)_4BEt_3H$	2	23	Deep red to black Cu almost completely dissolved	Ether	Ethanol	77.04	
15	$CuBr_2$	$N(octyl)_4BEt_3H$	2	23	Deep red to black Cu completely dissolved	Toluene	Ethanol	52.15	

Source: H. Bönnemann et al., *J. Mol. Catal.* 86, 129 (1994).

[a] Solvent: toluene.

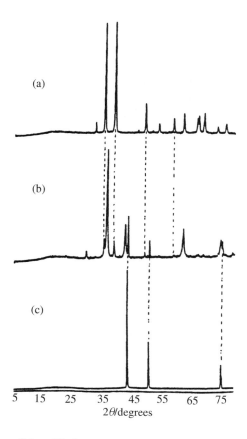

Fig. 9. XRD pattern of the solid phases present during the reduction of copper(II) oxide in ethylene glycol: (a) CuO precursor, (b) CuO + the intermediate Cu_2O + Cu, and (c) Cu obtained at the end of the reduction. Reprinted with permission from F. Fievet et al., *J. Mater. Chem.*, 3, 627 (© 1993 The Royal Society of Chemistry).

solid phases present during the course of the reaction that Cu_2O always exists as an intermediate crystalline phase. As a general rule, copper powders are made up of equiaxed particles whose size lies within the micrometer range. Some of these particles are single entities, but many of them are sintered with no definite shape. By similar reactions, the same group isolated cobalt and nickel in ethylene glycol by the reduction of $Co(OH)_2$ and $Ni(OH)_2$. The metals are recovered as a finely divided powder that appears in SEM as made up of particles with an isotropic shape and a size in the micrometer range. Almost perfectly spherical particles (Fig. 10a) can even be obtained in a mixture of diethylene glycol and ethylene glycol. The particles are usually homogeneous in size as shown in Figure 10b. The isotropically shaped particles have a narrow size distribution, owing to an effective separation between homogeneous nucleation and growth steps. Submicrometer-sized particles can be obtained either by raising the reaction temperature or by seeding the reactive medium with foreign metal nuclei, in order to induce a heterogeneous nucleation. The latter method is more convenient because the increase in temperature leads to degradation of the polyol and the heterogeneous nucleation allows, to some extent, control of the particle size in the submicrometer range.

A relatively new method for the synthesis of nanoscale metal/oxidized metal particles is described by Tsai and Dye [56] that utilizes homogeneous reduction of metal salts by dissolved alkalides or electrides in an aprotic solvent such as dimethyl ether or THF. Alkalides and electrides are crystalline ionic salts that contain either alkali–metal anions or trapped electrons. These compounds crystallize from solution to yield shiny bronze-

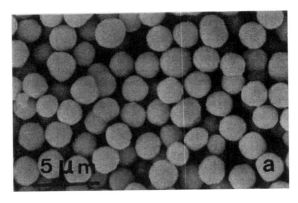

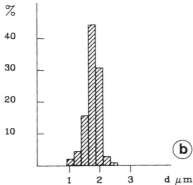

Fig. 10. (a) Cobalt powder obtained from $Co(OH)_2$ in a mixture of diethylene glycol and ethylene glycol. (b) Particle size distribution of the particles ($d_m = 1.75$ μm, $\sigma = 0.24$ μm). Reprinted from *Solid State Ionics*, F. Fievet et al., 32/33, 198 (© 1989), with kind permission of Elsevier Science-NL, Sara Burgerhartstraat 25, 105b KV Amsterdam, The Netherlands.

colored crystals (alkalides) or black crystals (electrides), which are all reactive toward air and moisture and thermally unstable at room temperature. Alkalides and electrides form M^- and e_{solv}^- when dissolved in a nonreducible solvent. These species are the strongest reducing agents that can exist in solution [57].

Soluble compounds of transition metals and post-transition metals in dimethyl ether or THF were rapidly reduced at $-50\,°C$ by dissolved alkalides or electrides to produce metal particles with crystallite sizes from less than 3 to 15 nm. Salts of Au, Cu, Te, and Pt formed metallic particles with little or no oxidation even when washed with degassed methanol. The reduction of salts of Ni, Zn, Ga, Mo, Sn, and Sb yielded surface oxidation over a metallic core. Stoichiometric amounts of the alkalide or electride were used and these were prepared either separately or *in situ*.

A typical reduction reaction follows the scheme [58]:

$$2AuCl_3 + 3K^+(15C5)_2e^- \rightarrow 2Au + 3K^+(15C5)_2Cl^- + 3KCl \tag{7}$$

or

$$AuCl_3 + 3K^+(15C5)_2e^- \rightarrow Au + 3K^+(15C5)_2Cl^- \tag{8}$$

where C5 = 15-crown-5 ether.

Only metallic gold peaks were detected by XRD from the precipitates after washing away the byproducts, $K^+(15C5)_2Cl^-$ and KCl, with water or methanol. The micrograph in Figure 11 agrees well with the particle size of about 100 Å as measured by the X-ray line broadening of gold produced in this reaction. The selected area diffraction (SAD) pattern

Fig. 11.   Electron micrograph of Au particles (bar = 330 Å). The SAD pattern in the upper right corner shows the cubic structure of gold. Reprinted with permission from K. L. Tsai and J. L. Dye, *Chem. Mater.*, 5, 540 (© 1993 American Chemical Society).

was also obtained from the same area and is shown at the top of the micrograph. The ring pattern confirms all of the *d* spacings of gold. In most cases, with all metals, a colloidal suspension is first produced as indicated by light scattering and color, followed by slow aggregation of the colloid to a precipitate that can be separated by centrifugation. The average particle sizes obtained were in most cases less than 3 nm.

Whereas the noble metals and some others such as Cu and Te can be separated from the byproducts, K^+X^- and $K^+(15C5)_2Cl^-$ ($X = Cl, Br, I$), by washing with methanol, more oxophilic metals were oxidized by this procedure. Also the inclusion of organic residues on the surface of the highly reactive metals could be removed in some metals by heating the samples to 100–150 °C under vacuum. Such procedures coupled with washing should improve the purity. With highly oxophilic metals such as Ti, it might not be possible to prevent oxidation.

A major advantage of this reduction method is that these reactions occur rapidly with homogeneous solutions. This has permitted the formation of intermetallic compounds or alloys when two metal salts are reduced simultaneously [59]. The strong reducing power (~ -3 V) of alkalides and electrides means that practically any soluble metal salt can be reduced to the metallic state.

Certain noble metals such as palladium and rhodium [60] have been prepared by the reduction of their respective inorganic salts in the presence of a polymer such as poy(vinylpyrrolidone) (PVP) or poly(vinyl alcohol) (PVA) and methanol as a reducing agent. The polymers were added because they can act as a protective agent to prevent the coagulation and precipitation of metallic particles [61]. More specifically, palladium acetate and rhodium(III) chloride were used as starting materials. Formaldehyde is produced quantitatively with the reduction of rhodium(III) chloride to metallic rhodium. The rhodium particles in the colloidal dispersion are found to be of two types, about 8 and 40 Å in diameter, by electron microscopy. The number of small particles, which form the large majority of particles at the early stage of refluxing, gradually decreases; concurrently, the number of large particles increases on prolonged refluxing. An absorption peak [by ultraviolet–visible (UV–vis) spectroscopy] appears at 260 nm at the early stage of refluxing and indicates that the coordination of poly(vinyl alcohol) to the rhodium(III) ion is indispensable for the formation of a homogeneous colloidal dispersion of rhodium. With the reduction of the rhodium(III) chloride to rhodium(0), methanol is oxidized to formaldehyde [62].

Similarly, when palladium acetate was refluxed with PVP in methanol, Bradley and co-workers produced colloidal palladium particles in the form of well-shaped microcrystallites with a mean diameter of 70 Å. This colloid shows no tendency to precipitate on standing for at least several weeks. X-ray diffraction shows the presence of fcc palladium, with line widths consistent with the particle size shown by TEM. CO adsorbs readily on the colloid in methanol, occupying only bridging sites (ν_{CO} 1944 cm^{-1}), similarly to CO on Pd (111) surfaces, and consistent with the crystalline nature of the colloid particles [63].

3.2. Synthesis of Intermetallics

Intermetallics are defined as solid solutions of two or more metals in varying proportions. The properties of the intermetallics are unique [64]. The general chemical synthesis methods used to make intermetallics are very similar to those used to produce individual metals. Therefore, this section will focus on the respective techniques, characterizations, and properties of the intermetallic systems. The synthesis and microstructural study of a four-component intermetallic, known as M50 steel [65] in our labs, will be discussed in detail.

3.2.1. Two-Component Intermetallics

Most commonly, the intermetallics are prepared by reduction reactions. For example, Buhro et al. [47] synthesized nanocrystalline powders of TiAl, TiAl$_3$, NiAl, and Ni$_3$Al by the reductions of TiCl$_3$ or NiCl$_2$ with LiAlH$_4$ in a mesitylene slurry followed by heating in the solid state ($\leqslant$550 °C). In the course of the reactions, Al and Ti or Al and Ni are initially precipitated in a segregated component phase, which, on subsequent heating, undergo an exothermic reactive-sintering process to give the nanocrystalline intermetallic. The microstructure of the air-sensitive aluminide powders consists of porous sintered aggregates of nanocrystallites, in which the primary nanocrystalline sizes are in the range of 25–35 nm. MoSi$_2$ was also prepared by the same group by sonochemical coreduction as described in the following reaction:

$$MoCl_5 + 2SiCl_4 + (13/2)NaK \rightarrow MoSi_2 + (13/2)NaCl + (13/2)KCl \qquad (9)$$

The ultrasonic irradiation formed a fine emulsion of the liquid NaK alloy, which maximized interfacial contact of the heterogeneous reactant phases and afforded a rapid, uniform coreduction reaction.

Co–Cu and Fe–Cu powders with varying metal concentrations have been prepared by the conventional method using sodium borohydride reduction of metallic salts [66]. As-synthesized Co–Cu powders showed a face-centered cubic (fcc) structure and an amorphous phase. The amount of amorphous phase was found to increase with the ratio of Co/Cu and grain sizes of the alloys were typically 300 Å. SEM and TEM revealed that the Fe–Cu powders were apparently agglomerated, as is frequently observed in ultrafine powders, and the grain sizes varied between 300 and 400 Å. Magnetic measurements confirmed that both the alloys and the composites of Fe–Cu were ferromagnetic at room temperature. The magnetometry curves showed that the metastable alloy possesses a low coercivity (ranging from 10–40 Oe).

Various other intermetallics such as TiB_2, Ni_2B, WC–Co, Co–B, Fe–B, Ni–B, and Pd–B have been prepared by reduction reactions using $NaBH_4$ as a reducing agent on different inorganic precursors such as $TiCl_4$, $NiCl_2 \cdot 6H_2O$, etc. The residual byproducts were either sublimed out of the mixture at elevated temperatures or washed away by organic solvents or water. Particles of varying sizes and morphologies were obtained, depending on the reaction conditions and the type of intermetallic being synthesized [47].

Some novel methods to prepare Ni and Pd boride colloids with core diameters of 1.4 and 1.6 nm have been described by Schmid and co-workers [67]. The particles are formed by the reaction of the metal complexes $Cl_2M(PR_3)_2$ (M = Ni, Pd, PR_3 = PPr_3, PBu_3) with B_2H_6 in toluene at room temperature with 40–70% yield. From the elemental analyses of the four colloids, the lowest formula units could be concluded as $Ni_6B_{10}Cl_{1.5}(PPr_3)$, $Ni_6B_{10}Cl_{1.5}(PBu_3)$, $Pd_4B_6Cl(PPr_3)$, and $Pd_4B_6Cl(PBu_3)$. Particle sizes of 1.6 nm were obtained by HRTEM (Fig. 12). X-ray powder investigations support the assumption that the particles are amorphous. ^{31}P nuclear magnetic resonance (NMR) investigations of the four colloids in solution resulted in singlets of free PR_3 molecules at -33 ppm and broad multiplets at ≈ -5 ppm indicating coordinated phosphines. This means that, as could be expected, at least parts of the ligands are dissociated in solution.

A thermochemical processing method for preparing high-surface-area powders starting from homogeneous precursor compounds has been reported. The method has been applied successfully to the synthesis of nanophase WC–Co powders [68]. The method is known as "spray conversion processing" and consists of three sequential steps: (1) preparation and mixing of aqueous solutions of the precursor compounds to fix the composition of the starting solution, (2) spray drying of the starting solution to form a chemically homogeneous precursor powder, and (3) thermochemical conversion of the precursor powder to the desired nanostructured end-product powder (Scheme IV). In a typical reaction, WC–Co powders were formed by thermochemical processing of a single chemical precursor compound, cobalt tris(ethylenediamine)tungstate $[Co(en)_3WO_4]$. The precursor is crystallized from solution and reduced in flowing argon/hydrogen to yield nanoporous/nanophase W–Co. This high-surface-area reactive intermediate was then converted directly to nanostructured WC–Co powder by gas phase carburization in flowing CO/CO_2. The resulting powder particles had the same morphology as the original $Co(en)_3WO_4$ particles, but the size of each particle was reduced by about 50%. In this process, the scale of the powder particle structure may be controlled from nanometer up to micrometer dimensions by adjusting the temperature of the carburization reaction, the residence time at a particular temperature, and the carbon activity of the gas phase. Alternatively, precursors such as $(NH_4)_6(H_2W_{12}O_{40}) \cdot 4H_2O$ and $CoCl_2$, $Co(NO_3)_2$, or $Co(CH_3COO)_2$ can be used to produce the desired nanopowders by the spray technique.

The alkalide and electride method of reduction used earlier for the synthesis of single metals can also be used for the preparation of alloys or compounds by the simultaneous reduction of two or more metal salts [56]. All systems tested (Au–Zn, Au–Cu, Cu–Te, and Zn–Te) yielded binary alloys or compounds. The Au–Cu system has been the subject of many investigations, and its phase diagram has been extensively studied. The most ideal stoichiometry of Au–Cu consists of alternate layers of Au and Cu atoms parallel to a cube face. The symmetry is slightly distorted to tetragonal with $c/a = 0.93$. The structure transforms to orthorhombic with $b/a = 10.03$ at about $380\,^{\circ}C$ [69]. The particles produced by the reduction of stoichiometric mixtures of $AuCl_3$ and $CuCl_2$ at $-50\,^{\circ}C$ have an SAD pattern that corresponds to a simple cubic pattern with each Au atom at the center of a cube of Cu atoms and vice versa.

The polyol reduction method has been used by Chow et al. [70] to produce nanocrystalline Co_xCu_{100-x} ($4 \leqslant x \leqslant 49$ at%) powders by refluxing metal acetates in a polyol. Typically, the powders were produced by suspending different proportions of Co(II) acetate tetrahydrate $[Co(O_2CCH_3)_2 \cdot 4H_2O]$ with Cu(II) acetate hydrate $[Cu(O_2CCH_3)_2 \cdot 4H_2O]$ in

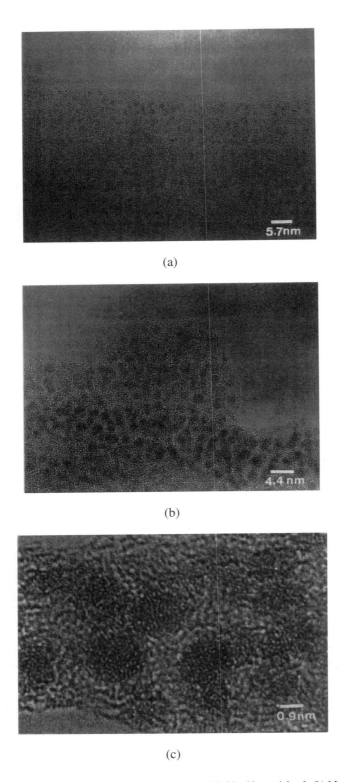

(a)

(b)

(c)

Fig. 12.   (a) Larger area covered with mainly uniform nickel boride particles **1**. (b) Magnified image of the same particles showing a size distribution of $\approx 1.6 \pm 0.3$ nm. (c) High-resolution image of a few particles of **1** indicating the amorphous structure. Reprinted with permission from G. Schmid et al., *Z. Anorg. Allg. Chem.*, **620**, 1170 (© 1994).

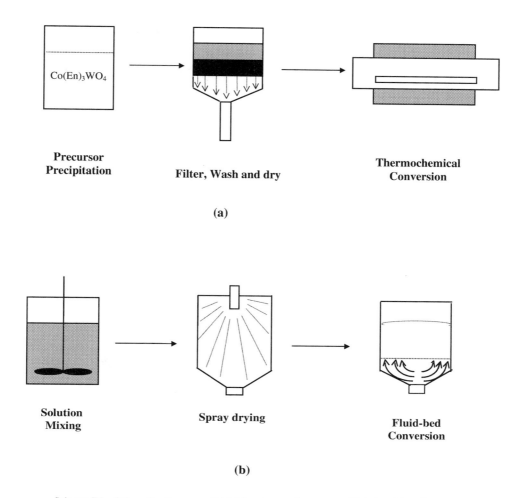

(a)

(b)

Scheme IV. Schematic diagrams of (a) laboratory scale process, (b) industrial scale process for the thermochemical processing of nanostructured powders, starting from aqueous solution mixtures. (Adapted from Nanostr. Mater. 3, 19–30 (1993), Kear et al.)

250 mL of ethylene glycol. The mixtures were refluxed at 180–190 °C for 2 h. During that time, Co–Cu particles precipitated out of solution. XRD data showed that the powders were crystalline. From the line broadening of the Cu(200) peaks, the crystallite size of Cu was in the range of 36 nm. The crystallite size of Co was estimated to be 18 nm from the Co(111) peaks for samples with $x \geqslant 30$. The morphology of the as-synthesized powders of Co_4Cu_{96} and $Co_{49}Cu_{51}$ is shown by TEM micrographs in Figure 13. All the powders appeared to be agglomerated, and the agglomerate size was typically 100 nm or larger. ^{59}Co spin-echo NMR spectra for the Co_xCu_{100-x} samples with $x = 4$, 19, and 49 have similar features in that there was a strong peak centered at 217 MHz and a broad shoulder in its high-frequency side with a small peak around 224–226 MHz. In addition, the spectra were very broad overall compared with the spectrum of pure Co metal. The strong 217-MHz peak and the absence of low-frequency resonances all indicated that the Co atoms existed as clusters rather than in a Co–Cu solid solution [71]. It is likely that the shoulder at the high-frequency side was due to either hexagonal closed-pack (hcp) Co grains or hcp stacking faults in fcc Co.

The magnetic properties of the samples were measured as a function of annealing temperature. The saturation magnetization (M_s) of as-synthesized powders with lower Co concentrations was much lower than those with higher Co content. Upon increasing the

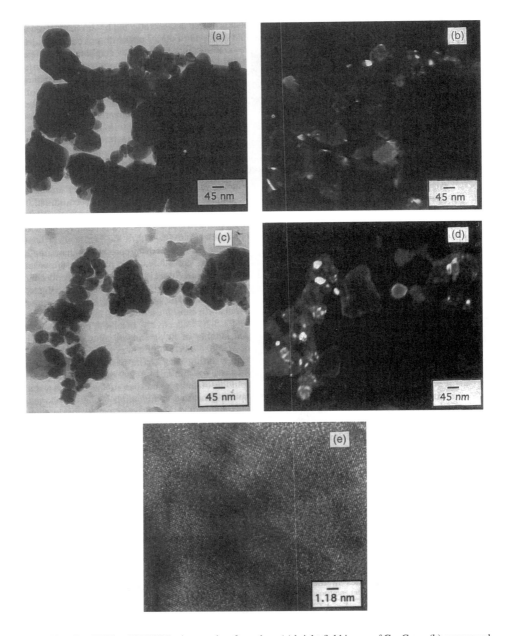

Fig. 13. TEM and HRTEM micrographs of powders: (a) bright-field image of Co_4Cu_{96}, (b) corresponding dark-field image of Co_4Cu_{96}, (c) bright-field image of $Co_{49}Cu_{51}$, (d) corresponding dark-field image of $Co_{49}Cu_{51}$, and (e) HRTEM showing the Co(111) lattice fringes in a $Co_{49}Cu_{51}$ sample. Reprinted with permission from G. M. Chow et al., *J. Mater. Res.*, 10, 1546 (© 1995 Materials Research Society).

annealing temperature, there was an increase in magnetization for samples with lower Co concentrations [72]. The increase in M_s could be due to precipitation of Co either from a metastable Co–Cu alloy or from the grain growth of superparamagnetic Co. Because HRTEM results of as-synthesized samples showed that alloying did not occur, the increase in M_s by annealing was attributed to the grain growth of Co clusters. It is notable that a coercivity as high as 370 Oe was observed for a nanocomposite with a low Co content (Co_4Cu_{96}) after annealing. $Co_{49}Cu_{51}$ showed a coercivity of 150 Oe and decreased to about 110 Oe at 650 °C.

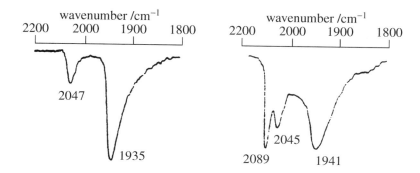

Fig. 14. Infrared spectra (CH$_2$Cl$_2$ solution) of carbon monoxide adsorbed on (a) PVP-stabilized colloidal palladium and (b) PVP-stabilized colloidal palladium–copper (Pd$_{37}$Cu$_{63}$). Reprinted with permission from J. S. Bradley et al., *Chem. Mater.*, 5, 254 (© 1993 American Chemical Society).

Bradley and co-workers [73] prepared Pd–Cu nanoscale colloids by heating mixtures of palladium acetate and copper acetate hydrate in 2-ethoxyethanol to reflux (135 °C) in the presence of poly(vinylpyrrolidone). The mean diameter of the particles varied in the range of 30–50 Å. EDAX analysis of the polymer films prepared with various PdCu ratios showed that the metal particles were bimetallic. Analysis of areas of the films between the colloid particles showed that no palladium(II) or copper(II) reamined unreduced. The observed electron diffraction rings were observed to be consistent with crystallinity, although the degree of crystallinity of the particles seemed to vary.

CO was adsorbed readily onto PdCu particles in dichloromethane at 25 °C, as shown by infrared (IR) spectroscopy (Fig. 14) on colloids with the composition Cu$_{63}$Pd$_{37}$ (particle size approximately 40 Å). On the bimetallic colloid, CO occupied both palladium and copper sites, as evidenced by the presence of IR bands at 2089, 2045, and 1941 cm^{-1}, demonstrating the presence of both metals at the surface of the particles [74].

Bönnemann and associates [48] have prepared several intermetallic colloids by the reduction of (1) two or more metal halides in the presence of hydrotriorganoborates, such as LiBEt$_3$H or NaBEt$_3$H (Table V summarizes the different products and some property characterizations of those powders), and (2) coreduction of inorganic salts such as metal halides in the presence of tetraalkylammonium-hydrotriorganoborates having alkyl groups preferentially of the chain lengths C$_6$–C$_{20}$. The NR$_4^+$ ions in these reactions stabilized the metal colloids by forming a protective shell around the metal ion and also helped in enabling the facile solvation of the colloids. The reactions occurring are similar to that described previously for the individual metal particles. A summary of the various intermetallics synthesized by this procedure is outlined in Table VI.

3.2.2. M50 Steel

M50 steel (4.0% Cr, 4.5% Mo, 1.0% V, and a balance of Fe) is widely used in the aircraft industry as the main-shaft bearings in gas turbine engines, because of its good resistance to tempering and wear and rolling contact fatigue. Conventional M50 steel consists of micrometer-sized carbide particles that can act as fatigue crack initiation sites in such bearing materials [75]. In contrast, an improvement in the mechanical properties of this material would be expected on reducing the grain size or, in other words, by making nanophase M50 steel.

Three different methods were used to synthesize the M50 nanopowders. The basic procedures are described in Scheme V and a detailed description of each procedure follows. After chemical synthesis, all the powders were subjected to an H$_2$ heat treatment at 420 °C to reduce the carbon and oxygen contents prior to compaction.

Table V. Preparation of Nanocrystalline Alloys by Coreduction of Metal Salts in THF

Number	Metal salt	Reducing agent	Condition		Product		DIF		Comments
			t (h)	T (°C)	Metal content (%)	Boron content (%)	2θ (°)	D (Å)	
1	Co(OH)$_2$ Ni(OH)$_2$	NaBEt$_3$H	7	65	Co: 48.3 Ni: 45.9	0.25	51.7	2.05	Single phase Nanocrystalline
2	FeCl$_3$ CoCl$_2$	LiH+10% BEt$_3$	6	65	Fe: 47.0 Co: 47.1	0.00	52.7	2.02	Single phase Grain size: 1–5 nm
3	FeCl$_3$ CoCl$_2$	LiBEt$_3$H	5	23	Fe: 54.8 Co: 24.5	0.00	52.5 99.9	2.02 1.17	Single phase Nanocrystalline
4	CoCl$_2$ PtCl$_2$	LiBEt$_3$H	7	65	Co: 21.6 Pt: 76.3	0.00	55.4 47.4	1.93 2.23	Single phase
5	RhCl$_3$ PtCl$_2$	LiBEt$_3$H	5	65	Rh: 26.5 Pt: 65.5	0.04	40.2 46.3	2.24 1.96	Single phase Grain size: 1–4 nm
6	RhCl$_3$ IrCl$_3$	LiBEt$_3$H	5	65	Rh: 33.5 Ir: 62.5	0.15	42.3	2.14	Single phase + Traces IrCl$_3$
7	PdCl$_2$ PtCl$_2$	LiBEt$_3$H	5	65	Pd: 33.6 Pt: 63.4	0.04	40.1 46.3	2.25 1.96	Single phase Grain size: 2–6 nm
8	PtCl$_2$ IrCl$_3$	NaBEt$_3$H	12	65	Pt: 50.2 Ir: 48.7	0.15	40.0 46.5	2.25 1.95	Single phase Nanocrystalline
9	CuCl$_2$ SnCl$_2$	LiBEt$_3$H	4	65	Cu: 49.6 Sn: 47.6	0.00	30.2 53.5	2.96 1.80	Cu$_6$Sn$_5$ + Cu + Sn
10	FeCl$_3$ CoCl$_2$ NiCl$_2$	LiBEt$_3$H	1.5	23	Fe: 30.1 Co: 31.4 Ni: 30.9	0.00	52.7 60.8 77.7 100.3	2.02 1.77 1.43 1.17	Single phase Nanocrystalline

Source: H. Bönnemann et al., *J. Mol. Catal.* 86, 129 (1994).

Table VI. Preparation of Colloidal Alloys

Number	Metal salt	Reducing agent	Condition t (h)	Condition T (°C)	Colloidal alloy solution colour	Workup solvent	Solvent added for precipitation	Metal content in isolated colloid (%)	Mean particle size (nm)
1	RhCl$_2$ PtCl$_2$	N(octyl)$_4$BEt$_3$H	18	50	Deep red to black Rh and Pt completely dissolved	Ether	Ethanol	Rh: 21.85 Pt: 45.96	2.3
2	PdCl$_2$ PtCl$_2$	N(octyl)$_4$BEt$_3$H	16	23	Deep brown to black Pd and Pt completely dissolved	Ether	Ethanol	Pd: 25.90 Pt: 33.60	2.8
3	CuCl$_2$ PtCl$_2$	N(octyl)$_4$BEt$_3$H	16	23	Deep red to black Cu and Pt completely dissolved	Ether	Ethanol	Cu: 15.60 Pt: 55.40	2.3
4	PtCl$_2$ CoBr$_2$	N(octyl)$_4$BEt$_3$H	18	23	Deep reddish brown to black Pt and Co completely dissolved	Toluene	Pentane/ethanol (25:1)	Pt: 25.40 Co: 6.47	—
5	NiBr$_2$ CoBr$_2$	N(octyl)$_4$BEt$_3$H	16	25	Dark red to black Ni and Co completely dissolved	ethanol	ether	Ni: 23.80 Co: 23.80	2.8
6	FeBr$_2$ CoBr$_2$	N(octyl)$_4$BEt$_3$H	18	50	Dark brown to black Fe and Co completely dissolved	ethanol	ether	Fe: 13.36 Co: 14.39	3.2

Source: H. Bönnemann et al., *J. Mol. Catal.* 86, 129 (1994).

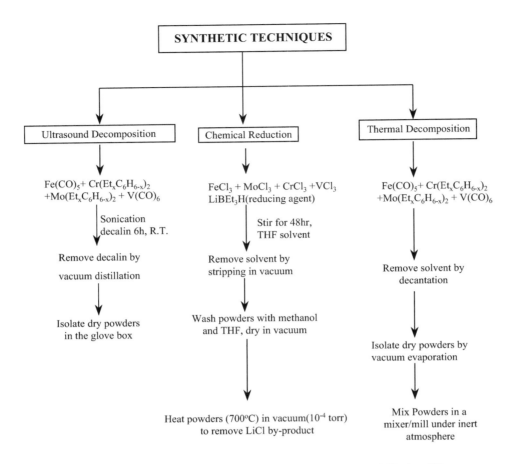

Scheme V. Schematic of synthetic procedures. (Adapted from Chemistry & Physics of Nanostructures and Related Non-Equilibrium Materials, TMS, p. 149, 1997, Gonsalves et al.)

3.2.2.1. Sonochemical Synthesis

A dispersion of 20 g $Fe(CO)_5$, 1.40 g $Cr(Et_xC_6H_{6-x})_2$, 1.03 g $Mo(Et_xC_6H_{6-x})_2$, and 0.063 g $V(CO)_6$ in dry decalin was sonicated at 50% of maximum vibration amplitude [sonochemical reactor horn (Sonics and Materials model VC-600, 20 kHz, 100 W cm^{-2})] for 6 h at room temperature in a sonochemical mercury bubbler. The color of the solution turned dark and then black within a few minutes and this reaction mixture was sonicated until the formation of shiny metallic particles was observed on the walls of the reaction vessel. The sonication was then stopped and the decalin solvent was removed from the reaction flask via vacuum distillation. Fine black powder (yield: 5.84 g) remained at the bottom of the reactor, which was then isolated and stored in a vial under nitrogen.

3.2.2.2. Coreduction Method

To a suspension of 10 g $FeCl_3$, 0.36 g $MoCl_3$, 0.46 g $CrCl_3$, and 0.11 g VCl_3 in 100 mL of THF, 202.98 mL of 1.0 M lithium triethyl borohydride in THF was added slowly while stirring at room temperature by a liquid addition funnel in the dry box. Slow efferves-cence was observed for a few minutes. After the reaction was stirred in the glove box for 48 h at room temperature, a black suspension was formed. The solvent THF was removed from the reaction flask by vacuum distillation and the black powders were washed with approximately100 mL of distilled degassed methanol until no further bubbling was ob-served. The fine black solid was washed with 50 mL of THF and dried under vacuum. The lithium chloride byproduct was removed from the preceding solid by vacuum sublimation in a tube furnace at $700\,^{\circ}C/10^{-4}$ torr. The yield of the powders after the sublimation was 3.38 g.

3.2.2.3. Thermal Decomposition

A dispersion of 20 g $Fe(CO)_5$, 1.40 g $Cr(Et_xC_6H_{6-x})_2$, 1.03 g $Mo(Et_xC_6H_{6-x})_2$, and 0.063 g of $V(CO)_6$ in dry decalin was refluxed for 6 h in a round-bottom flask fitted with a condenser and gas inlet and outlet tubes connected to a mercury bubbler. The color of the solution turned dark and then black within 3 h and this reaction mixture was left to reflux until the formation of shiny metallic particles was observed on the walls of the reaction vessel. The reaction was then stopped and the decalin solvent was decanted from the powders inside the glove box. A fine black powder (yield: 5.79 g) remained at the bottom of the flask, which was then thoroughly dried by gentle heating in vacuum, and then the dry powders were consequently mixed in a mixer/mill for 8 h under an inert atmosphere.

The morphology and the microstructure of the M50 steel powders produced by the three different techniques were examined by SEM, TEM, and XRD. The as-synthesized powders prepared by all the methods were amorphous and were agglomerated. Figure 15 shows the XRD spectra of a sonochemically synthesized M50 nanopowder. The broad peak [centered around 44–45° with full width at half-maximum (FWHM) of approximately 5° in 2θ units] is assigned to the main α-Fe reflection. The low processing temperatures involved in these methods prevented grain growth as well as full crystallization in the (Fig. 16) that they are highly agglomerated but consisted of smaller particles, approximately 4 nm in diameter.

M50 powders produced by the coreduction method, after workup of the reaction and washing with methanol and THF, were contaminated with the LiCl byproduct as shown in the EDAX spectrum (Fig. 17). The spectrum shows the presence of large amounts of LiCl as evidenced by the prominent chlorine peak. However, after sublimation of these powders at $700\,^{\circ}C/2$ h and a vacuum of 10^{-4} torr in a tube furnace, all the chloride byproduct is removed (Fig. 18) and the pure M50 nanopowders are isolated.

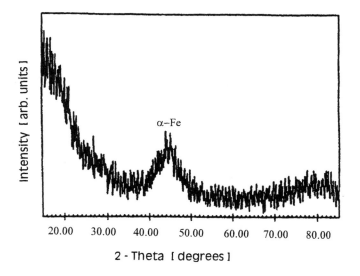

Fig. 15. XRD spectrum of as-synthesized steel powders (sonochemical method). Reprinted with permission from K. E. Gonsalves and S. P. Rangarajan, in "Chemistry and Physics of Nanostructures and Related Non-Equilibrium Materials" (E. Ma et al., eds.), p. 149. TMS Meeting, Orlando, FL (© 1997 Minerals, Metals & Materials Society).

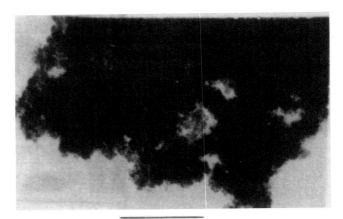

96 nm

Fig. 16. TEM micrograph of M50 steel powders (sonochemical method). Reprinted with permission from K. E. Gonsalves and S. P. Rangarajan, in "Chemistry and Physics of Nanostructures and Related Non-Equilibrium Materials" (E. Ma et al., eds.), p. 149. TMS Meeting, Orlando, FL (© 1997 Minerals, Metals & Materials Society).

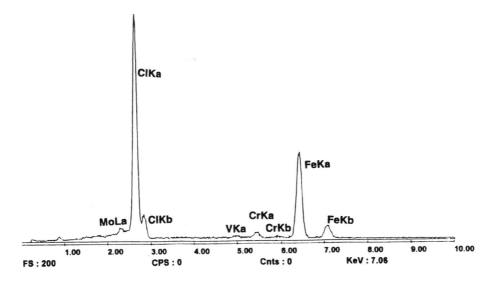

Fig. 17. EDAX spectrum of M50 steel powders (coreduction) before sublimation. Reprinted with permission from K. E. Gonsalves and S. P. Rangarajan, in "Chemistry and Physics of Nanostructures and Related Non-Equilibrium Materials" (E. Ma et al., eds.), p. 149. TMS Meeting, Orlando, FL (© 1997 Minerals, Metals & Materials Society).

The SEM micrograph of the coreduced powders taken after the heat treatment showed agglomeration and possibly sintering of the particles. This is indicated in Figure 19 by the presence of a chunky microstructure. Also the XRD spectrum of the powders taken after the heat treatment illustrated that the powders were crystallized and showed sharp peaks at the expected 2θ values of body-centered cubic (bcc) α-Fe, as well as the absence of any reflections from lithium chloride. This suggests that the sublimation process was effective in eliminating the powders of the chloride contaminant. The average grain size of these thermally treated coreduced powders as calculated from line broadening analysis was 34 nm ($\pm$3 nm).

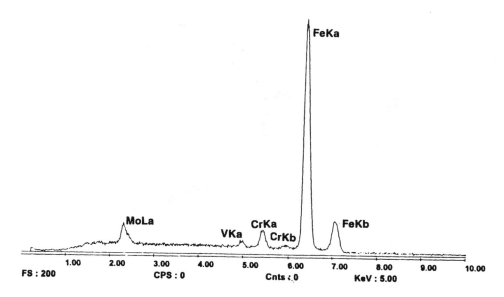

Fig. 18. EDAX spectrum of M50 steel powders (coreduction) after sublimation. Reprinted with permission from K. E. Gonsalves and S. P. Rangarajan, in "Chemistry and Physics of Nanostructures and Related Non-Equilibrium Materials" (E. Ma et al., eds.), p. 149. TMS Meeting, Orlando, FL (© 1997 Minerals, Metals & Materials Society).

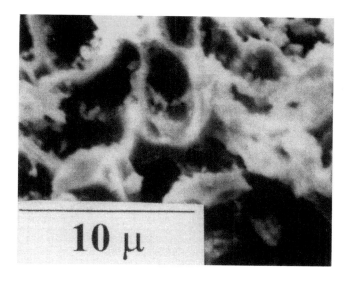

Fig. 19. SEM micrograph of powders (coreduction) taken after sublimation. Reprinted with permission from K. E. Gonsalves and S. P. Rangarajan, in "Chemistry and Physics of Nanostructures and Related Non-Equilibrium Materials" (E. Ma et al., eds.), p. 149. TMS Meeting, Orlando, FL (© 1997 Minerals, Metals & Materials Society).

The morphology of the powders produced by thermal decomposition was examined by SEM and their homogeneity was confirmed by EDAX analysis. These powders showed a porous coral-like morphology when observed in an SEM at 4400X as shown in Figure 20. Spot EDAX analysis at various locations on the micrograph (Fig. 21) showed the powders to be of homogeneous composition and of the expected stoichiometry for conventional M50 steel.

Fig. 20. SEM micrograph of powders (thermal decomposition). Reprinted with permission from K. E. Gonsalves and S. P. Rangarajan, in "Chemistry and Physics of Nanostructures and Related Non-Equilibrium Materials" (E. Ma et al., eds.), p. 149. TMS Meeting, Orlando, FL (© 1997 Minerals, Metals & Materials Society).

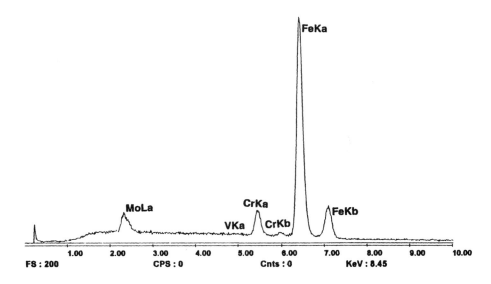

Fig. 21. EDAX spectrum of M50 steel powders (thermal decomposition). Reprinted with permission from K. E. Gonsalves and S. P. Rangarajan, in "Chemistry and Physics of Nanostructures and Related Non-Equilibrium Materials" (E. Ma et al., eds.), p. 149. TMS Meeting, Orlando, FL (© 1997 Minerals, Metals & Materials Society).

3.3. Synthesis of Semiconductors

The semiconductors being synthesized by chemical methods are mainly those of the group II–VI and III–V binary systems [76]. Frequently, these clusters are prepared in the form of dispersed colloids or can be trapped and stabilized within micelles, polymers, zeolites, or glasses [77]. The chemical synthesis is divided into categories according to the host material in which the semiconductor is created or embedded. In this section, we discuss the various methods that have been adopted to prepare single-size clusters, an important goal in this area.

3.3.1. Colloids/Micelles/Vesicles

There are several examples in the literature for the synthesis of semiconductor particles. The stabilization of a colloid in the small-cluster size regime requires an agent that can bind to the cluster surface and thereby prevent uncontrolled growth into larger particles. The simplest method to prepare these colloids involves using a solvent to act as a stabilizer of the small clusters; for example, base hydrolysis of a solution of a zinc salt results in the generation of ZnO in alcohol solvents [78]. On completion of this reaction, a transparent colloid is produced where the ZnO particles increase in size on standing. In a variation of this approach, that is, by a combined solvent/anion stabilization method followed by hydrolysis using LiOH, extremely high concentrations and very stable solutions of highly luminescent ZnO particles in ethanol have been produced.

Such colloids can also be made by the use of a polymeric surfactant/stabilizer that is added to a reaction designed to precipitate the bulk material. The polymer attaches to the surface of growing clusters and by either steric or electrostatic repulsion prevents the further growth of the nanoclusters. The most commonly used polymer is sodium polyphosphate (hexametaphosphate), and clusters of CdS, CdTe, and ZnTe with this surfactant have been studied [79].

The addition of anionic agents frequently called capping agents to the solutions of growing clusters prevents further growth of these materials by covalently binding to the cluster surface. Thiolates are the most commonly used capping agents and this method also forms the basis of synthesis of the monodispersed clusters described later on in this chapter. Figure 22 shows the synthetic strategies of producing clusters by this approach. This technique can be thought of as mimicking an organic polymerization reaction (initiation, propagation, and the termination phases) and is depicted for CdS in the Figure 22. Mixing the cadmium and sulfide ions initiates the polymerization, and the growth of the CdS clusters can be viewed as propagation steps that are sustained by the presence of additional cadmium and sulfide ions. The growth of the clusters can be terminated by providing a capping agent, in this case a thiophenolate ion that intercepts the growing clusters by binding to the cluster surface [80]. The average cluster size can then be controlled by adjusting the ratio of the sulfide to thiophenolate in the reaction solution. It has been found that thiophenolate-

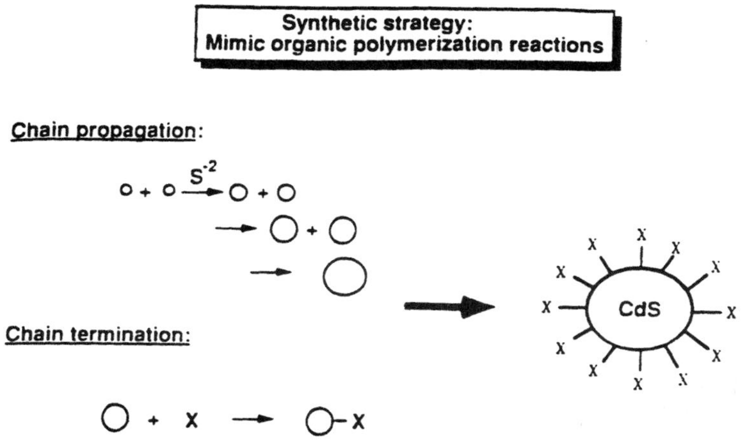

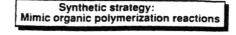

Fig. 22. Schematic diagram for the synthesis of the thiophenolate-capped CdS clusters (X represents the thiophenolate ion) drawing analogy to organic polymerizations. In the chain propagation step, the size of the circle represents the size of the CdS clusters. The growth of the CdS clusters is analogous to the growth of a polymer chain as long as the surfaces are not covered by the terminating agent, X. Reprinted with permission from N. Herron and Y. Wang, in "Nanomaterials: Synthesis, Properties and Applications" (A. S. Edelstein, ed.), p. 73 (© 1996 IOP Publishing, Bristol, U.K.).

capped CdS clusters act somewhat like living polymers that keep growing if fed on more sulfide ions [80].

Other capping agents that have been used are glutathione peptides produced by yeast for the capping of small CdS clusters. The generation of GaAs in THF solution or glycol ether may also be categorized as this type of a reaction [81]. The pentamethylcyclopentadienyl ligands on the Ga ions and the trimethylsilyl groups on the As ions regulate cluster growth and maintain GaAs colloids in solution. In another case, the acetylacetonate groups on the Ga ions and the trimethylsilyl groups on the As ions control cluster dimensions [82].

In similar approaches, micelle-forming reagents have been widely used as a method of controlling cluster growth. In this method, the semiconductor is precipitated in a small region of space defined by the micelle. In contrast to the colloidal approach, the micellar reagent acts as a physical boundary rather than a surface-capping agent. There are two approaches in the micellar technique: (1) normal micelles and (2) reverse micelles. Normal micelles using dihexadecyl phosphate or dioctadecyldimethylammonium chloride (DODAC) are generated in water to limit particle sizes between 150 and 300 nm [83]. The cadmium or zinc ions are dissolved in these clusters and precipitated with H_2S leading to clusters of up to 50 Å in diameter in these micelles. Reverse micelles using bis(2-ethylhexyl)sulfosuccinate salts (AOT) allow the formation of small water pools (<100 Å in radius) in heptane solvent, and again incorporation of metal ions followed by chalcogenide treatment can precipitate semiconductor clusters within these pools. Selenophenol-capped CdSe or CdSe/ZnS clusters have been developed by a combination of the reverse-micelle and the surface-capping approaches [84].

3.3.2. Polymers

The preceding methods have major problems of irreproducibility and colloidal instability, as well as the problem of good characterization. Also, it is often necessary to synthesize the semiconductor nanoclusters in solid thin-film form for use in practical applications. By using polymers as a matrix for the semiconductors, these problems can be alleviated. This method was first used with CdS in Nafion and PbS in Surlyn. These stable clusters were then identified by X-ray diffraction [85].

Composites of these clusters have been produced by several methods. The first is by ion exchange. Polymers such as Nafion or ethylene-methacrylic acid copolymer have cation exchange sites where Cd or Pb ions may be introduced into the polymer matrix (Fig. 23).

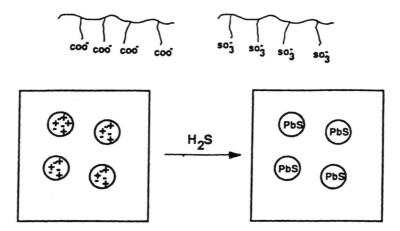

Fig. 23. Schematic showing the synthesis of PbS clusters in ionomers: Surlyn (top left) and Nafion (top right). Pb^{2+} irons are first exchanged into the ironic domains and then reacted with H_2S to form PbS clusters. Reprinted with permission from Y. Wang and N. Herron, *Res. Chem. Intermed.*, 15, 17 (© 1991 VSP BV).

Treatment of such ion-exchanged films with chalcogenide sources results in *in situ* precipitation of the compound semiconductor within the hydrophilic regions of the polymer. The ionic polymers can be considered as the solid-state analogs of micelles. The size of the trapped clusters is controlled by the appropriate control of the phase-separated hydrophilic and hydrophobic regions of the polymer [86]. Layered semiconductor clusters such as PbI_2, as well as magnetic particles such as Fe_2O_3, have been developed by this method [87].

Using ring opening methasesis polymerization (ROMP), Schrock et al. [88] incorporated the Pb^{2+} metal ion into monomer units that eventually became part of a norbornene-derived copolymer. In this case, the polymer is designed to phase separate into size-controlled regions of high and low hydrophobicity. The semiconductor is precipitated in these regions by gas phase treatment with H_2S.

Another approach that has been reported uses a combination of the polymer isolation and the surface-capped cluster approaches. A well-defined semiconductor cluster or colloid is prepared by the conventional capping technique described previously and is then dissolved in a solvent along with a soluble polymer. This mixed solution may then be spin coated onto a substrate and dried to produce a polymer film doped with the semiconductor particles. The CdS semiconductor has been incorporated into a photoconductive polymer such as polyvinylcarbazle using this method [89].

There are examples of the synthesis of CdS and CdSe nanoparticles stabilized by polymers in the literature. Yao et al. [90] first prepared CdS and CdSe particles by the reaction of $Cd(NO_3)_2 \cdot 4H_2O$ or $Cd(ClO_4)_2 \cdot 6H_2O$ with H_2S and H_2Se in the presence of PVP. In this case, PVP plays an essential role in preventing the flocculation of concentrated nanoparticles and controlling the particle size. The absorption spectra of CdS nanoparticles prepared with PVP (2.0 g/L) in acetonitrile showed a sharp peak at about 330 nm, and the particle diameters estimated from the absorption peak using tight binding approximation were about 2.0 nm. Similar blue-shifted absorption spectra of CdSe nanoparticles prepared in the presence of PVP in ethanol were also observed. The PVP-coated CdS and CdSe nanoparticles obtained were dispersed in acrylonitrile-styrene copolymer or in poly(2-hydroxyethyl methacrylate) (PHEMA) organic films, and the third-order nonlinear optical susceptibilities $\chi^{(3)}$ of these films were measured. A $\chi^{(3)}$ yield of 1.1×10^{-7} esu was obtained for a 4.0-vol% CdS-doped PHEMA film, and a value of 3.0×10^{-11} esu was obtained for a similar CdSe analog. CdS particles (2.5 nm in size) were also prepared by reactions of cadmium acetate, thiourea, and thioglycerol (which acts as a capping agent) by Chevreau et al. [91], who then dispersed them in polystyrene and cast films. The UV–vis spectrum of CdS thin films that were heated to $100\,^\circ C$ showed excitonic features at 375 nm. Luminescence spectra for the spin-coated films (Fig. 24) showed a maximum intensity around 540 nm. The transition of semiconductor chalcogenides from molecular CdX (X = Se, S) to bulk material has been observed in Nafion, a cation-exchanged membrane. The CdS and CdSe clusters were prepared by sonicating aqueous solutions of $CdCl_2 \cdot 4H_2O$ with the protonated Nafion. After drying, the samples were exposed to 1 atm of H_2S or H_2Se on a vacuum line for 10 h. TEM images indicate that CdS particle sizes varied between 2 and 4 nm. Exposure of the CdS and CdSe films to water caused a shift to longer wavelengths in the absorption spectra owing to Ostwald ripening and possible increase in grain size.

ZnO has many applications in luminescent devices, photocatalysis, and photoelectrochemistry [92]. A promising approach in nanocluster synthesis from a practical viewpoint is to use stabilizing agents to grow the particles in structured media. Highly stable, wurtzite quantum-sized ZnO colloids have been encapsulated in polymers such as PVP, PVA, and sodium hexa-meta phosphate (HMP) [93]. Freshly prepared uncapped ZnO clusters exhibited a blue shift in absorption, with the excitonic shoulder appearing at about 300 nm compared to the band edge for macrocrystalline ZnO at 365 nm. PVP-protected clusters showed an excitonic hump at about 310 nm. XRD revealed the particle sizes were 2.8 nm for the uncapped sample and 3.5 nm for the PVP-protected sample. It is, therefore, obvious

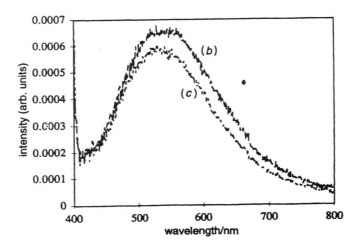

Fig. 24. Detail of luminescence spectra for 500 (b) and 1000 (c) rpm films showing the same emission linewidths. Reprinted with permission from A. Chevreau et al., *J. Mater. Chem.*, 6, 1643 (© 1996 The Royal Society of Chemistry).

that these clusters will exhibit quantum-sized effects as the Bohr exciton radius of ZnO is about 4.0 nm [94].

3.3.3. Glasses

Particles of CdS_xSe_{1-x} can be embedded in a borosilicate glass matrix and have traditionally been used as color filters [95]. These materials are prepared by traditional glass-making technology where Cd, S, and Se are added to a silicate or germanate glass melt at elevated temperatures. After casting, the glass is annealed at temperatures less than the melting point and the small semiconductors form within the dense glass matrix [96]. This technique is suitable for only a limited number of semiconductors as they must survive a very high temperature oxidizing environment during the glass-forming step. Thus, this method is not appropriate to make III–V semiconductor clusters of GaAs because of their thermal and oxidation sensitivity.

Rajh et al. [97] have reported size quantization effects in specially prepared silicate glasses containing small particles of CdS, CdSe, Bi_2S_3, PbS, HgSe, In_2Se_3, and AgI. The procedure involved mixing an aqueous colloidal solution of the semiconductor with tetramethoxysilane (TMOS) and accelerating the polymerization of SiO_2 with NH_4OH. In the starting colloidal solutions, the particle sizes were determined by electron microscopy and were in the 20–40-Å range. It was found that the luminescence intensity of colloidal semiconductor particles in silicate glasses is much higher compared to that of the same particles in aqueous solutions. The absorption and emission spectra for AgI ($\sim$100 Å) and CdS ($\sim$30 Å) are shown in Figure 25. The quantum yields for AgI and CdS in silicate glasses were 2% and 5%, respectively. The higher emission yields in silicate glasses could occur through passivation by the silicate polymer of surface states.

Minti et al. [98] have used a sol–gel method to prepare quantum dots of CdS in thin glass films, using tetraethoxysilane (TEOS), water, alcohol, and cadmium nitrate as precursors. Organically modified ceramic (ORMOCER) films were also prepared by the same group according to the procedures of Schmidt [99]. TEM micrographs of CdS particles in the glass films had sizes of 20–50 Å. Third-harmonic-generation (THG) signals from the CdS-doped thin glass and the ORMOCER films were easily observed. The strongest signals were about 1500–2000 times stronger than those generated by the glass substrate, and $\chi^{(3)}$ was measured to be 2×10^{-12} esu for the sol–gel CdS films and for the ORMOCER film it was approximately 10^{-12} esu. These values of $\chi^{(3)}$ are much lower than those of

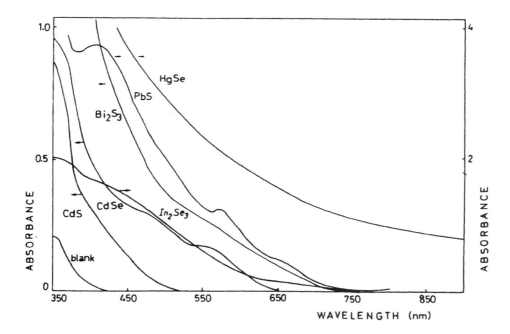

Fig. 25. Absorption spectra of various colloids incorporated in glasses. The monomer concentration of the starting colloids was 4×10^{-4} M and the stabilizing agents were: (a) 2×10^{-3} M polybrene for CdS and AgI, (b) 0.1% carbowax for CdSe and In_2Se_3, (c) 4×10^{-4} M hexametaphosphate for Bi_2S_3, (d) 0.1% PVA (MW 14,000) for PbS, and (e) 0.1% carbowax for HgSe. The absorption spectrum of the glass itself is also presented as blank. The optical path was 0.5 cm. Reprinted from *Chem. Phys. Lett.*, T. Rajh et al., 143, 305 (© 1988), with kind permission of Elsevier Science-NL, Sara Burgerhartstraat 25, 105b KV Amsterdam, The Netherlands.

pure CdS, but still quite high considering the low volume fraction of CdS particles in the film [100].

III–V semiconductors, because of their thermal and oxidation sensitivity, have been prepared by the selective acid leaching of borates from a borosilicate Vycor glass. The physical space restraint imposed by the pore sizes of the glass controls the size of the semiconductor that is precipitated *in situ* within the pores of this glass at low temperatures. Examples of II–VI and III–V semiconductors in the literature have been previously reported [101]. Sol–gel-derived porous silica glass has also been used for the production of these materials. In this case, the semiconductor particles were assembled within the pore structure of the host glass matrix and the residual porosity was then back filled with a polymer such as poly(methylmethacrylate) (PMMA) [102]. This construction strategy reduces light scattering, which is frequently observed from the empty pore network, and, therefore, makes the composite have more desirable optical properties.

A glassy thin film of TiO_2 nanoparticles has been reported by electrostatic spraying of an alkoxide solution. Radio frequency (rf) sputtering and gas evaporation techniques have also been used to produce small semiconductor clusters, which are sometimes doped in glasses. These methods are very useful for producing high-quality thin-film samples [103].

3.3.4. Crystalline and Zeolite Hosts

The use of a crystalline host lattice for the stabilization of semiconductor nanoclusters is an attempt to impose crystalline order and physical size constraints upon the included semiconductor particles. CuCl crystallites, ranging from 20 to 50 Å, have been prepared within single crystals of NaCl or KCl by a melt procedure and they exhibited exciton absorption features that were blue shifted from the bulk material [104]. This approach involves the use of a porous crystalline host, zeolite, into which semiconductor clusters are either directly

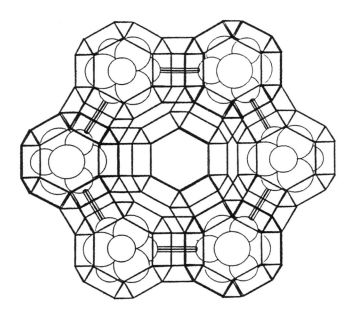

Fig. 26. Cd_4S_4 cluster array in the sodalite cages of zeolite Y. Circles represent CD_4S_4 clusters (smaller circles are Cd and larger circles are S). Six sodalite cages (occupied by Cd_4S_4) and one supercage (at the center, empty) are shown here. The sodalite cage has a diameter of ~ 5 Å and the supercage has a diameter of ~ 13 Å. Reprinted with permission from N. Herron and Y. Wang, in "Nanomaterials: Synthesis, Properties and Applications" (A. S. Edelstein, ed.), p. 73 (© 1996 IOP Publishing, Bristol, U.K.).

imbibed or created *in situ*. Examples of the former technique are the sublimation of Se or Te into aluminosilicate zeolites [105]. CdS and superclusters (cluster arrays) have been generated in zeolite Y by the second method, which is described next [106].

Zeolite Y occurs in nature as the mineral faujasite and consists of a porous network of aluminate and silicate tetrahedra linked through bridging oxygen atoms [107]. It has two types of cavities within the structure—the sodalite cavity of ~ 5 Å diameter with access through ~ 2.5-Å windows and the supercage of ~ 13 Å diameter with access through ~ 7.5-Å windows as shown in Figure 26. These well-defined and ordered cavities provide an ideal environment for the synthesis of single-sized clusters and cluster arrays, whereas the windows provide access for transporting reagents to the cavities. The ion-exchange properties of the zeolite are produced as a result of their chemical framework and are useful in producing the clusters. For example, CdS clusters were synthesized in zeolite Y by the ion-exchange method. Synchrotron X-ray, extended X-ray absorption fine structure (EXAFS), and optical absorption data reveal discrete $(CdS)_4$ cubes located within the small sodalite units of the structure (see Fig. 26). The Cd_4S_4 units are not isolated and they interact with each other via through-bond interactions to form a supercluster (or cluster array). The absorption spectrum shifts from approximately 290 nm for the isolated clusters to approximately 360 nm in the supercluster [108].

The zeolite confinement approach has been extended to include many other semiconductor guests and sodalite hosts [109]. AgI in zeolite mordenite has been shown to have unusual optical behavior in terms of its photosensitivity. Silver halides in the sodalites have been synthesized by Ozin et al. [110]. PbI_2 in X-, Y-, A-, and L-type zeolites shows evidence of a strong size effect on the exciton absorption. Cao et al. [111] investigated a variant of this approach and used layered crystalline hosts of the metal phosphonate series instead of zeolites.

The formation and the photophysical properties of CdS in zeolites with cages and channels such as the zeolite sodalites, A, X, chabazite, and offrretite, were investigated by Liu and Thomas [112]. The sizes of the clusters are constrained by the sizes of the cages. The results showed that the CdS clusters form in the biggest cages of the zeolites and in the main channel of the zeolites with channels. Absorption spectra of the CdS formed in these two types of systems show different features, indicating different particle sizes and distributions. The former covers a rather narrow range of wavelength and the latter covers a wider range, as the particles of CdS are smaller than and of a narrower size distribution in the former confined regions than in the latter.

Semiconductor superclusters of the preceding type represent a novel class of materials where the three-dimensional structure and electronic properties can be controlled by using different zeolites as the template [113]. The unique stability of the semiconductor clusters within these zeolite units is due to the coordination of Cd atoms with the framework oxygen atoms of the zeolites.

The zeolites offer the unique advantage of preparing three-dimensional arrays of mutually interacting clusters with geometric structures imposed by the zeolite internal pore structure. The electronic properties of the cluster array are controlled by the different spatial arrangements of the clusters, which, in turn, are controlled by using different zeolites as the template. However, imperfections still exist in this procedure, and the further success of this method will depend on the availability of high-quality zeolite single crystals as well as zeolites with larger pore sizes.

3.3.5. Synthesis of Single-Sized Clusters

It is imperative to produce clusters with a very narrow size distribution in order to realize their electrical and optical properties. Herron and Wang [26] have observed in their study of thiophenolate-capped CdS clusters that clusters can be grown simply by adding extra sulfide to the cluster solution. Based on this observation, it seems that if starting with a well-defined small molecular cluster, larger single-sized clusters could be obtained by the addition of reagents selected to cement the smaller units together. The $(NMe_4)_4Cd_{10}S_4SPh_{16}$ cluster synthesized by Dance and co-workers [114] provides an ideal candidate as a starting material. This compound belongs to a series of molecular clusters synthesized by Dance and co-workers, who produced crystalline, well-characterized semiconductor molecular clusters. The compounds $(NMe_4)_4Cd_{10}S_4SPh_{16}$ and $(NMe_4)_2Cd_{17}S_4SPh_{28}$ contain a crystalline core having the same atomic range as the bulk cubic phase of CdS and have particle dimensions in the 7–9-Å range [115]. These clusters are very soluble in polar organic solvents and have been used to synthesize single-sized clusters by two different approaches.

3.3.5.1. Controlled Cluster Fusion in Solution

In the work of Dance and associates [114], a molecular fragment of CdS containing a $Cd_{10}S_4$ core was capped by 16 thiophenolate groups. Although these syntheses are carried out in solution, the clusters can be collected as stable solids and redissolved. The reaction between the cadmium and the chalcogenide ions can be considered as an example of an inorganic polymerization with an initiation step forming a small nucleus, which then grows larger in a propagation step before precipitating in the termination step. It can be anticipated that the growth of the clusters can be controlled by the relative kinetics of attachment of the growing clusters by the propagation and termination steps, respectively.

The cluster fusion idea has been demonstrated by Dance et al., who synthesized the cluster $(Cd_{20}S_{13}(SPh)_{22})^{8-}$ just by the addition of extra sulfide ions to $(Cd_{10}S_4(SPh)_{16})^{4-}$:

$$2(Cd_{10}S_4(SPh)_{16})^{4-} + 5S^{2-} \rightarrow (Cd_{20}S_{13}(SPh)_{22})^{8-} + 10SPh^- \qquad (10)$$

The resultant 55-atom cluster showed a very sharp (FWHM $\sim$800 cm^{-1}) absorption peak at 351 nm (3.53 eV) and a weak shoulder at 330 nm (3.76 eV) at room temperature (Fig. 27). The clusters can be collected as stable solids and redissolved in solutions.

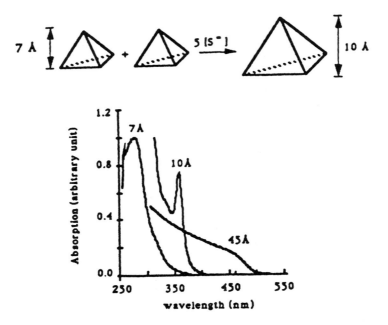

Fig. 27. The absorption of 7-Å $(Cd_{10}S_4(SPh)_{16})^{4-}$ clusters, 10-Å $(Cd_{20}S_{13}SPh_{22})^{8-}$ clusters, and 45-Å CdS clusters in Nafion film. Reprinted with permission from Y. Wang and N. Herron, *Res. Chem. Intermed.*, 15, 17 (© 1991 VSP BV).

By exercising care in controlling the mixing conditions (low concentrations and vigorous mixing) and component ratios during these types of syntheses, it is also possible to generate solutions of CdSe and CdTe clusters that are all capped by thiophenolate ion and that exhibit very sharp absorption features at the band edge in their electronic spectra. It is hoped that these controlled cluster fusion reactions coupled with various chemical separation techniques can eventually lead to the synthesis of single-sized clusters in the condensed phase.

3.3.5.2. Controlled Thermolysis

Thermolysis of the compound $(NMe_4)_4Cd_{10}S_4SPh_{16}$ in inert atmospheres leads to the loss of thiophenolate capping groups and the eventual formation of bulk CdS beginning at approximately 350 °C [116]. At an intermediate temperature of approximately 250 °C where a mass loss corresponding to 4 (NMe_4SPh) is observed, a material of stoichiometric $Cd_{10}S_4SPh_{12}$ can be isolated. This cluster has an 82-atom tetrahedral core of cubic phase CdS with an overall tetrahedral shape. This represents the largest crystallographically characterized semiconductor cluster and presents opportunities for the definitive assessment of optical and electronic properties for a monodispersed cluster size.

3.3.6. Nanostructured III–V Semiconductors

Binary III–V materials have recently attracted significant interest [117]. This is due to their wide band gap energy (3.1–3.8 eV), which makes them suitable for creating and processing blue light. Eventually, devices based on these materials will bring advances in the fields of communications and computing. However, most of the research on III–Vs has concentrated on vapor-deposited thin films.

Research has also been concentrated [118] on materials with large third-order nonlinear optical susceptibility ($\chi^{(3)}$). Third-order optical susceptibility ($\chi^{(3)}$) refers to the dependence of the polarizability on the third power of the optical electric vector (**E·E·E**).

One possible morphology for a $\chi^{(3)}$ material is a nanostructured composite with a polymer matrix. Optical nonlinearity arises in nanostructured systems because of the quantum confinement of the particle's electrons. Gallium nitride is a material known [119] to exhibit a significant $\chi^{(3)}$.

Nanostructured GaN has been chemically synthesized by Gladfelter's group by the pyrolysis of cyclotrigallazane ($[H_2GaNH_2]_3$) at 600 °C in argon [120]. The XRD of the heat-treated samples showed a diffraction pattern corresponding to a mixture of hexagonal and cubic gallium nitride. On the basis of X-ray line broadening and TEM, the particle size was found to be 60 Å. This nanocrystalline material, however, converted into the known wurtzite phase at 900 °C. The same group also synthesized nanocrystalline wurtzite GaN from the thermal decomposition of Ga_2O_3 and NH_3 at 600 °C.

Janik et al. [121] have synthesized gallium nitride powders by the pyrolysis at 450–500 °C under NH_3 or vacuum of a polymeric gallium imide $\{Ga(NH)_{3/2}\}_n$. This precursor is, in turn, prepared by the ammonolysis of $[Ga(NMe_2)_3]_2$ and NH_3 at ambient temperatures. The gallium imide precursor was shown to yield, upon pyrolysis, a rare cubic/hexagonal variety of GaN. On the basis of the relative broadness of the XRD powder patterns (Scherrer's equation) and the TEM results, the solid from the pyrolysis at 500 °C under NH_3 consisted of larger and less disordered GaN crystallites ($\sim$7 nm) than the solid from the pyrolysis at 450 °C under vacuum ($\sim$2 nm).

Another thermal synthetic route to nanocrystalline GaN was demonstrated by Xie et al. [122], who synthesized hexagonal GaN by the reaction of Li_3N with $GaCl_3$ in benzene at 280 °C under pressure in an autoclave. The XRD pattern obtained could be indexed to the hexagonal cell of GaN with lattice constants $a = 3.188$ Å and $c = 5.176$ Å, which are near the reported values. Some minor reflections can also be indexed to cubic GaN in a rock salt structure with $a = 4.006$ Å. The GaN crystallites have an average size of 32 nm and display a uniform shape. The photoluminescence (PL) spectrum of these nanoparticles shows one broad emission feature at 370 nm, which is in agreement with that of bulk GaN. Hexagonal nanocrystalline GaN, ranging in size from 5 to 10 nm, was produced by the controlled thermal decomposition of $[Ga(N_3)_3]_\infty$.

In our laboratory, we have synthesized nanostructured gallium nitride (GaN) and GaN/PMMA composites and have characterized them by HRTEM, chemical analysis, XRD, and image analysis. The deagglomeration of the initial GaN product, and its dispersion into a polymer matrix, yielding a composite with 3–8-nm-diameter particles, is also described.

The dimeric precursor, $Ga_2[N(Me)_2]_6$, was synthesized by the reaction of gallium chloride, $GaCl_3$, with lithium dimethylamide, $LiNMe_2$ [123]. A slight excess (6.44 g) of $LiNMe_2$ was reacted with 7.17 g $GaCl_3$ at room temperature for 2 days in dried, distilled hexane. After stirring, the solution was filtered through Celite and the volatiles were removed *in vacuo* to yield a colorless solid. The pale-yellow pure dimer was isolated by vacuum sublimation at 110 °C. The LiCl byproduct remained on the frit and was discarded. The yield was 3.82 g (71%).

The precursor dimer was placed in an alumina boat and decomposed using a glass tube with an inside diameter (i.d.) of 1 in. placed in a Thermolyne 21100 furnace with a 12-in. heating zone. The material was heated for 4 h at 600 °C. NH_3 flow at ambient pressure was maintained during heat treatment and while the material cooled to room temperature over several hours. The product, a grayish powder, was handled only under an inert atmosphere to prevent oxidation.

To form the nanostructured composite, 60 mg of this powder and 1 mL of methyl methacrylate were charged to a flask under argon. The mixture was sonicated in a cleaning bath for 2 h. The liquid was then decanted and placed in a sealed vial with an initiator (azobisisobutylnitrile, $\sim$5 mg). This mixture was polymerized thermally at 72 °C for 50 min. From the resulting solution, films were spin cast onto clean quartz substrates. The speed was 225 rpm and the time was 5 min. Film thicknesses, measured by profilometry, were approximately 10 μm.

Powder samples of GaN were examined by XRD under mineral oil to inhibit oxidation. The analysis was performed on a Scintag XDS-2000 diffractometer equipped with a Cu K$_\alpha$ source, $2\theta = 0$–70° at a rate of 2°/min. Duplicate chemical analyses were performed (Galbraith Labs) to determine gallium and nitrogen. The results indicate that the material is somewhat nitrogen poor, with an empirical formula of GaN$_{0.86}$. In a separate analysis (Microlytics), carbon and hydrogen contamination were found to be 0.8 and 0.5 wt%, respectively.

The powder sample was examined using a JEOL-4000EX electron microscope with an accelerating voltage of 400 kV and a point-to-point resolution of approximately 1.7 Å. HRTEM images were obtained at optimum (Scherzer) defocus. To prevent any contamination, the sample was prepared for HRTEM by simply grinding the powder between two glass plates and bringing the fine powder into contact with a carbon-coated copper grid under nitrogen atmosphere. The composite sample was embedded in epoxy resin, microtomed to a thickness of approximately 80 nm, picked up on a carbon-coated copper grid, and examined with a Phillips EM300 electron microscope. Subsequently, this sample was also examined using the JEOL instrument described previously.

The gallium nitride/PMMA composite film was examined using a Perkin-Elmer Lambda 6 spectrophotometer. The sample was scanned at a rate of 2 nm/min and the slit width was 0.5 nm.

The X-ray spectrum (Fig. 28) exhibits three relatively broadened reflections at $2\theta = $ 35.5, 58, and 69°. The broadening of the reflections is probably due to the effect of small domain size. These reflections were assigned to the (111), (220), and (311) planes, respectively, of zincblende gallium nitride (unit cell parameter $a = 4.5$ Å). The asymmetry of the (111) peak can be attributed to a confined, low-intensity (200) peak ($2\theta = 40.5°$), which is not explicitly observed. The low intensity of this latter peak may be caused by some nitrogen vacancies (as the chemical analysis indicates) and/or a high density of stacking faults. XRD calculations showed that a good match to experimental data was only obtained by introducing a high density of stacking faults.

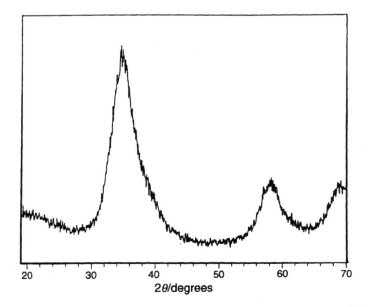

2θ/degrees

Fig. 28.   X-Ray spectrum of GaN powder. Reprinted with permission from K. E. Gonsalves et al., *J. Mater. Chem.*, 8, 1451 (© 1996 The Royal Society of Chemistry).

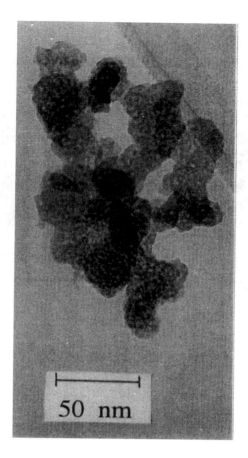

Fig. 29. Nanostructured GaN at high magnification. Reprinted with permission from K. E. Gonsalves et al., *J. Mater. Chem.*, 8, 1451 (© 1996 The Royal Society of Chemistry).

The low-magnification HRTEM image of the powdered GaN shows porous particles of relatively large size (around 50 nm major axis). Examination of the particles at higher magnification (Fig. 29) indicates that each of these large particles is an agglomeration of smaller particles, with nanostructured domains.

The mechanism of the particle deagglomeration is not fully characterized. Probably, the localized temperature and pressure transients introduced by the sonication broke the aggregates into their single-crystal domains, the amorphous material between the grains being weaker. However, there have been cases of stable nanoparticles produced by simple dissolution [124]. Similarly, the nature of the particle surface is unclear. The high curvature and roughness of the surface make it very reactive.

A close examination of a GaN particle (Fig. 30) shows evidence of a high density of stacking faults within the fcc structure. This particle is oriented along its ⟨110⟩ crystallographic axis. Usually, when a zincblende structure is oriented along this observation axis, the HRTEM image is composed of three atomic planes: two {111} and a {002}. This gives rise to a pseudo-hexagonal symmetry appearance. When a stacking fault is introduced, one of the {111} planes is displaced by a vector of 1/6⟨112⟩. In the case of our GaN nanoparticles, both {111} planes underwent this displacement, resulting in a bidimensional disorder. This behavior explains the asymmetry observed in the first X-ray peak and the low contribution of the {002} reflection.

The absorption of the composite film was examined using a Perkin-Elmer Lambda 6 spectrophotometer. The sample was scanned at a rate of 2 nm/s and the slit width was

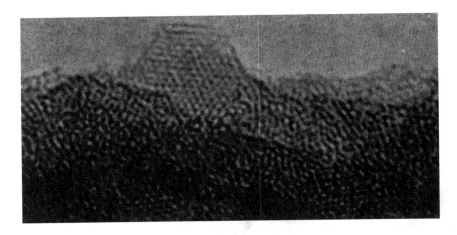

Fig. 30. HRTEM image of GaN. Reprinted with permission from K. E. Gonsalves et al., *J. Mater. Chem.*, 8, 1451 (© 1996 The Royal Society of Chemistry).

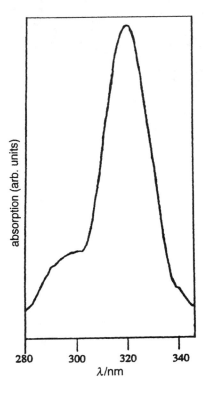

Fig. 31. Optical absorption spectrum. Reprinted with permission from K. E. Gonsalves et al., *J. Mater. Chem.*, 8, 1451 (© 1996 The Royal Society of Chemistry).

0.5 nm. The nanostructured GaN/poly(methyl methacrylate) composite exhibited a strong, symmetric optical absorption (Fig. 31) from 315 to 335 nm. This is probably the result of a surface-enhanced resonance that will lead to a large third-order nonlinear susceptibility ($\chi^{(3)}$) at this wavelength. The particle size distribution (Fig. 32), determined by analysis of TEM images of the composite, had a mean of 5.5 nm and a standard deviation of 2.6. High-resolution TEM of this composite (Fig. 33) confirms that the dispersed phase is zinc-blende GaN.

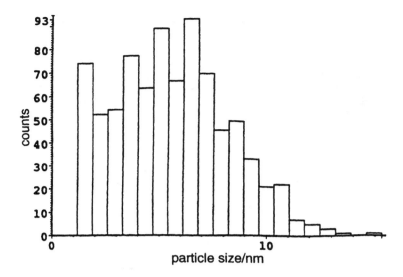

Fig. 32. Particle size distribution. Reprinted with permission from K. E. Gonsalves et al., *J. Mater. Chem.*, 8, 1451 (© 1996 The Royal Society of Chemistry).

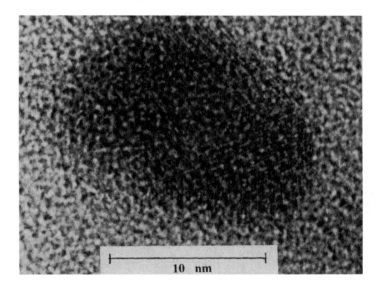

Fig. 33. HRTEM image of nanostructured GaN in PMMA. Reprinted with permission from K. E. Gonsalves et al., *J. Mater. Chem.*, 8, 1451 (© 1996 The Royal Society of Chemistry).

Photoluminescence was measured on a Perkin-Elmer LS 50B fluorescence spectrophotometer. The photoluminescence spectroscopy of GaN nanoparticles with a mean particle size of 5 nm has been carried out at room temperature in the UV excitation range. The excitation and emission features observed have not been reported in either epitaxially grown materials or nanoparticles reported elsewhere. The nanoparticles were enclosed in a quartz container that had been purged with nitrogen. During the measurement of photoluminescence, the sample compartment was continuously purged with nitrogen. The emission spectrum of the GaN nanoparticles was found to be extremely sensitive to the excitation wavelength. Figure 34a shows the excitation spectrum of GaN at room temperature when the emission was monitored at 350 nm. The quartz container has a cutoff around 200 nm

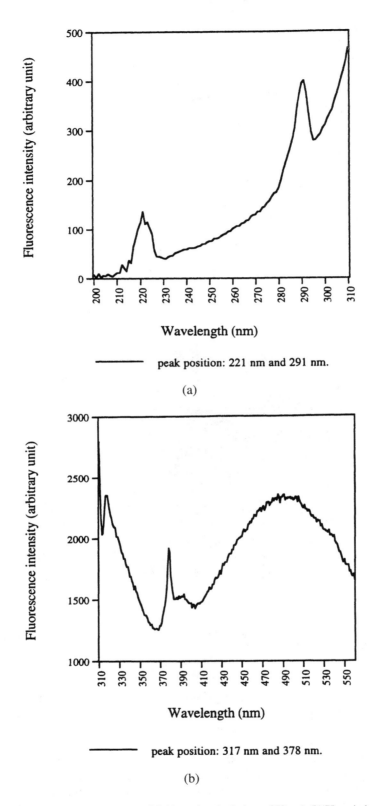

peak position: 221 nm and 291 nm.

(a)

peak position: 317 nm and 378 nm.

(b)

Fig. 34. (a) PL excitation spectrum of GaN powders (emission at 350 nm). (b) PL emission spectrum of GaN powders (emission at 293 nm).

and, therefore, no photoluminesence was observed at excitation below this wavelength. As the excitation wavelength is scanned to 300 nm, we see considerable structure in the strength of the fluorescence signal, suggesting the existence of certain resonances in the GaN nanoparticle electronic spectrum, possibly due to the quantum confinement effects. For a 5-nm particle size scale, we expect quantum confinement effects to modify the states in the conduction and valence bands considerably. Although the effective masses of electrons and holes in such small nanoparticles are not known, if we accept the bulk values we expect significant shifts to shorter wavelengths in the photoluminesence spectra. In addition, the large surface-to-volume ratio in such small particles may lead to the existence of surface states quite distinct from the states normally encountered in bulk or epitaxially grown GaN [125]. The photoluminesence emission spectrum obtained on excitation at 293 nm is shown in Figure 34b. There are several interesting features in the spectrum. A peak at 378 nm is observed, which is quite close to the 370-nm peak observed in 30-nm GaN particles and epitaxially grown material [122]. The observed 378-nm peak is very sharp, indicating that discrete energy states have taken the place of a continuous energy band. According to our proposed model, this is a transition from a discrete state situated at the bottom of the conduction band of bulk GaN to the discrete state situated at the top of the valence band of bulk GaN. The positions of these two discrete states are independent of the size and shape of the nanoparticle. Therefore, the distribution of the size and shape of the nanoparticle cannot appreciably broaden this peak, although it has broadened other peaks. According to the model, the allowed states are a series of sampling points on the energy band of bulk GaN. The transition between the discrete states situated on the conduction band of bulk GaN and those situated on the valence band should lead to the emission at wavelengths shorter than the band gap (378 nm) and this has been clearly observed by a peak at 317 nm. The excitation peaks at 221 and 291 (Fig. 34a) can be explained as the creation of electron–hole pairs in the discrete states (sampling points) with higher energy on the valence and conduction bands of bulk GaN. They are direct processes; that is, the created electron and hole have the same vector. Owing to the particle size and shape distribution, the emissions from various particle sizes would overlap and broaden the emission compared to a single particle size (except the 378-nm peak). A tighter control on the size can lead to narrower emissions. The emission at longer wavelengths is attributed to defect or impurity states. We also expect surface states to play an important role in the PL of these materials. PL in the visible range in GaN has been observed from epitaxial GaN as well and has been attributed to impurity and defect states [126].

The third-order nonlinear optical properties of nanoparticles in PMMA were studied using degenerate four-wave mixing (DFWM). The backward wave geometry was used, where the signal is a phase-conjugate replica of the probe, counterpropagating in the direction of the probe. The 532-nm, 30-ps pulses from a frequency-doubled, Q-switched neodymium-doped yttrium–aluminum–garnet (Nd:YAG) laser (Quantel) were used. The laser operated at a 10-Hz repetition rate. The average pulse energy was 25 mJ. A neutral density filter was placed in the probe beam to reduce its intensity to about 1% that of the pump beams. The phase-conjugate beam was separated from the signal using a beam splitter. The crossing angle was 6°. Fast silicon photodiodes calibrated against a laser energy meter were used to monitor the signal, the probe and the pump pulse energies. By comparison with a CS_2 standard, the $\chi^{(3)}$ was 2.6×10^{-11} esu, as compared to the $\chi^{(3)}$ value reported for gallium nitride films (1.5×10^{-11} esu) [119].

Several kinds of binary and ternary III–V semiconductor nanocrystals, such as GaAs or GaP, have been synthesized using dehalosilylation reactions of group III halides with $P[Si(CH_3)_3]_3$ or $As[Si(CH_3)_3]_3$ or in organic solvents. The byproduct, $(CH_3)_3SiCl$, is removed from the reaction mixture easily, as it is very soluble in commonly used organic solvents and can be isolated from the solid GaAs or GaP product [127].

3.3.7. Semiconductor Nanoparticle Films by Self-Assembly

The term self-assembly implies the spontaneous adsorption of molecules or nanoparticles onto a substrate. Self-assembled multilayer films are formed by the adsorption of subsequent monolayers of molecules or nanoparticles [128]. The evolution of self-assembled layers of molecularly nanostructured materials can be traced, from simple surfactant monolayers and multilayers, through their more complex particulate analogs to the self-assembly of simple molecules and larger particulates.

Self-assembly of nanoparticles to the oppositely charged substrate surface is governed by a delicate balance of the adsorption and desorption equilibria. The optimization of the self-assembly in terms of maximizing the adsorption of nanoparticles from their dispersions and minimizing their desorption on rinsing requires the judicious selection of stabilizers and the careful control of the kinetics of the process. The self-assembly of CdS and PbS nanoparticles was found to be most efficient, for example, if the semiconductor particles were coated by a 1:3 mixture of thiolactic acid and ethyl mercaptan [129]. The generation of size-quantized semiconductor nanoparticles as dispersed organoclay complexes and layered silicates illustrates the use of the nanophase reactors provided by adsorbed binary liquids. The binary liquid pairs, ethanol(1)–cyclohexane(2) and methanol(1)–cyclohexane(2), were selected because the polar component of the liquid mixture (1) preferentially adsorbed at the solid interface and the semiconductor precursors (Cd^{2+} and Zn^{2+}) were highly soluble in the liquid, which preferentially adsorbed at the interface (methanol and ethanol) but was insoluble in the bulk phase (predominantly cyclohexane). These conditions effectively limited the nucleation and growth of the semiconductors to the nanophase reactor provided by the adsorption layer at the solid interface. By varying the mole fraction of the polar liquid (1), it was possible to control the volume of the nanophase reactor and, hence, the size of the semiconductor particles grown therein.

3.3.8. Self-Assembly of Inorganic Nanoparticle Sandwich Films

3.3.8.1. Alternating Layers of Polyelectrolyte–Semiconductor Nanoparticles

The layer-by-layer self-assembly of polyelectrolyte–semiconductor nanoparticles onto substrates is simple. A cleaned substrate is primed by adsorbing a layer of surfactant or polyelectrolyte onto its surface. The primed substrate is then immersed into a dilute aqueous solution of a cationic polyelectrolyte, for adsorption of a monolayer, rinsed, and dried. Then the polyelectrolyte monolayer–covered substrate is immersed into a dilute dispersion of surfactant-coated negatively charged semiconductor nanoparticles, for a time optimized for adsorption of a monoparticulate layer, and subsequently rinsed and dried. These operations complete the self-assembly of a polyelectrolyte monolayer–monoparticulate layer of semiconductor nanoparticle sandwich onto the primed substrate. Subsequent sandwich units are deposited analogously. This method has been used to produce a self-assembly of a poly(diallylmethylammonium chloride), P, CdS nanoparticle onto different substrates (gold, silver, platinum, quartz, and Teflon) (Scheme VI) [129]. Similar methodologies were employed for the self-assembly of many other polyelectrolyte–semiconductor nanoparticle sandwich films.

3.3.8.2. Alternating Layers of Polyelectrolyte–Clay Platelet–Semiconductor Nanoparticles

The high lateral bond strength and aspect ratios have rendered clay organocomplexes such as Na^+-montmorillonite to be eminently suitable materials for nanoconstruction. Indeed, the nanocomposites prepared by mixing polymers and clay organocomplexes had superior mechanical properties.

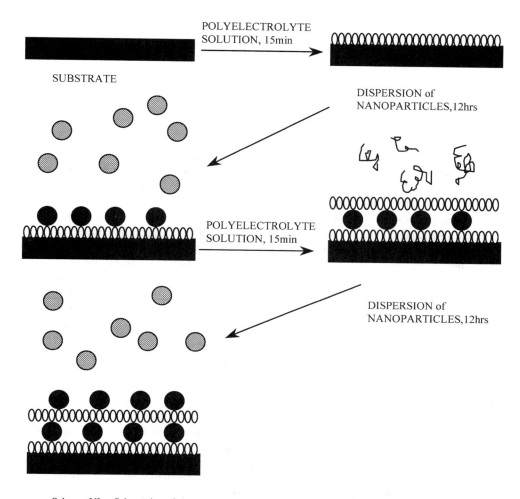

Scheme VI. Schematics of the self-assembly of an S-(P/CdS)$_n$ film. (Adapted from Chem. Mater. 8, 1616 (1996), Fendler et al.)

Alternating layers of polyelectrolyte(P)–clay platelet(M) sandwich films have self-assembled in a manner analogous to that shown for S-(P/CdS)$_n$ (Scheme VI). Significantly, the thickness of a given multilayer S-(P/M)$_n$ film was found to depend on the potential that was applied during the deposition(s) of M. Positive potentials increased the film thickness, whereas negative ones decreased it. A similar behavior has been noted in the electrophoretic deposition of metal particles onto conductive substrates.

3.3.8.3. Potential Applications of Self-Assembled Monolayers

Self-assembly has great advantages. This method is useful for the synthesis of composite films comprising different nanoparticles that can be layered in any desired order. Self-assembled films can function as membranes (barriers), with controllable levels of permeability, for gases, liquids, covalent molecules, ions, and, indeed, electrons (tunneling barriers). These properties have been exploited for the construction of insulators, passivators, sensors, and modified electrodes. Devices based on molecular recognition are candidates for synthesis by self-assembly. Supramolecular assemblies (nanoparticles) within a self-assembled layer can be aligned spontaneously or by changing the temperature, pressure, or pH, or by application of electric and magnetic fields. These properties permit the formation of superlattices with the desired symmetry, as well as the formation of a variety of photonic, electronic, magnetic, and nonlinear optical devices. Control of the sizes and agglomeration of the monodispersed nanoparticle within the self-assembled film can be exploited for construction of optical devices.

4. CONCLUSIONS

In this chapter, we have outlined several common and unique techniques for the synthesis of nanostructured metals, intermetallics, and semiconductors in powder and thin-film form. Considerable opportunities now exist for synthesizing nanophase materials with a variety of new architectures at nanometer length scales from atomic or molecular precursors via the assembly of atom clusters and several other chemical preparation methods. Solution chemistry can directly produce either the desired particles or the precursors that are further treated by various reaction methods to obtain the final products. It should be pointed out that detailed work on the characterization of these materials and their property evaluation is required. Concerns about the yield and the impurity content need to be critically addressed before a particular synthetic and processing method can be adopted for large-scale production. Although the problem of agglomeration of fine particles in solution can potentially be solved by controlling the interparticle forces, the subsequent processing must be designed carefully so that agglomeration can be minimized once the particles are removed from solution. The handling of fine particles after they are synthesized remains a serious problem. This problem is particularly acute for air-sensitive materials. The interesting and fast-developing area of membrane-mediated research, which addresses the stabilization and size control of nanoscale particles, is a potential practical approach to synthesize and process these materials.

The real future of nanophase materials, and other nanostructures as well, will depend on our ability to change significantly for the better, the properties of materials by structuring them artificially on nanometer scales and on developing economic and environmentally responsible methods for producing these materials in commercially viable quantities. Chemical synthesis of nanoparticles is a rapidly growing research area with a great potential to make technologically advanced and useful materials. The realization of this potential will require multidisciplinary interactions and collaborations between biologists, chemists, materials scientists, and engineers, in order to control and improve the properties of nanophase materials.

References

1. (a) H. Gleiter, *Adv. Mater.* 4, 474 (1992). (b) R. Birringer and H. Gleiter, in: "Encyclopedia of Materials Science and Engineering" (R. W. Cahn, ed.), Suppl. Vol. 1, p. 339. Pergamon Press, Oxford, U.K., 1988.
2. R. W. Seigel, *Nanostruct. Mater.* 3, 1 (1993).
3. R. W. Seigel, *Mater. Sci. Eng. B* 19, 37 (1993).
4. (a) R. W. Cahn, *Nature* 348, 389 (1990). (b) R. Dagani, *Chem. Eng. News.* 72, 18 (1992). (c) H. Gleiter, *Nanostruct. Mater.* 6, 3 (1995). (d) D. Chakravorty and A. K. Giri, "Chemistry for the 21st Century: Chemistry of Advanced Materials" (C. N. R. Rao, ed.), p. 217. Blackwell Scientific, London, 1993.
5. (a) Y. Wang and N. Herron, *J. Phys. Chem.* 95, 525 (1991). (b) J. P. Kuczynski, *J. Am. Chem. Soc.* 108, 2513 (1986). (c) E. E. Mendez and K. von Klitzing, *NATO Adv. Study Inst., Ser. B* 170, 1 (1989).
6. (a) Y. Yoshizawa, S. Oguma, and Y. Yamaguchi, *J. Appl. Phys.* 64, 6044 (1988). (b) H. J. de Witt, C. H. M. Wittmer, and F. W. A. Dirne, *Adv. Mater.* 3, 356 (1991).
7. (a) D. D. Beck and R. W. Seigel, *J. Mater. Res.* 7, 2840 (1992). (b) E. Boakye, *J. Colloid Interface Sci.* 163, 120 (1994).
8. (a) M. L. Steigerwald and L. E. Brus, *Acc. Chem. Res.* 23, 183 (1990). (b) A. I. Ekimov and A. A. Onushchenko, *Zh. Eskp. Teor. Fiz.* 40, 337 (1984). [English Translation, *JETP Lett.* 40, 1136 (1984)].
9. A. Henglein, *Top. Curr. Chem.* 143, 113 (1988).
10. (a) R. W. Seigel, *MRS Bull.* 15, 60 (1990). (b) H. Gleiter, *Prog. Mater. Sci.* 33, 223 (1990). (c) R. Herr, U. Birringer, and H. Gleiter, *Trans. Jpn. Inst. Met. Suppl.* 27, 43 (1986).

11. (a) B. W. Dodson, L. J. Schowalter, J. E. Cunningham, and F. E. Pollak, eds., *Mater. Res. Soc. Symp. Proc.* 160 (1989). (b) T. M. Bessman and B. M. Bessman, eds., *Mater. Res. Soc. Symp. Proc.* 168 (1990).

12. (a) L. M. Goldman, B. Blanpain, and F. Spaepen, *J. Appl. Phys.* 60, 1374 (1986). (b) D. S. Lashmore, R. Oberle, M. P. Dariel, L. H. Bennett, and L. Swartzendruber, *Mater. Res. Soc. Symp. Proc.* 132, 219 (1989).

13. (a) C. C. Koch, *Nanostruct. Mater.* 2, 109 (1993). (b) Materzzi et al., *Nanostruct. Mater.* 2, 217 (1993).

14. (a) G. Wasserman, in: "Proceedings of the 4th International Conference on Strength of Metals and Alloys," 1976, Vol. 3, p. 1343. (b) J. D. Embury, in: "Strengthening Methods in Crystals" (A. Kelly and R. B. Nicholson, eds.), p. 331. Applied Science Publishing, London, 1971.

15. (a) C. R. Aita, *Nanostruct. Mater.* 4, 257 (1994). (b) K. J. Balkus, Jr. et al., J. *Mater. Res. Soc. Symp. Proc.* 351, 437 (1994). (c) T. Kameyama et al., *J. Mater. Sci.* 25, 1058 (1994).

16. (a) J. Rivas et al., *J. Magn. Magn. Mater.* 122 (1993). (b) K. E. Gonsalves and T. D. Xiao, "Chemical Processing of Ceramics" (B. I. Lee and E. J. A. Pope, eds.), p. 359. Marcel Dekker, New York, 1994. (c) W. Chang et al., *Nanostruct. Mater.* 4, 345 (1994). (d) J. Phalippou, "Chemical Processing of Ceramics" (B. I. Lee and E. J. A. Pope, eds.), p. 265. Marcel Dekker, New York, 1994. (e) C. J. Brinker and J. Schener, "Sol–Gel Science, The Physics and Chemistry of Sol–Gel Processing." Academic Press, Boston, 1990.

17. R. W. Seigel, "Materials Science and Technology" (R. Cahn, ed.), Vol. 15, p. 583. VCH Publishers, Weinheim, Germany, 1991.

18. (a) S. K. Ganapathi and D. A. Rigney, *Scr. Metall. Mater.* 24, 1675 (1990). (b) H. J. Fecht, E. Hellstern, Z. Fu, and W. L. Johnson, *Adv. Powder Metall.* 1–2, 111 (1989). (c) E. Hellstern, H. J. Fecht, Z. Fu, and W. L. Johnson, *J. Appl. Phys.* 65, 305 (1989). (d) C. C. Kock, J. S. C. Jang, and S. S. Gross, *J. Mater. Res.* 4, 557 (1989).

19. P. A. Psaras and H. D. Langford, eds., "Advancing Materials Research, U.S. National Academy of Engineering and National Academy of Sciences," p. 203. National Academy Press, Washington, DC, 1987.

20. C. N. R. Rao, *Mater. Sci. Eng. B* 18, 1 (1993).

21. (a) V. K. LaMer and R. H. Dinegar, *J. Am. Chem. Soc.* 72, 4847 (1950). (b) J. T. G. Overbeek, *Adv. Colloid Interface Sci.* 15, 251 (1982).

22. F. Neilsen, *Manufacturing Chemist* 53, 38 (1982).

23. M. W. Real, *Proc. Br. Ceram. Soc.* 38, 59 (1986).

24. (a) W. Chang et al., *Nanostruct. Mater.* 4, 345 (1994). (b) H. Hahn and R. S. Averback, *J. Appl. Phys.* 67, 1113 (1990).

25. (a) K. E. Gonsalves et al., *J. Mater. Sci.* 27, 3231 (1992). (b) T. D. Xiao, et al., *J. Mater. Sci.* 28, 1334 (1993).

26. N. Herron and Y. Wang, in: "Nanomaterials: Synthesis, Properties and Applications" (A. S. Edelstein, ed.), p. 73. IOP Publishing, Bristol, U.K., 1996.

27. (a) N. Ichinose et al., "Superfine Particle Technology." Springer-Verlag, London, 1992. (b) Q. Xu and M. A. Anderson, *J. Am. Ceram. Soc.* 77, 1939 (1994). (c) K. E. Gonsalves et al., *Adv. Mater.* 6, 291 (1994). (d) B. H. Kear and L. F. McCandlish, *J. Adv. Mater.* 10, 11 (1993).

28. T. W. Smith and D. Wychick, *J. Phys. Chem.* 84, 1621 (1980).

29. (a) W. Heiber and R. Werner, *Chem. Ber.* 90, 286 (1957). (b) W. Heiber and N. Kahlen, *Chem. Ber.* 91, 2234 (1958).

30. (a) J. Dewar and H. O. Jones, *Proc. R. Soc. London, Ser. A* 76, 564 (1905). (b) H. G. Cutforth and P. W. Selwood, *J. Am. Chem. Soc.* 65, 2414 (1905).

31. (a) T. H. Whitesides and J. P. Neilan, *J. Am. Chem. Soc.* 95, 5811 (1973). (b) H. W. Whitlock, Jr. and Y. N. Chuah, *J. Am. Chem. Soc.* 87, 3605 (1965).

32. N. Smith, *J. Am. Chem. Soc.* 58, 173 (1936). (b) J. Kruger and J. P. Calvert, *J. Electrochem. Soc.* 114, 43 (1967).

33. I. Nakatani, M. Hijikata, and K. Ozawa, *J. Magn. Magn. Mater.* 122, 10 (1993).

34. J. Wonterghem, S. Morup, S. W. Charles, S. Wells, and J. Villadsen, *Phys. Rev. Lett.* 55, 410 (1985).

35. S. W. Charles and J. Popplewell, "Ferromagnetic Materials" (E. P. Wohlfarth, ed.), Vol. 2, p. 509. North-Holland, Amsterdam, 1980.

36. (a) B. S. Clausen, S. Morup, and H. Topsoe, *Surf. Sci.* 106, 438 (1981). (b) S. Morup, H. Topsoe, and B. S. Clausen, *Phys. Scr.* 25, 713 (1982).

37. (a) G. Le Caer, J. M. Dubois, M. Pijolat, V. Perrichon, and P. Bussiere, *J. Phys. Chem.* 86, 4799 (1982). (b) C. B. Ma, T. Ando, D. L. Williamson, and G. Krauss, *Metall. Trans. A* 14, 1033 (1983).

38. X. Cao, Y. Koltypin, G. Kataby, R. Prozorov, and A. Gedanken, *J. Mater. Res.* 10, 2952 (1995).

39. K. E. Gonsalves, S. P. Rangarajan, A. Garcia-Ruiz, and C. C. Law, *J. Mater. Sci. Lett.* 15, 1261 (1996).

40. J. Schneider, *Acta Crystallogr. Sect. A* 43, 295 (1987).

41. (a) D. M. McMillen, K. E. Lewis, G. P. Smith, and D. M. Golden, *J. Phys. Chem.* 8, 709 (1982). (b) W. M. Shaub and S. H. Bauer, *J. Chem. Kinet.* 7, 509 (1975).

42. J. R. Thomas, *Chem. Commun.* 2914 (1965).

43. F. E. Luborsky and T. O. Paine, *J. Appl. Phys.* 31, 68S (1960).

44. D. Zeng and M. J. Hampden-Smith, *Chem. Mater.* 5, 681 (1993).

45. H. Bönnemann, W. Brijoux, and T. Jousson, *Angew. Chem. Int. Ed. Engl.* 29, 273 (1990).

46. J. P. Collman, L. S. Hegedus, J. R. Norton, and R. G. Finke, "Principles and Applications of Organotransition Metal Chemistry." University Science Books, Berkeley, CA, 1987.

47. (a) W. E. Buhro, J. A. Haber, B. E. Waller, and T. J. Trentler, *Am. Chem. Soc. Symp. Ser.* 210, 20 (1995). (b) J. A. Haber, J. L. Crane, W. E. Buhro, C. A. Frey, S. M. L. Sastry, J. L. Balbach, and M. S. Conradi, *Adv. Mater.* 8, 163 (1996).

48. H. Bönnemann, W. Brijoux, R. Brinkmann, R. Fretzen, T. Joussen, R. Köpler, B. Korall, P. Neiteler, and J. Richter, *J. Mol. Catal.* 86, 129 (1994).

49. (a) A. Corrias, G. Ennas, G. Licheri, G. Marongin, and G. Paschina, *Chem. Mater.* 2, 363 (1990). (b) H. C. Brown and C. A. Brown, *J. Am. Chem. Soc.* 84, 1492 (1962).

50. H. Bönnemann, W. Brijoux, and T. Joussen, DE OS 3934351, Studiengesellschaft Kohle mbH, 1991.

51. G. Schmid, in: "Aspects of Homogeneous Catalysis" (R. Ugo, ed.), Vol. 7, p. 1. Kluwer Academic Publishers, Dordrecht, The Netherlands, 1990.

52. A. Duteil, G. Schmid, and W. Meyer-Zaika, *J. Chem. Soc., Chem. Commun.* 31 (1995).

53. (a) F. Fievet, J. P. Lagier, and M. Figlarz, *MRS Bull.* 14, 29 (1989). (b) M. Figlarz, F. Fievet, and J. P. Lagier, Eur. Patent, 0113281; U.S. Patent, 4,539,041.

54. (a) F. Fievet, F. Fievet-Vincent, J. P. Lagier, B. Dumont, and M. Figlarz, *J. Mater. Chem.* 3, 627 (1993). (b) F. Fievet, J. P. Lagier, B. Blin, B. Beaudoin, and M. Figlarz, *Solid State Ionics* 32/33, 198 (1989).

55. R. D. Nelson, Jr., "Dispersing Powders in Liquids," pp. 211. Elsevier, Amsterdam, 1988.

56. (a) K. L. Tsai and J. L. Dye, *J. Am. Chem. Soc.* 113, 1650 (1991). (b) K. L. Tsai and J. L. Dye, *Chem. Mater.* 5, 540 (1993).

57. (a) J. L. Dye, M. T. Lok, F. J. Tehan, R. B. Coolen, N. Papadakis, J. M. Ceraso, and M. G. DeBacker, *Ber. Bunsen-Ges. Phys. Chem.* 75, 3092 (1971). (b) J. L. Dye, M. G. DeBacker, and V. A. Nicely, *J. Am. Chem. Soc.* 92, 5226 (1970).

58. IUPAC name: 15-crown-5; 1,4,7,10,13-pentaoxacyclopentadecane.

59. J. L. Dye and K. L. Tsai, *Faraday Discuss. Chem. Soc.* 92, 45 (1991).

60. (a) J. S. Bradley, J. M. Millar, and E. W. Hill, *J. Am. Chem. Soc.* 113, 4016 (1991). (b) H. Hirai, Y. Nakao, and N. Toshima, *J. Macromol. Sci. Chem.* 12, 1117 (1978).

61. H. Thiele and H. S. von Levern, *J. Colloid Sci.* 3, 363 (1965).

62. J. H. Ross, *Anal. Chem.* 25, 1288 (1953).

63. A. M. Bradshaw and F. M. Hoffmann, *Surf. Sci.* 72, 513 (1978).

64. (a) M. J. Tracy and J. R. Groza, *Nanostruct. Mater.* 1, 369 (1992). (b) K. Higashi, T. Mukai, S. Tanimura, A. Inoue, K. Masumoto, K. Kita, K. Ohtera, and J. Nagahora, *Nanostruct. Mater.* 26, 191 (1992).

65. (a) K. E. Gonsalves, T. D. Xiao, G. M. Chow, and C. C. Law, *Nanostruct. Mater.* 4, 139 (1994). (b) K. E. Gonsalves, S. P. Rangarajan, C. C. Law, C. R. Feng, G.-M. Chow, and A. Garcia-Ruiz, *Am. Chem. Soc. Symp. Ser.* 622, 15. 220 (1996). (c) C. R. Feng, G. M. Chow, S. P. Rangarajan, X. Chen, K. E. Gonsalves, and C. C. Law, *Nanostruct. Mater.* 8, 45 (1997). (d) K. E. Gonsalves and S. P. Rangarajan, in: "Chemistry and Physics of Nanostructures and Related Non-Equilibrium Materials" (E. Ma, B. Fultz, R. Shull, J. Morral, and P. Nash, eds.), p. 149. TMS Meeting, Orlando, FL, 1997. (e) K. E. Gonsalves and S. P. Rangarajan, *J. Appl. Polym. Sci.* 64, 2667 (1997). (f) K. E. Gonsalves, U.S. Patent 389,778, 1996.

66. G. M. Chow, T. Ambrose, J. Xiao, F. Kaatz, and A. Ervin, *Nanostruct. Mater.* 2, 131 (1993).

67. G. Schmid, E. Schöps, J.-O. Malm, and J.-O. Bovin, *Z. Anorg. Allg. Chem.* 620, 1170 (1994).

68. (a) B. H. Kear and L. E. McCandlish, *Nanostruct. Mater.* 3, 19 (1993). (b) L. E. McCandlish and R. S. Polizzotti, *Solid State Ionics* 32/33, 795 (1989).

69. C. H. Johansson and J. O. Linde, *Ann. Phys.* 25, 1 (1936).

70. G. M. Chow, L. K. Kurihara, K. M. Kemner, P. E. Schoen, W. T. Elam, A. Ervin, S. Keller, Y. D. Zhang, J. Budnick, and T. Ambrose, *J. Mater. Res.* 10, 1546 (1995).

71. C. Meny, P. Panissod, and R. Loloee, *Phys. Rev. B* 45, 12269 (1992).

72. J. R. Childress, C. L. Chien, and M. Nathan, *J. Appl. Phys.* 70, 5885 (1991).

73. J. S. Bradley, E. W. Hill, C. Klein, B. Chaudret, and A. Duteil, *Chem. Mater.* 5, 254 (1993).

74. (a) H. A. C. M. Hendricks and V. Ponec, *Surf. Sci.* 192, 234 (1987). (b) J. S. Bradley, J. M. Millar, E. W. Hill, S. Behal, B. Chaudret, and A. Duteil, *Faraday Discuss. Chem. Soc.* 92, 255 (1991).

75. (a) F. Kayser and M. Cohen, *Metal. Progr.* 61, 79 (1952). (b) W. B. Pearson, "A Handbook of Lattice Spacing and Structures of Metals and Alloys," Vol. 1. Pergamon Press, Elmsford, NY, 1958.

76. (a) AI. L. Efros and A. L. Efros, *Fiz. Tekh. Poluprovodn.* 16, 1209 (1982) [English Translation, *Sov. Phys.-Semicond.* 16, 772 (1982)]. (b) M. L. Steigerwald and L. E. Brus, *Acc. Chem. Res.* 23, 183 (1990). (c) A. Henglein, *Top. Curr. Chem.* 143, 113 (1988). (d) Y. Wang, N. Herron, W. Mahler, and A. Suna, *J. Opt. Soc. Am. B* 6, 808 (1989).

77. N. Herron, J. C. Calabrese, W. E. Farneth, and Y. Wang, *Science* 259, 1426 (1993).

78. D. W. Bahnemann, C. Kormann, and M. R. Hoffmann, *J. Phys. Chem.* 91, 3789 (1987).

79. (a) L. Spanhel, M. Haase, H. Weller, and A. Henglein, *J. Am. Chem. Soc.* 109, 5649 (1987). (b) U. Resch, H. Weller, and A. Henglein, *Langmuir* 5, 1015 (1989).

80. N. Herron, Y. Wang, and H. Eckert, *J. Am. Chem. Soc.* 112, 1322 (1990).

81. C. T. Dameron, R. N. Reese, R. K. Mehra, A. R. Kortan, P. J. Carroll, M. L. Steigerwald, L. E. Brus, and D. R. Winge, *Nature* 338, 596 (1989).

82. H. Uchida, C. Curtis, P. V. Kamat, K. M. Jones, and A. J. Nozik, *J. Phys. Chem.* 96, 1156 (1992).
83. (a) Y. M. Tricot, A. Emeren, and J. H. Fendler, *J. Phys. Chem.* 89, 4721 (1985). (b) H.-C. Youn, S. Baral, and J. H. Fendler, *J. Phys. Chem.* 92, 6320 (1988).
84. (a) M. Meyer, C. Wallberg, K. Kurihara, and J. H. Fendler, *J. Chem. Soc. Chem. Commun.* 90 (1984). (b) C. Petit, P. Lixon, and M. P. Pileni, *J. Phys. Chem.* 94, 1598 (1990).
85. (a) Y. Wang, A. Suna, W. Mahler, and R. Kasowski, *J. Chem. Phys.* 87, 7315 (1987). (b) W. Mahler, *Inorg. Chem.* 27, 435 (1988). (c) E. Hilinski, P. Lucas, and Y. Wang, *J. Chem. Phys.* 89, 3435 (1988).
86. C. H. Fischer, H. Weler, L. Katsikas, and A. Henglein, *Langmuir* 5, 429 (1989).
87. (a) T. Goto, S. Saito, and M. Tanaka, *Solid State Commun.* 80, 331 (1991). (b) R. F. Ziolo, E. P. Giannelis, B. A. Weinstein, M. P. O'Horo, B. N. Ganguly, V. Mehrotra, M. W. Russell, and D. R. Huffman, *Science* 257, 219 (1992).
88. V. Sankaran, C. C. Cummins, R. R. Schrock, R. E. Cohen, and R. J. Silbey, *J. Am. Chem. Soc.* 112, 6858 (1990).
89. Y. Wang, N. Heron, M. Harmer, and A. Suna, "Proceedings of the MRS Spring Meeting," Pittsburgh, PA, 1992.
90. H. Yao, S. Takahara, H. Mizuma, T. Kozeki, and T. Hayashi, *Jpn. J. Appl. Phys.* 35, 4633 (1996).
91. A. Chevreau, B. Phillips, B. G. Higgins, and S. H. Risbud, *J. Mater. Chem.* 6, 1643 (1999).
92. (a) U. Koch, H. Fojtik, H. Weller, and A. Henglein, *Chem. Phys. Lett.* 122, 507 (1985). (b) P. Hoyer, R. Eichberger, and H. Weller, *Ber. Bunsen-Ges Phys. Chem.* 97, 630 (1993).
93. S. Mahamuni, B. S. Bendre, V. J. Leppert, C. A. Smith, D. Cooke, S. H. Risbud, and H. W. H. Lee, *Nanostruct. Mater.* 6, 659 (1996).
94. L. Spanhel and M. Anderson, *J. Am. Chem. Soc.* 113, 2826 (1990).
95. N. F. Borelli, D. W. Hall, H. J. Holland, and D. W. Smith, *J. Appl. Phys.* 61, 5399 (1987).
96. (a) B. G. Potter and J. H. Simmons, *Phys. Rev.* 37, 10838 (1988). (b) T. Arai, H. Fujumura, I. Umezu, T. Ogawa, and A. Fujii, *Jpn. J. Appl. Phys.* 28, 484 (1989).
97. T. Rajh, M. I. Vucemilovic, N. M. Dimitrijevic, and O. I. Micic, *Chem. Phys. Lett.* 143, 305 (1988).
98. H. Minti, M. Eyal, and R. Reisfeld, *Chem. Phys. Lett.* 183, 277 (1991).
99. H. Schmidt, "Proceedings of the Winter School on Glasses and Ceramics from Gels, Sol–Gel Science and Technology, Brazil" (M. A. Aegerter, M. Jafelicci, Jr., D. F. Souza, and E. D. Zanotto, eds.), p. 432. World Scientific, Singapore, 1989.
100. B. Buchalter and G. R. Meredith, *Appl. Opt.* 21, 3221 (1982).
101. (a) D. J. Scoberg, F. Greiser, and D. N. Furlong, *J. Chem. Soc., Chem. Commun.* 516 (1991). (b) J. C. Luong and N. F. Borelli, *Mater. Res. Soc. Symp. Proc.* 144, 695 (1989).
102. (a) Y. Wang and N. Herron, *Res. Chem. Intermed.* 15, 17 (1991). (b) R. Roy, S. Roy, and D. M. Komareni, *Mater. Res. Soc. Symp. Proc.* 32, 347 (1984). (c) Y. Kobayashi, S. Yamazaki, Y. Kurokawa, T. Miyakawa, and H. Kawaguchi, *J. Mater. Sci. Mater. Electron.* 2, 20 (1992).
103. (a) D. G. Park and J. M. Burlitch, *Chem. Mater.* 4, 500 (1992). (b) I. Tanahashi, A. Tsujimura, T. Mitsuyu, and A. Nishino, *Jpn. J. Appl. Phys.* 29, 2111 (1990). (c) M. Fujii, S. Hayashi, and K. Yamamoto, *Appl. Phys. Lett.* 57, 2962 (1990).
104. T. Itoh, Y. Iwabuchi, and M. Kataoka, *Phys. Status Solidi B* 145, 567 (1988).
105. (a) V. N. Bogomolov, V. V. Poborchii, and S. V. Kholodkevich, *JETP Lett.* 31, 434 (1980). (b) Y. Katayama, M. Yao, and Y. Ajiro, *J. Phys. Soc. Jpn.* 58, 1811 (1989). (c) J. B. Parise, J. E. MacDougall, N. Herron, R. Farlee, A. W. Sleight, Y. Wang, T. Bein, K. Moller, and L. M. Moroney, *Inorg. Chem.* 27, 210 (1988).
106. N. Herron, *Inorg. Chem.* 25, 4714 (1986).
107. (a) R. D. Stramel, T. Nakamura, and J. K. Thomas, *J. Chem. Soc., Faraday Trans. 1*, 84, 1287 (1988). (b) D. W. Breck, "Zeolite Molecular Sieves." Wiley, New York, 1974.
108. N. Herron, Y. Wang, M. Eddy, G. D. Stucky, D. E. Cox, T. Bein, and K. Moller, *J. Am. Chem. Soc.* 111, 530 (1989).
109. (a) T. Hirono, A. Kawana, and T. Yamada, *J. Appl. Phys.* 62, 1984 (1987). (b) Y. Nozue, Z. K. Tang, and T. Goto, *Solid State Commun.* 73, 31 (1990).
110. (a) G. A. Ozin, J. P. Godber, and A. Stein, U.S. Patent, August 1988. (b) A. Stein, Thesis, University of Toronto, 1988.
111. G. Cao, L. K. Rabenberg, C. M. Nunn, and T. E. Mallouk, *Chem. Mater.* 3, 149 (1991).
112. X. Liu and J. K. Thomas, *Langmuir* 5, 58 (1989).
113. V. Swayambunathan, D. Hayes, K. H. Schmidt, Y. X. Liao, and D. Meisel, *J. Am. Chem. Soc.* 112, 3831 (1990).
114. I. G. Dance, A. Choy, and M. L. Scudder, *J. Am. Chem. Soc.* 106, 6285 (1984).
115. G. S. H. Lee, D. C. Craig, I. Ma, M. L. Scudder, T. D. Bailey, and I. G. Dance, *J. Am. Chem. Soc.* 110, 4863 (1988).
116. W. E. Farneth, N. Herron, and Y. Wang, *Chem. Mater.* 4, 917 (1992).
117. (a) R. E. Treece, G. S. Macala, and R. B. Kaner, *Chem. Mater.* 4, 9 (1992). (b) B. K. Laurich, D. C. Smith, and M. D. Healy, *Mater. Res. Soc. Symp. Proc.* 351, 49 (1994). (c) R. L. Wells, S. R. Aubuchon, S. S. Kher, and M. S. Lube, *Chem. Mater.* 7, 793 (1995). (b) H. S. Nalwa and S. Miyata, eds., "Nonlinear Optics of Organic Molecules and Polymers." CRC Press, Boca Raton, FL, 1997.

118. (a) H. S. Nalwa and S. Miyata, eds., "Nonlinear Optics of Organic Molecules and Polymers," CRC Press, Boca Raton, FL, 1997. (b) J. W. Goodman, P. Chavel, and G. Roblin, eds., "Optical Computing," Vol. 963. SPIE, Washington, DC, 1989.

119. D. K. Wickenden, T. J. Kistenmacher, and J. Miragliotta, *J. Electron. Mater.* 23, 1209 (1994).

120. J. W. Hwang, J. P. Campbell, J. Kozubowski, S. A. Hanson, J. F. Evans, and W. L. Gladfelter, *Chem. Mater.* 7, 517 (1995).

121. J. F. Janik and R. L. Wells, *Chem. Mater.* 8, 2708 (1996).

122. Y. Xie, Y. Qian, W. Wang, S. Zhang, and Y. Zhang, *Science* 272, 1926 (1996).

123. H. Nöth and P. Konrad, *Z. Naturforsch. B: Chem. Sci.* 30, 681 (1975).

124. J. E. Turner, M. Hendewerk, J. Parmeter, D. Neiman, and G. A. Somorjai, *J. Electrochem. Soc.* 131, 1777 (1984).

125. H. P. Maruska and J. J. Tietjen, *Appl. Phys. Lett.* 15, 327 (1969).

126. S. Strite, *J. Vac. Sci. Technol. B* 9, 1924 (1991).

127. L. I. Halaoui, S. S. Kher, M. S. Lube, S. R. Aubuchon, C. R. S. Hagan, R. L. Wells, and L. A. Coury, *Am. Chem. Soc. Symp. Ser.* 622, 178 (1996).

128. J. H. Fendler, *Chem. Mater.* 8, 1616 (1996).

129. N. A. Kotov, I. Dekany, and J. H. Fendler, *J. Phys. Chem.* 99, 13065 (1995).

Chapter 2

NANOPARTICLES FROM LOW-PRESSURE, LOW-TEMPERATURE PLASMAS

Josep Costa

Grup de Recerca en Materials, Departament de Física, Universitat de Girona, Girona, Spain

Contents

1. INTRODUCTION

Low-pressure, low-temperature plasmas have been used extensively to manufacture materials and devices in the microelectronics industry since its beginning. As attempts were made to reduce the size of the devices and increase the scale of integration, it was soon realized that the formation of particles in the plasma itself could be the main source of wafer

Handbook of Nanostructured Materials and Nanotechnology, edited by H.S. Nalwa
Volume 1: Synthesis and Processing
Copyright © 2000 by Academic Press
All rights of reproduction in any form reserved.

ISBN 0-12-513761-3/$30.00

contamination and therefore loss of yield. The first reports on the observation of particles during thin-film deposition were published in the mid-1980s [1–3]. At that time, there was little knowledge of the physics involved in the formation of particles in a low-pressure, low-temperature plasma. For that reason, basic research programs on the understanding of particle formation were started with the ultimate purpose of avoiding particle contamination in semiconductor manufacturing processes. The first international program in this field was started in 1991 in Europe, with the title: "Powder Formation in Low-Pressure, Low-Temperature Plasmas." The author's activity in this field began within the framework of this project. During the next 3 years, similar national programs were started in the United States and Japan. Specific workshops and meetings on what were called "dusty plasmas" were organized [4, 5], as well as symposia on vacuum technology and materials research conferences.

The initial characterization of the powder generation dynamics allowed researchers to foresee that low-temperature plasmas could be a suitable, cost-effective source of nanostructured materials. In view of the potential use of low-temperature plasmas to produce and process nanometric powders, the European project cited previously was combined with another entitled: "Micropowder Processing Using Low-Pressure Plasma Technology." Therefore, an extensive study of the powder as a material was developed simultaneously to further research on particle generation and plasma–particle interactions.

In this chapter, it will be shown that low-pressure, low-temperature plasmas may generate a high particle nucleation rate and that they can allow for control of the composition, size, and atomic structure of the particles. Results on the formation of powders of different alloys will be presented. Moreover, these techniques have been used, more recently, to incorporate particles into a growing film to obtain a nanostructured thin film.

Although almost any plasma may generate particles, this chapter will not deal with etching or sputtering plasmas, where particles are introduced into the gas phase via removal from the substrate. In addition, most of the results presented have been obtained in radio frequency glow discharges of silane or silane mixtures, which are, by large, the discharges most widely studied.

This chapter is organized as follows. In Section 2, the scientific context of the formation of particles in low-temperature, low-pressure plasmas is reviewed. The basic studies of particles in interstellar space, the concern of the microelectronics industry about their contamination effect, the recent interest in nanostructured ceramics, and the new field of plasma crystals are described in this section.

In Section 3, we present the technological aspects of the formation and characterization of the powders. Particular attention is paid to the techniques for *in situ* characterization of powder suspended in the plasma.

Section 4 deals with research on the dynamics of particle formation and plasma-particle interactions. As will be discussed, knowledge acquired on the timing of particle development is the basis of the methods used both to avoid powder formation in semiconductor processes and to deposit particles of a chosen size on a growing film.

The material properties of nanometric powders of silicon and silicon alloys are presented in Section 5. Although few reports have been published on the *ex situ* characterization of powders generated in low-pressure plasmas, powders of silicon, silicon–carbon, silicon–nitrogen, and boron–nitrogen alloys have been studied to a different extent. The atomic structure, the relationship between the powder composition and the precursor gas mixture, the vibrational properties, and the hydrogenation of these alloys are described in this section.

Section 6 focuses on the applications of the particles generated in glow discharges that are currently under study. On the one hand, the light emission from silicon nanoparticles grown by plasma-enhanced chemical vapor deposition is described. This emission has been examined in detail and the main conclusions might be extended to the emission of related nanostructured materials such as porous silicon. On the other hand, the growth of nanostructured thin films, which consist of well-ordered particles of a few nanometers embedded in an amorphous matrix, is presented. The selective incorporation of particles into a growing silicon film has resulted in a new material with encouraging properties in the domain of amorphous and microcrystalline silicon-based devices.

This chapter concludes with a comprehensive summary of the results and a short discussion of the new applications that the peculiarities of this material promise.

2. SCIENTIFIC AND INDUSTRIAL CONTEXT

2.1. Occurrence of Particles in Plasmas

The presence of small particles in ionized media has been known for many years. As early as 1924, Langmuir observed a cloud of powder in a laboratory plasma [6], although astronomy is the science where most interest has been shown until the last decade. Hoyle and Wickramasinge [7] reviewed the insights into the weak galactic magnetic field and cosmic of chemical synthesis of small, charged dust particles. Gould and Salpeter [8] invoked the presence of small charged grains to explain the synthesis of molecular species in diffuse nebulae. The dust in planetary atmospheres and planetary ring structures has been studied extensively by Goertz [9], Whipple [10], Northrop [11], and others [12, 13]. Concerning laboratory plasmas, Emeleus and Breslin [14] studied the occurrence of dust in bounded positive column plasmas.

The general use of low-pressure, low-temperature plasmas to process materials for the microelectronics industry and the development of light scattering diagnostics triggered renewed scientific activity on the formation of particles in these plasmas. The particular features of laboratory plasmas changed the approach that astrophysics had been using to analyze the dust in interstellar space (colloidal suspensions, nucleation theory, and charged aerosols). The main distinctive features of laboratory discharges are, first, the fact that the plasmas have geometric boundaries whose properties influence the formation and transport of dust grains and, second, the spatiotemporal variations imposed on the dusty discharge by the external circuit that maintains the discharge.

It was soon realized that most laboratory plasmas are, to some extent, dusty. Not only those discharges from reactive gases but also etching or inert gas plasmas may contain a large amount of suspended particles. The mechanisms that cause the appearance of particles, however, may depend on the plasma.

An important consequence of the formation of dust in a low-pressure, low-temperature plasma, and, in fact, the initial motivation for the study of dusty plasmas, is its contamination effect during microelectronic device processing. For that reason, after discussing the potential of low-pressure plasmas to produce nanometric particles for advanced materials, we review the reports on particle formation in thin-film processing plasmas.

Figure 1 shows a scheme of the different branches of science that have an interest in the formation of particles in low-pressure, low-temperature plasmas.

2.2. Particles from Low-Pressure, Low-Temperature Plasmas in Nanostructured Materials

In recent years, there has been increased interest in nanostructured (or nanocrystalline, nanophase, or ultrafine grained) materials. Nanostructured materials can be one of four basic types: powder particles, laminated structures, polycrystalline coatings, or bulk (porous or dense). It is generally accepted that the largest grain size in nanostructured materials is about 100 nm. Such small grain sizes are responsible for the unique physical and mechanical properties of shed bodies made from nanometric particles. These materials exhibit dramatic changes in properties, such as enhanced sinterability, higher electrical conductivity in ceramics, enhanced ultraviolet (UV) light scattering, increased hardness and strength

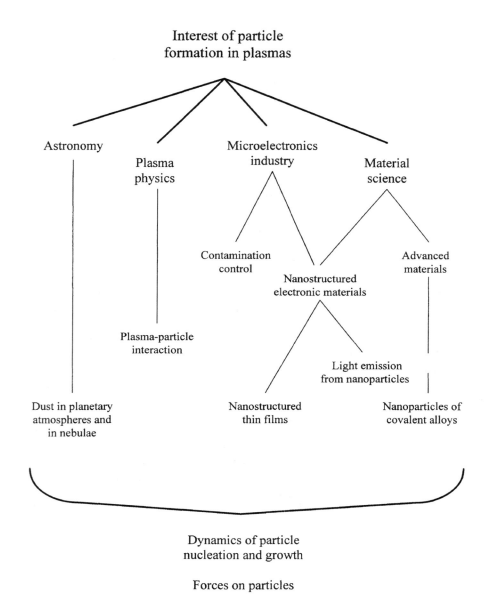

Fig. 1. Scheme of the different branches of science that have an interest in the formation of particles in low-pressure, low-temperature plasmas.

of metals and alloys, enhanced ductility and toughness in ceramics, increased magnetic coercivity, or increased luminescent efficiency of semiconductors. Table I summarizes the changes in properties at the nanocrystalline scale. Progress and development in nanocrystalline materials studies have been reported in the recent reviews of Gleiter [16, 17], Siegel [18], Andrievski [19], and Suryanarayana and Froes [20].

Much research activity has focused on the development of new synthesis techniques to produce nanocrystalline powders. It is still unclear which techniques will be the most cost effective. Few of them have been introduced on a large scale. Potential applications for nanopowders include structural ceramics, protective coatings, substrates for catalysts, ultrafine polishing, cosmetics, dielectrics, large area displays, and even biological components.

Table I.   Changes in Properties at the Nanocrystalline Size Scale [15]

Property	Change
Electrical	Higher electrical conductivity in ceramics and magnetic nanocomposites.
	Higher electrical resistivity in metals.
Magnetic	Increase in magnetic coercivity down to a critical size in the nanoscale size regime.
	Below critical crystalline size, decrease in coercivity leading to superparamagnetic behavior.
Mechanical	Increase in hardness and strength of metals and alloys.
	Enhanced ductility, toughness, and formability of ceramics.
	Superstrength and superplasticity.
Optical	Blue shift of optical spectra of quantum-confined crystallites.
	Increase in luminescent efficiency of semiconductors.

2.2.1. The Nanoparticles Industry [15]

Although much of the work being done in this field is developmental, there are, indeed, nanostructured products currently on the market. Nanosized particles are incorporated into commercially available abrasive polishing slurries, magnetic fluids, fire-retardant materials, sunscreens, magnetic recording tapes, among other products.

Nanostructured silica and iron oxide powders have a commercial history spanning about half a century, whereas nanocrystalline alumina, titania, antimony oxide, nonoxide ceramics, and other materials have entered the marketplace more recently. Besides powders of ceramic nanoparticles, metallic nanoparticles are extensively manufactured in Japan specifically for use in magnetic recording tapes.

More than 2775 nanomaterials-related publications have appeared in print since 1991, and, in 1996 alone, approximately 925 papers were published in the scientific literature. More than 300 U.S. patents have been awarded to corporations, research institutions, and individuals for developments in this area since 1990.

According to the technical-market research study "Opportunities in Nanostructured Materials" published recently by Business Communications Co., the overall U.S. market for nanostructured particles and coatings was valued at an estimated $42.3 million for 1996. This tabulation included ceramic, metallic, semiconducting, and diamond nanostructured materials produced in commercial quantities with the exception of nanoscale amorphous silica powder, which commands a market of several hundred million dollars. The market for nanostructured materials is projected to grow about 400% in 5 years. It is expected to reach $154.6 million in 2001, corresponding to an average annual growth rate of 29.6% from 1996 to 2001. Table II summarizes the overall U.S. markets for nanostructured materials.

2.2.2. Why Low-Pressure, Low-Temperature Plasmas?

Although powder is just one of the nanostructured materials that can be produced in a low-pressure, low-temperature plasma, the ability of this technique to produce ultrafine powders with respect to other methods is discussed in this section. The technological methods for producing ultrafine powders were reviewed recently in several reports [21–24]. Many

Table II. Overall U.S. Markets for Nanostructured Materials [15]

	1996		2001		
	$ (millions)	%	$ (millions)	%	Average annual growth rate (%)
Particles[a]	41.3	97.6	148.6	96.1	29.2
Coatings	1.0	2.4	6.0	3.9	43.1
Total	42.3	100.0	154.6	100.0	29.6

[a] Dry powders and liquid dispersions.

routes for ultrafine powder (UFP) preparation exist, which expands the technical possibilities but also creates competition.

Gas condensation techniques have long been known. Evaporation both into high vacuum and into inert or active gases applies to many metals, alloys, and compounds, using a variety of heating sources. A high level of material vapor oversaturation and the presence of a neutral gas are necessary in order to condense the nanoparticles [22]. It is difficult to obtain high purity and high consolidation levels with this technique. *Ball milling* is widely used for powder metallurgy although two main problems arise in the preparation of nanometric powders: It is difficult to prevent the contamination from the agitator mills (usually iron) [25], and the reduction in grain size below approximately 1 μm requires long milling times. Among the chemical methods, plasma synthesis and thermal decomposition are the most widely used [21, 24]. *Thermal decomposition* methods, such as thermolysis, pyrolysis, and calcination, are suitable for producing many ultrafine powders [23]. Organometallic precursors, such as diimide for Si_3N_4 [26] or polycarbosilanes [27] for SiC, are used in industry to prepare ultrafine powders. These techniques have a good yield of production although they require very high temperatures, so they are costly. Other gasphase synthesis techniques include the work of Prochazka and Greskovich [28], who pyrolyzed silane and ammonia, and the work of Haggerty and co-workers [29], who used a CO_2 laser to pyrolyze gaseous reactants for the synthesis of silicon, silicon nitride, and silicon carbide precursors. Plasma methods are very popular as they enable the preparation of two-component compounds as well as multicomponent powders. *Thermal plasmas* are often used to prepare ultrafine powders of almost any material: Direct-current, arc jet, radio frequency, or hybrid reactors are encountered in the literature [30–33]. In a thermal plasma, the temperatures of the gas and the electrons are comparable and are many thousands of degrees. Starting materials are typically atomized at these temperatures, and the powder synthesis occurs as condensation outside the plasma when the gases cool. In contrast, in a *low-pressure, low-temperature plasma*, the electron temperature is much higher than the gas or ion temperature, which is close to room temperature. The growth of particles may be completely determined by chemical kinetic factors. Therefore, thermal and low-temperature plasmas may produce materials with different structures and properties.

At this point, the question arises about which nanostructured materials can be produced in low-pressure, low-temperature plasmas and with which properties. To discuss the suitability of these plasmas to produce nanostructured materials, the following considerations should be taken into account:

1. *High nucleation rate*. The particle concentration can reach values in the range of 10^{17} m^{-3} in a fraction of a second depending on the discharge parameters. Measurements of particle size and particle concentration have been performed by laser light scattering (LLS) techniques (Section 4.1).

2. *Narrow size distribution function.* Both *in situ* LLS measurements and *ex situ* electron microscopy analyses revealed that the particle size distribution is very narrow in the first stages of particle growth. In fact, the wide size distribution observed after long periods of plasma-on is a consequence of the coagulation process described in Section 4.1.3. It can be prevented by controlling the duration of the plasma in a modulated discharge (Section 4.3.2).

3. *Control of particle size.* Mass spectrometry experiments have indicated that particle growth takes place because of a polymerization chain. Therefore, control of the duration of the plasma can determine the extent of the development of this polymerization pathway. Although very small clusters may not be stable outside the plasma, they might be incorporated into a growing film in the discharge in order to produce a composite film of a homogeneous matrix with nanometric particles (Section 6.1). Coagulation models such as those described in Section 4.1.3 are a very useful predictive tool to control particle size and the coagulation stage.

4. *Composition control and high purity.* On the one hand, the high vacuum attainable in low-pressure, low-temperature reactors guarantees a very low contamination level on the as-produced particles. On the other hand, the composition of the particles may be easily controlled through the gas mixture of the discharge (Section 5). Ideally, once the composition of the particle is chosen, hydrogen would be the only extra element.

5. *In situ deposition of composite thin films.* Besides the possibility of producing large amounts of ultrafine-sized powders of advanced materials such as nonoxide covalent ceramics, low-pressure, low-temperature plasmas permit the growth of thin films with a significant contribution of particles. This opens new possibilities to "classical" materials such as amorphous silicon thin films as it widens their structural characteristics (Section 6.1).

There are several unclear aspects of particle formation in cold plasmas that require more research. For instance, in terms of the structure of the particles, there is experimental evidence that the slow dynamics for particle formation lead to ordered atomic structures, whereas fast particle formation leads to amorphous (polymeric-like) structures. Another point that should be clarified is the stability of small clusters outside the plasma. There is no *ex situ* evidence of particles of less than approximately 3–4 nm, which, at some instant, must exist in the plasma.

2.2.3. Plasma Crystals

In 1994, it was reported that colloidal particles suspended in a low-power radio frequency (rf) plasma can undergo phase transitions and can behave as "plasma crystals" [34–36]. Although this effect is not directly related to nanostructured materials, it is worth briefly reviewing its theoretical basis.

Particles suspended in the plasma are negatively charged because of the high mobility of electrons relative to that of ions. Therefore, if particles approach each other, they will experience interparticle Coulomb forces. If the particle density is high, and thus Coulomb interactions are frequent, the plasma is called "strongly coupled". The parameter that is used to order the strength of this coupling is [37]:

$$\Gamma = \frac{Q^2 N_P^{1/3}}{k T_P} \tag{1}$$

where k is the Boltzmann constant, Q is the charge of the particle, N_P is the concentration of particles in the plasma, and T_P is their kinetic temperature defined from the velocity in the usual way. When Γ is larger than 2, a Coulomb liquid is predicted, and, when it exceeds 170, a Coulomb solid should result. From Eq. (1), one can see that as the density of the particles or their charge is increased, or their kinetic energy is lowered, the coupling constant increases.

The discovery of plasma crystals caused some excitement in the plasma physics community, as well as in certain sections of solid-state physics. In this latter field, the hope is that plasma crystals may be employed as a "model system" to study dynamic processes, in particular, in the emerging field of nanocrystals. Reports have even been published on the phase diagrams of plasma crystals [38].

Figure 2 [39] shows the trajectories of particles under different plasma conditions that determine the "structure" of the particle ordination. Particle disposition was monitored by a charge-coupled device (CCD) camera. Figure 2a corresponds to a crystalline particle arrangement that is hexagonal. In Figure 2b, the hexagonal structure is not as well established. The mobility of the particles has increased and some local changes in the structure have appeared. Further changes in the structure are shown in Figure 2c. A decrease of gas pressure in the discharge led to a liquid phase, where a scale length of one lattice constant was found, and to a gas phase, where particles moved randomly (the Coulomb interaction was so weak at this stage that particles only interacted when they came close together); see [39].

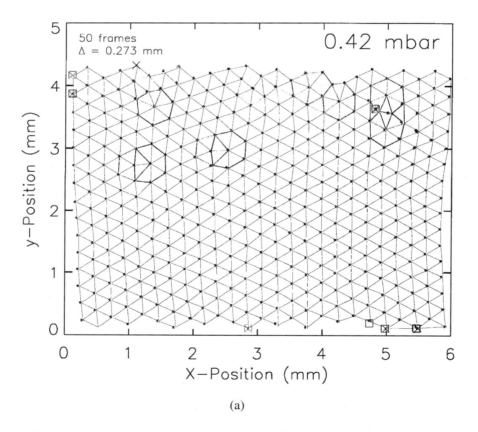

(a)

Fig. 2. Trajectories of melamine/formaldehyde spheres of 6.9 ± 0.2 μm suspended in a krypton rf plasma as observed over successive video frames (the number of frames and the mean lattice constant are indicated in the upper left-hand corner; the time between successive frames was 0.02 s) at different neutral gas pressures (in the upper right-hand corner) corresponding to different phases of the plasma crystal. (a) The crystalline phase and (b) and (c) a transition phase (possibly hexatic). Lower gas pressure led to liquid and gas phases (see [39]). Reprinted with permission from H. M. Thomas and G. E. Morfill, *J. Vac. Sci. Technol. A* 14, 501 (© 1996 American Vacuum Society).

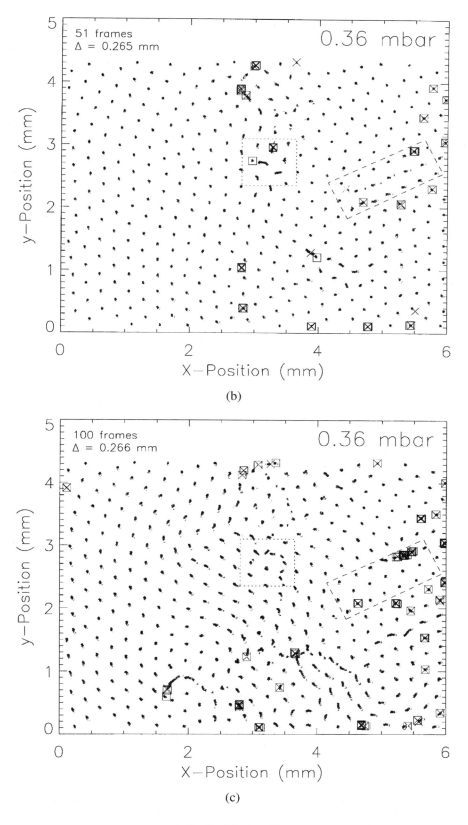

(b)

(c)

Fig. 2. (*Continued.*)

2.3. Powders in Thin-Film Processing Plasmas

The desire to maintain profitability motivates the semiconductor industry to improve manufacturing efficiency. These improvements typically include increasing the device speed and decreasing the cost per function. These require reduced device dimensions, increased wafer diameters, and increased device yields. One of the main causes of a reduction in the yield is the introduction of an unacceptable level of contamination in the course of the handling and processing of the wafer. Improvements in air filtration, clean room garments, and methods for wafer transport have dramatically reduced the contribution of the modern clean room environment to particle contamination. The largest source of contamination is now contributed by process-induced contamination [40, 41]. In high-volume manufacturing, approximately 75% of all yield losses are due to particles and as many as 90% of these particles are induced by the process itself [42].

The concern of the microelectronics industry for "process-inherent" particle contamination may be illustrated with the advertisement shown in Figure 3. The publicized apparatus contaminates the wafer with a controlled amount of particles in order to calibrate the wafer inspection systems of a device fabricator.

A review on the formation of particles in thin-film processing plasmas written by Steinbrüchel was published in 1994 [43].

Historically, plasma-generated particles during thin-film processing were first observed *in situ* in a silane–argon deposition plasma [1]. However, it became apparent that particles are also produced in etching plasmas, specifically in sputtering and in reactive ion etching

Fig. 3. The concern of the microelectronics industry about the contamination of devices from particles generated in the discharge is illustrated by this advertisement. This apparatus contaminates the silicon wafers with a selected amount of particles in order to calibrate wafer inspection systems.

(RIE) plasmas (see Section 2.3.1). Laser light scattering (LLS) techniques have been the most widely used to detect particles. The sensitivity of this technique to particle concentration and particle size depends on the geometrical configuration of the LLS setup, the laser intensity, and the detection method.

Since the earliest studies, LLS showed that particles accumulate in a quite localized region of the discharge, near the sheath edge at the powered electrode of a parallel-plate-type reactor [2]. This was the first indication that particles were negatively charged, and so they were suspended in the direction perpendicular to the electrodes. When the discharge was turned off, the particles fell onto the substrate or else they were swept out into the exhaust by the gas flow. The position of the particle cloud was strongly dependent on the discharge parameters, such as the reactor geometry, the gas pressure, the flow rate, the flow pattern across the electrode, and the temperature distribution in the reactor. We can briefly review some of the observations reported in sputtering, RIE, and deposition plasmas.

2.3.1. Particles in Sputtering Plasmas

Particles have been observed in systems chemically as simple as the sputtering of an elemental target in a noble-gas discharge. Particles have been observed in Ar plasmas sputtering Si [44–49], SiO_2 [50–52], graphite [53], Lexan and Teflon [50], as well as Al and Cu [52–55]. In sputtering plasmas, the particles consisted mainly of the target material [44, 45], although atoms coming from the electrodes were also reported [45, 46]. Whatever the cause, it was clear that the particles must have nucleated and grown in the gas phase from atoms removed by the sputtering process.

These studies showed that the electrical characteristics of the discharge have a marked effect on the appearance of a particle cloud [44]. Threshold behavior versus both rf power and pressure for the powder appearance was reported for sputtering of Si and SiO_2 in Ar by Yoo and Steinbrüchel [47–49]. At the onset of cloud appearance, particles were typically approximately 200 nm in diameter and quite monodisperse. Further particle development led to larger particles and wider particle size distributions.

Selwyn et al. [45, 51] reported the optical characterization of particle traps. In their reports, it was shown that particles were trapped not only on the plasma sheaths but also in a ring over the edge of the wafer. Figure 4 is a rastered LLS photograph showing trapped particle clouds over three Si wafers [56]. Selwyn et al. emphasized that any material or geometrical discontinuity on the wafer-holding electrode may give rise to a particle trap above it [57–59]. Further research on the particle traps induced by particular geometries of the electrodes was performed by other authors [60–64].

By the combination of LLS and a Langmuir probe, Carlile et al. [46] showed that the particle traps coincided with the localized maxima of the plasma potential above its surrounding value. Jellum et al. [55], in experiments on the rf sputtering of Al and Cu, investigated the effect of the electrode temperature on particle formation and particle cloud position between electrodes. These authors demonstrated a thermophoretic effect on particles, as they tended to move to the colder electrode.

2.3.2. Particles in Reactive Ion Etching

The first observations of particles on RIE plasmas were also reported by Selwyn et al. [57]. They reported particle clouds in CCl_2F_2/Ar, $O_2/CCl_2F_2/Ar$, or $SF_6/Cl_2/Ar$. For the same discharge conditions however, no particle cloud appeared in Cl_2/Ar, CF_4/Ar of CCl_2F_2/Ne. They combined LLS with laser-induced fluorescence (LIF) of Cl atoms and showed that both LLS and LIF signals were localized at the sheath edge. This led the authors to conclude that Si–halide etch products, with their propensity to form negative ions, may be involved in the formation or nucleation of particles.

Fig. 4. A photograph of a rastered laser light scattering image showing trapped particle clouds over three closely packed Si wafers on a graphite electrode. The particle clouds have a ring shape that reproduces the edge of the wafers. Reprinted with permission from G. S. Selwyn, *Plasma Sources Sci. Technol.* 3, 340 (© 1994 Institute of Physics Publishing Ltd.).

Yoo and Steinbrüchel [47] hypothesized, from observations of CCl_2/Ar etching of silicon, that the nucleating species must originate from the substrate as byproducts of the etching process. Stoffels and co-workers [65] reached similar conclusions, on the basis of infrared spectroscopy of 10% CCl_2F_2/Ar discharges.

SF_6/Ar etching plasmas have been shown to generate particle clouds [66]. Garrity and co-workers [67] proposed mechanisms to explain the formation processes of particles, including gas phase precursor formation, nucleation, and coagulation.

Kushner and collaborators developed a model for transport and agglomeration of particles in reactive ion etching plasma reactors [68].

2.3.3. Particles in Deposition Plasmas

Studies on the occurrence of particles in rf glow discharges of silane-based gas mixtures will be analyzed in detail in Section 4 because most of the basic knowledge of particle formation and plasma–particle interactions is based on them. However, in this section, we review preliminary studies mainly concerned with the particle contamination effect during the processing of microelectronic materials and devices.

The first detailed studies of particles on deposition plasmas were those of Spears and co-workers [1–3] starting in 1984. These authors investigated the position of the particle cloud in an Ar-diluted silane discharge and the influence of gas pressure, silane concentration, and flow rate on the appearance of the particle cloud. For the first time, they used LLS to gain information on the particle concentration and size distribution. Although later studies provided more accurate values, their measurements allowed them to argue that the particle size distribution was quite narrow. These authors, however, did not report the time course of the particle cloud development.

Around 1990, Watanabe [69–71], as well as Lloret [72], Verdeyen [73, 74], and others [75, 76], examined the modulation of the discharge as a method to control the appearance of particles in the discharge and to modify the thin film microstructure. Watanabe et al. [70] showed that with a modulated discharge it was possible to reach higher rf powers and, thus, obtain much higher deposition rates, without forming particles. Bertran and co-workers [72, 75, 76] studied the microstructure of a silicon thin film grown under different plasma modulation frequencies, and claimed that negatively charged species such as anions and particles contributed to film growth during the plasma-off times.

Later research on particle nucleation, growth dynamics, and plasma–particle interactions contributed to the present basic knowledge of the formation of particles in low-pressure, low-temperature plasmas and these will be discussed in Section 4.

3. TECHNOLOGY

3.1. Low-Pressure, Low-Temperature Plasmas

Plasma is a state of matter that consists of electrons, negatively and positively charged particles, and neutral atoms or molecules moving in random directions. Matter in this state is more highly activated than in the solid, liquid, or gas state. Most of the matter in the universe is in a plasma state. Occasionally, particles may nucleate and reside in the plasma. The plasma is electrically neutral. Therefore, in the absence of charged particles,

$$n_e + n_i^- = n_i^+ \tag{2}$$

where n_e is the electron density, n_i^- the anion density, and n_i^+ the cation density.

Low-temperature, low-pressure plasmas are the most often encountered, both in the microelectronics industries and in research laboratories. They are induced by applying an electric field to a low-pressure gas. This electrical excitation may be direct current (DC) or alternating current (AC). Commonly, the plasma is excited by a 13.56-MHz rf electrical field. This is the frequency allowed by the international authorities, because it does not interfere with communication signals. The electric field ionizes the gas and accelerates the electrons, which impact on neutral species and provoke their ionization. These new ionizations compensate for the loss of electrons and ions by mutual recombination or ambipolar diffusion to the walls. Laboratory plasmas are far from the equilibrium and the electron and ion temperature are markedly different. Although the electron temperature may be around 4×10^4 K (equivalent to 5 eV), ions are too heavy to follow the electric field and remain close to the gas temperature. For that reason, these discharges are referred to as low-temperature discharges or cold plasmas.

As indicative values of the external parameters for a low-pressure, low-temperature discharge, pressure ranges between 1 and 200 Pa and the rf electrical power typically lies between a few mW/cm^2 and 500 mW/cm^2. Concerning the internal parameters of the plasma, the plasma-bulk positive ion and electron density, n_i^+ and n_e, lie between 10^8 and 10^{10} cm^{-3}. However, in SiH$_4$ plasmas, the negative ion density, n_i^- (or negative charge density as powders), can exceed n_e by an order of magnitude [77–79]. The ratio n_i^+/N or ionized fraction of the gas ranges from 10^{-7} (low power and relatively high pressure) to 10^{-3} (high power and low pressure).

The plasma chemistry of the discharge is a consequence of the inelastic collisions between electrons and neutral or charged species, and of their recombination. Positive ions, anions, neutral radicals, excited molecules, and photons are products of these inelastic collisions. For instance, Table III shows the main dissociative reactions of the silane molecule resulting from electronic impact. To quantify this plasma chemistry, it is necessary to determine the energy distribution function (EDF) of the electrons in the discharge and the effective cross section for each reaction. Many reports have been devoted to the determination of these reaction cross sections [81–83].

Among rf plasmas, the capacitively coupled discharges are the most widely used. In this case, the electrical field is driven to the electrode through a blockage capacitor. Figure 5

Table III. Products of the Dissociative Collision of One Electron with the Silane Molecule [80]

	Products	Threshold energy (eV)
	$SiH_2 + 2H + e^-$	8(?)
	$SiH_3 + H + e^-$	(?)
	$SiH + H_2 + H + e^-$	10(?)
	$Si + 2H_2 + e^-$	12(?)
	$SiH^* + H_2 + H + e^-$	10.5
	$Si^* + 2H_2 + e^-$	11.5
$e^- + SiH_4 \rightarrow$	$SiH_2^+ + H_2 + 2e^-$	11.9
	$SiH_3^+ + H + 2e^-$	12.3
	$Si^+ + 2H_2 + 2e^-$	13.6
	$SiH^+ + H_2 + H + 2e^-$	15.3
	$SiH_3^- + H$	6.7
	$SiH_2^- + H_2$	7.7

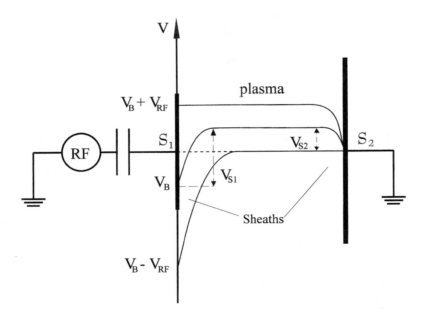

Fig. 5.   Scheme of the distribution of potentials between electrodes at different instants of the rf cycle in a capacitively coupled rf discharge. The plasma is confined between electrodes. The surfaces of the electrodes are S_1 and S_2. V_{S_1} and V_{S_2} are the time-averaged potentials between the plasma and each electrode. V_B is the resulting DC bias.

presents a scheme of the electric field between the electrodes at different instants of the rf cycle. The electrodes are usually placed inside the vacuum chamber.

Once the plasma is ignited, three different regions can be distinguished. In the central region, the plasma is electrically quasi-neutral and the negative charge density caused by electrons, anions, and charged particles equals the positive ion density. Between the plasma and the electrodes, there are space charge regions, the sheaths, that are mostly positive because of the difference in mobility between electrons and ions. The electric field in the sheaths tends to confine the negative species in the discharge and to accelerate the positive ions toward the walls.

In capacitively coupled reactors, a DC self-bias, V_B, may appear on the electrode connected to the rf generator, depending on the ratio between the area of the rf electrode and the effective area of the grounded walls. The time-averaged potential drops of each electrode through the sheaths, V_{S_1} and V_{S_2}, follow an inverse power law of S_1/S_2:

$$\frac{V_{S_1}}{V_{S_2}} = \left(\frac{S_1}{S_2}\right)^n \qquad 1 \leqslant n \leqslant 4 \tag{3}$$

where n depends on the discharge conditions [84].

It is commonly accepted that three different mechanisms of power dissipation in an rf discharge may dominate at different plasma conditions. These mechanisms are referred to as regimes [85–87].

Secondary Electron Emission (γ Regime). The bombardment of ions that have been accelerated by the sheaths on the electrodes can cause a strong emission of secondary electrons that will be dependent on the ion energy and the nature of the electrode. These secondary electrons are accelerated by the sheath electric field toward the plasma bulk, where they can cause ionization, dissociation, or excitation of neutral molecules [88]. This may be the main power dissipation mechanism at high power density.

Sheath Heating (α Regime). The movement of the plasma sheath, the successive contraction and expansion during the rf cycle, increases the energy of the electrons during sheath expansion. At high pressure, the sheath-heating mechanism is analogous to a surfer on a wave, and for that reason it is referred to as "wave riding" [78, 85].

Resistive Behavior (η Regime). If there is a depletion of the electron density in the discharge, an electric field builds up in the plasma zone, which compensates for the electron losses by increasing the ionization rate. These electron losses may be caused by electron attachment in electronegative gases such as C_2F_6 or SiH_4 or in the particles formed in the discharge. When this electric field builds up, the voltage distribution across the electrodes appears as indicated in Figure 6. This is referred to as the Joule-heating mechanism because the plasma impedance shifts from the capacitive behavior of the α regime to resistive behavior [88–90].

3.2. The Reactor

The widespread use of capacitively coupled rf reactors in the microelectronics industry and applied research centers has led to a diversity of reactor characteristics and, in particular, reactor geometries. However, all of them have features in common, which are discussed in this section. Special attention is paid to the particularities of rf reactors used for powder formation.

Gas System Management. Precursor gases provide the elements that the material grown in the discharge will contain. Inert gases influence plasma characteristics such as electron temperature and ion and neutral kinetic energy. The purity of the gases determines the presence of impurities in the material obtained. Materials of high purity can be produced in low-temperature, low-pressure plasmas because high vacuum can be achieved and precursor gases have low contaminant levels.

Special care should be taken when handling toxic, pyrophoric, or corrosive gases. For instance, silane, which is the gas most used in plasma-enhanced chemical vapor deposition (PECVD) processes, undergoes spontaneous combustion on contact with the atmosphere.

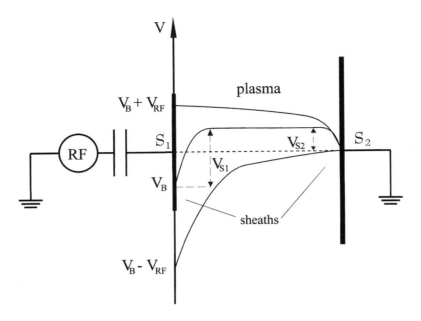

Fig. 6. Distribution of potentials in a capacitively coupled rf discharge in the η regime. Losses of charge caused by electron trapping in particles are compensated for by the electric field in the plasma bulk. The plasma impedance is mainly resistive.

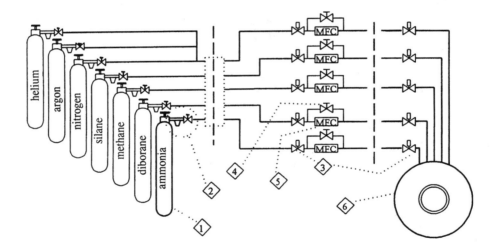

Fig. 7. Scheme of a typical arrangement of gas lines in a plasma reactor: (1) gases, (2) pressure reducers, (3) pneumatic valve, (4) manual valve, (5) mass flow controllers, and (6) reactor chamber.

Moreover, it is highly toxic (threshold value: 0.5 ppm). Diborane undergoes spontaneous combustion above 38 °C and it is extremely toxic (0.1 ppm). It is usually diluted in argon or hydrogen. Ammonia is also toxic (25 ppm) and corrosive, so it requires gas lines and sealing O-rings of chemically resistant materials such as Viton.

Concerning the toxicity of gases, powder generated in a plasma may have gases adsorbed on particle surfaces. Therefore, the reactor operator should take the toxicity of these gases into account when he or she opens the reaction chamber after a powdery discharge.

Bottles of reactive gases should be connected to independent lines in order to avoid a reaction between the gases before they enter the reaction chamber. Figure 7 shows a

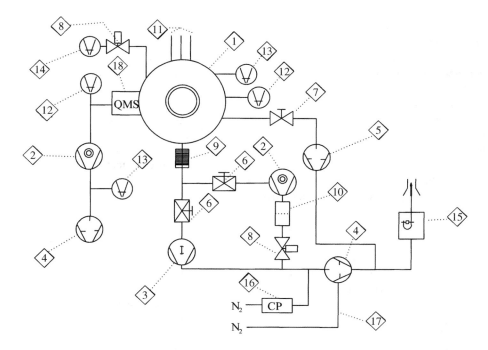

Fig. 8. Scheme of a vacuum pumping system. The turbomolecular pump is used to achieve the ultimate pressure before starting the discharge. During the discharge, gases are evacuated by means of a roots pump followed by a rotary pump. (1) reactor chamber, (2) turbomolecular pump, (3) roots vacuum pump, (4) rotary pump, (5) auxiliary rotary pump, (6) gate valve, (7) manual valve, (8) pneumatic valve, (9) flange connector to attenuate vibrations, (10) filter of alumina, (11) gas inlet, (12) Pirani vacuum gauge, (13) Pening vacuum gauge, (14) capacitive vacuum gauge, (15) exhaust, (16) pressure controller, (17) gas ballast, and (18) mass spectrometer.

representative scheme of the gas lines that connect each bottle with the reaction chamber. Every line is provided with a mass flow controller to regulate the gas flow. A bypass to the mass flow controller must be installed in order to evacuate lines and bottles in case it does not operate correctly.

Accessibility to the reaction chamber and simple cleaning routines should be foreseen when the reactor has to be used for powder generation.

Vacuum System Design. If the reactor has to be used for powder generation, some precautions should be taken against pumping system damage. First, powders should not be evacuated through particle-sensitive vacuum pumps such as turbomolecular pumps. Figure 8 illustrates a vacuum system design in which the turbomolecular pump is used only to attain the ultimate pressure in the reaction chamber. Gas evacuation during the discharge is provided by a roots vacuum pump followed by a rotary pump, which are quite insensitive to powders. A second precaution consists of preventing pumps that evacuate pyrophoric gases during the process from evacuating oxygen. Therefore, rough vacuum is provided by an auxiliary rotary pump independent of the rest of the vacuum system. If this precaution is not taken, there is a risk of violent reaction between atmospheric oxygen and pyrophoric gases trapped in the oil of the rotary pump.

The control of gas pressure during the discharge can be attained by several methods. It should be considered that powder generated in the plasma may stick to pipes and valves causing their conductance to diminish. If the flow of gas entering the chamber is constant, pressure in the reactor will increase. Therefore, control of pressure in real time is required. For this purpose, it is possible to inject a gas flow, controlled in real time, to the rotary pump inlet. Then, the evacuation rate of the pumping system is varied and the pressure in the chamber can be controlled without varying the gas flow that enters into the reactor.

The Reaction Chamber.   The shape of the reaction chamber determines the dynamics of the gas flow inside the chamber, the geometry of the electric potential between electrodes, and the DC bias of the rf electrode. Therefore, results from different authors concerning experiments or materials obtained with the same discharge parameters but different chamber geometries cannot be compared without precaution.

Two reaction chamber geometries are most often encountered. On the one hand, some researchers prefer closed boxes in order to obtain a laminar gas flow parallel to the electrodes (see Fig. 9b). In this chamber geometry, the large grounded surface compared to that of the rf electrode causes a strong DC bias in the rf electrode [see Eq. (3)]. This DC bias may have interesting effects in thin-film deposition as it enhances the ion bombardment on the film placed on the rf electrode. However, it causes a strong asymmetry in the distribution of the electric potential across electrodes.

Nevertheless, reaction chambers designed to produce symmetric discharges are reported more often. In this case, the shape and the arrangement of the electrodes are designed so that no self-bias appears on the rf electrode. Usually, both rf and grounded electrodes are cylindrical and placed a few centimeters apart, without lateral walls (Figs. 9a and 9c).

The gas dynamics in the discharge are determined by diffusion or convection. High pressure (which lowers diffusivity), high flow, or long distances between the gas inlet and the exhaust gas ports favor convection. In this case, the partial pressure of molecules varies in the direction of the flow and then there may exist spatial inhomogeneities in the discharge. In contrast, low pressure and low gas flow with well-distributed gas injection points (like shower-type injectors) lead to diffusive gas dynamics. Then, the partial pressure of the molecules is spatially homogeneous. Figure 9 shows that symmetric electrodes with shower-type gas injectors lead to homogeneous gas distribution, whereas gas dynamics in closed boxes are dominated by convection.

The distribution of temperatures in the reactor also influences the spatial distribution of the discharge. As is discussed in Section 4.2, nanometric particles are very sensitive to thermophoretic forces that appear if there is a temperature gradient. Particles tend to go away from heated walls and move toward the cold zones of the reactor. It is possible to gain advantage of this behavior to collect the particles efficiently. However, in basic studies of dusty plasmas, it is preferred that particle dynamics not be dominated by these forces. In any case, it should be decided whether a thermally homogeneous reaction chamber is desired or it is preferred to have cold zones that attract particles.

The Radio Frequency Circuit.   Usually, an rf power source, working at 13.56 MHz, excites the plasma. There are many generators available on the market, most of which can be

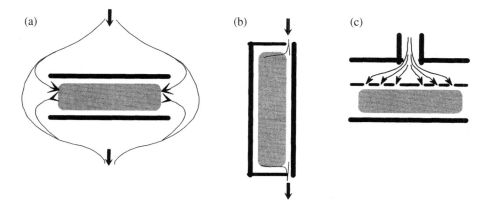

Fig. 9.   Different configurations of gas injection in a PECVD reactor. Shower-type injectors lead to diffusive gas dynamics and homogeneous discharges.

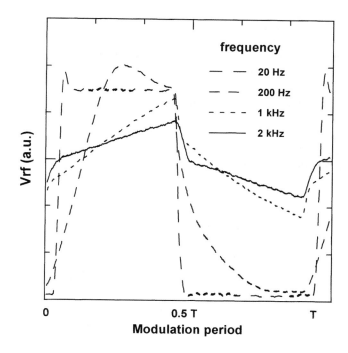

Fig. 10. Peak-to-peak rf voltage, V_{RF}, during one cycle of square-wave modulation at four frequencies (0.02, 0.2, 1, and 2 kHz). At 1 and 2 kHz, the rf generator cannot reproduce the square-wave modulation signal.

externally driven to generate a square-wave-modulated rf signal. The power modulation, the alternation of plasma-on and -off periods, is a method to control the formation of powder in low-pressure, low-temperature plasmas [69–76]. For that purpose, the modulation signal (usually square-wave) created by a function generator is sent to the rf generator, which then produces an rf wave with amplitude proportional to the modulation signal.

Some rf generators are not able to produce a square-wave-modulated signal if the modulation frequency is above approximately 1 kHz. Figure 10 illustrates this effect. It shows the rf voltage during one cycle of square-wave modulation for different modulation frequencies. Indeed, at low frequencies, the rf voltage is a square wave, but as the frequency increases the generator cannot reproduce the modulation signal and then the voltage tends to an undulation. The amplitude of the voltage does not reach its expected value during the plasma-on period and does not attain zero during the afterglow. This fact has serious consequences. The plasma modulation is performed so that negative species confined by the plasma sheaths leave the plasma during the afterglow. This requires the sheath to collapse in the afterglow, so the rf voltage must be zero during this period. Therefore, the ability of the rf power source to produce a square-wave-modulated signal at high frequencies should be checked either with a voltage probe or by monitoring the optical emission of the modulated discharge.

To transmit the rf power from the generator to the electrode, it is necessary not only to drive it through a line but to match the plasma impedance to that of the generator. For this purpose, an impedance matchbox is connected in between. Usually, the rf generator has an output impedance that is resistive and of 50 Ω, whereas the plasma impedance is mainly imaginary. The impedance matchbox guarantees that the load to the rf generator is always 50 Ω, irrespective of the plasma impedance. Ideally, there is no reflected power from the matchbox to the electrode. Nevertheless, between the matchbox and the electrode, with or without plasma, the addition of incident and reflected power leads to the establishment of a stationary wave. Care should be taken to ensure that electrical lines, connectors, and vacuum throughput have 50 Ω of characteristic impedance, in order to minimize the reflected power component.

When the plasma is ignited, the impedance resulting from the plasma itself plus the reaction chamber elements (electrodes, connectors, etc.) evolves with plasma characteristics. It is necessary to know the power actually dissipated in the discharge in order to compare results from different reactors and to correlate the plasma characteristics with external discharge parameters. This measurement is not obvious. Experimental evidence indicates that up to 90% of the total power delivered by the generator is lost in the circuit, electrical connections, and impedance matchbox, depending on the plasma conditions [91].

Several methods allow the correct measurement of the rf power dissipated in the plasma. For instance, the integral method relies on the determination of the instantaneous voltage and intensity close to the electrode [92]. An alternative method, less demanding with respect to the instrumentation required, is the subtractive method. In this case, the power dissipated in the discharge, P_{PL}, is estimated from the difference between the incident power with plasma, P_{TOT}, and without it, P_{VAC}, for the same rf voltage, V_{RF} [91]:

$$P_{PL} = P_{TOT} - P_{VAC}|_{V_{RF}=ct} \qquad (4)$$

Therefore, it is required only to monitor the incident power and the rf voltage close to the electrode. This method assumes that the plasma impedance, Z_{PL}, is in parallel with that of the reaction chamber and impedance matchbox. Figure 11 shows an electrical scheme of these loads. When the plasma is not ignited, the power delivered, P_{VAC}, at an electrode voltage, V_{RF}, corresponds to the circuit losses. This method assumes that, for a given rf voltage, the power losses will not depend on whether the plasma is ignited. Figure 12 shows the experimental estimation of the power dissipated in the discharge.

Once the actual power dissipated in the discharge can be measured, it is possible to determine the efficiency of the discharge, $\eta = P_{PL}/P_{TOT}$. Experimental evidence indicates that the discharge efficiency depends on parameters such as the pressure (Fig. 13), the rf power, and the nature of the gas (Fig. 14).

3.3. *In Situ* Particle Characterization Techniques

Since the pioneering studies of Spears et al. [1–3], special techniques have been developed to monitor the particle concentration and the size of the dust suspended in the plasma. Among these techniques, commonly referred to as diagnostics, laser light scattering (LLS) has been the most widely used. Moreover, instrumentation for both characterization of materials, such as infrared spectroscopy, and gas monitoring, such as mass spectrometry, have been adapted to investigate the precursors of particles.

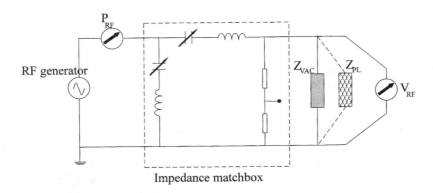

Fig. 11.   Equivalent electrical circuit assumed in the subtractive method to determine the rf power dissipated in the discharge. The figure shows the arrangement of the rf generator, impedance matchbox, voltage probe, impedance of the reactor in vacuum, and plasma impedance.

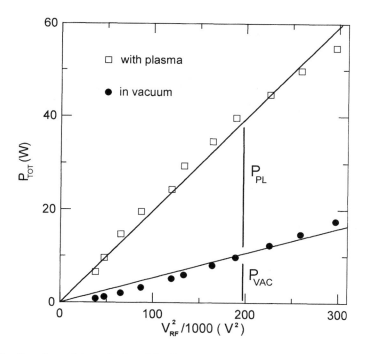

Fig. 12. Experimental determination of the rf power dissipated in the discharge. The dots correspond to the rf power dissipated when the reactor is in vacuum, whereas the squares correspond to the dissipated power when the plasma is ignited.

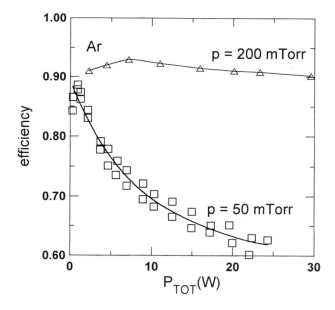

Fig. 13. Dependence of the discharge efficiency of argon discharges on the delivered power. In discharges at low pressure, the percentage of power losses can be very important.

In this section, we review briefly the most widely used diagnostics of particles in dusty plasmas. Each technique is especially suited to monitor a particle size range, from particle precursors of a few atoms to coagulated particles of around 100 nm.

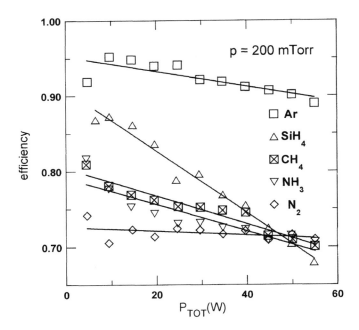

Fig. 14. Dependence on the delivered power of the efficiency of the discharge for different gases. There is a clear relationship between the nature of the gas and the discharge efficiency.

3.3.1. Laser Light Scattering

Elastic light scattering by spherical homogeneous particles is described by the Rayleigh–Gans theory for particles smaller than about $R_P < 0.1\lambda$ and by the Lorenz–Mie theory for larger particles [93, 94]. Particles below approximately 2 nm are not observable experimentally by LLS. The experimental arrangement allows measurement of the scattered light at a single angle from the incident beam (usually 90°) [95] or at multiple angles [45, 96–98].

The objective is to determine the particle concentration, N_P, the particle size, R_P, and the particle refractive index, $m = m_r + im_i$. Occasionally, LLS is performed simply to monitor the onset of powder appearance [99]. The measured magnitudes are the scattered-light intensity polarized both perpendicular and parallel to the polarization of the incident beam, and the normalized transmitted intensity ($W, W_n, T/T_o$). With these three measurements it is only possible to find the four unknown quantities (R_P, N_P, m_r, m_i) by an iterative algorithm (if single-angle LLS is used) [95]. Alternatively, one of the unknown quantities should be estimated from *ex situ* analyses. Indeed, some reports estimate the refractive index of particles in silane discharges as that of amorphous silicon thin films. Nevertheless, several authors use the scattering angular dissymmetry to solve this problem. The scattered light at two [45, 97] or even three angles [96] has been measured. Figure 15 shows the experimental arrangement for a three-angle laser light scattering experiment.

The laser light scattering method is nonperturbative and easy to install. However, it is necessary that in the analyzed plasma zone the particles do not suffer any drift in order to monitor the particle evolution [95]. Therefore, only plasma conditions that lead to stable powder development can be investigated by these methods. Moreover, it must be assumed that the particles are spherical and monosized.

Hayashi and Tachibana [100] proposed a sophisticated LLS method to derive the size, size dispersion, density, and refractive index of the particles by measuring both the phase difference and the intensity ratio between polarized components of light scattered by particles.

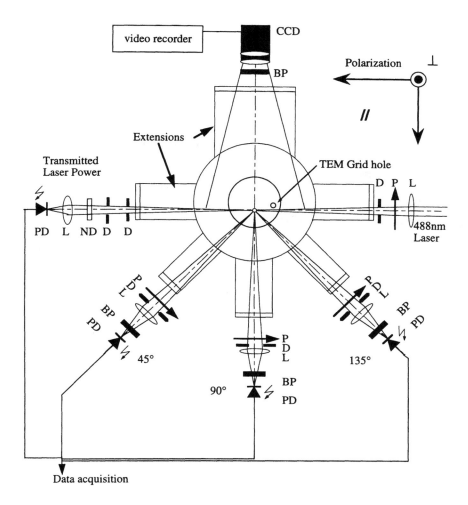

Fig. 15. Schematic top view of the experimental arrangement for a three-angle LLS. The figure shows the plasma reactor, the laser transmission measurement, and the three-angle detection of the laser beam scattering. P: polarizer, L: lens, D: diaphragm, BP: 488-nm bandpass filter, ND: neutral density filter, and PD: photodetector. Reprinted with permission from C. Courteille, Ch. Hollenstein, J.-L. Dorier, P. Gay, W. Schwarzenbach, A. A. Howling, E. Bertran, G. Viera, R. Martins, and A. Macarico, *J. Appl. Phys.* 80, 2069 (© 1996 American Institute of Physics).

3.3.2. Mass Spectrometry

Among the diagnostics that are sensitive to particles below 2 nm, mass spectrometry can provide the most complete data. Mass spectrometry monitors positive and negative ions as well as neutral molecules and radicals, and it is able to resolve an atom in the cluster. The experimental arrangement used most is quadrupole mass spectrometry [95, 101–103], although results obtained with time-of-flight [103, 104] or Fourier transform mass spectrometry [105] have also been reported. A differential pumping between the plasma chamber (~1 mbar) and the mass spectrometer housing is required. This technique allowed researchers to monitor the time evolution of positive and negative ions in dusty and nondusty discharges. Section 4.1 discusses the mass spectrometry results obtained in modulated discharges that allowed the determination of the polymerization pathway that leads to particle formation in silane plasmas.

The localization of the ion extractor head of the mass spectrometer in the discharge has an influence on which plasma species are monitored. Usually, the ion extractor head is positioned with its axis in the electrode plane, beyond one electrode.

To the author's knowledge, published results on mass spectrometry applied to dusty discharges reported masses up to 1300 amu, which corresponded to clusters of around 1 nm [103].

3.3.3. Other Techniques

Several diagnostics have been developed to detect particles on the nanometer scale on the basis of the interaction between a highly energetic laser beam and the particles. On the one hand, Stoffels et al. [106] showed that particles could be detected with high sensitivity by laser heating of particles and recording blackbody-like emission. Another detection method for small particles was based on laser-induced particle explosive evaporation (LIPEE). Particles were irradiated with an excimer laser beam and then the emission resulting from bremsstrahlung was monitored [107].

Laser-induced photodetachment combined with a microwave resonance technique was specifically developed to detect negative ions. A high-power pulsed laser was used to irradiate the discharge and the extra shift in the resonance frequency caused by the photodetached electrons was measured [108].

4. DEVELOPMENT OF PARTICLES IN SILANE PLASMAS

4.1. From Molecules to Particles in Silane Plasmas

Discharges of silane and mixtures of an inert gas with silane are the systems most widely studied to identify the reaction pathways that lead to the formation of particles. Although the earliest reports on the formation of powder in a low-pressure, low-temperature plasma claimed that particle growth was a heterogeneous nucleation process on a cluster sputtered from the reaction chamber walls by electron bombardment [109], it is well established nowadays that particle nucleation is a homogeneous process in the gas phase.

The chemical mechanisms of particle nucleation and growth are not completely understood. However, there is general agreement on the role of negative ions in the polymerization pathway [74, 101, 110–113]. In spite of that, the great differences in the time for particle formation between low-power discharges (few or 10–100 mW/cm^2), where it can take several minutes, and high-power discharges, where particles appear in a few milliseconds, suggest that the polymerization pathway may vary depending on the discharge conditions [103, 111].

4.1.1. Clusters Below the Nanometer Scale

At the first stages of growth, the particles are too small to be monitored by direct laser light scattering techniques. Therefore, mass spectrometry, with different mass ranges, has been used by several research groups to analyze the ionic and neutral species [74, 101, 114–117]. The analysis of neutral and positive ions in the plasma does not present particular problems. However, negative ions are electrostatically confined in the discharge by the plasma sheaths and they cannot be detected by the mass spectrometer during continuous discharge. Therefore, negative ions can be detected only in the afterglow. During the afterglow of a silane plasma, the electron density decays in 30–50 μs, the sheath potential collapses, and negative ions can then escape [101, 114]. Therefore, the discharge should be modulated so as to permit the detection of negative ions. As discussed later, the alternation of short periods of plasma-on and plasma-off (the modulation of the rf power) is also a common technique to control powder development. Consequently, mass spectrometry can monitor negative ions provided the modulation frequency is low enough to allow the sheaths to collapse (~below 10 KHz) [101, 102, 115].

Detailed research on the presence of positive and negative ions and neutral radicals has been performed in modulated discharges by quadrupole mass spectrometry. In particular, the reports of Howling, Hollenstein, and co-workers are especially relevant [95, 101–103, 117, 118]. The main conclusion of their studies was that the polymerization that leads to particle formation is caused by anion polymerization. This conclusion is supported by the following evidence:

1. There is an anticorrelation between the detection of negative ions and the presence of powder in the discharge. When the modulation frequency is low enough to allow the anions in the discharge to escape during the afterglow, no powder was observed. In contrast, modulation frequencies that did not permit the complete depletion of anions during the afterglow led to dusty discharges because the anion polymerization continued during the following plasma-on period [95, 101, 102]. A dynamical model of polymerization was proposed, which showed good agreement with the experimental results [101].

   Figure 16 illustrates the fact that long plasma-off times are needed to allow the anions to escape from the discharge and thus prevent them from reaching high masses. It shows the dependence of anion flux on time during the afterglow of a 50-μs plasma pulse. When the afterglow lasted more than 150 μs, the anions detected were mainly monosilicon hydride anions, which were diffused in around 200 μs after each plasma pulse. No powder was formed in this situation. However, the relative contribution of heavy anions increased significantly when the afterglow period was reduced to 150 μs, whereupon the anions did not completely diffuse and thus were retrapped in successive plasma-on periods. This new situation gave rise to the formation of powders, demonstrating that this is a consequence of the polymerization of anions trapped in the discharge.

2. They showed that silane anions up to mass 1300 amu (the maximum of their experimental setup) could be found in the discharge, whereas the positive or neutral species never exceeded 200 amu for the same plasma parameters [101, 103]. These results had already been reported by other authors before these studies [115, 116].

3. By partial-depth modulation experiments, Howling et al. [103] provided further evidence of the anion polymerization pathway. They studied a modulated discharge at 1 kHz and observed that it did not produce powders because of its long afterglow. Moreover, they detected negative anions by mass spectrometry in these conditions. They then set the rf voltage during the plasma-off period to 15% of the rf voltage during plasma-on. This slight change, which was not expected to modify the dynamics of neutral radicals or positive ions, caused powder formation in the discharge. Besides, the negative ion flux detected during the total modulation was no longer measurable during the partial-depth modulation. In this second case, the low rf voltage during the afterglow was nevertheless high enough to prevent the sheath from collapsing and, therefore, negative ions remained trapped in the plasma. The continuation of anion polymerization led to particle formation.

Other theoretical studies based on rate coefficients for the various reactions in the silane plasma also support the hypothesis of the anion polymerization pathway [113].

Other studies, however, indicate that this anion trapping in the plasma is too slow to explain the fast particle growth that occurs in relatively high-power discharges, where powder is already evident after a few milliseconds [119, 120]. In this case, the positive ions are also disqualified because bottlenecks occur in cation nucleation [116] and, in addition, they are rapidly lost by evacuation across the plasma sheaths. Therefore, it is proposed that a rapid neutral pathway, such as SiH_2 insertion into higher silanes, may account for the fast particle growth [119, 120]. If these neutral clusters can grow to a critical size to start the

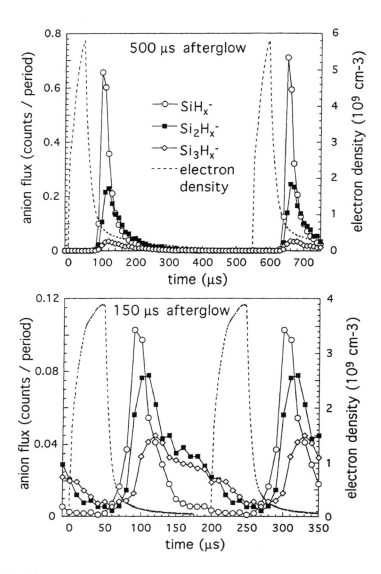

Fig. 16. Time-resolved measurements of electron density (dashed curve) and low-mass anions using 50-μs plasma pulses for two different afterglow intervals. Powder formed only for afterglow intervals less than or equal to 150 μs, which were not long enough to cut the polymerization chain as revealed by the detection of heavy anions. Reprinted with permission from Ch. Hollenstein, W. Schwarzenbach, A. A. Howling, C. Courteille, J.-L. Dorier, and L. Sansonnens, *J. Vac. Sci. Technol. A* 14, 535 (© 1996 American Vacuum Society).

coagulation or to become negatively charged by the rf plasma [121, 122] before escaping from the discharge, there is no need to invoke an anionic pathway.

4.1.1.1. The Structure of These Clusters

Very little is known about the atomic structure of these clusters or particle precursors. Concerning the hydrogen content, mass spectrometry studies revealed that only for anions with n silicon atoms ($n < 4$) was there a clear maximum signal that corresponded to $Si_nH_{2n+1}^-$, which is isoelectronic with its equivalent saturated neutral Si_nH_{2n+2} [102]. Heavier clusters showed a broad intensity distribution with respect to their hydrogen content. This hydrogen content on the cluster was dependent on the cluster size for small clusters ($n < 10$) [102]. It decreased strongly with the cluster size from an [H]:[Si] ratio of approximately 5 to a size-independent [H]:[Si] ratio of 4/3. These findings indicate that these clusters cannot be understood as silicon cores covered with hydrogen. In contrast, the monotonic decrease of the [H]:[Si] ratio reveals that these clusters are tridimensional, cross-linked structures [123] and that this cross-linking increases with cluster size.

The hydrogenation of these clusters at these initial moments has been studied by *in situ* infrared spectroscopy by Kroesen et al. [124]. Their measurements showed that no solid-state vibrational absorption of SiH or SiH$_2$ bands was observed until the nucleation and coalescence of particles had been completed. This indicated the polymeric character of the particles in the first stages of their development and the increase in cross-linking as the particles grow. *In situ* infrared absorption had previously been performed to study the growth of particles in an rf plasma of mixtures of CF$_4$, CF$_2$Cl$_2$, and CHF$_3$ with Ar by the same research team [125].

4.1.2. Particles of Approximately 1 to 2 Nanometers

The maximum cluster, or particle, analyzed by mass spectrometry ($\sim$1300 amu) corresponds to a particle with approximately 45 silicon atoms with a diameter of about 1 nm. Such particles are not observable by LLS. Other, less conventional laser-based techniques permitted the detection of particles in the 1–2-nm range with high sensitivity. Stoffels et al. [108] used laser-induced photodetachment in combination with a microwave resonance technique to measure the spatially averaged electron density, the spatially resolved negative ion density, and the charge on small clusters in an SiH$_4$/Ar plasma.

Another laser-based method to detect these small particles consisted of irradiating the plasma with a high-power excimer laser, which caused the evaporation of the particles suspended in the plasma [107]. Particle size was then determined from the consequent bremsstrahlung emission. This method was called laser-induced particle explosive evaporation (LIPEE).

These techniques showed that particle size development could be divided into two stages. In the first 10–50 ms, crystallites approximately 2 nm in size formed; they then coalesced into larger structures. The photodetachment signal was low before the coagulation, indicating that the total negative charge on the crystallites was very small [108]. Afterwards, the particles formed during the coalescence of the crystallites acquired a nonzero average negative charge, sufficient to trap them in the glow.

These studies revealed the influence of the gas temperature on the time for powder development. The higher the gas temperature, the longer was the delay in particle formation [107]. Figure 17 illustrates the dependence of the bremsstrahlung emission as a function of the duration of the discharge.

Particles of 2 nm and above are measurable by laser light scattering (LLS) techniques. The light scattering of small particles (below $\sim$40 nm for a 514-nm laser) is described by the Rayleigh theory (see Section 3.3.1). Laser scattering techniques are the most powerful methods to determine the particle concentration, N_P, and particle radius, R_P, *in situ*. Once the particles became detectable, the particle concentration grew up to a critical value in the range of 10^{17} m^{-3}. At this stage, the particle radius was about 2 nm [96]. Then the particle concentration dropped and the particle radius increased rapidly because of particle coagulation, which is described in detail in the next section.

At a size of 1–2 nm, each particle undergoes charge fluctuations, although the mean charge is negative because electrons have higher mobility than ions. Therefore, at any instant there is a fraction of not-charged particles. This fraction of neutral particles depends on the particle size. The smaller the size, the higher is the fraction of neutral particles. Therefore, these neutral particles can escape from the plasma even in a continuous discharge. As discussed in Section 6.1, this fact allowed Roca i Cabarrocas et al. [99] to deposit nanometric particles in silicon thin films in a continuous rf discharge.

The high particle concentration in the plasma before the coagulation dramatically affects the electrical characteristics of the discharge. The fact that particles act as electron traps

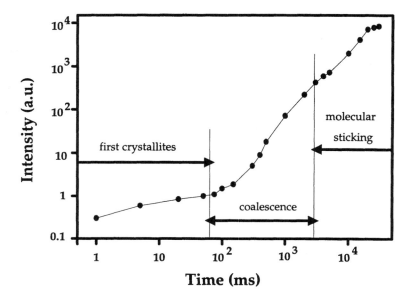

Fig. 17. LIPEE emission intensity as a function of rf plasma discharge duration for 1.2-sccm silane flow at ambient temperature. Reprinted with permission from L. Boufendi, J. Hermann, A. Bouchoule, B. Dubreuil, E. Stoffels, W. W. Stoffels, and M. L. de Giorgi, *J. Appl. Phys.* 76, 148 (© 1994 American Institute of Physics).

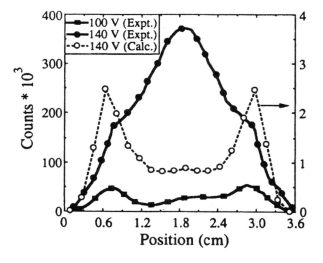

Fig. 18. Spatial profile of the measured optical emission intensity of SiH and calculated excitation rate of SiH (Particle-in-Cell Monte Carlo model). The discrepancy between the experiment and the model was solved by assuming a significant electron attachment on existing particles. Reprinted with permission from J. P. Boeuf and Ph. Belenguer, *J. Appl. Phys.* 71, 4751 (© 1992 American Institute of Physics).

causes an increase in the plasma electrical field, which tends to compensate for these electron losses. The discharge impedance shifts from capacitive to resistive [86, 89, 126]. This impedance transition is clearly shown by the measurement of the spatial time-averaged distribution of the optical emission of the discharge. During the transition, the emission profile changes from two peaks (maximum of the emission at the sheath edge, lower emission in the plasma) to a profile with a much more intense emission in the plasma volume (Fig. 18). This change is due to the fact that the distribution of the electric field in the plasma has to increase substantially to balance the electron losses by means of electron impact ionization.

4.1.3. Coagulation of Particles

Experimental evidence indicates that, after the establishment of this high particle concentration. there is a fast coagulation process during which the particle radius increases rapidly and the particle concentration drops by about two orders of magnitude [96, 107, 108, 127]. The particle diameter increases from about 2 nm to a fraction of a micrometer in a short time, which depends on the plasma conditions. Laser light scattering has been applied successfully to monitor this coagulation process [96, 98, 128]. It has also been shown by this technique that during coagulation the refractive index of particles decreases, which indicated evolution of the particle microstructure [95].

Several studies focused on an *ex situ* analysis of the size evolution of silicon particles after the onset of coagulation [129, 130]. Good agreement was obtained between the size determination by *ex situ* transmission electron microscopy and laser light scattering techniques [96]. As discussed in Section 4.1.1, the polymerization pathway leading to the formation of particles depends on the discharge parameters. Therefore, the temporal evolution of the particle size depends on the plasma conditions. As an example, Figure 19 shows the particle diameter versus time for an Ar-diluted silane discharge at low rf power [129] and for a silane–methane–ammonia discharge at relatively high rf power [130]. The coagulation process is similar in both cases. It appears that high rf powers and the lack of dilution in an inert gas favors fast particle formation. One of the general conclusions of these studies was that, for any plasma conditions, the diameter of the largest particles never exceeded a critical value in the range of 100 to 200 nm, irrespective of the duration of the plasma.

Various mechanisms have been proposed to explain this coagulation. Courteille et al. [96] described it as Brownian free molecule coagulation (BFMC), which assumes that the particles are neutral during the process. Good agreement with experimental results was

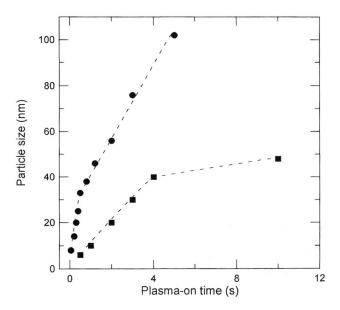

Fig. 19. Dependence of the particle size on time for two different discharge conditions. Squares correspond to a low-rf-power SiH_4/Ar discharge [129], whereas circles correspond to an $SiH_4/NH_3/CH_4$ discharge under conditions for high yield of powder formation [130]. Both dependences are qualitatively similar because of the coagulation process.

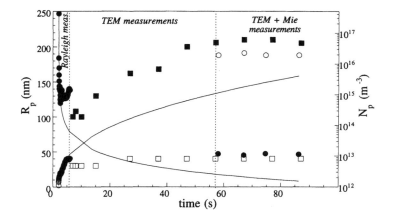

Fig. 20. Time development of the particle radius R_P (o) and number density (•) for early discharge times (Rayleigh scattering). The solid line shows the best fit of the Brownian free molecule coagulation model. The inset presents the first second of the scattered light time evolution (□). Reprinted with permission from C. Courteille, Ch. Hollenstein, J.-L. Dorier, P. Gay, W. Schwarzenbach, A. A. Howling, E. Bertran, G. Viera, R. Martins, and A. Macarico, *J. Appl. Phys.* 80, 2069 (© 1996 American Institute of Physics).

obtained for a wide time interval (~60 s). Figure 20 shows the time evolution of the particle concentration and particle radius during coagulation, measured by Rayleigh and Mie laser light scattering, compared with the expected evolution of a BFMC process.

Fridman et al. [127] proposed a k-body collision process in which they supposed that the particle interaction mean radius was proportional to the particle physical radius and that all direct collisions result in particle aggregation. According to this model, the coagulation rate increases strongly when the particle number density exceeds a threshold, causing phase transition–like behavior. This model is consistent with the particle concentration evolution during the first milliseconds of the discharge but would overestimate the coagulation rate for longer times.

Watanabe et al. [128] suggested that the collision of high-energy electrons of the discharge on small neutral particles may cause them to emit secondary electrons. The resulting positively charged particles would rapidly coagulate with other negatively charged particles. However, V. A. Schweigert and I. V. Schweigert [131] developed a more accurate theoretical study of the coagulation in a low-temperature plasma. By introducing the frequencies of charging by electrons and ions, they calculated the particle charge distribution and the fraction of particles with k charges. They deduced analytical approximations for the rate of coagulation, the particle radius distribution function, and the mean particle radius. The coagulation rate was considered to be a consequence of binary collisions, taking into account the electrostatic interaction between particles. Their calculations could satisfactorily explain most of the previous experimental results in Ar-diluted discharges. Indeed, besides reproducing the mean radius development over time, their model explained both the increase in the dispersion in the particle size and the depletion for particles of small radius when the mean radius increases. Finally, they deduced a logarithmic increase in particle size for long times, which accounted for the limited size of the particles observed. Figure 21 shows the evolution of the particle size distribution over time in an SiH_4/Ar discharge calculated according to the analytical expressions deduced in [131].

4.2. Forces on Particles

The earliest studies on the presence of particles in low-pressure, low-temperature plasmas indicated that dust is formed and then suspended in the plasma sheaths near the chamber walls [1–3]. Dust is suspended in layers near the electrodes in a pattern that depends on the

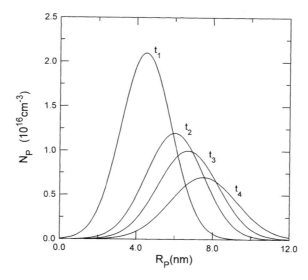

Fig. 21. Evolution of particle size distribution over time in an SiH$_4$/Ar discharge calculated from the analytical expressions deduced in [131]. Note that the size dispersion increases with time.

shape of the electrodes and the plasma characteristics. The spatial distribution of the dust particles in the discharge depends on the forces acting on them. These forces have been summarized by several authors [132–135] and include gravitational, neutral, and ion drag, electrostatic, thermophoretic, and pressure gradient forces. Which of theses forces is the largest depends on the plasma conditions.

There is a concern in the microelectronics industry to understand and control these forces, as they determine which zones of the reaction chamber the particles contaminate. On the other hand, they determine which fraction of the particles generated contributes to the film growth and which fraction is lost in the exhaust gas or impinges on the walls during the deposition of a nanostructured thin film. Some authors have attempted to qualify these forces. For this purpose, particle radii, r_P, particle concentration, N_P, particle charge, Z_P, and the electric potential around the particle, V_P, must be estimated. This requires the estimation of the ion and electron flux impinging on the particle and the particle capacitance. Charge fluxes are dependent on the local plasma parameters, that is, plasma density, electron temperature, ion kinetic energy, electric field, and ion mass [135].

Consider a particle suspended in a plasma. The plasma is characterized by neutral gas density, n_0, at temperature T, and electron, positive ion, and negative ion densities (n_e, n_i^+, and n_i^-, respectively) at temperatures T_e, $T_+ \approx T_- \approx T$. When particles are present in the plasma, the charge neutrality of Eq. (2), should be rewritten as

$$n_+ = n_e + n_- + Z_P N_P \qquad (5)$$

Daugherty et al. [136] showed that the particle potential could be approximated by a Debye–Hückel potential:

$$V(r) = \frac{V_P(r_P)}{r} \exp\left(-\frac{(r - r_P)}{\lambda}\right) \qquad (6)$$

where λ is

$$\lambda = \left[\frac{e^2}{\varepsilon_0} n_\infty \left(\frac{1}{kT_e} + \frac{1}{2E_+}\right)\right]^{-1/2} \qquad (7)$$

$V_P(r_P)$ is the potential of the particle surface with respect to the plasma, ε_0 is the vacuum permittivity, e is the electron charge, n_∞ is the plasma density far from the particle, k is the Boltzmann constant, and E_+ is the kinetic energy of positive ions. According to this approach, the particle capacitance, C_P, is

$$C_P = 4\pi\varepsilon_0 \frac{r_P}{1 + r_P/\lambda} \tag{8}$$

4.2.1. Electrostatic Forces

From Eqs. (6) and (8) and the expressions for the ion and electron fluxes impinging on the particles [137, 138], the monopolar electrostatic force caused by an electric field E_0 on a particle suspended in the plasma can be written as [139]:

$$F_m = -eZ_P E_0 \left(1 + \frac{(r_P/\lambda)^2}{3(1 + r_P/\lambda)}\right) \tag{9}$$

A similar calculation for the dipolar moment leads to a result that was identical to the dipolar moment of a charged sphere in vacuum ($p = 4\pi\varepsilon_0 r_P^3$) provided that terms of r_P/λ and smaller are dropped. Because the dipole moment, p, is proportional to r_P^3, the resulting dipole force on the particle from a gradient in the electric field is generally smaller than the monopole force [135].

4.2.2. Ion Drag

Ion drag is also referred to as ion wind force. When the positive ions accelerated by the electric field in the sheath move toward walls, they interact with the particle sheaths, transferring momentum in what are essentially Coulomb interactions. Particle dynamics may be dominated by this force depending on the particle size and plasma characteristics. This force depends on the momentum transfer rate coefficient, K_{mt}, which can be calculated from the cross section and the ion speed distribution function. Then, the ion drag force is calculated according to [135]:

$$F_{\text{ion drag}} = K_{mt} n_+ m_+ v_{+P} \tag{10}$$

where m_+ is the mass of the ion and v_{+P} the ion speed relative to the particle.

4.2.3. Thermophoretic Forces

A dust particle of radius r_P in a gas or liquid with a temperature gradient dT/dx at the position of the dust will be acted upon by a so-called thermophoretic force along the direction of x equal to [140]:

$$F_x = -8r_P^2 n_0 lk_B \left(\frac{dT}{dx}\right) \tag{11}$$

where k is the Boltzmann constant, n_0 is the neutral gas density, and l is the average collision length in the neutral gas.

If the walls in the reaction chamber are not isothermic [141] or simply a result of gas heating [133], temperature gradients may appear in the discharge. Refinements in the calculation of thermophoretic forces in situations such as particles close to the walls have been performed recently [142].

4.2.4. Neutral Drag

Additional forces on the particles include gravity and drag caused by the relative velocity between the particle and the neutral gas. Gravity is normally not relevant when powders are suspended in the plasma. The exact expression for the force resulting from the neutral gas is deduced from gas kinetic theory and can be found in [143]. However, when the

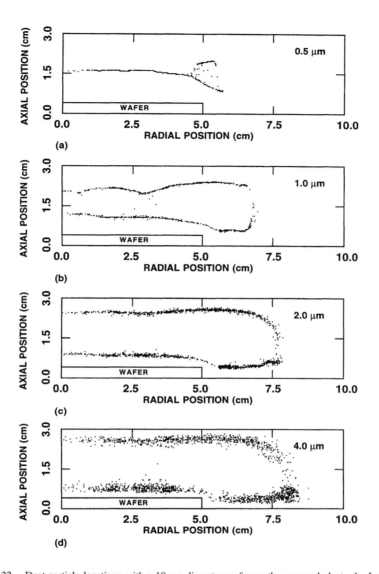

Fig. 22. Dust particle locations with a 10-cm-diameter wafer on the powered electrode. Locations are shown for (a) 0.5-, (b) 1.0-, (c) 2.0-, and (d) 4.0-μm particles. Small particles are more sensitive to the electric potential and to domes and rings around local maxima in the potential. Large particles have larger ion drag forces. Reprinted with permission from S. J. Choi, P. L. G. Ventzek, R. J. Hoekstra, and M. J. Kushner, *Plasma Sources Sci. Technol.* 3, 418 (© 1994 Institute of Physics Publishing Ltd.).

average relative velocity between the neutral gas and the particle, v_{nP}, is much smaller than the thermal speed of the neutrals, c_n, the neutral drag force reduces to the Epstein expression [135]:

$$F_{\text{neutral drag}} = \frac{4}{3} \pi r_P^2 m_n n_0 c_n v_{nP} \qquad (12)$$

Several models of dusty rf discharges concluded that the main force on the particles is due to the averaged electric field in the plasma. Therefore, the negatively charged particles are confined between the plasma sheaths. Figure 22 shows a theoretical calculation of the dust location in a reaction chamber with a 10-cm-diameter wafer on the powered electrode for four particle sizes. This Figure indicates that small particles are sensitive to the electric potential and may form domes and rings around the local maxima of the electric potential.

Large particles, however, suffer ion drag forces, which push them toward the boundaries. Figure 4 presents further evidence of the sensitivity of small particles to the electric potential.

Several reports have been devoted to the observation of the spatial distribution of the powder in the discharge by optical methods, which ranged from the direct visual observation [144] to the more elaborate rastered laser light scattering techniques [59, 98, 145]. Of particular relevance is the study of Dorier et al. [145], who analyzed the temporal evolution of particle size and particle concentration in two dimensions between electrodes, with a two-dimensional Mie scattering setup. Similar two-dimensional analyses were performed by Shiratani et al. [98]. For instance, Figure 23 shows the spatial distribution of particles between electrodes in a pure silane discharge. These photographs were obtained by illuminating the discharge volume from one side with a diverged laser beam and taking pictures from the front.

The localization of particles in well-defined regions of the discharge gave rise to what was referred to as particle traps. Several reports studied particle traps by means of the visualization of particles on them [60, 61] or by direct mapping of the electric potential [62].

4.3. Plasma Parameters for Powder Formation

In this section, we summarize the effect of the plasma parameters on powder formation in a silane-based discharge. Particles formed in sputtering or RIE discharges will not be considered in this section as they enter into the gas phase via removal from the substrate or walls. The effect of some of these parameters has been mentioned previously but, for clarity, we group them together in this section. Parameters such as the frequency of the rf signal [146] and the effect of the gas drag on the particle [147] also affect the temporal evolution of particle formation but they will not be considered as their influence is unclear.

4.3.1. Gas Mixture, Temperature, and Radio Frequency Power

Gas Mixture. We should distinguish between the influence of inert and reactive gases on particle formation. Silane dilution in inert gases such as Ar or He affects particle formation in different ways. It has been claimed that argon dilution advances whereas helium dilution retards the time for powder appearance with respect to pure silane plasmas [148]. Other authors reported that silane dilution in hydrogen inhibited or delayed powder formation [149].

Concerning the discharges of mixtures of reactive gases that are used to produce particles with a chosen composition, there are several factors that determine whether the particle formation is enhanced with respect to a pure silane plasma, for instance, the ionization energy of the gas. Mixtures of $SiH_4 + N_2$ have less tendency to produce particles than $SiH_4 + NH_3$ mixtures because of the higher energy required to ionize the N_2 molecule. Obviously, the reactivity of the molecule and its fragments will also affect powder formation. In general, the dilution of gases such as methane, ammonia, or diborane tends to diminish the powder yield with respect to a pure silane plasma.

Gas and Reaction Chamber Temperature. Empirical evidence indicates that heating the electrodes leads to a drastic reduction in powder formation [107, 146, 147, 149, 150]. In addition, it has been observed that the power needed to trigger the transition of the plasma impedance associated with powder formation scales with temperature [146]. These findings are attributed to changes in the plasma chemistry and not to the effect of thermophoretic forces [118, 148].

Therefore, the reactor is usually kept at room temperature in discharges producing large amounts of powder. In contrast, in the microelectronics industry, where the rf power is raised in order to increase the deposition rate, heating the electrodes might prevent the

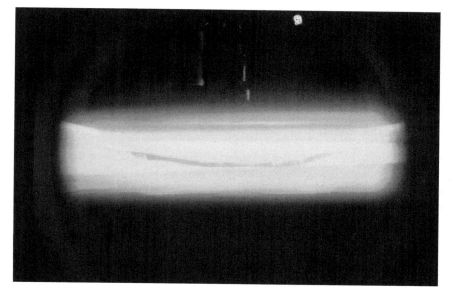

(a)

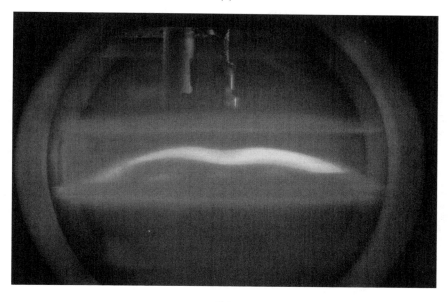

(b)

Fig. 23. Spatial distribution of particles between the electrodes in a pure silane rf plasma (a) and just after the plasma has been extinguished (b). The images were obtained by illuminating the interelectrode space with a diverged laser beam and taking the picture from the front.

formation of particles. Finally, the input gas can be cooled, or heated, before it enters the reaction chamber, which would enhance, or prevent, powder formation.

As particles are very sensitive to thermophoretic forces, the reaction chamber can be designed in such a way that it is possible to introduce temperature gradients either to reduce powder contamination in thin-film processing or to collect the powder suspended in the discharge.

Radio Frequency Power.   Increasing the rf power in the discharge enhances the formation of powder and advances its appearance, irrespective of the other discharge parameters. In silane, as well as other plasma-processing gases, achievement of large deposition rates and high-quality films is limited by the formation of powders or particles when the power dissipation is increased.

4.3.2. Plasma Modulation

Among the discharge parameters, modulation of the plasma has been the most often invoked method to control powder formation [70–77, 101]. For any given set of plasma conditions, it is possible to find a plasma modulation that inhibits the generation of powders. Plasma modulation consists of alternating periods of plasma-on (duration: T_{ON}) with afterglow periods (T_{OFF}). In the framework of thin-film deposition, this technique is referred to as square-wave modulation (SQWM) and it has been used to control the microstructure of the films [70–77]. Several effects can be pursued by controlling T_{ON} and T_{OFF}. Let us summarize them:

1. *Interruption of the polymerization chain.* To inhibit the generation of particles, T_{ON} should be chosen so that no particle is formed during one single T_{ON} period. Besides, it is necessary to set T_{OFF} long enough to deplete completely the heavy anion population during the afterglow in order to avoid their further polymerization in the next T_{ON}. In this case, the polymerization pathway is completely interrupted at each afterglow and no particles are formed.

2. *Increase in thin-film deposition rate and control of film microstructure.* In microelectronic processes, the discharge modulation is used to control the microstructural characteristics of the deposited films (mainly amorphous silicon thin films). The effect of the SQWM frequency on the film microstructure has been explained in terms of the contribution to film growth, during the afterglow, of negative species confined in the plasma. The dependence of roughness, void fraction, and deposition rate on the modulation frequency has been investigated [75].

3. *Size-selective particle generation.* In a dusty plasma, a long T_{ON} (~1 min or more) leads to a broad particle size distribution, which can diminish interest in particles for applications that require monodisperse powders. Plasma modulation is effective in controlling the particle size and size distribution. T_{ON} is set to permit particle growth up to the chosen size, and T_{OFF} is taken long enough to eliminate anions and particles completely. Therefore, a new particle generation initiates in the next T_{ON}. Usually, T_{OFF} must be larger than T_{ON} (several seconds in contrast to a fraction of second, depending on the particle size required). T_{ON} is chosen on the basis of theoretical coagulation models or previous experimental studies such as *ex situ* transmission electron microscopy (TEM) analysis.

4. *Enhancement of yield of powder formation.* In nonmodulated dusty discharges, the powder development reaches a quasi-steady-state size distribution and particle concentration. At this moment, the nucleation and growth of new particles is lower than at the beginning of the discharge. Therefore, it is convenient to modulate the discharge, in order to deplete the powder during the afterglow, and then initiate the formation of particles in the next plasma-on. This enhances the powder yield. The modulation is set so as to reach this quasi-steady-state particle concentration during T_{ON} (a fraction of a minute) and then particles can be removed during T_{OFF} (several seconds). This kind of low-frequency modulation is used to produce large amounts of powders.

5. MATERIALS

In this section, it is shown that controlling the dynamics of particle formation and the duration of particle growth allows the production of particles with different size and size distribution function. Moreover, the structure of these particles can range from totally amorphous to crystallite, depending on the plasma characteristics. The ability of low-pressure, low-temperature plasmas to produce particles of different size, size distribution, and structure permits the production of diverse materials such as powders or nanostructured thin films.

In Section 5.1, we show how silane-based plasmas with different discharge conditions may produce either small, well-ordered, isolated particles or powders consisting of agglomerated amorphous particles. Then, the microstructural characteristics of particles produced in low-pressure, low-temperature plasmas of different gas mixtures are described. Although few reports about the characteristics of powders generated in cold plasmas have been published, papers concerning silicon (Section 5.2), silicon–carbon, silicon–nitrogen, and boron–nitrogen (Section 5.3) are reviewed.

5.1. Control of Particle Coagulation in Order to Obtain Powders or Nanostructured Films

Here, we summarize a report that compared particles from discharges for powder production and for thin-film deposition [151]. This paper showed that particle size and structure were dependent on the discharge parameters, and it highlighted particle coagulation as the key step in powder development.

Two different discharges were chosen to illustrate the possibility of obtaining powders and nanostructured thin films in low-pressure, low-temperature discharges: Discharge A was an argon-diluted discharge where the appearance of powder takes several seconds, whereas discharge B was a pure silane plasma at relatively higher rf power and pressure that causes a fast powder appearance ($\sim$10 ms). The discharge parameters were 30 sccm of Ar flow, 1 sccm of SiH_4, a total pressure of 20 Pa, and 20 W of rf power for discharge A; 20 sccm of SiH_4, 80 Pa, and 70 W for discharge B.

Both discharges were modulated to control the particle size and structure. Experiments with two different plasma-on periods ($T_{ON} = 240$ ms and $T_{ON} = 1$ s) were performed with discharge A parameters. T_{OFF} was chosen long enough to assure that all particles formed during T_{ON} leave the plasma. Discharge B was modulated at lower frequencies ($T_{ON} = 2.5$ s, $T_{OFF} = T_{ON}$). Discharge B produced large amounts of powder that deposited on vacuum chamber walls. Plasma conditions for discharge A led to the deposition of a thin film on a substrate located in the reaction chamber. As is discussed later, small particles participated in the film growth in discharge A.

Discharge A. The technological parameters for discharge A were adopted from previous experimental reports devoted to the nucleation and coalescence of silicon particles in Ar-diluted discharges at room temperature [147]. These reports concluded that the time for the coagulation onset was about 200 ms for these particular discharge parameters. In the report that is being summarized, two different T_{ON}'s close to this coagulation onset were selected (240 ms and 1 s).

Figure 24 shows a high-resolution micrograph of a particle generated in the discharge with $T_{ON} = 240$ ms. The discharge was maintained during 10 modulation periods in order to limit film deposition on the TEM grid, which could mask the structure of the particle. Under these conditions, a few monosized particles of about 4 nm in diameter could be observed on the grid. Figure 24 shows that these particles had an ordered atomic structure so they were referred to as crystallites. However, the existence of amorphous particles of similar size to those shown in Figure 24 cannot be excluded because the small contrast between them and the amorphous carbon membrane would make them practically unobservable.

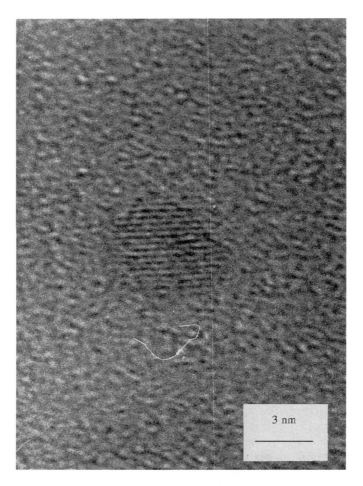

Fig. 24. High-resolution TEM photograph of one particle generated in an Ar-diluted discharge (discharge A) with T_{ON} = 240 ms.

Longer discharge intervals led to a different set of particles. Figures 25 and 26 show high-resolution micrographs of the grid after 10 periods of discharge A with T_{ON} = 1 s. Figure 25 shows a particle, 10 nm in diameter, that resulted from the coalescence of two crystallites similar to those shown in Figure 24. However, Figure 26 shows a particle with a different shape. In this case, the atomic order spreads all over the particle; thus, no smaller particles are identifiable. It is presumed, however, that the particle resulted from the coagulation of smaller particles.

Discharge B. In this case, the rf power and the total gas pressure were higher than those commonly selected for stable and reproducible studies of particle formation. Yes, indeed, instabilities on the plasma optic emission were evident simply by direct visual observation. This discharge produced a large amount of powder, which covered the inner reactor walls. No film was deposited under these plasma conditions.

The electron micrographs revealed spherical particles whose diameter ranged from 15 to 100 nm (Fig. 27). These particles agglomerated to form irregularly shaped aggregates of sizes up to a few micrometers. TEM analyses showed that the particle size distribution was broad.

The differences in particle size and particle size distribution between discharges A and B may be understood by considering that the time evolution of particle growth was quite different in both discharges. That is, particles from each discharge came from a different stage

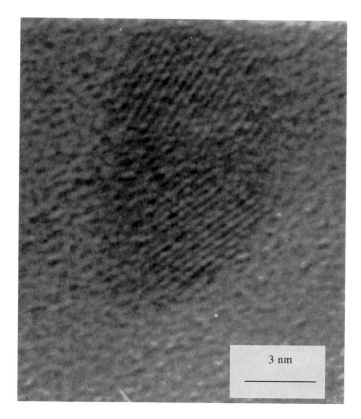

Fig. 25. HRTEM of one particle produced in discharge A ($T_{ON} = 1$ s) that resulted from the coagulation of two smaller particles.

in the coagulation process. Particles from discharge A, $T_{ON} = 0.240$ s, correspond to particles from a just-initiated coagulation. Therefore, their size was so small (<5 nm) and they appeared monodisperse. In contrast, those from discharge B, where a fast appearance and coagulation of particles occurred, correspond to a completed coagulation, which motivated the wide distribution of particle sizes observed. The smallest particles on Figure 27 are likely a generation of particles that started their growth later in the plasma-on period.

However, the explanation of why the structural characteristics of particles from discharges A (crystallites) and B (amorphous) are completely different is not as evident. Concerning particles from discharge A, the presence of small ordered domains of a few nanometers embedded in larger particles was reported beginning with the early studies on Ar-diluted silane discharges [152, 153]. These crystallites were claimed to be the first generation of particles from which the coagulation started. It was argued that the electrostatic trapping of these particles in the plasma could explain their well-ordered structure, although no definite physical reasons were suggested [127]. Two different physical mechanisms appear to explain this phenomenon. On the one hand, the crystallization of the particle may be a consequence of the temperature peak produced in the collision of two smaller particles during coagulation. Previous theoretical calculations concluded that this temperature may be considerable (~1000 K), depending on the size of the particles and their kinetic energy [154]. As the particle size increases, the temperature reached during the collision would be lower, in good agreement with the fact that only small particles appear as crystallites. On the other hand, the particle crystallization might be the result of the impingement of highly energetic species of the plasma (electrons and ions) on the particles suspended in the plasma. Because of the small particle size, collisions of approximately

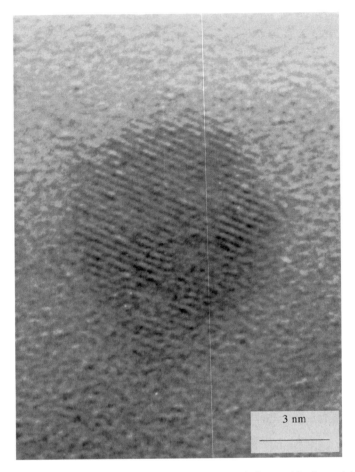

Fig. 26. HRTEM of one particle from discharge A (T_{ON} = 1 s). Presumably, the particle resulted from the coagulation of smaller particles and posterior crystallization.

Fig. 27. TEM images of particles obtained in a powdery discharge (discharge B). Particles appeared highly agglomerated and the distribution of particle sizes was wide. Particles were amorphous.

15-eV electrons may provide enough energy to alter its atomic structure. As a rough indicative value, the energy required to melt a silicon particle of 2 nm is 85 eV (calculated with bulk silicon density and enthalpy of fusion).

The main conclusion of that report was that it is possible to obtain particles of different size and structure by low-pressure, low-temperature plasmas by controlling both the discharge parameters, which determine the dynamics of particle formation, and the duration of the plasma-on, which limits the coagulation. Just-coagulated particles grown under discharges of slow particle formation, such as those shown in Figures 24 to 26, can be incorporated into a growing thin film to produce a nanostructured film with a new structure, that is, properties, as will be discussed in Section 6.1. Plasma conditions for fast particle development, such as that of discharge B, permits the formation of considerable amounts of powders with potential uses as raw material for advanced ceramics.

There remain open questions regarding the atomic structure of the crystallites (they are closer in size to a cluster than to a bulk material) or why isolated particles of sizes smaller than 4–5 nm have never been observed by TEM. This fact suggests that these particles are stable in the plasma but cannot survive outside it. Some of these questions will be discussed later in this chapter, although most remain unresolved.

Once the possibilities of low-pressure, low-temperature plasmas to produce different nanostructured materials have been pointed out, the microstructural characterization of silicon powders and silicon alloy powders will be presented in the next sections.

The material characterization does not include the description of the atomic structure of small particles (<10 nm) because they will be discussed in Section 6.1, which is devoted to the deposition of nanostructured thin films consisting of these nanometric particles embedded in an amorphous matrix.

5.2. Silicon Powders

5.2.1. Microstructural Characterization

Plasmas of silane and mixtures of silane and an inert gas have been the plasmas most studied both to investigate the temporal development of particle growth and to generate significant amounts of silicon powders.

When not specified otherwise, the results presented in this section were obtained from the analyses of silicon powders produced under the following technological parameters: pure silane plasma at 80 Pa of pressure, 30 sccm of silane gas flow, and 50 W of rf power square-wave-modulated at 0.2 Hz.

5.2.1.1. Powder Aspect

Figure 28 shows the inner walls of the vacuum chamber and the reaction box after 2 h of a powdery discharge. The photograph shows a frontal view of the cylindrical vacuum chamber and the reaction chamber (a parallelipedic box described in Section 3.2) inside it. The squared rf electrode is mounted vertically inside this reaction chamber. This photograph shows that the color of the silicon powder varies depending on the location in the vacuum chamber where it is deposited. The color ranges from bright yellow to dark brown. The powder inside the reaction box tends to be darker than the powder outside it. The darkest colors are found at the corners of the electrodes.

These differences in color, which reveal some differences in particle structure, may be attributed to the effect of the plasma on the particles once they have deposited on the walls. The analyses of the zones where the particles appear darker suggest this hypothesis. The yellowish powder on the cylindrical walls would correspond to unaltered powder dragged outside the reaction chamber by the gas flow. In contrast, the powder deposited inside the reaction box would have suffered plasma ion bombardment, which would have altered the particle structure. Section 5.2.2 provides further evidence of the plasma effect on the

Fig. 28. Photograph of the reaction chamber after a powdery discharge. The squared rf electrodes are placed vertically inside the cylindrical vacuum chamber. The differences in color of silicon powder should be attributed to the plasma effect on its microstructure.

microstructure of the silicon powders. This hypothesis is in good agreement with the fact that particles on the rf electrode, where an important DC bias exists, appear darker. Other authors have reported variations in the color of silicon powder in different locations of the reactor [155]. Ion bombardment on silicon powder would cause Si–H bond breaking and, therefore, the loss of hydrogen and the increase in Si–Si bonds, which absorb in the visible region of the spectra.

On the other hand, silicon powder presents the usual characteristics of ultrafine powders, which makes its handling difficult. Indeed, the powder sticks to the closest surfaces as if it were electrostatically charged, exhibits collective movements, and its apparent density is very low (~ 0.1 g/cm^3). These facts are motivated by the high surface forces relative to gravity and the large amount of adsorbed gases in the particle surfaces. Both features are a consequence of the high surface-to-volume ratio in ultrafine materials.

Silicon powders exhibited a very large specific surface area (~ 162 m^{2}g^{-1}) [156].

Several authors have studied the structure of silicon powder by transmission electron microscopy (TEM) [152, 153, 155–159]. The shape of the coagulated particles is strongly dependent on the plasma characteristics (the dependence of particle size with time before the coagulation was discussed in Section 4.1.3).

Examination of particles found in Ar-diluted silane plasmas provided evidence of the agglomeration that led to their present size. Indeed, particles of about 100 nm appeared to have been formed by the agglomeration of 5–10-nm particles. These particles appeared isolated in the TEM grid [152, 153]. Dark-field analysis of these particles revealed inten-

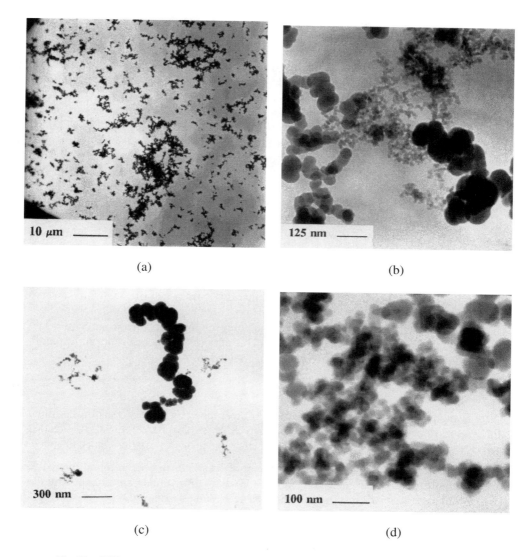

Fig. 29.   Different TEM micrographs of silicon particles obtained in a discharge for powder production. The discharge parameters were 80 Pa of pressure, 30 sccm of pure silane, and 50 W of rf power modulated at 0.2 Hz.

sity rings with a dense distribution of bright points that were attributed to nanocrystallites of about 2 nm. Therefore, it was claimed that the nanocrystallites were formed during the first milliseconds of the discharge and that their agglomeration gave rise to larger particles [153].

The powder structure is different in pure silane discharges. Particles are homogeneous amorphous spheres, and they are rarely isolated [155–159]. However, relatively low power discharges and short T_{ON} (<10 s) produced monodisperse isolated particles [96]. Usually, particles appear highly agglomerated in discharges with a high yield of powder production (pure silane and high rf power and pressure). Moreover, the size distribution is wide (Fig. 29), which can be attributed to the fact that new generations of particles start their growth at different instants of the plasma-on period.

Figure 29 shows representative photographs of these agglomerates. It should be taken into account that these agglomerates suspended in the plasma act as individual nonspherical particles in laser light scattering experiments [96].

5.2.1.2. Crystallinity

As discussed previously, the atomic structure of the particles depends on the discharge characteristics and the duration of the plasma-on periods. Particles formed in Ar-diluted discharges appear as small crystallites for short T_{ON} ($\sim$250 ms), whereas these crystallites agglomerate at larger T_{ON} times (Section 5.1).

In contrast, those particles generated in pure silane plasmas under discharge parameters with a high yield of powder formation do not exhibit crystalline features. Dutta et al. [160], however, showed that these particles contained ring structures in a detailed high resolution transmission electron microscopy analysis (Fig. 30).

At this point, it is worth discussing the atomic structure of both the crystallites of Ar-diluted discharges and the nonordered particles of pure silane plasmas. In relation to the structure of the crystallites, it is commonly accepted that the most stable structure of small silicon clusters with few atoms (<2 nm) is not the diamond structure [161]. Many proposals of cage or cage–core structures appear in the literature [162–164]. In addition, clusters formed in the plasma are highly hydrogenated and the saturation of silicon bonds with hydrogen can influence the most stable structure of the cluster. The homogeneous nucleation of the hydrogenated silicon clusters in the gas phase indicates that these clusters might initially adopt geometric arrangements that differ from that of the diamond structure. Moreover, bond geometry in cagelike clusters is not dissimilar to the tetrahedral geometry of bulk silicon. Therefore, it is reasonable to suppose that silicon particles might contain cagelike clusters in a metastable state. This possibility will be discussed further in the section devoted to nanostructured thin films (Section 6.1).

5.2.1.3. Vibrational Properties

The vibrational properties of silicon powders as determined by Fourier transform infrared (FTIR) and Raman spectroscopies are summarized in this section. From these detailed analyses, relevant data concerning particle structure and growing kinetics can be deduced.

Infrared Spectroscopy. The vibrational properties of silicon powders are compared to those of thin films of amorphous silicon (a-Si:H) as both materials are obtained by the same technique and a-Si:H is a material that has been studied extensively.

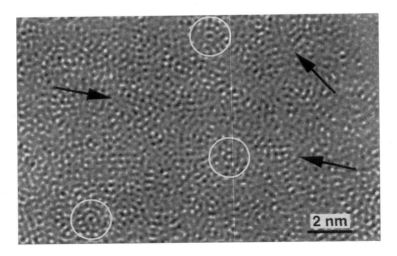

Fig. 30. HRTEM image of part of an amorphous Si particle that exhibits ringlike (marked by circles) and fringelike (marked by arrows) contrast features, respectively. Reprinted from *Nanostruct. Mater.*, J. Dutta, R. Houriet, H. Hofmann, and H. Hofmeister, 9, 359 (© 1997), with kind permission from Elsevier Science Ltd., The Boulevard, Langford Lane, Kidlington OX5 1GB, UK.

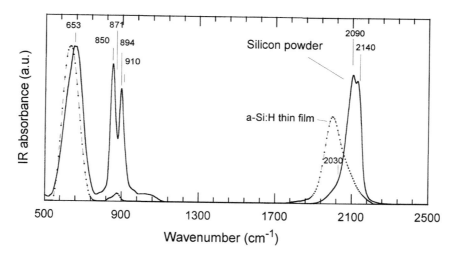

Fig. 31. IR spectra of silicon powder compared to that of standard amorphous silicon thin films.

Figure 31 shows the infrared (IR) spectra of both materials. The IR spectrum is usually divided into three absorption zones (stretching, bending, and wagging bands).

Stretching Band. In a-Si:H thin films, the stretching band comprises one single peak at 2000 cm^{-1}, which is attributable to monohydride arrangements ($\equiv$Si–H) [165]. Poor-quality films show a shoulder in the high-energy side of this peak ($\sim$2100 cm^{-1}), which is usually assigned to both dihydride groups ($=$SiH$_2$) [166] and monohydride groups in surfaces or in internal cavities where collective vibrations are expected [165, 167].

This absorption band presents a number of significant differences in silicon powders. It shifts toward higher wavenumbers and it comprises three distinct absorption contributions. Figure 32 shows the deconvolution of the stretching band for silicon powder in three gaussian peaks. The absorption peak centered at around 2030 cm^{-1} is attributable to the monohydride arrangement ($\equiv$SiH) inside the particle. This shift in wavenumber may be explained as follows. The Si–H vibration appears close to 2180 cm^{-1} where the bond can freely oscillate as in the silane molecule [165, 168]. This vibration energy falls as the density of the medium surrounding the bond increases as a result of a depolarizing effect [165]. In a relatively dense material, such as a-Si:H films, this bond absorbs at 2000 cm^{-1}. Therefore, the shift can be explained by the effect of the small particle size and its lower density relative to that of films. Thus, the position of this peak can be taken as being indicative of particle density (Section 5.2.2).

The second contribution is centered at about 2100 cm^{-1} and is usually assigned to dihydride groups (SiH$_2$) or polymeric-like arrangements, (SiH$_2$)$_n$ [166]. Certain authors claim that monohydride arrangements in internal cavities or in surfaces can contribute to the absorption in this band [169]. This band is the most prominent in silicon powders. Finally, the contribution at 2140 cm^{-1} has been assigned to polymeric-like chain terminations (SiH$_3$).

Bending Band. This absorption band, between 840 and 910 cm^{-1}, is indicative of the presence of dihydride arrangements ($=$SiH$_2$) and polymeric chains, (SiH$_2$)$_n$ [169]. In a-Si:H thin films, this peak correlates with film quality. In electronic grade (high-quality) a-Si:H films, this absorption band does not appear. This band is the most prominent in silicon powders, revealing the polymeric character of this material.

Wagging Band. The integrated absorption of this band is commonly used to quantify the hydrogen content in a thin film because all Si–H$_n$ arrangements contribute to this

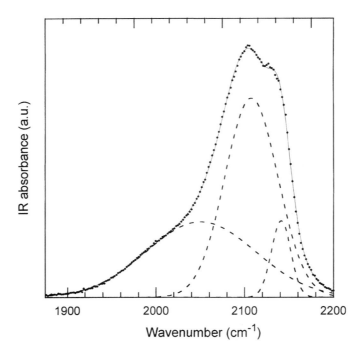

Fig. 32. Stretching band of the absorption spectra in the infrared of silicon powders and its deconvolution into three Gaussian peaks. Peaks centered at 2030, 2100, and 2140 cm^{-1} correspond to $\equiv$SiH, $=$SiH$_2$, and $-$SiH$_3$ arrangements.

peak [165]. In silicon powders, it appears at 653 cm^{-1} shifted from the position in a-Si:H thin films (630 cm^{-1}).

Oxidation-Related Peaks. Exposure to atmosphere oxidizes the powders, as can be seen by the presence of a prominent silicon–oxygen absorption band centered around 1050 cm^{-1}. Section 5.2.3 contains a detailed study of the oxidation of silicon powder.

Raman Spectroscopy. The Raman spectra of silicon powders have been studied by Dutta et al. [155]. They studied two samples of powder collected in different zones of the reactor. These samples were different in color: That collected on the electrodes was reddish brown, whereas that deposited on a probe 3 cm apart from the electrodes was yellow. The structural origin of these differences has already been discussed in Section 5.2.1.1. The Raman spectrum of the reddish-brown powder was highly nonuniform. The spectra of both amorphous silicon and spectra close to that of crystalline silicon were measured. These findings indicated that both amorphous and crystalline phases were nonuniformly distributed in the powder sample. In contrast, the spectrum of the yellow powder was uniform throughout the sample. It consisted of a broad structure between 430 and 530 cm^{-1} with several distinct peaks superimposed. The broad band at 480 cm^{-1} was attributed to amorphous silicon. Distinct superimposed weaker peaks had not previously been reported. The authors related the low-energy position of these peaks to very small crystalline particles, which could be highly strained because of the large fraction of atoms at the surface.

Unpublished Raman results on silicon powder indicate that the spectrum consists of only bands associated with silicon–hydrogen vibrational modes when low excitation laser intensities are used [170]. Only when the laser intensity is increased are the vibrational modes attributed to the silicon network seen [170].

These findings suggest that the atomic structure of the yellow powder, when powders are neither exposed to the plasma nor heated by a laser beam, is close to a polymer-like structure. Whether this structure contains well-ordered cagelike clusters or is mainly a disordered polymeric chain remains unclear.

5.2.1.4. Hydrogen in Silicon Powders

Infrared spectroscopy analyses showed that silicon powder contains hydrogen bonded to silicon in monohydride and dihydride arrangements and that most hydrogen is bonded in dihydride groups. The hydrogen bonding configuration was also analyzed by means of thermal desorption spectrometry [159, 171]. This technique consists of heating the sample in a vacuum furnace with a controlled temperature ramp and simultaneously monitoring the hydrogen flow evolving from the sample with a quadrupole mass spectrometer. Figure 33 shows the experimental apparatus.

This technique has been used extensively to study the hydrogenation of a-Si:H thin films [172–174]. In this material, hydrogen evolution is dominated not only by the bond breaking, but also by the diffusion of atomic hydrogen through the film and the formation of an H_2 molecule, which eventually evolves from the sample [174]. The use of this technique for the study of nanometric silicon powder is of particular interest because in this material there are no significant diffusion effects and the thermal evolution of hydrogen can be understood as a single desorption process, the kinetics of which are dominated by hydrogen–silicon bond breaking. Therefore, direct information about hydrogen–silicon bond energy in silicon powders can be inferred from the thermodynamical analysis of the desorption peaks.

Let us summarize the thermodynamics theory behind a desorption process [175, 176]. In a desorption process, there is no possibility that the gas that evolves from the sample will flow back and be reabsorbed in the sample. Therefore, the desorption rate is proportional to the nth power of the hydrogen concentration in the sample (C_H):

$$-\frac{dC_H}{dt} = K C_H^n \tag{13}$$

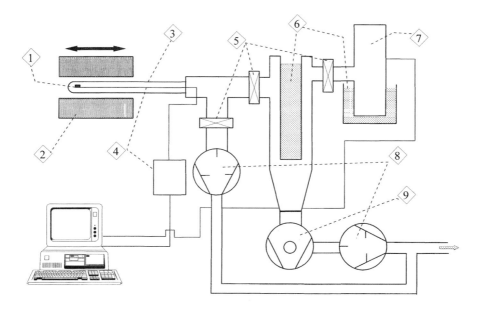

Fig. 33. Experimental arrangement for the experiment of thermal desorption of hydrogen: (1) sample, (2) tubular furnace, (3) quartz tube, (4) thermocouple, (5) gate valves, (6) cold trap of liquid nitrogen, (7) mass spectrometer, and (8) and (9) vacuum pumps.

where n depends on the order of the reaction ($n = 1$ for a first-order process) and K is the dynamic coefficient, which is thermally activated:

$$K = K_0 \exp\left(-\frac{\Delta E}{kT}\right) \tag{14}$$

where ΔE is the activation energy and k is the Boltzmann constant.

In a detailed thermodynamic analysis performed by Oguz et al. [175], the dynamic coefficient K was related to the variation of Gibbs energy of the local arrangement when the hydrogen evolves:

$$K = r\left(\frac{1}{\tau_0}\right)\exp\left(-\frac{\Delta G}{kT}\right) \tag{15}$$

where r is the probability of overcoming the energy barrier and τ_0^{-1} is the frequency of collision with this energy barrier. $\tau_0^{-1} \sim h/KT$, where h is Planck's constant. Taking into account that $\Delta G = \Delta H - T\,\Delta S$ and defining the experimental entropy as $\Delta S^* = \Delta S + k\ln(r)$, the dynamic coefficient can be written as:

$$K = \frac{1}{\tau_0}\exp\left(\frac{\Delta S^*}{k}\right)\exp\left(-\frac{\Delta H}{kT}\right) = \frac{1}{\tau_0}\exp\left(-\frac{\Delta G^*}{kT}\right) \tag{16}$$

Therefore, the preexponential factor of Eq. (14) corresponds to

$$K = K_0 \exp\left(-\frac{\Delta H}{kT}\right) \tag{17}$$

where

$$K_0 = \frac{1}{\tau_0}\exp\left(\frac{\Delta S^*}{k}\right) \tag{18}$$

and the activation energy can be directly related to the enthalpy of the process.

In a thermal desorption experiment, the partial pressure of hydrogen is monitored and related to the hydrogen flow evolving from the sample. Therefore, if the total hydrogen content in the sample is N_0 and N is the actual evolved hydrogen, Eq. (13) can be rewritten as

$$\frac{d(N/N_0)}{dt} = \left(\frac{kT}{h}\right)\left(1 - \frac{N}{N_0}\right)^n\exp\left(\frac{\Delta S^*}{k}\right)\exp\left(-\frac{\Delta H}{kT}\right) \tag{19}$$

and, taking logarithms,

$$\ln\left[\frac{\frac{d(N_{jT}/N_{0jT})}{dt}}{\left(1 - \frac{N_{jT}}{N_{0jT}}\right)^n}\frac{h}{kT}\right] = \frac{\Delta S_{jT}^*}{k} - \frac{\Delta H_{jT}}{kT} \tag{20}$$

where the index j accounts for each peak of the desorption curve.

The thermal desorption of hydrogen of the silicon powder produced in pure silane plasmas was analyzed according to this theoretical background [171].

Figure 34 shows the results of the experiment. The hydrogen evolution starts at 490 K and two different peaks are clearly distinguishable: The low-temperature (LT) peak has its maximum at around 600 K, whereas the high-temperature (HT) peak is less intense and is centered at about 720 K.

The mathematical treatment presented in Eq. (20) allowed the determination of ΔS^*, ΔH, and ΔG^* ($\Delta G^* = \Delta H - T\Delta S^*$) for each peak. Figures 35 and 36 show the left-hand side of Eq. (20) versus $1000/T$ assuming a first-order process ($n = 1$). The linear dependence confirms the hypothesis of a first-order process.

If these results are compared to those obtained for amorphous silicon films, it can be seen that the temperature of the maximum of the hydrogen evolutions in silicon powder

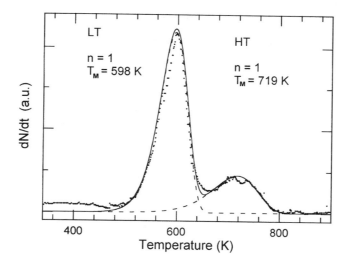

Fig. 34. Evolution of the effusion rate of hydrogen in silicon powder during thermal annealing. The experimental points are indicated by dots. Dashed curves correspond to the calculated best fit to a first-order desorption process.

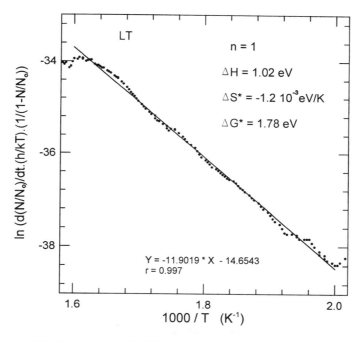

Fig. 35. Analysis of the LT peak of Figure 34 in terms of Eq. (20).

is significantly lower than in a-Si:H (650 and 770–800 K, respectively). Besides, the predominant hydrogen evolution is found at low temperatures for silicon powders, whereas the high temperature peak is the most prominent in a-Si:H films.

In relation to the thermodynamic parameters for hydrogen desorption in silicon powders, the fact that the activation energy, ΔE, is much lower than ΔG^* ($\Delta E_{LT} = 1.1$ eV for $\Delta G^*_{LT} = 1.82$ eV and $\Delta E_{HT} = 0.9$ eV for $\Delta G^*_{HT} = 2.22$ eV) has been explained for thin films [176]. The activation energy is lower than the energy required to break the Si–H bond even if the energy delivered in the formation of H_2 is taken into account. This finding

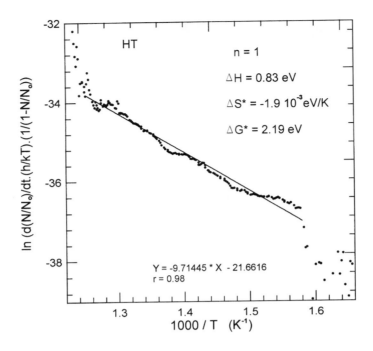

Fig. 36.   Analysis of the HT peak of Figure 34 in terms of Eq. (20).

was explained satisfactorily by a desorption mechanism on the basis of short-lived energy fluctuations of the silicon–hydrogen bond [176].

It is worth noting that the Gibbs energies for desorption found in silicon powder, $\Delta G_{LT}^* = 1.82$ and $\Delta G_{HT}^* = 2.22$ eV, are in very good agreement with those obtained for the desorption from the 111 silicon surface with high hydrogen coverage (1.86 and 2.2 eV, respectively) [177]. Each peak of the hydrogen evolution can be tentatively attributed to a particular Si–H arrangement as follows. The HT peak can be assigned to breaking two Si–H bonds to form a hydrogen molecule. In this case, the energy balance for the desorption would be given by

$$\Delta G_{HT}^* = 2E(\text{Si–H}) + E(\text{H}_2) \tag{21}$$

As $E(\text{H}_2)$ is known, 4.52 eV, and ΔG_{HT}^* has been determined experimentally (2.22 eV), the Si–H bond energy of monohydride groups in silicon powder may be estimated to be about 3.38 eV.

Similarly, the LT peak can be associated with Si–H bonds in dihydride arrangements. An equivalent energy balance to that described previously, with $\Delta G_{LT}^* = 1.82$ eV, gives an energy of 3.17 eV for the Si–H bond in SiH_2 arrangements.

These estimated values of the Si–H bond in SiH and SiH_2 (3.38 and 3.17 eV) are close to those reported for porous silicon (3.58 and 3.10 eV) [173] and gas phase silane (between 2.8 and 3.9 eV) [178].

5.2.2. Influence of the Modulation Frequency on the Structure of Silicon Powders

A study of the influence of plasma modulation frequency on the structure of silicon powder in discharges for high powder yield provided information about the evolution of particle structure during the development of these particles in the plasma [159]. These results are summarized next.

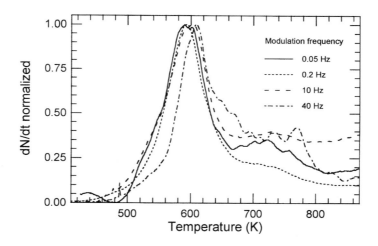

Fig. 37. Normalized hydrogen thermal desorption spectra corresponding to silicon powders grown at different square-wave modulation frequencies.

Figure 37 shows the thermal evolution of hydrogen for four samples of silicon powder obtained at four plasma modulation frequencies (0.05, 0.20, 10, and 40 Hz). The parameters for these discharges were 15 sccm of pure silane, 65 Pa of pressure, and 200 mW/cm^2 power density. All discharges were performed at room temperature. For comparison, the four curves were normalized to their maximum. These evolutions do not show significant differences in the low-temperature peak. However, the high-temperature evolution of samples produced under the 40-Hz modulation frequency exhibited a peculiar behavior. Indeed, this evolution revealed a set of superimposed peaks whose maximum ranged between 640 and 800 K. This evolution corresponds to the sample grown under shorter plasma-on periods. As the thermal evolution of hydrogen in silicon powder should be assigned only to a desorption process, the presence of distinguishable peaks suggests that hydrogen is bonded to silicon at particular sites with different bond energies. This could be a consequence of the presence of clusters, such as cagelike structures, in silicon powder. Further investigations are required to confirm this hypothesis.

On the other hand, infrared analysis of these samples showed that the silicon–hydrogen configuration varied with the modulation frequency. The stretching band of each sample was normalized to its integrated absorption in order to compare the vibrational characteristics of the samples. Then, this band was deconvoluted into the three Gaussian contributions described previously and the integrated absorption of each Gaussian peak was calculated. This method allowed the relative comparison of Si–H bonds (peaks at 2030 cm^{-1}), SiH$_2$ (2100 cm^{-1}), and SiH$_3$ (2140 cm^{-1}) to be performed. Figure 38 shows the dependence on the plasma-on period (T_{ON}) of the integrated absorption of each contribution in the stretching band. The sample grown with shorter T_{ON} exhibited a higher relative contribution of hydrogen bonded in SiH$_2$ arrangements. The relative contribution of SiH bonds increased monotonically with T_{ON}.

The evolution of SiH$_n$ arrangements with T_{ON} suggests that the atomic structure of the silicon powders grown in discharges for large powder yield evolves from an initial polymeric-like structure to a denser material more similar to that of amorphous silicon. This structural evolution should be regarded as a consequence of cross-linking inside the particle during its permanence in the plasma. Cross-linking may be caused by ion and electron bombardment on the growing particle. As cross-linking increases, the relative amount of hydrogen bonded in the SiH$_2$ groups diminishes and, moreover, the hydrogen concentration in the sample also decreases. The color of the samples was also indicative of the evolution in the atomic structure. As the cross-linking increases, there are more Si–Si bonds,

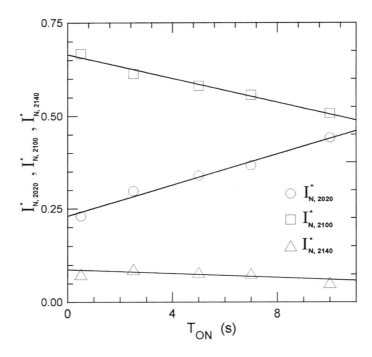

Fig. 38. Representation of the relative concentration of different Si–H bonds, calculated from the IR absorption band, and corresponding to the stretching vibration modes, against the plasma-on period.

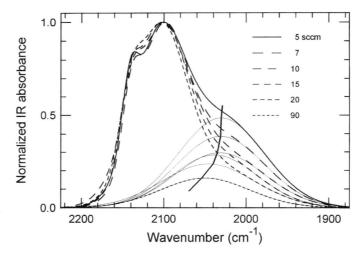

Fig. 39. Stretching band of the IR spectra of silicon powder samples grown in discharges with different gas flows. As the gas flow increases (i.e., the residence time of particles in the plasma decreases), the IR spectra indicate that the atomic structure approaches that of a polymeric-like structure.

which absorb in the visible spectrum, and, thus the powder is darker. Powders grown at 40 Hz were yellow, whereas samples grown for longer T_{ON} presented a darker color.

The effect of plasma-on on the particle structure was also shown by a series of experiments in which the silane gas flow entering the reaction chamber was varied [159]. Gas flow determines the residence time of the gas in the reactor, besides controlling the drag forces on the particles. For both these reasons, particles grown under discharges with higher gas flows are expected to remain in the plasma for shorter periods. Figure 39 shows the stretching band for the samples of the series. Curves have been normalized to their maximum for comparison. This figure shows two main features. First, the relative contribution of Si–H arrangements increases as gas flow diminishes (the residence time in the plasma increases). Second, the density of the material also increases with residence time. The density can be inferred from the position of the low wavenumber peak at around 2020 cm^{-1}; the material is more dense as the position of this peak is shifted to lower wavenumbers, as discussed in Section 5.2.1.3.

5.2.3. Oxidation of the Silicon Powder

The high specific surface and large hydrogenation of nanometric silicon powder suggest that it is highly reactive. It has been shown that these particles oxidize spontaneously in the atmosphere. This section describes an infrared study of the oxidation of silicon powders both spontaneously and thermally activated [171].

5.2.3.1. Spontaneous Oxidation

Silicon samples that had been in the atmosphere for a few minutes already presented oxidation-related absorption bands in the infrared spectrum. Figure 40 compares the IR spectrum of a sample as deposited with that of one that had been in air for 1 week. Table IV shows the vibrational modes of the oxygen-related peaks [179]. In particular, the absorption bands caused by silicon–hydrogen bonds shift their position toward higher wavenumbers as a consequence of the electronegative bond environment caused by the presence of oxygen. This allows researchers to distinguish the Si–H arrangements back-bonded to one, two, or three oxygens (2150, 2190, and 2260 cm^{-1}, respectively) [180]. Figure 41 shows the deconvolution into five different Gaussian peaks of the stretching band of an oxidized sample of silicon powder.

Moreover, the presence of the bending absorption band associated with O_2–SiH_2 (980 cm^{-1}) in Figure 40 reveals that oxygen is incorporated in an interstitial position between two silicon atoms close to the particle surface where dihydride arrangements are more abundant.

Further studies to clarify the mechanism of spontaneous oxidation of silicon powder are currently being conducted. Mass spectrometry of gases evolved during thermal annealing of powder in dry and wet air is foreseen as a suitable method for studying the oxidation process.

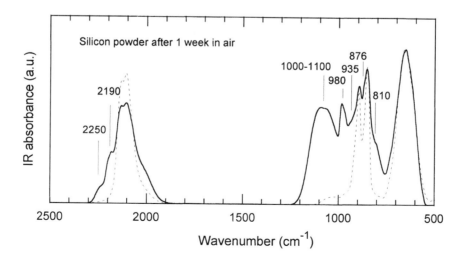

Fig. 40.   IR spectra of one sample of silicon powder that had been in air for several weeks, compared to that of one as-deposited powder sample (dashed curve).

Table IV. Vibrational Modes in the Infrared Related to the Oxidation of
Hydrogenated Silicon [179]

Arrangement	Position (cm^{-1})	Mode
Silicon–hydrogen bond		
O_3–SiH	2,260	Stretching
O_3–SiH	876	Bending
O_2–SiH$_2$	976,935	Bending
O_2Si–SiH	2,190	Stretching
OSi$_2$–SiH	2,150	Stretching
Silicon–oxygen bond		
Si–O–Si	1,000–1,100	Stretching
Si–O–Si	810	Bending

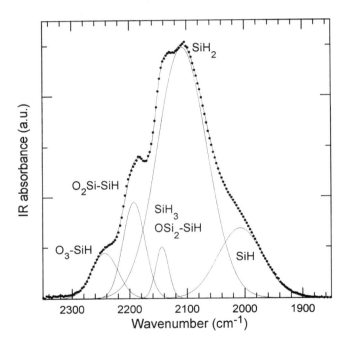

Fig. 41. Deconvolution of the stretching band of the IR spectra shown in Figure 40. Besides vibrational modes related to SiH$_n$ (2140, 2100, and 2030 cm^{-1}, $n = 3, 2, 1$), the absorption of peaks from the O_jSi$_{3-j}$–SiH modes is indicated (2250, 2190, and 2150 cm^{-1}, $j = 3, 2, 1$). Contributions at 2140 and 2150 cm^{-1} are grouped into one Gaussian peak.

5.2.3.2. Thermal Oxidation

Thermal annealing of silicon powder in air was studied in a small open furnace adapted to the FTIR interferometer. Silicon powder is an ideal material to use in studying all possible O–Si–H arrangements because of its large specific surface and large hydrogen content.

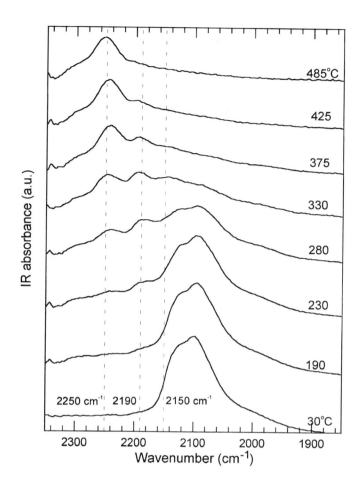

Fig. 42. Thermal oxidation experiment. The evolution with temperature of the IR absorption of silicon powder in the spectral region corresponded to the stretching band of SiH_n.

There was a remarkably good agreement between the results obtained experimentally and those predicted theoretically by Lucovsky and collaborators [180]. Figures 42 and 43 show the dependence of the IR absorbance in two spectral ranges on the heating temperature. These figures indicated the hydrogen loss process and the simultaneous oxygenation of the sample. This evolution can be summarized as follows. At the onset of heating, there was an exchange from the $Si–H_n$-related absorption bands to those related to $O–Si–H_n$. At around 200 °C, the effect of the thermal desorption of hydrogen in the IR spectrum added to the oxidation process. At 330 °C, there was no trace of SiH_2 without back-bonded oxygen. However, absorptions related to dihydride arrangements (980 cm^{-1} for $O_2–SiH_2$) were still observed at 375 °C. At this temperature, all hydrogen in SiH_2 was expected to have evolved (as indicated by the results of thermal desorption discussed previously). The presence of $Si–H_2$ at this temperature was a consequence of oxygen in the neighborhood of the $Si–H_2$ bonds. Oxygen is electronegative and, therefore, it attracts the electron cloud, diminishes its length, and increases its strength [180, 181]. In theoretical reports, it has been estimated that each oxygen back-bonded to the $Si–H$ group increases in 0.24 eV the hydrogen–silicon bond energy [180]. At the higher temperature available during the experiment (485 °C), the only remaining hydrogen-related band was that at 2250 cm^{-1}, which is attributed to $O_3–SiH$.

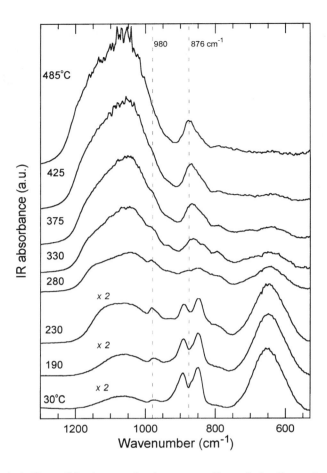

Fig. 43.   As in Figure 42 for the spectral region corresponding to the bending modes of SiH$_n$.

5.2.4. Annealing of the Silicon Powder

The results presented in Section 5.2.1 for the characterization techniques of IR and Raman spectroscopies, as well as the thermal desorption of hydrogen, indicate the possible presence of small clusters in the amorphous particles. It is known that clusters having a cage or cage–core configuration may play an important role in the crystallization of tetrahedrally bonded amorphous materials [182, 183]. Therefore, thermal annealing of silicon powders was expected to reveal these clusters.

Hofmeister et al. [184] performed a detailed analysis of the atomic structure of silicon powders upon thermal treatment by high-resolution transmission electron microscopy (HRTEM). This treatment was carried out in a reducing atmosphere to avoid oxidation of the powders.

Upon mild annealing for 1 h, between 300 and 600 °C, no apparent structural changes were observed. However, a careful inspection of the micrographs revealed circular contrast features extending from 1.5 to 2.5 nm in size, as denoted by the circles in Figure 30. Diffractograms, obtained by Fourier transformation of such images, exhibited a spotty appearance indicating deviations from the homogeneous atomic distribution of amorphous materials. It was claimed that these circular features were due to partially ordered regions, which were probably already present in the as-prepared powder and which became more evident upon annealing. It was thought that the attachment of atoms from the surrounding amorphous phase to a 15-atom cluster might lead to a fivefold twinned crystallite.

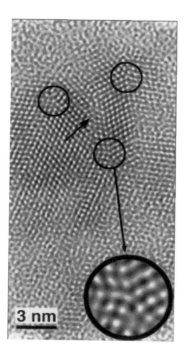

Fig. 44. Fivefold twin junction (circle with arrow) surrounded by parallel and azimuthally rotated twin lamellas with additional multiple twin junctions (encircled) formed by growth twinning. Reprinted with permission from H. Hofmeister, J. Dutta, and H. Hofmann, *Phys. Rev. B* 54, 2856 (© 1996 American Physical Society).

The onset of crystallization was observed during the prolonged annealing of powders at 600 °C (5.5 h). Numerous circular contrast features, up to a medium-range order, were observed. Upon 1-h annealing at 700 °C, however, crystallization was distinctly seen to set in. Selected area electron diffraction (SAED) patterns exhibited dotted rings superimposed on diffuse broad rings indicating a low-dimensional crystalline phase in an amorphous matrix. Fast Fourier transform (FFT) diffractograms of the images exhibited {111} spots. In addition, fast-grown fivefold twinned crystallites occurred. At this stage of crystallization, the common growth twinning in diamond cubic semiconductors was already observed. Typical structures formed by growth twinning are shown in detail in Figure 44.

Upon annealing at 800–900 °C, an extended crystal lattice with the characteristics of a diamond cubic structure appeared. The extensive growth twinning led to a heavily faulted structure. Even at this annealing temperature, no sintering of the particles or change in particle size was observed. The thin layer of oxide (~2 nm) observed on the particle surface impeded sintering.

5.3. Powders of Silicon Alloys

5.3.1. Powders of Silicon–Nitrogen Alloys

To the author's knowledge, the first report dedicated to the properties of nanometric powders produced in a low-pressure, low-temperature rf discharge was that of Ho and co-workers in 1989 [185], which was concerned with the formation of silicon nitride precursors in silane–ammonia discharges. Their work was followed promptly by that of Anderson and co-workers [109], who examined particulate generation mechanisms and dynamics in discharges of similar gas mixtures. In the mid-1990s, Buss and Babu [186] took the study of Si–N powders a stage further. On the basis of the the prevailing understanding of powder

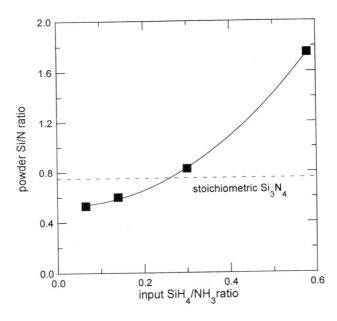

Fig. 45. Dependence of the Si/N ratio in the powder on the SiH$_4$/NH$_3$ ratio in the input gas mixture. Data taken from [185].

development, they improved the controllability of particle size and showed that heat treatment could prevent the oxidation of the powder.

Ho and co-workers [185] determined the powder stoichiometry by neutron activation analysis. The N and H content were also determined by microcombustion elemental analysis. The powder stoichiometry dependence on the silane–ammonia gas mixture was investigated. It was observed that an excess of ammonia was required to produce a stoichiometric silicon nitride precursor (Si/N ratio of 0.75). Figure 45 reproduces this measured dependence.

Ho et al. [185] described a powder color that ranged from yellow to pure white, varying with the Si/N ratio (the higher the nitrogen content, the closer the color was to white). Elemental analysis showed a hydrogen content of about 40 at%.

The IR spectra of these samples showed that hydrogen was bonded to silicon (Si–H at 2100 cm^{-1}) and nitrogen (N–H at 3400 cm^{-1}). The relative concentration of the Si–H and N–H bonds was dependent on the gas mixture. The gas mixture with a low SiH$_4$/NH$_3$ ratio had the Si–H and N–H stretching vibrations shifted to 2174 and 3312 cm^{-1}, respectively. This fact was explained as a consequence of the electronegativity of nitrogen. It was claimed that the small oxygen content revealed by IR analysis was a result of powder hydrolysis on exposure to atmosphere during handling and analysis. Indeed, Ho et al. [185] compared the oxidation of powders stored in different environments by IR spectroscopy. One sample was placed in a dry box containing air and a desiccator, and the other was placed in a wet box with 100% relative humidity, both at room temperature. The powder exposed to water vapor rapidly hydrolyzed with the complete loss of the Si–H feature and with the strong development of the band at approximately 1100 cm^{-1} (Si–O). These results showed that water vapor, rather than O$_2$, was responsible for the oxidation of these powders. In a later report [186], it was shown that, once the powders had been annealed in ammonia atmosphere at 800 °C for 20 min, they were largely insensitive to air exposure (Fig. 46). After this heat treatment, powders changed from white to dark brown and the hydrogen–nitrogen absorption bands disappeared from the IR spectrum.

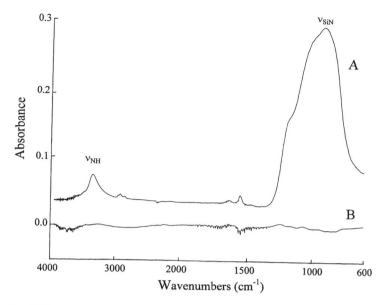

Fig. 46. FTIR spectra of heat-treated silicon nitride particles obtained in an rf discharge of silane and ammonia ($T_{ON} = 0.2$ s). The heat treatment was performed at 800 °C for 20 min in ammonia atmosphere. (A) Freshly heat-treated powder. (B) Difference spectrum of powder after 24-h exposure to ambient air. Reprinted with permission from R. J. Buss and S. V. Babu, *J. Vac. Sci. Technol., A* 14, 577 (© 1996 American Vacuum Society).

Recent results from other authors compared the formation of Si–N powders in $SiH_4 + N_2$ and $SiH_4 + NH_3$ gas mixtures [171]. In $SiH_4 + N_2$ discharges, the gas flow ratio SiH_4/N_2 was chosen to be 1/40 in order to incorporate a significant amount of nitrogen into the sample. Figure 47 presents the IR spectra of powders of both discharges. It shows that hydrogen is mainly bonded to silicon in powder from $N_2 + SiH_4$ discharges, whereas the N–H absorption band is the most prominent in powder from $SiH_4 + NH_3$ ($SiH_4/NH_3 = 0.8$). Both samples show a relevant absorption from silicon–nitrogen bonds.

Ho et al. [185] observed highly agglomerated particles with diameters ranging from 10 to 200 nm. Nitrogen Brunauer–Emmett–Teller (BET) measurements gave a surface area of 35 m^2/g. The density of the material was estimated to be about 1.8 g/cm^3 [185]. They estimated the yield for the production process as the percentage of silicon atoms delivered to the plasma that were incorporated into the powder, assuming an average molecular weight of 58 amu (SiN_2H_2). Maximum yields of 40% were obtained. According to these authors, one of the main drawbacks of this technique for producing nanometric powders was the lack of control provided over size and size distribution. However, Buss and Babu [186] showed that these parameters could be controlled, to some extent, by adjusting the plasma-on time; even so, the size dispersion was large.

5.3.1.1. Annealing

Silicon nitride powders produced in the rf discharges were heat treated in closed molybdenum crucibles under flowing N_2 in a tungsten-element resistance furnace [185]. The crystallinity of the powders after the treatment was investigated by X-ray diffraction. Heat treatment for 1 h up to 1600 °C did not lead to the crystallization of powder. After thermal annealing for 8 h at 1600 °C, the X-ray spectrum showed lines corresponding to α-Si_3N_4. Electron diffraction, however, indicated that the amorphous particles were transformed to α-Si_3N_4, present both as roughly equiaxed grains and needles, and β-Si_3N_4, which was present only as needles. It was found that the crystallization temperature could be lowered by increasing the silicon content.

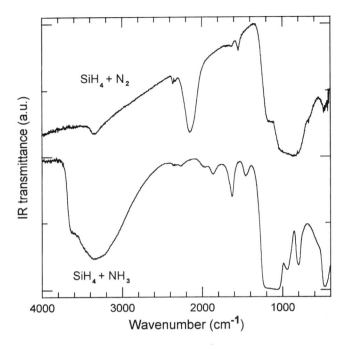

Fig. 47.   IR spectra of Si–N powders obtained in silane-ammonia and silane nitrogen discharges.

Ho et al. [185] claimed that their as-synthesized powders were similar to silicon diimide, $[Si(NH)_2]_n$, although their material contained SiH bonds that are not present in silicon diimide (diimide is known to convert to α-Si_3N_4 at high temperatures and to be easily hydrolyzed [187]).

5.3.2. Powders of Boron–Nitrogen Alloys

The capacity of low-pressure, low-temperature rf discharges to produce powders of B–N alloys was demonstrated in a preliminary work on particle formation in a diborane–hydrogen–ammonia discharge [171, 188]. In this report, 20 sccm of 1% B_2H_6 in H_2 and 10 sccm of NH_3 were introduced in an 80-W rf discharge at 70 Pa. The discharge was modulated at low frequency ($T_{ON} = 5$ s).

In this discharge, the powder formation was significantly lower than in a pure silane plasma. The size of the observed particles ranged from 20 to 100 nm and they were highly agglomerated. The electron diffraction of this powder consisted of three diffused rings, which demonstrated that short-range order existed in these samples. For instance, HRTEM showed ordered domains of less than 2.5 nm (Fig. 48). The interplanar distance of 3.4 Å was close to the (002) planes of the hexagonal structure of BN (3.33 Å). Therefore, the structure of these samples was qualified as turbostatic (short-range ordered domains).

The atomic structure of B–N powders was also investigated by Raman spectroscopy. The Raman spectrum had an intense continuous luminescence background for all excitation laser powers. Figure 49 shows that it was necessary to reach high laser intensities to observe the characteristic peak of hexagonal B–N centered at 1366.2 cm^{-1}. The full width at half maximum of the Raman peak was higher than in bulk boron nitride. To clarify whether this was a consequence of the phonon confinement in small ordered domains, a phonon confinement model was used to simulate this peak [171]. The expression for the phonon dispersion curve in BN was taken from Nemanich [189] and a Gaussian confinement was supposed. Figure 50 shows the experimental peak (once the background was subtracted) and two different fits corresponding to two sizes of crystallite ordered domains.

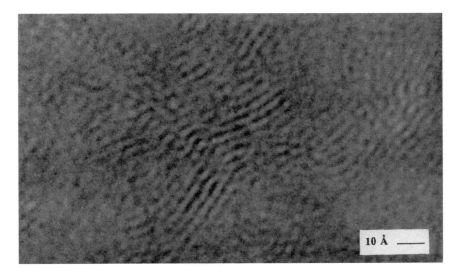

Fig. 48. High-resolution TEM image of BN powder obtained in a silane–ammonia discharge.

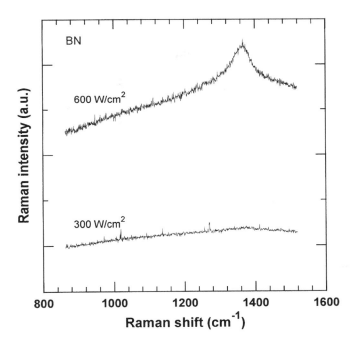

Fig. 49. Raman spectra of nanometric B–N powders using two laser intensities. Only for high laser fluences did the spectra show the characteristic features of hexagonal BN. A continuous luminescent background appeared irrespective of the laser intensity.

None of the calculated curves was able to reproduce the experimental points. The Raman peak observed, therefore, was attributed either to thermal effects resulting from the high laser intensities or to a large dispersion in size of the ordered domains.

The thermal desorption of hydrogen of these samples was examined [188]. Two evolutions appeared, one centered at around 500 °C (LT) and a second one that started at about 700 °C and increased up to the temperature limit of the experiment (HT). The LT evolution was attributed to the B–H bond, whereas the HT evolution was attributed to the N–H bond.

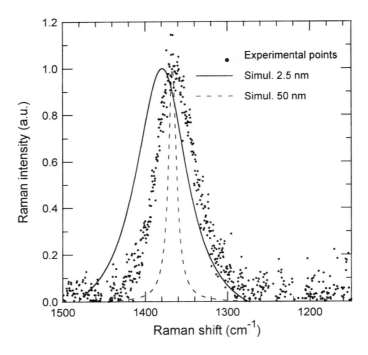

Fig. 50. Raman spectra measured experimentally (dots) and two simulated spectra for h-BN with crystalline domains of 2.5 nm (continuous line) and 50 nm (dashed line). From this comparison, it appeared that the spectral shape was not a consequence of phonon confinement.

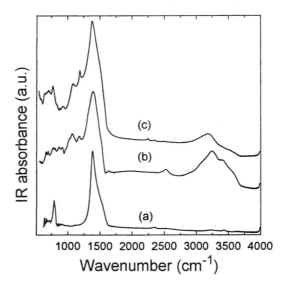

Fig. 51. FTIR spectra of B–N powder produced in an rf discharge of silane and ammonia. For comparison, the IR spectra of a PECVD BN film is also shown (a). (b) As-deposited powder at room temperature and (c) powder after thermal treatment at 800 °C.

A heat treatment of the B–N powder was carried out in order to remove the hydrogen content and to improve the B–N bonding. Figure 51 shows the IR spectra of the B–H samples before and after the heat treatment. The spectra showed the absorption band centered at 1385 cm^{-1}, corresponding to the in-plane B–N stretching vibration mode, characteristic of hexagonal BN. Besides, the as-prepared powder showed small absorption peaks in the 700–1200 cm^{-1} region (1180, 1070, 918, 780, and 695 cm^{-1}), which were related to those reported to appear in chemical vapor deposition (CVD) boron nitride films [190, 191] and in cyclic boron–nitrogen compounds [192].

Furthermore, the as-deposited BN powder exhibited a broad absorption related to hydrogen bonded to nitrogen in the region 3000–3500 cm^{-1}. Finally, a weak absorption peak at 2520 cm^{-1} was ascribed to the stretching vibration of B–H. Following heat treatment up to 800 °C, the B–H-related band disappeared, whereas that of N–H decreased but was still present. The absorption associated with the B–N stretching and B–N–B bending modes (1385 and 780 cm^{-1}, respectively) became more evident.

5.3.3. Powders of Silicon–Carbon Alloys

Nanometric particles of different silicon–carbon stoichiometries have been produced in rf glow discharges of silane (SiH_4) and methane (CH_4) mixtures [193–198]. In relation to the choice of methane as the gas precursor containing carbon, Hollenstein et al. [117] showed that acetylene plasmas yield more highly polymerized ions than CH_4 and have a very strong tendency to form powder in rf plasmas.

The dependence of the carbon content, x ($S_{1-x}C_x$:H), on the methane fraction of the gas precursor, R (R = [CH_4]/([CH_4] + [SiH_4])), was examined in order to determine the gas mixtures that lead to stoichiometric powders [193, 194].

The discharges were modulated at low frequencies to improve the yield of powder formation. It was realized that the time evolution of the rf voltage during the first milliseconds of the discharge was strongly dependent on the methane fraction. Figure 52 shows the peak-to-peak rf voltage during a period of plasma-on [194]. When the plasma was ignited, there was a narrow voltage peak of a few milliseconds. After reaching its maximum, the rf voltage decreased smoothly to a stable value. Figure 53 shows that the relative height of this peak increased with the methane concentration. These transient phenomena were associated with the losses of electrons involved with the formation of particles. Particles

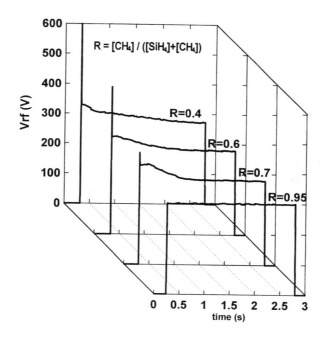

Fig. 52. Radio frequency peak-to-peak voltage during a plasma-on period for different methane fractions in a silane–methane discharge.

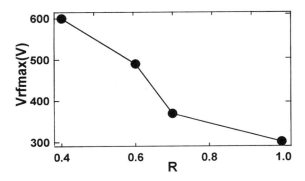

Fig. 53. Dependence of the voltage peak shown in Figure 53 on methane concentration.

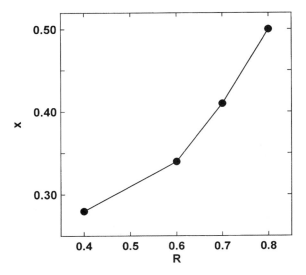

Fig. 54. Dependence of the carbon concentration x ($Si_{1-x}C_x$) on the precursor gas mixture R, R = $[CH_4]/([CH_4]+[SiH_4])$.

act as traps of negative charges and the electric field rises so as to balance the lost charges. The fact that this peak scaled with the silane concentration was in good agreement with the experimental evidence that the formation of powder diminished with the methane content on the precursor gas.

The dependence of the carbon content, x, on the methane fraction was studied by means of elemental analysis [193, 194]. Figure 54 shows this dependence. To obtain a stoichiometric powder, $x = 0.5$, a methane-rich gas (R = 0.8) was necessary. This is related to the higher energy required to ionize the methane molecule compared to that needed for silane. In addition to carbon content, the elemental analysis measurements revealed that the hydrogen content was about 45%.

X-ray photoelectron spectroscopy (XPS) analyses also provided a compositional factor, x [197]. The 1-s carbon peak of an $Si_{1-x}C_x$ sample with $x = 0.56$ is shown in Figure 55, where it is compared to the experimental spectrum of commercial β-SiC. The deconvolution of this spectrum showed that it was composed of peaks arising from carbon bonded to silicon, at 283.5 eV, and carbon bonded to carbon, at 284.6 eV. The dominance of Si–C bonds over C–C bonds was evident from their relative peak intensities. No C–O was detected in spite of the presence of a significant amount of oxygen in the sample revealed by

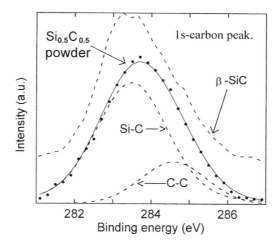

Fig. 55.   XPS spectra of C (1 s) core levels of $Si_{0.44}C_{0.54}$ and commercial β-SiC powders.

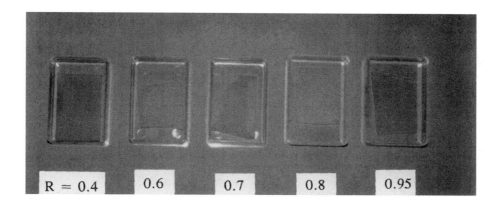

Fig. 56.   A series of $Si_{1-x}C_x$ powder samples obtained in rf discharges of methane and silane mixtures. Sample color ranges from brown yellow for silicon-rich samples to white for carbon-rich samples.

IR and elemental analysis. Therefore, this oxygen content was assumed to be bonded to only silicon atoms.

The color of the samples was dependent on the carbon content [171]. Figure 56 shows a series of powder samples deposited on a glass substrate (R = 0.4, 0.6, 0.7, 0.8, 0.95). The color ranged from red yellow for the silicon-rich samples to transparent white for the carbon-rich powders.

These studies were mainly devoted to the structural characterization of the material but no effort was made to control the size of these powders. Therefore, to the author's knowledge, there is no report concerning the development of particle size in silane–methane mixtures.

TEM investigations on the $Si_{1-x}C_x$ powder samples revealed a wide size distribution of particles, which ranged from 10 to 300 nm [193–198]. As explained for particles produced in silane discharges, specification of the plasma-on duration and the plasma conditions should allow for control of the particle size and size distribution.

Raman measurements, as well as those of SAED and HRTEM, showed that the particles were amorphous. Moreover, Raman analysis provided evidence of the presence of Si–H and C–H vibrational modes, whereas those of Si–C could not be detected.

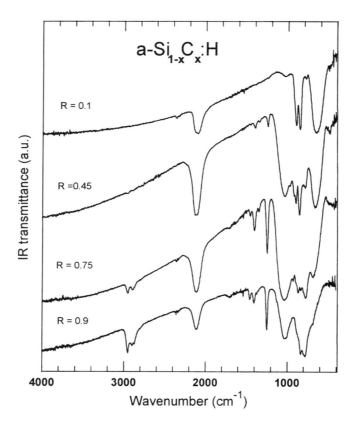

Fig. 57. IR transmission spectra of a series of samples produced in silane/methane discharges with different methane fractions.

The IR spectra of these samples, however, provided more detailed information on the Si–H, C–H, and Si–C bond configurations. Figure 57 shows the IR spectra for a series of samples, from silicon-rich to carbon-rich samples.

The vibrational modes related to silicon-hydrogen bonds appeared in each spectrum, the wagging modes, at 650 cm^{-1}, and the bending modes related to $(SiH_2)_n$ and SiH_2, between 870 and 910 cm^{-1} [199, 200]. The stretching band, at around 2100 cm^{-1}, did not shift toward higher wavenumbers as the carbon concentration increased as expected from the electronegativity of carbon.

Absorption contributions from the carbon–hydrogen groups did not appear in the silicon-rich samples. Stretching modes of the C–H arrangements (between 2875 and 2955 cm^{-1}) as well as bending modes (around 1400 cm^{-1}) were evident in samples with R = 0.75 and R = 0.9. The absorption band at 1250 cm^{-1} was attributed to C–C skeletal vibrations or to C–H$_3$ bending vibrations in the Si–C$_3$ groups.

In the spectral region between 700 and 1100 cm^{-1}, several absorption peaks were superimposed on each other. Besides the bending modes of Si–H, the absorption at around 780 cm^{-1} was related to Si–C bonds, whereas that at around 1000 cm^{-1} might be due to Si–CH$_2$ bonding arrangements or to silicon–oxygen bonds. Figure 58 shows the IR absorbance in the region between 425 and 1500 cm^{-1}.

Figure 59 describes the thermal desorption of hydrogen from the $Si_{1-x}C_x$ sample with $x = 0.28$. Hydrogen evolution started at around 473 K and rose continuously to its maximum at around 930 K. The shape of the peak revealed that the evolution had two distinct contributions: The contribution of the low-temperature peak increased with the silicon content in the sample, as shown in the inset to Figure 59. Therefore, these two contributions,

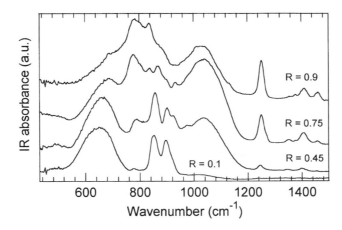

Fig. 58.   IR absorption spectra in the range between 450 and 1450 cm^{-1} for the series of samples shown in Figure 57.

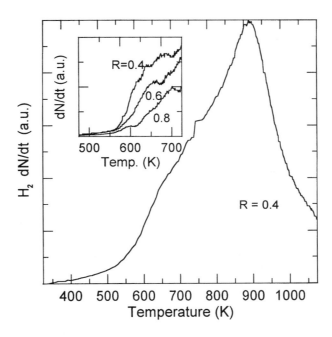

Fig. 59.   Thermal desorption spectra of hydrogen for the sample produced in a discharge with a fraction of methane of R = 0.4. The figure inset shows that the low-temperature evolution scales with the silicon concentration in the sample.

at low and high temperature, could be assigned to silicon–hydrogen bonds and to carbon–hydrogen bonds, respectively. From the temperature of the maximum of the evolution, it was deduced that the free energy for the desorption of C–H was $\Delta G = 2.94$ eV [171]. Although this value was slightly lower than that in amorphous carbon thin films [174], an energy balance equivalent to that presented in Section 5.2.1.4 allowed the determination of the C–H bond energy (3.73 eV), which was in very good agreement with that reported in the literature (3.7 eV) [201]. The free energy related to the evolution of Si–H was $\Delta G = 2.08$ eV and therefore greater than that determined in silicon powder. This could be due to the increase in the Si–H bond strength as the carbon content increases.

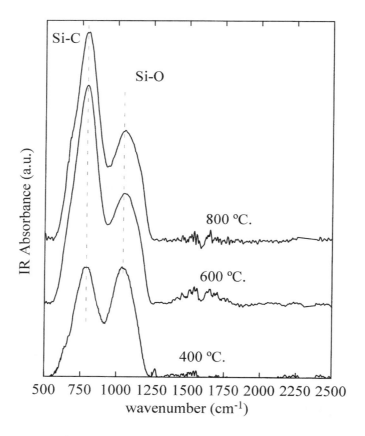

Fig. 60. IR vibrational spectra of $Si_{0.63}C_{0.37}$ powder at different annealing temperatures. The spectra were normalized to the Si–O peak area. The thermal annealing was performed under vacuum.

5.3.3.1. Annealing

The objectives of annealing $Si_{1-x}C_x$ powders were to study the crystallization of the as-prepared amorphous samples and the stabilization of the samples against their spontaneous oxidation on exposure to the atmosphere.

To investigate the stabilization against spontaneous oxidation, Viera et al. [198] compared the IR spectra of two samples that had been in air for several months. One was as prepared and the other sample was thermally annealed up to 800 °C in vacuum. After 1 month, the as-prepared sample exhibited the spectral features of silicon–oxygen-related bonds and its oxygen content had increased by a factor of 2. However, the IR spectra of the annealed sample did not show any change after 6 months, and its oxygen did not vary, either.

In addition, differential scanning calorimetry (DSC) showed that oxidation at room temperature can be avoided by annealing the powder in an inert atmosphere at a relatively low temperature [196].

To investigate the structural changes induced by thermal annealing, Viera et al. [197] performed thermal treatments on a silicon-rich $Si_{1-x}C_x$ sample in vacuum at 400, 600, and 800 °C. Figure 60 shows the IR spectra at different temperatures. At 400 °C, SiH stretching vibrations were found to be absent and only weak signals of C–H vibrations could be observed. One interesting finding was that the Si–C-related band (at around 775 cm^{-1}) increased relative to the band at around 1040 cm^{-1}. This occurred because of the transformation of $Si(CH_2)_n$ Si groups (which, as well as Si–O modes, contribute to the 1040-cm^{-1} band) into Si–C bonds, as hydrogen was removed. The structural changes produced in the

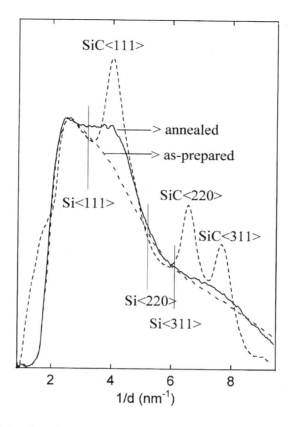

Fig. 61. Radial profiles of the electron diffraction patterns for as-deposited (a) and annealed (b) nanometric SiC powder compared with that obtained for standard β-SiC microcrystalline powder. d represents the interplanar distance.

powder by heat treatment under vacuum at 800 °C were analyzed by HRTEM and SAED. No changes in particle size or shape were observed after the treatment, which indicated that no sintering was initiated at this temperature. The SAED pattern of the annealed powder revealed diffuse rings and no spots. The short-range order of these samples was analyzed by digitalizing the electron diffraction pattern and plotting the diffraction intensity versus the corresponding interplanar distance. Figure 61 shows these curves and compares them to those of commercial cubic SiC microcrystalline powder. The figure shows that the short-range order of the annealed SiC powder is closer to that in cubic SiC than in crystalline silicon or carbon.

The SiC powder was heated at high temperatures (∼2000 °C) [171] and was irradiated with a pulsed excimer laser, for a preliminariy study of the crystallization of the samples [197].

The excimer laser irradiation of a silicon-rich sample produced Si and SiC crystalline domains of around 20 nm [197].

However, heat treatment at 2000 °C in a graphite furnace with an Ar atmosphere of stoichiometric SiC powders effectively crystallized the samples. Figure 62 shows scanning electron microscope (SEM) images of SiC powder after the heat treatment. The micrographs revealed SiC crystals of up to 50 μm together with small structures whose sizes were in the nanometer range. The huge size of the crystals, in comparison with the nanometer size of the initial particles, indicated that the heat treatment should be performed at lower temperatures in order to crystallize the particles without leading to this abnormal grain growth. Further studies are in progress.

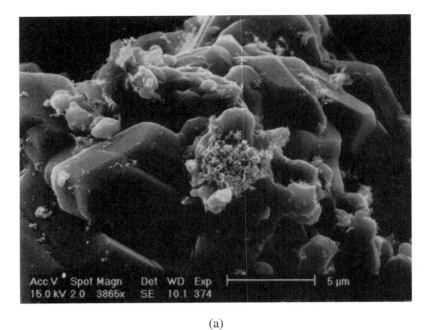

(a)

(b)

Fig. 62. SEM images of SiC powder annealed in a graphite furnace under vacuum at 2000 °C for 30 min.

Raman analysis of heat-treated samples showed that, besides the well-crystallized SiC, they contained some crystalline carbon (Fig. 63). This feature of powders that initially were stoichiometric could be attributed to the fact that all of the inner elements of the furnace were graphite. However, the main result of this preliminary study was the evidence that well-crystallized SiC from powders produced in a glow discharge of silane and methane mixtures can be obtained. The control of the crystal size and particle sintering is yet to be performed.

(c)

Fig. 62. (*Continued.*)

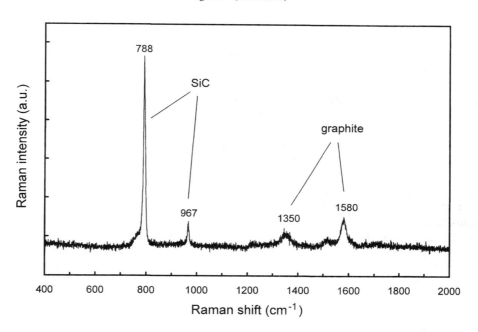

Fig. 63. Raman spectra of SiC powder heated to 2000 °C.

5.4. Summary

The main conclusions of the review presented in this section can be summarized as follows.

Low-pressure, low-temperature plasmas can produce particles of alloys whose composition is determined by the gas mixture. The properties of each gas, such as the ionization energy, determine its efficiency to contribute to particle growth.

Particle development can be divided into two stages: nucleation and coagulation. Therefore, theoretical models of these processes, or *ex situ* microscopy measurements, enable one to control the particle size and size distribution by adjusting the discharge conditions and the plasma-on duration.

Two different kinds of nanoparticles can be produced in these plasmas. Quasimonodisperse particles of a few nanometers (<10 nm) can be grown in discharges with slow kinetics of particle development, such as low-power argon-diluted discharges, provided that the plasma-on is short enough to avoid coagulation. Laser light scattering methods do not detect these particles because of their small size. These particles have an ordered atomic structure that is not completely understood. On the other hand, fast particle development, such as in pure-silane high-power discharges and long plasma-on periods, leads to large quantities of powders with sizes up to approximately 200 nm with a wide size distribution. Their atomic structure is amorphous. They are observable by laser light scattering.

The silicon powders obtained under discharge conditions for a large yield of powder production are highly hydrogenated. The atom content approaches 40%, and the atomic structure is amorphous. The color ranges from yellow to brown. Particles appear highly agglomerated. The vibrational characterization indicates that silicon powder has a polymer-like structure, which tends to become more compact during its exposure to the plasma. Raman measurements also indicate that there is no Si network in the as-deposited particles.

Differences in the color of silicon powders depending on the zone of the reactor where they deposit reveal that their microstructure is affected by the plasma. This plasma effect was further demonstrated by the IR study of particles that were been suspended in the plasma over different lengths of time. This was performed in two experiments in which the plasma duration and the gas residence time were varied. Ion and electron bombardment diminishes the hydrogen content and enhances cross-linking in the particle. Therefore, Si–Si bonds are more abundant as the time for particle suspension in the plasma increases. Silicon powder oxidizes spontaneously when it is in air.

Powders of silicon–nitrogen alloys can be produced in both silane–ammonia and silane–nitrogen gas mixtures. As-deposited powders have strong similarities to silicon diimide, $[Si(NH)_2]_n$. These powders suffer spontaneous oxidation as a result of powder hydrolysis on exposure to atmosphere. This oxidation is prevented by thermal annealing at relatively low temperature. When silicon–nitrogen powders are annealed at 1600 °C for 1 h, the amorphous particles transform to α-Si_3N_4 and β-Si_3N_4.

Discharges of silane and diborane diluted in N_2 or H_2 produce powders of boron and nitrogen. In this case, the atomic structure is turbostatic. Annealing at 800 °C removes the hydrogen content and enhances the boron–nitrogen bonds.

Powders of silicon–carbon alloys can be produced in discharges of silane–methane mixtures. Other carbon-containing gases such as acetylene are thought to be more efficient for producing powders. Low-temperature annealing prevents the oxidation of these samples in air at room temperature; otherwise, it is a spontaneous process. High-temperature annealing, at 2000 °C, leads to crystals of SiC of grain size around 40 μm, which suggests that a lower temperature is required to avoid this abnormal grain growth.

6. APPLICATIONS

Nanoparticles produced in low-pressure, low-temperature plasmas can have several applications. Although this material is currently under research and development, some uses of these particles are reviewed in this chapter. The structural properties of powders of different alloys were described in Section 5. Moreover, the ability to deposit nanostructured thin films consisting of nanoparticles embedded in a homogeneous matrix is reviewed in detail in Section 6.1. Finally, the detailed analyses of light emission in silicon nanoparticles are reviewed in Section 6.2. These are the applications that have received the most attention until now. However, nanoparticles of covalent, nonoxide ceramics are expected to have specific advanced applications in the near future.

6.1. Nanostructured Silicon Thin Films

Initial studies of the formation of particles in low-pressure, low-temperature plasmas focused on their contamination effect during the deposition of amorphous silicon thin films (Section 2.3). Industrial applications of silicon-based microelectronic devices require very high deposition rates and film qualities. Attempts to improve the deposition rate by increasing the rf power also causes polymerization of the silane radicals, which leads to the formation of particles. Particles are the main source of contamination in these processes.

A better understanding of the phenomena under different discharge conditions was achieved through research into powder development. Indeed, it is now possible to calculate the particle size and size distribution dependence on time for a given set of plasma characteristics (Section 4.1.3). On the one hand, contamination of a discharge can be avoided by controlling the plasma-on duration in a modulated discharge. As was discussed in Section 4.3.2, plasma modulation can interrupt effectively the reaction pathway to the formation of particles.

On the other hand, and this will be the subject of this section, knowledge of the coagulation dynamics permits the incorporation of particles of a chosen size into a growing thin film. Therefore, the ratio of atoms contributing to film growth as particles to the total number of silicon atoms contributing to film growth, and the structure of these particles, largely determine the properties of the thin film. The discussion will show how films growing with a large number of small particles have optical and electronic properties that promise to be useful for device applications. These new kinds of thin films are referred to as nanostructured thin films.

To the author's knowledge, the selective incorporation of particles into a growing film has been used only for the deposition of silicon films [99, 202–205], although it is expected that films of silicon alloys and other compounds can be deposited using the same technique. It should be mentioned that many reports have been devoted to investigating the discharge conditions that prevent particle formation and thus contamination on the wafer during amorphous (a-Si:H), microcrystalline (μc-Si:H), or polycrystalline (pc-Si:H) silicon thin-film deposition (see Section 2.3 and [43] for a review). This section does not discuss these reports but others that use the nanometric particles to obtain silicon thin films with new atomic structures and properties.

Hydrogenated amorphous silicon has been extensively studied during the last 30 years. The most widely used technique to deposit this material is plasma-enhanced chemical vapor deposition (PECVD), using an rf electric signal as the source of plasma excitation. The device-quality a-Si:H is usually obtained at low rf power and substrate temperatures around 200 °C. The deposition rate under these discharge conditions is very low ($\sim$1 Å/s).

The atomic structure of a-Si:H can be described as a disordered network of silicon atoms tetrahedrally bonded with some distortion in bond length and angle compared to that on crystalline silicon (diamond structure) [206, 207]. The short-range order is expected to reach only the first or second neighbors.

Microcrystalline silicon (μc-Si:H) in thin-film technology appears promising because it combines the high optical absorption of amorphous silicon with higher and more stable electrical properties. In this case, the atomic structure consists of well-crystallized silicon domains that are easily shown by HRTEM or X-ray diffraction. The atomic order spreads over more than approximately 100 nm [208–210].

Films with crystalline domains of a fraction of a micrometer are referred to as polycrystalline silicon (pc-Si:H). On the basis of theoretical calculations, some authors claim that no crystalline domains smaller than 2–3 nm can be stable in a silicon network [211], which seems to determine the borderline between amorphous and microcrystalline silicon thin films. Many experimental observations, however, support the existence of ordered domains with sizes below 3 nm [212, 213].

This section shows that the incorporation of ordered particles of 1–2 nm into the growing film may lead to a short-range ordered structure of about the particle size. These ordered regions do not need to be understood as small crystals but, rather, as clusters with particular well-defined geometries.

6.1.1. Nanostructured Silicon Thin Films in Continuous Discharges

Discharge conditions that ensure a moderate fraction of atoms contributing to film growth as particles, as well as small sizes, are required to produce a good-quality thin film. Excessive particle sizes or particle contribution would lead to unwanted roughness and void fraction for a device-quality material [72, 75, 76]. Therefore, discharge conditions, including plasma modulation, for the formation of small particles should be chosen.

From the discussion of particle coagulation in Section 4.1.3, it appears that the easiest way to control the size of the particles to be incorporated into the film is by adjusting the plasma-on duration in a modulated discharge. Roca i Cabarrocas et al. [99], however, gave experimental evidence of particle deposition under a continuous argon–silane plasma. Using laser light scattering measurements to monitor powder formation, they determined the discharge conditions that led to the formation of small particles (not detectable by LLS) but not of powders. For a given argon flow, they studied how the silane gas flow and total gas pressure affected the formation of powders (observable by LLS). They showed that, as the silane gas flow decreased, the LLS signal diminished. In addition, the scattered light intensity exhibited periodic fluctuations, whose frequency diminished as the silane flow decreased. These fluctuations were explained as follows: Silicon powders grow up to a critical size and/or concentration above which the plasma cannot confine them any longer, which leads to particle evacuation and the start of a new cycle. Under these conditions, the resulting particles are too big and the films deposited may have structural and electronic defects. Therefore, the silane gas flow was decreased until no powder could be observed by LLS in the discharge, which meant that no particles with diameters above approximately 60 nm were present in the discharge. Nevertheless, *ex situ* TEM measurements showed that particles of about 8 nm were produced under these plasma conditions after 1 s of discharge.

Particles in this size range experience charge fluctuations in the plasma. At the moment when the particles are electrically neutral, they are not electrostatically confined by the plasma sheaths and they are likely to leave the plasma. Therefore, there is an "escape rate" as well as the particle nucleation rate. If the escape rate is higher, then particles grow for a relatively short time up to a limited size before escaping the plasma, which prevents the formation of larger particles. When the plasma conditions lead to a high nucleation rate (higher than the escape rate), however, the powder formation develops until it reaches a combination of particle size and particle concentration that cannot be maintained in the plasma any longer. Then, large particles are formed and oscillations in the laser light scattered signal or in the optical emission are detected. This escape rate is very sensitive to particle concentration and size and to other plasma characteristics such as electron temperature. The escape rate is higher during the first stages of particle formation when the instantaneous fraction of neutral particles is higher [131].

In their report, Roca i Cabarrocas et al. [99] showed that nanometric particles did contribute to film growth in a continuous-wave discharge. They compared spectroscopic ellipsometry measurements of films deposited under the same discharge conditions but with different plasma-on times: 0.24 s, 1 s, and continuous-wave (cw) conditions. Figure 64 shows the imaginary part of the pseudoelectric function $\langle \varepsilon_2 \rangle$ for each sample. In all samples, $\langle \varepsilon_2 \rangle$ was very low because of their high porosity. The key result in this discussion was that the film grown under the continuous discharge had the same $\langle \varepsilon_2 \rangle$ as the film deposited

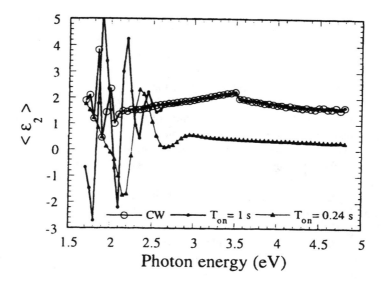

Fig. 64. Imaginary part of the pseudoelectric function of films deposited under a continuous-wave discharge and with $T_{ON} = 1$ s and $T_{ON} = 0.24$ s. Note that the films deposited under cw and under modulation with $T_{ON} = 1$ s have the same $\langle \varepsilon_2 \rangle$ in the high-energy part of the spectra, which indicates that the films have the same optical properties. Reprinted with permission from P. Roca i Cabarrocas, P. Gay, and A. Hadjadj, *J. Vac. Sci. Technol., A* 14, 655 (© 1996 American Vacuum Society).

with a modulation of 1 s of plasma-on. Moreover, the same deposition rate for both discharges (4 Å/s) was found, which confirmed that the deposition during the plasma-off time was negligible.

The imaginary part of the pseudoelectric function exhibited a characteristic step at 3.5 eV for the films grown under continuous plasma and under modulation with 1 s of plasma-on (Fig. 64). This step was claimed to be a characteristic feature of the nanoparticles deposited on the film. These authors argued that the film grown with a plasma-on duration of 0.24 s did not show this characteristic step in $\langle \varepsilon_2 \rangle$ because the plasma duration was too short to allow particle growth up to the requisite size. Using effective medium theories, they calculated that a nanoparticle fraction of about 10% on the film was needed to account for the observed step. This percentage was consistent with their calculation of the ratio of the number of silicon atoms impinging as 8-nm particles to the total number of silicon atoms contributing to film growth. They based their calculation on the observed deposition rate and on TEM measurements.

The same authors suggested that the interesting properties of other silicon films grown previously under silane–helium discharges, where powder was produced, might be a consequence of the incorporation of particles during their growth, although this had not been directly investigated by the research team [214–216]. Indeed, even though powders are confined in the discharge and do not contribute to the deposition, some fraction of the smaller particles (which are more likely to escape) could contribute to film growth. The structural analyses of film surface by tunnelling microscopy suggested this idea [217].

6.1.2. Characterization of Nanostructured Films

6.1.2.1. Structural Properties

Although these authors produced films that can be regarded as an amorphous matrix with embedded nanoparticles, they referred to these films as polymorphous silicon (pm-Si) [203, 205, 218]. These films were deposited under a wide range of discharge conditions [203]:

1. Pure silane at low temperature, low pressure, and low rf power. These conditions were thought to produce a relatively poor quality a-Si:H, although they demonstrated device quality a-Si:H even at 50 °C [219].
2. High hydrogen dilution at 100 °C at low rf power.
3. High hydrogen dilution and high rf power at 200 °C and high pressure.

Roca i Cabarrocas et al. observed that, in spite of the very different deposition conditions, the films had similar properties. They compared these nanostructured thin films with silicon microcrystalline films grown under similar conditions. In fact, they presented the nanostructured films as the material bridging the gap between amorphous and microcrystalline silicon [203]. For that purpose, they compared the properties of films grown in a hydrogen–2% silane discharge under the same conditions except for the total gas pressure. Discharges at 400, 800, and 1200 mtorr produced a-Si:H, μc-Si:H, and pm-Si, respectively [203].

The imaginary part of the pseudoelectric function of the pm-Si:H was quite similar to that of standard a-Si:H. However, the spectrum of the pm-Si film shifted toward higher energies, indicating a slightly higher gap, confirming optical transmission and photothermal desorption spectroscopy (PDS) measurements. Despite the similarity of both spectra, the pm-Si film spectrum could not be simulated by a mixture of an amorphous phase with a fraction of voids, as was done with the a-Si:H film spectrum.

Later they reported the modeling of these spectra by means of Bruggeman effective medium approximation (BEMA) theory [220]. This modeling showed that the $\langle \varepsilon_2 \rangle$ of the film grown at 800 mtorr was that of μc-Si:H because the $\langle \varepsilon_2 \rangle$ could be described by a mixture of amorphous silicon, polycrystalline silicon (89%), and voids. To describe the $\langle \varepsilon_2 \rangle$ of the pm-Si:H, however, a mixture of amorphous silicon, voids, and polycrystalline silicon in a multilayer system had to be assumed. This multilayer system consisted of an amorphous top layer and a bulk layer with a high crystalline fraction. This description was not arbitrary, as it was supported by HRTEM measurements. Nevertheless, no explanation could be provided for this lack of homogeneity in the film composition.

The absorbance in the infrared of these samples showed further structural differences between a-Si:H and pm-Si:H [203]. The absorption in the stretching region, which consists of one single peak at around 2000 cm^{-1} in a-Si:H, could be deconvoluted into two Gaussian peaks in pm-Si:H. One peak was centered at 2000 cm^{-1}, as in a-Si:H, and the other one at about 2030 cm^{-1}. The latter had also been found in μc-Si and was reported to appear in silicon powders (Section 5.2.1).

The hydrogen content was 14% in the pm-Si and 15% in the μc-Si film, both values being much higher than the typical hydrogen content in standard a-Si:H films deposited at the same temperature ($\sim$7%).

To characterize the structure of these nanostructured samples, Viera et al. [221] performed an HRTEM analysis of a cross section of a similar film and showed that it consisted of small crystalline domains (1–2 nm in diameter) embedded in an amorphous matrix. The assignation of these ordered domains to particles formed in the plasma was corroborated by another report in which it was shown that the individual particles contributing to the film growth had a well-ordered atomic structure [222]. The detailed structure of particles of 4–5 nm or above can be easily understood as small silicon crystals with a diamond structure. Clusters of smaller diameter, however, might have other atomic arrangements. It is commonly accepted that the diamond structure is not the most stable structure for small silicon clusters. Several authors have proposed different fullerene-like or cage–core structures [162–164].

The crystallization of the nanostructured films grown with the incorporation of a fraction of nanoparticles induced by thermal treatment has been investigated both by Raman spectroscopy and by monitoring the electrical conductivity of the sample during heat treatment [222, 223]. The Raman experimental setup allowed *in situ* monitoring of the crystal-

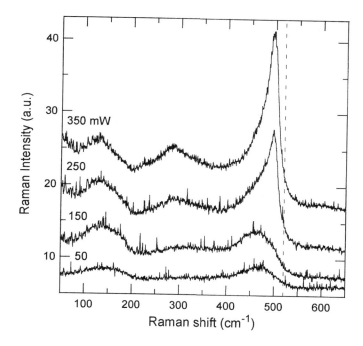

Fig. 65. Raman spectra of a nanostructured silicon thin film at increasing laser intensities. Note the position of the TO peak ($\sim$495 cm^{-1}) shifted from the position in c-Si (520 cm^{-1}). The laser intensity for the crystallization onset was lower than in amorphous silicon.

lization of a nanostructured sample deposited under a silane–argon discharge ($T_{ON} = 1$ s). The Raman spectra were measured at increasing laser intensities and the scattered signal was integrated for 240 s [222]. Because of this measurement method, each spectrum was the result of integrated signals coming from the different microstructures of the sample during the interval of detection. Although this is a drawback of the method, it ensures that the Raman signal comes from the annealed point of the sample. Moreover, it can provide a good picture of the crystallization dynamics.

The crystallization of this sample, as well as of other nanostructured silicon films, was characterized by the appearance of a sharp peak at around 495–498 cm^{-1} and by the crystallization onset starting at lower laser fluences than in standard amorphous silicon films. Figure 65 shows the Raman spectra of the sample at increasing laser fluences. Note the shift of the crystalline peak from the position of the transverse optical (TO) peak in bulk crystalline silicon (520 cm^{-1}) [224].

The most prominent peak in crystalline silicon is the TO peak (at 520 cm^{-1}), which can shift its position owing to several effects, such as the size of the crystalline domain. When the crystalline domain size diminishes below approximately 10 nm, the TO peak shifts to lower energies because of phonon confinement, which causes an indetermination on the wavevector [208, 209, 225].

Figure 66 shows the TO peak for silicon crystalline domains between 1.5 and 20 nm, which were calculated from Gaussian phonon confinement in a spherical crystal [225]. The comparison between Figures 65 and 66 shows clearly that phonon confinement does not account for the observed position of the peak observed experimentally.

Another possible cause of the displacement of the peak is the temperature reached by the sample during laser irradiation. A point often overlooked in recent papers is that there is linear dependence of the frequency shift on the temperature because of the disturbance of the interatomic potential [226, 227]. However, the temperature needed to account for a frequency shift of 25 cm^{-1} is close to 900 °C, which is unlikely to occur at the onset of

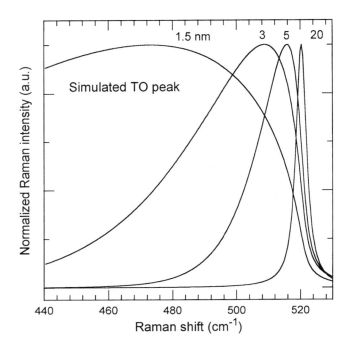

Fig. 66. Simulated Raman intensity of the TO peak of silicon for different sizes of the crystalline domains. These peaks have been calculated according to a Gaussian confinement model of the phonon in a spherical crystalline domain.

crystallization, when the applied laser fluence is still quite low. As is explained later, the temperature for the onset of crystallization in these films is around 500–600 °C. Although this tends to rule out a temperature effect, the position shift toward lower energies as the laser fluence increases reveals that temperature does affect the peak position to some extent.

These findings led to the comparison between nanostructured silicon thin films and silicon clathrates [222]. The crystallized sample in Figure 65 has similar Raman spectra to silicon clathrates [228–230]. Despite caution in linking silicon clathrate and nanostructured silicon films, nevertheless, the clathrate structure does consist of silicon polyhedra (12-, 14-, and 16-hedral silicon cages) arranged to form a periodic structure [230]. In the center of some of these cages, there is a metal atom (Na, Ba) that does not bond to any silicon atom, but prevents the collapse of the cage and consequent rearrangement into a diamond cubic structure. Given that the growth of the nanostructured film is partly due to the incorporation of small clusters of unknown structure, these clusters could be similar in structure to the polyhedra that form the clathrate. In this case, hydrogen in nanostructured films could play the role of the metal atoms in silicon clathrate. Further structural analyses are required to assess this hypothesis.

Monitoring the electrical conductivity of the sample during heat treatment showed that the crystallization dynamics in nanostructured silicon films were faster than in standard amorphous silicon [223]. Although the kinetic parameters for the crystallization were not provided, isothermal heat treatment of both samples showed that the crystallization of the nanostructured film occurred at lower temperatures than in a-Si:H. The onset of crystallization was detected by a sudden increase in the electrical conductivity. This was assumed to be due to ordered nanoparticles inside the amorphous matrix acting as seeds for heterogeneous nucleation process. Figure 67 compares the evolution over time of the electrical conductivity of the nanostructured and the amorphous films during isothermal heat treatment.

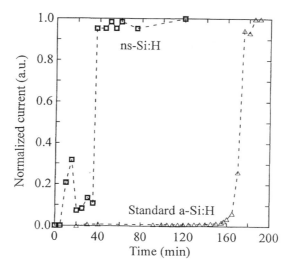

Fig. 67. Time evolution of the normalized current of the nanostructured silicon and standard amorphous silicon thin films during isothermal annealing at 590 °C.

6.1.2.2. Electrical Properties

Although nanostructured thin films have still not been fully characterized electrically, some insights into their transport properties have already been published [205, 218]. Some of these findings, such as the electronic mobility–lifetime product ($\mu\tau$), seem very encouraging in their device applications. There is no well-based understanding of the origin of the peculiar properties of nanostructured silicon thin films. Moreover, it may be that silicon films grown under unconventional discharge conditions, which were understood to be standard amorphous silicon films even though they exhibited improved electrical properties, have a significant contribution of small clusters. For instance, the Energy Research Unit (ERU) of Calcutta showed that silicon films grown under helium–silane mixtures at high pressure ($\sim$240 Pa) had their $\mu\tau$ product at a factor of 50 higher than standard amorphous silicon [231, 232]. Although the evidence is not conclusive, the high pressure used to deposit these films points to the presence of small clusters in the discharge. Longeaud et al. [218] compared the transport properties and the density of defects, N_d, of a nanostructured thin film produced in a hydrogen-diluted silane discharge with a standard amorphous film. The $\mu\tau$ product for the nanostructured silicon film was 2.9×10^{-4} cm^2 V^{-1}, whereas that of standard amorphous silicon was 1–3×10^{-6} cm^2 V^{-1}. These authors observed a strong correlation between N_d and $\mu\tau$, the lower N_d values corresponding to higher $\mu\tau$ values. Constant photocurrent (CP) and modulated photocurrent (MPC) techniques revealed that the deep defect density in nanostructured films was approximately 20 times lower than that of standard samples, whereas both conduction tails were quite similar. The high $\mu\tau$ values obtained for the nanostructured film could not be attributed to an n-doping caused by any contamination because the deep defect density was lower than that in standard samples (in the case of n-doping, the reverse would have been expected). Moreover, the photosensitivity under AM1 ranged from 8×10^6 to 10^8 (compared to 10^5–10^6 for standard samples).

The high values of the $\mu\tau$ product can be attributed to an increase in either the electronic mobility or the lifetime. Longeaud et al. [218] observed that the electronic mobility of extended states was increased by only a factor of 2 at the most. Thus, the high values of $\mu\tau$ were attributed to an increase in τ, which can be linked to a decrease in either the density of the deep defects or their capture cross section or both. The observed decrease in the deep defect density by a factor of 20 could not account for the increase in $\mu\tau$ (a factor of 200). They concluded that in nanostructured silicon films the capture cross sections of the deep states, or at least part of them, have been reduced compared to standard amorphous silicon films.

Finally, the light soaking for both nanostructured and amorphous samples was studied. Apart from the high density of states in the gap that limits transport, the major drawback of amorphous silicon films is the degradation of their transport properties after an exposure to high-intensity light: the well-known Staebler–Wronski effect [233]. This study showed that the nanostructured film reached saturation much faster than the amorphous silicon sample. The light-soaking kinetics was approximately 10 times faster and the value of the $\mu\tau$ product in the "light-soaked state" was equivalent to that of the standard sample in the "annealed state" [218].

St'ahel et al. [205] performed further studies on the light-soaking kinetics and metastability of amorphous, nanostructured, and microcrystalline silicon thin films as representative materials of short-, medium-, and long-range atomic order. Besides showing that the nanostructured material had a higher gap than a-Si:H (probably because of its higher hydrogen content or because of the presence of nanometer-sized ordered domains), they confirmed the results of Longeaud et al. [218] of higher dark conductivity and faster light soaking for the nanostructured films than for the amorphous films. Moreover, they compared the absorption coefficients of the following samples: (i) as prepared, (ii) once the steady state of the light soaking was reached, and (iii) after subsequent low-temperature annealing. They concluded that light soaking induced some irreversible microstructural changes in the nanostructured films, which increased the subgap absorption significantly. In contrast, the light soaking of the microcrystalline sample caused a decrease in the absorption coefficient at low energies.

6.1.3. Summary

It has been shown that it is possible to incorporate small clusters into a growing silicon film and that this cluster participation can modify the structural and electrical properties of standard amorphous silicon. The size of the nanometric particles can be controlled by the plasma modulation and discharge conditions. *Prior* knowledge of particle size development can be inferred from theoretical coagulation models. The discharge parameters for the slow-particle-formation dynamics needed to deposit these nanostructured films (a fast particle formation would lead to the undesirable appearance of powders) cause the clusters to have an ordered atomic structure.

Therefore, these films consist of nanometric ordered domains, whose structure is not necessarily the diamond structure, embedded in an amorphous matrix. Optical techniques such as ellipsometry, IR absorbance, or PDS revealed the structural differences between nanostructured and amorphous silicon thin films. X-ray diffraction or Raman spectroscopy gave amorphous signals.

Less energy was required to crystallize these nanostructured samples than required to crystallize amorphous silicon, because of the presence of nanometer-sized domains, which act as nucleation seeds for a heterogeneous nucleation process.

The electrical properties of these films are encouraging for device applications. Nanostructured films have a higher energy gap, lower density of defects, higher dark conductivity, and faster light-soaking kinetics than amorphous silicon. In particular, the $\mu\tau$ product is 200 times higher than in standard a-Si:H.

Further research is required to clarify the specific structure of these films. The structure of the clusters has to be determined and then related to the transport properties of the nanostructured thin films.

6.2. Light Emission from Nanoparticles

Although the photoluminescence from silicon nanostructured materials, such as porous silicon and silicon nanocrystals, is discussed extensively in other chapters of this book, the light emission of the silicon nanoparticles produced by low-pressure, low-temperature plasmas is discussed in this section. The peculiarities of this system, the detailed studies that have been performed on the emission dynamics, and the fact that some results might be extrapolated to other nanostructured systems justify discussing this topic in this chapter.

6.2.1. Preliminary Studies

Hollenstein and co-workers monitored the laser-induced light emission from silicon particles suspended in the plasma as a diagnostic tool for particle development [95, 234]. They irradiated the particles suspended in the plasma with a 488-nm argon laser and detected the emission intensity at 90° via a 530-nm high-pass filter and an optical multichannel analyzer. They suggested that the light emission observed could originate from a quantum confinement effect in nanocrystallites. They observed a monotonical shift toward longer wavelengths of the emission spectrum as the time of plasma-on increased. They associated this spectral shift to changes in the particles' size. Further studies by the same research team ruled out the possibility that oxygen bonding to these nanoparticles could significantly affect the light emission, and reinforced the hypothesis of a quantum confinement effect [234].

The *ex situ* studies on the emission of silicon particles, however, led to a more complete understanding of the phenomena. The investigations of Roura et al. [235–241] into the transient behavior of the emission led to their elaboration of a detailed dynamic model (a multistep–multiphoton excitation process) capable of accounting for all the observed features of the emission [236]. Although, in a later report, the same authors proved that the physical origin of the emission was thermal emission and that a model based on blackbody equations could describe all the observed features [241], their description of a multistep–multiphoton excitation process established the transient behavior of such a process. Moreover, their model based on blackbody emission suggested that not only the photoluminescence observed in silicon powder, but also that observed on porous silicon and similar materials, could, in some cases, be blackbody radiation. The *ex situ* analyses of light emission of silicon powders grown by PECVD can be summarized as follows.

Nanometric particles produced in a pure silane discharge under plasma conditions that led to considerable powder production (20 sccm of SiH_4, 65 Pa of pressure, and an rf power density of 200 mW/cm^2 modulated at 0.1 Hz) were selected to study their photoluminescence [235–241]. It was thought that the high surface area relative to the volume of these particles, their oxidation, and their amorphous atomic structure could help to clarify the origin of photoluminescence in other materials such as porous silicon or silicon nanocrystals.

The photoluminescence was excited by the 488-nm line of an argon gas laser, which produced a spot about 4 mm^2 on the sample surface. The incident laser power ranged from 1 to 50 mW. The light emission spectra were analyzed by a 0.5-m monochromator and detected with an InAs photodiode cooled to 70 K or a GaAs photomultiplier cooled at -30 °C, depending on the wavelength range studied. Samples were mounted inside a closed-circuit helium cryostat where the surrounding gas pressure (air or another inert gas) could be controlled. A mechanical shutter was used to analyze the transient behavior of the emission whose commutation time was 2 ms. Figure 68 shows a model of the experimental setup.

The main observable features of the light emission from silicon nanoparticles produced by PECVD were reported in a preliminary study. The light emission of two different samples was investigated [235]. One sample consisted of as-deposited Si powder after 2 months

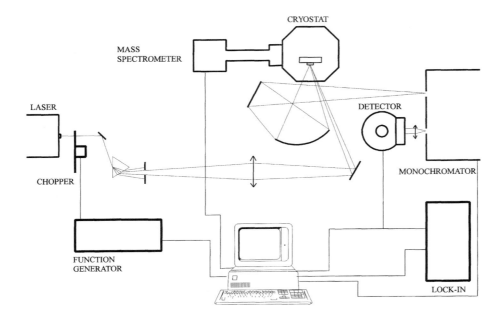

Fig. 68. Experimental setup for the measurement of light emission of silicon powder at different temperatures and gas pressures.

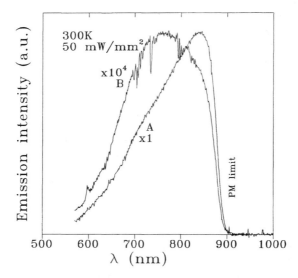

Fig. 69. Typical emission spectra of silicon powder obtained at atmospheric pressure (B) and under low-pressure conditions (A). The spectral range was limited by the detector response.

of exposure to the atmosphere, whereas the other sample was a pellet obtained from compressed as-deposited powder and exposed to the atmosphere for 2 months.

The emission spectra were of two types, corresponding to spectra A and B in Figure 69, both with ranges in the very near IR region. At atmospheric pressure, the type-B band ranged from 550 nm to the detector limit (880 nm) and showed an intensity maximum at 760 nm. The shape and intensity level were similar in both the powder and the pellet samples. When the air pressure was reduced below 5 Pa, the pellet continued to exhibit type-B emission, whereas for the powder sample the emission appeared as a new band,

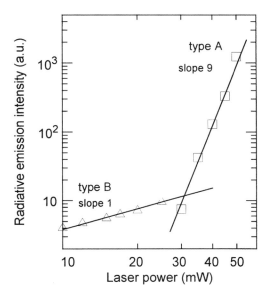

Fig. 70. Dependence on laser power of the type-A and type-B emissions. Note the logarithmic scale on the emission intensity axis.

type A, shown in Figure 69. The type-A emission was several orders of magnitude more intense than the type-B emission and its spectrum shifted to longer wavelengths.

The A and B emissions corresponded to independent processes because of their different characteristics. For instance, the dependence of the type-A emission intensity on laser power was extremely supralinear:

$$I_A \propto I_L^r \tag{22}$$

whereas that of type B was sublinear. As Figure 70 shows, an increase in the laser power by a factor of 2 resulted in an increase of nearly three orders of magnitude in the emission intensity.

Another striking result was the behavior of Si powder emission with gas pressure inside the cryostat. The type-A emission intensity showed an exponential decrease with pressure below 15 Pa:

$$I_A = I_0 \exp\left(-\frac{p}{p_0}\right) \tag{23}$$

The type-B emission under vacuum could be observed only in the pellet sample and did not exhibit the pressure dependence shown by the type-A emission. Because of its high intensity and unusual features, the type-A emission was studied in detail in further reports [236–241].

The structural origin of this emission was related to the laser annealing of the powder. Figure 71 demonstrated that the type-A emission appeared after an irreversible structural change of the sample caused by relatively high-power laser irradiation [239]. Further evidence of the structural change was provided by the simultaneous monitoring of the hydrogen desorbed from the sample and its light emission [239]. In this experiment, the powder sample was mounted in a vacuum chamber connected to a quadrupole mass spectrometer, which detected the evolved molecular hydrogen. After a few seconds of laser irradiation, both H_2 effusion and light emission began. Whether this laser annealing caused the crystallization of the sample is not clear. However, this was not an important question, as the emission was not related to quantum confinement effects in nanometric crystalline domains but to the heating of the sample.

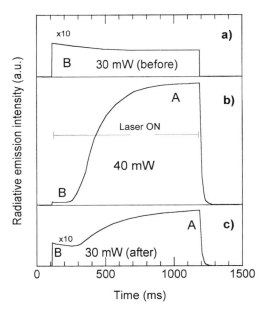

Fig. 71. Plots of the complete emission transient caused by a laser excitation pulse. (a) The emission of the as-grown sample before laser annealing; (b) associated with the structural change, a new emission appeared. The irreversibility of this light emission was evidenced with transient (c) taken at the same laser intensity as (a).

Because the origin of the observed emission was the thermal emission of the sample, the laser-induced structural change leading to its luminescent structure may be easily understood. Laser irradiation caused the loss of hydrogen from the sample and the consequent reorganization of the particle structure. The increase in the number of Si–Si bonds increased the absorption coefficient of the sample in the visible region. Therefore, the light absorption in the annealed sample was more efficient than in the as-prepared powder. Higher absorption implies more intense heating.

As has been explained, the type-A emission from silicon powders was first interpreted in terms of a particular energy level structure and specific excitation dynamics. To explain the supralinear dependence of the emission intensity on laser power in terms of photoluminescence, a multistep–multiphoton excitation process was proposed [236]. This model could satisfactorily explain both the complicated dynamics of the emission and the supralinear dependence of the emission intensity on laser power. Within the framework of this model, the quenching of the emission intensity with gas pressure was simply the result of a similar pressure dependence on the dynamic parameters (lifetime and optical cross section) of each level [237]. With this phenomenological model, all the observed features of the emission could be explained, although its physical origin remained obscure.

The first insight into the physical origin of the light emission was provided by the experimental evidence that the effect of the gas on the emission intensity was linked to the energy that its molecules were able to extract from the powder after every collision with the particles [238, 240]. This was made clear by the dependence of the characteristic pressure, p_0 in Eq. (23), on different gases [238].

6.2.2. Blackbody Emission

In a final paper, it was demonstrated that the emission was simply due to the blackbody emission of the particles [241]. The energy absorbed from the laser beam can be dissipated through the surrounding gas by radiation. Under vacuum, energy dissipation by radiation is not efficient enough to avoid the heating of the particles, whereas at higher pressures the energy released by the gas molecules quenches the emission.

All the dynamic characteristics of the emission that had been previously explained in terms of multistep–multiphoton excitation dynamics were reinterpreted in this final paper as a consequence of the blackbody emission of these particles. Therefore, a theoretical study of the emission intensity dependence on pressure and laser intensity, and of its dynamic behavior was presented. New correlations were established that both highlighted the predictive power of the theory and permitted the unambiguous identification of a blackbody emission. The most relevant results concerning blackbody emission intensity in nanometric particles can be summarized as follows [241].

Consider the energy balance in an isolated particle suspended in a gas that is irradiated by a laser beam. The particle temperature will depend on the balance between the energy absorbed and that dissipated through the gas or by radiation. It will be assumed that there is no thermal conductivity between the particles. The particle is also assumed to be small enough to be at a homogeneous temperature. Therefore, the particle temperature would be governed by the following equation:

$$\frac{4}{3}\pi r^3 \rho c \frac{dT_t}{dt} = \pi r^2 Q_{abs} I_L - 4\pi r^2 (q_R + q_K) \tag{24}$$

where r is the particle radius, ρ its density, and c the specific heat; Q_{abs} is the absorption efficiency and I_L is the laser intensity. Finally, q_R and q_K are the heat fluxes caused by radiative emission and thermal conduction through the gas, respectively.

The radiative emission is given by the Stephan–Boltzmann law:

$$q_R = \varepsilon_i \sigma \left(T_p^4 - T_R^4\right) \tag{25}$$

where σ is the Stephan–Boltzmann constant and ε_i is the integrated emissivity of the particle; T_R is the radiation temperature around the particle.

To find the heat dissipation caused by the gas, q_K, we will consider the limiting case where the pressure is so low that the mean free path is greater than the particle size. Then, q_K can be calculated approximately by multiplying the number of collisions by the mean energy per molecule in the gas. This procedure gives

$$q_K = \frac{1}{4\sqrt{\pi}} \frac{c_V}{R} \sqrt{\frac{8k_B}{m}} \frac{P}{\sqrt{T_G}} \alpha (T_p - T_G) \tag{26}$$

where k_B is the Boltzmann constant, m is the molecular mass, T_G is the gas temperature, c_V is its specific heat at constant volume, and R is the universal constant of gases. The parameter α is the "accommodation coefficient" that accounts for the fact that after one collision the gas molecules will not be thermalized to the particle temperature ($\alpha < 1$).

Once the temperature of the particle has been determined, the radiative emission at any wavelength can be calculated according to Planck's distribution:

$$I_e(\lambda) = \varepsilon(\lambda) \frac{8\pi hc}{\lambda^5} \exp\left(-\frac{hc}{\lambda k_B T_p}\right) \tag{27}$$

where it is assumed that $\exp(hc/\lambda k_B T_p) \gg 1$.

A similar model was used to determine the temperature of silicon particles suspended in the plasma [242], where the heating power was delivered by the collisions of ions existing in the plasma. In this case, the suspended particles were clearly independent. It is doubtful, however, whether the particles in the powder are completely independent. In fact, following the usual classification of ceramic raw materials, as the particle size of our powder is well below 1 μm, the system is colloidal [243]. In such a system, the inertial forces on the particles are insignificant and the surface forces are dominant. This is the case here, given the low density of the powder (10 mg/cm^3 [236]). This value means that the particles occupy only about 1/200th of the powder volume. Consequently, the thermal conductivity of the powder is very small and its contribution to the energy balance is negligible. So, it has not been considered in Eq. (23). Another proof leading to this conclusion comes from

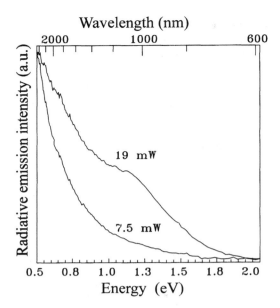

Fig. 72. Emission spectra of the silicon powder at two laser intensities.

the dependence of the radiative intensity on the cryostat temperature. From 14 to 300 K, the intensity experiences only a slight increase of a factor of 2. So, the particle temperature at the excitation conditions does not follow the variations of the cold finger to which it is glued, which indicates negligible thermal conduction.

The comparison between the experimental results and the prediction of this simple model demonstrated the validity of this explanation of the physical origin of the type-A emission of silicon nanoparticles. To simplify the analysis, it was assumed that $T_R^4 \ll T_G^4$. Particular care was taken to thermalize the gas with the cryostat walls for the measurements at cryogenic temperatures.

6.2.2.1. Spectral Shape of the Emission

First, it was shown that the experimental spectrum of the emission had the expected characteristics of a Planck distribution. Figure 72 shows the spectra measured at two laser intensities. As expected, the spectra were continuous. The spectrum obtained at lower exciting laser intensities was displaced to longer wavelengths and its relative intensity near the visible diminished. Unfortunately, the measured spectra were limited by the detector response and the positions of their maxima were not visible. Furthermore, the feature observed around 1 μm at the highest intensity was claimed to be due to the effect of the band gap on the emissivity of the samples [236].

6.2.2.2. Steady-State Intensity

According to this model, the steady-state intensity at vacuum conditions is

$$I_e(\lambda) = \varepsilon(\lambda) \frac{8\pi hc}{\lambda^5} \exp\left(-\frac{hc}{\lambda k_B}\left(\frac{4\sigma\varepsilon_i}{I_L Q_{abs}}\right)^{1/4}\right) \tag{28}$$

which indicates that for a given wavelength the logarithm of the intensity should be proportional to $I_L^{-1/4}$ and the slope should be proportional to λ^{-1}.

The intensity of the emission at two wavelengths ($\lambda = 2.1$ and 3.0 μm) for a range of laser powers was measured in order to test Eq. (28). The results are plotted in Figure 73.

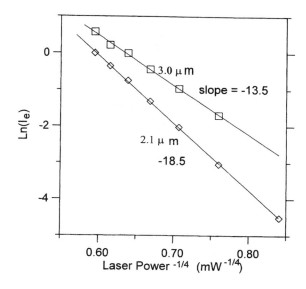

Fig. 73. Dependence of the radiative intensity on the laser power at different wavelengths.

The experimental points appeared well aligned and the product of the slope and the wavelength was constant.

6.2.2.3. Excitation and Deexcitation Transients

As well as the analysis of the steady-state emission intensity, the excitation and deexcitation transients predicted for a thermal radiative emission were compared with the transients observed experimentally. The main characteristics of the measured excitation and decay transients were as follows:

1. The excitation transient began with zero slope.
2. The time constant for the excitation is much longer than the decay lifetime.
3. The decay transient is nonexponential.

Figure 74 shows the measured response of the emission intensity to a laser pulse. The computed simulation of the emission evolution reproduced all the characteristic features detailed previously [241].

A more detailed analysis of the model of thermal emission revealed that the value of the slope for the decay transient just after the laser has been turned off is

$$\tau = \frac{2}{3} \frac{\lambda k_B \rho C}{hc \sqrt{\sigma \varepsilon_i Q_{abs}}} I_L^{-1/2} \tag{29}$$

The measured time constants of the decay transient at $t = 0$ were plotted versus $I_L^{-1/2}$. Indeed, Figure 75 shows that the time constants were proportional to $I_L^{-1/2}$, as expected from Eq. (29).

6.2.2.4. Pressure Quenching of the Radiative Emission

However, the most striking characteristic of the measured emission was its intensity quenching with pressure [Eq. (23)]. Within the framework of the thermal radiation, the effect of the gas is straightforward. The temperature of the particles drops, as does the intensity of the emission, as the dissipation through the gas increases.

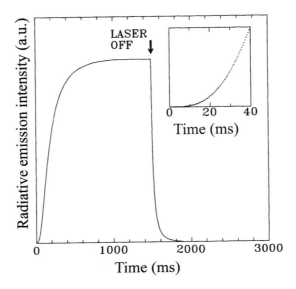

Fig. 74. Excitation and decay transients when a laser pulse 2 s long was applied. In the inset, note the zero slope at the beginning of the excitation.

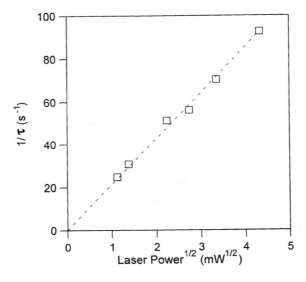

Fig. 75. Dependence of the deexcitation time constant at $t = 0$ on the laser power.

From Eq. (24) to (27), the characteristic pressure p_0 of Eq. (23) is given by

$$p_0 = \frac{4\lambda Q_{abs}}{\alpha hc}\sqrt{\frac{\pi k_B}{8}}\frac{R}{c_V}\sqrt{mT_G}\frac{T_p}{T_P - T_G}I_L \qquad (30)$$

According to this expression, p_0 must be proportional to I_L and, moreover, must depend on the gas species through their molecular mass ($m^{-1/2}$) and their ability to transport energy (R/c_V). Finally, p_0 must be proportional to $T_G^{1/2}$ times $T_p/(T_p - T_G)$.

The proportionality between p_0 and the laser intensity is shown in Figure 76. Unexpectedly, however, p_0 was proportional to T_G, suggesting that the capacity of the gas molecules to dissipate the heat is proportional to the number of collisions and the temperature of the particle, but independent of their energy before the collision.

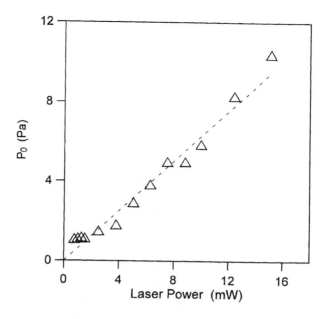

Fig. 76.   Dependence of the characteristic pressure p_0 on the laser power.

6.2.2.5. The Temperature of the Particles

The temperature of the particles near the powder surface at steady-state conditions and under vacuum can be easily derived from Eqs. (24) and (25):

$$T_p = \left(\frac{Q_{abs}}{4\sigma\varepsilon_i}\right)^{1/4} I_L^{1/4} \tag{31}$$

If the optical parameters are known, then the temperature can be calculated. However, as stated before, this cannot be done with sufficient accuracy. Anyway, if the powder is taken as a whole and because the radiative emission diminishes very quickly with laser power, the emission intensity can be expressed as

$$\ln\frac{I_e}{I_L^{1/4}} = A - \frac{hc}{\lambda k_B}\left(\frac{4\sigma\varepsilon_i}{Q_{abs}}\right)^{1/4} I_L^{-1/4} \tag{32}$$

We see that the temperature can be calculated from the slope of the corresponding curve. The result is plotted in Figure 77, which shows the temperature corresponding to the measurements of Figure 73. In view of the excellent linearity of the points in Figure 73, the temperature values can be taken with great confidence.

An independent experimental measurement of the particle temperature and the cooling effect caused by the gas was provided by Raman measurements [241]. The sample was located in a vacuum chamber and was illuminated by the laser beam. Figure 78 shows the experimental equipment for this experiment. It is known that the TO peak of pc-Si shifts from 520.5 cm^{-1} at 20 °C to lower wavenumbers as the temperature increases, in a linear relationship [226, 227]:

$$\omega_{TO}(T) = \omega_{TO}(273\,K) - \left(2.81 \times 10^{-2}\,cm^{-1}/K\right)(T - 273) \tag{33}$$

Figure 79 shows the TO peak of silicon powder at different gas pressures. The calculated temperatures of the Si particles from the TO peak shift are shown in Figure 80. The values are higher than those reached during luminescence experiments because of the higher laser intensity obtained in the Raman microscope.

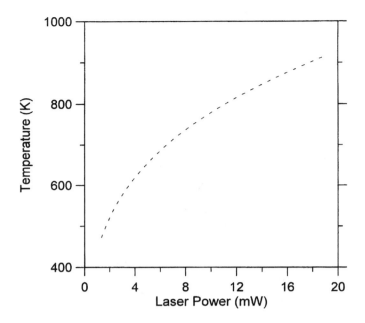

Fig. 77.   Temperature of the particles during laser excitation.

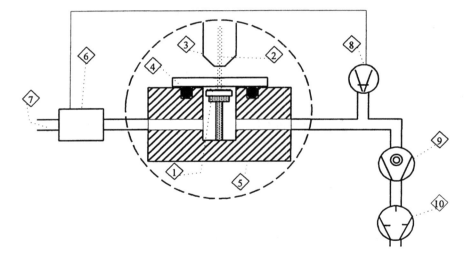

Fig. 78.   Experimental setup for Raman spectroscopy measurements at different gas pressures: (1) sample, (2) laser beam, (3) microscope lens, (4) sealing O-ring, (5) vacuum chamber, (6) mass flow controller, (7) gas inlet, (8) vacuum gauge, and (9) and (10) vacuum pumps.

6.2.2.6. Why Blackbody Emission?

In this section, we discuss the conditions that led to the observation of blackbody emission in nanometric silicon particles at the exciting laser intensities usual in photoluminescence experiments. This will allow us to predict whether such behavior can be expected or not for other kind of nanoparticles. The only particle parameters that determine the temperature are the fraction of the incident laser beam intensity that is absorbed by the particle, Q_{abs}, and the integrated emissivity, ε_i. These parameters can be related to the absorption coefficients

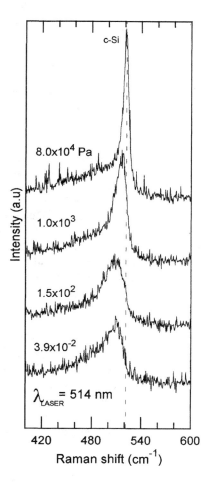

Fig. 79.   Raman spectra of silicon powder measured at different gas pressures. The shift of the TO peak is interpreted as being due to the heating of the powder.

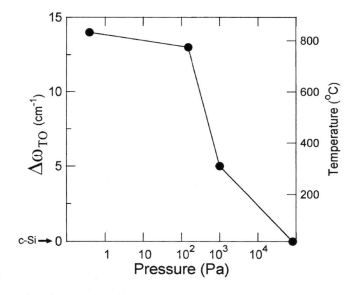

Fig. 80.   Temperatures of the particles at the gas pressures of the Raman measurements presented in Figure 79.

at the laser wavelength, α_L, and in the infrared, α_i. If we neglect the dispersion of light, then we can state approximately that

$$Q_{abs} \approx \alpha_L r \qquad \varepsilon_i \approx \alpha_i r \qquad (34)$$

So, in view of Eq. (31), the temperature is proportional to $(\alpha_L/\alpha_i)^{1/4}$. In the case of silicon, we can calculate a priori the temperature expected at, say, 20 mW/4 mm^2. If we take $\alpha_L = 18 \times 10^3$ cm^{-1} [244] and for α_i the value of a highly doped silicon at elevated temperatures, $\alpha_i \simeq 100$ cm^{-1} [245], we obtain $T_p \simeq 1400$ K. Although this value is far from the experiment, it tells us that the radiative thermal emission could have been predicted a priori. A high value of $\alpha_L/\alpha_i \simeq 10^2$ is obtained in silicon because the excitation is at a photon energy higher than the band gap energy.

This enables us to understand why the nanopowder emits thermal radiation only after it has been dehydrogenated [239]. This is because before dehydrogenation the predominant bond is Si–H, which has an energy in the ultraviolet region, $h\nu_L \simeq 3.5$ eV. So, α_L is very low before the hydrogen has evolved from the powder.

A simple general criterion for thermal emission could lie in the comparison between the band gap energy E_g and the exciting photon energy, $h\nu_L$. Under our experimental conditions, we did not observe any radiation from BN or Si$_3$N$_4$ because, in this case, $E_g > h\nu$. However, we think that it may be detectable in SiC once dehydrogenated ($E_g = 2.35$ eV for the cubic polytype).

Although Eq. (34) could lead one to think that particle heating is independent of the particle dimensions, this is not the case. In fact, the smaller the particles, the more intense in the heating for three reasons: (a) When $\alpha_L > \alpha_i$, as the radius increases, Q_{abs} reaches its limiting value ($\sim$1) before ε_i; (b) the dispersion of the radiation makes Q_{abs} and ε_i smaller than in Eq. (34), an effect more pronounced at longer wavelengths (that is for ε_i); and (c) particles with radii greater than about 1 μm tend to agglomerate because of their own weight [243], making the thermal radiation irrelevant in comparison to the thermal conduction.

Given the conditions described previously for blackbody emission and once the theoretical characteristics for such a process have been established, the question arises whether the photoluminescence of low-conductivity materials such as porous silicon may be thermal emission rather than luminescence. This hypothesis should not be overlooked, especially in reports on the supralinear dependence of emission intensity on laser power [246] or on pressure quenching emission [247].

6.2.3. Summary

The laser-stimulated light emission of silicon nanoparticles has been studied both on powder suspended in the discharge and by *ex situ* measurements. Two different types of emissions were observed. At atmospheric pressure, the emission (type B) was rather weak, whereas that observed under vacuum (type A) was more intense and with unusual features: It exhibited an intensity quenching with pressure and the dependence on laser power was highly supralinear. The type-A emission was interpreted in terms of photoluminescence in preliminary reports. However, detailed analysis showed that the physical reason for the emission under vacuum is blackbody emission. The inefficient heat dissipation by conduction causes particles to heat up under laser irradiation. A model based on blackbody emission could explain all the observed features of this emission.

6.3. Other Possibilities

Pioneering work on nanocrystals has shown that materials with improved fracture strength and toughness can be obtained when their microstructure approaches nanoscale dimensions [19]. Materials with superior strength can be obtained by embedding a high volume

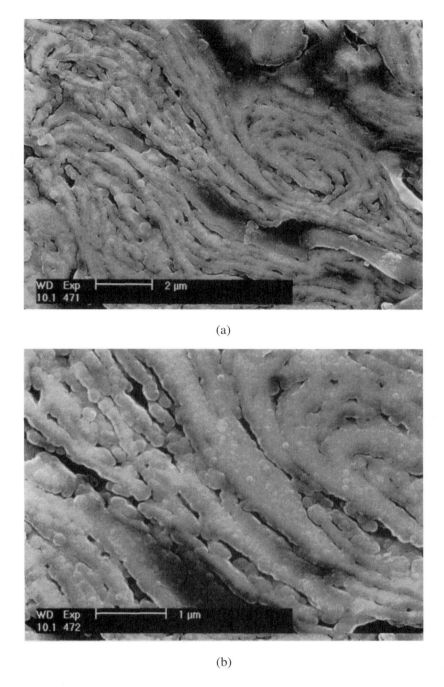

(a)

(b)

Fig. 81. SEM images of powder obtained by mechanical alloying of 40-μm aluminum particles with SiC nanometric powders (10–100 nm in diameter).

fraction (greater than 50%) of ultrafine particles in a ductile matrix. Currently, research is being carried out on metal matrix composites reinforced with nanoparticles of ceramic materials. These composite materials are usually produced by mechanical alloying in a ball milling device. As an example, Figure 81 shows a series of SEM pictures at different magnifications of a mechanically alloyed powder consisting of aluminum particles and nanometric powders of SiC produced in an rf discharge. The resulting powder has a grain

(c)

Fig. 81. (*Continued.*)

size in the nanometer range, which is difficult to obtain by mechanical milling of pure aluminum. This is an encouraging result as the small grain size is thought to lead to a high-strength, high-toughness material. Further studies on the mechanical properties of these alloys are underway.

Another application of powders produced in rf discharges that has been investigated preliminarily is the production of nanoporous membranes. Membranes for gas filtration require a porous size in the nanometer domain with a narrow size distribution. Figure 82 shows a series of SEM pictures of a porous graphite cylinder covered by SiC particles produced in an rf discharge. The deposition of particles on the graphite tube was performed by means of an electrophoretic technique described elsewhere [248]. This method requires further refinement in order to avoid the formation of cracks in the particle coverage such as those illustrated in Figure 82. The presence of these cracks makes these membranes useless.

The fact that the nanoparticles are grown in a reactive medium such as the plasma opens up the possibilities for producing new nanostructured materials. For example, one can foresee the deposition of particles of one material covered by another material. This can be achieved by changing the gas mixture during particle growth. As a result, it could be possible to obtain a particle core resistant to high temperatures covered by a material with specific surface properties such as catalytic qualities.

In the field of nanostructured thin films, there are many possibilities, for example, the deposition of nanostructured thin films of alloyed materials such as SiC, SiN, or BN, and the deposition of multilayered films in which layers with and without nanoparticles are alternated.

7. FINAL SUMMARY

This chapter presented a review of the reports that have been dedicated to the study of low-pressure, low-temperature plasmas, such as rf discharges, as sources for different nano-

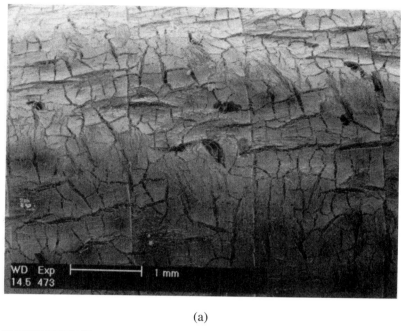

(a)

(b)

Fig. 82.   SEM images of a porous graphite tube covered by SiC nanometric particles. SiC powder was deposited by an electrophoretic method.

structured materials. These plasmas can produce nanometric particles of different composition, size, size distribution, and microstructure, depending on the discharge conditions. Besides the plasma parameters, modulation of the discharge and control of the duration of the plasma-on time can determine particle features. The onset of particle coagulation in the plasma triggers the change from mainly isolated small particles to agglomerates of larger particles (the size at each stage depends on the discharge conditions).

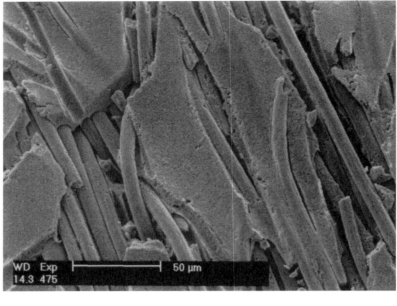

(c)

(d)

Fig. 82. (*Continued.*)

Plasma conditions that lead to fast kinetics of particle formation are suitable for the production of ultrafine powders of alloys (SiC, SiN, BN) that may be used as a raw material for advanced ceramics. However, as-deposited powders have atomic structures close to that of polymers and they oxidize spontaneously in air. Therefore, heat treatment is required to improve the atomic structure and to passivate them against oxidation. Much technological development is required to collect the powder generated in the discharge more efficiently.

Discharge conditions for slow growth kinetics lead to isolated particles of a few nanometers in size with an ordered atomic structure. Discharge conditions for such slow particle growth enable the deposition of thin films consisting of nanoparticles embedded in a homogeneous matrix. The fraction of particles in the film along with their size and atomic structure determines the properties of the film. Nanostructured silicon thin films have encouraging properties in terms of device applications.

Detailed reports on the strong light emission observed in silicon nanoparticles produced in an rf discharge at low pressures showed that its origin is blackbody emission. In spite of using low laser intensities, nanoparticles heat up as a result of their low efficiency for dissipating heat through conduction between particles. This effect may be, in some cases, responsible for the light emission in other nanostructured materials.

Although fundamental knowledge of particle development in low-pressure, low-temperature plasmas has been gained since the 1990s, questions still remain such as the following: What is the polymerization pathway that leads to particle formation under different discharge conditions, what is the atomic structure of particles below 5 nm? What is the origin of the optical and electrical properties of nanostructured thin films? Why have isolated particles below 5 nm never been observed by TEM?

Much effort is still required to find out the suitability of these particles for applications, although this question is, nowadays, a common feature of nanostructured materials.

Economic costs compared to those of other methods will largely determine the potential of this technique for production of powders for advanced ceramics. However, the ability of these plasmas to incorporate particles of a chosen size and structure into a film during its growth is of unpredictable technological importance.

Acknowledgments

The author's activity on nanoparticles has been supported by BRITE-EURAM Contracts CT91-01411 and CT94-0944 and by CICYT from Spain under Conract MAT96-1194-C02-02. The author thanks Professors E. Bertran and P. Roura for their guidance in this research. The author is grateful to Dr. P. Roca i Cabarrocas from LPICM (France) for providing results and discussion on nanostructured thin films; Dr. A. A. Howling from CRPP (Switzerland) for useful discussions on particle development; Professor Ludo Froyen from MTM (Belgium) where preliminary investigations on composites of nanoparticles in a metal matrix were performed; Dr. V. Paillard from the University Paul Sabatier (France) for Raman measurements of annealed SiC; G. Viera and J. A. Mayugo for preparing several figures; Robin Rycroft from SAL of the University of Barcelona for English language corrections; and to Mrs. Veronique Neyrand for her help in preparing the manuscript.

References

1. R. M. Roth, K. G. Spears, and G. Wong, *Appl. Phys. Lett.* 45, 28 (1984).
2. R. M. Roth, K. G. Spears, G. D. Stein, and G. Wong, *Appl. Phys. Lett.* 46, 253 (1985).
3. K. G. Spears, T. J. Robinson, and R. M. Roth, and *IEEE. Trans. Plasma. Sci.* PS-14, 179 (1986).
4. "Nato Advanced Research Workshop on the Formation, Transport and Consequences of Particles in Plasmas," Chateau de Bonas, Castera-Verduzan, France, 1993.
5. "Dusty Plasmas '95 Workshop on Generation, Transport, and Removal of Particles in Plasmas," Wickenburg, AZ, 1995.
6. I. Langmuir, G. Found, and A. F. Dittmer, *Science* 60, 392 (1924).
7. F. Hoyle and N. C. Wicksramasinge, "The Theory of Cosmic Grains." Kluwer Academic Publishers, Dordrecht, The Netherlands, 1991.
8. R. J. Gould and E. E. Salpeter, *J. Astrophys.* 138, 393 (1963).
9. C. K. Goertz, *Rev. Geophys.* 27, 271 (1989).
10. E. C. Whipple, *Rep. Prog. Phys.* 44, 1197 (1981).
11. T. G. Northrop, *Phys. Scr.* 45, 475 (1992).
12. U. de Angelis, *Phys. Scr.* 45, 465 (1992).
13. T. W. Hartquist, W. Pilipp, and O. Havnes, *Astrophys. Space Sci.* 246, 243 (1997).
14. K. G. Emeleus and A. C. Breslin, *Int. J. Electron.* 29, 1 (1970).

15. M. N. Rittner and T. Abraham, *Am. Ceram. Soc. Bull.* 76, 51 (1997).

16. H. Gleiter, *Prog. Mater. Sci.* 33, 223 (1989).

17. H. Gleiter, *Nanostruct. Mater.* 1, 1 (1992).

18. R. W. Siegel, *Ann. Rev. Mater. Sci.* 21, 559 (1991).

19. R. A. Andrievski, *J. Mater. Sci.* 29, 614 (1994).

20. C. Suryanarayana and F. H. Froes, *Metall. Trans. A* 23, 1071 (1992).

21. R. W. Chorley and P. W. Lednor, *Adv. Mater.* 4, 474 (1991).

22. R. Uyeda, *Prog. Mater. Sci.* 35, 1 (1991).

23. G. Franz and G. Schwier, in "Raw Materials for New Technologies" (M. Kursten, ed.), p. 139. Nagele and Obermiller, Stuttgart, Germany, 1990.

24. S. J. Savage and O. Grinder, in "Advances in Powder Metallurgy and Particulate Materials—1992" (J. M. Capus and R. M. German, eds.). MPIF, Princeton, NJ, 1992.

25. Y. Ogino, M. Miki, and T. Yamasaki, *Mater. Sci. Forum* 88–90, 795 (1992).

26. K. S. Mazdiyasni and C. M. Cooke, *J. Am. Ceram. Soc.* 56, 189 (1984).

27. S. Schwier, G. Nietfeld, and G. Franz, *Mater. Sci. Forum* 47, 1 (1989).

28. S. Prochazka and C. Greskovich, *Am. Ceram. Soc. Bull.* 57, 579 (1978).

29. J. S. Haggerty and W. R. Cannon, in "Laser-Induced Chemical Processes" (J. I. Steinfield, ed.), p. 165. Plenum Press, New York, 1981.

30. K. Kijima, H. Noguchi, and M. Konishi, *J. Mater. Sci.* 24, 2929 (1989).

31. F. Allaire and S. Dallaire, *J. Mater. Sci.* 26, 6736 (1991).

32. H. J. Lee, K. Eguchi, and T. Yoshida, *J. Am. Ceram. Soc.* 73, 3356 (1990).

33. G. P. Vissokov, K. D. Manolova, and L. B. Brakalov, *J. Mater. Sci. Lett.* 16, 1716 (1981).

34. J. Chu and I. Lin, *Phys. Rev. Lett.* 72, 4009 (1994).

35. H. Thomas, G. E. Morfill, V. Demmel, J. Goree, B. Feuerbacher, and D. Mohlmann, *Phys. Rev. Lett.* 73, 652 (1994).

36. Y. Hayashi and K. Tachibana, *Appl. Phys. Lett.* 33, L476 (1994).

37. H. Ikezi, *Phys. Fluids* 29, 1764 (1986).

38. H. C. Lee and B. Rosenstein, *Phys. Rev. E* 55, 7805 (1997).

39. H. M. Thomas and G. E. Morfill, *J. Vac. Sci. Technol., A* 14, 501 (1996).

40. J. F. O'Hanlon, *J. Vac. Sci. Technol., A* 7, 2500 (1989).

41. R. A. Bowling, G. B. Larrabee, and W. G. Fisher, *J. Environ Sci. (USA)* 32, 22 (1989).

42. B. Van Eck and T. P. Schneider, in "Dusty Plasmas '95 Workshop on Generation, Transport, and Removal of Particles in Plasmas," Wickenburg, AZ, 1995.

43. Ch. Steinbrüchel, in "Physics of Thin Films" (M. H. Francombe and J. L. Vossen, eds.), Vol. 18, p. 289. Academic Press, New York, 1994.

44. J. L. Petrucci, Jr. and Ch. Steinbrüchel, in "Proceedings of the 8th Symposium on Plasma Processes" (G. S. Mathad and D. W. Hess, eds.), p. 219. Electrochemical Society, Pennington, NJ, 1990.

45. G. S. Selwyn, J. E. Heidenreich, and K. L. Haller, *J. Vac. Sci. Technol., A* 9, 2817 (1991).

46. R. N. Carlie, S. Geha, J. F. O'Hanlon, and J. C. Stewart, *Appl. Phys. Lett.* 59, 1167 (1991).

47. W. J. Yoo and Ch. Steinbrüchel, *Appl. Phys. Lett.* 60, 1073 (1992).

48. W. J. Yoo and Ch. Steinbrüchel, *J. Vac. Sci. Technol., A* 10, 1041 (1992).

49. W. J. Yoo and Ch. Steinbrüchel, *J. Vac. Sci. Technol., A* 11, 1258 (1993).

50. J. A. Durham and Ch. Steinbrüchel, in "Proceedings of the 8th Symposium on Plasma Processes" (G. S. Mathad and D. W. Hess, eds.), p. 207. Electrochemical Society, Pennington, NJ, 1990.

51. G. S. Selwyn, J. S. McKillop, and K. L. Haller, *J. Vac. Sci. Technol., A* 8, 1726 (1990).

52. G. M. Jellum and D. B. Graves, *J. Appl. Phys.* 67, 6490 (1990).

53. B. Ganguly, A. Garscadden, J. Williams, and P. Haaland, *J. Vac. Sci. Technol., A* 11, 1119 (1993).

54. G. M. Jellum and D. B. Graves, *Appl. Phys. Lett.* 57, 2077 (1990).

55. G. M. Jellum, J. E. Daugherty, and D. B. Graves, *J. Appl. Phys.* 69, 6923 (1991).

56. G. S. Selwyn, *Plasma Sources Sci. Technol.* 3, 340 (1994).

57. G. S. Selwyn, J. Singh, and R. S. Bennet, *J. Vac. Sci. Technol., A* 7, 2758 (1989).

58. G. S. Selwyn *J. Vac. Sci. Technol., B* 9, 3487 (1991).

59. G. S. Selwyn and E. F. Patterson, *J. Vac. Sci. Technol., A* 10, 1053 (1992).

60. R. N. Carlile, J. F. O'Hanlon, L. M. Hong, M. P. Garrity, and S. M. Collins, *Plasma Sources Sci. Technol.* 3, 334 (1994).

61. S. M. Collins, D. A. Brown, J. F. O'Hanlon, and R. N. Carlile, *J. Vac. Sci. Technol., A* 14, 634 (1996).

62. J. Kang, R. N. Carlile, J. F. O'Hanlon, and S. M. Collins, *J. Vac. Sci. Technol., A* 14, 639 (1996).

63. M. Dalvie, M. Surendra, G. S. Selwyn, and C. R. Guarnieri, *Plasma Sources Sci. Technol.* 3, 442 (1994).

64. S. G. Geha, R. N. Carlile, J. F. O'Hanlon, and G. S. Selwyn, *J. Appl. Phys.* 72, 374 (1992).

65. W. W. Stoffels, E. Stoffels, G. M. W. Kroesen, M. Haverlag, J. H. W. G. den Boer, and F. J. de Hoog, *Plasma Sources Sci. Technol.* 3, 320 (1994).

66. M. M. Smadi, G. Y. Kong, R. N. Carlile, and S. E. Beck, *J. Vac. Sci. Technol., B* 10, 30 (1992).

67. M. P. Garrity, T. W. Peterson, and J. F. O'Hanlon, *J. Vac. Sci. Technol., A* 14, 550 (1996).

68. S. J. Choi, P. L. G. Ventzek, R. J. Hoekstra, and M. J. Kushner, *Plasma Sources Sci. Technol.* 3, 418 (1994).

69. Y. Watanabe, M. Shiratani, Y. Kubo, I. Ogawa, and S. Ogi, *Appl. Phys. Lett.* 53, 1263 (1988).
70. Y. Watanabe, M. Shiratani, and H. Makino, *Appl. Phys. Lett.* 57, 1616 (1990).
71. Y. Watanabe, M. Shiratani, and M. Yamashita, *Appl. Phys. Lett.* 61, 1510 (1992).
72. A. Lloret, E. Bertran, J. L. Andújar, A. Canillas, and J. L. Morenza, *J. Appl. Phys.* 69, 632 (1991).
73. L. J. Overzet and J. T. Verdeyen, *Appl. Phys. Lett.* 48, 695 (1986).
74. J. T. Verdeyen, J. Beberman, and L. Overzet, *J. Vac. Sci. Technol., A* 8, 1851 (1990).
75. A. Canillas, E. Bertran, J. L. Andújar, and B. Drévillon, *J. Appl. Phys.* 68, 2752 (1990).
76. J. L. Andújar, E. Bertran, A. Canillas, J. Campmany, J. Serra, and C. Roch, *J. Appl. Phys.* 71, 1546 (1992).
77. Y. Yamaguchi, A. Sumiyama, R. Hattori, Y. Morokuma, and T. Makabe, *J. Phys. D: Appl. Phys.* 22, 505 (1989).
78. M. J. Kushner, *IEEE Trans. Plasma Sci.* PS-14, 188 (1986).
79. F. Tochikubo, A. Suzuki, S. Kakuta, Y. Terazono, and T. Makabe, *J. Appl. Phys.* 68, 5532 (1990).
80. J. Perrin and J. F. M. Aarts, *Chem. Phys.* 80, 351 (1983).
81. S. K. Srivastava and E. Krishnakumar, *Int. J. Mass Spectrom. Ion Processes* 107, 83 (1991).
82. Y. Ohmori, M. Shimouza, and H. Tagashira, *J. Phys. D: Appl. Phys.* 19, 1029 (1986).
83. E. Krishnakumar and S. K. Srivastava, *Int. J. Mass Spectrom. Ion Processes* 103, 107 (1991).
84. M. A. Lieberman and A. L. Lichtenberg, "Principles of Plasma Discharges and Materials Processing." Wiley, New York, 1994.
85. V. A. Godyak and A. S. Kanneh, *IEEE Trans. Plasma Sci.* PS-14, 112 (1986).
86. C. Böhm and J. Perrin, *J. Phys. D: Appl. Phys.* 24, 865 (1991).
87. J. Perrin, P. Roca i Cabarrocas, B. Allain, and J. M. Friedt, *Jpn. J. Appl. Phys.* 27, 2041 (1988).
88. Ph. Belenguer and J. P. Boeuf, *Phys. Rev. B* 41, 4447 (1990).
89. Ph. Belenguer, J. Ph. Blondeau, L. Boufendi, M. Toogood, A. Plain, A. Bouchoule, C. Laure, and J. P. Boeuf, *Phys. Rev. A* 46, 7923 (1992).
90. J. P. Boeuf, *Phys. Rev. A* 46, 7910 (1992).
91. C. M. Horwitz, *J. Vac. Sci. Technol., A* 1, 1795 (1983).
92. C. Böhm and J. Perrin, *J. Phys. D: Appl. Phys.* 24, 865 (1991).
93. C. F. Bohren and D. R. Huffman, "Absorption and Scattering of Light by Small Particles." Wiley, New York, 1983.
94. H. C. V. d. Hulst, "Light scattering by small particles." Wiley, New York, 1975.
95. Ch. Hollenstein, J. L. Dorier, J. Dutta, L. Sansonnens, and A. A. Howling, *Plasma Sources Sci. Technol.* 3, 278 (1994).
96. C. Courteille, Ch. Hollenstein, J.-L. Dorier, P. Gay, W. Schwarzenbach, A. A. Howling, E. Bertran, G. Viera, R. Martins, and A. Macarico, *J. Appl. Phys.* 80, 2069 (1996).
97. U. I. Schmidt and D. B. Graves, *J. Vac. Sci. Technol., A* 14, 595 (1996).
98. M. Shiratani, H. Kawasaki, T. Fukuzawa, and Y. Watanabe, *J. Vac. Sci. Technol., A* 14, 603 (1996).
99. P. Roca i Cabarrocas, P. Gay, and A. Hadjadj, *J. Vac. Sci. Technol., A* 14, 655 (1996).
100. Y. Hayashi and K. Tachibana, *Jpn. J. Appl. Phys.* 33, L476 (1994).
101. A. A. Howling, L. Sansonnens, J. L. Dorier, and Ch. Hollenstein, *J. Appl. Phys.* 75, 1340 (1994).
102. A. A. Howling, J. L. Dorier, and Ch. Hollenstein, *Appl. Phys. Lett.* 62, 1341 (1993).
103. A. A. Howling, C. Courteille, J.-L. Dorier, L. Sansonnens, and Ch. Hollenstein, *Pure Appl. Chem.* 68, 1017 (1996).
104. W. D. Reents, S. W. Downey, A. B. Emerson, A. M. Mujsce, A. J. Muller, D. J. Siconolfi, J. D. Sinclair, and A. G. Swanson, *Plasma Sources Sci. Technol.* 3, 369 (1994).
105. W. D. Reents and M. L. Mandich, *Plasma Sources Sci. Technol.* 3, 373 (1994).
106. E. Stoffels, W. W. Stoffels, G. M. W. Kroesen, and F. J. de Hoog, in "Proceedings of the 9th Symposium E.P.C.R.L.T.P.," Pila, Czechoslovakia, 1992 (unpublished), p. 131.
107. L. Boufendi, J. Hermann, A. Bouchoule, B. Dubreuil, E. Stoffels, W. W. Stoffels, and M. L. de Giorgi, *J. Appl. Phys.* 76, 148 (1994).
108. W. W. Stoffels, E. Stoffels, G. M. W. Kroesen, and F. J. de Hoog, *J. Vac. Sci. Technol., A* 14, 588 (1996).
109. H. M. Anderson, R. Jairath, and J. L. Mock, *J. Appl. Phys.* 67, 3999 (1990).
110. A. Garscadden, B. N. Ganguly, P. D. Haland, and J. Williams, *Plasma Sources Sci. Technol.* 3, 239 (1994).
111. S. J. Choi and M. J. Kushner, *J. Appl. Phys.* 74, 853 (1993).
112. P. Haaland, and *J. Chem. Phys.* 93, 4066 (1990).
113. J. Perrin, Ch. Bhöm, R. Etemadi, and A. Lloret, *Plasma Sources Sci. Technol.* 3, 252 (1994).
114. L. J. Overzet, J. H. Beberman, and J. T. Verdeyen, *J. Appl. Phys.* 66, 1622 (1989).
115. I. Haller, *Appl. Phys. Lett.* 37, 282 (1980).
116. M. L. Mandich, W. D. Reents, and K. D. Kolenbrander, *Pure Appl. Chem.* 62, 1653 (1990).
117. Ch. Hollenstein, W. Schwarzenbach, A. A. Howling, C. Courteille, J.-L. Dorier, and L. Sansonnens, *J. Vac. Sci. Technol., A* 14, 535 (1996).
118. A. A. Howling, L. Sansonnens, J. L. Dorier, and Ch. Hollenstein, *J. Phys. D: Appl. Phys.* 26, 1003 (1993).
119. S. Veprek, K. Schopper, O. Ambacher, W. Rieger, and M. G. J. Veprek-Heijman, *J. Electrochem. Soc.* 140, 1935 (1993).
120. Y. Watanabe, M. Shiratani, and H. Makino, *Appl. Phys. Lett.* 57, 1616 (1990).

121. A. C. Breslin and K. G. Emeleus, *Int. J. Electron.* 31, 189 (1971).

122. J. Goree, *Plasma Sources Sci. Technol.* 3, 400 (1994).

123. T. P. Martin and H. Schaber, *J. Chem. Phys.* 83, 855 (1985).

124. G. M. W. Kroesen, J. H. W. G. den Boer, L. Boufendi, F. Vivet, M. Khouli, A. Bouchoule, and F. J. de Hoog, *J. Vac. Sci. Technol., A* 14, 546 (1996).

125. M. Haverlag, E. Stoffels, W. Stoffels, H. den Boer, G. Kroesen, and F. de Hoog, *Jpn. J. Appl. Phys.* 33, 4202 (1994).

126. J. P. Boeuf and Ph. Belenguer, *J. Appl. Phys.* 71, 4751 (1992).

127. A. A. Fridman, L. Boufendi, T. Hbid, B. V. Potapkin, and A. Bouchoule, *J. Appl. Phys.* 79, 1303 (1996).

128. Y. Watanabe, M. Shiratani, H. Kawasaki, S. Singh, T. Fukuzawa, Y. Ueda, and H. Ohkura, *J. Vac. Sci. Technol., A* 14, 540 (1996).

129. L. Boufendi, A. Plain, J. Ph. Blondea, A. Bouchoule, C. Laure, and M. Toogood, *Appl. Phys. Lett.* 60, 169 (1992).

130. J. L. Andújar, G. Viera, M. C. Polo, Y. Maniette, and E. Bertran, *Vacuum* 52, 153 (1999).

131. V. A. Schweigert and I. V. Schweigert, *J. Phys. D: Appl. Phys.* 29, 655 (1996).

132. M. S. Barnes, J. H. Keller, J. C. Foster, J. A. O'Neill, and D. K. Coultas, *Phys. Rev. Lett.* 68, 313 (1992).

133. J. Perrin, P. Molinàs-Mata, and Ph. Belenguer, *J. Phys. D: Appl. Phys.* 27, 2499 (1994).

134. J. Y. Liu and J. X. Ma, *Phys. Plasmas* 4, 2798 (1997).

135. D. B. Graves, J. E. Daugherty, M. D. Kilgore, and R. K. Porteous, *Plasma Sources Sci. Technol.* 3, 433 (1994).

136. J. E. Daugherty, R. K. Porteus, M. D. Kilgore, and D. B. Graves, *J. Appl. Phys.* 72, 3934 (1992).

137. A. A. Uglov and A. G. Gnedovets, *Plasma Chem. Plasma Proc.* 11, 251 (1991).

138. X. Chen and X. Chen, *Plasma Chem. Plasma Proc.* 9, 387 (1989).

139. J. E. Daugherty, R. K. Porteus, and D. B. Graves, *J. Appl. Phys.* 73, 1617 (1993).

140. L. Talbot, R. K. Cheng, R. W. Schefer, and D. R. Willis, *J. Fluid. Mech.* 101, 737 (1980).

141. G. M. Jellum, J. E. Daugherty, and D. B. Graves, *J. Appl. Phys.* 69, 6923 (1991).

142. O. Havnes, T. Nitter, V. Tsytovich, G. E. Morfill, and T. Hartquist, *Plasma Sources Sci. Technol.* 3, 448 (1994).

143. M. J. Baines, I. P. Williams, and A. S. Asebiomo, *Mon. Not. R. Astron. Soc.* 130, 63 (1965).

144. A. A. Howling, Ch. Hollenstein, and P. J. Paris, *Appl. Phys. Lett.* 59, 1409 (1991).

145. J. L. Dorier, Ch. Hollenstein, and A. A. Howling, *J. Vac. Sci. Technol., A* 13, 918 (1995).

146. J. L. Dorier, Ch. Hollenstein, A. A. Howling, and U. Kroll, *J. Vac. Sci. Technol., A* 10, 1048 (1992).

147. A. Bouchoule, A. Plain, L. Boufendi, J. Ph. Blondeau, and C. Laure, *J. Appl. Phys.* 70, 1991 (1991).

148. C. Courteille, L. Sansonnens, J. Dutta, J.-L. Dorier, Ch. Hollenstein, A. A. Howling, and U. Kroll, in "Proceedings of the 12th European Photovoltaic Solar Energy Conference," Amsterdam, 1994, p. 319.

149. R. Banerjee, S. N. Sharma, S. Chattopadhyay, A. K. Batabyal, and K. Barua, *J. Appl. Phys.* 74, 4540 (1993).

150. S. Yokoyama, M. Tanaka, T. Yamaoka, M. Ueda, and A. Matsuda, in "Proceedings of the 5th Plasma Processing Meeting," Japan, 1988, p. 219.

151. J. Costa, P. Roca i Cabarrocas, et al., unpublished manuscript.

152. A. Bouchoule and L. Boufendi, *Plasma Sources Sci. Technol.* 2, 204 (1993).

153. L. Boufendi and A. Bouchoule, *Plasma Sources Sci. Technol.* 3, 262 (1994).

154. E. S. Machlin, in "Materials Science in Microelectronics," p. 7. Giro Press, New York, 1995.

155. J. Dutta, W. Bacsa, and Ch. Hollenstein, *J. Appl. Phys.* 77, 3729 (1995).

156. J. Dutta, I. M. Reaney, C. Bossel, R. Houriet, and H. Hofmann, *Nanostruct. Mater.* 4, 121 (1994).

157. J. Costa, G. Sardin, J. Campmany, J. L. Andújar, and E. Bertran, *Mater. Res. Soc. Symp. Proc.* 286, 155 (1993).

158. J. Costa, G. Sardin, J. Campmany, J. L. Andújar, A. Canillas, and E. Bertran, *Mater. Res. Soc. Symp. Proc.* 351, 1031 (1993).

159. E. Bertran, J. Costa, G. Sardin, J. Campmany, J. L. Andújar, and A. Canillas, *Plasma Sources Sci. Technol.* 3, 348 (1994).

160. J. Dutta, R. Houriet, H. Hofmann, and H. Hofmeister, *Nanostruct. Mater.* 9, 359 (1997).

161. J. R. Chelikowsky and J. C. Phillips, *Phys. Rev. Lett.* 63, 1653 (1989).

162. M. V. Ramakrishna and J. Pan, *J. Chem. Phys.* 101, 8108 (1994).

163. J. Pan and M. V. Ramakrishna, *Phys. Rev. B* 50, 15431 (1994).

164. U. R. Röthlisberger, W. Andreoni, and M. Parinello, *Phys. Rev. Lett.* 52, 665 (1994).

165. M. Cardona, *Phys. Status Solidi B* 118, 463 (1983).

166. E. C. Freeman and W. Paul, *Phys. Rev. B* 41, 4288 (1978).

167. H. Shanks, C. J. Fang, L. Ley, M. Cardona, F. J. Demond, and S. Kalbitzer, *Phys. Status Solidi B* 100, 43 (1980).

168. L. J. Bellamy, "The Infrared Spectra of Complex Molecules." Chapman & Hall, London, 1975.

169. G. Lucovsky, *J. Non-Cryst. Solids* 141, 241 (1992).

170. V. Paillard, private communications.

171. J. Costa, J. J. Suñol, J. Fort, and P. Roura, *Mater. Res. Soc. Symp. Proc.* 513, 427 (1998).

172. A. Triska, D. Dennison, and H. Fritzsche, *Bull. Am. Phys. Soc.* 20, 392 (1975).

173. P. Gupta, V. L. Colvin, and S. M. George, *Phys. Rev. B* 37, 8234 (1988).

174. W. Beyer, *Physica B* 170, 105 (1991).

175. S. Oguz and M. A. Paesler, *Phys. Rev. B* 22, 6213 (1980).

176. Yu. L. Khait, R. Weil, R. Beserman, W. Beyer, and H. Wagner, *Phys. Rev. B* 42, 9000 (1990).

177. G. Schulze and M. Henzler, *Surf. Sci.* 124, 336 (1983).

178. R. Walsh, *Acc. Chem. Res.* 14, 246 (1981).

179. D. V. Tsu, G. Lucovsky, and B. N. Davidson, *Phys. Rev. B* 40, 1795 (1989).

180. Z. Jing, J. L. Whitten, and G. Lucovsky, *Phys. Rev. B* 45, 13978 (1992).

181. G. Lucovsky, R. J. Nemanich, and J. C. Knights, *Phys. Rev. B* 19, 2064 (1979).

182. S. Matsumoto and Y. Matsui, *J. Mater. Sci.* 18, 1785 (1983).

183. T. Okabe, Y. Kagawa, and S. Takai, *Philos. Mag. Lett.* 63, 233 (1991).

184. H. Hofmeister, J. Dutta, and H. Hofmann, *Phys. Rev. B* 54, 2856 (1996).

185. P. Ho, R. J. Buss, and R. E. Loeman, *J. Mater. Res.* 4, 873 (1989).

186. R. J. Buss and S. V. Babu, *J. Vac. Sci. Technol., A* 14, 577 (1996).

187. E. G. Rochow, "An Introduction to the Chemistry of Silicones," 2nd ed., p. 16. Wiley, New York, 1951.

188. J. Costa, E. Bertran, and J. L. Andújar, *Diamond Relat. Mater.* 5, 544 (1996).

189. R. J. Nemanich, S. A. Solin, and R. M. Martin, *Phys. Rev. B* 23, 6348 (1981).

190. A. C. Adams and C. D. Capio, *J. Electrochem. Soc.* 127, 399 (1980).

191. S. V. Nguyen, T. Nguyen, H. Treichel, and O. Spindler, *J. Electrochem. Soc.* 141, 1633 (1994).

192. K. W. Bödekker, S. G. Shore, and R. K. Bunting, *J. Am. Chem. Soc.* 88, 4396 (1966).

193. J. Costa, G. Viera, R. Q. Zhang, J. L. Andújar, E. Pascual, and E. Bertran, *Mater. Res. Soc. Symp. Proc.* 410, 173 (1996).

194. E. Bertran, J. Costa, G. Viera, and R. Q. Zhang, *J. Vac. Sci. Technol., A* 14, 567 (1996).

195. G. Viera, S. N. Sharma, J. Costa, R. Q. Zhang, J. L. Andújar, and E. Bertran, *Vacuum* 48, 665 (1997).

196. J. Costa, J. J. Suñol, J. Saurina, G. Viera, S. Martinez, E. Bertran, and P. Roura, *Key Eng. Mater.* 132–136, 145 (1997).

197. G. Viera, S. N. Sharma, J. Costa, R. Q. Zhang, J. L. Andújar, and E. Bertran, *Diamond Relat. Mater.* 6, 1559 (1997).

198. G. Viera, S. N. Sharma, J. L. Andújar, R. Q. Zhang, J. Costa, and E. Bertran, *Vacuum*, to appear.

199. H. Wieder, M. Cardona, and C. R. Guarnieri, *Phys. Status Solidi B* 92, 99 (1979).

200. A. H. Mahan, P. Raboison, D. L. Williamson, and R. Tsu, *Solar Cells* 21, 117 (1987).

201. K. P. Huber, in "AIP Handbook of Physics" (D. E. Grey, ed.), p. 168. McGraw–Hill, New York, 1972.

202. D. M. Tanenbaum, A. L. Laracuente, and A. Gallagher, *Appl. Phys. Lett.* 68, 1705 (1996).

203. P. Roca i Cabarrocas, S. Hamma, S. N. Sharma, G. Viera, E. Bertran, and J. Costa, *J. Non-Cryst. Solids* 227, 871 (1998).

204. M. Ehrbrecht, B. Kohn, F. Huisken, M. A. Laguna, and V. Paillard, *Phys. Rev. B* 56, 6958 (1997).

205. P. St'ahel, S. Hamma, P. Sládek, and P. Roca i Cabarrocas, *J. Non-Cryst. Solids* 227, 276 (1998).

206. D. Beeman, R. Tsu, and M. F. Thorpe, *Phys. Rev. B,* 32 874 (1985).

207. D. M. Bhusari, A. S. Khumbar, and T. Kshirsagar, *Phys. Rev. B* 47, 6460 (1993).

208. J. Kanellis, J. F. Morhange, and M. Balkanski, *Phys. Rev. B* 21, 1543 (1980).

209. M. Cardona, *Superlattices Microstruct.* 5, 34 (1989).

210. N. Beck, J. Meier, J. Fric, Z. Rennes, A. Poruba, R. Flükiger, J. Pohl, A. Shah, and M. Vanecek, *J. Non-Cryst. Solids* 198–200, 238 (1996).

211. S. Veprek, Z. Iqbal, and F.-A, Sarott, *Philos. Mag. B* 45, 137 (1982).

212. A. Ourmazd, J. C. Bean, and J. C. Phillips, *Phys. Rev. Lett.* 55, 1599 (1985).

213. Y. He, Ch. Yin, G. Cheng, L. Wang, and X. Liu, *J. Appl. Phys.* 75, 797 (1994).

214. M. Cuniot, J. Dixmier, and P. Roca i Cabarrocas, *J. Non-Cryst. Solids* 198, 540 (1996).

215. K. Zellama, J. H. von Bardelebe, V. Quillet, Y. Bouizem, P. Sládek, M. L. Thèye, and P. Roca i Cabarrocas, *J. Non-Cryst. Solids* 164–166, 285 (1993).

216. P. Morin and P. Roca i Cabarrocas, *Mater. Res. Soc. Symp. Proc.* 336, 281 (1994).

217. G. C. Stutzin, R. M. Ostrom, A. Gallagher, and D. M. Tanenbaum, *J. Appl. Phys.* 74, 91 (1993).

218. C. Longeaud, J. P. Kleider, P. Roca i Cabarrocas, S. Hamma, R. Meaudre, and M. Meaudre, *J. Non-Cryst. Solids* 227, 96 (1998).

219. P. Roca i Cabarrocas, *Appl. Phys. Lett.* 65, 1674 (1994).

220. P. Roca i Cabarrocas and S. Hamma, in "Colloque International sur les Procédés Plasama," Le Mans, France, 1997, p. 172.

221. G. Viera, P. Roca i Cabarrocas, S. Hamma, S. N. Sharma, J. Costa, and E. Bertran, MRS Spring Meeting 1997, to be published.

222. J. Costa, P. Roura, P. Roca i Cabarrocas, S. Hamma, and E. Bertran, submitted to *J. Appl. Phys.*

223. E. Bertran, S. N. Sharma, G. Viera, J. Costa, P. St'ahel, and P. Roca i Cabarrocas, *J. Mater. Res.* 13, 2476 (1998).

224. A. Zwick and R. Carles, *Phys. Rev. B* 48, 6024 (1993).

225. P. M. Fauchet and I. H. Campbell, *Crit. Rev. Solid State Mater. Sci.* 14, S79 (1988).

226. T. R. Hart, R. L. Aggarwal, and B. Lax, *Phys. Rev. B* 1, 638 (1970).

227. R. Tsu and J. G. Hernandez, *Appl. Phys. Lett.* 41, 1016 (1982).
228. J. S. Kasper, P. Hagenmuller, M. Pouchard, and C. Cros, *Science* 150, 1713 (1965).
229. C. Cros, M. Pouchard, and P. Hagenmuller, *J. Solid State Chem.* 2, 570 (1970).
230. H. Kawaji, H. Horie, S. Yamanaka, and M. Ishikawa, *Phys. Rev. B* 74, 1427 (1995).
231. S. Ray, S. Hazra, A. R. Middya, and A. K. Barua, Proc. of First World Conf. on Photovoltaic Energy Conversion, Hawaï, USA 5–9 Dec. 1994.
232. A. R. Middya, S. Harza, S. Ray, A. K. Barua, and C. Longeaud, *J. Non-Cryst. Solids* 198–200, 1067 (1996).
233. D. L. Staebler and C. R. Wronski, *Appl. Phys. Lett.* 31, 292 (1977).
234. C. Courteil, J.-L. Dorier, J. Dutta, Ch. Hollenstein, and A. A. Howling, *J. Appl. Phys.* 78, 61 (1995).
235. J. Costa, P. Roura, G. Sardin, J. R. Morante, and E. Bertran, *Appl. Phys. Lett.* 64, 463 (1994).
236. P. Roura, J. Costa, G. Sardin, J. R. Morante, and E. Bertran, *Phys. Rev. B* 50, 18124 (1994).
237. J. Costa, P. Roura, N. A. Sulimov, G. Sardin, J. Campmany, J. R. Morante, and E. Bertran, *Mater. Sci. Technol.* 11, 707 (1995).
238. P. Roura, J. Costa, J. R. Morante, and E. Bertran, *J. Appl. Phys.* 81, 3290 (1997).
239. J. Costa, P. Roura, A. Canillas, E. Pascual, J. R. Morante, and E. Bertran, *Thin Solid Films* 276, 96 (1996).
240. P. Roura, J. Costa, N. A. Sulimov, J. R. Morante, and E. Bertran, *Appl. Phys. Lett.* 67, 19 (1995).
241. J. Costa, P. Roura, J. R. Morante, and E. Bertran, *J. Appl. Phys.* 83, 7879 (1998).
242. J. E. Daugherty and D. B. Graves, *J. Vac. Sci. Technol., A* 11, 1126 (1993).
243. J. S. Reed, "Principles of Ceramic Processing," 2nd ed., p. 70. Interscience, New York, 1994.
244. O. Madelung, ed., "Data in Science and Technology, Semiconductors, Group IV Elements and III–V compounds." Springer-Verlag, Berlin, 1991.
245. P. J. Timans, *J. Appl. Phys.* 74, 6353 (1993).
246. D. P. Savin, Ya. O. Roizin, and D. A. Demchenko, *Appl. Phys. Lett.* 69, 3048 (1996).
247. T. D. Shen, I. Shmagin, C. C. Coch, R. M. Kolbas, Y. Fahmy, C. Bergman, R. J. Nemanich, M. T. McLure, Z. Sitar, and M. X. Quan, *Phys. Rev. B* 55, 7615 (1997).
248. J. H. Kennedy and A. Foissy, *J. Electrochem. Soc.* 122, 482 (1975).

Chapter 3

SYNTHESIS OF NANOSTRUCTURED COATINGS BY HIGH-VELOCITY OXYGEN–FUEL THERMAL SPRAYING

Honggang Jiang

Materials Science & Technology Division, Los Alamos National Laboratory, Los Alamos, New Mexico, USA

Maggy Lau, Victoria L. Tellkamp, Enrique J. Lavernia

Department of Chemical and Biochemical Engineering and Materials Science, University of California, Irvine, Irvine, California, USA

Contents

1. INTRODUCTION

Technological progress in many domains of modern society is intimately linked to the materials science and engineering community's ability to conceive novel materials with extraordinary combinations of physical and mechanical properties. In the transportation industry, for example, the ever-increasing demand to manufacture lighter vehicles that can travel at higher speeds, and can withstand a higher payload capacity, has motivated the development of high-strength/low-density materials with improved damage tolerance and enhanced temperature capabilities. It is precisely the unusual combinations of physical attributes that are required of these materials that often preclude their fabrication by conventional processing techniques. As a result, the successful implementation of such materials depends on the development of radically different processing schemes. Research in materials science and engineering, driven in part by this critical need, has progressively shifted toward the study and application of nonequilibrium processes. The significant departure from thermodynamic equilibrium associated with these types of processes allows

Handbook of Nanostructured Materials and Nanotechnology, edited by H.S. Nalwa
Volume 1: Synthesis and Processing
Copyright © 2000 by Academic Press

material scientists and engineers to exercise a degree of microstructural control heretofore unattainable, and thereby develop materials with unusual combinations of microstructure and physical attributes.

Significant interest has been generated recently in the field of nanoscale (also described as nanocrystalline or nanophase) materials, in which the grain size is usually in the range of 1–100 nm. The sudden burst of enthusiasm stems not only from the outstanding properties that can be obtained in such materials, but also from the realization that early skepticism about the ability to produce high-quality, unagglomerated nanoscale powder was unfounded. There are literally dozens of methods utilized by over 60 companies involved in nanocrystalline materials in the United States alone, some of which are fully commercialized [1]. Accordingly, the focus is shifting from synthesis to processing, that is, the manufacture of useful coatings and structures from these powders. The potential applications span the entire spectrum of technology, from thermal barrier coatings for turbine blades to wear-resistant rotating parts. The potential economic impact is several billions of dollars per year [1]. The importance of the field (and widespread interest in it) can also be seen by the fact that among the 25 most highly cited authors in the field of materials science and engineering, seven are engaged in research on nanoscale materials [2].

Significant progress has been made in various aspects of processing on nanoscale materials. Most of this work has been focused on the fabrication of bulk structures [3–5]. However, the process most likely to have the earliest (and perhaps the greatest) major technological impact is deposition of coatings by thermally activated processes. This includes the so-called "thermal spray processes" such as the high-velocity oxy–fuel (HVOF) process, plasma spray process, and others, but also includes new innovations such as the chemical vapor condensation (CVC) process and a number of exciting new combustion processes. For example, a modified HVOF process has been used to fabricate dense coatings of Co/WC with remarkable properties [6]. Moreover, recent modeling work of the HVOF process, motivated by the problem of depositing nanoscale materials, promises to revolutionize the thermal spray industry as a whole.

Inspection of the available scientific literature reveals that a great deal of effort has gone into enhancing our understanding of the synthesis and structural characteristics of nanocrystals. More recently, greater scientific emphasis is being placed on the physical and mechanical characteristics of nanocrystalline ceramics and metals, because it is evident that it is possible to achieve combinations of properties that are otherwise unachievable with equilibrium materials. For example, it is possible to sinter nanophase ceramics at temperatures that are substantially lower than those required by coarse-grained ceramics, owing to their fine microstructures, small diffusion scales, and high-grain-boundary purity [7, 8]. Nanophase ceramics are reported to exhibit unusually high ductility, whereas nanophase metals are noted to exhibit ultrahigh hardness values [7, 8]. In addition, what is perhaps most unusual about nanocrystalline materials is the fact that, despite being classified as nonequilibrium materials, recent work shows that their grain size may, in some cases, remain metastable during exposure to elevated temperatures [8, 9]. Although this phenomenon is not clearly understood, it has been suggested that the unusual resistance of the nanocrystals to coarsening may be due to their narrow distribution of grain size [8, 9].

To assess our current understanding of the synthesis and characterization of nanocrystalline coatings prepared by thermal spraying, the following topics will be described in the present chapter:

- A brief overview of thermal spraying processes
- The dynamic and thermal processes associated with HVOF thermal spraying
- The synthesis and characterization of coarse-grained or fine-grained HVOF thermally sprayed coatings
- The examination of various properties (density, hardness, wear resistance, corrosion resistance, mechanical properties, and thermal stability) of HVOF thermally sprayed coatings

- The synthesis and characterization of nanocrystalline coatings obtained by HVOF spraying

2. OVERVIEW OF THERMAL SPRAYING

Significant progress has been made in various aspects of the synthesis on nanoscale materials, as evidenced by the large amount of scientific literature available [7, 8]. Most of this work is focused on the fabrication of bulk structures. However, the progress most likely to have a short-term major technological impact is deposition of coatings by thermally activated processes. This includes the so-called "thermal spray" such as HVOF and plasma spray, as well as new innovations such as chemical vapor condensation and a number of new exciting new combustion processes. Thermal spraying combines particle melting, quenching, and consolidation in a single operation. This spraying technology, facilitated by an achievement of metallurgical and chemical homogeneity, is used to fabricate a variety of simple preform shapes in addition to coatings. Thermal spraying was originally developed to produce corrosion-resistant zinc coatings, as well as coatings for other refractory metals [10, 11]. Today, thermal spraying technology has created many coating applications including:

- Arc plasma spray (APS) coating of Cr_2O_3 on hardened-steel drilling components used in petroleum mining to improve the service lifetime [10]
- High-velocity oxy–fuel (HVOF) coating of stainless-steel 316L to provide protection against sulfur and ammonia corrosion in chemical refinery vessels [10, 12]
- WC–cermet coatings produced by HVOF onto the contact surface of steel rolls to increase their abrasion and friction resistance in steel-rolling applications [13]
- Al_2O_3 coating by APS on aluminum midplates for diode assembly in automotive alternators to provide resistance against salt corrosion and moisture absorption [10, 14]
- Thermal barrier coatings (TBCs) of ZrO_2–Y_2O_3 (outer layer)/CoCrAlY (bond layer) by plasma spray to reduce heat transfer and thus increase engine efficiency in an adiabatic diesel engine [10, 15]
- Wear-resistant coatings of WC–M (M = Ni, Co, or Co–Cr) by HVOF or APS on compressor fans and disc midspan stiffeners in aeroengines [16, 17]

Thermally sprayed coatings are also widely used in the electronics industries [18], power generation plants [19], marine gas turbine engines [20], ceramics industries, and printing industries.

In principle, powder, rods, and wires can be used as precursor materials (or feedstock) in the thermal spray process. Metals and alloys in the form of rods or wires are commonly used in arc spraying (AS) and flame spraying (FS) [10]. Powders of metals, alloys, ceramic oxides, cermets, and carbides are often used in thermal spraying to produce a homogeneous microstructure in the resultant coating. In most cases, the sprayed surface is degreased, masked, and roughened prior to spraying to maximize the bonding strength between the coating and the substrate material. Various techniques for presprayed treatment are described in detail elsewhere [21].

Various thermal spraying techniques have been developed since the 1900s. Today, flame spraying (FS), atmospheric plasma spraying (APS), arc spraying (AS), detonation gun spraying (DGS), high-velocity oxy–fuel (HVOF) spraying, vacuum plasma spraying (VPS), and controlled-atmosphere plasma spraying (CAPS) are widely used to produce various coatings for industrial applications. The typical process parameters of the various thermal spraying techniques mentioned previously are listed in Table I [10, 22–27].

Flame spraying (FS), sometimes referred to as combustion flame spraying, involves the combustion of fuel gas in oxygen (1:1 to 1.1:1 in volume ratio) to heat the feedstock [28]. The flame gases are introduced axially, and the particles are introduced in a direction

Table I.   Process Parameters of Various Thermal Spraying Techniques

Thermal spraying technique	Working flame	Flame temperature (K)	Flame velocity (m/s)	Powder particle size (μm)	Powder injection feed rate (g/min)	Spraying distance (mm)
FS	Fuel + O_2 (g)	3000–3350	80–100	5–100	50–100	120–250
APS	Ar or mixture of Ar + H_2, Ar + He, and Ar + N_2 (g)	Up to 14,000	800	5–100	50–100	60–130
AS	Various electrically conductive wires (Zn, Al)	Arc temperature up to 6100 K by an arc current of 280 A [22]	Velocity of molten particles up to 150 m/s	—	50–300	50–170
DGS	Detonation wave from a mixture of acetylene + O_2	Up to 4500 K with 45% acetylene	2930 [23]	5–60	16–40 [24]	100 [25]
HVOF	Fuel gases (acetylene, kerosene, propane, propylene, or H_2) with O_2	Up to 3440 K at ratio O_2:acetylene (1.5:1) by volume [26]	2000	5–45	20–80	150–300
VPS	Ar mixed with H_2, He, or N_2	In electron temperature of 1000 to 1500 K	Velocity of plasma between 1500 and 3000 K	5–20	50–100 (spray in vacuum)	300–400
CAPS	Same as APS	Same as APS	Same as APS	Same as APS	Same as APS	100–130 mm in SPS [27]

perpendicular to the flame gases. The particles are thereby heated and accelerated toward the target substrate. The coating thickness produced by FS is typically 100–2500 μm and porosity ranges from 10 to 20%. The bond strength for FS ceramic coatings is approximately 15 MPa. Other materials may have bond strength values of up to 30 MPa. For example, the bond strength of a NiAl coating produced by FS can reach 60 MPa [10].

In atmospheric plasma spraying (APS), the flame gas (Ar or a mixture of $Ar + H_2$, $Ar + He$, or $Ar + N_2$) is heated by a plasma generator (60 kW or more), which produces an electric arc. The advantages of the plasma processing include: a clean reaction atmosphere, which is needed to produce a high-purity material; and a high enthalpy to enhance the reaction kinetics by several orders of magnitude; and high temperature gradients to provide the possibility of rapid quenching and the generation of fine particles [29]. Because of the high temperature of the flame gases (up to 14,000 K), APS is commonly used to produce ceramic TBCs of Y_2O_3-stabilized ZrO_2 [15, 30], Al_2O_3–ZrO_2, and other cermet coatings. The bond strength of typical ceramic coatings produced by APS is between 15 and 25 MPa. Some bonding alloys (NiAl, NiCrAl) and metals (Mo) may reach bond strength values of 70 MPa. The porosity of APS coatings is generally lower (1–7%) than those produced by FS and the thickness of the coating ranges from 50 to 500 μm [10].

Arc spraying (AS) involves two electrically conductive wires (Zn or Al) that are arc melted, and the molten particles propelled by a compressed gas. The high-velocity gas (flow rate of 1–80 m^3/h) acts to atomize the melted wires and to accelerate the fine particles to the substrate. Alloy coatings can be produced if the wires are composed of different materials [10]. The thickness of the coating produced by AS is between 100 and 1500 μm, and the bond strength is in the range of 10–30 MPa for Zn and Al coatings [31].

Detonation gun (D-gun) spraying is commonly used to produce WC–Co and Al_2O_3 coatings owing to the resultant low porosity ($\sim$0.5% for WC–Co) and high bond strength (83 MPa for WC–Co) [10, 32]. In D-gun spraying, a mixture of flame gas (oxygen and acetylene) is fed to a long barrel with a charge of powder. Upon ignition, a detonation wave is produced (1–15 detonations/s) that delivers the powder particles at a velocity up to 750 m/s to the substrate [10].

High-velocity oxy–fuel (HVOF) spraying is the most significant development in the thermal spray industry since the development of the original plasma spray. HVOF is characterized by high particle velocities and relatively low thermal energy when compared to plasma spraying. The applications of HVOF have expanded from the initial use of tungsten carbide coatings to include different coatings that provide resistance to wear or erosion/corrosion [33]. HVOF uses an internal-combustion jet fuel (propylene, acetylene, propane, and hydrogen gases) to generate a hypersonic gas velocity of approximately 2000 m/s, more than five times the speed of sound. When burned in conjunction with pure oxygen, these fuels can produce a gas temperature greater than 3029 K [10]. The powder particles are injected axially into the jet gas and simultaneously heated and propelled toward the substrate. With the relatively low temperature of the flame gas associated with the HVOF system as compared to plasma spraying, the particles are made highly plastic by convective heat transfer and superheating or vaporization of individual particles is prevented [34]. Furthermore, the lower particle temperatures experienced in carbide coatings lead to less carbide depletion than plasma-sprayed coatings. In effect, the advantages of the HVOF process over conventional plasma spraying include higher coating bond strengths, lower oxide contents, and improved wear resistances owing to a homogeneous distribution of carbides [11, 35].

Vacuum plasma spraying (VPS), sometimes referred to as low-pressure plasma spraying (LPPS), consists of a plasma jet stream produced by heating an inert gas with an electric-arc generator (requiring more power than that for APS) [10]. The powders are introduced into the plasma jet in vacuum, undergo melting, and are accelerated toward the substrate material [10, 11]. An important parameter in VPS is the position of the injection port in the nozzle. The pressure of the powder injector must be greater than the pressure in the nozzle in order to propel the powders properly [36]. The advantages of utilizing VPS over conventional APS include lower porosity, fewer oxides in the resulting coating, and denser deposits [11, 35]. The advantages are a result of the decrease in incomplete melting, less fusing together of deposited particles, and the higher particle velocities associated with VPS.

Any thermal plasma spraying technique enclosed in a controlled atmosphere other than air or vacuum can be classified as controlled-atmosphere plasma spraying (CAPS). Inert plasma spraying (IPS) involves plasma spraying into an inert-gas (He, N_2) chamber. Shrouded plasma spraying is often used to produce TBCs in which the plasma jet is protected from the atmosphere. The shielding nozzle is connected to the anode of the plasma torch, and the nozzle is in close proximity (100–130 mm) to the substrate [10].

3. HIGH-VELOCITY OXY–FUEL THERMAL SPRAYING

Among the various thermal spraying techniques, the HVOF process is emerging as a novel and unique technology for producing high-performance coatings with greater thicknesses, stronger bonds, a higher hardness, and improved durability than currently possible with other thermal spray processes. More recently, HVOF spraying has been successfully used as a means of producing nanocrystalline coatings [6, 37–44]. The extremely brief exposure of the precursor nanocrystalline particles to the high temperatures of the HVOF process appears to preserve the nanocrystalline structure in most of the particles deposited on the substrate. In general, the quality of the HVOF-sprayed coating is determined by a combination of various processes. These are described in some detail next.

3.1. Dynamic and Thermal Processes

The HVOF thermal spraying process (Fig. 1) [45] results in numerous dynamic and thermal processes that can drastically affect the microstructures of the final coatings. These processes may be broadly classified into the following categories:

- Combustion flame and spraying gas dynamics
- In-flight mechanical and thermal behavior of powder particles
- In-flight mass transfer in composite coatings

3.1.1. Combustion Flame and Spraying Gas Dynamics

3.1.1.1. Combustion Reaction of Oxygen–Fuel Mixture

As mentioned previously, the HVOF thermal spraying process starts with a combustion reaction of fuel gas and oxygen inside a gun barrel to achieve high temperatures and high pressures, which can subsequently heat the particles and drive the particles toward the substrate at a high speed. Because the combustion process is similar to that in a rocket engine, a single one-dimensional rocket performance model can be utilized to simulate this process [46–48]. In the case where kerosene and oxygen are used as the combustion gas mixture, the following results can be obtained [49, 50]:

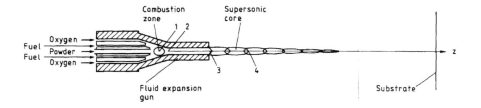

Fig. 1. Schematic diagram of HVOF spraying. Reprinted with permission from V. V. Sobolev et al., *J. Mater. Process. Manuf. Sci.* 4, 25 (1995). © 1995 Technomic Publishing.

- The flame temperature is a nonuniform function of the equivalence ratio, ϕ, and the maximum attainable value for most hydrocarbons with a slightly oxygen-rich mixture is $\phi = 1.1–1.2$.
- The gas temperature at the nozzle exit is lower than the combustion chamber flame temperature because of the expansion and acceleration of the gases in the convergent–divergent nozzle; the frozen composition model calculation provides an exit temperature about 500 K lower than that for the equilibrium case [49].
- The gas velocity at the nozzle exit increases with the equivalence ratio, whereas the gas momentum flux, serving as the accelerating force for the powder particles, changes slightly with ϕ; both parameters are higher in the case of the frozen composition model.
- The gas pressure and gas velocity at the nozzle exit vary slightly with the combustion chamber pressure, but the gas momentum flux at the exit increases with the pressure.

Swank et al. [48] conducted experiments to study gas combustion behavior. The results from this study revealed much higher values of enthalpy, gas temperature, and gas velocity at the gun exit than those obtained from numerical simulation. This significant deviation indicates that the combustion products are far from equilibrium compositions. Hence, in an effort to improve the simulations, a single one-dimensional model for the frozen combustion flow has been employed, which proved to be the most suitable model in predicting the HVOF combustion process [48]. By incorporating both combustion kinetics and gas dynamics in the combustion chamber into this model, quantitative results of fluid temperature (at point 1 of Fig. 1), T_{f1}, and fluid velocity (at point 1 of Fig. 1), v_{f1}, can be more accurately obtained [51]:

$$T_{f1} = T_{f0} + b_1 \left[2500\phi - 700(\phi - 1) \right] \qquad \phi > 1 \tag{1}$$

$$v_{f1} = b_2 v_{f1} = \frac{0.78 \times 10^4}{\left(\frac{10^4}{T_f} + \frac{900}{T_{f0}} \right)^{4.938}} P_{f1}^{-0.09876} \tag{2}$$

where T_{f0} is the fluid temperature at point 0, P_{f1} is the fluid pressure at point 1 (see Fig. 1), and b_1 and b_2 are coefficients. The preceding set of equations can be precisely used in predicting the fluid temperature and velocity [45].

3.1.1.2. Flame Gas Dynamics Inside the Gun

In the case of an adiabatic compressible gas flow in a gun, the Hugoniot theorem yields the following [52]:

- The velocity of flow changes inversely with the cross-sectional area, F, of the pipe for subsonic (Mach number <1) flow.
- The velocity of flow increases with F for supersonic flow.
- In the preceding two cases, the gas pressure P varies inversely with the flow velocity.
- The flow velocity can be equal to the local speed of sound ($Ma = 1$) only in a place where the pipe cross section has its upper or lower limits (maximum or minimum).

In the case of various types of HVOF thermal spray guns, including the Diamond Jet, the Jet-Kote, and the CDS gun, Fanno curves (enthalpy or temperature vs temperature) are usually utilized to study the particular features of a constant adiabatic irreversible flow in a straight gun during HVOF spraying. Representative studies are available in the literature [46, 52–54].

Once the combustion gas jet approaches the exit of the nozzle, it starts to expand with a corresponding increase in the Mach number ($Ma > 1$). Simultaneously, the so-called "shock diamonds" form [50], as illustrated in Figure 1. The pressure at the boundaries decreases to atmospheric pressure rapidly; however, along the gas flow axis the pressure drops more slowly. Moreover, a central supersonic core is formed inside the jet, separating supersonic and subsonic flow regions, in which Mach waves are developed. The gas velocity reaches its maximum value at the surface of these conical waves, and the gas pressure and temperature drop to minimum. The quantitative descriptions of the jet velocity and temperature along the jet axis and with increasing radial distance can be found elsewhere [46, 48, 55, 56].

3.1.2. In-Flight Mechanical and Thermal Behavior of Powder Particles

The optimization of the experimental parameters requires an understanding of the mechanical and thermal behavior of the powder particles during HVOF spraying [50]. Mathematical modeling is an effective means capable of providing this understanding. Models describing the in-flight behavior of the powder particles and gas dynamics during plasma spraying have been developed and well documented [50, 52, 57]. Modifications of these models have been adapted for the HVOF process, in which the following factors have been taken into consideration:

- The fluid–particle interactions and the effect of particle density on the fluid velocity
- The structure of the composite particles (carbides + binding phases) (in the case of composite particles)
- The comparatively small influence of the Knudsen discontinuum effect on heat and momentum transfer
- The influence of the particle prehistory on the initial values of the particle temperature and velocity (particularly in the combustion chamber)
- The effects of particle form, the particle thermal state such as its heating or cooling, and the processes of melting and solidification
- The effects of chemical reactions between the fluid and the particle materials (oxidation, decarburization), as well as the processes of the diffusive mass transfer inside the powder particle (diffusion and dissolution)

The velocity and temperature of the powder particles are noted as important parameters characterizing the in-flight behavior of the powder particles. Generally, the following factors are thought to govern the movement of the powder particles [58]: (i) drag force, (ii) force due to pressure gradients, (iii) force due to added mass, (iv) Basset history term, and (v) external potential forces (gravitational, electrical, among others). In principle, among the factors that affect the movement of particles during the HVOF process, only the drag force plays a dominant role; other factors can be neglected in most cases [10]. Therefore, the equation describing the velocity of the particle and the gas is expressed as

$$\frac{1}{6}\rho_p \pi d_p^3 \frac{dv_p}{dt} = \frac{1}{8}C_D \pi d_p^2 \rho_g (v_g - v_p)^2 \tag{3}$$

where v_p is the particle velocity, v_g is the gas velocity, d_p is the particle diameter (spherical), ρ_p and ρ_g are the particle and gas densities, respectively, and C_D is the drag coefficient. The dependence of the drag coefficient on the particle velocity relative to the flame velocity (referred to as the Reynolds number) is summarized as follows:

$$C_D = \frac{24}{Re} \qquad Re < 0.2 \tag{4}$$

$$C_D = \frac{24}{Re}\left(1 + \frac{3}{16}Re\right) \qquad 0.2 \leqslant Re \leqslant 2 \tag{5}$$

$$C_{\mathrm{D}} = \frac{24}{Re}\left(1 + 0.11 Re^{0.81}\right) \qquad 2 \leqslant Re \leqslant 21 \tag{6}$$

$$C_{\mathrm{D}} = \left(1 + 0.189 Re^{0.632}\right) \qquad 21 \leqslant Re \leqslant 200 \tag{7}$$

where the Reynolds number is $Re = \rho_{\mathrm{g}} d_{\mathrm{p}}(v_{\mathrm{p}} - v_{\mathrm{g}})/\eta_{\mathrm{g}}$ (η_{g} is the dynamic viscosity of gas; other symbols have been described earlier). It should be kept in mind that in a specified flow the appropriate selection of the drag coefficient is still one of the unresolved problems in the field of gas–particle flow [59].

Regarding the thermal behavior of the powder particles during the HVOF process, the powder particle temperature, T_{p}, as a function of time, t, and radial coordinate, x, can be described by a heat conduction equation:

$$\rho_{\mathrm{p}} c_{\mathrm{p}} \frac{\partial T_{\mathrm{p}}}{\partial t} = \frac{1}{x^n} \frac{\partial}{\partial x}\left(x^n \lambda_{\mathrm{p}} \frac{\partial T_{\mathrm{p}}}{\partial x}\right) \qquad 0 \leqslant x \leqslant R_{\mathrm{p}} \qquad t > 0 \tag{8}$$

where c_{p} is the specific heat of the particles and λ_{p} is the heat conductivity of the particles. The boundary conditions describing the center and the surface of the particles are as follows:

$$\frac{\partial T_{\mathrm{p}}}{\partial x}(0, t) = 0 \tag{9}$$

$$\lambda_{\mathrm{p}} \frac{\partial T_{\mathrm{p}}}{\partial t}(R_{\mathrm{p}}, t) = \alpha_{\mathrm{h}}\left(T_{\mathrm{f}} - T_{\mathrm{p}}(R_{\mathrm{p}}, t)\right) \tag{10}$$

where $T_{\mathrm{p}}(x, 0) = T_{\mathrm{p}0}$, and the coefficient of heat transfer, α_{h}, can be determined by the Ranz–Marshall semiempirical equation [60]:

$$N_{\mathrm{u}} = \frac{\alpha_{\mathrm{h}} d_{\mathrm{p}}}{\bar{\lambda}_{\mathrm{f}}} = 2 + 0.6 Re^{1/2} Pr^{1/3} \qquad Pr = \frac{\bar{c}_{\mathrm{f}} \bar{\eta}_{\mathrm{f}}}{\bar{\lambda}_{\mathrm{f}}} \tag{11}$$

It is thought that during the HVOF spraying process the surface temperature of a homogeneous particle can become as high as the melting temperature of that material [50]. Hence, subsequent propagation of the melting front toward the particle center is controlled by the Stefan heat balance condition [61]. Inserting the effective specific heat into the heat conductivity Eq. (8), Eq. (8) then becomes

$$\rho_{\mathrm{p}} c_{\mathrm{p}} \Psi(T_{\mathrm{p}}) \frac{\partial T_{\mathrm{p}}}{\partial t} = \frac{1}{x^n} \frac{\partial}{\partial x}\left(x^n \lambda_{\mathrm{p}} \frac{\partial T_{\mathrm{p}}}{\partial x}\right) \qquad 0 \leqslant x \leqslant R_{\mathrm{p}} \qquad t > 0 \tag{12}$$

$$\Psi_{\mathrm{p}}(T_{\mathrm{p}}) = 1 + q^{\mathrm{p}} c_{\mathrm{p}}^{-1}(1 - k)^{-1}(T_{\mathrm{k}} - T_{\mathrm{l}})^{-1}\left(\frac{T_{\mathrm{k}} - T_{\mathrm{l}}}{T_{\mathrm{p}} - T_{\mathrm{l}}}\right)^{(2-k)/(1-k)} \qquad T_{\mathrm{s}} \leqslant T_{\mathrm{p}} \leqslant T_{\mathrm{l}} \tag{13}$$

$$\Psi(T_{\mathrm{p}}) = 1 \qquad T_{\mathrm{p}} > T_{\mathrm{l}} \qquad T_{\mathrm{p}} < T_{\mathrm{s}}$$

The boundary and initial conditions for Eq. (12) are the same as in Eq. (8).

3.1.3. In-flight Mass Transfer in Composite Coatings

An interesting and important phenomenon that must be addressed in the preparation of composite coatings during the HVOF process is the mass transfer between different phase components. This may result in an undesirable microstructure in the final coating, thereby degrading its final properties [50]. Previous work has revealed that a typical mass transfer situation can be exemplified in the case of WC–Ni particles [55, 62, 63]. During the in-flight motion, WC dissolution and the subsequent diffusion of the W, C, and Ni occur immediately following the melting of the Ni. As a result, a W–C–Ni phase may form, in which the concentration of W and C in the liquid metallic phase decreases in the direction of the particle surface and the Ni concentration diminishes in the direction of the particle center. Therefore, a particle impinging onto a substrate will have a liquid zone rich in W,

C, and Ni, respectively, indicating the presence of a chemical inhomogeneity. The mass transfer mainly occurs during the time interval (Δt), between the melting of the metallic binder and the impingement of the particle onto the substrate surface. Hence, mass transfer increases with an increase in this time interval. The time interval is a function of the particle size and density. Mass transfer may also be enhanced by an increase in the diffusion rate as a result of an increase in the overall temperature.

In addition to the aforementioned mass transfer mechanism that occurs in the particle during the in-flight motion following the melting of the metallic binder, mass transfer may also take place in the liquid splat formed after the droplet deforms at the substrate surface. The duration of this process does not exceed the characteristic time, t_s, of splat solidification. Usually, t_s is much less than Δt, implying that the main contribution to the chemical inhomogeneity of the particle liquid phase arises from the in-flight mass transfer. In the case of HVOF-sprayed coatings, the mass transfer may induce chemical segregation in the growing deposit. In general, the kinetics of chemical segregation depend significantly on the crystallization Péclet number, Pe_k. Pe_k is equal to the ratio of the solidification velocity, V_k, to the characteristic velocity of diffusion [62, 64]. For crystals with a semithickness of radius R and an intercrystalline distance of $2L$, Pe_k can be expressed as $V_k L/2R$. It is worth noting that chemical segregation can only be accounted for when Pe_k approaches unity [62, 64]. In the case of ordinary solidification, if the velocity of solidification is relatively high (e.g., $V_k = 0.01$ m/s), this segregation is negligible because of the larger values of Pe_k ($\gg 1$) [62, 65]. In the case of HVOF spraying, the parameter Pe_k may approach unity [62]. Hence, the chemical segregation may be distinct.

An example of this chemical segregation may be demonstrated in the case of a spherical particle of WC surrounded by a binder metal of Ni, creating a two-zoned particle prior to HVOF spraying (Fig. 2a). When the particle is heated, the Ni will melt, causing dissolution of the WC in the liquid Ni. As a result, three zones are formed in the moving particle: WC, $W + Ni + C$, and Ni (Fig. 2b). The continuing dissolution will eventually lead to the disappearance of the Ni zone owing to the diffusion of W and C into the Ni. In this case, only two zones will remain: WC and $W + Ni + C$ (Fig. 2c). The composition of the $W + Ni + C$ zone will vary along the radius of the particle according to the diffusion rates of W and C in Ni.

In an effort to study the mass transfer process, related diffusion equations were selected [50, 66, 67]. Further approximation of these equations was completed to obtain practical solutions [50, 68]. In the case described in Figure 2, the decrease in the diameter (R_1) of the original WC particle can be expressed as [50, 67]:

$$R_1 = R_{10}(1 - B_1 t)^{1/2} \tag{14}$$

where R_{10} is the initial radius of the WC. For simplification, the diffusion of W and C into Ni is assumed to occur at the same rate. Similarly, an equation describing the variation of R_2, the radius of the $W + Ni + C$ zone, can be written as

$$R_2 = R_{10}(1 + B_2 t)^{1/2} \tag{15}$$

The parameters B_1 and B_2 are associated with the diffusion coefficients of W and C in liquid Ni.

The diffusion causing chemical segregation takes place primarily during the time interval, Δt, the time between the melting of the metallic binder and the impingement of the particle onto the substrate surface. By applying Eq. (14), it follows that

$$B_1 = (\Delta t)^{-1}\left(1 - R_1^2 R_{10}^{-2}\right) \tag{16}$$

B_2 can be obtained in a similar manner. The analysis of the coating structure shows that the volume fraction of WC at the moment of impingement on the substrate is about 25–35% of its initial value [50, 69].

Experimental results demonstrated that pure Ni is absent in the coatings, indicating that chemical segregation was present. The rates of WC dissolution and W and C diffusion in

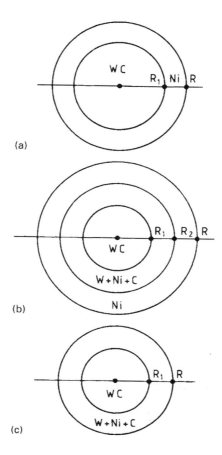

Fig. 2. Variation of composition of WC–Ni composite powder particles during the HVOF process. Reprinted from V. V. Sobolev et al., *Surf. Coat. Technol.* 81, 136 (© 1996), with permission from Elsevier Science, The Boulevard, Langford Lane, Kidlington 0X5 1GB, UK.

the liquid Ni can be assumed to be the same, that is, $B_1 = B_2$, which is in good agreement with the experimentally observed results.

3.2. Spray Parameters

Control of the spraying parameters is essential to obtain optimal combinations of tensile strength, superficial hardness, microhardness, and microstructure in a coating. Crawmer et al. [70], for example, conducted experiments to investigate the fundamental parameters determining the physical properties of WC–17 at% Co, CrC–25 at% NiCr, CrC–20 at% NiCr, and Tribaloy 800 coatings sprayed by the Miller Thermal HVOF system. Results from this study indicated that using a 12-mm-diameter nozzle is the dominant factor for increasing the tensile strength of a CrC–25 at% NiCr coating, whereas the spray distance is the most critical factor for the tensile strength of the CrC–20 at% NiCr and Tribaloy 800 coatings [70].

The critical spray parameters for HVOF processes include the following [10, 70]:

- Fuel and oxygen flow
- Fuel and oxygen ratio
- Carrier gas flow
- Combustion nozzle type
- Powder feed rate

- Particle diameter
- Torch linear speed
- Gun water temperature
- Gun-to-substrate spray distance

3.2.1. Fuel and Oxygen Flow

A constant flow of oxygen and fuel is an important factor in the operation of HVOF spraying processes to produce consistent quality in coatings. The total flow of fuel and oxygen affects the microstructure, oxide content, microhardness, and abrasive wear resistance of a coating [71]. The temperature of the combustion flame depends strongly on the stoichiometry of the oxygen–fuel mixtures. For instance, the ideal stoichiometric ratio for the combustion of propylene to produce a neutral flame is 4.5:1 of O_2 to C_3H_6 as expressed in the following equation:

$$2C_3H_6 + 9O_2 \rightarrow 6H_2O + CO_2 \qquad (17)$$

A lean combustion will yield an excess of O_2 that will enhance the formation of oxides in the coatings. However, excessive fuel introduced during combustion will lower the flame temperature owing to the depletion of O_2. Consequently, a high concentration of unmelted particles may result, which will lead to an increase in porosity [34].

Creffield et al. [71] studied the effects of total gas flow rate and different oxygen-to-propane ratios on the properties of WC–Co coatings sprayed using a Miller Thermal HV2000 system. Liquid propane with two different compositions was selected as the fuel in this study to determine the effects on coating properties of the WC–Co coatings. A greater amount of oxygen is required for complete combustion when the liquid propane composed of higher hydrocarbons is selected. The incomplete combustion leads to the deposition of carbon on the interior walls of the gun, which results in a buildup of powders in the gun barrel.

Oxidation is a serious problem for corrosion-resistant coatings and thermal barrier coatings. For example, oxidation was reported to be the dominant factor affecting the wear performance and microhardness of WC–Co coatings [71]. Low-oxide coatings have been reported to improve the performance of the coatings in selected corrosion-resistant applications, although these improvements are not as significant as those offered by carbides and silicides [72]. In thermal barrier coating applications, a coefficient of thermal expansion mismatch between the oxides and the metal often leads to degradation in the coatings [72]. Therefore, controlling the oxide content becomes a critical factor in obtaining a desirable coating.

In an effort to elucidate the mechanisms of coating oxidation that result from the gas dynamics of the HVOF process, the phenomenon of gas shrouding has been studied extensively [72, 73]. Gas shrouding refers to the interaction between the jet gas carrying the spray particles and the surrounding atmosphere. Various laser scattering visualization devices have been used by Hackett and Settles to study the flow field of the entrainment to distinguish between the jet gas and the atmosphere [72]. In this study, the microsized particles introduced in the HVOF process are visible by laser illumination, and the particle path lines yield the flow field of the jet gas in the recorded image. In particular, measurements from the planar laser scattering image of the HVOF jet mixing show that the atmospheric entrainment rapidly increases the mass flux at the exit of the nozzle [72].

3.2.2. Spray Distance

The distance from the gun to the substrate is critical in determining the thermal stress behavior of the coating. A shorter spray distance increases the temperature of the coating surface, which leads to undesirable thermal stress fractures. Thermal stress fractures can be reduced by increasing the total flow of cooling air onto the coating surface [70]. Varacalle et al. [34] assessed different processing parameters (oxygen and fuel flow, air flow, powder feed rate, standoff distance) on the structure and mechanical properties of Inconel 718 coatings [34]. Results from this study showed that shorter spray distances increased the deposition efficiency and superficial hardness in the Inconel 718 coatings [34].

3.2.3. Deposition

The formation of a thermal spray coating results from the buildup of individual molten or semimolten particles that strike the surface of a substrate. Particle impingement at the substrate surface is a dynamic process that combines particle deformation and solidification simultaneously. The temperature of the particle at the moment of impact influences the grain size and phase composition of the coatings [10]. Therefore, the phenomena related to the impingement of melted or semimelted particles on a substrate are crucial in determining the coating characteristics such as porosity, inclusions, and chemical segregation [62].

3.2.3.1. Particle Deformation at Impact

The flattening process of individual particles has been studied analytically, numerically, and experimentally [50, 62, 74–77]. Results from these studies show that the kinetics of particle flattening depend on particle size and impact velocity [50]. Inertia dominates at the initial moment after impact and the effects of viscous flow proceeds, influencing particle spreading. The kinetic energy is dissipated to overcome the viscous forces of the flowing particles [78, 79].

The presence of porosity in HVOF coatings is likely to emerge as a critical issue in the application of nanocrystalline coatings, as confirmed by preliminary results that will be presented later in this chapter. Porosity in thermally sprayed coatings is closely related to the thermal, fluid flow, and solidification conditions. Consequently, the deformation and solidification of one or multiple droplets impinging on a substrate should be addressed in detail. Work in this area is currently underway at the University of California, Irvine (UCI), using numerical simulation. The numerical simulations have been conducted on the basis of the full Navier–Stokes equations and the volume of fluid (VOF) function by using a two-domain method for the thermal field and solidification problem and a new two-phase flow continuum model for the flow problem with a growing solid layer [80–83]. On the basis of this mechanism, some fundamental trends and effects of important processing parameters on microporosity may be reasonably explained, and optimal processing conditions for dense coatings may be determined. The simulations reveal that the spreading liquid separates from the solidified splat [84]. This mechanism is believed to be responsible for the formation of micropores in the solidified layer [85, 86]. It has also been shown that a fully liquid droplet impinging on a solid substrate may generate good contact between the splat and the substrate, whereas it will produce liquid ejection if it strikes onto another splat. Simulations of deforming molten-metal droplets as they interact with other droplets and/or a nonflat substrate provided further insight concerning the fundamental mechanisms governing pore formation in spray-processed materials [87]. If the roughness width is larger than the initial droplet diameter, the droplet undergoes a succession of accelerations and decelerations and eventually breaks apart. If the droplet is larger than the roughness width, the spreading process is hindered. In the cases mentioned previously, droplet impact velocities are high and the effect of the surrounding gas can be neglected [87]. The viscous and surface tension effects at the interface (free surface when gas is neglected) are relatively less important than the inertia force owing to high Reynolds and Weber numbers [87, 88]. The results obtained with W, Ni, and Ti demonstrate that a droplet spreads uniformly in the radial direction during impingement onto a flat substrate and eventually forms a thin splat [87]. The flattening rate is fast initially and decreases asymptotically. The spreading process of droplets under thermal spray conditions is essentially governed by the inertia dynamics. Therefore, increasing impact velocity, droplet diameter, and material density, or decreasing material viscosity, leads to an increase in the final splat diameter and to a larger spreading time [87]. The inherent dependence of the final splat diameter and the spreading time on the inertia and viscous effects may be approximated by the correlation derived from the regression analyses of the calculated results. A fully liquid droplet impinging onto a flat solid substrate may lead to good contact and adhesion between the splat and the substrate, whereas a fully liquid droplet striking onto the flattening, fully liquid splat produces ejection, rebound, and breakup of the liquid. These phenomena may reduce the deposit rate and deteriorate the bonding and deposit integrity. Except for the complete axisymmetric case, a fully liquid droplet colliding with the flattening, fully liquid splat also causes formation of voids within the liquid, which may become inner pores when solidification and subsequent contraction occur [87]. The validity of the droplet deformation simulations has been verified by a test calculation and by examining the viscous and surface tension effects in earlier work on Ti droplets [86, 87].

3.2.3.2. Residual Stresses

Residual stresses in coatings originate from the misfit strains that arise during spray deposition and from differential thermal contraction stresses during cooling. Residual stresses ultimately affect the electrical, mechanical, and optical properties of the coatings [89]. Extensive studies were conducted by Clyne [89] on the relationship between stress distributions and release rate of interfacial strain energy in various types of coatings.

Stresses induced from differential thermal contraction result from the differences in thermal expansivities between the coating and the substrate. The misfit in strain can be calculated by the following equation [89]:

$$\Delta\varepsilon = \int_{T_2}^{T_1} \left(\alpha_s(T) - \alpha_d(T)\right) dT \tag{18}$$

where $\Delta\varepsilon$ is the misfit strain (unitless), T_1 and T_2 are cooling temperatures (K), and α_s and α_d are the thermal expansivities of the substrate and coating deposit, respectively (K^{-1}).

Because thermal expansivity increases with rising temperature, large errors associated with the value of the misfit strain can occur if thermal expansivities are assumed constant [89].

3.3. Experimental Synthesis and Analysis

3.3.1. Feedstock Materials for High-Velocity Oxy–Fuel Thermal Spraying

Materials eligible for thermal spraying may take various forms, such as powders, rods, and wires. In the case of HVOF thermal spraying, however, the feedstock materials usually take the form of powders. Presently, there are several routes, namely, atomization, fusing or sintering, spray drying and cladding, and sol–gel, that have been documented in manufacturing powders for HVOF thermal spraying [10]. Nevertheless, the selection of a specific method of preparation depends primarily on the types of materials and desired coating characteristics. For example, the atomization method is commonly considered to be suitable for preparing metals and alloy powders. The preparation of ceramic powders, such as oxides and carbides, has been effective utilizing fusing (or sintering) followed by crushing, sol–gel, and thermal pyrolysis or by calcination routes [10].

More recently, as a result of an ongoing research program aimed at fabricating nanocrystalline coatings by HVOF spraying, Tellkamp et al. [37] reported that cryogenic attritor milling in liquid nitrogen can be successfully employed to prepare nanocrystalline feedstock metal powders suitable for HVOF spraying. This method promises more flexibility in synthesizing a wide variety of HVOF thermal spray materials, including metals, alloys, ceramics, and cermets.

The morphology of the feedstock powders has been reported to play a critical role in controlling the quality of the final coatings fabricated by HVOF thermal spraying [10]. This is thought to be due to the difference in the impingement mechanisms with which the spherical or pancake particle powders strike the substrate; however, a set of criteria defining the role of each factor on the structure of the final coatings remains to be established. Usually, the morphology of powders is dependent on the types of materials, as well as the technical routes, employed to synthesize these powders. For example, gas-atomized powders tend to have a spherical morphology, which is free of porosity (see Fig. 3a), whereas mechanical alloyed/milled metal powders take the form of a pancake (see Fig. 3b). The shape of the powder particles is thought to be a crucial factor influencing the impingement mechanism. The powders fabricated by fusing, followed by crushing, generally take an angular, blocky shape, differing significantly from those of the atomized powders [10]. The sol–gel method may be employed for synthesizing ceramic powders, in particular, ultra-fine nanoscale ceramic powders. These powders usually exhibit a perfect, spherical shape, analogous to that produced by the atomization method. Recently, as a result of a fruitful cooperative work between Rutgers University (Piscataway, NJ) and the University of Connecticut (Storrs, CT) [6], advances in the manufacturing of industrial scale-up nanopowders have been made. It has been demonstrated that (i) nanocrystalline metal powders can be produced by an aqueous solution reaction method, (ii) nanocrystalline cermet powders can be produced by the spray conversion processing method, and (iii) nanocrystalline ceramic powders can be produced by the chemical vapor condensation method. Notably, some of these nanocrystalline powders have been successfully used for HVOF spraying [6].

The size distribution of the feedstock powders also plays a critical role in the quality of the HVOF-sprayed coatings [10]. The optimal range of powder size for HVOF spraying usually lies between 20 and 60 μm [10]. Smaller powders, with sizes below 20 μm, can be subjected to a variety of pretreatments, such as sintering or agglomeration, to increase the size of the agglomerates, thereby making them suitable for HVOF spraying [6]. This advance has significantly improved the prospects of using fine powders, particularly the nanometer metal or ceramic powders, for HVOF spraying.

3.3.2. Synthesis of Coatings

HVOF thermal spraying has evolved into a desirable method of fabricating coatings covering a broad spectrum of materials, including metals, alloys, ceramics, and composites [6, 10]. These coatings provide wear resistance, corrosion resistance, and thermal resistance for the sprayed components to extend their service lifetime as required by the aerospace, petrochemical, and automotive industries [53].

3.3.2.1. Pure Metal Coatings

The synthesis of pure metal coatings by HVOF spraying has been reported in the current literature [6, 90–92]. Molybdenum coatings are well noted for their low coefficient of friction and excellent resistance to scuffing [90, 91]. These unusual properties may significantly enhance the production of piston rings. In related studies, Kreye et al. [90] fabricated Mo coatings by HVOF spraying. Evaluation of the mechanical properties revealed that the HVOF-sprayed Mo coatings exhibited very high hardnesses owing to the formation of MoO_2 during the HVOF process [91]. This high hardness is thought to provide additional benefits for the production of piston rings. Chromium is another pure metal that has attracted extensive attention in the form of coatings [92]. Generally, Cr coatings may be fabricated via various routes, such as plating, vacuum deposition, and HVOF spraying [92]. An assessment of these methods reveals that HVOF thermal spraying appears to be the

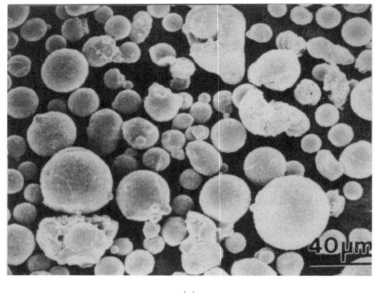

(a)

(b)

Fig. 3. Scanning electron microscopy (SEM) images of (a) gas-atomized powders of Inconel 718 and (b) cryomilled Inconel 718 powders for HVOF spraying. Reprinted from V. L. Tellkamp et al., *Nanostruct. Mater.* 9, 489 (© 1996), with permission from Elsevier Science.

most viable method in terms of cost and environmental protection. Knotek et al. [92] compared the properties of the HVOF-sprayed chromium coatings with those of galvanic hard chromium platings, which are generally applied to reduce friction and corrosion but are environmentally hazardous during their production. They pointed out that if the oxide content is well controlled during the HVOF spraying, the HVOF-sprayed chromium coatings, which are environmentally friendly, can be expected to replace the galvanic hard chromium coatings. Tantalum coatings have never been a cost-effective item for the petrochemical industry; however, corrosion-resistant tantalum coatings have been and continue to be a very cost-effective solution for many complex metallurgical applications [93]. HVOF thermal spraying of tantalum coatings offers an attractive alternative for corrosion-resistant applications. Optimal tantalum coatings can be achieved by an appropriate implementation of the experimental parameters during the HVOF process [93].

3.3.2.2. Alloy Coatings

Alloy coatings fabricated by HVOF thermal spraying mainly consist of superalloys [94, 95], stainless-steels [96, 97], Tribaloy [98], NiCr alloy [99], aluminum bronze [100], and others [6, 16, 85]. Among these alloy coatings, superalloy coatings have been extensively studied. For example, Arvidsson [94] prepared superalloy coatings (Anval 625 and Anval 718) by both HVOF and APS techniques. Subsequent characterization revealed that the Anval 625 HVOF coating exhibits better corrosion resistance than its plasma-sprayed counterparts owing to a very high density and the subsequent absence of large pores. Moreover, bond strength is higher for HVOF coatings. Noting the benefits that the high particle velocity associated with HVOF spraying would bring to improving the quality of the coatings, Irons [95] characterized the HVOF-sprayed coatings of Inconel 718 alloy. The results from this study demonstrated that the high particle velocity inherent to the HVOF spraying process indeed improves the quality of the coating density, hardness, oxide content, and thickness significantly. More recently, Edris et al. [101] deposited Inconel 625 alloy on a mild steel substrate using HVOF spraying and investigated the evolution of the microstructures of the coatings as a function of the oxygen-to-fuel-gas ratio, total gas flow rate in the gun, and combustion chamber length. The coatings consist of high- and low-alloy Ni-based metallic regions together with oxides exhibiting Cr_2O_3 and $NiCr_2O_4$ (spinel) crystal structures, and the relative amount of each phase present in the coatings depends significantly on the three processing variables—oxygen-to-fuel-gas ratio, combustion chamber length, and total gas flow rate [101].

Another class of alloys that has received a great deal of interest are the stainless-steels. HVOF spraying is noted to produce low-oxide, dense, thick stainless-steel coatings with superior mechanical properties [96, 97]. Excellent mechanical properties were observed in the coating formed by spraying 316L stainless-steel powder with high particle velocities, within a particle temperature range between the solidus and liquidus temperatures of the 316L alloy [102]. Recently, Siitonen et al. [103] studied the corrosion resistance of Fe–20% Cr–18% Ni–6.2% Mo–0.8% Cu–0.2N coatings fabricated by HVOF, APS, and others. It was noted that the HVOF coatings exhibited the best corrosion resistance owing to a reduction in porosity. To seek an alternative usage for cast-iron liners in aluminum engine blocks in the automotive industry, a rotating HVOF spray gun was used to produce alloy coatings of aluminum bronze (Cu–9 Al–1 Fe) and 1025 and 1049 steels on the cylinder bore [100]. Comparisons among the HVOF coatings, the gray cast iron, and 390 aluminum alloy were conducted. The results revealed that the splat delamination wear mechanism controls all of these coatings; the oxide particles present in all thermally sprayed coatings do not appear to act as hard particles that bear the load. In addition to the aforementioned examples representing, the HVOF spraying of alloy coatings, related work also includes amorphous (or amorphous–crystalline) Fe–Cr–P–C (or Ni–Cr–Si–B–C) and quasicrystalline Al–Cu–Fe coatings produced by HVOF spraying [104, 105].

3.3.2.3. Ceramic Coatings

In principle, ceramic coatings can be classified into two categories: (i) single-phase ceramic coatings and (ii) multiphase ceramic coatings. Because the highest temperatures that HVOF spraying can attain are lower than those of plasma spraying, the application of HVOF spraying in the synthesis of ceramic coatings, which usually have high melting points and low thermal conductivities, is relatively limited. Vuoristo et al. [106] prepared

Al_2O_3, Al_2O_3–3–40% TiO_2, Al_2O_3–30% MgO, and Cr_2O_3 coatings by both HVOF spraying and atmospheric plasma spraying (APS), and examination of various properties was subsequently conducted. The results revealed that the HVOF-sprayed coatings exhibited superior properties than the APS coatings in terms of compositional homogeneity, microhardness, and wear resistance [106].

A representative example of multiphase ceramic coating is the $ZrO_2 + Y_2O_3$ coating, commonly known as a thermal barrier coating. In the past, extensive efforts have been directed toward the fabrication and characterization of this material because of its superior thermal resistance, corrosion resistance, and wear resistance properties. Inspection of the available literature; however, revealed that most of the thermally sprayed $ZrO_2 + Y_2O_3$ ceramic coatings were fabricated by plasma spraying. In recent years, with the advancement of new routes to synthesize nanometer-sized ceramic powders, in which the melting point can be significantly lowered relative to those of their micrometer counterparts, it is expected that HVOF spraying of ceramic coatings such as $ZrO_2 + Y_2O_3$ may be achieved [6].

Additionally, HVOF has proved to be useful in fabricating superconducting ceramic materials [107] and biocompatible coatings on implants such as artificial joints [108]. Research activities in these fields are also increasing.

3.3.2.4. Composite Ceramic Coatings

The class of composite coatings that has seen the most research is the ceramic–metal (cermet) coating. The idea behind the creation of cermet coatings using HVOF spraying originated from the conjecture that the incorporation of metal binders into the ceramic coatings would create coatings combining the advantages of both ceramics and metals. Early work led to the development of a wide variety of cermets, as summarized in Table II, by a selective combination of ceramics and metal binders. Cemented carbides, especially those based on tungsten carbide, have been widely used in applications requiring high hardness and good wear resistance. The metal binder materials usually include pure Ni, Co, CrNi alloy, and NiAl intermetallic compounds.

Chromium carbides, nitrides, and oxides are well characterized as being hard and highly corrosion resistant [112]. This combination of properties makes them excellent candidates for the fabrication of cermet coatings having unusual properties. For instance, Cr_3C_2-based cermets such as Cr_3C_2–NiCr can be used in relatively high temperature applications [112]. In recent years, intermetallic compounds such as NiAl have been commonly used as metal binders in both WC- and Cr_3C_2-based cermets [112]. The use of intermetallic compounds has led to an increase in service temperatures [112]. More recently, the application of ceramic coatings has expanded, and ceramic composites such as WB, CrB, SiC, WC/VC, and WC/CrC have been developed to meet the increasing demands of coatings with improved properties.

3.3.3. Possible Nonequilibrium Structures of High-Velocity Oxy–Fuel Sprayed Coatings

HVOF thermal spraying has been used extensively as a means to synthesize coatings with coarse-grained polycrystalline structures. In many instances, however, varying degrees of nonequilibrium microstructures can also be achieved using HVOF thermal spraying. These nonequilibrium microstructures are dependent on the thickness of the coating, as well as the specific experimental parameters such as cooling rate, substrate temperature, and thermal conductivity. The nonequilibrium structures can range from metallic glass to quasicrystalline and nanocrystalline materials [6, 37, 104, 105]. For example, in the iron-based (Fe–Cr–P–C) and nickel-based (Ni–Cr–Si–B–C) eutectic alloys, it has been reported that a mixture of amorphous and crystalline or purely amorphous coatings can be obtained using an appropriate set of parameters for HVOF spraying [104]. Sordelet et al. [105] attempted

Table II. Typical Cermets Produced by HVOF Spraying

Ceramic	Cermet	Author	Reference
WC	WC + Ni	Sobolev and Guilemany	[62]
	WC + Co	Iwanoto et al.	[109]
	WC + Co	Guilemany et al.	[110]
	WC + Co	Creffield et al.	[71]
	WC + CrNi (or Co)	Nakahira et al.	[111]
	WC + NiAl	Shaw et al.	[112]
	WC + Co + Cr	Niemi et al.	[17]
WC/VC	WC/VC + Co	Kear and Strutt	[6]
WC/CrC	WC/CrC + Ni	Froning and Keller	[113]
WB	WB + Co	Froning and Keller	[113]
(W, Ti)C	(W, Ti)C + Ni	Guilemany et al.	[114]
WC/TiC	WC/TiC + Ni	Niemi et al.	[17]
Cr_3C_2	Cr_3C_2 + NiCr	Shaw et al.	[112]
	Cr_3C_2 + NiCr	Irons et al.	[115]
	Cr_3C_2 + NiAl	Shaw et al.	[112]
	Cr_3C_2 + NiCr	Froning and Keller	[113]
CrB	CrB + NiCr	Froning and Keller	[113]
(Ti, Mo)C	(Ti, Mo)C + NiCo	Vuoristo et al.	[116]
SiC	SiC + Al	Steffens et al.	[117]
SiC	SiC + NiCr8020	Steffens et al.	[117]
$TiC–ZrO_2$	NiCrAl–11%, TiC–16% ZrO_2	Bernard et al.	[118]

to prepare Al–Cu–Fe coatings using HVOF spraying. Characterization of the as-sprayed coatings revealed a quasicrystalline structure. This quasicrystalline Al–Cu–Fe coating is being considered as a potential replacement for the electrodeposited chromium on various components of the space shuttle main engine. More recently, it has been found that HVOF spraying is capable of synthesizing dense coatings with nanoscale grain sizes (the so-called nanocrystalline materials) [6, 37].

3.3.4. Characterization of High-Velocity Oxy–Fuel Coatings

The requirement of achieving optimum properties for HVOF coatings necessitates an understanding of the spraying parameters that control the morphology, microstructure, and composition of the coating. Inspection of the available literature reveals that the characterization of the HVOF coatings can be primarily accomplished using the following techniques.

3.3.4.1. X-Ray Diffraction

X-ray diffraction (XRD) measurements can be conveniently performed on the slightly polished as-sprayed coatings. It is not necessary to separate the coatings from the substrates because the thickness of the coatings is usually larger than the penetration depth of the X-rays. Typically, XRD measurements are carried out to identify the phase components present in the coatings, especially the oxides that may have formed during HVOF spraying. Tellkamp et al. [37] prepared Inconel 718 coatings by HVOF spraying of cryomilled Inconel powders. Subsequent XRD analysis enabled the determination of the oxides present in the coatings as $(Cr, Fe)_2O_3$. Similar work is also available elsewhere [119]. More recently, XRD has been widely utilized for quantitative analysis to verify the successful creation of nanocrystalline coatings utilizing HVOF spraying [6, 37, 38]. Tellkamp et al. [37] succeeded in extracting the grain size (D) and microstrain (e) of Inconel 718 HVOF coatings using the integral breadth analysis [120]:

$$\frac{(\delta 2\theta)^2}{\tan^2 \theta_0} = \frac{K\lambda}{D(\delta 2\theta / \tan \theta_0 \sin \theta_0) + 16e^2} \tag{19}$$

where θ_0 is the position of peak maximum, K is a constant factor, being taken as 0.9, and λ is the wavelength of the X-rays. By plotting $(\delta 2\theta)^2 / \tan^2 \theta_0$ against $\delta 2\theta / \tan \theta_0 \sin \theta_0$ and applying the results from a few broadened XRD peaks, a linear relationship can be obtained. The grain size and microstrain may be determined from the slope, $K\lambda/D$, and ordinate intercept, $16e^2$. The XRD results, indicating nanocrystalline structures in the HVOF-sprayed Inconel 718 coatings, were recently confirmed by transmission electron microscopy observations [39].

Other methods, such as the Scherrer equation, the single-line approximation, and the Warren–Averbach method, may also be used to extract the grain size as well as the microstrain from the XRD analyses [121]. Assuming that the physical origin of the XRD peak broadening is due to the small grain size alone, the relationship between grain size, D, and the full width of peak at half maximum (FWHM), $\delta 2\theta$, in radians, is given by the Scherrer equation [120]:

$$D = \frac{0.9\lambda}{\delta 2\theta \cos \theta} \tag{20}$$

where λ is the wavelength and θ is the diffraction angle. The grain size obtained from this equation is the volume-averaged grain size in the direction perpendicular to the plane of diffraction.

The single-line approximation method has the ability to determine simultaneously the grain size and microstrain [122]. This method was developed by de Keijser et al. [122] by assuming that the XRD profile can be matched by a Voigt function. In fact, the experimentally measured line profile, h, is a combination of the structurally broadened profile, f, and the standard profile, g. The g profile is usually used for correcting the instrumental broadening. If h, f and g are assumed to be Voigt functions, then [123]:

$$h_C = g_C * f_C \qquad \text{and} \qquad h_G = g_G * f_G \tag{21}$$

where the subscripts C and G denote the Cauchy and Gaussian components of the respective Voigt profiles, and the operational symbol "$*$" denotes convolution. From Eq. (21), it follows that the integral widths (for the f profile) of β_f^C and β_f^G (after the correction of instrumental broadening) are given by

$$\beta_C^f = \beta_C^h - \beta_C^g \qquad \text{and} \qquad \left(\beta_G^f\right)^2 = \left(\beta_G^h\right)^2 - \left(\beta_G^g\right)^2 \tag{22}$$

The constituent Cauchy and Gaussian components can be obtained from the β and the ratio of $2w/\beta$ for the h and g profiles, where $2w$ is the FWHM of the XRD peak and β is the integrated peak width. The following empirical equations have been derived to determine the Cauchy and Gaussian components [122]:

$$\beta_C = \beta\left(2.0207 - 0.4803(2w/\beta) - 1.7756(2w/\beta)^2\right) \tag{23}$$

$$\beta_G = \beta\left(0.6420 + 1.4187\left((2w/\beta) - 2/\pi\right)^{1/2} - 2.2043(2w/\beta) + 1.8706(2w/\beta)^2\right) \tag{24}$$

The grain size (D) and heterogeneous strain (e) can then be estimated by

$$D = \frac{K\lambda}{\beta_C^f \cos\theta} \tag{25}$$

$$e = \frac{\beta_G^f}{4\tan\theta} \tag{26}$$

where K is a constant factor, being taken as 0.9.

3.3.4.2. Transmission Electron Microscopy

Transmission electron microscopy (TEM) observation offers a unique technique for imaging submicrostructures of materials. This technique primarily consists of (i) regular TEM (plane-view and cross-sectional TEM depending on the sample preparation methods) and (ii) high-resolution TEM (HREM). In the case of plane-view TEM, HVOF coatings should be separated from the substrate in order to prepare TEM samples. Plane-view TEM observation has been widely utilized in providing the following information:

1. Morphology and size distribution of grains (the grain size measurements range from several nanometers to several tens of micrometers)
2. Observation of porosity
3. Distribution of ceramic phases, including carbides, oxides, and others
4. Phase identification by means of selected area diffraction (SAD)
5. Distribution of elements by means of energy-dispersive spectrometry (EDS)

Cross-sectional TEM is specifically suitable for determining interfacial reactions between the HVOF coatings and the substrates that may have occurred during the HVOF process. Knowledge regarding the interfacial state can be used to assess the quality of bonding between the coatings and the substrates in terms of porosity and formation of interfacial phases. Owing to the difficulty in preparing cross-sectional TEM samples, limited work has been available in this area to date [124].

High-resolution TEM (HREM) distinguishes itself from the conventional TEM in its ability to produce high-resolution microstructure images on the atomic scale [7]. The detection of entities as small as a few angstroms in the HVOF-sprayed coatings is possible with this technique. The HREM technique is especially suitable for the characterization of nanocrystalline HVOF-sprayed coatings, because the details of the ultrafine grains can be hardly observed otherwise.

3.3.4.3. Scanning Electron Microscopy

Scanning electron microscopy (SEM) is another powerful technique used to characterize the HVOF-sprayed coatings [10]. Its functions primarily include (i) secondary electron imaging (SE), (ii) backscattered electron imaging (BS), (iii) energy-dispersive spectrometry (EDS), and (iv) electron microprobe analysis (EMPA). Based on SEM observations, the following information may be obtained: (i) morphology and microstructure of the coatings (SE and BS), (ii) porosity present in the coatings as well as at the bonding interfaces (SE and BS), (iii) elemental distribution of the coatings (EDS and EMPA), and (iv) interface bonding between the coatings and the substrates (SE and BS).

3.3.4.4. Optical Microscopy

Optical microscopy (OM) is a conventional technique capable of characterizing the microstructure of coating materials [10]. Presently, it is widely used in nearly all the studies of HVOF-sprayed coatings. Some of the functions of OM are similar to those of SEM, the simplicity and cost effectiveness of optical microscopy, however, determine its dominant role in characterizing the HVOF-sprayed coatings. Routine analysis with OM consists of studying (i) the fraction and size of voids in the coatings, (ii) the fraction and size of un-melted particles in the coatings, (iii) the microstructures of various phases in the coatings, and (iv) the bonding interface between the coatings and the substrates.

3.3.4.5. Other Methods

Additionally, in some cases, Auger electron spectroscopy (AES), X-ray photoelectron spectroscopy (XPS), and scanning acoustic microscopy (SAM) may also be used to study the compositions, valence states, and microstructures of HVOF-sprayed coatings [10, 125].

3.3.5. Properties of High-Velocity Oxy–Fuel Coatings

HVOF spraying is described as a thermal spraying process, in which the powder particle temperature is thought to be relatively lower, and the particle velocity higher, than those of other thermal spraying processes, such as plasma spraying [126]. As a result, the HVOF-sprayed coatings exhibit a variety of superior properties, such as a high coating density, a greater thickness, an improvement in hardness and wear resistance, strong interfacial bonds, and a reduction of unwanted oxides.

3.3.5.1. Porosity (Density)

Typical values of porosity in coatings fabricated by a variety of thermal spraying techniques are listed in Table III [10]. These results indicate that HVOF spraying is the technique that produces the densest coatings among all of the thermal spraying techniques. In related work, Voggenreiter et al. [102] reported that in the case of HVOF spraying of alloy 316L, high-density coatings with a minimum porosity of 0.1–0.2% could be obtained with the high particle velocities associated with the HVOF process. Creffield et al. [71] have attempted to study the effect of gas flow rate on the density of the HVOF coatings. It was determined that decreasing the gas flow rate can significantly increase the porosity, which, in turn, affects the oxidation and hardness of the coatings.

D-gun spraying (DGS) can be regarded as a type of HVOF spraying because of the similarities in the spraying processes. The residual porosity of DGS, atmospheric plasma spraying (APS), and flame spraying (FS) coatings are compared in Table IV. This table indicates that DGS produces the densest coatings [127].

3.3.5.2. Hardness

The hardness of thermally sprayed coatings can be evaluated in terms of the Vickers mi-crohardness and Rockwell hardness measurements. In general, an increase in density of the coating results in an increase in hardness. For example, the mean values of hardness for D-gun-sprayed Al_2O_3, $Al_2O_3 + 3\%$ TiO_2, $Al_2O_3 + 13\%$ TiO_2, $Al_2O_3 + 40\%$ TiO_2, $Al_2O_3 + 30\%$ MgO, and Cr_2O_3 are higher than those of plasma-sprayed coatings, as shown in Table V [106]. In many cases, however, the hardness is not a simple function of density, but a function of a number of factors, including oxide or carbide content and

Table III.   Values of Porosity of Thermally Sprayed Coatings [10]

HVOF	FS	APS	AS	DGS	VPS
<1%	10–20%	1–7%	10–20%	<2%	<1–2%

Table IV. Typical Values of Residual Porosity of Sprayed Coatings [127]

	Coating material	Porosity
DGS	NiAlSi	0.5 ± 0.4
DGS	NiFeAlSi	0.6 ± 0.4
APS	NiAlSi	4.7 ± 0.6
APS	NiFeAlSi	5.2 ± 0.7
FS	NiAlSi	10 ± 2.0
FS	NiFeAlSi	9 ± 2.0

Table V. Mean Microhardness Values of the Plasma-Sprayed and Detonation-Gun-Sprayed Alumina, Alumina–Titania, Alumina–Magnesia, and Chromia Coatings [106]

Coating composition	Mean microhardness ($Hv_{0.2}$)	
	Plasma sprayed	Detonation gun sprayed
Al_2O_3	780 (98)[a]	1020 (55)
$Al_2O_3 + 3\% \ TiO_2$	881 (125)	992 (71)
$Al_2O_3 + 13\% \ TiO_2$	843 (59)	1053 (99)
$Al_2O_3 + 40\% \ TiO_2$	827 (72)	916 (67)
$Al_2O_3 + 30\% \ MgO$	726 (102)	952 (58)
Cr_2O_3	1894 (127)	2082 (125)

[a] Standard deviation values are given in parentheses.

elemental segregation. Therefore, anomalies in the correlation between hardness and density may occur. In related work, Sordelet et al. [105] prepared Al–Cu–Fe coatings using atmospheric plasma spraying (APS), vacuum plasma spraying (VPS), and HVOF spraying, and conducted a study of hardness for the different coating techniques. Their results, as illustrated in Table VI, show explicitly that the relatively dense coatings prepared by VPS and HVOF do not exhibit higher hardnesses than those prepared by APS. Quantitative energy-dispersive spectroscopy of these coatings indicated that localized oxidation occurred during the HVOF process [105]. This may have caused elemental segregation, which might explain the decrease in coating hardness.

3.3.5.3. Wear Resistance

In most cases, the term "wear resistance" of thermally sprayed coatings refers to the following: (i) adhesive wear resistance (frictional wear resistance), (ii) abrasive wear resistance, and (iii) erosive wear resistance (including cavitation erosion wear). It has been noted that the HVOF coatings usually exhibit superior wear resistance than other thermally sprayed coatings [17, 53, 106]. This is especially true in tungsten carbide/cobalt (WC/Co) and chromium-carbide nickel-chromium (CrC-NiCr) coatings. In some cases; however, the microhardness behavior of HVOF-sprayed coatings fail to follow this trend. For example,

Table VI. Hardness of Al–Cu–Fe Coatings
Prepared via APS, VPS, and HVOF [105]

Spraying method	Hardness ($Hv_{0.03}$)
APS1	876 ± 74
APS2	773 ± 154
VPS1	803 ± 161
VPS2	876 ± 144
HVOF	610 ± 80

Parker and Kutner [53] reported that the HVOF-sprayed WC/Co and CrC–NiCr coatings show only two thirds of the wear resistance of coatings prepared by other methods.

Owing to the complexities of the coating methods, coating materials, and testing methods, a detailed comparison of the wear resistance of HVOF coatings relative to that achievable by other thermally sprayed coatings is beyond the scope of this chapter. Hence, the interested reader is encouraged to consult the available literature. Recently, Pawlowsky [10] reviewed the friction and adhesive wear of thermally sprayed coatings, in which the following issues have been taken into consideration: (i) materials categories, including ceramics, metals, alloys, and composites; and (ii) testing methods, involving two-body abrasive wear, three-body abrasive wear, erosive wear, fretting wear, corrosion wear, and impact wear.

3.3.5.4. Corrosion Resistance

HVOF thermal spraying can produce coatings with extremely high densities and very low oxide contents. This is especially beneficial to corrosion applications where a dense barrier is usually needed. Typical materials for HVOF spraying in the case of corrosion resistance include the following: stainless-steel (FeCrNi), Hastelloy C-276, Anval 254 SMO, WC + Co, and oxide ceramics. Amoco Oil Company has routinely employed HVOF-sprayed AISI 316 stainless-steel and Hastelloy C-276 coatings for the corrosion protection of various machine parts that are subjected to severe erosive wear and corrosion conditions [12, 97]. It has been demonstrated that refinements in HVOF spraying have significantly improved the quality of metallic alloy coatings [37]. These coatings behave as true corrosion barriers, in which the corrosion resistance is as good as the wrought materials. In related work, it has been reported that the HVOF coatings may be used on many shipboard items that traditionally are chrome plated [97]. Almost no hazardous waste was generated from this process as compared to other methods of producing chrome coatings. Siitonen et al. [103] attempted to study the corrosion resistance of the coatings produced by various thermal spraying methods, specifically atmospheric plasma spraying (APS), atmospheric plasma spraying using gas shielding around the plasma (APS/S), low-pressure plasma spraying (LPPS), detonation gun spraying (DGS), and HVOF spraying. The results of this study showed that the best coating quality can be achieved by the LPPS and HVOF techniques, and that oxidation and porosity restrict the use of APS coatings. In another study on Anval 625 alloy coatings, better corrosion resistance was found in the HVOF coatings over their plasma-sprayed counterparts [94].

3.3.5.5. Bond Strength

The strength of the interfacial bond between the thermally sprayed coating and the substrate depends primarily on the coating parameters, the coating materials, and the substrate materials. For numerous types of thermal spray coating materials (carbides, oxides, metals, and alloys) and substrate materials (stainless-steel, mild steel, aluminum, and nickel alloys), the available results of bond strengths were collected, tabulated, and systematically compared by Pawlowsky [10]. It was noted that, for carbide coatings, the bond strength in most cases was greater than 50 MPa. Thus, the carbide coatings adhere very well to the metal and alloy substrates. In the HVOF-sprayed carbides, the bond strength can reach as high as 90 MPa [10]. The same study demonstrated that the bond strength of carbides sprayed with the APS technique was lower than those sprayed with the HVOF processes [10]. The bond strength of oxide coatings is usually less than that of carbides, because the adhesion of the oxide coatings to the metal substrate is mostly by mechanical bonding [10].

3.3.5.6. Fatigue, Erosion Resistance, and Modulus

Pawlowsky [10] published a general overview covering the elastic modulus, strength, and fracture toughness of a variety of thermally sprayed coatings. It is apparent from this work that limited information on the fatigue, erosion resistance, and modulus of these coatings is available. Tipton [128] studied the effect of HVOF spraying on the elevated temperature high cycle fatigue behavior of a martensitic stainless-steel. They concluded that HVOF-sprayed coatings can significantly influence the high-cycle fatigue behavior of a martensitic 12Cr stainless-steel at elevated temperatures. In this case, the dominant factor is the thermal expansion mismatch between the coating and the substrate. In related work, Steffens et al. [129] attempted to elucidate the influence of HVOF-sprayed WC/Co coatings on the high-cycle fatigue strength of mild steel. It was demonstrated that the HVOF-sprayed WC/Co coating did not cause a decrease of the high-cycle fatigue resistance of the mild steel St 52. In some cases, a significant increase of the fatigue strength was observed.

A study of the erosion resistance of chromium carbide coatings (both HVOF- and plasma-sprayed coatings) has been conducted by Irons et al. [115]. Eight different chromium carbide materials were studied, and the erosion resistance of the HVOF chromium carbide coatings was found to be 1.5 times better than their plasma-sprayed counterparts, as measured by a room temperature erosion test. Moreover, it was found that the powder size is the most important factor affecting the erosion resistance. A larger powder size yielded improved high-angle erosion resistance.

Young's modulus is defined as the ratio of stress to strain in the elastic range of a material. In an attempt to study the modulus of the HVOF-sprayed WC–12 wt% Co coating, Nakahira et al. [111] investigated the elastic properties parallel and perpendicular to the coating plane. The results of this study revealed the presence of anisotropy in the coating. The Young's modulus value parallel to the coating plane was twice as large as that in the perpendicular direction.

3.4. High-Velocity Oxy–Fuel Spraying of Nanocrystalline Coatings

3.4.1. Background

More recently, there has been increasing interest in the manufacture of nanocrystalline coatings by HVOF spraying. Since the pioneering work of Kear and Strutt [6] and Tellkamp et al. [37], advances have been made toward:

- Synthesis of nanocrystalline feedstock powders suitable for HVOF thermal spraying
- Development of new material systems having the potential to produce nanocrystalline coatings with superior mechanical properties
- Synthesis of nanocrystalline coatings from nanocrystalline feedstock powders

- Increased fundamental understanding of the formation of nanocrystalline coatings during HVOF spraying
- Modeling of the HVOF process to optimize the experimental parameters

The development of nanocrystalline HVOF coatings is driven by the need to enhance the physical characteristics of engineered coatings (e.g., hardness, bond strength, corrosive behavior, etc.), as well as the potential to develop coatings with heretofore unobtainable microstructures (e.g., with grain sizes < 100 nm) and the accompanying improvements in performance [7, 8]. Preliminary research results [6, 37] are encouraging and potential industrial applications are rapidly expanding. It is also apparent, however, that, to exploit fully the potential of this technology, a firm understanding of the relevant fundamental physical phenomena involved will have to be established. This will be a challenge to the scientific community, owing in part to the complex interactions of the fluid, thermal, and solidification phenomena that are present during the thermal spraying of nanocrystalline materials. Fortunately, exciting preliminary results have fueled the rapid development of this technology, as end users seek to develop coatings with unusual combinations of microstructure and physical attributes.

3.4.2. Formation Mechanisms

The formation of metastable materials such as amorphous and quasicrystalline materials by HVOF spraying has been documented [104, 105]. The mechanism that is thought to be responsible for the formation of these metastable materials is the combined effects of nucleation and grain growth under rapid cooling conditions, such as those present during thermal spraying. In the case of the formation of nanocrystalline coatings, however, there may be other operative mechanisms depending on the various types of precursors.

- In the case of a liquid precursor (i.e., the feedstock powders experience complete melting prior to impingement onto the substrate), the formation of a nanocrystalline microstructure will be strongly influenced by the cooling rate. A high cooling rate creates a high nucleation rate and a sluggish grain growth rate favoring the formation a nanocrystalline structure. This mechanism is similar to that associated with the formation of amorphous and quasicrystalline materials [104, 105].
- In the case of a solid precursor (i.e., the feedstock powders experience a high temperature exposure but no melting occurs prior to impingement), the formation of a nanocrystalline structure will depend on the thermal stability of the nanocrystalline feedstock powders. To that effect, it has been argued that the thermal stability of nanocrystalline feedstock powders is critical in the retention of a nanometric microstructure in the final coatings [37].
- In the case of a mixture of liquid and solid precursors, both the cooling rate and the thermal stability of the feedstock powders play dominant roles. This is the case that most frequently occurs in practice because of the wide range in size distribution of feedstock powders.

3.4.3. Fundamental Factors Influencing the Behavior of High-Velocity Oxy–Fuel Nanocrystalline Coatings

The complexity of the HVOF process poses a challenge in predicting the behavior of HVOF-sprayed nanocrystalline coatings, both experimentally and theoretically. In principle, however, it is evident that the behavior of the nanocrystalline powders during HVOF will be influenced by (i) the inherent thermal stability of the sprayed materials, (ii) the thermal environment present during the HVOF process, and (iii) the thermal history of the particle powders (liquid or solid) following impingement, initially on a substrate and subsequently on each other.

3.4.3.1. Thermal Stability of Nanocrystalline Materials against Grain Growth

The thermal stability of nanocrystalline materials, in particular, the resistance to grain growth, is of vital importance to both the preparation of nanocrystalline coatings and the retention of the nanocrystalline structure during service. Nanocrystalline materials inherently possess a significant fraction of high-energy, disordered grain boundary regions that provide a strong driving force for grain growth. The ability to retain ultrafine grain sizes during elevated temperature consolidation or HVOF spraying, however, is critical because it is precisely the fine grain size and large grain boundary volume that provide the unique properties to the bulk material. Reviews on the subject by Suryanarayana [8] and Malow and Koch [9] have demonstrated that a number of nanocrystalline materials experience significant grain growth at room temperature, including Pd, which has a melting point of 1552 °C. Conversely, other nanocrystalline alloys have been found to exhibit an inherent grain size stability, which has been explained on the basis of narrow grain size distributions, equiaxed grain morphology, low-energy grain boundary structures, relatively flat grain boundary configurations, and porosity [8]. In many cases, abnormal grain growth is observed, which may indicate the inhomogeneous distribution of grain growth inhibitors, such as pores or impurities. Although these observations are numerous and well documented, direct evidence linking them to the proposed mechanisms is scarce.

As a general observation, Malow and Koch [9] noted that those nanocrystalline materials that have exhibited significant thermal stability tend to share one common feature: They are multicomponent; that is, either they are alloys or they contain impurities. As an example, significant improvements in the thermal stability of nanocrystalline Al alloys have been achieved through the *in situ* formation of fine dispersoids during cryogenic high-energy milling [130]. In this work, Luton et al. [130] milled Al powder in an attritor under a liquid nitrogen slurry. Fine particles 2–10 nm in diameter spaced 50–100 nm apart formed as a result of this "cryomilling" process were found to impede grain growth at high temperature [130]. Using electron energy loss spectroscopy (EELS), these were shown to contain equal amounts of O and N, and were henceforth identified as aluminum oxynitride particles [131]. The dispersoids effectively maintained a stable 50–300-nm grain size in the cryomilled Al after heat treatment for 5 h at 783 K, or 84% of the melting point of Al. In addition, the liquid nitrogen cryomilling process has been shown to be effective in producing fine-grained, stable microstructures in Al-containing alloys, including NiAl [132] and Fe–10 wt% Al [133].

The theoretical foundation for particle pinning of grain boundaries has been well established for coarse-grained materials [134]. This phenomenon was first addressed by Zener [135] who considered the minimum radius of spherical grains, R, that could be effectively pinned by a homogeneously distributed volume fraction, f, of particles of radius r. The criterion for the limiting case is described by the equation

$$R = \frac{4r}{3f} \tag{27}$$

This was later expanded by Gladman [136] to take into account the grain size inhomogeneity, Z, defined as the size of the largest grain divided by the size of the average grain. The resultant equation relates the volume fraction, f, of particles of radius r, capable of pinning grains of average radius R_o, with size inhomogeneity Z:

$$r = \frac{6R_o f}{\pi}\left(\frac{3}{2} - \frac{2}{Z}\right)^{-1} \tag{28}$$

Moreover, the segregation of solute atoms to grain boundaries may also impede the grain growth in nanocrystalline materials. This phenomenon has been studied extensively in conventional materials [137–139]. This process is commonly associated with the heat treatment of supersaturated solid solutions, in which excess solute atoms are rejected from the grain interior. The resultant high concentration of insoluble atoms residing in the grain boundaries has been reported to impede the grain growth kinetics of coarse-grained [140, 141], as well as nanocrystalline [142, 143], materials.

Finally, it has also been observed that the rate of grain growth depends on the chemical short- and long-range order in an alloy. In $(Fe, Mn)_3Si$ [144], a reduction of the rate of growth is reported as the order parameter increases, indicating reduced mobility and/or driving force for migration of the grain boundaries or the interfaces between ordered and disordered grains.

3.4.3.2. In-Flight Thermal History of Particles During High-Velocity Oxy–Fuel Spraying

The thermal history of the feedstock powders during HVOF spraying is a necessary input for the prediction of the evolution of a nanocrystalline structure. The thermal history depends primarily on the particular HVOF experimental arrangement, as well as the experimental parameters employed. The development and application of novel diagnostic techniques has enabled the experimental measurement of the variation of velocity and temperature of sprayed particle powders as a function of time during HVOF spraying [10]. In recent work [145], it was reported that the velocity and temperature of the HVOF-sprayed particle powders can be controlled independently. These results indicate that accurate measurement and control of the thermal history of the HVOF-sprayed particle powders is possible. In addition to the previously mentioned experimental efforts, numerical simulations have also been extensively used to extract the temperature and velocity of the powders as a function of time [50].

3.4.3.3. Solidification Behavior during Thermal Spraying

The results reviewed in previous sections reveal that the conditions of the particles during impingement are influenced by many factors. Under some conditions, the upper layer of thermally sprayed materials may be in one of two possible states: liquid [146–151] and semiliquid [148, 150, 152–158]. Figure 4 illustrates these two typical upper-layer condi-

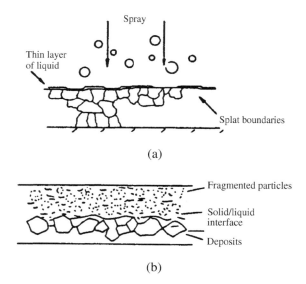

Fig. 4. Schematic diagrams illustrating the upper-layer conditions during spray deposition: (a) liquid and (b) semiliquid. (a) Reprinted with permission from A. R. E. Singer and R. W. Evans, *Metall. Technol.* 10, 61 (1983). © 1983 Institute of Materials. (b) Reprinted with permission from J. L. Estrada and J. Duszezyk, *Mater. Sci.* 25, 1381 (1990). © 1990 Kluwer Academic Publishers.

tions as suggested by the thermal spraying literature [147, 157]. In a related study, Liang and Lavernia [159] defined this upper layer as an interaction domain on the basis that its microstructure evolution depends on the interactions that occur among gas, droplets, and deposited material. Compared to the concept of a "semiliquid layer," the concept of an interaction domain is defined in such a way as to ensure its applicability to extreme spray conditions, such as the case of an extremely low liquid fraction in the impinging droplets with a concomitant absence of a semiliquid layer.

The microstructure of thermally sprayed materials is typically characterized by the absence of macrosegregation and the presence of droplet boundaries with interdispersed, micrometer-sized pores. Macrosegregation is typically absent as a result of the fact that thermally deposited materials are formed by the gradual deposition of individual droplets, and, consequently, the segregation distance of alloying elements is limited by droplet size. In some cases, the grain morphology of thermally sprayed materials is consistently reported in the literature to be equiaxed, regardless of alloy composition [146, 150, 152–155, 158–165]. Despite numerous investigations, a quantitative understanding of the relevant thermal, momentum, and solidification mechanisms that govern microstructure evolution during deposition has yet to be established. Nevertheless, a number of physical models have been proposed in these studies that offer useful insight into important phenomena, such as fluid flow, deformation, fracture of the solid phase, and cooling, on microstructural development. For example, there are various mechanisms that have been proposed to rationalize the formation of equiaxed grains during deposition. These may be generally grouped into two categories: nucleation limited mechanisms [146, 150, 160, 166, 167] and growth-limited mechanisms [153–155, 168].

Nucleation-Limited Mechanisms. The proposed nucleation-limited mechanisms for equiaxed grain formation are primarily qualitative. For example, in early spray-deposition studies, it was argued that the formation of equiaxed grains could be rationalized on the basis of dendritic arm fragmentation and grain multiplication during deposition [146, 150, 160, 166, 167]. Such processes occur in the liquid or semiliquid interaction domain at the upper surface of spray-deposited materials as shown in Figure 4. A schematic illustration of dendrite arm fragmentation is shown in Figure 5 [146]. In these early models, it was argued that extensive fragmentation of the solid phases that are present in partially solidified

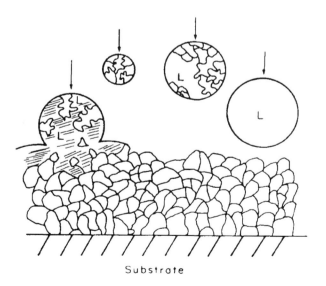

Substrate

Fig. 5. Schematic illustration of dendrite arm fragmentation during impingement. Reprinted with permission from E. J. Lavernia, *Int. J. Rapid Solid.* 5, 47 (1989). © 1989 A B Academic Publishers.

droplets may provide a high density of dendrite fragments in the interaction domain. Such dendrite fragments become potent nuclei for solidification as a result of these particles being chemically and crystallographically similar or identical to the nucleating solid phase. Subsequently, solid phases grow from such nucleation centers. The growth fronts from different nuclei will eventually converge, providing an equiaxed grain structure upon the completion of solidification [146, 150, 160, 166, 167]. Under ideal conditions, when dendrite fragment remelting is absent, each dendrite fragment will ultimately develop into a grain. Accordingly, the grain size in spray deposited materials should be less than the average spacing of the fragments. On the basis of this mechanism, the smallest grain size will be limited by the number of available nuclei in the semiliquid layer.

Regarding the phenomena of dendrite arm fragmentation during deposition, it has been suggested that the fracture of dendrite arms occurs as a result of (a) direct mechanical deformation and (b) the shear forces that are induced by the turbulent fluid convection in the interaction domain [146, 150, 151, 157, 159, 160, 166, 167]. Regarding droplet deformation during impingement, available experimental results show that the impingement velocities range from several tenths of a meter per second [163, 169, 170] to values in excess of 10^2 m/s [159, 171]. Accordingly, the energetic impingement of the droplets on the deposition surface will lead to the deformation and fracture of not only dendrites that are present in the atomized droplets, but also dendrites that are present in the interaction domain. Although direct experimental evidence providing support to the phenomenon of dendrite arm fragmentation is not currently available, some results reported in the literature suggest that there is extensive redistribution of solid phases during impingement.

Growth-Limited Mechanisms. Growth-limited mechanisms have been proposed in an effort to correlate quantitatively grain size and thermal history during rapid solidification [153, 155, 156, 172]. Such mechanisms recognize the effects of grain growth and coarsening during solidification and during post solidification cooling on the final grain size. Grain size may be considered to be the result of a competition between nucleation and grain growth kinetics. A general equation for the grain size during solidification in spray deposition can be written as [156]:

$$D = A\left(\frac{V_{sl}}{I}\right)^b \tag{29}$$

where D is the grain size, V_{sl} is the solid–liquid interfacial velocity, I is the nucleation rate, and A and b are constants. A general expression for interfacial velocity can be written as

$$V_{sl} = (G\dot{T})^{-1} \tag{30}$$

where G is the thermal gradient across the solid–liquid interface and $\dot{T}$ is the cooling rate. Under steady-state spray deposition conditions, the thermal gradient and the nucleation rate are constant. Combining Eqs. (29) and (30) yields expressions that relate grain size and cooling conditions:

$$D = K\dot{T}^{-b} \tag{31a}$$

and

$$D = K't_f^b \tag{31b}$$

where K and K' are constants, and t_f is the local solidification time. Equation (31) was used as an empirical relationship in various spray deposition studies to estimate grain size from the calculated cooling rate and local solidification time [153, 155, 156, 172].

Published results in the literature regarding the relationship between grain size and cooling rate, however, are typically qualitative. One approach that has been used in estimating grain size is to invoke the relationship between segregation distance, that is, dendrite arm spacing, and cooling rate. The rationale for this approach is based on the limitation of the smallest possible grain size by the segregation distance. An example of this approach is provided by the work of Annavarapu et al. [154] in which the cooling rate present in a spray-deposited steel strip was determined numerically on the basis of heat transfer considerations. The grain size was then estimated from the calculated cooling rate. In this study, the following relationship between secondary dendrite arm spacing and cooling rate, as originally determined by Suzuki et al. [173] for low-carbon steels, was used to predict grain size in spray-deposited steel:

$$\text{SDAS} = 709\dot{T}^{-0.387} \tag{32}$$

where SDAS is the secondary dendrite arm spacing. As shown in Table VII, the grain size predicted from Eq. (32) is much greater than the experimentally measured grain size in the spray-deposited steel strip. For example, as shown in Table VII, the grain size calculated by Annavarapu et al. [154] for a steady-state cooling rate of 95 K/s was 122 μm, whereas the measured grain size was 94 μm. Moreover, the predicted and measured grain sizes in the strip corresponding to a distance of 0.5 mm from the substrate surface were 39 and 11 μm, respectively. As is evident from Table VII, a similar trend of overestimating as-spray-deposited grain sizes by using established dendrite arm–cooling rate relationships has been reported by other investigators [155, 156, 174, 175].

The reasons for the aforementioned discrepancy between measured and predicted grain sizes on the basis of cooling rate conditions are not presently well understood. Available experimental results, however, suggest that this phenomenon may be partially attributed to an alteration of grain growth kinetics as a result of the elevated fraction of solid phases that is present in the interaction domain [176, 177]. In an effort to understand the mechanisms that govern grain size during spray deposition, Grant et al. [176] and Annavarapu and Doherty [177] studied grain-coarsening behavior in spray-deposited 2014 Al, 2618 Al, Al–4 wt% Cu, N707 (Al–10 Zn–2 Mg–1 Cu–0.2 Zr), and Cu–4 wt% Ti alloys by reheating these materials into the two-phase region of the corresponding phase diagram. Figure 6, for example, illustrates the microstructure of a spray-deposited Cu–4 wt% Ti alloy following heat treatment in the two-phase region for 15 min [177]. The selected heat treatment

Table VII. Comparison of Measured Grain Size and Predicted Grain Size from Dendritic Arm Spacing–Cooling Rate Relationships

Material (wt%)	Measured size (μm)	Cooling rate (K/s)	Relationship used	Calculated grain grain size (μm)	Reference
Low-carbon steel	11	1800	$\text{SDAS} = 709\dot{T}^{-0.387}$	39	[154]
	73	155	$\text{SDAS} = 709\dot{T}^{-0.387}$	101	[154]
	94	95	$\text{SDAS} = 709\dot{T}^{-0.387}$	122	[154]
Fe–20Mn	13	7.4	$\text{DAS} = 150\dot{T}^{-0.25}$	91	[167]
	40	2.9	$\text{DAS} = 150\dot{T}^{-0.25}$	115	[167]
	57	1.3	$\text{DAS} = 150\dot{T}^{-0.25}$	140	[167]
	79	0.5	$\text{DAS} = 150\dot{T}^{-0.25}$	175	[167]
7075 Al	6–25	1–10	$\text{DAS} = 50\dot{T}^{-0.3}$	25–50	[164]

DAS, dendrite arm spacing; SDAS, secondary dendrite arm spacing; $\dot{T}$, cooling rate.

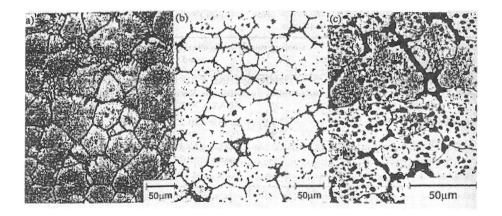

Fig. 6. Microstructure of spray-deposited Cu–4Ti (wt%) quenched following isothermal holding for 15 min in the two-phase regions corresponding to various solid fractions, f_s: (a) 1015 °C ($f_s = 0.62$), (b) 1000 °C ($f_s = 0.73$), and (c) 975 °C ($f_s = 0.87$). Reprinted from S. Annavarapu and R. D. Doherty, *Acta Metall. Mater.* 43, 3207 (1994), with permission of Elsevier Science.

temperatures were 1015, 1000, and 975 °C corresponding to solid fractions of 0.62, 0.73, and 0.86, respectively. As shown in Figure 6, grain morphology remains equiaxed following heat treatment in the semiliquid state. However, there is clear evidence that the extent of grain boundary melting increases with increasing heating temperature. Regarding grain growth, the results of Grant et al. [176] indicated that the coarsening rate is much greater in the semiliquid state than in the solid state. Figure 7 illustrates the grain size as a function of heating time for spray-deposited 2618 Al and N707. For example, the results for 2618 Al indicate that heat treatment in the solid state at 530 °C for 360 min only affects grain size marginally. During heat treatment in the semiliquid state (in the temperature range of 592 to 610 °C), however, the grain size increases progressively with increasing annealing time and temperature. Grant et al. [176] argued that grain coarsening may follow parabolic growth kinetics if it is governed by the reduction in grain boundary area:

$$D^2 - D_o^2 = k_2 t \tag{33}$$

where D_o is the original grain size, D is the grain size, k_2 is a coarsening rate constant, and t is time. When grain boundary migration is diffusion controlled, the coarsening behavior may follow cubic-law kinetics:

$$D^3 - D_o^3 = k_3 t \tag{34}$$

where k_3 is the coarsening rate constant for a cube law. Regarding Eqs. (33) and (34), grain size regressions conducted by Grant et al. [176] on 2618 Al, Al–4 wt% Cu, and N707 reveal that cubic-law kinetics offer a slightly better fit with experimental data than a parabolic law.

Annavarapu and Doherty [177] noted that in spray-deposited materials the grain growth rate in the solid state is considerably retarded relative to that in the semiliquid state. In the semiliquid state, the grain growth rate was observed to increase with a decreasing fraction of solid phases [177]. Moreover, they observed that the solid fraction affects the grain growth rate primarily by influencing the growth rate constant. In this study, the grain size was noted to increase heating time according to the cubic-law relationship given by Eq. (34). Annavarapu and Doherty [177] reported that in the semiliquid state the growth rate constant decreases rapidly with an increasing solid fraction. For example, for spray-deposited Inconel625, when the solid fraction increases from 0.75 to 0.85, the growth rate constant decreases from 200 μm^3 s^{-1} to 120 μm^3 s^{-1} (Fig. 8).

Recently, it has been reported that nanocrystalline coatings may be obtained by HVOF thermal spraying of various feedstock powders [6, 38–44, 178]. In this case, the nucleation

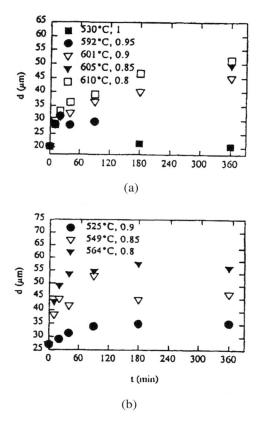

(a)

(b)

Fig. 7. Grain size in spray-deposited materials as a function of isothermal heat treatment time in the two-phase regions: (a) 2618 Al and (b) N707. Reprinted with permission from P. S. Grant et al., in "Spray Forming 2, Proceedings of the Second International Conference on Spray Forming," p. 45. Woodhead, 1993. © 1993 American Educational Systems.

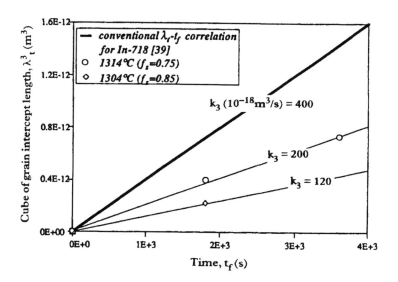

Fig. 8. Dependence of grain size on isothermal heat treatment time in the two-phase regions for a spray-deposited Inconel 625 superalloy where k_3 is the grain-coarsening rate constant for the cubic law. Also shown is the relationship between the segregation spacing, λ, and the solidification time, t_f (thick line). Reprinted from S. Annavarapu and R. D. Doherty, *Acta Metall. Mater.* 43, 3207 (1994), with permission of Elsevier Science.

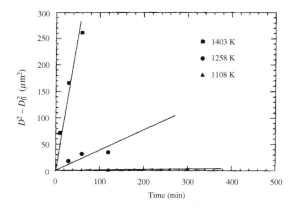

Fig. 9. Grain size in spray-deposited Ni_3Al as a function of solid-state annealing time where D is the grain size and D_0 is the initial grain size. Reprinted from X. Liang et al., *Acta Metall. Mater.* 40, 3003 (1992), with permission of Elsevier Science.

rate must be extremely high, while the grain growth rate is maintained extremely low during the deposition of any liquid precursors onto the substrates. In addition, the thermal stability of the spray-deposited materials must be high enough to limit grain growth in the sprayed nanocrystalline materials. Research in this area is continuing.

Solid-State Coarsening. Grain coarsening during solid-state cooling has also been proposed as a mechanism that influences grain size in thermally sprayed materials [168]. This, in some instances, represents the conditions typically present during HVOF thermal spraying. It has been argued that the relatively slow cooling rate that is present during deposition $(1–10^2$ K/s$)$ will allow significant grain growth and coalescence. This phenomenon may be particularly significant during the spray deposition of high-temperature materials, such as intermetallics. To verify the proposed mechanism, Liang et al. [168] conducted solid-state annealing experiments on a spray-deposited Ni_3Al. The experiments revealed that the grain morphology changed from a deformed dendritic microstructure to a spheroidal grain microstructure during annealing, and that the grain size increased with increasing annealing temperature and time; see Figure 9. The relationship between grain size, D, isothermal annealing temperature, T, and time, t, was noted to obey parabolic coarsening kinetics as

$$D^2 - D_o^2 = 1.44 \times 10^{11}\left\{\exp\left(-\frac{338,000}{T}\right)\right\}t \tag{35}$$

where D_0 is the initial grain size or dendrite size. Liang et al. [168] further argued that a spray-deposited material is effectively exposed to an elevated temperature annealing as a result of the slow cooling rate that is present in the deposition stage. Accordingly, grain growth during this thermal exposure will influence the final grain size and morphology.

3.4.3.4. *Prediction of Grain Growth Behavior of Nanocrystalline Feedstock Powders during High-Velocity Oxy–Fuel Spraying*

Over the past two decades, thermal spray technology has matured and it is now possible to manufacture high-quality coatings in a reproducible manner for various applications of industrial relevance, such as wear and thermal protection [10]. In the case of synthesis of nanocrystalline coatings, the grain growth behavior during HVOF thermal spraying must be controlled to preserve the nanocrystalline structure in the resultant coating. In this chapter, a combination of isothermal and nonisothermal grain growth studies will be used to predict the grain growth behavior of nanocrystalline powders.

In the case of isothermal annealing, the variation of grain size as a function of time can be described by [179]:

$$D^{1/n} - D_0^{1/n} = kt \qquad \text{or} \qquad D = \left(kt + D_0^{1/n}\right)^n \tag{36}$$

where D is the mean grain size after an annealing time of t, D_0 is the mean grain size at $t = 0$, and k is the temperature (T)-dependent rate constant, which can be expressed as

$$k = k_0 \exp\left(-\frac{Q}{RT}\right) \tag{37}$$

where Q is the activation energy for isothermal grain growth and R is the molar gas constant. The activation energy Q is often used to infer the microscopic mechanism that dominates the grain growth process.

A theoretical framework that may be used to describe grain growth during nonisothermal annealing has been derived by Bourell and Kaysser [80]. From this work, it follows that derivation of the isothermal grain growth equation Eq. (36) with respect to time t yields

$$nD^{n-1} dD = A \exp\left(-\frac{Q}{RT}\right) dt \tag{38}$$

For nonisothermal conditions, a linear change of temperature with time can be expressed as

$$dt = \frac{1}{v_0} dT \tag{39}$$

where v_0 is a constant heating (positive) or cooling (negative) rate. Inserting Eq. (39) into Eq. (38) and integrating, then

$$D^n - D_0^n = \frac{A}{v_0} \int_{T_i}^{T_f} \exp\left(-\frac{Q}{RT}\right) dt \tag{40}$$

where T_i and T_f are the initial and final temperatures (in absolute K), respectively. Following further mathematical simplification, a final equation describing the thermal dependence of grain size during linear heating or cooling can be expressed as

$$D^n - D_0^n = \frac{AR}{v_0 Q}\left(T_f^2 \exp\left(-\frac{Q}{RT_f}\right) - T_i^2 \exp\left(-\frac{Q}{RT_i}\right)\right) \tag{41}$$

In this equation, the effects of heating (positive v_0) and cooling (negative v_0) on coarsening are mathematically equivalent. Moreover, assessment of the error associated with Eq. (41) shows that if the T_i and T_f are not very close, the error can be neglected [180]. Because n, Q, T_i, and T_f can be extracted from isothermal annealing experiments, as described in the last section, the dependence of grain size on nonisothermal annealing can be predicted on the basis of Eq. (41).

3.4.4. Modeling of High-Velocity Oxy–Fuel Synthesis of Nanocrystalline Coatings

In reference to the development of nanocrystalline coatings, mathematical modeling may be used for the following:

- To provide guidance in the establishment of HVOF experimental parameters that will be required to generate optimal combinations of microstructure and physical behavior
- To verify the recommended designs for a HVOF facility

In related studies, Eidelman and Yang [181] used a three-dimensional computational fluid dynamics (CFD) model, consisting of a conservative equation and constitutive relations for both gas and particle phases, to simulate the HVOF spraying of nanocrystalline coatings. As a result, process optimization, new process and coating tool design, and control design for the thermal spraying system were evaluated. In the case of thermal spraying of WC/Co, the results of this study show that the deposition of a high-density coating requires the preparation of 10–20-μm WC/Co agglomerates, which should be stable and not disintegrate in supersonic flow. In addition, the results show that low-temperature exposure tends to prevent grain growth, particle disintegration resulting from Co melting, and substrate overheat [181].

Lau et al. [40] selected a different route to simulate the HVOF spraying process for producing nanocrystalline coatings from cryomilled nanocrystalline feedstock powders, in which a number of issues relative to the specific characteristics of the cryomilled nanocrystalline feedstock powders (not spherical), such as powder aspect ratio, were addressed. During the HVOF process, the powder particles that are propelled into the flame undergo acceleration and a significant amount of heating before contacting the substrate. Therefore, the microstructural evolution of the sprayed coating and the resulting properties of the coating are influenced by the momentum and thermal transport between the flame gas and the powder particles during flight [10]. Ramm et al. [182] investigated the relationship between spraying conditions and the resulting microstructure for Al_2O_3 coatings, and reported that porosity is closely related to the impact velocities of the impinging particles on the substrate. From this work, it was argued that a low impact velocity causes the formation of coarse pores (3–10 μm) owing to the incomplete melting of the particles. This leads to incomplete filling of the interstices between prior layers of deposited particles. On the other hand, fine pores (smaller than 0.1 μm), which are present at the higher temperature of the injected particles, are caused by incomplete contact between lamellae [182]. Therefore, the study of the momentum and thermal history of the impinging particles is required to optimize the spraying parameters.

Although the HVOF spraying process is considered to be inherently simpler than the plasma spraying process, the spraying parameters are still very complex because of the characteristic processes associated with the HVOF spraying gun. These processes are related to the thermodynamic laws of compressible fluid flow, heat transfer principles, and coating formation [183]. In developing a mathematical model for the velocity and temperature profile of the impinging particles, the following parameters are considered:

- Composition and stoichiometry of the flame gas
- Temperatures of the flame gas and the particles
- Velocities of the flame gas and particles
- Flight distance

The following assumptions are commonly made in HVOF thermal spray modeling [183]:

- The flame gas obeys the ideal behavior of a perfect gas (changes of thermophysical fluid properties, mixing, etc.)
- Area changes are considered to be adiabatic and frictional effects are neglected; that is, the laws of isentropic flow of compressible fluids may be applied
- The combustion of the fuel gas takes place instantaneously within the combustion chamber
- The rise of the gas temperature owing to combustion follows the laws of heat transfer to a perfect gas flowing within a constant area duct without friction (Rayleigh)
- The spray powder particles are spheres
- The chemical reactions of the exhaust gases with the surrounding media and the particles with the gases are not considered

- Vaporization and evaporation of the particles
- The principles of heat transfer apply to heat exchange between the hot exhaust gases and the spray powder particles, the hot exhaust gases and the nozzle wall, the impinging hot gas stream onto substrate and the coating, and the impinging molten/heated particles and the solidified particles or substrate.

The velocity profile of the impinging particles can be expressed by Newton's second law [184, 185]:

$$V_p \rho_p \frac{dV_p}{dt} = A_p C_D \rho_g |v_g - v_p|(v_g - v_p) \tag{42}$$

where V_p is the volume of the spherical particles (m^3)

ρ_p, ρ_g is the density of the particle and flame gas, respectively (kg/m^3)

v_p, v_g is the particle velocity and flame gas velocity, respectively (m/s)

t is the time (s)

A_p is the particle area (m^2)

C_D is the drag coefficient

The velocity profile of the particles along the flame gas can be plotted as a function of the time or flight distance. Because of the small size of the particles (5–45 μm in diameter) used in the HVOF spraying, the drag coefficient must account for the viscous flow owing to the Basset history term, variable property effects (property variation within the boundary layer of the particle/gas interface), and noncontinuum effects [184, 185] influencing the dynamic interaction between the injecting particles and the flame gas. To account for these factors, the modified C_D is expressed as follows [184]:

$$C_D = \left[\frac{24}{Re} + \frac{6}{1 + \sqrt{Re}} + 4 \right] \left(\frac{\rho_g \eta_g}{\rho_w \eta_w} \right)^{-0.45} \left[1 + \frac{c_p(T_g - T_w)}{\Delta H_{vap}} \right] \tag{43}$$

where R_e is the Reynolds number of the particle ($Re < 100$)

ρ_w is the density of the particle/gas interface (kg/m^3)

η_g, η_w is the dynamic viscosity of the flame gas and the particle/gas interface (kg/m·s)

c_p is the specific heat of the particle (kJ/kg·K)

T_g, T_w is the temperature of the flame gas and particle/gas interface (K)

ΔH_{vap} is the heat of vaporization of the particle (kJ/kg)

The heat transfer between the flame gas (including the effects of convection and radiation) and the particles strongly influences the quality of the resulting coatings. The energy balance, following Newtonian conditions, is expressed in the following equation [184]:

$$\rho_p V_p c_p \frac{dT_p}{dt} = h(T_g - T_p) A_p + \varepsilon \sigma \left(T_g^4 - T_p^4 \right) A_p \tag{44}$$

where c_p is the heat capacity of the particle (kJ/kg·K)

T_p is the temperature of the particle (K)

h is the modified heat transfer coefficient (kJ/s·m^2·K)

ε is the emissivity of the surface

σ is the Stefan–Boltzmann constant

The convective heat transfer coefficient, adapted from the Ranz–Marstall correlation [186–188], which includes the effects of viscous flow of the particles and the vaporization, is expressed as

$$h = \frac{\lambda_g}{d}\left[2 + 0.6\sqrt{Re^3}\,\sqrt{Pr}\left(\frac{\rho_g\eta_g}{\rho_w\eta_w}\right)^{0.6}\left(\frac{c_g}{c_w}\right)^{0.38}\left(1 + \frac{c_p(T_g - T_w)}{\Delta H_{vap}}\right)\right] \quad (45)$$

where λ_g is the thermal conductivity of the gas (W/m·K)

Pr is the Prandtl number of the plasma gas

c_g is the specific heat of the flame gas

3.4.5. Experimental Analysis and Results

In view of the fact that published experimental and numerical studies on the synthesis of nanocrystalline coatings through HVOF spraying are very scarce (this field is at a nascent stage), emphasis will be placed on research work carried out at UCI in this section. At UCI, nanocrystalline Ni, Inconel 718, and 316 stainless-steel powders were prepared by mechanical milling (both cryomilling and methanol milling), followed by thermal spraying using the HVOF process. Subsequent structural characterization and property examination of the nanocrystalline powders and coatings were also conducted. Although preliminary results have been encouraging, there exist a series of technical and scientific challenges that must be addressed in order for this technology to achieve its full potential. These issues are discussed in some detail in the following sections.

3.4.5.1. Preparation of Nanocrystalline Feedstock Materials

Nanocrystalline materials are characterized by a microstructural length scale in the 1–100-nm regime [7]. More than 50 vol% of the atoms are associated with grain boundaries or interfacial boundaries when the grain size is small enough (i.e., 5 nm) [189]. Thus, a significant amount of interfacial component between neighboring atoms contributes to the physical properties of nanocrystalline materials [189]. The available techniques capable of producing nanocrystalline materials include gas condensation, mechanical alloying/milling, crystallization of amorphous alloys, chemical precipitation, spray conversion processing, vapor deposition, sputtering, electrodeposition, and the sol–gel processing technique [8]. Mechanical alloying/milling techniques have been used to produce large quantities of nanocrystalline materials for possible commercial use [8].

Mechanical alloying or ball milling is a process, in which powder particles are welded, fractured, and rewelded in a dry (or wet) environmental high-energy ball chamber. During the process, elemental mixtures or prealloyed powders are ground under a protective atmosphere in equipment capable of high-energy compressive impact forces such as attrition mills, vibrating ball mills, and shaker mills. The typical ball mill experiment utilizes a ball-to-powder mass ratio ranging from 3:1 to 50:1 [190]. Sometimes a lubricant fluid is used [191]. The milling time may range from minutes to several hundred hours and, in some cases, the grain size is found to reach a steady-state minimum value [192]. Periodically, a distinction is made between "low energy" and "high energy" ball milling processes. This is based on the vial volume and the number and diameter of balls utilized [193].

The HVOF spraying process has been documented as an effective means in fabricating nanocrystalline coatings having a high potential of industrial applications [6, 37]. In many instances, the synthesis of nanocrystalline coatings through HVOF spraying necessitates the effective preparation of the feedstock materials with a nanocrystalline structure. Recently, investigations have been successfully carried out in producing nanocrystalline coatings through HVOF thermal spraying using nanocrystalline particles and clusters of particles [6, 37, 39]. Preliminary results are encouraging, and presently several research groups, funded by the Office of Naval Research (Arlington, VA), and supported by various industries, are actively pursuing this technology. The anticipated benefits of nanocrystalline coatings include:

- Extended part performance
- Improved energy conservation
- Increased useful life of parts
- Reduced manufacturing costs
- Improved bonding structure in the coating/substrate interface

A number of techniques are currently available in producing nanocrystalline feedstock powders for HVOF spraying synthesis of nanocrystalline coatings [6, 37, 39]. In the case of preparing ultrafine nanopowders, the aqueous solution reaction, spray conversion processing, and chemical vapor deposition methods [194] are very often reported. Because of the smaller size of the nanopowders, additional agglomeration processes are required to increase the size of the agglomerates to that suitable for HVOF spraying. In the case of nanograined powders, Lavernia et al. [37, 39] have demonstrated the capability of producing metal nanograined powders using methanol or cryogenic attritor milling. Cryogenic milling is a mechanical milling process, as shown in Figure 10, in which the metal powders are immersed in an inert atmosphere of liquid nitrogen or liquid argon. The purpose behind the cryogenic method is: (i) to minimize oxygen contamination, (ii) to decrease the amount of heat generated during the milling process, and (iii) to increase of the effectiveness of plastic deformation, which induces nanometer grain size formation. In addition, recent data points to the formation of nanodispersed phases, for example, oxynitride phases in Al-based alloys, which enhances the thermal stability of the nanocrystalline materials [133, 195]. Moreover, the powders thus synthesized can be directly used for HVOF spraying; no further processing is required to adjust the particle size. For example, at UCI, the cryomilled Ni, Inconel 718, and 316 stainless-steel agglomerated powders have been successfully used to prepare nanocrystalline coatings by means of HVOF thermal spraying [43, 44].

3.4.5.2. Feedstock Powder Processing at the University of California, Irvine

In recent studies [43, 44], inert-gas-atomized Ni, Inconel 718, and 316 stainless-steel powders Sulzer Metco (US) Inc. with a nominal size of 45 ± 11 μm were selected. Mechanical milling in a methanol environment was conducted in a modified Union Process 01-ST attritor mill with a grinding tank capacity of 0.0057/m^3. Stainless-steel balls (0.635 cm

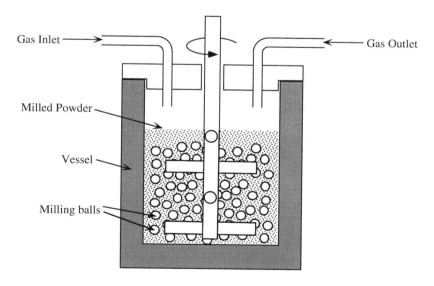

Fig. 10.   Attritor-type ball mill.

in diameter) served as the grinding media with a powder-to-ball mass ratio of 1:20. The drive shaft operated at a speed of 180 rpm, and, to minimize powder accumulation, a shaft clearance of 0.635 cm from the bottom of the grinding tank was maintained. Cryogenic milling of the atomized powders was performed with the same parameters as those used in the methanol milling experiments. To ensure complete immersion of the powders, liquid nitrogen was continually introduced into the mill. The temperature was maintained in the range of 83–98 K throughout the experiment by using a globe valve to control the amount of liquid nitrogen introduced into the mill manually.

The powders were then removed from the grinding tank of the attritor frame and placed in a glove box filled with argon. The mixture containing the balls and powders was removed from the grinding tank after all of the liquid nitrogen evaporated; sieving proceeded to separate the powders from the milling balls. The particle size distributions of the milled powders were determined by the Microtrac Standard Range Particle Analyzer.

3.4.5.3. Thermal Spray of Nanocrystalline Powders

The mechanically milled powders were thermally sprayed onto a 1020 stainless-steel substrate using a Sulzer Metco DJ 2600 HVOF spray system. Hydrogen gas (H_2) was used to generate a hypersonic gas velocity of approximately 2000 m/s and a pressure of 0.28 MPa. The combustion of the fuel gas with oxygen (O_2) produced a nominal flame temperature of 2755 K. The powders were injected axially through a fluidized bed into the jet gas, in which the powders are heated and propelled toward the substrate to produce a coating. Each material was sprayed with two different carrier gases. Table VIII lists the parameters that characterize the spraying conditions.

3.4.5.4. Methods of Analysis

X-Ray Diffraction Analysis. The milled powders and the thermally sprayed coatings were analyzed in a Siemens D5000 diffractometer using Mo $K\alpha$ radiation and a monochrometer to determine the phases present. The average grain size of the milled powders was determined by the Klug–Alexander method [120].

Scanning Electron Microscopy Analysis. Scanning electron microscopy (SEM) analysis was performed on the powder samples obtained directly from the milling processes, as well as those from the thermally sprayed coatings. The cross-sectional coating samples were mounted in a conductive mold and mechanically ground. The agglomerate size of the milled powders was determined by SEM analysis performed in a Philip XL30 FEG scanning electron microscope equipped with EDAX analysis.

Transmission Electron Microscopy Analysis. The milled powder particles to be analyzed by transmission electron microscopy (TEM) were dispersed in methanol, deposited on carbon film substrates, and allowed to dry in air. The coatings were sectioned into 3 mm × 3 mm TEM samples, mechanically ground, and jet-polished. Table IX lists the jet-polishing solutions used to prepare various coating materials.

Table VIII.   Spraying Parameters for the Ni, Inconel 718, and 316 Stainless-Steel Powders

	Pressure (MPa)				Gas flow (m^3/h)			
	O_2	H_2	Air	N_2	O_2	H_2	Air	N_2
Standard	1.17	0.96	0.69	0	13.8	41.0	22.3	0
Low oxide	1.17	0.96	0	0.76	13	46.4	0	31.1

Table IX. Jet-Polishing Solutions for the Preparations of TEM Samples

Coating material	Jet-polishing solution
Ni	20% perchloric acid
	80% ethanol
Inconel 718	10% perchloric acid
	20% ethanol
	70% butanol
316 stainless steel	20% sulfuric acid
	80% methanol

TEM analysis was performed in a Philips 200 transmission electron microscope operated at 200 keV. The grain sizes of the milled powders and the coatings were determined from measurements obtained from dark-field images.

Microhardness. Microhardness measurements were performed on the coatings using a Buehler microhardness tester with a diamond indenter and a 300-g load. The coatings were mechanically polished to provide a smooth surface prior to the measurements. At least 10 measurements were taken and averaged for each sample.

3.4.6. Results

3.4.6.1. Powder Characterization

The milling of powders conducted in methanol or liquid nitrogen produced flake-shaped agglomerates. The aspect ratio, the ratio of length to width of the agglomerates, increases with increasing milling time as a result of continuous welding and fracturing during mechanical milling [43]. Parts a and b of Figure 11 show the morphological changes of 316 stainless-steel powders milling for 5 and 10 h in methanol. Table X lists the agglomerate size, average grain size determined by TEM dark-field imaging, and aspect ratio of Ni, Inconel 718, and 316 stainless-steel powders under different milling conditions. Results obtained from TEM dark-field imaging indicate that the average grain size decreases with increasing milling time for Ni, Inconel 718, and 316 stainless-steel powders [43].

3.4.6.2. Thermally Sprayed Nanocrystalline Coatings

X-ray diffraction analysis was performed on the as-deposited Ni, Inconel 718, and 316 stainless-steel coatings. Figure 12 compares the X-ray diffraction spectra of the as-received 316 stainless-steel powders, the methanol-milled powders for 5 and 10 h, and the thermally sprayed coatings. The X-ray diffraction spectra indicated the presence of characteristic diffraction peaks of Fe–Cr in all five samples. After 5 and 10 h of mechanical milling, the diffraction peaks were broadened. Furthermore, characteristic peaks of Fe_3O_4 were observed in the thermally sprayed coating when air was used as a carrier gas, indicating the possibility of various chemical reactions taking place during thermal spraying [43].

Results from the backscattered electron images of SEM analysis performed on the as-sprayed nanocrystalline coatings indicated a higher porosity than those of conventional coatings with identical spraying parameters [43]. Parts a–d of Figure 13 compare the coating structures of conventional 316 stainless-steel coatings and nanocrystalline

(a)

(b)

Fig. 11.   Scanning electron microscopy (SEM) images of (a) methanol-milled 316 stainless-steel powders for 5 h, and (b) methanol-milled 316 stainless-steel powders for 10 h.

Table X.   Characteristic Powder Properties of Mechanical-Milled Ni, Inconel 718, and 316 Stainless-Steel under Different Milling Conditions

Material	Milling condition (h/medium)	Agglomerate size, D_{50} (μm)	Grain size (nm)	Aspect ratio
Ni	5/methanol	80	90	1.40
Ni	10/methanol	35	82	1.42
Ni	5/liquid N_2	25	26	1.55
Ni	10/liquid N_2	43	28	1.50
Inconel 718	10/liquid N_2	65	16	1.34
316 stainless-steel	5/methanol	67	67	1.43
316 stainless-steel	10/methanol	54	24	1.68
316 stainless-steel	5/liquid N_2	50	38	1.38
316 stainless-steel	10/liquid N_2	31	21	1.68

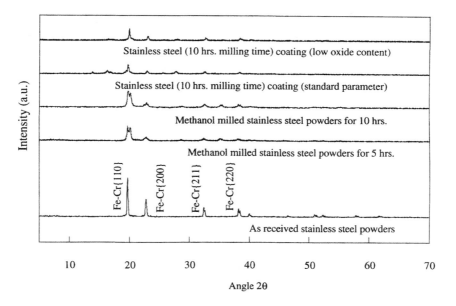

Fig. 12. X-ray diffraction patterns comparing the as-received 316 stainless-steel powders, methanol-milled powders for 5 and 10 h, and thermally sprayed coatings using different spraying parameters.

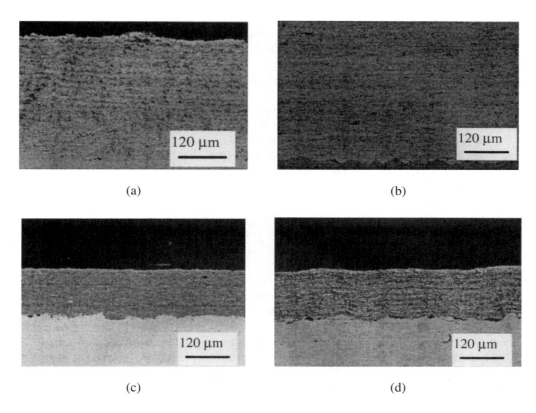

(a)

(b)

(c)

(d)

Fig. 13. Scanning electron microscopy (SEM) cross-sectional images of (a) conventional 316 stainless-steel coating, standard spray; (b) conventional 316 stainless-steel coating, low-oxide content spray; (c) methanol-milled 316 stainless-steel (10 h) coating, standard spray; and (d) methanol-milled 316 stainless-steel (10 h) coating, low-oxide content spray.

Table XI. Coating Characteristics and Properties of Ni, Inconel 718, and 316 Stainless Steel

Material	Milling (h/medium)	Spraying parameter	Porosity (%)	Microhardness 300-g load (DPH)
C-Ni	Not applicable	Standard	1.09	447
C-Ni	Not applicable	Low oxide content	<0.5	390
N-Ni	10/methanol	Standard	1.42	535
N-Ni	10/methanol	Low oxide content	1.42	484
C-Inconel 718	Not applicable	Standard	<0.5	440
N-Inconel 718	30/liquid N_2	Standard	1.41	712
C-stainless steel	Not applicable	Standard	<0.5	451
C-stainless steel	Not applicable	Low oxide content	<0.5	411
N-stainless steel	10/methanol	Standard	1.5	613
N-stainless steel	10/methanol	Low oxide content	2.4	503
N-stainless steel	10/liquid N_2	Standard	2.1	460
N-stainless steel	10/liquid N_2	Low oxide content	1.9	486

C, conventional; N, nanocrystalline.

316 stainless-steel coatings using different spraying parameters. Table XI lists the coating characteristics of Ni, Inconel 718, and stainless-steel 316, which include porosity and microhardness.

Interestingly, the nanocrystalline coatings exhibited a larger percentage of porosity than that of the conventional coatings [43]. The pancake shape of the cryomilled powders defined by the aspect ratio is thought to play a role in the mechanisms responsible for porosity; work in this area is continuing. Despite the high porosity, the microhardness values corresponding to the nanocrystalline coatings were higher than those of the conventional coatings for all the Ni, stainless-steel, and Inconel 718 alloys studied. The Inconel 718 alloy exhibited a maximum hardness for the nanocrystalline coating that was 60% higher than the conventional polycrystalline coating. Consequently, it seems reasonable to assume that if the porosity of the nanocrystalline coating can be reduced to the level of the conventional coating, the microhardness would be much higher [43]. It is noteworthy that it is often reported that a higher microhardness corresponds to a superior wear resistance, although this trend remains to be confirmed in nanocrystalline materials.

TEM analysis was performed on the as-sprayed Ni, Inconel 718, and 316 stainless-steel coatings [43, 44]. Results from the TEM analysis of methanol-milled Ni powders for 10 h demonstrated the presence of nanocrystalline grains with an average grain size of 15 nm. Furthermore, nanocrystalline grains were also observed in the TEM analysis of the methanol-milled Ni (10 h) coating, shown in Figure 14a, as well as the cross-sectional coating, shown in Figure 14b, c. Areas exhibiting elongated grains with an aspect ratio of 2–3 and a grain size range of 100–200 nm were also observed. Elongated grains, resulting from the process of mechanical milling, suggest that fractions of the nanocrystalline agglomerates did not melt during HVOF spraying [43].

3.4.6.3. Thermal Stability

Figure 15 shows the XRD patterns of the as-received, attritor-milled and HVOF-sprayed Inconel 718 alloys [37, 41]. Evidently, the XRD peaks of the attritor-milled and the

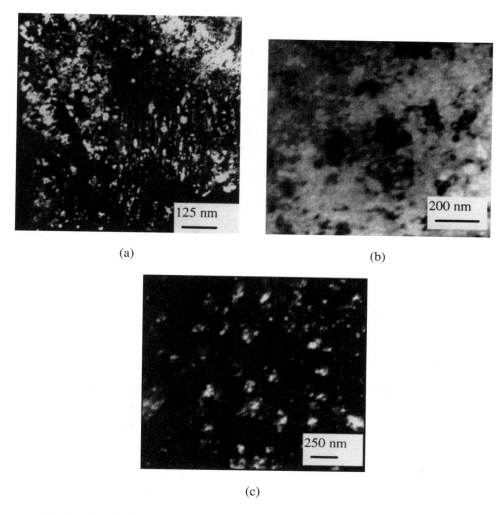

Fig. 14. Transmission electron microscopy (TEM) images of (a) dark-field image of methanol-milled Ni (10 h) coating, (b) bright-field image of cross-sectional methanol-milled Ni (10 h) coating, and (c) dark-field image of cross-sectional methanol-milled Ni (10 h) coating.

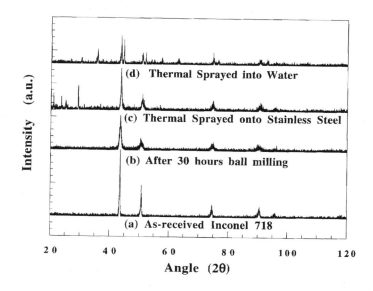

Fig. 15. X-ray diffraction patterns of the as-received, attritor-milled, and HVOF-sprayed Inconel 718 alloys.

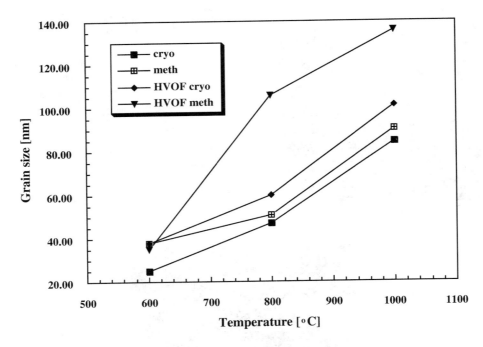

Fig. 16. Relationship between grain size and annealing temperature (annealing time, 60 min) for various Inconel 718 alloys.

HVOF-sprayed Inconel alloys are broadened in comparison to those of the as-received powders as a result of the fine-grained structures. The grain growth behavior of the nanocrystalline Inconel 718 powders and coatings, under isothermal annealing, is shown in Figure 16. It is noted that all these nanocrystalline powders and coatings exhibited thermal stability against grain growth up to 800 °C. Following annealing at 1000 °C for 1 h, the average grain sizes of the methanol-milled powders, the cryomilled powders, the HVOF coatings of the methanol-milled powders, and the HVOF coatings of the cryomilled powders were 91, 84, 137, and 102 nm, respectively [41].

The mechanisms that are responsible for the enhanced thermal stability against grain growth in nanocrystalline materials are typically described as [196]: (i) solute drag, (ii) Zener pinning of secondary phases, and (iii) chemical ordering. In the present case of various nanocrystalline Inconel 718 alloy powders and coatings, as there are no available reports claiming the observation of solute segregation or chemical ordering, it is speculated that the Zener-pinned nanometer-scale oxides and/or carbides revealed by the XRD analysis [37] may play a dominant role in impeding the grain growth during thermal annealing. The coarsening of the nanometer-scale oxides and/or carbides at higher temperatures causes subsequent grain growth of the matrix materials, thereby eliminating the original nanocrystalline characteristics. This mechanism of Zener pinning in impeding grain growth has also been found in numerous other nanocrystalline materials [42, 130, 133, 197–201].

To provide insight into the grain growth behavior of nanocrystalline materials during HVOF thermal spraying, the cryomilled nanocrystalline Ni was investigated under isothermal and nonisothermal conditions, and relevant theories [Eqs. (36) and (41)] were used to rationalize the experimental results.

The variations of the grain size of nanocrystalline Ni as a function of annealing time at different temperatures are shown in Figure 17. It is apparent that significant grain growth occurred in the case of cryomilled nanocrystalline Ni powders even when annealing at lower temperatures (equivalent to about 0.17 of the melting temperature), indicating the poor thermal stability of these powders. Mathematical fitting of the data plotted in

Figure 17 using Eq. (36) yielded a variety of k and n values corresponding to different annealing temperatures. By plotting $\ln(k)$ vs $1/T$, a linear relationship between $\ln(k)$ and $1/T$ can be established (Fig. 18). From the slope $(-Q/R)$ of the plot, the activation energy for grain growth, Q, was extracted to be 146.2 kJ/mol.

Table XII lists the collected grain growth data for various nanocrystalline Ni. The activation energy values for both the as-plated nanocrystalline Ni coatings (131.5 kJ/mol [202]) and the cryomilled nanocrystalline powders (146.2 kJ/mol) are close to the grain boundary diffusion value for Ni (188 kJ/mol) [196]. Therefore, it is thought that grain growth in the present study was primarily controlled by grain boundary diffusion. In addition, the time exponent, n, values are very close to 4.0 (see Fig. 19), which also implies that a

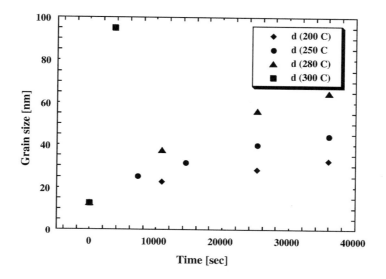

Fig. 17. Grain size of nanocrystalline Ni as a function of annealing time at different temperatures.

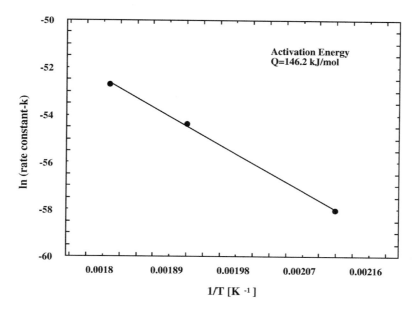

Fig. 18. Extraction of activation energy of grain growth.

Table XII. Comparison of Various Nanocrystalline Ni (*n*-Ni) Prepared via Different Methods

	As-plated *n*-Ni	Thin-film *n*-Ni	Cryomilled *n*-Ni
Q (kJ/mol)	131.5	220	146.2
DSC heat signal	Yes	Yes	No
Time exponent, *n*	—	—	3.6–3.9
Reference	[202]	[203]	Present work

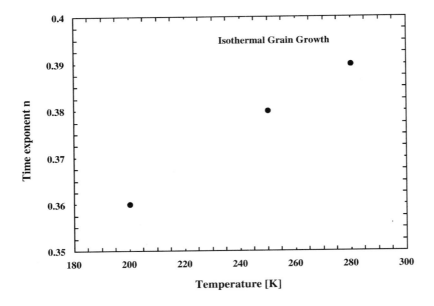

Fig. 19. Variation of time exponent vs temperature.

grain boundary diffusion mechanism controls the grain growth [204]. Exothermic peaks corresponding to grain growth were found in the differential scanning calorimetry (DSC) diagrams for both the thin-film and the as-plated nanocrystalline Ni; thus, extraction of the grain growth activation energy could be achieved through the Kissinger method [205]. However, in the case of the cryomilled Ni, there was no exothermic peak that would be attributed to grain growth, and, hence, an isothermal annealing approach was employed to study the kinetics of grain growth.

TEM images of the cryomilled nanocrystalline Ni, following thermal annealing at various temperatures, are shown in Figure 20. The observed uniform distribution of grain size for the powders annealed at lower temperatures such as 250 °C suggests that during low-temperature annealing grain growth occurs primarily through normal grain growth. When the annealing temperature approaches higher temperatures such as 300 °C, abnormal grain growth occurs. As a result, a small proportion of large grains, as shown in Figure 20, is found to coexist with the fine-grained matrix. Thermal annealing at even higher temperatures above 600 °C promotes recrystallization of the nanocrystalline matrix, resulting in a large-grained structure (see Fig. 20c).

In principle, the thermal history that the powders experience during HVOF spraying can be experimentally measured and mathematically described by a number of nonisothermal annealing techniques [41, 205]. If the grain growth of nanocrystalline materials during

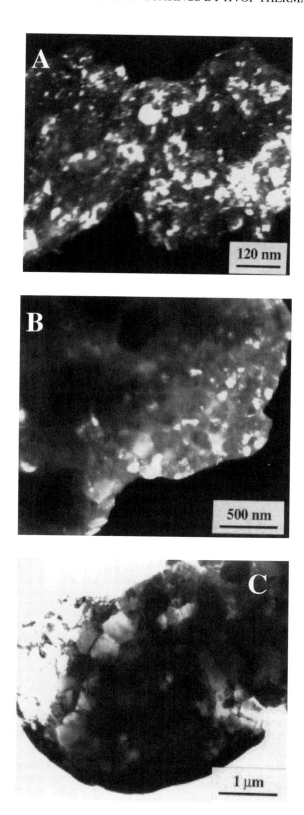

Fig. 20.   Transmission electron microscopy (TEM) images of the cryomilled nanocrystalline Ni follow-ing thermal annealing at various temperatures.

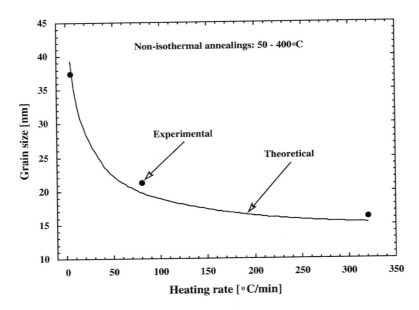

Fig. 21. Experimental and simulation results of grain growth for the Ni nanocrystalline powders under nonisothermal annealing (heating, 50–400 °C; cooling, 400 °C to room temperature, 180 °C/min).

nonisothermal annealing is theoretically predictable, then the grain growth during HVOF spraying can be estimated. Figure 21 shows an example of the grain growth behavior of Ni nanocrystalline powders under nonisothermal annealing conditions [41, 206]. Evidently, there is a good correspondence between the experimental results and the theoretical simulation results using Eq. (41). As the heating rate increases, the increment of grain size decreases significantly, implying that the fine-grained structure can be retained at higher temperatures if the heating rate is high enough, similar to that of HVOF spraying.

4. FUTURE PERSPECTIVES

The development of nanocrystalline coatings by means of thermal spraying, including HVOF spraying and reactive plasma spraying, has the potential to improve the structure and quality of thermally sprayed coatings. The successful exploitation of this technology will depend on:

- The development of novel material systems, particularly new cermet materials, with superior physical properties and damage tolerance
- The development of new, cost-effective routes to produce nanocrystalline feed-stock powders having high chemical homogeneity and enhanced thermal stability, suitable for thermal spraying
- The development of a theoretical formulation capable of predicting microstructural evolution, particularly grain growth behavior, during thermal spraying
- The formulation and application of robust models to optimize the experimental parameters that are necessary for the thermal spraying of nanocrystalline systems
- The development and application of diagnostic tools to assure reproducibility of results

Additionally, the continued commercial success that is currently enjoyed by standard (i.e., nonnanocrystalline) thermally sprayed coatings is critical to the development of nanocrystalline coating technology [39, 43]. The potential technological benefits of thermally sprayed nanocrystalline coatings are manifold. For example, there are approximately 1500 weld overlays in a single ship. The anticipated life cycle of these welds could be significantly extended if a nanocrystalline coating, with the associated improvements in hardness and wear characteristics, could be used. Moreover, it has been estimated that a significant proportion of the valve stems that fail in ships are due to steam erosion. The improved wear properties of nanocrystalline coatings are ideally suited for this particular application. This and other examples suggest that the applications of thermally sprayed nanocrystalline coatings can extend to a broad range of industries (e.g., the U.S. Navy and the aerospace and automotive, industries) and cover a wide spectrum of materials that could be used to fabricate diverse components [39, 43].

Acknowledgments

The authors would like to acknowledge the financial support provided by the Office of Naval Research (Grant N00014-94-10017 and N00014-98-1-0569). The authors would also like to thank Dr. L. T. Kabacoff for his valuable suggestions. Experimental help from Dr. Robert J. Perez was highly appreciated. In addition, Professor E. J. Lavernia would like to acknowledge the Alexander von Humboldt Foundation in Germany for support of his sabbatical visit at the Max-Planck-Institut für Metallforschung, in Stuttgart, Germany.

References

1. C. Cheung, D. Wood, and U. Erb, in "Processing and Properties of Nanocrystalline Materials" (C. Suryanarayana, J. Singh, and F. H. Froes, eds.), p. 479. Minerals, Metals and Materials Society, Warrendale, PA, 1996.
2. *Science Watch*, 1995.
3. E. Y. Gutmanas, L. I. Trusov, and I. Gotman, *Nanostruct. Mater.* 8, 893 (1994).
4. G. E. Fougere, L. Riester, M. Ferber, J. R. Weertman, and R. W. Siegel, *Mater. Sci. Eng.*, *A* 204, 1 (1995).
5. G. E. Korth, and R. L. Williamson, *Metall. Mater. Trans. A* 26, 2571 (1995).
6. B. H. Kear and P. R. Strutt, *Naval Res. Rev.* 4, 4 (1995).
7. H. Gleiter, *Prog. Mater. Sci.* 33, 223 (1989).
8. C. Suryanarayana, *Int. Mater. Rev.* 40, 41 (1995).
9. T. R. Malow, and C. C. Koch, in "Synthesis and Processing of Nanocrystalline Powder" (D. L. Bourell, ed.), p. 33. Minerals, Metals and Materials Society, Warrendale, PA, 1996.
10. L. Pawlowski, "The Science and Engineering of Thermal Spray Coatings." Wiley, Chichester, U.K., 1995.
11. T. S. Srivatsan and E. J. Lavernia, *J. Mater. Sci.* 27, 5965 (1992).
12. L. N. Moskowitz, in "Thermal Spray: International Advances in Coatings Technology" (C. C. Berndt, ed.), p. 611. ASM International, Materials Park, OH, 1992.
13. Y. Matsubara and A. Tomiguchi, in "Thermal Spray: International Advances in Coatings Technology" (C. C. Berndt, ed.), p. 637. ASM International, Materials Park, OH, 1992.
14. L. Byrnes and M. Kraser, in "1994 Thermal Spray Industrial Applications" (C. C. Berndt and S. Sampath, eds.), p. 39. ASM International, Materials Park, OH, 1994.
15. H. Chen, Z. Liu, Y. Zhuang, and L. Xu, *Chin. J. Mech. Eng.* 5, 183 (1992).
16. A. R. Nicoll, A. Bachmann, J. R. Moens, and G. Loewe, in "Thermal Spray: International Advances in Coatings Technology" (C. C. Berndt, ed.), p. 149. ASM International, Materials Park, OH, 1992.
17. K. Niemi, P. Vuoristo, T. Mantyla, G. Barbezat, and A. R. Nicoll, in "Thermal Spray: International Advances in Coatings Technology" (C. C. Berndt, ed.), p. 685. ASM International, Materials Park, OH, 1992.
18. C. W. Smith, in "Science and Technology of Surface Coatings" (B. N. Chapman and J. C. Anderson, eds.), p. 262. Academic Press, London, 1974.
19. P. R. Taylor and M. Manrique, *JOM* 48, 43 (1996).
20. H. C. Chen, Z. Y. Liu, and Y. C. Chuang, *Thin Solid Films* 223, 56 (1992).
21. H. Schmidt and D. Matthaus, in "9th International Thermal Spray Conference," Netherlands Instituut voor Lastechniek, The Hague, The Netherlands, 1980, p. 225.
22. R. C. Tucker, in "Deposition Technologies for Films and Coatings" (R. F. Bunshah, ed.), p. 454. Noyes Publications, Park Ridge, NJ, 1982.
23. D. R. Marantz, in "Science and Technology of Surface Coatings" (B. N. Chapman and J. C. Anderson, eds.), p. 308. Academic Press, London, 1974.

24. R. G. Smith, in "Science and Technology of Surface Coating: A NATO Advanced Study Institute on the Science and Technology of Surface Coating" (B. N. Chapman and J. C. Anderson, eds.), p. 271. Academic Press, London, 1972.

25. Y. S. Borisov, E. A. Astachov, and V. S. Klimenko, "Detonation Spraying: Equipment, Materials and Applications," Thermische Spritzkonferenz, Essen, Germany, 1990.

26. E. Schwarz, in "9th International Thermal Spray Conference," Netherlands Institut voor Lastechniek, The Hague, The Nertherlands, 1980, p. 91.

27. K. Niederberger and B. Schiffer, "Eigenshaften Verschiedener Gase und Deren Einfluss Beim Thermischen Spritzen," Thermische Spritzkonferenz, Essen, Germany, 1990.

28. M. Okada and H. Maruo, *Bri. Welding J.* 15, 371 (1968).

29. R. T. Smyth, F. J. Dittrich, and J. D. Weir, in "International Conference on Advances in Surface Coating Technology," London, 1978, p. 233.

30. A. P. Bennett and M. B. C. Quigley, *Welding Metal Fabrication* 485 (1990).

31. D. J. Wortman, *J. Vac. Sci. Technol. A* 3, 2532 (1985).

32. T. J. Roseberry and F. W. Boulger, A Plasma Flame Spray Handbook, U.S. Department of Commerce Report MT-043, National Technical Information Service, Springfield, VA, 1977.

33. D. W. Parker and G. L. Kutner, *Adv. Mater. Process.* 140, 68 (1991).

34. D. J. Varacalle, M. G. Ortiz, C. S. Miller, T. J. Steeper, A. J. Rotolico, J. Nerz, and W. L. R. Riggs, in "Thermal Spray: International Advances in Coatings Technology" (C. C. Berndt, ed.), p. 181. ASM International, Materials Park, OH, 1992.

35. D. Apelian, D. Wei and B. Farouk, *Metall. Trans. B* 20, 251 (1989).

36. M. E. Vinayo, L. Gaide, F. Kassabji, and P. Fauchais, in "7th International Symposium on Plasma Chemistry," Eindoven, The Netherlands, 1985, p. 1161.

37. V. L. Tellkamp, M. L. Lau, A. Fabel, and E. J. Lavernia, *Nanostruct. Mater.* 9, 489 (1996).

38. B. H. Kear and G. Skandan, *Nanostruct. Mater.* 7, 913 (1996).

39. E. J. Lavernia, M. L. Lau, and H. G. Jiang, *J. Therm. Spray Technol.*, to appear.

40. M. L. Lau, H. G. Jiang, and E. J. Lavernia, in "15th International Thermal Spray Conference and Exhibition," Nice, France, 1998, to appear.

41. H. G. Jiang, M. L. Lau, and E. J. Lavernia, in "15th International Thermal Spray Conference and Exhibition," Nice, France, 1998, to appear.

42. H. G. Jiang, M. L. Lau, and E. J. Lavernia, *J. Therm. Spray Technol.*, to appear.

43. E. J. Lavernia, M. L. Lau, and H. G. Jiang, in "NATO ASI Conference," St. Petersburg, Russia, 1997, in press.

44. M. L. Lau, H. G. Jiang, R. Schweinfest, W. Nuchter, and E. J. Lavernia, *Phys. Status Solidi A* 166, 257 (1998).

45. V. V. Sobolev, J. M. Guilemany, and J. A. Calero, *J. Mater. Process. Manuf. Sci.* 4, 25 (1995).

46. M. L. Thorpe and H. J. Richter, *J. Therm. Spray Technol.* 1, 161 (1992).

47. W. L. Oberkampf and M. Talpalikar, in "Thermal Spray Industrial Applications," p. 381. ASM International, Boston, 1994.

48. W. D. Swank, J. R. Fincke, D. C. Haggard, and G. Irons, in "Thermal Spray Industrial Applications," p. 313. ASM International, Boston, 1994.

49. S. Gordon and B. McBride, NASA SP-273, Lewis Research Center, 1976.

50. V. V. Sobolev and J. M. Guilemany, *Int. Mater. Rev.* 41, 13 (1996).

51. J. H. Horlock and D. E. Winterbone, "The Thermodynamics and Gas Dynamics of Internal Combustion Engines," Vol. 2. Clarendon Press, Oxford, U.K., 1980.

52. R. Ouziaux and J. Perrier, Mecanique des fluides appliquée, Vol. 2. Dunod, Paris, 1967.

53. D. W. Parker and G. L. Kutner, *Adv. Mater. Process.* 7, 31 (1994).

54. A. J. Sturgeon, *Metal. Mater.* 8, 547 (1992).

55. V. V. Sobolev, J. M. Guilemany, J. C. Garmier, and J. A. Calero, *Surf. Coat. Technol.* 63, 181 (1994).

56. J. Szekely and N. J. Themelis, "Rate Phenomena in Progress Metallurgy," Wiley, New York, 1971.

57. S. V. Joshi, *Int. Powder Metall.* 24, 373 (1992).

58. J. A. Lewis and W. H. Gauvin, *AIChE J.* 19, 982 (1973).

59. R. Clift, J. R. Grace, and M. E. Weber, "Bubbles, Drops and Particles." Academic Press, New York, 1978.

60. W. E. Ranz and W. R. Marshall, *Chem. Eng. Prog.* 48, 141, 173 (1952).

61. M. C. Flemings, "Solidification Processing." McGraw–Hill, New York, 1974.

62. V. V. Sobolev and J. M. Guilemany, *Mater. Lett.* 25, 285 (1995).

63. V. V. Sobolev and J. M. Guilemany, *Mater. Lett.* 18, 304 (1994).

64. V. V. Sobolev, J. M. Guilemany, and J. A. Calero, *Mater. Sci. Technol.* 11, 810 (1995).

65. J. M. Guilemany, V. V. Sobolev, J. Nutting, Z. Dong, and J. A. Calero, *Scr. Metall. Mater.* 31, 915 (1994).

66. J. Crank, "Mathematics of Diffusion," 2nd ed. Oxford University Press, New York, 1975.

67. R. H. Doremus, "Rates of Phase Transformations." Academic Press, London, 1985.

68. V. V. Sobolev, J. M. Guilemany, J. R. Miguel, and J. A. Calero, *Surf. Coat. Technol.* 81, 136 (1996).

69. J. M. Guilemany, J. R. Miguel, and Z. Dong, *Powder Metall.* 37, 219 (1994).

70. D. C. Crawmer, J. D. Krebsbach, and W. L. Riggs, in "Thermal Spray: International Advances in Coatings Technology" (C. C. Berndt, ed.), p. 127. ASM International, Materials Park, OH, 1992.

71. G. K. Creffield, M. A. Cole, and G. R. White, in "Proceedings of the 8th National Thermal Spray Conference" (C. C. Berndt and S. Sampath, eds.), p. 291. Houston, TX, 1995.

72. C. M. Hackett and G. S. Settles, in "Advances in Thermal Spray Science and Technology," (C. C. Berndt, ed.). ASM International, Materials Park, OH, 1995.

73. C. M. Hackett and G. S. Settles, in "Thermal Spray Industrial Applications" (C. C. Berndt and S. Sampath, eds.), p. 21. ASM International, Materials Park, OH, 1994.

74. V. V. Sobolev and J. M. Guilemany, *Mater. Lett.* 25, 71 (1996).

75. V. V. Sobolev, J. M. Guilemany, and A. J. Martin, *J. Therm. Spray Technol.* 5, 207 (1996).

76. V. V. Sobolev, J. M. Guilemany, and A. J. Martin, *Mat. Lett.* 29, 185 (1996).

77. H. Fukanuma, *J. Therm. Spray Technol.* 3, 33 (1994).

78. G. Trapaga and J. Szekely, *Metall. Trans. B* 22, 904 (1991).

79. G. Trapaga, E. F. Matthys, J. J. Valencia, and J. Szekely, *Metall. Trans. B* 23, 710 (1992).

80. J.-P. Delplanque, E. J. Lavernia, and R. H. Rangel, *Int. J Nonequilib. Process.* 10, 185 (1997).

81. J.-P. Delplanque, S. Dai, R. H. Rangel, and E. J. Lavernia, *J. Mater. Synth. Process.* 5 (1997).

82. J.-P. Delplanque, W. D. Cai, R. H. Rangel, and E. J. Lavernia, *Acta Mater.* 45, 5233 (1997).

83. J.-P. Delplanque, E. J. Lavernia, and R. H. Rangel, *Int. J. Numer. Heat Transfer, A* 30, 1 (1996).

84. H. Liu, E. J. Lavernia, and R. H. Rangel, *J. Phys. D: Appl. Phys.* 26, 1900 (1993).

85. J.-P. Delplanque, E. J. Lavernia, and R. H. Rangel, Paper Presented at the 1995 ASME Winter Annual Meeting, San Francisco, CA, 1995.

86. H. Liu, R. H. Rangel, and E. J. Lavernia, *Acta Metall. Mater.* 42, 3277 (1993).

87. H. Liu, E. J. Lavernia, and R. H. Rangel, *Acta Metall. Mater.* 43, 2053 (1995).

88. G. Trapaa and J. Szekely, *Metall. Trans. B* 22, 901 (1991).

89. T. W. Clyne, *Key Eng. Mater.* 116–117, 307 (1996).

90. H. Kreye and D. Blume, in "Proceedings of the International Thermal Spray Conference and Exposition," Orlando, FL, 1992, p. 177.

91. S. Zimmermann and H. Kreye, in "Proceedings of the 8th National Thermal Spray Conference" (C. C. Berndt and S. Sampath, eds.). Houston, TX, 1995, p. 297.

92. O. Knotek, E. Lugscheider, P. Jokiel, U. Schnaut, and A. Wiemers, in "Proceedings of the 7th National Thermal Spray Conference" (C. C. Berndt and S. Sampath, eds.). Boston, 1994, p. 179.

93. C. Hays, J. L. Watson, Sr., and J. P. Walker, Jr., in "Proceedings of the 8th National Thermal Spray Conference" (C. C. Berndt and S. Sampath, eds.). Houston, TX, 1995, p. 589.

94. P. E. Arvidsson, *pmi* 24, 176 (1992).

95. G. Irons, in "28th Annual Aerospace/Airline Plating and Metal Finishing Forum and Exposition," San Diego, CA, 1992, p. 1.

96. L. N. Moskowitz, in "Proceedings of the International Thermal Spray Conference and Exposition" (C. C. Berndt, ed.). Orlando, FL, 1992, p. 611.

97. R. Shah, W.-C. Wang, K. Sampath, S. Parthasaathi, J. Jo, and E. J. Onesto, in "Proceedings of the 7th National Thermal Spray Conference" (C. C. Berndt and S. Sampath, eds.). Boston, 1994, p. 675.

98. X. X. Guo and H. Zhang, in "Proceedings of the International Thermal Spray Conference and Exposition" (C. C. Berndt, ed.). Orlando, FL, 1992, p. 729.

99. R. Knight and R. W. Smith, in "Proceedings of the International Thermal Spray Conference and Exposition" (C. C. Berndt, ed.). Orlando, FL, 1992, p. 159.

100. S. E. Hartfield-Wünsch and S. C. Tung, in "Proceedings of the 7th National Thermal Spray Conference" (C. C. Berndt and S. Sampath, eds.). Boston, 1994, p. 19.

101. H. Edris, D. G. McCartney, and A. J. Sturgeon, *J. Mater. Sci.* 32, 863 (1997).

102. H. Voggenreiter, H. Huber, S. Beyer, and H.-J. Spies, in "Proceedings of the 8th National Thermal Spray Conference" (C. C. Berndt and S. Sampath, eds.). Houston, TX, 1995, p. 303.

103. P. Siitonen, T. Konos, and P. O. Kettunen, in "Proceedings of the 7th National Thermal Spray Conference" (C. C. Berndt and S. Sampath, eds.). Boston, 1994, p. 105.

104. T. Shmyreva, A. Mukhin, and L. Mukhina, in "Proceedings of the 8th National Thermal Spray Conference" (C. C. Berndt and S. Sampath, eds.). Houston, TX, 1995, p. 243.

105. D. J. Sordelet, P. D. Krotz, R. L. Daniel, Jr., and M. F. Smith, in "Proceedings of the 8th National Thermal Spray Conference" (C. C. Berndt and S. Sampath, eds.). Houston, TX, 1995, p. 627.

106. P. M. J. Vuoristo, K. J. Niemi, and T. A. Mantyla, in "Proceedings of the International Thermal Spraying Conference and Exposition" (C. C. Berndt and S. Sampath, eds.). Orlando, FL, 1992, p. 171.

107. M. Vanolo, F. Pavese, D. Giraudi, and M. Bianco, in "7th Congresso SATT," Turin, Italy, 1994, p. 2119.

108. E. Lugscheider, P. Remer, A. Nyland, and R. Sicking, in "Proceedings of the 8th National Thermal Spray Conference" (C. C. Berndt and S. Sampath, eds.). Houston, TX, 1994, p. 583.

109. N. Iwanoto, M. Kamai, and G. Ueno, in "Proceedings of the International Thermal Spraying Conference and Exposition" (C. C. Berndt and S. Sampath, eds.). Orlando, FL, 1992, p. 259.

110. J. M. Guilemany, J. Nutting, Z. Dong, and J. M. de Paco, *Scr. Metall. Mater.* 33, 1055 (1995).

111. H. Nakahira, K. Tani, K. Miyajima, and Y. Harada, in "Proceedings of the International Thermal Spray Conference and Exposition" (C. C. Berndt and S. Sampath, eds.). Orlando, FL, 1992, p. 1011.

112. K. G. Shaw, M. F. Gruninger, and W. J. Jarosinski, in "Proceedings of the 7th National Thermal Spray Conference and Exposition" (C. C. Berndt and S. Sampath, eds.). Boston, 1994, p. 185.

113. M. J. Froning and H. Keller, in "Proceedings of the 8th National Thermal Spray Conference" (C. C. Berndt and S. Sampath, eds.). Houston, TX, 1995, p. 549.

114. J. M. Guilemany, N. Llorca-lsern, and J. Nutting, *pmi* 25, 176 (1993).

115. G. Irons, W. R. Kratochvil, W. R. Bullock, and A. Roy, in "Proceedings of the 7th National Thermal Spray Conference" (C. C. Berndt and S. Sampath, eds.). Boston, 1994, p. 127.

116. P. Vuoristo, K. Niemi, T. Mäntylä, L.-M. Berger, and M. Nebelung, in "Proceedings of the 8th National Thermal Spray Conference" (C. C. Berndt and S. Sampath, eds.). Houston, TX, 1995, p. 309.

117. H.-D. Steffens, J. Wilden, and K. Nassenstein, in "Proceedings of the 8th National Thermal Spray Conference" (C. C. Berndt and S. Sampath, eds.). Houston, TX, 1995, p. 689.

118. D. Bernard, O. Yokota, A. Grimaud, P. Fauchais, S. Usmani, Z. J. Chen, C. C. Berndt, and H. Herman, in "Proceedings of 7th National Thermal Spray Conference" (C. C. Berndt and S. Sampath, eds.). Boston, 1994, p. 171.

119. M. S. Seehra and V. S. Babu, *J. Mater. Res.* 11, 1133 (1995).

120. H. P. Klug and L. E. Alexander, "X-ray Diffraction Procedures," p. 643. Wiley, New York, 1974.

121. H. G. Jiang, M. Rühle, and E. J. Lavernia, *J. Mater. Res.*, to appear.

122. Th. H. de Keijser, J. I. Langford, E. J. Mittemeijer, and A. B. P. Vogels, *J. Appl. Crystallogr.* 15, 308 (1982).

123. J. I. Langford, *J. Appl. Crystallogr.* 11, 10 (1978).

124. J. M. Guilemany, J. Nutting, V. V. Sobolev, Z. Dong, J. M. de Paco, J. A. Calero, and J. Fernandez, *Mater. Sci. Eng., A* 232, 119 (1997).

125. S. Parthasarathi, K. Sampath, and B. R. Tittmann, in "Proceedings of the 8th National Thermal Spray Conference" (C. C. Berndt and S. Sampath, eds.). Houston, TX, 1995, p. 505.

126. E. B. Smith, G. D. Power, T. J. Barber, and L. M. Chiappetta, United Technologies Research Center Report 91-8, East Hartford, CT, 1991.

127. V. H. Kadyrov, V. B. Brik, F. J. Worzala, and C. Florey, in "Proceedings of the 7th National Thermal Spray Conference" (C. C. Berndt and S. Sampath, eds.). Boston, 1994, p. 269.

128. A. A. Tipton, in "Proceedings of the 8th National Thermal Spray Conference" (C. C. Berndt and S. Sampath, eds.). Houston, TX, 1995, p. 463.

129. H.-D. Steffens, J. Wilden, K. Nassenstein, and S. Möbus, in "Proceedings of the 8th National Thermal Spray Conference" (C. C. Berndt and S. Sampath, eds.). Houston, TX, 1995, p. 469.

130. M. J. Luton, C. S. Jayanth, M. M. Disko, S. Matras, and J. Vallone, in "Multicomponent Ultrafine Microstructures" (L. E. McCandlish et al., eds.), Materials Research Society Symposium Proceedings, Pittsburgh, PA, 1989, p. 79.

131. M. M. Disko, M. J. Luton, and H. Shuman, *Ultramicroscopy* 37, 202 (1991).

132. B. Huang, J. Vallone, C. F. Klein, and M. J. Luton, *Mater. Res. Soc. Symp. Proc.* 273, 171 (1992).

133. R. J. Perez, B. Huang, and E. J. Lavernia, *Nanostruct. Mater.* 9, 71 (1997).

134. M. F. Ashby, in "Recrystallization and Grain Growth of Multi-Phase and Particle Containing Material" (N. Hansen, A. R. Jones, and T. Leffers, eds.), p. 325. Fyens Stiftsbogtrykkeri, Denmark, 1980.

135. C. S. Smith, *Trans. AIME* 9, 15 (1949).

136. T. Gladman, *Proc. R. Soc. London, Ser. A* 294, 298 (1966).

137. C. L. Briant, *Philos. Mag. Lett.* 73, 345 (1996).

138. P. Lejcek and S. Hofmann, *Crit. Rev. Solid State Mater. Sci.* 20, 1 (1995).

139. K. Ishida, *J. Alloys Compd.* 235, 244 (1996).

140. B. Ralph, *Mater. Sci. Technol.* 6, 1139 (1990).

141. H. C. Fielder, *Metall. Trans. A* 8 1307 (1977).

142. B. Fultz, L. B. Hong, Z. Q. Gao, and C. Bansal, in "Synthesis and Processing of Nanocrystalline Powder" (D. L. Bourell, ed.), p. 249. Minerals, Metals and Materials Society, Warrendale, PA, 1996.

143. G. J. Fan, W. N. Gao, M. X. Quan, and Z. Q. Hu, *Mater. Lett.* 23, 33 (1995).

144. C. Bansal, Z. Q. Gao, and B. Fultz, *Nanostruct. Mater.* 5, 327 (1995).

145. C. M. Hackett and G. S. Settles, in "Thermal Spray: Practical Solutions for Engineering Problems" (C. C. Berndt, ed.), p. 665. ASM International, Materials Park, OH, 1996.

146. E. J. Lavernia, *Int. J. Rapid Solid.* 5, 47 (1989).

147. A. R. E. Singer and R. W. Evans, *Metall. Technol.* 10, 61 (1983).

148. R. H. Bricknell, *Metall. Trans. A* 17, 583 (1986).

149. B. P. Bewlay and B. Cantor, in "Rapidly Solidified Materials" (P. Lee and R. Carbonara, eds.), p. 15. ASM International, Materials Park, OH, 1986.

150. S. Annavarapu, D. Apelian, and A. Lawley, *Metall. Trans. A* 19, 3077 (1988).

151. Z. Feng, J. Duszczyk, and A. G. Leatham, *J. Mater. Sci.* 27, 4511 (1992).

152. S. Annavarapu and R. Doherty, *Int. J. Powder Metall.* 29, 331 (1993).

153. E. Gutierrez-Miravete, E. J. Lavernia, G. M. Trapaga, and J. Szekely, *Int. J. Rapid Solid.* 4, 125 (1988).

154. S. Annavarapu, D. Apelian, and A. Lawley, *Metall. Trans. A* 21, 3237 (1990).

155. P. Mathur, S. Annavarapu, D. Apelian, and A. Lawley, *Mater. Sci. Eng. A* 142, 261 (1991).
156. P. Mathur, D. Apelian, and A. Lawley, *Acta Metall.* 37, 429 (1989).
157. J. L. Estrada and J. Duszezyk, *Mater. Sci.* 25, 1381 (1990).
158. A. Kahveci, in "Science and Technology of Rapid Solidification and Processing" (M. A. Otooni, ed.), p. 271. Kluwer Academic Publishers, Dordrecht, The Netherlands, 1995.
159. X. Liang and E. J. Lavernia, *Metall. and Mater. Trans. A* 25, 2341 (1994).
160. P. S. Grant, W. T. Kim, and B. Cantor, *Mater. Sci. Eng. A* 134, 1111 (1991).
161. E. J. Lavernia and N. J. Grant, *Int. J. Rapid Solid.* 2, 93 (1986).
162. J. Marinkovich, F. A. Mohamed, J. R. Pickens, and E. J. Lavernia, *J. Metall.* 41, 36 (1989).
163. P. S. Grant and B. Cantor, *Cast Metals* 4, 140 (1991).
164. E. Gutierrez-Miravete, E. J. Lavernia, G. M. Trapaga, J. Szekely, and N. J. Grant, *Metall. Trans. A* 20, 71 (1989).
165. J. Baram, *Metall. Trans. A* 22, 2515 (1991).
166. B. P. Bewlay and B. Cantor, *J. Mater. Res.* 6, 1433 (1991).
167. B. P. Bewlay and B. Cantor, *Mater. Sci. Eng. A* 118, 207 (1989).
168. X. Liang, J. C. Earthman, and E. J. Lavernia, *Acta Metall. Mater.* 40, 3003 (1992).
169. B. P. Bewlay and B. Cantor, *Metall. Trans. B* 21, 899 (1990).
170. P. S. Grant, B. Cantor, and L. Katgerman, *Acta Metall. Mater.* 41, 3109 (1993).
171. E. J. Lavernia, E. Gutierrez-Miravete, J. Szekely, and N. J. Grant, *Int. J. Rapid Solid.* 4, 89 (1988).
172. S. N. Ojha, J. N. Jha, and S. N. Singh, *Scr. Metall.* 25, 443 (1991).
173. A. Suzuki, T. Suzuki, Y. Nagaoka, and Y. Iwata, *J. Jpn. Inst. Metall.* 32, 1301 (1968).
174. S. Ashok., *Int. J. Rapid Solid.* 7, 283 (1993).
175. E. Gutierrez-Miravete, G. M. Trapaga, and J. Szekely, in "Casting of Near Net Shape Products" (Y. Sahai, J. E. Battles, R. S. Carbonara, and C. E. Mobley, eds.), p. 133. TMS, Warrendale, PA, 1988.
176. P. S. Grant, R. P. Underhill, W. T. Kim, K. P. Mingard, and B. Cantor, in "Spray Forming 2, Proceedings of the Second International Conference on Spray Forming" (J. V. Wood, ed.), p. 45. Woodhead, 1993.
177. S. Annavarapu and R. D. Doherty, *Acta Metall. Mater.* 43, 3207 (1994).
178. H. Edris and D. G. McCartney, *J. Mater. Sci.* 32, 863 (1997).
179. J. W. Christian, "The Theory of Transformations in Metals and Alloys." Pergamon Press, Oxford, U.K., 1975.
180. D. L. Bourell and W. Kaysser, *Acta Metall. Mater.* 41, 2933 (1993).
181. S. Eidelman and X. Yang, *Nanostruct. Mater.* 9, 79 (1997).
182. D. A. J. Ramm, T. W. Clyne, A. J. Sturgeon, and S. Dunkerton, in "1994 Thermal Spray Industrial Applications" (C. C. Berndt and S. Sampath, eds.), p. 239. ASM International, Materials Park, OH, 1994.
183. O. Knotek and U. Schnaut, in "Thermal Spray: International Advances in Coatings Technology" (C. C. Berndt, ed.), p. 811. ASM International, Materials Park, OH, 1992.
184. X. Liang, E. J. Lavernia, J. Wolfenstine, and A. Sickinger, *J. Therm. Spray Technol.* 4, 252 (1995).
185. E. Pfender and Y. C. Lee, *Plasma Chem. Plasma Proc.* 5, 211 (1985).
186. D. Apelian, M. Paliwal, R. W. Smith, and W. F. Schilling, *Int. Metall. Rev.* 271 (1983).
187. E. Pfender, *Thin Solid Films* 238, 228 (1994).
188. Y. C. Lee, Y. P. Chyou, and E. Pfender, *Plasma Chem. Plasma Proc.* 5, 391 (1985).
189. R. Birringer, *Mater. Sci. Eng., A* 117, 33 (1989).
190. J. S. Benjamin, in "Materials Science Forum" (P. H. Shingu, ed.), Vols. 88–90, p. 1. Trans Tech Publications, Zürich, Switzerland, 1992.
191. R. B. Schwarz, S. Srinivasan, and P. B. Desch, in "Materials Science Forum" (P. H. Shingu, ed.), Vols. 88–90, p. 595. Trans Tech Publications, Zürich, Switzerland, 1992.
192. J. Eckert, J. C. Holzer, C. E. Krill III, and W. L. Johnson, *J. Mater. Res.* 7, 1751 (1992).
193. M. L. Trudeau, R. Schulz, L. Zaluski, S. Hosatte, D. H. Ryan, C. B. Doner, P. Tessier, J. O. Strom-Olsen, and A. Van Neste, in "Materials Science Forum" (P. H. Shingu, ed.), Vols. 88–90, p. 537. Trans Tech Publications, Zürich, Switzerland, 1992.
194. B. H. Kear and P. R. Strutt, *Nanostructured Mater.* 6, 227 (1995).
195. R. J. Perez, H. G. Jiang, and E. J. Lavernia, *Nanostructured Mater.* 9, 71 (1997).
196. T. R. Malow and C. C. Koch, *Acta Mater.* 45, 2177 (1997).
197. R. J. Perez, H. G. Jiang, and E. J. Lavernia, *Metall. Trans.* 29A, 2469 (1998).
198. H. G. Jiang and E. J. Lavernia, *J. Mater. Res.*, to appear.
199. M. L. Lau, H. G. Jiang, R. J. Perez, J. Juarez-Islas, and E. J. Lavernia, *Nanostruct. Mater.* 7, 847 (1997).
200. B.-L. Huang, J. Vallone, M. J. Luton, *Nanostruct. Mater.* 5, 411 (1995).
201. S. Seshan and J. Kaneko, *Adv. Compos. Mater.* 2, 153 (1992).
202. N. Wang, Z. K. Wang, T. Aust, and U. Erb, *Acta Mater.* 45, 1655 (1997).
203. P. Knauth, A. Charaï, and P. Gas, *Scr. Metall. Mater.* 28, 325 (1993).
204. U. Köster, *Mater. Sci. Forum* 235–238, 377 (1997).
205. H. E. Kissinger, *Anal. Chem.* 29, 1702 (1957).
206. H. G. Jiang, M. L. Lau, and E. J. Lavernia, *Nanostruct. Mater.* 10, 169 (1998).

Chapter 4

LOW-TEMPERATURE COMPACTION OF NANOSIZE POWDERS

E. J. Gonzalez, G. J. Piermarini

Ceramics Division, National Institute of Standards and Technology, Gaithersburg, Maryland, USA

Contents

1. INTRODUCTION

Recently, there has been a significant increase in interest in fabricating ceramic materials from ultrafine powders that consist of nanosize primary particles ranging in mean diameter from 1 to 100 nm. Theoretical predictions by Frenkel [1] and Herring [2] clearly indicate that the rate of densification varies inversely as a function of particle size. Thus, based on this prediction, as particle size decreases from micrometers to nanometers, a substantial decrease in sintering time can be expected at a given temperature. Indeed, many experimental investigations support this theoretical prediction. For example, Rhodes [3] produced densely packed compacts of nanosize zirconia particles and observed sintering of the compacts to near theoretical density at much lower temperatures than are used for sintering coarse zirconia particles. Recently, Skandan et al. [4] sintered nanosize titania at 800 °C, well below the sintering temperature for conventional titania powders.

These results suggest that nanosize particles as starting materials might offer considerable advantages for fabricating ceramics, especially because the reduced sintering

Handbook of Nanostructured Materials and Nanotechnology, edited by H.S. Nalwa
Volume 1: Synthesis and Processing
This contribution is a U.S. government work
and is not subject to copyright.

temperatures and sintering times required may inhibit undesired grain growth, one of the most important microstructural parameters because it is directly related to materials properties. It has been substantiated in many polycrystalline metallic and nonmetallic materials that the finer the grain size the higher the yield strength [5]. The expected reduced grain size (possibly on a nanoscale dimension) in a ceramic fabricated from nanosize particles results in a significant increase in grain interfaces, so that the number of interfacial atoms becomes comparable to the number of lattice site atoms. Thus, the overall properties of the solid will be influenced strongly by the local atomic arrangements in the interfaces, which are quite different from the atoms in the ordered crystalline state. Under these conditions, where significant contributions are made by the interfacial atoms, novel and possibly improved physical properties in ceramic materials fabricated from nanosize particles are anticipated.

One of the more important challenges in the fabrication of nanostructured materials is how to achieve full densification of the powder while simultaneously retaining a nanoscale microstructure. At present, relatively high densities (>95%) can be achieved with 10–20-nm particles, but the high densities invariably are accompanied by significant grain growth (grain sizes >50 nm) [6]. To reduce grain growth during sintering (where most of the densification process occurs), a high-density (random close packed) homogeneous green body with minimum pore size is desired, because large-size pores begin to close only after substantial grain growth has occurred and most of the small pores have been eliminated. Thus, prior to sintering a body composed of nanosize powders, it is important to try to achieve high densities, that is, full random close packing, of the nanosize particles in the green body. One way to achieve this condition is by compaction of the powder at high pressures. However, it is now known that forming densely packed green compacts from nanosize ceramic particles is very difficult to achieve by the application of pressure, because strong aggregation forces, such as the van der Waals attraction, increase dramatically as particle size decreases. For nanosize particles, the van der Waals attraction forces can essentially prevent the particles from sliding by one another, thereby causing agglomeration by diffusion during compaction [7]. As a result of aggregation and subsequent agglomeration of primary particles, green compacts of nanosize particles usually have relatively low densities after room temperature (RT) pressing or compaction.

The low green density of nanosize particle compacts is due primarily to two factors: (1) the presence of large voids formed by agglomerates of irregular shapes that cannot be easily broken down using conventional compaction techniques because of the presence of very strong bridging forces and (2) inefficient packing of individual particles because of large interparticle (interagglomerate) friction forces [8]. To obtain the desired nanoscale microstructure in the green body, agglomeration of the particles during synthesis and subsequent processing must be minimized. Recently, Duran et al. [9] sintered 3 mol% Y_2O_3–ZrO_2 powder to full density at relatively low temperatures. The authors emphasized that it is important to avoid the formation of gradients in the green-body density, inhomogeneities, and agglomerations, to produce defect-free sintered specimens. Their results also indicated that a change in sintering temperature of only 70 °C can have a significant effect on the densification kinetics. Based on these results, it is evident that special criteria must be met when selecting the processing conditions for nanosize powders.

High-pressure compaction of nanosize particles has been used in an attempt to achieve the desired nanoscale microstructure and high density in the green body. Typical compaction pressures usually exceed 1.0 GPa and, therefore, require specialized equipment. For example, to attain high green density by uniaxial compaction of a nanosize powder, a high-strength tungsten carbide–cobalt (WC/Co) piston–cylinder die has been used, and, for even higher pressures, a diamond anvil-type pressure cell (DAC) has been utilized. Because of the demanding strength requirements of apparatus capable of producing pressures in the gigapascal range, the compaction equipment necessarily must be designed to produce relatively small samples. For these reasons, the DAC was one of the first devices to be utilized in compaction of nanosize particle experiments. Moreover, it is relatively easy to use and cost effective when compared with large press equipment.

Recently, it was discovered that the densification of nanosize powders can be improved in the green body by the use of lubricants that allow particles to slide easily over one another, permitting their rearrangement to produce a minimum interstice structure, that is, random close packing [10]. A good lubricant can improve significantly the packing properties of the nanosize particles. However, selection of a suitable lubricant is severely limited because of the existence of strongly reacting interfaces between nanosize particles, which the lubricant must overcome, as well as the presence of nanoscale voids in the structure formed by densely packed nanosize particles, which the lubricant must be able to penetrate to be effective. Thus, the lubricant must be necessarily a low-molecular-weight substance, small enough to fit into these nanoscale voids. High-density green-body compacts of nanosize particles are typically transparent, indicating that the voids contained within them are also of nanoscale dimensions, that is, less than the wavelengths associated with white light. Thus, only a few substances are commonly available that can be used as lubricants meeting this dimensional requirement.

Pechenik et al. [10] demonstrated the concept that simple molecular liquids can provide efficient lubrication to enhance the densification of nanosize particles when subjected to high pressures. They compacted nanosize silicon nitride particles at liquid nitrogen temperatures and also used LN_2 as a lubricant. The first compactions were performed in a DAC because that was the only pressure equipment conveniently available to them at that time. This cryogenic compaction process was successful in producing densely packed transparent (light brownish in color) compacts of nanosize particles that sintered close to full density at a much lower temperature (1200 °C) than that required for the conventional processing of silicon nitride. As a follow-up to this original research, Chen et al. [8] studied the cryogenic powder compaction process in much greater detail. To do this, these authors designed and constructed a special scaled-up WC/Co piston–cylinder die apparatus for fabricating much larger samples than those that were made in the DAC. The compaction equipment they developed is capable of producing 3-mm disk-shaped samples under vacuum or in a variety of controlled conditions, such as temperatures in the range from 77 to 1000 K and pressures up to 3 GPa. In addition, throughout the compaction process, continuous measurements of the sample volume, applied force, and frictional force between the sample and the die walls were performed. This novel system, because of its uniqueness and its applicability to understanding the rheology and compaction of nanosize powders, is described in great detail later in this chapter, along with a number of experimental studies on the compaction of nanosize silicon nitride and γ-alumina powders.

Unfortunately, published reports on the effects of lubricants on nanosize powder compaction processes are very limited. However, the authors are aware of unpublished current activity in private industry related to the use of lubrication techniques to achieve the high green densities necessary for low-temperature sintering of nanosize powders. Unfortunately, results related to such work have not been reported publicly by the different ceramic and metal processing industries because the information is regarded as being very exclusive and proprietary. In the following sections, we will describe the more important published techniques used to compact nanosize ceramic powders at low temperatures. The advantages and disadvantages of these techniques will be discussed also. Because of their prominence in this field, the high-pressure piston–cylinder compaction system developed by Chen et al. [8] and the diamond anvil method used by Pechenik et al. [10] will be described in detail in conjunction with the different lubrication experiments that were conducted with these systems.

2. LOW-TEMPERATURE–HIGH-PRESSURE POWDER COMPACTION

2.1. Diamond Anvil Pressure Cell

The diamond anvil pressure cell has been used for many years to investigate a variety of physical phenomena too numerous to list here. For the purposes of this discussion, it is sufficient to note that one of the more recent and less well known applications of the DAC has been in ceramic processing, particularly of nanosize powders at high pressures and moderate temperatures ($<600\,°C$). Because the DAC is relatively easy to use and can typically achieve pressures in excess of 5 GPa, it is an attractive alternative for routinely compacting small samples of nanosize powders for exploratory research purposes. Because much of the early results in this area of ceramic processing were obtained by utilizing the DAC, we consider the device important enough to describe in detail.

A schematic diagram of the typical design of a DAC for use at room temperature is shown in Figure 1 [11]. For the sake of clarity, there are no features incorporated in this particular design for specifically heating this cell. However, cells designed for simple resistance heating have been developed that can be routinely heated to temperatures in the range of $500\,°C$ with occasional excursions to $600\,°C$. The sample, located between two opposed diamond anvils shown in detail on the left side of Figure 1, is confined by a metallic (usually Inconel X750)* gasket with a small circular hole about 250 μm in diameter for confining the nanosize powder sample. The initial thickness of the gasket is also about 250 μm. Pressure is applied by rotation of the large screw, which compresses the Belleville spring washers, generating a load transmitted to the gasket/sample via the lever arms. The design has $180°$ optical access (see Fig. 1), which permits the ruby fluorescence method of pressure measurement as well as other optical and X-ray spectroscopic techniques to be used [12, 13]. The fluorescence pressure measurement method utilizes a small ruby sphere (≈ 10 μm in diameter), which is included with the powder to be pressed and serves as the internal pressure sensor.

Heating an anvil-type pressure cell of this design is typically done with a cylindrical resistance coil heating furnace positioned either around the anvils themselves internally or around the anvil–piston assembly [13]. The latter method, shown in Figure 2, is a modification of the design shown in Figure 1. This static method of heating was the first developed for this type of pressure cell and is limited to routine temperatures of about $500\,°C$, because at higher temperatures the metal components of the pressure cell lose their required strength and hardness and, thus, may lead to anvil failure [10]. Utilizing the resistance coil heating method with the DAC is not useful for processing ceramic powders at typical sintering temperatures, usually well above $500\,°C$. The DAC is more readily adaptable to low-temperature or cryogenic compaction experiments, which is accomplished by total immersion of the pressure cell in LN_2 as pressure is applied to the sample.

The main disadvantages of using the DAC for ceramic processing experiments is that the sample's volume is very small, making characterization of the resulting sintered compact very difficult (a disk ≈ 0.2 mm in diameter and 0.2 mm in height). Thus, although the DAC has some serious limitations, it has enough advantages to justify its use as a device for relatively simple and rapid exploratory research. Its role in ceramic processing has evolved into a complementary tool to a more useful compaction technique, that is, the piston–cylinder die developed by Chen et al. [8], which is described in the following section.

---

* Certain trade names and company products are mentioned in the text or identified in illustrations in order to specify adequately the experimental procedure and equipment used. In no case does such identification imply recommendation or endorsement by the National Institute of Standards and Technology, nor does it imply that the products are necessarily the best available for the purpose.

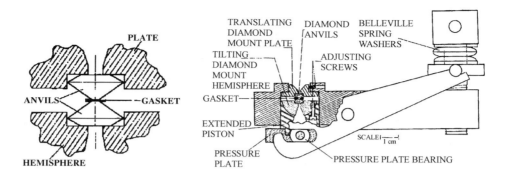

Fig. 1. Typical high-pressure optical diamond cell (DAC) designed for use at room temperature. Details of the anvil assembly are shown in the insert on the left side of the figure. (Source: Piermarini and Block [11].)

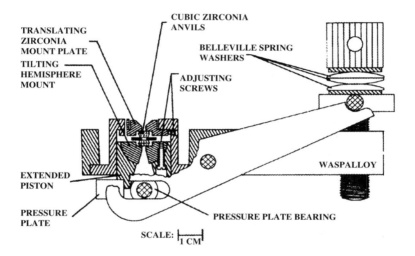

Fig. 2. A schematic diagram of an Inconel 718 diamond anvil cell used for heating a sample to temperatures in the range of 500 °C. The heating is provided by a cylindrical coil furnace located in the recess surrounding the anvil assembly. (Source: Barnett et al. [13].)

2.2. High-Pressure Compaction with the Piston–Cylinder Device

The equipment for the compaction of nanosize powders, originally developed by Chen et al. [8], was designed to perform at least four functions. The first, and most important, was to produce a 3-mm-diameter disk-shaped sample so that accurate measurements of density, hardness, fracture toughness, optical properties, and pore-size distribution could be carried out. Prior DAC samples, because of their miniature size, did not permit many of these measurements to be made easily or with the desired accuracy. The second function was to provide a means of measuring the compaction rheology of powders in the form of the volume (or density) of the compacting sample as a function of the applied pressure, the rate of pressure application, the processing temperature, and the environment. To provide complete information about the compaction process, the authors designed the system so that the previously mentioned parameters were measured continuously and simultaneously. Also, recognizing that nanosize particles are extremely reactive chemically, the third function was to design the compaction chamber to support a vacuum, as well as a positive gas pressure to permit various environments to be used. With such a system, the authors had the capability to experiment with different gases and fluids under selected pressures for protection and lubrication of the surfaces of the nanosize particles prior to compaction. From a theoretical standpoint, it is reasonable to expect that the thickness of the adsorbed coating on the particle surfaces and properties of the coating should depend strongly on two parameters: (1) temperature and (2) gas pressure inside of the environmental chamber. Because of the large amount of data acquired during an experimental run with their device, the authors built in a real-time computer control of all measurements and experimental procedures. To detect and measure accurately phase transitions accompanied by a volume change, they designed the equipment to provide accurate and precise data for the development of pressure–volume (PV) curves. Finally, the fourth function was to design the equipment to withstand temperatures of up to 1000 K for short periods of time. Owing to the importance of this equipment in the high-pressure compaction of nanosize ceramic powders, a detailed description of the compaction apparatus developed by Chen et al. [8] and later used by Gonzalez et al. [14] follows.

3. PISTON–CYLINDER DIE

3.1. Equipment Configuration

A 10,000-kg-capacity screw-driven Instron press with two large parallel platens was used to apply load to the die. The servocontrolled drive system permitted the selection of platen displacement rates ranging from 0.85 to 850 μm/s, a feature that is useful for the study of stress relaxation in a powder during compaction. In this apparatus, a 2100-kg load is required to generate a pressure of 3 GPa on a 3-mm-diameter sample.

The authors designed their equipment to be used under two different modes of operation for the powder compaction experiments: (1) gas lubrication and (2) liquid lubrication. In the first mode, a small amount of lubricating cover gas is introduced into the environmental chamber under controlled pressure. The introduced gas condenses on the surface of the particles and the powder is compacted under dry or semidry conditions. The basic design for this first experimental setup is shown in Figure 3. The environmental chamber (14) is designed for isolating the sample in order to control compaction conditions while the movable piston (13), equipped with two O-ring seals, applies force to the compacting piston (2). Lubricant gas can be introduced from a gas tank into the vacuum chamber (14), and the gas pressure is controlled by a vacuum needle valve (not shown). A thermocouple gauge located outside of the chamber on the supply line measures the gas pressure in the chamber. A single-acting (bottom punch fixed) piston–cylinder-type device (3) is used for the application of variable pressures up to 3 GPa. The piston (2) and the die (3) are made of tungsten carbide and ground to about a 0.1-μm finish on the contact surfaces. The bottom end of the cylindrical hole of the die has a small conical shape for easy removal of the compacted samples. The temperatures can be varied from that of liquid nitrogen, 77 K, to room temperature. A cooling cylinder (4) and a heater (5) are designed with a coil wound on a frame surrounding the die. A cooling loop of LN$_2$ can pass through a heat-insulating flange into the environmental chamber.

For cryogenic compaction, the applied load is adjusted continually at low-temperatures to maintain the desired pressure. One difficulty is to measure the pressure in nonhydrostatic and low-temperature environments. The frictional force on the wall of the die, the punch, and the sample must be taken into account for pressure calibration. As depicted in Figure 3, two load cells are used for pressure measurement. One, the main load cell, measures the primary force applied by the press and is located on the bottom platen of the press. The other load cell (11) measures the frictional load between the punch and the die. The authors are not aware of any commercially available load cells calibrated for use at extreme temperatures. Therefore, they incorporated into the design for low-temperature compaction a special thermally insulating mechanical structure to maintain the frictional load cell at room temperature. To solve this temperature problem of the load cell, they used

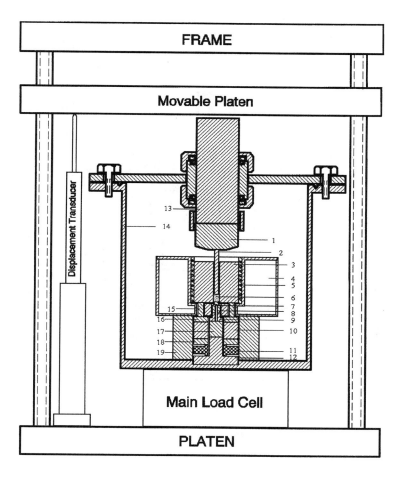

Fig. 3. Configuration of "gas lubrication"–type powder compaction assembly: (1) Bridgman anvil, (2) moving punch, (3) die, (4) cooling cylinder, (5) heater, (6) powder, (7) thermocouple, (8) fixed punch, (9) hardening disk, (10) heat-insulating ring disk, (11) frictional load cell, (12) positioning and heat-conducting disk, (13) pressure transmission piston, (14) vacuum chamber, (15) frictional load transmission pistons, (16) hardening ring disk, (17) heat-insulating disk, (18) hardening ring disk, and (19) plastic alignment ring. (Source: Chen et al. [8].)

frictional force transmission pistons (15) and heat-insulating glass–ceramic (Macor) ring disks (10). The pistons (15) transmit frictional force from the die to the load cell that operates at room temperature. For low-temperature compaction, two ceramic disks (10, 17) prevent heat transfer from the room temperature region to the cold region. The average pressure at the bottom of the sample is calculated from the total and frictional load data.

A K-type thermocouple (7) was placed at the center of the fixed punch as near the sample as possible to provide measurement and control of the temperature. A displacement transducer was positioned between the two platens of the press, as shown in Figure 3, to measure the thickness of the sample versus the force on the sample for the rheological investigations.

An alternative design of their experimental setup to study the effects of lubricants on the compaction of powders is shown in Figure 4. In this design, a cryogenic container (3) holds the die (2) and powder sample (4), and a metal (Inconel 600) foil (6) prevents leakage of LN_2 or other lubricating liquids. The load of the sample (total load minus friction) on the fixed punch (5) is transmitted via the metal foil (6), the hardened-steel load transmission disk (8), the heat-insulating disk (9), and the positioning and heat-conducting disks (13)

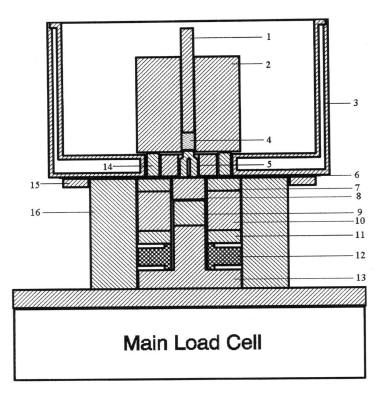

Fig. 4. Configuration of "liquid lubrication"–type powder compaction assembly: (1) moving piston, (2) die, (3) cryogenic container, (4) powder, (5) fixed punch, (6) metal foil, (7) hardening ring disk, (8) hardening disk, (9) heat-insulating disk, (10) heat-insulating ring disk, (11) hardening ring disk, (12) frictional load cell, (13) positioning and heat-conducting disk, (14) frictional load transmission pistons, (15) flange, and (16) plastic alignment ring. (Source: Chen et al. [8].)

to the main load cell as shown in Figure 4. The frictional load on the die (2) is transmitted to the frictional load cell via frictional transmission pistons (14), the metal foil (6), the hardened-steel ring disk (7), the heat-insulating ring disk (10), and the hardened-steel frictional load transmission ring disk.

3.2. Computer Control and Software Development

The authors implemented real-time computer control to measure and/or control (1) the temperature of the chamber, in the case of hot pressing; (2) the total applied and frictional forces; and (3) piston displacement. The applied pressure and rate of pressure application were calculated from measured forces, times, and areas. The configuration of the computer control setup is shown in Figure 5. The computer controls the temperature from remote points with a temperature control unit by means of a digital-to-analog converter card. A digital multimeter combined with a scanner performs the voltage measurements on the load cells, the displacement transducer, and the thermocouples. An IEEE-488 general-purpose interface bus card was used for communicating between the computer and the digital multimeter. The computer controls the sequence of scanning measurements, presets the temperature in the environmental chamber, and conducts autodetermination of steady-state temperature.

The control program uses a built-in clock as the time counter for the measurement of pressure as a function of time, that is, the rate of pressure application. The code of the program for controlling and monitoring the powder processing operation was written in

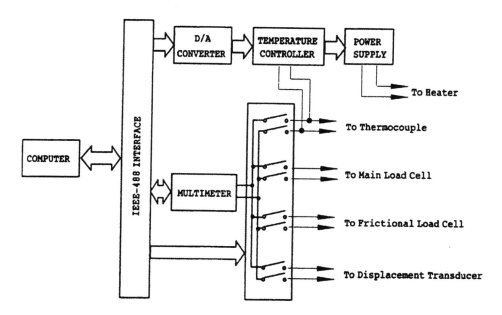

Fig. 5. Block diagram of computer control system. (Source: Chen et al. [8].)

a GFA-BASIC language for MS-Windows on a 486 IBM-compatible personal computer. Eight windows are opened to plot and display experimental data. The following data are displayed graphically in six windows: (1) sample volume vs applied pressure, (2) calculated density vs pressure, (3) calculated density vs time, (4) pressure vs time, (5) temperature vs time, and (6) load fraction (real load on sample or total load minus friction/total load) vs the applied pressure. The numerical data output and picture simulation of the piston position in the die and change of temperature are displayed in windows (7) and (8). During an experimental run, the program estimates the range for each measurement according to the initial scanning measurements and some preset values, and replots all diagrams using an autoscale subroutine to give an appropriate range and boundary for each quantity. The replotting is performed at a preset time interval.

A typical curve of the load fraction (real load/total load) vs pressure for dry compaction of nanosize γ-Al$_2$O$_3$ is shown in Figure 6. The load fraction depends on the type of powder, the magnitude of the gap between the punches and the die, the roughness of the contact surfaces, the sample thickness, and the lubrication conditions. The figure shows that the contribution of friction for a real load calculation is larger in the lower-pressure region.

To determine the accuracy of the PV curves measured by this apparatus, two well-studied materials with known pressure–volume characteristics, rubidium bromide and potassium chloride, were selected by the authors. Both alkali halides (RbBr and KCl) at room temperature exhibit face-centered cubic (fcc) structures that undergo a polymorphic transition to a denser simple cubic (sc) form at elevated pressures [15–17]. Both materials were in the form of powders with primary particle size in the range of 1 μm. X-ray powder diffraction analysis indicated that both materials were in the fcc phase in the form of granular crystalline powders and that the degree of hydration was minimal. Hydration is known to cause erroneous results in determining accurate transition pressures for these materials [17]. The results of the compaction of RbBr and KCl at room temperature are presented as pressure vs volume curves in Figures 7 and 8, respectively. These experiments demonstrated the ability and sensitivity of the instrument in measuring volume changes as a function of pressure.

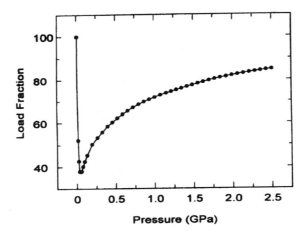

Fig. 6.   Fraction of real load (total load minus friction) and total load versus pressure for compaction of nanosize γ-Al$_2$O$_3$ powder. (Source: Chen et al. [8].)

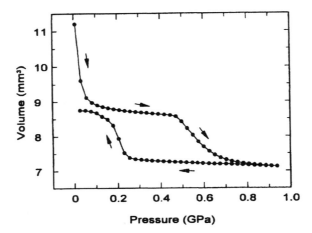

Fig. 7.   Hysteresis behavior and volume phase transition of RbBr granular crystalline powder. (Source: Chen et al. [8].)

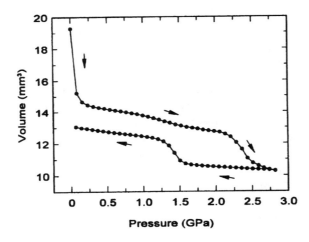

Fig. 8.   Hysteresis behavior and volume phase transition of KCl granular crystalline powders. (Source: Chen et al. [8].)

4. COMPACTION AND LUBRICANTS

4.1. Compaction of Si_3N_4 Powder

As previously mentioned, for nanosize particles, the van der Waals attraction forces can prevent the particles from sliding past each other, and, thereby, promote agglomeration. As a result, compacts of nanosize particles usually have low green densities after cold pressing. As demonstrated using several examples in a later section of this chapter, the use of suitable lubricants can improve the packing properties of the nanosize particles, but, in selecting a lubricant, one is severely limited because of the strongly reacting interfaces between particles and the small size of voids in the resulting green microstructure formed by densely packed nanosize particles [10]. Generally speaking, the diameter of the channels that exist in the structure of densely packed particles is approximately 1/10th that of the particle size. Therefore, for a 10-nm-diameter particle, the channel diameter could be approximately 1 nm. Not many molecules exist that can penetrate such small channels. Because the lubricant must be eliminated from the densely packed structure prior to sintering, one must select a lubricant that consists of very small molecules. Moreover, the selected molecule must adsorb on the surface of the nanosize particles strongly enough to provide good sliding properties during cold compaction, but it must also be eliminated easily after achieving a densely packed structure. This could be accomplished by heating in vacuum for an appropriate length of time.

As an example of the effects of lubricants, we will discuss the work done by Pechenik et al. [10] on the compaction of nanosize Si_3N_4. Pechenik et al. used amorphous stoichiometric Si_3N_4 powder produced via a laser-driven gas phase reaction between silane and ammonia. More details about the preparation of this powder can be found elsewhere [18]. The average particle size of this powder was estimated to be between 16 and 17 nm by transmission electron microscopy, and the starting powder had an oxygen mass fraction of approximately 6%.

The initial experiments by Pechenik et al. [10] were conducted in the DAC under three different environmental conditions: (1) RT, (2) under LN_2, and (3) after outgassing at 200 °C. The samples were compacted according to the following procedures. The starting powder was precompacted in a WC/Co die under 0.1 GPa to facilitate the handling and loading of the nanosize powder in the DAC. The precompacted powder achieved a density of approximately 30% of theoretical random close packing. For the room temperature compaction, the powder was loaded in the DAC and pressed to 5 GPa. A molybdenum gasket was used to confine the sample between the two opposed anvils in the DAC and served as the sample container. The maximum size of the compacts fabricated by the DAC in this work was 0.2 mm in diameter by 0.15 mm in thickness. Pressures were measured by the calibrated shift of the ruby fluorescence R_1 line. For this purpose, a small ruby sphere (about 10 μm in diameter) serving as the pressure sensor was included with the sample. In some cases, the samples were hot-pressed for 3 h under a pressure of 5 GPa to further study the effects of compaction.

Cryogenic compaction (LN_2) was performed following a more complex procedure. The gasketed DAC, containing both the Si_3N_4 precompacted sample and the small ruby pressure sensor, was first sealed and isolated from the external environment at RT by applying a nominal load to the gasket. The DAC was then immersed in an LN_2 bath. After about 10 min, it reached equilibrium with the LN_2. When temperature equilibrium was attained, the load on the gasket in the DAC was reduced to zero, thereby breaking the seal between the gasket and the diamond surface and exposing the precompacted Si_3N_4 powder to LN_2. After 1 min, the cell was resealed by applying the load to produce pressures ranging from 1 to 3 GPa on the sample, which now included some LN_2. The pressurized DAC was then removed from the LN_2 bath and warmed to RT. At RT, the sample pressure was measured and defined as the initial compaction pressure. Pressure cycling at RT was then carried out and consisted of reducing the pressure to ambient and then increasing it again to the desired maximum pressure defined as the final compaction pressure (always 5 GPa for these LN_2 experiments). The pressure cycling procedure was carried out usually two or three times until the LN_2 had been eliminated from the sample.

The third group of samples was prepared in a specially designed vacuum chamber that accommodated the DAC and also permitted access to rotation of the load-transmitting screw in the DAC. The loaded cell was heated to 200 °C for 2 h inside the vacuum chamber. The sample was also sealed under vacuum by applying pressures of 0.5–1.0 GPa.

To evaluate the effects of the different compaction conditions, the authors used the hardness of the sample as a measure of compaction efficiency. The authors measured the hardness using a Vickers indenter. Figure 9 summarizes the results for green bodies and sintered samples. It is clear from the plot that the samples processed under LN_2 were always harder. On the contrary, the outgassed samples had the lowest hardness, suggesting that vacuum heating can change the flow properties of the powders during compaction. Pechenik et al. [10] speculated that an adsorbed substance on the surface, either a gas or a liquid, acts as a lubricant during compaction, and that this layer is eliminated during vacuum heat treating. The authors emphasized that samples processed in LN_2 were mostly transparent to visible light and the outgassed samples and samples compacted at RT were typically opaque.

The authors also presented data for compaction at different pressures to show that hardness increases with compaction pressure (Fig. 10). Because these powders are not expected to deform plastically during compaction like metals, the reported increments in hardness would be expected to be a result of a tighter and more efficient packing density. Again, their results indicate that the LN_2-compacted samples were always harder, even after heat treating at high temperatures as indicated in Figure 9.

Chen et al. [8] also studied these powders using the piston–cylinder compaction apparatus. Figure 11 shows a maximum random packing density of approximately 64% for the nanosize amorphous silicon nitride compacted at less than 2.5 GPa pressure under LN_2 using the piston–cylinder apparatus described earlier. At the same pressure, for a dry compaction of the same powder, the green density was about 57% of theoretical. It is important to note that the rate of volume change with pressure is higher for the sample compacted under an LN_2 environment. This suggests that friction between particles has decreased and particle rearrangement is occurring at lower pressures.

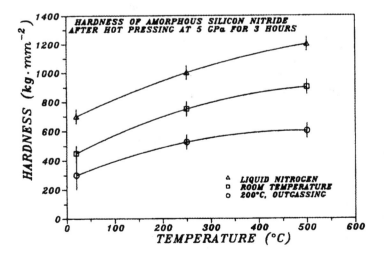

Fig. 9. Comparison of hardness measurements of amorphous Si_3N_4 specimens as a function of hot-pressing temperature. All samples were pressed in the DAC and hot-pressed at 5 GPa for 3 h. Each curve represents a different initial compaction technique; liquid nitrogen, room temperature, and 200 °C outgassing during initial compaction. (Source: Pechenik et al. [10].)

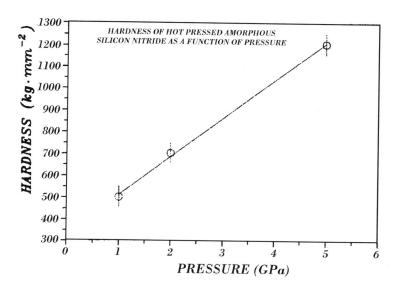

Fig. 10. Hardness of hot-pressed amorphous Si_3N_4 specimens as a function of compaction pressure for samples compacted under liquid nitrogen and hot-pressed in the DAC to the respective pressures at 500 °C for 3 h. (Source: Pechenik et al. [10].)

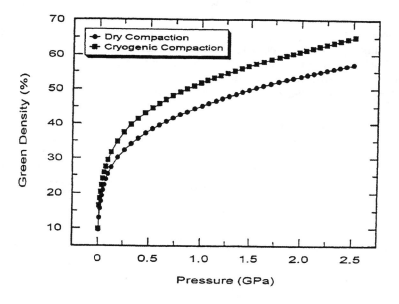

Fig. 11. Variation of density of green compacts of nanosize amorphous Si_3N_4 as a function of compaction pressure. (Source: Chen et al. [8].)

These results clearly indicate that cryogenic compaction is an efficient technique to reduce the compaction pressure necessary to obtain high packing densities and small-scale porosity. The green body produced by LN_2 processing exhibits transparency under visible light, which is an indication of nanoscale, uniform porosity. The actual mechanism of lubrication has not been studied in detail, or at least we are not aware of any report in the literature. Stronger evidence of the effects of lubricants has been reported in the compaction of γ-Al_2O_3 powders.

4.2. Compaction of γ-Al_2O_3 Powder

We have discussed several times throughout this chapter the difficulty of compacting nanosize powders into green bodies that can be easily examined without crumbling. Nanosize γ-Al_2O_3 powder is also very difficult to compact, and high pressures are typically required to obtain structurally sound green bodies. The following discussion presents results reported by Gallas et al. [19] and Gonzalez et al. [14] on a number of compaction attempts of nanosize particles of γ-Al_2O_3. Samples were prepared at RT, and using LN_2 as a lubricant. The main objective of the authors was to produce a compact of γ-Al_2O_3 with uniform and relatively high green density (60%–70%). Without first producing a homogeneous compact of random-closed-packed particles, the green-state body cannot be sintered to high density because coarse porosity is not easily eliminated via surface diffusion, the predominant mechanism for sintering compacts of nanosize particles [20].

As presented in the earlier discussion on the compaction of nanosize particles of silicon nitride, it was shown that dense packing was not achieved by using high pressures alone [10, 21]. It was found that LN_2 provided an excellent lubricant for moving the silicon nitride particles closer together. The hardest compacts were those that were fabricated under LN_2. In light of this previous work, LN_2 also was used to compact nanosize particles of γ-Al_2O_3 and the results were compared with room temperature compactions. Both methods produced transparent green-state compacts, which were then heat treated under vacuum. Hardness was measured before and after heat treatment. In addition, the compacts were analyzed by energy-dispersive X-ray diffraction to identify the crystallographic phases present, and by transmission electron microscopy (TEM) to characterize the microstructure.

The starting nanosize γ-Al_2O_3 powder used was a commercially available material, Aluminum Oxide C, supplied by Degussa AG, Geschäftsbereich Anorganische Chemieprodukt, Frankfurt, Germany. It had an average particle diameter of 20 nm. Angle-dispersive X-ray powder diffraction measurements on the starting powder indicated the presence of primarily γ phase with some δ phase, estimated from diffraction intensities to be less than 10% by volume. The powder was used as received without any prior treatment and was handled in laboratory air.

The experimental procedures used were very similar to the steps followed for the compaction of Si_3N_4 reported earlier. The starting powder was precompacted in a WC/Co die under 0.1 GPa and then subsequently compacted in a DAC under pressures ranging from 1 to 3 GPa. The initial compactions were carried out at two different temperatures: (1) RT and (2) LN_2 temperature. In the case of room temperature compaction, the precompacted starting material was initially pressed to transparency. Typically, transparency was reached at about 1 GPa. However, final sample pressures exceeded this value and ranged from 2 to 4 GPa. The final sample pressure was reached in two ways: (1) by one pressure cycle and (2) by multiple (normally two) pressure cycles. It was found that the hardness of the green-body compact was independent of the number of cycles used to reach the final sample pressure at room temperature. This is an important detail because, as will be discussed in the following paragraph, the procedure for LN_2 compaction requires cycling, and confirms that the increase in hardness is exclusively a lubrication effect.

The compaction of the γ-Al_2O_3 sample under LN_2 (cryogenic compaction) was performed by immersing the DAC in LN_2 as was done with Si_3N_4. Pressure cycling at RT was then carried out and consisted of reducing the pressure to that of the ambient atmosphere and then increasing it again to the desired maximum pressure defined as the final compaction pressure (always 3 GPa for these LN_2 experiments). The pressure cycling procedure was carried out usually two or three times until the LN_2 had been eliminated from the γ-Al_2O_3 sample.

After final compaction and removal from the DAC, the transparent green-body compact of nanosize particles of γ-Al_2O_3 was evaluated by measuring its hardness using a Vickers microhardness indentor at a fixed load of 50 g. Optical clarity was evaluated by microscopy examination. The energy-dispersive X-ray powder diffraction technique was used to determine the polymorphic phases present.

In the final step, the transparent green-body compacts produced by both RT and LN_2 compaction procedures were heat treated in a tube furnace under 100 Pa vacuum at 800 °C for 10 h. Heating and cooling rates of 5 °C/min were used in the heat treatment procedure. The molybdenum gasket was found to be chemically inert to γ-Al_2O_3 at the 800 °C heat treatment condition. After heat treatment, the optical transparency, microhardness, and X-ray characterization were again evaluated. The microstructure of the heat-treated and green-state compacts were also characterized by TEM.

It is important to mention that no phase change occurred during the compaction procedures, and, therefore, the possibility of a different phase of alumina contributing to changes in hardness was ruled out. The authors confirmed this by studying X-ray diffraction patterns collected from both the RT and the LN_2 compacted samples. The authors pointed out that, while optical transparency in the green-state compact is necessary, it is not a sufficient condition for achieving dense packing of nanosize particles [21]. Optical transparency indicates that particles are packed such that the pores are reduced in size to below 100 nm and suggests sample uniformity. Thus, optical clarity alone cannot be used as an indicator for characterizing density or fine-scale homogeneity of particle packing. Similar to the Si_3N_4 study, the shear strength of a compact, characterized by measuring the Vickers hardness, is thought to be a reliable parameter to characterize the density of particle packing because it depends strongly on the coordination number of the particle arrangement. Therefore, the authors used Vickers microhardness numbers as a measure of the quality and efficiency of particle packing.

Figures 12 and 13 summarize hardness measurements done before and after heat treatment (HT) as a function of the compaction pressure applied at RT and at LN_2 temperature, respectively. The HT procedure was the same for all the γ-Al_2O_3 samples (800 °C for 10 h under 100 Pa vacuum) so that relative differences are strictly due to the compaction conditions, for example, RT and LN_2 compaction. The general features of both Figures 12 and 13 are similar. The hardness after heat treatment (H_a) appears to be strongly dependent on the hardness before heat treatment (H_b). In both cases, the ratio H_a/H_b is roughly constant at about 1.5. Furthermore, the hardness before heat treatment appears to be directly dependent on the pressure and temperature of compaction.

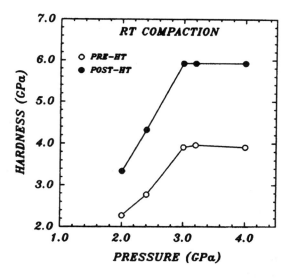

Fig. 12. Hardness of γ-Al_2O_3 compacts as a function of final pressure used to compact the powder at room temperature (RT) before and after a heat treatment of 800 °C for 10 h. (Source: Gallas et al. [19].)

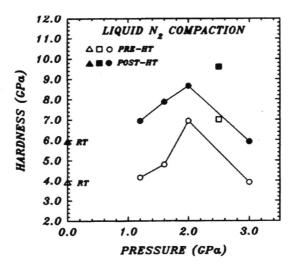

Fig. 13. Hardness of γ-Al$_2$O$_3$ compacts as a function of final pressure used to compact the powder at liquid nitrogen temperatures before and after a heat treatment of 800 °C for 10 h. The two triangles in the hardness axis designate the measured limiting hardness for samples pressed at room temperature (Fig. 12). The solid and open square symbols designate measured hardness values for the samples heated at 100 °C to avoid the β phase of N$_2$ prior to retrieval of the green-state compact. (Source: Gallas et al. [19].)

For samples compacted at RT for pre- and post-HT, the hardnesses were not as high as the values measured in the cryogenic samples, indicating less dense compacts, even though they were also optically transparent (Figs. 12 and 13). For 2 GPa, the hardness of the RT sample is much lower than the hardness measured for the LN$_2$ sample at this same initial pressure, under both conditions, before and after heat treatment. Another interesting result shown in Figure 12 is that the hardness reaches a limiting value for both the green body and the heat-treated compact and is nearly constant for the compaction pressure between 3 and 4 GPa.

In Figure 13, for the pressure range up to 2 GPa, the hardness increases with the applied initial pressure and then decreases abruptly at 3 GPa. The highest hardness (approximately 9 GPa) was obtained with the initial compaction pressure under LN$_2$ between 2.0 and 2.5 GPa. When pressures exceeding this range were used, lower hardnesses were obtained. Note that precompaction under LN$_2$ at 3 GPa followed by ambient compaction at 3 GPa (standard procedure) resulted in a sample hardness comparable to that obtained for RT compaction identified on the hardness axis in Figure 13 as triangles. This result indicates that all beneficial effects of cryogenic compaction are lost when the pressure exceeds roughly 2.5 GPa.

The unexpected decrease in hardness observed for the powder compacted initially at 3 GPa under LN$_2$ may be explained by referring to the equilibrium pressure/temperature phase diagram for nitrogen (Fig. 14). Of interest here are the successive phase transformations that occur in N$_2$ with increasing pressure at different temperatures. The temperature of concern here is approximately 77 K, the equilibrium temperature at which the γ-Al$_2$O$_3$ powders were initially compacted. According to this pressure/temperature phase diagram for N$_2$, there are three solid phases, β, γ, and δ, which exist between ambient pressure and 2.5 GPa at 77 K [22–24]. As the temperature increases in the pressure range of concern here, both the γ and the δ phases transform to the β polymorph, which is the form that is in equilibrium with the liquid phase. At RT and 2 GPa, only the liquid phase of nitrogen is stable and is, therefore, present with the powder compact. However, if the powder is pressed initially to 3 GPa at 77 K, then we are on or very near to the phase boundary between the ε and δ polymorph so that either polymorph or a mixture of both solid phases of

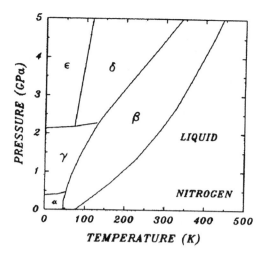

Fig. 14. Equilibrium pressure–temperature phase diagram for nitrogen showing pressure and temperature range pertinent to this work. Reprinted with permission from D. A. Young, "Phase Diagrams of the Elements," pp. 113–117 (© 1991 University of California Press).

nitrogen may exist. When the temperature increases to ambient, the alumina sample may be subjected to the effects of several nitrogen phase transformations, each involving an expansion to a less dense phase. For example, if the ε phase is produced at low temperature by applying a pressure of 3 GPa, then, on warming to RT, the ε phase transforms to the δ phase, which, in turn, transforms to the β phase. At 3 GPa, the stable phase at RT is the β phase rather than the liquid as is the case at lower pressures. Thus, the initial dense packing of the alumina powder achieved by compacting to 3 GPa at low temperatures is significantly altered when the temperature is increased to ambient. This is a consequence of the successive expansions that the solid nitrogen, trapped in the interstices of the powder, undergoes when it transforms from the ε phase, to the δ phase, and, ultimately, to the β phase. However, this may not completely explain the experimental observations. An additional factor must be considered when the solid β phase, trapped in the interstices of the alumina powder, melts.

At RT, LN$_2$ transforms to solid β-N$_2$ at 2.44 GPa. The β phase of nitrogen has a molar volume that is about 20% less than the molar volume of the liquid phase [22–24]. When β-N$_2$ melts on lowering the pressure on the compact at RT, the nitrogen pressure-transmitting medium experiences a 20% expansion. Because most of the solid nitrogen occupies interstices in the powder compact, the resulting expansion upon melting tends to reopen the structure of the compacted powder, thereby decreasing its bulk density. Thus, the mechanism that permits an efficient particle packing under LN$_2$ must also include the phenomenon of phase transformations that the surrounding nitrogen medium experiences. The solid-to-liquid phase transformation, in particular, involves a significant increase in molar volume of the N$_2$. However, the manner in which the phase boundary is traversed to produce the melt is also important because it will have a greater or lesser effect on reopening the compressed structure. If the boundary is traversed by lowering the pressure at constant temperature to produce the liquid, as in the case of the 3-GPa run, then there will be a larger net expansion than when the melt is produced by increasing the temperature at constant pressure. In the former case, although the transition volume is the same, to produce the liquid, the pressure has to be lowered and, because of considerable hysteresis in the transition pressure, the final state can be at a substantially lower pressure and, thus, have a larger sample volume than in the latter case. Thus, the particles are not as well packed when the solid/liquid boundary is crossed by lowering the pressure.

To test this theory of the roles that the various phases of nitrogen play in the cryogenic compaction process, the authors fabricated a compact of γ-Al_2O_3 following a procedure that avoided the presence of the solid β-N_2 phase at the final compaction pressure of 3.0 GPa. An initial cryogenic compaction of 2.5 GPa was used. At the final RT compaction pressure of 3.0 GPa, the solid β-N_2 phase was melted by increasing the temperature to 373 K. At this temperature and pressure, only liquid nitrogen exists in the γ-Al_2O_3 sample. On retrieval of the green body from these conditions, the hardness of the transparent compact (H_b) was measured to be 7.0 GPa (open square symbol in Fig. 13). This value is comparable to the hardness found for the green-state sample processed at 2.0 GPa shown also in Figure 13. The green body was then heat treated according to the procedure followed in their earlier experiments. The hardness of the heat-treated transparent compact (H_a) was then measured and found to be 9.6 GPa (solid square symbol in Fig. 13), a value significantly larger than the highest hardness found earlier for the heat-treated sample processed at 2.0 GPa and plotted in Figure 13. It is of interest to compare the H_a/H_b ratio for these two compacts with the other values. The lowest ratio, 1.25, is for the sample with highest hardness, 8.7 GPa, cryogenically compacted at 2.0 GPa. It also has the least difference between the hardness before and after heat treatment. The ratio, 1.37, for the sample of concern here, indicates that increased packing efficiency, and, therefore, higher hardness, can be achieved by avoiding the presence of the solid β phase of nitrogen at the final compaction pressure. Thus, it appears that it is the liquid state of nitrogen that plays the important role in processing nanosize powders of γ-Al_2O_3 by cryogenic compaction.

Comparison of the results plotted in Figures 12 and 13 revealed that cryogenic compaction produces harder samples at much lower pressures. Because all samples were heat treated under the same conditions, the only difference was in the processing before heat treatment, so the microhardness results must be related to the process of compaction. In the earlier work with Si_3N_4 [8], it was clear that LN_2 played an important role as a lubricant in the compaction process. These present results on γ-Al_2O_3 support this earlier conclusion and have also extended the understanding of the mechanism involved in this cryogenic lubrication process. However, it is interesting to note that, in the earlier work with Si_3N_4 [8], the effects of the LN_2 phase transformation on the packing density were not investigated by the authors.

To determine whether the observed differences in hardness with processing conditions can be related to microstructural differences, selected samples were examined by TEM. These samples included one green-body compact, pressed under LN_2 ($H \approx 4$ GPa), and five heat-treated samples. Three of the heat-treated samples were compacted in LN_2 and had hardnesses of approximately 7, 8, and 10 GPa. The other two were compacted at ambient temperature and had hardnesses of approximately 4 and 6 GPa. With the exception of one sample, which was ion milled to electron transparency, TEM sample preparation simply involved fragmenting the compacts and then containing the fragments between two grids with predeposited, holey carbon films. Representative views, illustrating the microstructure of the nanoscale aluminas, are shown in Figures 15 and 16a–c. A typical large-area electron diffraction pattern is included as an insert in Figure 15.

TEM examination of these samples, including the ion-milled sample where there was no fragmentation of the specimen, revealed no significant differences in microstructure. With reference to Figure 16, each sample can be similarly described as a uniform aggregate of nanoscale particles. Selected area electron diffraction gave "spotted" ring patterns, largely consistent with cubic γ-Al_2O_3. The presence of the tetragonal δ-Al_2O_3 phase, established as a minor constituent by angle-dispersive X-ray diffraction, could only be inferred from the excessive widths of individual rings in these patterns.

As illustrated in Figure 16, both green-state (part a) and heat-treated compacts (parts b and c) consist of densely packed particles of irregular, but nearly spherical shape. Particle sizes ranged from approximately 5 to 40 nm, but were most commonly 15 to 20 nm in size, corresponding to the quoted starting powder size. In all cases, the surfaces of individual

Fig. 15. Fragment from compact exhibiting highest hardness (nearly 10 GPa) attained in this study. Note uniform distribution of particles and interconnected porosity. Electron diffraction (insert) confirmed major particle phase is cubic γ-Al_2O_3. (Source: Gallas et al. [19].)

particles appear rough because of fine-scale surface asperities. These asperities are evident not only along the edges of particles, but also along the junctions separating adjacent contacting particles. Although it was not possible by the authors to characterize further the nature of the interparticle junctions, it is important to note that heat treatment did not result in any resolvable changes in these junctions. Finally, all samples, including the ion-milled specimen, appeared to contain a similar, appreciable volume fraction of open or interconnected porosity. Thus, the observed porosity cannot be considered an artifact related to the fracture process used in TEM sample preparation. Distinct differences in porosity could not be reliably established, regardless of differences in processing conditions or measured hardness. Here, however, it should be noted that the void structure can only be defined with reasonable certainty within small regions at the fragment edges, which are about three particle diameters or less in thickness. At best, comparative results from such regions (as in Fig. 16) showed that void dimensions never exceeded the average particle size (15–20 nm) and were typically much smaller. Accordingly, it appears that, for even the lowest compaction pressures, the level of particle packing achieved can be classified as "random dense packing." The significance of this result is that, at such levels of particle packing, the network of interconnected particles becomes load bearing, and stress-induced changes in the structure and density of the compact with increased pressure require relative displacements of the particles, for example, interparticle sliding. Here, structural changes to the particles, resulting from either plastic deformation or fracture of the particles, or from heat treatment (as will be discussed), are not considered relevant based on the observations by the authors.

The absence of discernible differences in porosity in the TEM results complicates discussion of the effect of compaction pressure on the structure and hardness of these compacts. On the one hand, the observed increase in green-state hardness with increased pressure may be due to small, but significant, increases in particle packing density, defined only by correspondingly small decreases in the sizes of distributed voids. On the other hand, it is also possible that the nature of the stresses acting on the compact changes as dense packing is achieved and results in constrained compression rather than pure hydrostatic compression. Under such conditions, deformation of the compact can occur with little change in density. As deformation involves particle motion, realignment occurs and leads to an increased resistance to further deformation because of particle interlocking. This increased resistance to deformation is ultimately reflected in the measured hardness. In any case, it appears that any change in the structure of particles with increased pressure occurs by particle sliding and the associated particle realignment results in more efficient packing. Moreover, because changes in structure occur by interparticle sliding, they will

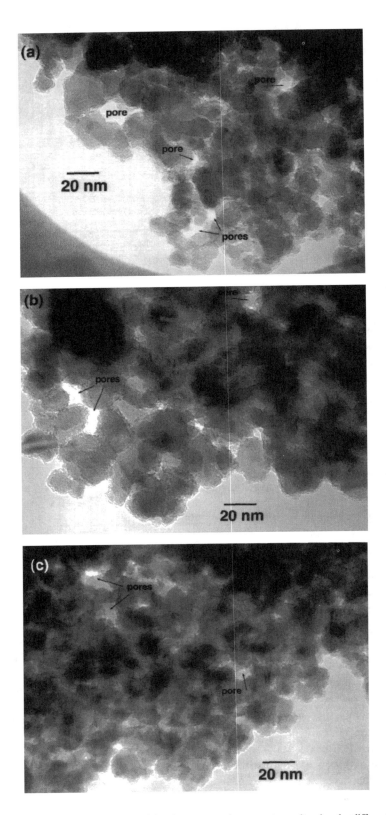

Fig. 16. Comparative views of particle microstructure in compacts produced under different conditions and exhibiting different hardnesses: (a) green body, LN_2 compacted, $H = 4$ GPa; (b) heat treated, ambient temperature compacted, $H = 4$ GPa; and (c) heat treated, LN_2 compacted, $H = 10$ GPa. Note microstructural similarity and dimensions of large pores (P) relative to particle sizes. (Source: Gallas et al. [19].)

be governed largely by the elastic/plastic properties of the particles, the nature of forces binding the particles, and the properties of the interparticle or "void space" phase. On this basis, the pronounced effect of compaction under LN_2 appears directly related to the increase in elastic modulus of the particles at low temperatures and to the presence of a liquid intergranular phase. As has already been suggested, LN_2 appears to act as an effective lubricant, facilitating particle rearrangement to give a densely packed microstructure at low pressures.

The effect of heat treatment on hardness cannot be similarly explained in terms of particle rearrangement or densification. In fact, comparative observations on green-state compacts and heat-treated compacts (e.g., Fig. 16a vs Fig. 16b and c) revealed no apparent changes in particle size, shape, and surface texture or in the appearance of the particle contact junctions. This leads to the conclusion that only extremely limited transport, whether by intrinsic (bulk), surface, or interfacial diffusion, occurred during heat treatment at 800 °C. Thus, the roughly 50% increase in hardness universally realized by this heat treatment cannot be attributed to various microstructural changes that normally occur by diffusion during sintering. Instead, the authors speculated that pressureless heat treatment brings about a change in the nature of the forces acting across the contact junctions of these nanosize particles. Conceivably, heat treatment results in a change from strictly surface force attraction (e.g., van der Waals) to one that includes interatomic bonding (e.g., covalent interactions). Considering the irregular structure of the junctions between particles, interatomic bonding would appear to be limited to the small junctions formed by surface asperity contact. The authors admit that this conclusion is speculative and can only be resolved when complete characterization of the particles, particularly in the surface regions, is carried out.

Finally, we should point out that the authors failed to consider the possibility of any solubility of the γ-Al_2O_3 particles in LN_2 that might promote a solution–reprecipitation mechanism. A solution–reprecipitation process would result in the formation of necks between particles and, therefore, an increase in hardness. In fact, we are aware of claims that water-based solutions can dissolve small amounts of different phases of nanosize aluminas, resulting in the formation of strong agglomerates [25]. However, we must say that TEM observations by the authors do not indicate that this mechanism of sintering is occurring during compaction.

More evidence of the lubricating ability of LN_2 during compaction has been reported by Gonzalez et al. [14]. The authors were studying the transformation of γ-Al_2O_3 to α-Al_2O_3 using the high-pressure DAC to compact the nanosize powder. They observed that, when the γ-Al_2O_3 powder was compacted to high compaction pressures using LN_2 as a lubricating medium and pressureless sintered at high temperatures, the final microstructures were more uniform and the specimens had a smaller average grain size than the same powder compacted under dry conditions. The authors claimed that, when the sample is compacted under LN_2, the load distribution across all particle contact surfaces is more homogeneous because there is less interparticle friction, and, therefore, particle rearrangement and redistribution to a uniform packing occurs more readily. This example is once again clear evidence of the benefits of using LN_2 as a lubricating fluid during the compaction of nanosize γ-Al_2O_3.

It is important to recognize that a lubricating fluid that works for one material does not necessarily work for other powders. For instance, Hahn et al. [26] compacted nanosize TiO_2 using high-pressure and high-vacuum environments to obtain densities greater than 70%. On the other hand, Pechenik et al. [10] established that, in the case of the nanosize amorphous Si_3N_4 powder used in his work, the green density of the compacts was the lowest when the samples were compacted at 200 °C under vacuum, in comparison with RT compaction and LN_2 compaction, with the latter producing the highest green densities. According to Nilsen et al. [27], the Si_3N_4 powder used by Pechenik et al. [10] contained physically absorbed water, nitrogen, and other chemisorbed groups on the surface of the particles. Because these absorbed species can be, in many cases, partially or fully removed by exposing the powder to high temperatures and vacuum, the authors argued that this could explain the different compaction behaviors observed under the three different compaction environments studied by Pechenik et al. [10]. Thus, these results suggest that the presence of these species on the surface of the particles reduces interparticle friction. In fact, it is not hard to visualize these surface modifications having a significant effect on the flow properties of the powder and, therefore, on a powder compaction process. Putting it all together, we have presented clear evidence from the literature that shows that the nature of the particle surface and the ability of the surface to absorb the lubricating fluid, either a liquid, a gas, or the formation and growth of a thermodynamically stable surface layer, is very important. We must say that, even though particle morphology was not discussed in detail by the authors, the morphology of the powder particles will have a strong effect on the flow properties of the powder and will surely have a significant effect on the compaction efficiency when a lubricant is used.

The major emphasis in this chapter up to now has been on how to achieve high random packing densities of nanosize powders in the green body, with the expectation that the resulting sintered material will have a nanoscale dimension microstructure. We now present experimental results on the compaction and sintering of nanosize γ-Al$_2$O$_3$ powder that demonstrate that achieving a high-density, random-close-packed green body in the starting material is by no means the ultimate answer to producing a fine-grain nanoscale microstructure in the final sintered material.

4.3. Nanosize γ-Al$_2$O$_3$ Powder Processing

Sintering of alumina powders at relatively low temperatures, 1000–1150 °C, has been reported in the literature [28–30]. In some cases, a solid-state phase transformation has been exploited to aid sintering at these temperatures to produce high-density polycrystalline alumina with submicrometer grain size. For example, Kumagai et al. [28] have studied the effects of seeding transitional oxides of alumina (boehmite) with α-Al$_2$O$_3$ to lower the sintering temperature and increase the kinetics of the transformation to α-Al$_2$O$_3$. Sol–gel techniques were employed in order to obtain homogeneous mixtures of the transitional oxides and the α-Al$_2$O$_3$ seeds. Sol–gel methods, however, suffer from severe cracking during the drying or "curing" period. Yeh and Sacks [30] made slurries of fine-grain α-Al$_2$O$_3$ by conventional suspension techniques for alumina that were used to slip-cast samples to relatively high bulk densities (69%). The samples, sintered at 1150 °C in air, had relative densities greater than 99.5% and an average grain size of 0.25 μm. In this case, a phase transformation was not used, yet dense samples with submicrometer grain size were obtained. Thus, the phase transition appears to be unnecessary to obtain this result. It also appears that it is important to have a uniform density in the green body to reduce internal stresses that occur during inhomogeneous densification. However, making homogeneous suspensions (mixtures) with submicrometer-sized powders is very difficult, particularly when nanosize particles ($\approx$20 nm) are used. The authors recognized that an alternative approach to achieving high-density green compacts with a nanoscale microstructure is to utilize high-pressure compaction techniques.

Earlier in this chapter, it was demonstrated that, by increasing the compacting pressure on nanosize particles, an increase in green-body bulk density can be achieved. However, it has yet to be demonstrated that the gain in green-body density has a significant effect on the densification process that should take place during sintering. To investigate this problem, Gonzalez et al. [14] studied nanosize γ-Al$_2$O$_3$ as a model material. Although the thermodynamically stable phase of alumina, α-Al$_2$O$_3$, is obviously the better choice, these authors were unable to obtain nanosize particles of the α phase commercially. This is a common problem with nanosize particles of ceramic materials, because these powders usually exist as thermodynamically unstable amorphous phases (silicon nitride) or polymorphic crystalline phases (γ-Al$_2$O$_3$) at ambient conditions of pressure and temperature.

These unstable nanophases readily transform to a stable phase, for instance, by elevating the temperature.

For example, nanosize γ-Al_2O_3 particles transform to the α-Al_2O_3 phase at approximately 1150 °C accompanied by rapid grain growth. The kinetics of the transformation have been studied by Dynys and Halloran [31], who reported that the rate of transition is strongly dependent on the mechanical pretreatment of the powder. They showed that mechanical milling of the powder can reduce the time for complete transformation at 1150 °C. In addition, the authors reported that compaction under different loads changed the transformation kinetics. From their work, it appears that both compaction and milling involve the creation of large contact stresses, which can cause an increase in the internal energy of the material. For example, lattice defects can be created, and these can act as heterogeneous nucleation sites for the transformation. This variability in the transformation kinetics complicates the interpretation and understanding of the compaction and sintering results on γ alumina that will now be presented. Nevertheless, the results on γ-Al_2O_3 reported by Gonzalez et al. [14] give us a closer look at the possible effects that high-pressure compaction and high green-body density have on the sintering of nanosize powders in general.

The preliminary work conducted by the authors consisted of preparing green compacts of γ-Al_2O_3 made using the piston–cylinder device described earlier in this chapter. Two sets of green compacts were produced at room temperature using no lubricant. One set was pressed to 1.0 GPa, the other to 2.5 GPa. Samples from both compaction pressures were sintered in an alumina tube furnace under the rough vacuum of a mechanical pump at 1000, 1100, 1200, or 1300 °C for 5 h, employing heating and cooling rates of 300 °C/h. After the heat treatments, the weight and volume of the samples were remeasured to determine the extent of densification. To further understand the densification process, scanning electron microscopy (SEM) and TEM were used to characterize the samples. In addition to the microscopy work, X-ray diffraction was used to identify phase composition.

4.3.1. Results and Discussion

The average densities for the samples compacted at 1 and 2.5 GPa were 2.00 g/cm^3 and 2.37 g/cm^3, respectively, which correspond to 54% and 65% of the theoretical density (3.67 g/cm^3). The average density at each sintering temperature is summarized in Figure 17. The densities increase monotonically for both compaction pressures and approach each other at a density of approximately 3.3 g/cm^3, or 83% of the theoretical density of α-Al_2O_3. It is also of interest to note that the 2.5-GPa compacted sample sintered at

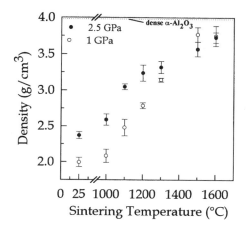

Fig. 17. Density of γ-Al_2O_3 samples compacted to 1 or 2.5 GPa at RT as a function of sintering temperature. (Source: Gonzalez et al. [14].)

$1000\,°C$ showed a slight increase in density, which was not as significant in the 1.0-GPa samples. The authors determined the microstructure and phase fraction of the α and γ phases to gain further understanding of the densification process occurring in their specimens.

4.3.2. Microstructure

Figure 18 shows a TEM micrograph of a green body compacted at 2.5 GPa. Both X-ray diffraction and electron diffraction results on these samples confirmed that they are predominantly γ-Al_2O_3. As seen in Figure 18, the γ particles are equiaxed with an average equivalent spherical diameter of approximately 20 nm. More generally, observations on 1.0 and 2.5-GPa compacts indicate a random-dense-packed particle structure with uniform interconnected porosity. For both compaction pressures, the pore dimensions were less than the particle size. As a consequence, the 16% difference in packing density between the samples compacted at 1.0 and 2.5 GPa was not resolved in TEM.

The samples sintered at $1000\,°C$ have a completely different microstructure. Parts a and b of Figure 19 show, in comparison, TEM micrographs of the samples sintered at $1000\,°C$ for the 1.0- and 2.5-GPa compaction pressures. The sample compacted at 1.0 GPa consists of a mixture of equiaxed particles of γ phase and isolated clusters of predominantly α phase. Even at this temperature, necking between γ particles, indicative of γ phase sintering, was not resolved in TEM. The isolated α clusters grow in a wormy or spongy structure with continuous porosity. X-ray diffraction results corroborate that both the γ and the α phases of Al_2O_3 are present in the 1.0-GPa pressed samples. In contrast, the samples pressed at 2.5 GPa are made up of all α phase. The microstructure is spongy or wormy with continuous porosity similar to the isolated α clusters in the 1.0-GPa samples. Again, the authors confirmed by X-ray diffraction that the 2.5-GPa samples contained α-Al_2O_3 exclusively.

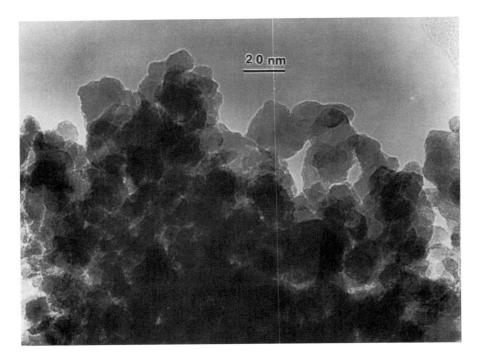

Fig. 18.   TEM bright field image shows the microstructure of a green body of γ-Al_2O_3 compacted at 2.5 GPa. (Source: Gonzalez et al. [14].)

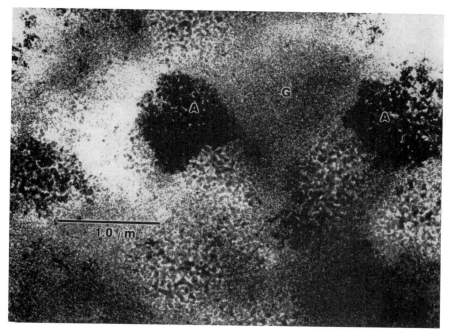

Fig. 19.   Bright-field image in TEM shows the typical microstructure of samples sintered at 1000 °C for (a) 1 GPa where A and G identify regions of α-Al$_2$O$_3$ and γ-Al$_2$O$_3$ respectively and (b) 2.5 GPa. (Source: Gonzalez et al. [14].)

It should be noted that the spongy α-phase structure actually consists of interconnected individual grains, ranging in size from about 100 nm to nearly 1 μm. As illustrated in Figure 19b, where regions of dark contrast define areas of common orientation, the grains have a complex morphology, and the boundaries connecting adjacent grains are typically 50–100 nm in length.

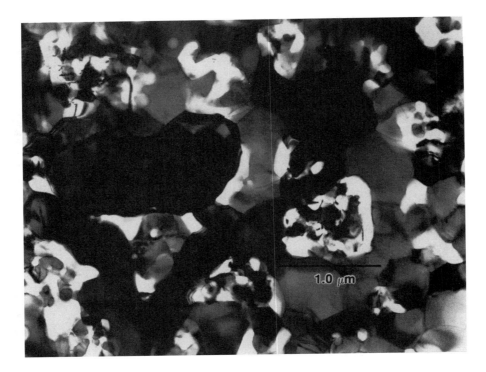

Fig. 20.   TEM bright-field image of 1-GPa compacted sample sintered at 1300 °C. (Source: Gonzalez et al. [14].)

The authors also studied the surface area and pore size distributions by the Brunauer–Emmett–Teller (BET) adsorption isotherm on a green, 1000 °C sintered body (2.5-GPa compacted sample). The average pore size for the green body was 5 nm, consistent with a packing coordination number of 4. However, the porosity determined by TEM ranges in size from 100 to 300 nm, which is an order of magnitude larger than the pore size measured in the BET. The average surface area also decreased during sintering, suggesting neck formation and some densification in agreement with the observed microstructure.

As mentioned earlier, the 2.5-GPa samples showed some densification at 1000 °C, whereas the 1.0-GPa samples did not. It is evident that the samples have different grain morphologies, which contribute to the differences in densities. Another reason for this difference in density may be because the theoretical density of α-Al$_2$O$_3$ (3.987 g/cm^3) is larger than the density of γ-Al$_2$O$_3$ (3.67 g/cm^3). The samples pressed to 2.5 GPa experienced significant shrinkage during the transformation from the γ to the α phase (given the same mass) and, as a result, yielded higher bulk densities. The amount of α phase produced in the 1.0 GPa samples, however, is so small that the change in volume associated with the transition is insignificant and could not be determined by the technique employed.

The microstructural changes in the samples sintered at 1300 °C were also reported. The density of these samples did not exceed 83% of theoretical. X-ray diffraction results indicated that both the 1.0- and the 2.5-GPa compacted samples were 100% α-Al$_2$O$_3$. The 1.0-GPa samples are made up of a combination of the wormy grain structure and equiaxed micrometer-sized grains of alumina (Fig. 20). Significant porosity is visible, and necking between grains is evident. In contrast, the 2.5-GPa compact exhibited only an equiaxed grain morphology (Fig. 21). No evidence for the spongy α-phase structure was observed in these samples.

As originally anticipated, the processing of alumina from γ-phase particle compacts is strongly influenced by the γ-to-α transformation. The results reported by Gonzalez

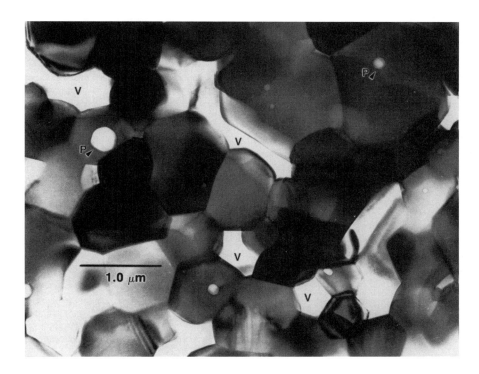

Fig. 21. TEM bright-field image of 2.5-GPa compacted sample sintered at 1300 °C where V and P identify voids and pores respectively. (Source: Gonzalez et al. [14].)

et al. [14] confirm the earlier observations of Dynys and Halloran [31], namely, that agglomerates of nanosize γ particles do not simply transform into agglomerates of similarly sized α particles. Instead, transformation appears to involve heterogeneous nucleation followed by a period of rapid, almost explosive growth. This growth involves the rearrangement of hundreds to many thousands of 20-nm γ particles into single-crystal grains of α phase, ranging in size from 0.1 to 1 mm. With this consolidation of the solid phase, there is a corresponding rearrangement of porosity into pores of micrometer dimensions. This large-scale change in microstructure, brought about by the transformation, clearly complicates the discussion of the effect of initial compaction pressure on the densification that occurs during heat treatment. Nevertheless, the results reported so far do indicate a twofold effect of compaction pressure on the sintering of alumina from nanosize γ particles.

First, increased compaction pressure enhances the γ-to-α transition, reducing the transition temperature or, alternatively, increasing the nucleation rate at temperatures ranging from 1000 to 1150 °C. Evidence for this is based on the observations that, for compacts pressed to 2.5 GPa, complete conversion to the α phase occurs within 5 h at 1000 °C, whereas those pressed to 1.0 GPa showed less than 50% conversion to the α phase. These compacts were completely converted to the α phase at 1100 °C.

Second, increased compaction pressure leads to increased densification upon heat treatment at still higher temperatures. The results show a significant difference in the microstructure, density, and hardness of 1.0- and 2.5-GPa compacts sintered at 1300 °C. For 2.5-GPa compacts, this heat treatment results in equilibrium or faceted α-phase grains of 1 μm or larger dimensions. Adjacent grains tend to be connected by straight grain boundaries of nearly grain size dimensions. Although considerable porosity is retained, it is in the form of distributed voids at multigrained junctions and isolated pores within the grains. By contrast, 1300 °C heat treatment of 1.0-GPa compacts results in limited formation of faceted grains and retention of large (1 μm and larger) porous structures, reminiscent of the

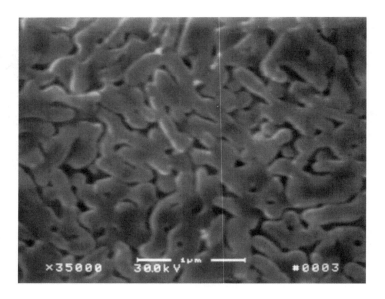

Fig. 22.   SEM micrograph of an alumina sample compacted in a DAC to 2.1 GPa and pressureless sintered at 1150 °C for 2 h. (Source: Gonzalez et al. [14].)

initial spongy α structure. The change from what has been termed the spongy α structure to a faceted grain structure must involve diffusive transport as there is a change in both grain shape and size. Accordingly, the observed differences in microstructures for the 1.0- and 2.5-GPa compacts, after heat treatment at 1300 °C, strongly suggests a relative increase in the transport kinetics associated with the increased compaction pressure. Further study, however, will be required to elucidate the underlying reasons for these observed effects.

In the case of γ alumina, the compaction and sintering results are complicated by the large microstructural changes that occur during the γ-to-α phase transformation, which takes place at about 1000–1100 °C, depending on the initial compaction pressure. The authors realized the relationship between the compaction pressure and the transformation. Their results also agreed with other investigators in that the transformation is nucleation limited and that presumably the higher the compaction pressure the more nucleation sites are created. For this reason, they explored even higher compaction pressures. The samples were compacted in the DAC to 3, 4, or 5 GPa. Because an Inconel X750 gasket is used in the DAC technique, the sintering temperatures were limited to 1150 °C for 5 h. In agreement with the piston–cylinder results, the authors observed that samples compacted at 2.1 GPa developed an α-phase vermicular microstructure as shown in Figure 22. In contrast, the formation of this vermicular structure was completely eliminated in samples compacted at 3 GPa and higher. The 3-GPa samples were sintered at 1150 °C for 1 h (Fig. 23). The authors estimated an average grain size of 223 ± 37 nm (one standard deviation) using the line intercept method. The relative density of the 3-GPa samples was 89% and most of the resolved porosity appeared to be isolated at grain interstices. The authors noticed that, as the compaction pressure was increased, the average grain size decreased and the resulting microstructure was more homogeneous after sintering. Evidence of these observations was also seen in the sample precompacted to 4 GPa. The authors estimated an average grain size of 192 ± 22 nm and a relative density of 93% (Fig. 24). In a similar analysis of sintered 5-GPa compacted samples, the average grain size was determined to be 189 ± 26 nm and the relative density was also 93% (Fig. 25). The pores in these samples were exclusively accommodated at grain interstices. To study the effects of time, the authors sintered the 5-GPa samples at 1150 °C for 1 h. The authors explained that the results from these samples support the contention that, once the a phase nucleates, the grains

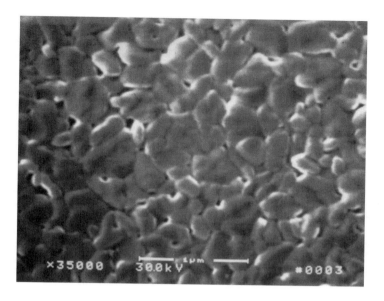

Fig. 23. SEM micrograph of an alumina sample compacted in a DAC to 3.0 GPa and pressureless sintered at 1150 °C for 2 h. (Source: Gonzalez et al. [14].)

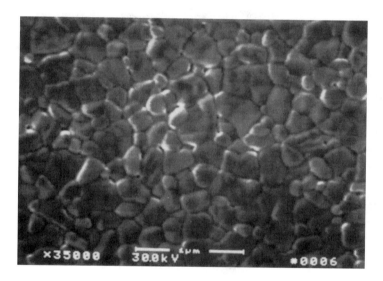

Fig. 24. SEM micrograph of an alumina sample compacted in a DAC to 4.0 GPa and pressureless sintered at 1150 °C for 2 h. (Source: Gonzalez et al. [14].)

grow rapidly. The microstructure, however, was very inhomogeneous. The average grain at the center of the sample was estimated to be 181 ± 21 nm, and 144 ± 16 nm near the edges of the sample. The authors speculate that the nonuniform microstructure is a result of the nonuniform distribution of nucleation sites in the compact that were created by a nonuniform distribution of stress during compaction. Because the authors were aware of the benefits of LN_2, they proceeded to compact specimens to 5 GPa using LN_2 as a lubricant, hoping for a more uniform stress distribution during compaction and a homogeneous distribution of nucleation sites. The microstructure that develops after sintering is shown in Figure 26. In contrast with the samples compacted under dry conditions at room

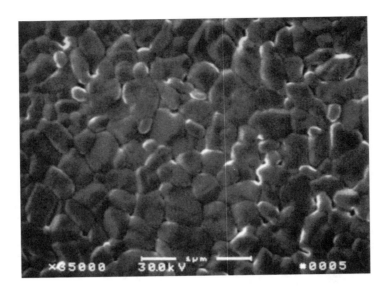

Fig. 25. SEM micrograph of an alumina sample compacted in a DAC to 5.0 GPa and pressureless sintered at 1150 °C for 1 h in air. (Source: Gonzalez et al. [14].)

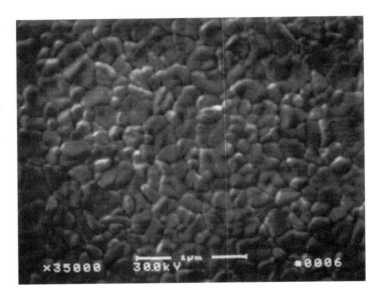

Fig. 26. SEM micrograph of an alumina sample compacted in a DAC to 5.0 GPa and pressureless sintered at 1150 °C for 1 h in liquid nitrogen. (Source: Gonzalez et al. [14].)

temperature, the average grain size (146 ± 15 nm) is uniform throughout the entire cross section of the sample. This result provides additional evidence of the lubricating properties of LN_2 on the compaction of nanosize powders. For a more detailed description of this work, we refer the reader to Gonzalez et al. [14].

4.3.3. Conclusions

Gonzalez et al. [14] were able to list eight conclusions concerning the compaction and sintering of γ alumina nanosize powder relevant to the compaction process. They are as follows:

1. Random dense packing of γ alumina nanosize powder can be achieved at high compaction pressures ($\geqslant 1$ GPa).
2. Increased compaction pressure results in a corresponding increase in green-body density, specifically from 54% theoretical at 1.0 GPa to 65% theoretical at 2.5 GPa.
3. Green-body hardness values reflect the final compaction pressure.
4. Increased compaction pressure enhances the γ-to-α phase transformation, effectively increasing the nucleation rate, or, alternatively, lowering the transition temperature.
5. Heterogeneous nucleation of the α phase in nanoscale γ alumina compacts is invariably followed by a period of rapid growth, which results in a coarsening of both the solid phase and the interconnected porosity. Although this microstructural coarsening decreases the internal surface area, reduction of the volume fraction of porosity is not apparent. As a consequence, densification is limited to the solid phase component and corresponds to the relative increase in density, accompanying the γ-to-α phase transition, 3.67 g/cm^3 to 3.99 g/cm^3, respectively.
6. Despite the enhancement of sintering kinetics with increasing green-body density at temperatures below 1300 °C and pressures below 3 GPa, the results, based on density measurement and microstructural evaluation, indicate that the final density is independent of the initial density. The consequence of this is that full densification is not achieved by pressureless sintering. It appears that a similar volume fraction of closed porosity is retained.
7. Increased compaction pressure below 3 GPa (green-body density) appears to have no effect on the final density of the product sintered at temperatures above approximately 1400 °C. Increased compaction density, however, does affect how fast one can achieve 83% of theoretical density when the sintering temperatures are below 1400 °C.
8. Compaction pressures in the DAC over 3 GPa are necessary to avoid the formation of the vermicular structure during sintering.

These are very important conclusions because they have basically changed our views concerning the processing of nanosize particles to form dense nanoscale microstructures. As a consequence, it appears that the notion of achieving full packing density in the green body of a nanosize powder will result in a dense nanoscale grain-sized sintered ceramic is not necessarily a correct one. From what we now know, this is particularly so when a phase transformation occurs during the sintering process, as in the cases of nanosize γ alumina and amorphous silicon nitride powders. In both materials, high-density green bodies did not produce dense nanoscale microstructures when the sintering temperature employed exceeded the phase transformation temperature. Amorphous silicon nitride crystallized to the α phase with a significant increase in density and grain size. As we saw earlier, γ alumina transformed to the α phase also accompanied by an increase in density and grain size. Unfortunately, there are few ceramic powders that are thermodynamically stable as nanosize particles at room temperature and pressure. Thus, until such materials become available and can be tested by these procedures, the notion of sintering fully packed green bodies of single-component nanosize particles to dense nanoscale microstructures remains an open issue.

5. COMPACTION EQUATIONS FOR POWDERS

Considering all the information that has been presented so far in this chapter, the question that now needs to be addressed is how can we explain, in a more detailed or quantitative manner, the effects of lubricants on the compaction efficiency of nanosize powders. Intuition suggests that this should be addressed by studying the process of compaction and the compaction equations that attempt to model these processes. Unfortunately, most mathematical models are exclusively phenomenological and, therefore, offer little insight into the mechanics of compaction. Most analytical problems require information that, in the case of powder compaction, is not available or is difficult to obtain experimentally. For example, many expressions call for information such as (1) the distribution of powder particle-to-particle contact stresses, (2) the flow properties of the powder, (3) the load distribution throughout the compact, and (4) the compressibility and strength of the individual particles [32]. When some of these critical parameters are known and are used in compaction equations, the information gained during compaction is considered to be useful [32]. Chen and Malghan [32] suggest that the compressibility of the material, compaction ratio, and average particle size could be determined from such model equations. Most equations appear to focus on factors such as (1) the different possible stages of the compaction process and (2) the different mechanisms involved, which also can provide useful information. The difficulty lies in putting all these parameters together in one general expression applicable to the general powder system. In the following section, we will briefly discuss the basic ideas behind compaction equations, with special emphasis on the equation derived by Chen and Malghan, which is the most familiar to the authors.

One of the goals in the study of the compaction process in powders is to estimate the pressure necessary for achieving a desired bulk density. The process of compaction of a ceramic powder, however, is very complex, especially for nanosize particulates. Ceramic powders can exhibit a high compressibility during the initial or low-pressure stages of compaction, and a low compressibility at higher pressures because the particles are closer together and the coordination number of individual particles has increased significantly [32]. The former is more pronounced during the compaction of nanosize powders, because the tap densities of these powders can be as low as 30% of theoretical. Because of the absence of a sharp or well-defined transition between these two stages of compaction, the modeling of volume changes over the entire pressure range is difficult. Furthermore, even if an equation describes the compaction behavior of a particular powder, it may not be directly applicable to other powders with different compaction properties.

Chen and Malghan [32] developed a new approach to deduce a compaction equation. The authors conducted a number of compaction experiments with ceramic powders and fitted the experimental data using modified nonlinear and linear least-squares techniques. As they pointed out, even the most basic parameters in compaction equations, for example, initial and final volume, are difficult to obtain. The experimental determination of initial volume is not accurate because factors such as particle size distribution, particle agglomeration (especially for nanosize particles), and flow properties, to name a few, will affect the ability of the powder to pack efficiently without the application of pressure. Similarly, the prediction of the final volume based on theoretical considerations can be incorrect because it is typically measured in the absence of an external load. Chen and Malghan avoided dealing with these errors by choosing these values as fitting parameters in their equations.

Their new compaction equation suggests a classification of voids into two types with the resulting equation containing only five parameters:

$$V = a_1 + a_2 \exp\left(\frac{1}{1 + a_3 P}\right) + \frac{a_4}{1 + a_5 P} \tag{1}$$

where a_i, $i = 1, \ldots, 5$, are parameters, $V_0 = a_1 + a_2 e + a_4$, $V_\infty = a_1 + a_2$, and P is the applied pressure. V_0 represents the initial volume and V_∞ is the net particle volume under extremely high pressure ($P \rightarrow \infty$). The authors compared their equation with that of Cooper and Eaton [33], and found that their equation provided a better fit to the experimental data. It is important to recognize that powder compacts are made of many voids of different sizes and shapes that contribute to the complexity of the modeling. Moreover,

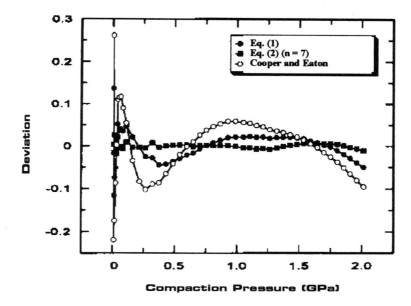

Fig. 27. Deviation of compaction equations and experimental data for α-phase Si_3N_4 powder. (Source: Chen and Malghan [32].)

the size distribution of these voids is changing continuously as pressure increases and particle rearrangement occurs. According to Chen and Malghan, if a compaction equation could consider all possible types of voids, the equation would approach a real compaction process. However, this would require an infinite number of parameters and would make mathematical manipulation impractical.

Chen and Malghan proposed another compaction equation to achieve an even better representation of the different types of voids. The final expression they propose is

$$V = \sum_{i=0}^{n} \frac{a_i}{(1 + u_i \alpha P)} \qquad (2)$$

where α is a factor with a dimension of inverse pressure, $a_i = (V_0 - V_\infty)a_i'$, and u_i is the ith zero of the nth-order Laguerre polynomial. The number of terms in the compaction equation is $n + 1$. All parameters a_i in this equation are linear and can be determined by linear least-squares fits, which simplifies the fitting process. For specific details concerning the derivation of this equation, the reader is referred to [32]. Chen and Malghan claim that the advantage of this equation is that the variety of void types is proportional to the number n of parameters, and, therefore, by adjusting n, a better representation of the compaction process is achieved.

Figure 27 shows the divergence of the two equations, one derived by Chen and Malghan [32] and the other by Cooper and Eaton [33], from the experimental data for α-Si_3N_4 powder. It is clear from this figure that the mathematical treatment developed by Chen and Malghan gives a better fit to these data. However, these equations appear to fall into the same trap as all the other equations of powder compaction, because they fail to provide information on transferable physical parameters. For example, there are no individual parameters that provide information on flow properties, lubricating conditions, time dependence, particle size distribution, shape of the powder particles, and, for example, the state of agglomeration of the powder. All these characteristics are basically convoluted in the compaction ratio, defined as

$$Compaction\ ratio = \frac{V_\infty}{V_0} \qquad (3)$$

Compaction ratios are useful to some extent because they can provide overall information on compaction characteristics. For example, Chen and Malghan indicated that the compaction ratio of powders with an average particle size between 100 nm and 1.0 mm is approximately 0.4 and approximately 0.1 for a powder with an average particle size below 100 nm. Although compaction ratios are qualitatively useful, they are severely limiting if the intention is to understand fully the compaction process in powders.

In summary, the compaction equations developed by Chen and Malghan are very effective in curve fitting pressure–volume data over the entire pressure range, probably better than other available models. The mathematical manipulation and fitting routines are relatively simple, because the a_i fitting parameters can be determined by linear least squares. It appears that the most important value that can be obtained from these compaction equations is the compaction ratio. Because the compaction ratio is, in essence, a measure of the packing efficiency of the powder, it can provide qualitative information about the compaction process, and is especially relevant to compaction under different lubricating conditions. However, the compaction ratio can only help to create an overall picture of the compaction performance of the powder, because it cannot provide details on the most relevant powder characteristics to the compaction process. Although interpretation of its physical meaning is difficult, it would be useful to study changes in the compaction ratio of powders under different lubrication conditions, including the use of LN_2. Such experiments have not been reported yet.

6. CONCLUSIONS

In recent years, there has been a strong interest in the processing of nanosize ceramic powders because of their promising low-temperature sintering capabilities and because it has been suggested that ceramic pieces made up of nanosize grain structures may exhibit superior mechanical properties. In our understanding, a true nanosize polycrystalline microstructure is defined as a material with an average grain size well below 100 nm. At first glance, to achieve this fine-grain structure might appear trivial, because ceramic powders are readily available with an average particle size of 20 nm or less. However, processing nanosize powders into dense bulk specimens that retain their original nanosize microstructure can be very difficult.

The driving force for densification of nanosize powders is very high because of their small particle size. However, in many situations, their relatively low green-body densities require longer sintering times or higher temperatures than expected to achieve full densification. The longer sintering times and higher temperatures required promote excessive grain growth and/or irregular grain growth, resulting in a microstructure that does not exhibit the desired nanosize grain structure. The homogeneity and density of the green compact is critical in many situations, especially because the cold compaction density of most ceramic nanosize powders is extremely low (<45% of theoretical). As discussed in this chapter, high-pressure compaction can be a means of increasing the green-body density and structural integrity of the compact. Green densities in excess of 70% were achieved with nanosize amorphous silicon nitride powders. Similar densities were also obtained with γ-Al_2O_3. Even higher densities were achieved when LN_2 was used as a lubricant during compaction. Liquid nitrogen appears to be an excellent lubricant for nanosize particles. As far as we know, a detailed description of the lubrication mechanism of LN_2 for nanosize particles has not been reported yet. However, evidence of its effect on improving the packing density of nanosize particles has been amply demonstrated by several authors and their results are very convincing. Care must be taken when using LN_2 to compact nanosize particles, because it can experience a series of phase transformations, for example, liquid $\rightarrow$ solid, under pressure at low temperatures, which can have undesirable molar volume changes during compaction. These aspects of LN_2 have been discussed in detail here, and are relevant to the researcher interested in using LN_2 as a lubricant.

Finally, as mentioned earlier, obtaining a high green density is not the ultimate accomplishment in the processing of nanosize powders. The reason for this is that, in many cases, ceramic powders exist as metastable phases when they are prepared as very fine powders ($\approx$20 nm, nanosize). These materials are metastable and may transform, typically involving significant volume changes, during high-temperature sintering. Examples are the two materials discussed in this chapter. γ-Al$_2$O$_3$ is a metastable phase that transforms to the thermodynamically stable α-Al$_2$O$_3$ phase when heated from around 900 to 1100 °C. As far as we know, nanosize Si$_3$N$_4$ powder is available only in the amorphous state. It crystallizes upon heating to temperatures in the 1300 °C range. How large an effect such phase transformations have on the microstructure of the sintered product is difficult to predict. Certainly for the two materials presented here, γ-Al$_2$O$_3$ and Si$_3$N$_4$, the effect is great. Only with further experimental work in this area can we answer this question.

References

1. J. Frenkel, *J. Phys. (USSR)* 8, 386 (1945).
2. C. Herring, *J. Appl. Phys.* 21, 301 (1950).
3. W. H. Rhodes, *J. Am. Ceram. Soc.* 64, 19 (1981).
4. G. Skandan, H. Hahn, and J. C. Parker, *Scr. Metall.* 25, 2389 (1991).
5. T. H. Courtney, "Mechanical Behavior of Materials." McGraw–Hill, New York, 1990.
6. G. Skandan, "Processing of Nanostructured Zirconia Ceramics, Nanostructured Materials," Vol. 5, pp. 111–126. 1995.
7. M. D. Sacks and J. A. Pask, *J. Am. Ceram. Soc.* 65, 70 (1982).
8. W. Chen, A. Pechenik, S. J. Dapkunas, G. J. Piermarini, and S. G. Malghan, *J. Am. Ceram. Soc.* 77, 1005 (1994).
9. P. Duran, M. Villegas, F. Capel, J. F. Fernandez, and C. Moure, *J. Mater. Sci.* 32, 4507 (1997).
10. A. Pechenik, G. J. Piermarini, and S. C. Danforth, *J. Am. Ceram. Soc.* 75, 3283 (1992).
11. G. J. Piermarini and S. Block, *Rev. Sci. Instrum.* 46, 973 (1975).
12. R. A. Forman, G. J. Piermarini, J. D. Barnett, and S. Block, *Science* 176, 284 (1972).
13. J. D. Barnett, S. Block, and G. J. Piermarini, *Rev. Sci. Instrum.* 44, 1 (1973).
14. E. J. Gonzalez, B. Hockey, and G. J. Piermarini, *Mater. Manufacturing Process.* 11, 951 (1996).
15. P. W. Bridgman, *Z. Kristallogr.* 67, 363 (1928).
16. P. W. Bridgman, *Proc. Am. Acad. Arts Sci.* 76, 1 (1945).
17. C. E. Weir and G. J. Piermarini, *J. Res. Natl. Bur. Stand. (U.S.) Phys. Chem.* 68A, 105 (1964).
18. W. Symons and S. C. Danforth, "Ceramic Materials and Components for Engines" (V. J. Tennery, ed.), pp. 67–75. American Ceramic Society, Westerville, OH, 1989.
19. M. R. Gallas, B. Hockey, A. Pechenik, and G. J. Piermarini, *J. Am. Ceram. Soc.* 77, 2107 (1994).
20. J. E. Bonevich and L. D. Marks, *J. Mater. Res.* 7, 1489 (1992).
21. A. Pechenik, G. J. Piermarini, and S. C. Danforth, *Nanostruct. Mater.* 2, 479 (1993).
22. D. A. Young, "Phase Diagrams of the Elements," pp. 113–117. University of California Press, Berkeley, CA, 1991.
23. R. L. Mills, D. H. Liebenberg, and J. C. Bronson, *J. Chem. Phys.* 63, 4026 (1975).
24. W. L. Vos and J. A. Schouten, *J. Chem. Phys.* 91, 6302 (1989).
25. S. Kwon, C. S. Nordahl, and G. L. Messing, *Mater. Manufacturing Process.* 11, 969 (1996).
26. H. Hahn, J. Logas, and R. S. Averback, *J. Mater. Res.* 5, 609 (1990).
27. K. J. Nilsen, R. E. Riman, and S. C. Danforth, *Ceram. Trans.* 1, 469 (1988).
28. M. Kumagai and G. L. Messing, *J. Am. Ceram. Soc.* 68, 500 (1985).
29. G. L. Messing and M. Kumagai, *J. Am. Ceram. Soc.* 72, 40 (1989).
30. T.-S. Yeh and M. D. Sacks, *J. Am. Ceram. Soc.* 71, 841 (1988).
31. F. W. Dynys and J. W. Halloran, *J. Am. Ceram. Soc.* 65, 442 (1982).
32. W. Chen and S. G. Malghan, *Powder Technol.* 81, 75 (1994).
33. A. R. Cooper, Jr., and L. E. Eaton, *J. Am. Ceram. Soc.* 45, 97 (1962).

Chapter 5

KINETIC CONTROL OF INORGANIC SOLID-STATE REACTIONS RESULTING FROM MECHANISTIC STUDIES USING ELEMENTALLY MODULATED REACTANTS

Christopher D. Johnson, Myungkeun Noh, Heike Sellinschegg, Robert Schneidmiller, David C. Johnson

Department of Chemistry and Materials Science Institute, University of Oregon, Eugene, Oregon, USA

Contents

1. INTRODUCTION

The synthesis of inorganic solid-state compounds has a long history, extending at least back to the smelting of ores in the Bronze Age or even earlier as clay pots were fired in

Handbook of Nanostructured Materials and Nanotechnology, edited by H.S. Nalwa
Volume 1: Synthesis and Processing
Copyright © 2000 by Academic Press

ISBN 0-12-513761-3/$30.00

prehistoric times. These early practitioners empirically discovered that high temperatures were required for successful preparations. Subsequent experiments have shown that the high temperatures are usually necessary to overcome the high activation energy required for solid-state diffusion. As a consequence, inorganic solid-state chemistry has mainly been focused on thermodynamically stable compounds, which can be prepared by direct, high-temperature reaction of simple binary compounds or elements. While solid-state chemists have been very clever in controlling the reaction conditions to produce many new compounds, the synthesis of new compounds has remained as much an art as a science [1].

In contrast, the development of molecular synthetic chemistry was dramatically different from that of extended inorganic solids. A major achievement of molecular chemists has been the ability to design rationally a sequence of synthetic steps to prepare very complex molecules. This synthetic control results from an understanding of reaction mechanisms and the ability to use catalysts to control the relative kinetics of competing reactions. Complex transformations are broken down into discrete reaction steps on individual components of the reacting molecule. After each step, the desired product is separated from byproducts and unreacted starting materials before beginning the next step. Molecular reactions can be controlled using catalysts and reaction conditions because the rate limiting step of molecular reactions is the making or breaking of bonds.

In the synthesis of extended inorganic solids, the rate limiting step in the traditional "heat and beat" approach is solid-state diffusion. Solid-state diffusion, in thermodynamic terms, decreases free-energy gradients. Even though atoms are typically tightly bound, thermal vibrations permit some atoms to move. The rate of movement of atoms through many materials has been found to obey an Arrhenius expression:

$$D = D_0 \exp\left(-\frac{Q}{RT}\right) \tag{1}$$

where the prefactor, D_0, is related to the number of vacancies and defects, Q is the activation energy, R is the gas constant, and T is the absolute temperature [2].

Several methods have been used to increase solid-state reaction rates by overcoming the inherently slow interdiffusion rates for atoms through solids. Given the Arrhenius expression for the diffusion rate, the most straightforward way to increase diffusion rates is to raise reaction temperatures. Indeed, the vast majority of inorganic solid-state compounds have been prepared by direct high-temperature reaction of either elements or binary precursors. This approach typically results in the formation of the thermodynamically most stable mix of products. This method of synthesis has the obvious advantage of being experimentally simple. One challenge, however, is finding a suitable container for the reactants that will not itself react at the elevated temperatures of the reaction. In addition, those compounds that are stable only at low temperatures or are only kinetically stable are not accessible via this approach [3].

An often used alternative strategy has been to reduce the activation energy by having the dominant diffusion path proceed through either the gas phase or the liquid phase. In the late 1960s, Schäfer [4] developed vapor phase transport reactions, in which selected chemical reagents enable otherwise nonvolatile products or reactants to be moved along an activity (or temperature) gradient at temperatures well below that required for direct volatilization. More recently, there has been a marked increase in the use of low-temperature fluxes and supercritical solvents as media for the synthesis of new solids at low temperatures [5]. The advantage of these growth techniques is the ability to prepare diffraction-quality single crystals that permit the structures of the resulting new compounds to be easily solved. A drawback of using these approaches is that the complexity of the reacting system usually prevents any predictive ability in terms of the structure or even the composition of the new compounds formed.

A third approach to decrease the synthetic difficulties associated with slow solid-state diffusion rates has been to shorten the necessary diffusion distances by preparing molecular or solid-state precursors that contain an intimate mix of the desired elements. This "soft" chemistry approach has been shown to yield new crystalline forms with compositions identical to existing compounds at surprisingly low temperatures [6]. Long-range diffusion path lengths still remain for the removal of byproducts (ligands and solvent molecules) that were used to stabilize the precursor. This can lead to long annealing times as well as elevated reaction temperatures. This technique also requires the development of a new precursor chemistry for each system under study.

One of the distinct challenges associated with each of the preceding synthetic approaches is the difficulty in following the progress of the reactions. As recently pointed out by Schollhorn [7], developing an understanding of the reaction mechanisms is a necessary precursor to predictive ability in these solid-state synthesis routes. There is little understanding of the reaction mechanisms in the preceding systems, because most of the reactions either begin as heterogeneous reactions or progress through heterogeneous intermediates. This problem is compounded by the lack of *in situ* probes, which results in a scarcity of information regarding the sequence of chemical events occurring in the reactions [2]. Unfortunately, without an understanding of the reaction mechanism, it is impossible to use the reaction conditions in a rational manner to control the structure of the products. This leads directly to the present situation, where the synthesis of new compounds is still as much an art as a science.

Given the present state of solid-state inorganic synthesis, elementally modulated multilayer reactants present several unique features that can be exploited. The most important feature is the creation of a new synthetic variable, the diffusion length of which is controlled by the period of the deposited multilayer. If the diffusion lengths are short (on the order of tens of angstroms), the resulting exothermic reactions are very rapid. This permits the reaction progress to be easily monitored using differential scanning calorimetry. The regular structure of the superlattice also results in a low-angle diffraction pattern in which the positions of the diffraction maxima are directly related to the thickness of the repeating unit of the modulated reactant. The intensities of these diffraction maxima are related to the composition profiles through the multilayer. This provides a probe to follow the course of the interdiffusion reactions by monitoring the intensity of the various diffraction orders as a function of reaction time and temperature. We have shown that if the diffusion distances are short enough, the overall composition of the elementally modulated reactant becomes an important factor in controlling the subsequent reaction mechanism and kinetics. This synthesis method has the advantage of fast sample throughput—an ideal trait for an exploratory synthetic method. The main disadvantage of this approach is the small amount of sample produced (typically 10 mg per sample). Although half gram samples have been prepared to evaluate physical properties, the production of large quantities of product via this approach remains a challenge.

2. BACKGROUND

Before looking at the reactivity of elementally modulated reactants, it is useful to first review several areas that have established important precedence for using multilayers to control solid-state reactivity. We will first review low-angle X-ray diffraction on artificial compositionally modulated materials as a method to measure exceedingly small diffusion coefficients. At the end of this section, we will also review the ability to follow interfacial roughness by tracing the evolution of both the spectral and the diffuse X-ray scattering. This will be followed by a brief review of thin-film reaction studies. The sequential evolution of compounds forming at the reacting interface show the importance of kinetics in these systems. A review of solid-state amorphization reactions concludes this section.

2.1. Measuring Diffusion Rates Using Multilayers

In 1940, DuMond and Youtz [8] prepared a Cu/Au superlattice with a period of approximately 100 Å to calibrate X-ray wavelengths. Although the film was not of sufficient quality for their original purpose, they did observe a Bragg reflection resulting from the compositional modulation. They found that the intensity of the diffraction maxima decreased with time while the sample was at room temperature. Half the initial intensity had disappeared 2 days after being prepared. They determined that the decay of the diffraction signal resulted from interdiffusion of the sample and analyzed the data according to the simple diffusion equation:

$$\frac{\partial c}{\partial t} = D \frac{\partial^2 c}{\partial x^2} \tag{2}$$

where c is the atom fraction of one of the components, x is the spatial coordinate normal to the film plane, t is time, and D is a composition-independent diffusion coefficient. DuMond and Youtz showed, by expanding the composition modulation in a Fourier series, that the higher harmonics of the series decay rapidly, resulting in a sinusoidal composition modulation. The relative intensities of the Bragg diffraction maxima are determined by the electron density distributions within the unit cell. For a sinusoidal composition modulation of wavelength d, they derived that the intensity of the resulting Bragg reflection should decrease according to

$$\frac{d}{dt}\left[\ln\left(\frac{I}{I_0}\right)\right] = -\frac{8\pi^2}{d^2} D \tag{3}$$

where I_0 is the initial intensity observed. With this simple analysis, they showed that it is possible to determine lower diffusivities than with any other technique. In the subsequent nearly 60 years, several factors either ignored or oversimplified in this original study have been included. In particular, the diffusivity has been found to be a function of d because the sharp concentration gradients found in the multilayers require corrections to the simple diffusion equation. In addition, DuMond and Youtz ignored the effects of strain in the artificial structure, which can arise from differential thermal contraction and the fact that the diffusivity is typically composition dependent. These last two effects can be minimized by using samples that are nearly compositionally homogeneous [9]. For those interested in a more thorough treatment of this area, a review by Greer and Spaepen [10] develops the theory required to analyze the data and summarizes experimental studies on both amorphous and crystalline systems.

Experimentally, the small diffraction angles make alignment of the diffraction system crucial for measuring accurate diffraction intensities and angles. Data collected by Novet (Fig. 1) clearly shows the dramatic changes in intensity and position as a function of alignment errors in either the height or the tilt of the sample relative to the goiniometer circle. A simple check of the sample alignment is to measure the rocking curve at several different diffraction angles. If the sample is aligned, the maximum intensity corresponding to the specular signal will occur when the incident and exit angles are equal. If there is a constant offset between the incident and exit angles, this indicates that the sample height is correct but there is an angular offset between the surface of the sample and the zero angle of the goiniometer. If the offset between the incident and exit angles changes as a function of diffraction angle, the sample is not located on the center of the goiniometer [11].

The iron–silicon system is an example of a diffraction study used to determine diffusivities as well as follow changes in interfacial structure as a function of annealing time. Figure 2 shows the decay of the diffraction signal as a function of annealing time at 150 °C. During this anneal, the diffraction peak also shifts to a higher angle as the size of the repeating unit decreases. These diffraction data can be interpreted using a more general form

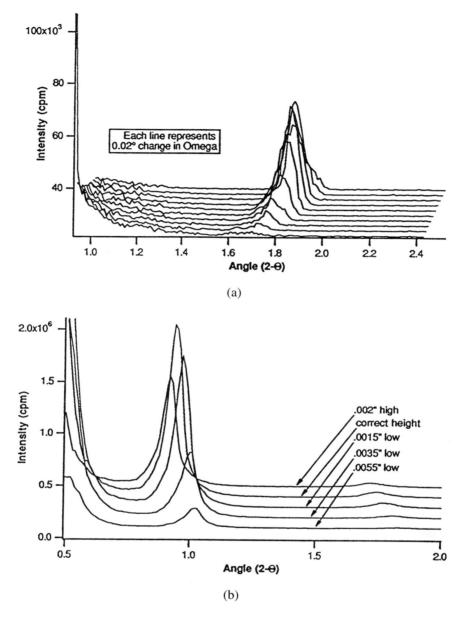

Fig. 1. X-ray diffraction scans demonstrating the sensitivity of diffraction intensity to errors in alignment of the sample: (a) the effect of varying the incident angle relative to the exit angle of the diffractometer and (b) the effect of changing the height of the sample.

of Eq. (3):

$$\frac{d}{dt}\left[\ln\left(\frac{I}{I_0}\right)\right] = -\frac{8\pi^2 n^2}{d^2}D \tag{4}$$

where n is the order of the diffraction peak [10]. Figure 3 shows the data from Figure 2 now plotted as indicated by Eq. (4). The data from both the first and the second diffraction orders fall on the same line, indicating that the diffusion coefficient is roughly independent of concentration. The diffusivity obtained from the slope is 1.3×10^{-19} cm^2/s. Using this composition-independent diffusivity, the decay of the composition modulation within the

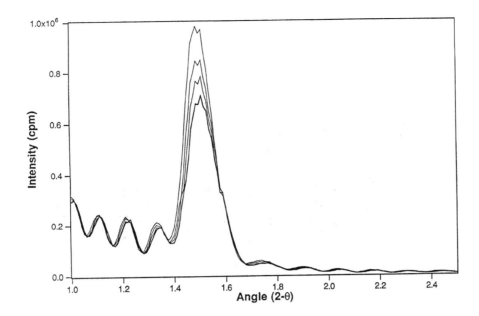

Fig. 2.   The change in diffraction intensity of the first-order diffraction peak of an iron–silicon multilayer reactant as a function of annealing time at 150 °C. The multilayer has a repeat spacing of 66 Å and an iron:silicon ratio of 3:1.

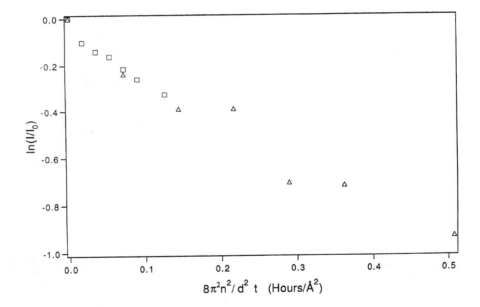

Fig. 3.   The change in the intensity of the first and second Bragg diffraction maxima of an iron–silicon multilayer reactant as a function of annealing time at 150 °C. The slope of the line through these points results in a diffusion coefficient of 1.3×10^{-19} cm^2/s.

Fe–Si elemental multilayer can be modeled and agrees well with an analysis of the heat evolved as measured with differential scanning calorimetry.

Further studies of this type are needed to follow the changes in structure and composition at reacting interfaces. Unfortunately, most diffusivities are not independent of composition, leading to a more complex analysis of the diffraction data. To overcome these difficulties, one can prepare a sequence of samples with varying average composition that are nearly compositionally homogeneous. By measuring the diffusivity in each of these samples, the composition dependence of the diffusivity can be experimentally determined. These data can then be used to analyze the more complex interdiffusion behavior in multilayers prepared with large composition gradients at the reacting interfaces.

2.2. Interfacial Roughness at Reacting Interfaces

In addition to being able to follow compositional changes at reacting interfaces, X-ray diffraction can also be used to obtain information about roughness and changes in interfacial roughness as a function of annealing conditions by looking at the changes in diffuse scattering in addition to the specular light discussed previously. Interest in extracting information about interfacial roughness from X-ray data has been increasing, as researchers have found that a range of unique optical and electrical properties of multilayers depends on the structural perfection of the superlattice. As an example, indirect evidence has shown that large-scale substrate roughness can be propagated through soft X-ray multilayer mirrors and degrades their efficiency [12].

Many papers have discussed the scattering of light (both visible and X-ray wavelengths) from nonideal surfaces and interfaces, focusing on the loss of specular reflectivity as a consequence of surface or interface disorder [13–18]. The most common approach has been to describe the distribution of roughness at the interfaces in terms of a Gaussian, leading to an exponential attenuation of the reflectivity if multiple scattering can be ignored. This approach is analogous to the treatment of the dynamic Debye–Waller factor resulting from thermal vibrations, except that the disorder (the roughness) is static. If the roughness is propagated perfectly from layer to layer, then multiple scattering analysis gives the same result as the kinematic approximation, an exponential attenuation of the specular scattering. Choosing a roughness distribution other than Gaussian will give rise to a different decay of the specular reflectivity as a function of diffraction order [13, 16].

There have been fewer studies measuring and modeling the diffusely scattered radiation. The measurement of the diffuse component of scattering in a wide angular range around the specularly reflected beam gives information on the correlation length ξ, with which the roughness decays laterally [19]. Sinha et al. [14] used the first distorted-wave Born approximation, treating the surface roughness as a small perturbation of the smooth surface to describe the diffuse scattering of a rough surface. This theory was then extended by Weber and Lengeler [20] to treat systems in which the roughness is no longer a small perturbation. These authors were able to characterize a rough surface in terms of a root mean square (rms) roughness σ, the height–height correlation length ξ, and the roughness exponent h. The value for the rms roughness obtained from the fitting of the diffuse scattering agreed with that calculated from the decrease in specular scattering. The values for the height–height correlation length and the roughness exponent obtained from diffuse scattering measurements were in excellent agreement with those obtained from an atomic force microscopy study [20].

Multilayer films present several variables that are not present in treatments of rough surfaces. In particular, the interfacial roughness in multilayer films can be random or correlated [19]. Correlated roughness occurs when a feature, for example, a surface defect, is replicated from layer to layer. The shape of the diffuse intensity scattered from a multilayer and measured in a direction parallel to the interfaces gives information on the lateral correlation of the interfacial roughness. Measuring the distribution of the diffuse intensity as a function of the angle of the incident radiation will provide information about the vertical correlation of roughness from layer to layer. Current research in this area has focused on developing a straightforward analysis of the data to yield lateral correlation lengths, rms roughness, and roughness exponents [19, 20]. The use of diffuse scattering from multilayers to explore the evolution of reacting interfaces will be a growing area of interest.

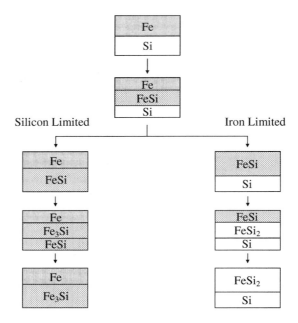

Fig. 4. The sequence of binary compounds observed forming at the interface of an iron–silicon thin-film diffusion couple.

2.3. Kinetics of Thin-Film Reactions

In the 1980s, several groups began to explore the reactivity of thin-films of transition metals deposited on silicon substrates [21–25]. Much of the resulting early interest in the reactivity of thin films grew out of the need to understand reactions between thin polycrystalline metal films and single crystal silicon, which are used in integrated-circuit contacts and interconnection schemes. Issues such as reaction temperature, first phase formation, and morphology were found to be important in making reproducible and reliable products. As work proceeded in this area of study, it became evident that the reaction between each pair of elements evolved through a unique sequence of phases. Understanding the mechanism responsible for the sequence of phase formation became a focus of research, as it offered the possibility of using previously unexplored parameters to control the outcome of solid-state reactions.

Early work focused on thin-film reactions in which one or both of the reacting layers were on the order of hundreds to thousands of angstroms [21–25]. A typical study involved depositing one or more layers (a multilayer) on a substrate, typically silicon. Annealing the samples at several hundred degrees Celsius resulted in interdiffusion and the growth of compounds. Below a critical thickness, a single compound was formed at the interfaces. It was found to grow until one of the reagents was exhausted. On further annealing, a second compound eventually nucleated at the interface of the initial compound and the remaining reagent. This second phase grew until either the initial compound layer or the initial reagent was exhausted. This process continued until the equilibrium combination of products was obtained. In general, the reactions proceeded stepwise from the initial phase to the next phase richer in the nonlimited reagent. Not all compounds found in the bulk system are necessarily formed. Some compounds are occasionally omitted in the sequence of phase formation. Figures 4 and 5 contain examples of the sequence of phases observed for representative binary systems.

These observations first demonstrated that kinetics were important in direct solid-state reactions between elements. The length scale of these reactions proved to be very im-

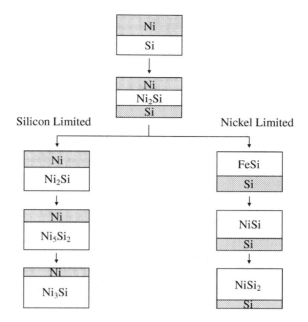

Fig. 5. The sequence of binary compounds observed forming at the interface of a nickel–silicon thin-film diffusion couple.

portant, revealing the competition between diffusion and nucleation that occurs at reacting interfaces. In these thin-film reactions, diffusion and nucleation alternately are the rate limiting steps in the reaction. Diffusion controls the rate at which the phases grow until one of the reactants is exhausted. Nucleation then limits the reaction and determines which phase forms next. Several groups developed rules to predict the sequence of phase formation.

The empirical rules developed by Walser and Bené provide insight into the mechanism of the phase sequence. They correlated the available data on the sequence of phase formation with the respective equilibrium phase diagrams. For metal–metal systems, they proposed that [26]:

- The first phase nucleated in metal–metal thin-film reactions is the phase immediately adjacent to the low-temperature eutectic in binary phase diagrams.

For metal–group IV element systems, they proposed [27]:

- The first phase nucleated in a metal–covalent semiconductor thin-film reaction is the highest-temperature congruently melting phase adjacent to the low-temperature eutectic.

These rules can be used to predict the first phase formed in most metal/silicon and metal/metal thin-film reactions. As an example, in the iron/silicon system, the lowest-temperature eutectic is at 34 at% silicon, and the highest congruently melting phase adjacent to this eutectic is FeSi. Experimentally, FeSi is the first phase formed in the thin-film diffusion couple.

In developing these rules, Walser and Bené assumed the formation of an amorphous interfacial layer. The composition of this layer was proposed to be near that of the lowest-temperature eutectic in the binary system because it is the most stable liquid composition in the phase diagram. They assumed the phase that would nucleate first would be the one closest in composition to that of the amorphous layer. This requires the smallest changes in composition and local bonding arrangement in forming the nuclei. In those cases where more than one congruently melting phase existed next to the low-temperature eutectic, the phase with the highest melting point was thought to be the first phase to form. This implicitly assumes that melting points scale with the driving force for nucleation.

Walser and Bené assumed the formation of an amorphous interlayer as the initial step in the reaction between the elemental layers. Since the publication of their rules, transmission electron microscopy (TEM) studies of several binary systems have confirmed the formation of amorphous layers at reacting interfaces [28–31]. The formation of these amorphous layers has been rationalized from a thermodynamic standpoint as a result of lowering interfacial energies [32]. Essentially, the idea is that a reduction of the interface energies occurs by replacing a metal–metal interface with a metal–amorphous-metal interface. In some systems, defects or grain boundaries at the interface are thought to be required to form the amorphous interfacial layers [33].

The work in the area of thin-film reactions demonstrated the importance of kinetics in solid-state reactions. Diffusion and nucleation were alternately found to be the rate limiting step in product formation. It was suggested that amorphous interface layers were a key reaction intermediate. The composition of the amorphous layers was thought to control the sequence of phase formation by influencing the relative nucleation energy of potential products.

2.4. Solid-State Amorphization Reactions

Solid-state amorphization reactions were first reported by Schwarz and Johnson in 1983 [34]. They observed the formation of a homogeneous amorphous alloy by low-temperature annealing of Au/La thin films, which were 100–600 Å thick. The surprising lack of crystallization was attributed to the anomalously fast diffusion of Au through La. The low mobility of one element relative to the other prevented the local rearrangements necessary for crystallization.

Since this initial report, numerous other systems have been found in which thin films do not react to form compounds, but instead interdiffuse to form stable amorphous alloys [35–39]. The formation of an amorphous phase generally requires two factors: a large heat of mixing of the two elements and a kinetic factor inhibiting crystallization. The ratio of the atomic radii has been used to estimate the differential rate of interstitial diffusion of reacting species. A fast rate of formation of the amorphous phase has also been used to rationalize solid-state amorphization reactions [40]. Although it was found that these factors can explain most cases of the formation of amorphous alloys from multilayer films, examples exist where either the atomic radii ratio is close to 1 or the heat of mixing is a very small negative or even a positive quantity [41].

One interesting aspect of the solid-state amorphization of multilayers is the stability of the amorphous alloys with respect to both the unreacted crystalline starting materials and the nucleation of binary crystalline compounds. A large heat of mixing is thought to supply the driving force for the mixing and to make the amorphous alloy more stable than the starting materials. In many cases, the heat of mixing of multilayers to amorphous alloys accounts for the majority of the heat of formation of the final crystalline compounds. For example, in the Ni/Zr system, over 90% of the heat of formation of the intermetallic compound evolves during mixing [42]. The remaining heat of formation is released when crystallization of the amorphous alloy occurs.

The stability of the amorphous alloy with respect to crystalline compounds is thought to be due to the presence of an activation barrier for nucleation. The compound with the lowest activation barrier will be the one that nucleates. Because the compound that crystallizes is not necessarily the thermodynamically most stable compound, crystallization of amorphous alloys potentially provides access to compounds that cannot be prepared via more traditional synthetic approaches. Determining which experimental parameters affect nucleation and how these parameters can be used to control which compound crystallizes or to prevent crystallization entirely is an important area of current research [43].

2.5. Nucleation

Phase transformations involve both the nucleation of the new compound (or phase) and its subsequent growth. The rate limiting step in a phase transformation may be either nucleation or growth, depending on the relative activation energies. In solid-state amorphization reactions where the initial layers within a multilayer reactant are thin enough, the initial reactant will interdiffuse completely before nucleating a crystalline compound. In this situation, nucleation is clearly the rate limiting step in the formation of the thermodynamically more stable crystalline state. The constrained nature of the solid-state complicates nucleation. Stresses and strains can build up because of volume differences, impurities may aid or hinder nucleation, and the limited atomic mobilities may favor a compound with a simple structure over a compound with a more complicated atomic arrangement.

Homogeneous nucleation from a fluid phase, a much simpler system, provides a reasonable starting point to discuss the factors that influence nucleation in the solid state. In crystallization from a homogeneous liquid, the formation of embryos having the structure and composition of the nucleating crystalline compound is thought to occur via fluctuations in local energies and atomic densities. In the simplest picture, these embryos are assumed to have sharp interfaces with the surrounding medium. The energy of formation of the crystallizing compound from the fluid is assumed to be negative. The energy associated with creating the interface or surface between the fluid and the nucleating solid is taken to be positive. For small embryos, the surface free energy dominates the total free-energy change. These small embryos are unstable and redissolve into the surrounding fluid. At some size, the decrease in free energy resulting from the increase in the volume of the particle will become larger than the increase in the free energy resulting from the increase in surface area. Particles larger than this size will spontaneously grow, driven by the decrease in free energy. Figure 6 illustrates the competition between these two energy terms as a function of radius, assuming spherical particles and an isotropic system. The value of the radius (r) for the embryo with the maximum free energy is called the critical radius (r^*). The energy associated with this critical radius is the activation barrier for nucleation. This is the magnitude of the fluctuation in local free energy that must occur for the sample to crystallize [44].

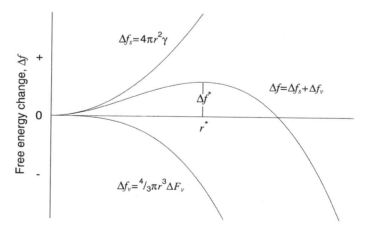

Fig. 6.   The competition between surface energy $4\pi r^2 \gamma$ versus volume energy $\frac{4}{3}\pi r^3 \Delta F_v$ as a function of the size of a growing particle. r^* is the size of a critical nucleus. Particles smaller than r^* spontaneously dissolve; those larger spontaneously grow. ΔF is the size of the energy barrier for nucleation of the more stable phase.

Understanding the experimental factors that control the relative nucleation energies of different compounds is crucial to either preventing nucleation, if one is interested in using the properties of the amorphous alloy, or controlling nucleation, if one is interested in using an amorphous precursor to prepare desired crystalline compounds. To study the effect of experimental parameters on nucleation, it is necessary to have a method to measure the activation energy for nucleation. A convenient method of measuring the activation energy for nucleation is by Kissinger analysis of nonisothermal differential scanning calorimetry (DSC) data, as discussed in the following section.

2.6. Kissinger Analysis of Nonisothermal Differential Scanning Calorimetry Data

Kissinger analysis of nonisothermal DSC data takes advantage of the fact that the temperature at which a compound nucleates is dependent on the rate at which the sample is heated. Figure 7 shows this effect for a multilayer sample of molybdenum and silicon heated at rates of 2, 5, 10, and 20 °C/min. This shift in the nucleation temperature can be explained in terms of the activation energy for nucleation, assuming there is a minimum temperature required before fluctuations in local energies and atomic densities become large enough for nucleation to occur. The probability that a sample will nucleate at each temperature above the minimum temperature is proportional to the time the sample spends at each temperature because the number of nucleation attempts is proportional to time. The probability that a sample will nucleate per unit time increases with increasing temperature above the minimum temperature because the magnitude of the energy fluctuations will become larger. The time it takes for a sample to nucleate, therefore, depends on the scan rate. Suppose a sample is observed to nucleate at a temperature of 300 °C when scanned at 10 °C/min. At lower heating rates, nucleation occurs at a lower temperature because the sample spends more time at each temperature above the minimum temperature. At higher heating rates, the sample spends less time at each temperature, resulting in a higher nucleation temperature.

This phenomenon was used by Kissinger to develop the following equation [45]:

$$\frac{d \ln[Q/T_p^2]}{d[1/T_p]} = \frac{-E_{\text{crystallization}}}{R} \tag{5}$$

where T_p is the temperature of the maximum heat flow during the exotherm, Q is the scan rate, R is the ideal gas constant, and E is the activation energy. By plotting $\ln[Q/T_p^2]$

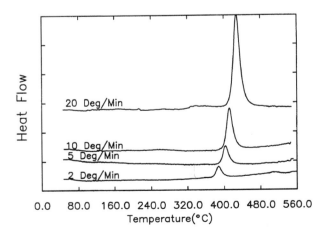

Fig. 7. Differential scanning calorimetry data of a Mo–Si multilayer as a function of scan rate. The exotherm for the nucleation of MoSi$_2$ shifts to a higher temperature as the scan rate increases.

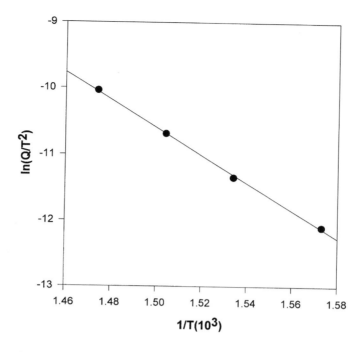

Fig. 8.   Kissinger analysis for the nucleation exotherm for the formation of $MoSi_2$. The data are plotted as suggested by Eq. (5), yielding a straight line. The slope of the straight line yields a nucleation energy of 2.4 eV.

versus $1/T_p$, a straight line is obtained with slope $-E/R$. Figure 8 shows such a graph for DSC data collected on a molybdenum and silicon multilayer sample that had a sharp exotherm at approximately $400\,^{\circ}C$ for the formation of $MoSi_2$. The slope of the straight line yields an activation energy of 2.4 eV. The straight line obtained through the data points suggests that the assumptions made in obtaining Eq. (5) were met. In deriving this equation, Kissinger assumed that nucleation and growth can be described by the Johnson–Mehl–Avrami equation [46]. The derivation of the Johnson–Mehl–Avrami equation assumes that there are isothermal transformation conditions, that the nucleation is spatially random, and that the growth kinetics are linear. A further approximation made in deriving this equation is that both the nucleation rate and the growth rate may be described by Arrhenius expressions over the range of temperature in which the peak temperature varies with the scan rate. Also, the initial and the final states have the same composition and the nucleation and growth rates are considered constant at constant temperature.

3. MULTILAYERS AS REACTANTS

The concepts and ideas discussed in the preceding sections led us to propose that multilayer reactants might provide a general route to the amorphous state. Nucleation would then be the rate controlling step in the formation of crystalline compounds. Subsequent studies with multilayers have shown that there are several ways to control a reaction sequence based on the design of the repeat unit of the initial multilayer. The following sections discuss the concepts and present key experimental results demonstrating the ability to control the reaction pathway. We begin by discussing the concept of a critical layer thickness and examine the nucleation-limited behavior found for samples with repeat layer thicknesses below this critical value. Examples are presented that demonstrate the ability of this approach to make thermodynamically unstable compounds. We then discuss the reaction pathway observed in samples with repeat layer thickness above this critical value. In this thickness regime, interfacial nucleation occurs before diffusion is complete. The composition modulations present in the initial reactant can persist during subsequent crystal growth, forming a crystalline superlattice.

3.1. Critical Layer Thickness

A key concept that came out of the studies of solid-state amorphization reactions was the idea of a critical thickness of the multilayer repeat unit below which interdiffusion of the elements occurs before interfacial nucleation. Several ideas were proposed in the literature that give insight into the factors that determine a critical thickness [47–49]. The nucleation time approach presented by Meng et al. [49] provides a useful framework for understanding the origin of this phenomenon. In this proposal, the authors assumed competition between nucleation and diffusion at the interfaces. They proposed that the critical nucleus of the crystalline phase needs to exceed a critical length in the direction of the moving interface to grow spontaneously. The typical time for the amorphous phase to move over this distance and therefore destroy any nucleating crystalline phase will be proportional to the velocity of the diffusion front. The competing time scale is that required for nucleation, which will be inversely proportional to the nucleation rate. Assuming diffusion-controlled growth, the rate of growth of the amorphous layer will slow down as the thickness of the amorphous phase increases. The rate of movement of the interface will be proportional to the square of the thickness of the amorphous phase. At some thickness, the time scales will be equal and nucleation will occur. If the multilayer repeat thickness is below this critical thickness, the sample will interdiffuse to form a bulk amorphous intermediate.

Several groups have determined the critical layer thickness in a number of metal–metal systems that were known to undergo solid-state amorphization reactions. For systems with either a large driving force for mixing or a large difference in size between the metal atoms, the critical thicknesses are on the order of hundreds of angstroms. The Ni–Zr system is an example of such a system, where bilayers 450 Å thick interdiffuse to form a homogeneous amorphous alloy [42]. To probe how general this idea of a critical thickness was, we studied the dependence of the reaction pathway as a function of repeat layer thickness in several metal–selenium and metal–silicon systems that were not expected to undergo solid-state amorphization reactions.

The Mo–Se system is typical of a number of the systems investigated. For multilayers containing Mo–Se bilayers greater than 40 Å thick and with a 1:2 ratio of molybdenum to selenium, a single broad exotherm below 250 °C is observed in the DSC experiment. This exotherm, shown in Figure 9, appears to contain at least two overlapping components. Diffraction data collected as a function of annealing temperature show that nucleation of $MoSe_2$ occurs during the first exotherm while the sample is still layered. Only the 00l Bragg diffraction maxima are observed, indicating that there is a significant preferred alignment of the crystallites. After the second exotherm, more 00l diffraction orders are observed and the line widths of the peaks have narrowed, indicating growth of the crystallites. For multilayers containing Mo–Se bilayers less than 30 Å thick, two well-separated exotherms are observed in the DSC data as shown in Figure 10. There is a broad low-temperature exotherm with a maximum heat flow at 150 °C and a second sharp exotherm at 550 °C. Diffraction data collected as a function of annealing temperature suggest that, during the first, broad, low-temperature exotherm, the layers interdiffuse without the formation of any crystalline compound. After the second exotherm, the diffraction data show that crystalline $MoSe_2$ has formed. This suggests that the second exotherm results from the nucleation and growth of $MoSe_2$. The large shift in the nucleation temperature for $MoSe_2$ as a function of layer thickness shows the ability of interfaces to reduce nucleation energy. In thick bilayers, nucleation occurs at the interfaces. In the samples with thin bilayers, interdiffusion eliminates the composition gradients before the sample has a chance to nucleate. Nucleation in the relatively homogeneous amorphous intermediate is more difficult because of the lack of interfaces [50].

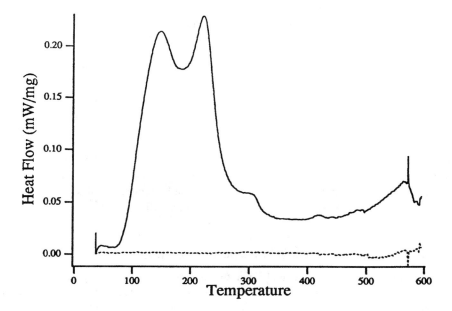

Fig. 9. Differential scanning calorimetry data collected on a molybdenum–selenium multilayer containing twice as much selenium as molybdenum with a repeat bilayer thickness of 60 Å. The low-temperature exotherm is the superposition of two overlapping thermal events.

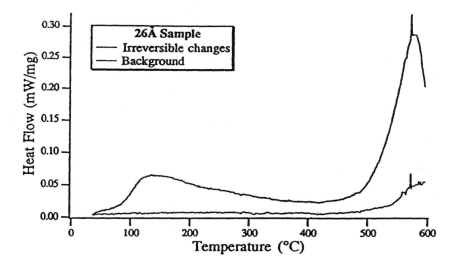

Fig. 10. Differential scanning calorimetry data collected on a molybdenum–selenium multilayer containing twice as much selenium as molybdenum with a bilayer thickness of 27 Å. A broad low-temperature exotherm with a maximum at 150 °C caused by interdiffusion of the multilayer and a sharp high-temperature exotherm resulting from the nucleation and growth of MoSe$_2$ at 550 °C are observed.

The competition between diffusion and nucleation is expected to depend on the overall composition of the interdiffusing multilayers because both the diffusion coefficient and the nucleation energies depend on composition. A study of the niobium–selenium system clearly demonstrates this composition dependence [51]. Multilayer films that were niobium rich and had bilayer thicknesses less than 60 Å formed an amorphous reaction intermediate on low-temperature annealing. Further annealing led to the crystallization of Nb$_2$Se,

265

Nb_5Se_4, or Nb_3Se_4, depending on the composition of the amorphous intermediate. Multilayer films that were selenium rich were always observed to nucleate heterogeneously $NbSe_2$ at the reacting interfaces regardless of the composition or bilayer thickness (layer thicknesses from 15 to 100 Å were investigated). The nucleation of more selenium-rich phases, $NbSe_3$, $NbSe_4$, or Nb_2Se_9, does not occur.

As can be seen by these examples, the critical thickness in different binary systems can vary over a considerable range. On the thick side, on the order of hundreds of angstroms, one has systems that undergo solid-state amorphization reactions. On the other extreme, some binary systems cannot be interdiffused to the amorphous state regardless of how thin the bilayer thicknesses are. The Fe–Al system is an example of this latter extreme, with the binary compound FeAl interfacially nucleating as soon as diffusion begins [52]. The magnitude of the critical thickness depends on the bilayer repeat thickness, the overall composition of the bilayers, and the nucleation energetics of the possible crystalline compounds.

3.2. Controlling Crystallization of Amorphous Intermediates

The preceding studies on the critical layer thicknesses of some binary systems demonstrated that elementally modulated reactants could be used to access in general the amorphous state, provided the bilayers in the initial multilayer are thin enough. The usefulness of the amorphous state as a synthetic intermediate, however, depends on the ability to direct the nucleation process so the desired product crystallizes. The nucleation step involves the assembly of atoms in the correct ratio and in the three-dimensional structure of the nucleating compound. Intuitively, the difficulty of this nucleation step will depend on the composition of the amorphous intermediate. If the amorphous intermediate has the composition of the nucleating phase, only short-range rearrangements will be necessary to form the structure. If the amorphous intermediate is different in composition from the nucleating compound, however, longer-range diffusion will be necessary to disproportionate the amorphous intermediate in forming a nuclei of the compound. This requires a larger fluctuation in energy and is correspondingly more difficult and less probable. Therefore, the composition of the amorphous intermediate should control which compound nucleates.

The iron–silicon system illustrates the ability to control crystalline compound formation via the composition of the amorphous intermediate [53]. Binary multilayers containing alternating layers of Fe and Si in which the binary repeat thickness is less than 70 Å interdiffuse, forming amorphous intermediates on extended annealing at temperatures below 250 °C. Amorphous intermediates were prepared with compositions of the known compounds in the equilibrium phase diagram—$FeSi_2$, FeSi, Fe_5Si_3, and Fe_3Si. The amorphous intermediate with an iron:silicon composition of 1:2 nucleated $FeSi_2$ at 485 °C. FeSi nucleated at 290 °C from the amorphous intermediate of composition 1:1. Fe_5Si_3 nucleated at 455 °C from the amorphous intermediate of composition 5:3. The amorphous intermediate with an iron:silicon ratio of 3:1 nucleated Fe_3Si at 540 °C. If the amorphous intermediate was not close to the composition of a binary compound, nucleation did not occur below 600 °C. Significantly, Fe_5Si_3 is not thermodynamically stable with respect to a mixture of FeSi and Fe_3Si at the temperature at which it nucleated from the amorphous intermediate. Presumably, Fe_5Si_3 nucleates from the amorphous intermediate because of the diffusional constraints imposed by the low reaction temperatures. Essentially, it is easier to nucleate Fe_5Si_3 than to disproportionate the amorphous intermediate and nucleate a mixture of FeSi and Fe_3Si.

The iron–silicon study clearly demonstrates that the composition of the amorphous intermediate controls the nucleation process. Kissinger analysis of nonisothermal DSC data as a function of the composition of the amorphous intermediate would provide the composition dependence of the nucleation energy. In 1996, such a study was conducted, which measured the nucleation energy of InSe [54]. The authors observed a minimum in the nucleation energy as a function of composition when the composition was close to 1:1 as

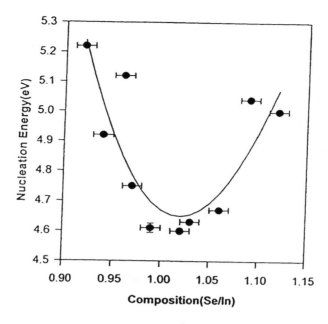

Fig. 11. The nucleation energy of InSe, measured using Kissinger analysis, as a function of composition of the amorphous reaction intermediate.

shown in Figure 11. This minimum reflects the minimal long-range rearrangements required for nucleation of InSe at this composition. Samples more indium rich than Se/In = 0.90 formed a mixture of InSe and In_4Se_3 as expected from the phase diagram. Samples more selenium rich than Se/In = 1.12 formed a mixture of InSe and In_2Se_3, skipping the intermediate phase In_6Se_7 expected from the published phase diagram. This study also suggests that composition of the amorphous phase can be used to control nucleation. As discussed in the following section, this control can be used to avoid stable compounds and prepare new compounds that are thermodynamically unstable with respect to mixtures of known compounds.

3.3. Application of Multilayer Reactants to the Synthesis of New Binary Compounds

Multilayer reactants provide a unique reaction pathway that can be used to discover new compounds in well-studied binary systems. This technique circumvents the difficulties present using more traditional synthetic techniques. Three examples are given in this section. We first discuss the synthesis of a compound that peritectically decomposes at low temperature. We then describe the synthesis of a binary compound in a binary system with limited mutual solubility of the elements and conclude with an example of the synthesis of a metastable binary compound.

Consider the traditional ceramic synthesis of a compound A_2B that is predicted to be thermodynamically stable but peritectically decomposes at a low temperature as shown in Figure 12. If the formation temperature is very low relative to that required for diffusion, the compound is very difficult, if not impossible, to prepare by cooling from high temperatures. Above the peritectic decomposition temperature, the compound does not exist. Below, but near the decomposition temperature, the driving force for the formation of the new compound will be small, making nucleation difficult, if not impossible. The driving force for nucleation can be increased by lowering the temperature; however, this further decreases the rate.

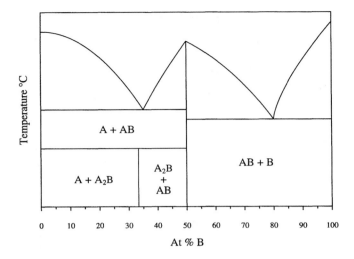

Fig. 12. A schematic phase diagram showing the compound A_2B that decomposes peritectically into a mixture of AB and A.

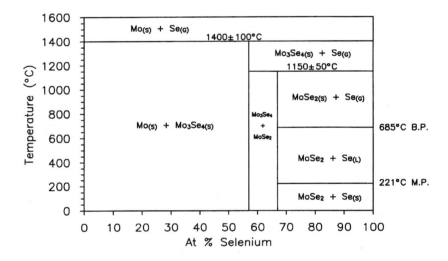

Fig. 13. The binary Mo–Se phase diagram.

Multilayer reactants provide a reaction route to an intimately mixed amorphous alloy. The composition of the amorphous alloy can then be used to avoid the nucleation of known compounds and favor the nucleation of the desired compound. To increase the rate, the temperature can be raised to just below the peritectic decomposition temperature. There will still be a significant driving force for nucleation of the new compound because of the relatively high energy of the amorphous intermediate. An example of the use of multilayer reactants to prepare a compound that peritectically decomposes is the preparation of Mo_3Se. This compound was not observed in previous investigations of the binary Mo–Se phase diagram, shown in Figure 13. Schneidmiller et al. found that Mo_3Se formed below 300 °C at the reacting interfaces of a Mo–Se multilayer reactant if its bilayer thickness was less than 15 Å and its composition was close to a 3:1 ratio of molybdenum to selenium. Annealing the reactant below 300 °C resulted in crystal growth of this new compound, which has the A-15 structure. Increasing the temperature significantly above 300 °C results in the peritectic decomposition of this compound. Surprisingly, the thermodynamically stable bi-

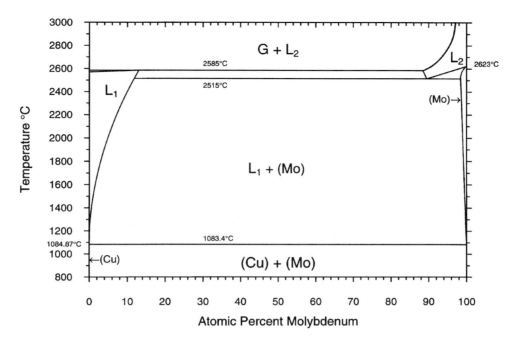

Fig. 14.   The binary Mo–Cu phase diagram.

nary compound Mo_6Se_8, which contains Mo_6 octahedra capped with selenium atoms, is not observed to form directly from a multilayer reactant. Mo_6Se_8 was only formed as the product of a reaction between $MoSe_2$ and either Mo_3Se or Mo. Schneidmiller et al. suggested that the nucleation behavior of this system reflects the complexity of the crystal structures, proposing the nucleation of Mo_3Se involves the assembly of a relatively small number of atoms relative to that required to nucleate Mo_6Se_8.

The second situation in which multilayer reactants can provide access to new compounds is in binary phase diagrams in which the elements have little, if any, mutual solubility in the liquid or solid state. An example of such a situation is the molybdenum–copper phase diagram shown in Figure 14. There are no binary compounds in this system. Investigating Mo–Cu multilayer reactants, Fister observed a reversible phase transformation (shown in Fig. 15) at 530 °C in a DSC experiment. This behavior was observed in multilayer reactants with bilayer thicknesses less than 40 Å and copper-rich compositions. Presumably, the reaction that results in the formation of this compound is driven by the increased energy of the system as a result of the high density of unstable interfaces. The system eliminates this interfacial energy through interdiffusion of the layers. The area of the reversible transition of the DSC decreased with each additional cycle through the transition, indicating that the amount of the substance causing the phase transition decreased with each cycle. A diffraction study as a function of annealing temperature showed the growth and subsequent disappearance of a diffraction maximum midway between the 111 diffraction maximum of Cu and the 110 diffraction maximum of Mo. Fister proposed that the reversible transition was an order–disorder transition of a new binary compound [55]. However, further studies are needed to determine the structure of this new phase and confirm this hypothesis.

Multilayer reactants also provide access to new binary compounds where the new compound is thermodynamically unstable with respect to known compounds. In this situation, the new binary compound can only be prepared if one can avoid the formation of any other binary compounds. Annealing a modulated reactant with a bilayer thickness less than the critical thickness results in the formation of a homogenous amorphous alloy. The

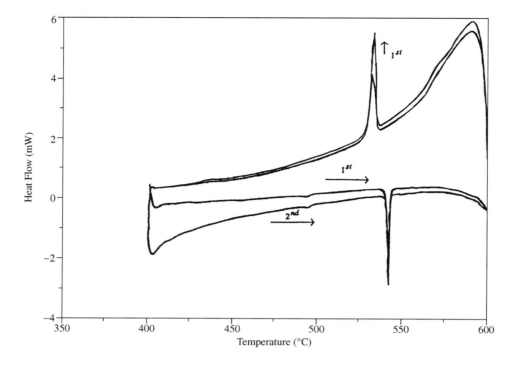

Fig. 15. Differential scanning calorimetry trace of a Mo–Cu multilayer showing the presence of a reversible phase transition at approximately 530 °C.

compound formed from the amorphous alloy depends on the relative nucleation energies of potential compounds, not on their absolute thermodynamic stability. As discussed earlier, the composition of the amorphous intermediate can be used to control the relative nucleation energies. The formation of the binary skutterudite, $FeSb_3$, is an example of this situation. This binary phase system has recently been investigated by Richter and Ipser [56], who showed that, at equilibrium, a sample containing 25 at% iron and 75 at% antimony consists of a mixture of $FeSb_2$ and antimony below 624 °C (see Fig. 16). Binary multilayer reactants of this composition with bilayer thicknesses less than 40 Å, however, evolve into a homogeneous amorphous alloy below 100 °C. Annealing these amorphous alloys above 150 °C results in the nucleation and growth of a new compound, $FeSb_3$ [57]. Further annealing of $FeSb_3$ above 350 °C results in the exothermic decomposition of this new compound. The exothermic decomposition implies that $FeSb_3$ is thermodynamically unstable with respect to a mixture of $FeSb_2$ and Sb. If the bilayer thickness is greater than 40 Å, $FeSb_2$ nucleates at the reacting interfaces. If the bilayer thickness is less than 40 Å and the composition of the initial multilayer is more iron rich than a 3:1 ratio of antimony to iron, $FeSb_2$ is observed to nucleate from the amorphous intermediate. The targeted compound, $FeSb_3$, can only be prepared by avoiding the more stable binary compounds.

3.4. Application of Multilayer Reactants to the Synthesis of New Ternary Compounds

In the synthesis of ternary and higher-order compounds, the need to avoid stable binary compounds as reaction intermediates is well recognized. As stated by Brewer many years ago, there are a multitude of undiscovered compounds that are thermodynamically stable with respect to the elements yet metastable with respect to a mixture of known binary compounds [58]. Discovering synthetic conditions and procedures to make these unknown compounds is an ongoing challenge.

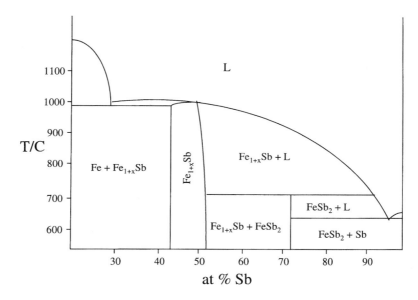

Fig. 16. The binary Fe–Sb phase diagram.

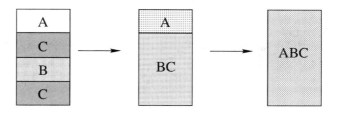

Fig. 17. A schematic of the expected reaction pathways of ternary reactants with different layer sequences. In a binary A–B multilayer, the phase AB is assumed to form at the reacting interfaces.

Multilayer reactants offer some significant advantages for the preparation of ternary and higher-order compounds by providing access to an amorphous reaction intermediate. A new experimental parameter, the order of the elemental layers within the repeating unit, can be used to control the reaction pathway. As an example, consider a ternary system ABC in which the elements A and B react to form the compound AB at the interfaces of binary A–B multilayers. In a ternary multilayer reactant with layer order ABC within the repeating unit, one would expect the compound AB to form at the reacting A–B interfaces. In a ternary multilayer with layer order ACBC within the repeating unit, there are no A–B interfaces where the binary compound AB can form. These various reaction sequences are illustrated in Figure 17.

The different diffusion rates of the elements through the different layers also create opportunities to control the reaction sequence by adjusting diffusion lengths to control time and/or the sequence of layer mixing. For example, consider the ternary system ABC in which A diffuses into B 10 times faster than it diffuses into C. A ternary multilayer reactant with a simple ABC sequence in the repeating unit on annealing will form a mixed AB layer that will then react with the C layer. By preparing a multilayer reactant with a more complex repeating unit, for example ACBCBC, one can force C and B to interdiffuse before A can react with B. This ability to design the initial structure of the multilayer reactant provides several options to select and control reaction intermediates. One consequence of the differences in relative diffusion rates of elements through different layers is that the "critical thicknesses" are not transferable from the binary systems. The different diffusion rates can be used to mix sequentially the reacting layers or force the layers to interdiffuse simultaneously, depending on the design of the initial reactant structure.

The ternary metal–molybdenum–selenium system provides a convenient platform to illustrate the principles discussed previously. In a study of the binary Mo–Se system, the thermodynamically stable binary compound Mo_6Se_8 was not observed to nucleate from binary multilayer reactants, regardless of the repeat spacing or composition [59]. Schneidmiller et al. explored the evolution of ternary reactants, M–Mo–Se, as a function of concentration and identity of the M atom. When the M atom was nickel, a slow diffusing species relative to the rate of Mo and Se, the binary compound $MoSe_2$ was observed to nucleate interfacially at the Mo–Se interfaces. When the M constituent was a fast diffusing species relative to that of Mo and Se, for example, Zn, In, Sn, or Cu, a change in reaction pathway was observed as a function of M atom concentration. When M was below a critical concentration, $MoSe_2$ was observed to nucleate interfacially. When M was above this critical concentration, the multilayer was observed to interdiffuse and form an intermixed amorphous intermediate. This is summarized in Figure 18. The low-angle diffraction pattern was observed to decay as a function of annealing temperature until the sample was no longer modulated in composition. The authors have proposed that this is due to the initial interdiffusion of the M and Se layers of the multilayer reactant to form an intermixed M–Se amorphous alloy, which then interdiffuses with the Mo layers. The interfacial nucleation of $MoSe_2$ is inhibited by the large concentration of the M cation at the reacting interface. Further annealing of the resulting amorphous alloys in these systems leads to crystallization of a number of different compounds, depending on the ternary metal. When M was tin, a layered dichalcogenide compound was formed. When M was indium, ternary molybdenum selenides containing larger clusters than the Mo_6Se_8 phase were observed. When M was copper, the desired compound, $Cu_xMo_6Se_8$, exothermically nucleated at 250 °C [60]. Further studies are required to understand the factors that control the nucleation behavior in these systems. We rationalize the difficulty in nucleating the desired cluster compounds as resulting from the large difference between the structure of the amorphous state and

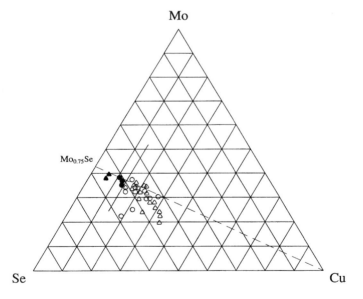

Fig. 18. A summary of the reaction behavior observed in the ternary Cu–Mo–Se system. For samples below 13% Cu, the binary phase $MoSe_2$ was observed to nucleate interfacially, denoted by filled triangles and circles. Above this composition, the ternary reactant was observed to form an amorphous intermediate, denoted by the empty triangles and circles. The samples denoted by triangles had an exotherm in the DSC data corresponding to crystallization, whereas samples denoted by circles did not have an exotherm in their DSC scans.

that of the desired cluster compounds. Although this idea will be explored in future experiments, the ternary amorphous state is clearly accessible via ternary elementally modulated reactants.

More straightforward nucleation tendencies were found in the ternary iron–antimony systems. The ternary antimonides, known as "filled" skutterudites [61], with formulas $M_xM'_4Sb_{12}$ where M is a lanthanide and M' is either Fe or Co, have been recently touted as potentially useful thermoelectric materials [62]. The antimonides with this filled structure are promising thermoelectric materials because of an unusual structural property (see Fig. 19). They can be formed with cations that are significantly smaller than their interstitial site. As a result, these ions have unusually large thermal vibration amplitudes and, therefore, are strong phonon scatterers. The conduction, however, occurs in the transition metal and antimony framework and is not effected by the motion of the cations. The result is a suppression of the phonon thermal conductivity without adverse affects on the electrical properties [63].

The weak bonding of the ternary metal in the "filled" skutterudites makes many of the potential ternary skutterudites containing small ternary metal atoms thermodynamically unstable with respect to disproportionation to a mix of binary compounds. The size mismatches cause a reduction in the Madelung energy, which is important for the stability of the crystalline structure. Consequently, only the early rare-earth skutterudites from $LaFe_4Sb_{12}$ to $NdFe_4Sb_{12}$ have been prepared via traditional synthesis techniques.

The basic structure of the skutterudites suggests that their nucleation from the amorphous state should be relatively straightforward. It consists of iron atoms octahedrally coordinated by antimony. Adjacent octahedra share corners to prepare a rather open structure, as shown in Figure 19. The ternary metal atom resides in the asymmetric cavities in this structure. An amorphous alloy with a 3:1 ratio of antimony to iron is likely to contain some of these structural features, aiding the formation of the critical nuclei of the skutterudite phase.

The DSC of a La–Sb–Fe multilayer reactant, shown in Figure 20, contains two sharp exotherms. Low-angle X-ray diffraction indicates that the multilayer interdiffused before the first sharp exotherm and high-angle diffraction indicates that the sample is amorphous. Diffraction data collected after the first exotherm are consistent with the formation of a

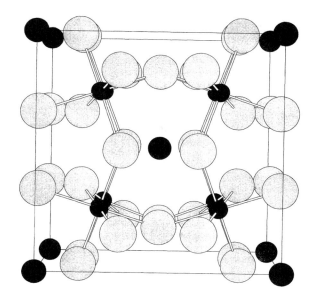

Fig. 19.   The filled skutterudite structure. The ternary M cation sits in a large and asymmetric site.

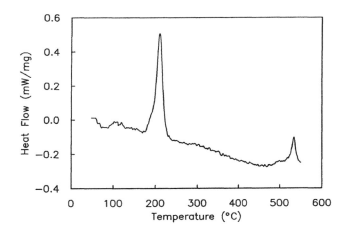

Fig. 20.   Calorimetry data of the La–Sb–Fe ternary reactant. Two sharp exotherms are observed.

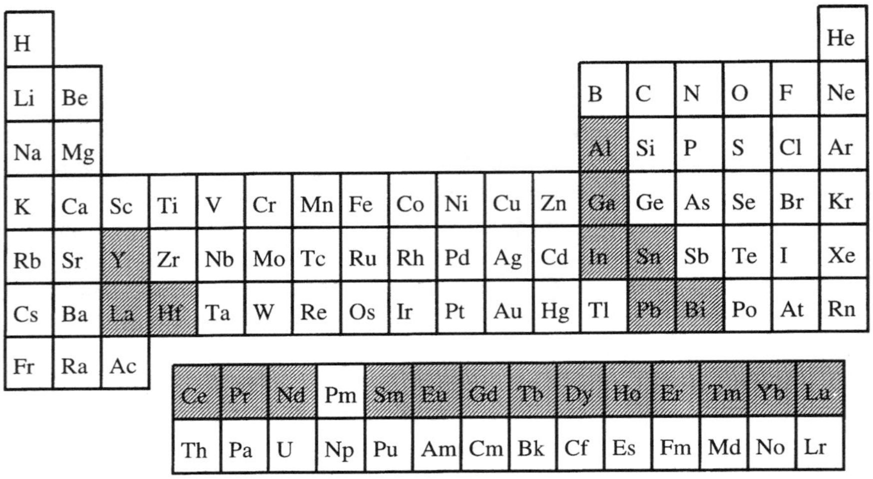

Fig. 21.   The shaded elements have been successfully inserted into the ternary site of FeSb₃, which has the skutterudite structure.

cubic skutterudite and a small amount of an impurity phase. After the second exotherm, the original cubic phase has decomposed into a mix of binary compounds and another cubic skutterudite with a smaller unit cell. The lattice parameters of the low-temperature skutterudite are larger than those previously reported. The lattice parameters of the high-temperature skutterudite agree with those of skutterudites prepared using conventional high-temperature synthesis. We suspect the difference between these two compounds is a rotation or tilt of the iron octahedra [57, 64].

Ternary M–Sb–Fe multilayer reactants have been prepared with approximately 20 different ternary metals, as shown in Figure 21 [57, 64, 65]. All interdiffuse to the amorphous state and nucleate the filled skutterudite structure at low temperature. Rare-earth–Sb–Fe multilayer reactants nucleate the "filled" skutterudite structure around 150 °C. The low-temperature skutterudite structure decomposes exothermically at approximately 450 °C for all of the rare-earth cations. The heavier, later rare earths all decompose to a mixture of binary compounds. The early rare earths through gadolinium all show at least traces of a small unit-celled skutterudite on decomposition of the low-temperature phase. Ba, Y, and Hf containing multilayers also interdiffuse and nucleate the filled skutterudite structure. Posttransition metal (Al, Ga, In, Zn, Bi, Sn, and Pb)–containing multilayers form the filled skutterudite structure, but decompose at lower temperatures than observed for the rare earths. In all of the ternary systems studied, the multilayer reactant was observed to interdiffuse to an amorphous state below 150 °C. The nucleation temperature of the skutterudite phase was very low in all of the systems studied, varying from 120 to 250 °C depending on the ternary cation. This suggests that all of the amorphous intermediates must be structurally similar to each other and contain the essential structural building blocks of the skutterudite structure.

The ternary skutterudite system demonstrates the importance of being able to vary the diffusion length in the multilayer reactant. For multilayers with repeating trilayers less than 20 Å thick, the formation exotherm of the skutterudite compound is sharp. Doubling this thickness to 40 Å broadens and adds considerable structure to the exotherm as shown in Figure 22. The compound nucleated is still the filled skutterudite. Tripling the trilayer thickness results in the interfacial nucleation of the binary compound $FeSb_2$. Because a mixture of binary compounds is more stable than the ternary skutterudite, the formation of $FeSb_2$ prevents the formation of the skutterudite. This highlights the importance of controlling reaction intermediates. Multilayer reactants provide a simple and systematic way to vary the diffusion distances to form an amorphous intermediate. The composition of the amorphous phase can then be used to control the subsequent nucleation. The design of the initial reactant avoids binary intermediates and permits the preparation of undiscovered compounds that are thermodynamically stable with respect to the elements yet metastable with respect to a mixture of known binary compounds.

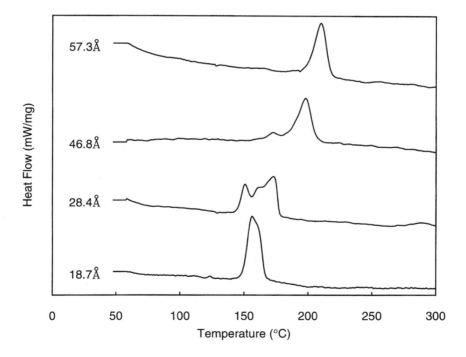

Fig. 22. The change in differential scanning calorimetry traces as a function of the thickness of the repeating trilayer in the Sn–Fe–Sb system. In the thinnest sample, $SnFe_4Sb_{12}$ nucleates at the exotherm. In the thickest sample, the binary compound $FeSb_2$ is observed to form at the reacting interfaces.

4. CRYSTALLINE SUPERLATTICES FROM MULTILAYER REACTANTS: CONTROL OF INTERFACIAL NUCLEATION

4.1. Background

The key to the preparation of metastable compounds using multilayer reactants is the control of composition on an angstrom length scale. During the course of the preceding studies, we pondered what would happen to a multilayer reactant with a long (50 to several hundred angstroms) compositional period. There are at least three reaction pathways that would lead to a crystalline superlattice product as shown in Figure 23. In the pathway to the lower left,

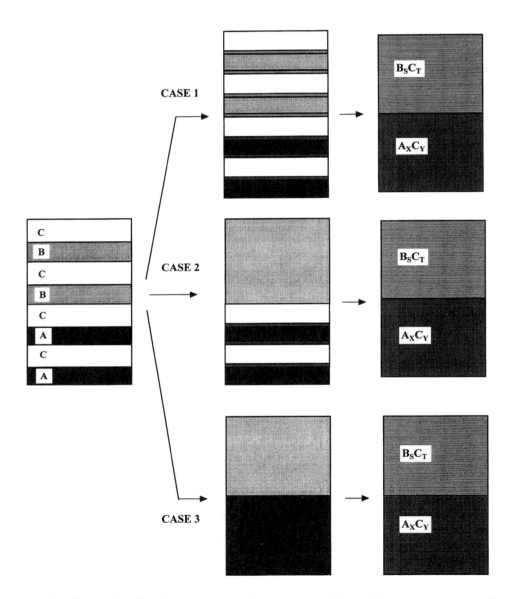

Fig. 23. A schematic of three proposed reaction pathways in which a multilayer reactant with a modulated composition could evolve into a crystalline superlattice. The pathway to the upper right interfacially nucleates both components of the final superlattice at the reacting interfaces. The pathway to the lower left shows the initial reactant interdiffusion that forms an amorphous intermediate while still maintaining a long length scale composition modulation. The middle pathway is a combination of these two extremes.

the initial reactant interdiffuses the elemental layers but maintains the long-range compositional period. Nucleation of this amorphous intermediate results in a crystalline product with the built-in compositional period. In the reaction pathway shown to the upper left, the initial reactant interfacially nucleates binary compounds at the reacting interfaces. Annealing at low temperature leads to the formation of the crystalline superlattice because insufficient time or energy is available to interdiffuse the components. The center pathway is a combination of these two extremes, with one compound interfacially nucleating while the other interdiffuses. Low-temperature annealing again may produce a crystalline superlattice.

The ability to fabricate crystalline superlattices with controlled superstructure is a prime example of fundamental research leading to new technology. Several synthetic approaches based on epitaxial growth, including molecular beam epitaxy (MBE), chemical vapor deposition, and liquid phase epitaxy, have been developed to prepare these materials. In these techniques, the deposition rates and substrate temperature are carefully controlled such that epitaxial growth of the growing sample occurs in a layer-by-layer manner. MBE has emerged as the growth technique with the most control, able to prepare samples with atomic-scale control of composition and nearly ideal interfaces [66]. These synthetic advances have led to new physical phenomena, including the quantum Hall effect [67] and the fractional quantum Hall effect [68], as well as new high-performance devices through modulation doping [69] and band gap engineering [70].

Although most of the early efforts in MBE focused on semiconducting materials, more recently researchers have explored other systems in an effort to manipulate properties and discover new phenomena. In the early 1980s, researchers prepared crystalline superlattices containing two metals with large differences in their lattice parameters and different crystal structures [71]. More recently, superlattice structures containing high-temperature superconductor components such as $SrCuO_2$–$BaCuO_2$ and $BaCuO_2$–$CaCuO_2$ have been prepared using pulsed laser deposition onto heated substrates [72, 73]. Another recent synthetic development is called van der Waals epitaxy (VDWE). This technique is used to grow structures containing van der Waals gaps, resulting in interfaces with no dangling bonds. VDWE growth has produced high-quality epitaxial films on substrates that have both large lattice mismatches and different crystal structures than that of the deposited films [74, 75].

In spite of this success, epitaxial-based techniques have several drawbacks for the exploratory synthesis of new materials. Determining epitaxial growth conditions for new materials can be a daunting task, especially when the material being grown contains elements with a wide range of vapor pressures, surface mobilities, and surface residency times. This challenge increases as the number of elemental components in a material increases. If deposition conditions for two compounds are incompatible, it is difficult to toggle between deposition of these components in building the desired superstructure. The preparation of crystalline superlattices using modulated reactants, if successful, would permit a rapid survey of new systems for unusual properties, providing targets for future MBE studies.

Noh et al. [76] decided to initiate their studies using transition metal dichalcogenides. The dichalcogenides were chosen because they were known to nucleate interfacially and grow with the $00l$ direction perpendicular to the interfaces. The basic structure of the layered transition metal dichalcogenides contain hexagonal sheets of transition metal atoms with each sheet sandwiched between two hexagonal sheets of chalcogen. The transition metal bonds covalently to the six nearest-neighbor chalcogen atoms in the adjacent chalcogen layers, forming a tightly bound XMX trilayer sandwich. The XMX sandwiches are coupled together by weak van der Waals bonding. This two-dimensional structure results in anisotropy in many physical properties. The electrical properties of these layered materials vary from insulators to true metals, depending on the coordination of the metal atom and the degree of filling of the bands. The strong dependence of physical properties on stoichiometry results from excess transition metal atoms donating additional electrons to the nonbonding d bands [77].

4.2. The Growth of Crystalline Superlattices on Annealing Multilayer Reactants

To explore the feasibility of preparing superlattices using interfacial nucleation and growth, a multilayer reactant with a composition modulation designed to yield three $TiSe_2$ and three $NbSe_2$ layers in the unit cell of the final superlattice was prepared [78]. The structure of the initial multilayer reactant is shown in Figure 24.

The reactant shown in Figure 24 was annealed in a tube furnace in a nitrogen atmosphere. The evolution of its structure as a function of annealing time and temperature was studied using X-ray diffraction. Figure 25 shows the changes in the low-angle

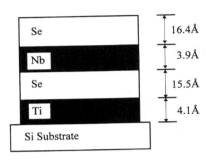

Fig. 24. The structure of an initial multilayer designed to evolve into a superlattice containing three $NbSe_2$ slabs and three $TiSe_2$ slabs in the repeating unit.

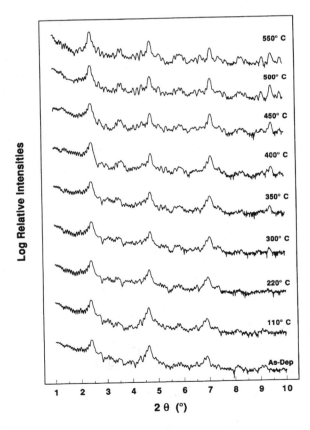

Fig. 25. The evolution of the low-angle diffraction pattern of the reactant shown in Figure 24 as a function of annealing temperature. The sample was annealed for 8 h at each temperature.

diffraction pattern as the sample is consecutively annealed for 8 h at each of the indicated temperatures. The Bragg diffraction maxima in the low-angle diffraction patterns confirm the compositionally modulated nature of the as-deposited sample and clearly indicate that the modulated structure persists throughout the annealing process. There is little change in the intensity of the low-order superlattice diffraction maxima during annealing below 220 °C. At higher annealing temperatures, the fourth-order Bragg diffraction peak increases with time, indicating that the sample is developing sharper concentration gradients.

In addition to developing sharper concentration gradients, the sample becomes smoother, as indicated by the increase in the intensity, regularity, and persistence with increasing angle of the subsidiary maxima. The subsidiary maxima result from a combination of incomplete destructive interference from the layers as well as interference between the front and back surface of the multilayer. The persistence of the subsidiary maxima is correlated to the roughness of the multilayer. Further evidence for the smoothing of the layers comes from rocking angle scans on the low-angle diffraction maxima. As shown in Figure 26, the low-angle rocking curve about the 002 Bragg maxima consists of a sharp specular peak on a broad nonspecular background. The intensity of this diffuse background depends only on the magnitude of the rms roughness of the interfaces. During low-temperature annealing, the intensity of this diffuse scattering decreases. This is shown in Figure 27, which plots the integrated area of the nonspecular scattering as a function of annealing temperature. The decrease in the roughness of the interfaces is probably related to the elimination of voids and defects in the structure, as the sample contracts by 3% during annealing.

High-angle diffraction data collected during the annealing are shown in Figure 28. The diffraction patterns obtained on the as-deposited sample indicate that dichalcogenide nuclei

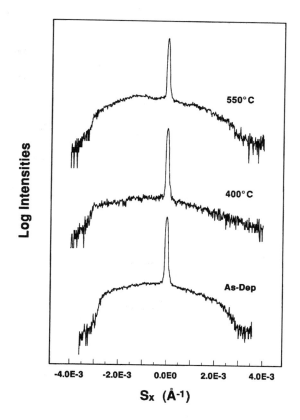

Fig. 26.   The low-angle rocking curve of the 002 Bragg diffraction maxima. The sharp spike in the center of the curve is the specular peak. The broad background is the diffuse scattering resulting from roughness.

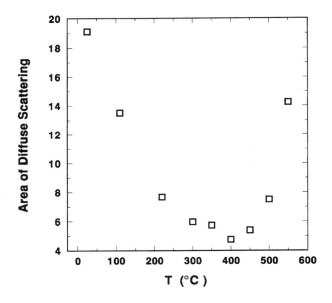

Fig. 27.   The integrated area of the diffuse scattering around the 002 diffraction maxima as a function of annealing temperature.

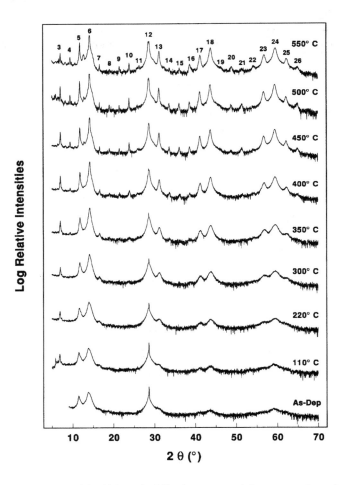

Fig. 28.   The evolution of the high-angle diffraction pattern of the reactant shown in Figure 24 as a function of annealing temperature. The sample was annealed for 8 h at each temperature.

are formed during the deposition process. Annealing below 300 °C increases the intensity of the diffraction maxima without decreasing the line width, suggesting that the amount of material crystallized increases without increasing the domain size perpendicular to the layers. This combined with the decreased roughness of the interfaces suggests that the crystallites grow along the reacting metal–selenium interfaces rather than perpendicular to them. After annealing above 350 °C, crystal growth perpendicular to the interfacial planes is clearly evident through the sharpening of the diffraction peaks as a function of annealing temperature and time, as shown in Figure 29.

Rocking curve measurements show a narrowing with increasing annealing time and temperature about the developing high-angle diffraction maxima. The half widths of the rocking curves about the 006 Bragg diffraction maxima decrease from 1.41° to 0.07° as annealing temperature is increased, as shown in Figure 30. The observed narrowing of the rocking curve widths indicates that the degree of alignment of the superlattice increases

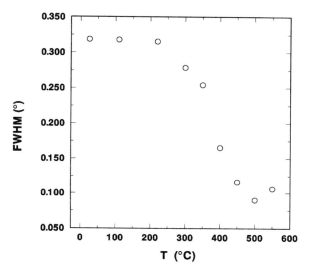

Fig. 29.   The full width at half maxima of the 005 diffraction maxima as a function of annealing temperture.

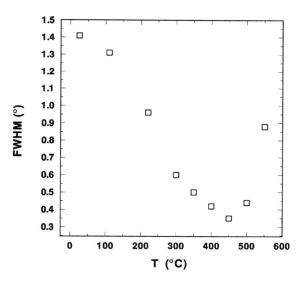

Fig. 30.   The half widths of the rocking curve diffraction scans about the 006 Bragg diffraction order as a function of annealing temperature.

as the sample evolves into the desired, kinetically trapped crystalline superlattice. The 00l diffraction planes become increasingly aligned parallel to the substrate. This also suggests the preferred growth of crystals along the interdiffusing interfaces. Dichalcogenide crystals grown from vapor phase transport reactions are flat platelets up to a square centimeter in size while less than a tenth of a millimeter in thickness. The increase in preferred orientation on annealing suggests that the basal ab plane is also the preferred growth direction in the multilayer reactant.

4.3. Proposed Mechanism

The information from the preceding diffraction studies is consistent with the four-step reaction mechanism shown schematically in Figure 31. On low-temperature annealing, the as-deposited layers (a) begin to interdiffuse. Continued annealing results in interfacial nucleation (b) and lateral growth of the respective binary components (c). Continued low-temperature annealing results in the kinetic trapping of the desired superstructure as the final product (d). The systematic annealing experiments discussed previously suggest that the optimum growth conditions for obtaining crystalline superlattices begin with a low-temperature anneal to nucleate and grow the desired component compounds at and along the interfaces [76]. Because of the low temperatures, the reaction is restricted to the interfaces, preventing long-range diffusion between the different component regions. The interfaces are the only regions with the correct compositions for the desired compounds to grow. We suspect that "pancake-like" domains formed at the interfaces act as diffusion barriers, preventing intermixing of the different metal layers. Subsequent higher-temperature annealing leads to grain growth perpendicular to the layers and eliminates defects and misoriented crystallites. This proposed mechanism was used as a guide to optimize the annealing conditions of a series of superlattices containing different numbers of each of the component layers in the repeating unit of the superlattice. The X-ray patterns collected on the final crystalline superlattices, $(TiSe_2)_m(NbSe_2)_n$, contain many well-resolved diffraction maxima, as shown in Figure 32 [79]. Significantly, this mechanism suggests that this growth technique can produce heterostructures with no epitaxial relationship between the constituents.

4.4. Determining Superlattice Structure Using X-Ray Diffraction

Standard analysis of X-ray diffraction patterns from superlattice materials treats the superlattice as a perturbation on top of the original structure of the components. The superlattice peaks are indexed as $\pm n$ relative to the main maxima of the component materials. The observation of satellite peaks around the main peak is proof of the existence of a superlattice structure.

The high degree of preferred orientation along with the quality of the diffraction patterns collected on the $(TiSe_2)_m(NbSe_2)_n$ superlattices allows the structure to be treated as a single large unit cell containing multiple layers of each of the parent components. All of the diffraction maxima in each of the diffraction patterns can be indexed as 00l reflections. The c-axis lattice parameters obtained using a least-squares refinement of the positions of the 00l reflections are consistent with the designed number of $TiSe_2$ and $NbSe_2$ layers in the repeating unit of the superlattice structure as summarized in Table I. A plot of the refined unit cell parameters versus the desired number of $TiSe_2$ monolayers in the unit cell of the superlattice is shown in Figure 33. From the calculated slopes of these four linear fits, the average $TiSe_2$ c-axis lattice parameter is 6.005(28) Å, very close to the value observed in the binary compound $TiSe_2$, 6.008 Å. The average $NbSe_2$ lattice parameter calculated from the intercepts of these lines is 6.349(16) Å, close to the range of values reported for the different polytypes of the binary compound $NbSe_2$.

For a more quantitative evaluation of the structure along the c axis, the observed diffraction patterns can be refined using Rietveld analysis. The refined structure of the crystalline

Table I.   Layer thicknesses d (Å) and lattice parameters c (Å) of selected titanium–niobium–selenides[a]

Compound	d_{Ti}[b]	d_{Se}[c]	d_{Nb}[d]	d_{Se}[e]	$\sum d$[f]	d_w[g]	c_{exp}[h]	c_{calcd}[i]
[TiSe$_2$]$_3$[NbSe$_2$]$_{12}$	1 × 4.07	1 × 15.52	4 × 3.93	4 × 16.39	(100.87)	100.09(56)	94.23(96)	94.72
[TiSe$_2$]$_6$[NbSe$_2$]$_{12}$	2 × 4.07	2 × 15.52	4 × 3.93	4 × 16.39	(120.46)	119.98(93)	112.1(17)	112.87
[TiSe$_2$]$_9$[NbSe$_2$]$_{12}$	3 × 4.07	3 × 15.52	4 × 3.93	4 × 16.39	(140.05)	138.8(10)	130.2(20)	131.03
[TiSe$_2$]$_{12}$[NbSe$_2$]$_{12}$	4 × 4.07	4 × 15.52	4 × 3.93	4 × 16.39	(159.64)	154.6(30)	148.1(34)	149.18
[TiSe$_2$]$_6$[NbSe$_2$]$_{15}$	2 × 4.07	2 × 15.52	5 × 3.93	5 × 16.39	(140.78)	138.3(11)	131.2(23)	132.01
[TiSe$_2$]$_9$[NbSe$_2$]$_{15}$	3 × 4.07	3 × 15.52	5 × 3.93	5 × 16.39	(160.37)	158.19(49)	149.2(29)	150.17
[TiSe$_2$]$_{12}$[NbSe$_2$]$_{15}$	4 × 4.07	4 × 15.52	5 × 3.93	5 × 16.39	(179.96)	177.6(10)	166.9(27)	168.32
[TiSe$_2$]$_{15}$[NbSe$_2$]$_{15}$	5 × 4.07	5 × 15.52	5 × 3.93	5 × 16.39	(199.55)	188.6(40)	185.6(33)	186.48
[TiSe$_2$]$_1$[NbSe$_2$]$_9$	1 × 1.36	1 × 5.17	3 × 3.93	3 × 16.39	(67.49)	66.69(39)	63.35(35)	63.47
[TiSe$_2$]$_3$[NbSe$_2$]$_9$	1 × 4.07	1 × 15.52	3 × 3.93	3 × 16.39	(80.55)	78.42(15)	75.16(92)	75.58
[TiSe$_2$]$_6$[NbSe$_2$]$_9$	2 × 4.07	2 × 15.52	3 × 3.93	3 × 16.39	(100.14)	98.36(42)	93.2(15)	93.73
[TiSe$_2$]$_9$[NbSe$_2$]$_9$	3 × 4.07	3 × 15.52	3 × 3.93	3 × 16.39	(119.73)	120.62(74)	110.9(17)	111.89
[TiSe$_2$]$_{12}$[NbSe$_2$]$_9$	4 × 4.07	4 × 15.52	3 × 3.93	3 × 16.39	(139.32)	136.48(28)	129.1(27)	130.04
[TiSe$_2$]$_1$[NbSe$_2$]$_6$	1 × 1.36	1 × 5.17	2 × 3.93	2 × 16.39	(47.17)	47.10(11)	44.09(22)	44.33
[TiSe$_2$]$_2$[NbSe$_2$]$_6$	1 × 2.71	1 × 10.35	2 × 3.93	2 × 16.39	(53.70)	53.39(13)	50.23(31)	50.38
[TiSe$_2$]$_3$[NbSe$_2$]$_6$	1 × 4.07	1 × 15.52	2 × 3.93	2 × 16.39	(60.23)	59.40(16)	56.15(40)	56.44
[TiSe$_2$]$_6$[NbSe$_2$]$_6$	2 × 4.07	2 × 15.52	2 × 3.93	2 × 16.39	(79.82)	78.89(31)	74.16(85)	74.59
[TiSe$_2$]$_9$[NbSe$_2$]$_6$	3 × 4.07	3 × 15.52	2 × 3.93	2 × 16.39	(99.41)	95.07(11)	92.4(11)	92.75

[a]The factor in front of the layer thickness indicates the number of times a layer was deposited. [b]Intended thickness of the titanium layer. [c]Intended thickness of the selenium layer. [d]Intended thickness of the niobium layer. [e]Intended thickness of the selenium layer. [f]Thickness of the total layer. [g]Measured thickness of the repeat unit after deposition. [h]Calculated lattice parameter c of the superlattice of the product, based on the lattice parameters of the binary constituents. [i]Calculated lattice parameter c of the superlattice of the reactant.

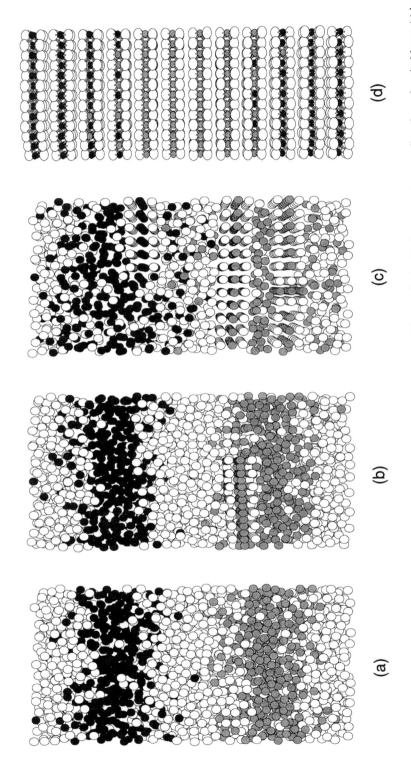

Fig. 31. Proposed reaction mechanism for the evolution of an initial multilayer reactant into a crystalline superlattice. On low-temperature annealing, the as-deposited layers (a) begin to interdiffuse. Continued annealing results in interfacial nucleation (b) and lateral growth of the respective binary components (c). Continued low-temperature annealing results in the kinetic trapping of the desired superstructure as the final product (d).

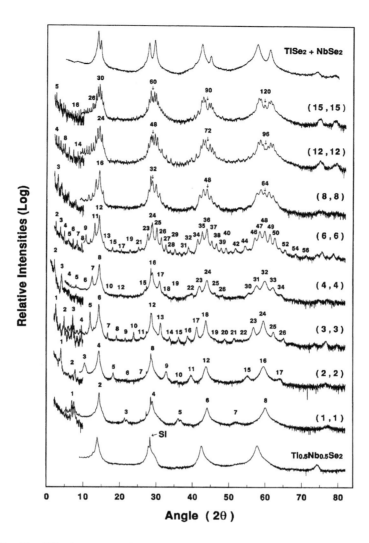

Fig. 32. The diffraction patterns of a number of $[NbSe_2]_m[TiSe_2]_n$ superlattices obtained by annealing multilayer reactants with designed initial structure. The numbers above the diffraction maxima correspond to the $00l$ index of the superlattice unit cell.

superlattice $(TiSe_2)_6(NbSe_2)_6$ based on the observed X-ray diffraction pattern is shown in Figure 34. The basic structure is as designed, containing six unit cells of each of the dichalcogenide components. The refined structure revealed that there are four niobium diselenide layers with little titanium content and one layer next to the titanium layers that is approximately 40% titanium and 60% niobium. The titanium diselenide layers all contain approximately 7% niobium substituted for titanium. The van der Waals gaps in both the $TiSe_2$ and $NbSe_2$ blocks are comparable to those of the pure dichalcogenides. The van der Waals gaps on either side of the 40% titanium and 60% niobium mixed layers are slightly larger because of the a-axis mismatch between the niobium and the titanium dichalcogenides [76].

The diffraction pattern normal to the surface contains no information about the bonding in the layers resulting from preferred orientation. To obtain information on the structure along the planes, a number of experiments were performed. To determine if the superlattice samples were crystalline in the plane and aligned between planes, off-specular diffraction data were collected. Figure 35 contains the diffraction map obtained on a $(TiSe_2)_{12}(NbSe_2)_9$ superlattice. In addition to the $00l$ Bragg diffraction maxima, the

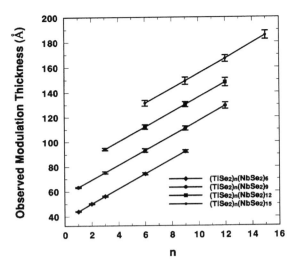

Fig. 33. The change in the unit cell parameters of various $[NbSe_2]_m[TiSe_2]_n$ superlattices as a function of the number of $TiSe_2$ units of the unit cell. The straight lines go through points of superlattices having the same number of $NbSe_2$ units in the unit cell of the superlattice as indicated.

complete family of $10l$ diffraction planes are observed, implying that this superlattice is crystallographically ordered in the ab plane. Pole figure measurements of the $10l$ diffraction planes show that these planes are completely isotropic. This suggests that the samples are composed of microcrystalline domains that are highly oriented along the c axis, but each domain has a random orientation in the ab plane. To confirm this hypothesis, Noh et al. [80] collected scanning electron microscopy (SEM) images of the $(TiSe_2)_{12}(NbSe_2)_9$ superlattice. The SEM image of this sample formed using secondary electrons, shown in Figure 36, reveals that most areas of the sample surface are very flat without any topographic structure. Cracks are observed on the annealed samples that are not observed on the as-deposited samples. The authors suggested that the cracks result from shrinkage of the sample during annealing, leading to stress and fracture during cooling. The backscattered image of this sample, which is much more sensitive to the average atomic number, is shown in Figure 37. The significant contrast observed in this image suggests that the mean atomic number varies across domains of the sample. To confirm this result, electron microprobe data were collected. The light areas had an average composition of $Nb_{0.88}Ti_{1.02}Se_4$, while the dark areas had average compositions of $Nb_{0.93}Ti_{1.03}Se_4$ [80].

4.5. Metastability of the Crystalline Superlattices

The interdiffused layers observed in the crystal structure at the interface between the $NbSe_2$ and $TiSe_2$ blocks suggest that the crystalline superlattices are only kinetically trapped. To test the stability of these compounds, a $(TiSe_2)_6(NbSe_2)_6$ superlattice was annealed at elevated temperatures and diffraction data were collected as a function of annealing, as shown in Figure 38. The data in Figure 38 show that, as the annealing temperature is increased, the superlattice diffraction pattern decays as a result of the intermixing of the $TiSe_2$ and $NbSe_2$ layers. The decay of the superlattice pattern is not that expected from a simple picture. If Fick's law for diffusion for a composition-independent diffusion coefficient holds, the nth-order diffraction peaks should decay n^2 times faster than the first-order diffraction peak as the composition profile approaches a sinusoidal modulation. The data in Figure 38 indicate a different picture, however, because both low- and high-order diffraction peaks decay at about the same rates. This suggests that the square-wave composition profile in the initial structure remains throughout intermixing. One explanation of these data is that

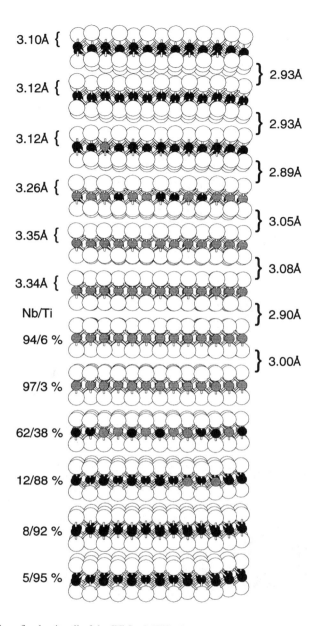

Fig. 34.   The refined unit cell of the $[NbSe_2]_6[TiSe_2]_6$ superlattice determined by Rietveld refinement of its diffraction pattern.

the electron density difference between the two regions decreases with annealing, as shown in Figure 39. The composition profile within each of these regions remains flat, however, because the rate limiting step in mixing is the motion of the transition metals across the boundary between them.

4.6. Preparation of Superlattices Containing Other Component Phases Using Multilayer Reactants

To probe the generality of the reaction pathway observed for the evolution of multilayer reactants containing Nb/Se/Ti/Se layers into $(TiSe_2)_m(NbSe_2)_n$ superlattices, several additional systems have been investigated. These are briefly discussed next.

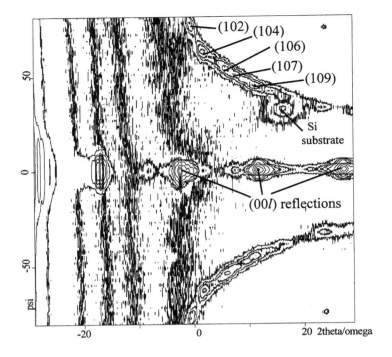

Fig. 35. The diffraction map obtained on the $[NbSe_2]_9[TiSe_2]_{12}$ superlattice. The numbers on the map correspond to the indices of the $00l$ and $10l$ reflections.

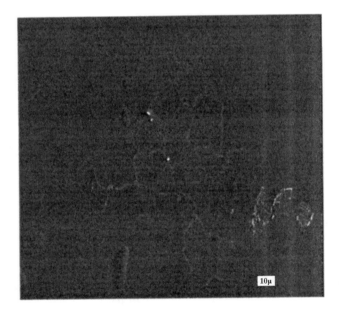

Fig. 36. SEM image of the $[NbSe_2]_9[TiSe_2]_{12}$ superlattice formed using secondary electrons.

4.6.1. *(TiSe₂)₆(TaSe₂)₆ Superlattice Formation from Ti/Se/Ta/Se* *Multilayer Reactants*

To determine whether it is possible to prepare other dichalcogenide superlattices using multilayer reactants, a Ti/Se/Ta/Se multilayer reactant was prepared. The composition of each

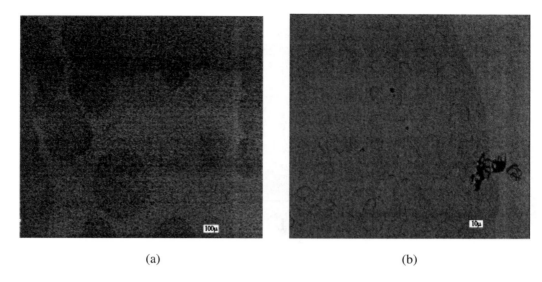

(a)   (b)

Fig. 37.   SEM image of the $[NbSe_2]_9[TiSe_2]_{12}$ superlattice formed using backscattered electrons.

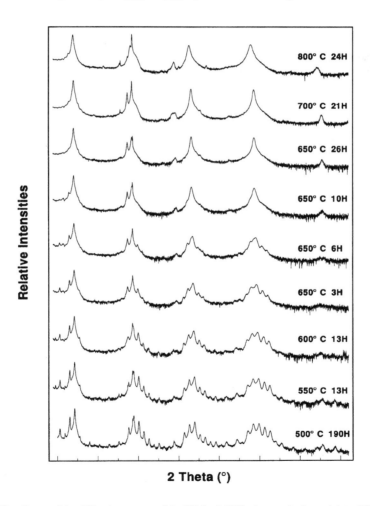

Fig. 38.   Decay of the diffraction pattern of the $[NbSe_2]_6[TiSe_2]_6$ superlattice as it interdiffused to form a uniform solid solution of composition $NbTiSe_4$.

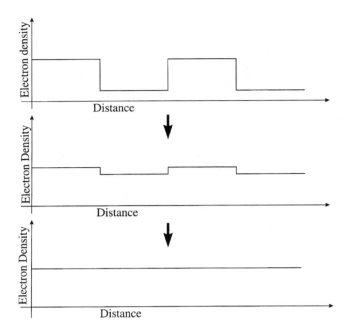

Fig. 39. A schematic of the proposed interdiffusion of the $[NbSe_2]_6[TiSe_2]_6$ superlattice as it forms a uniform solid solution of composition $NbTiSe_4$. The rate limiting step is the transfer of the transition metal cations across the boundary region between the $NbSe_2$ and $TiSe_2$ blocks.

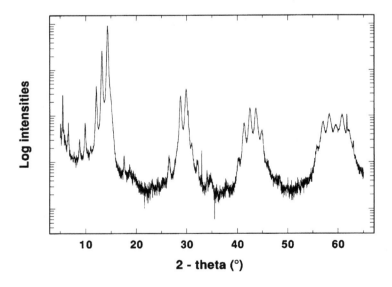

Fig. 40. The diffraction patterns of the $[TaSe_2]_6[TiSe_2]_6$ superlattice obtained by annealing a multilayer reactant. The numbers above the diffraction maxima correspond to the 00l index of the superlattice unit cell.

Ti–Se and Ta–Se period was chosen to be that of the desired dichalcogenide compound and the thicknesses of each elemental layer in the initial multilayer were chosen to form integral multiples of the dichalcogenide unit cells after annealing. Because the chemical properties of tantalum are similar to that of niobium, a similar superlattice formation was expected. The X-ray diffraction pattern collected on this sample after annealing is shown in Figure 40. The data contain all of the (00l) diffraction maxima expected from the designed

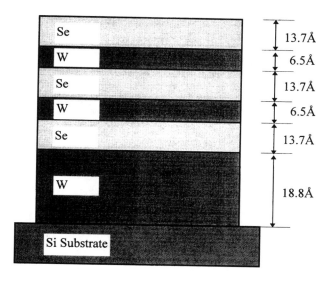

Fig. 41. A schematic of the initial multilayer reactant used to form the $W_n[WSe_2]_8$ superlattice.

repeat unit containing six $TiSe_2$ layers and six $TaSe_2$ layers. This result suggests that multilayer reactants can be designed to prepare superlattices with most of the dichalcogenide compounds as components.

4.6.2. W/WSe₂ Superlattice Formation from W/Se/W/Se/W/Se Multilayer Reactants

In the previous experiments, two layered transition metal dichalcogenide species were used as the components of the superlattice. To test the extension of the observed reaction mechanism to systems without both components having van der Waals gaps, the tungsten–tungsten diselenide system was explored. Tungsten is a hard metal with a body-centered cubic structure, whereas tungsten diselenide is a semiconductor with a layered structure of hexagonal symmetry. The lack of epitaxial relationships and structural similarities between the components of the proposed superlattice made it doubtful that long-range structural coherence would develop across many layers of the material.

Previous research in a binary tungsten–selenium system revealed that WSe_2 nucleated and grew at the internal W–Se interfaces. This information was used to prepare an initial reactant designed to evolve into a tungsten and tungsten diselenide superlattice. This reactant consisted of a thick tungsten and several thin tungsten and selenium layers in the repeating unit, as shown in Figure 41. The sample was cut into several pieces and each piece was annealed under different annealing procedures and examined using X-ray diffraction. The best result was observed when the sample was annealed at 250 °C for more than 12 h at which point WSe_2 nucleated. After the 600 °C annealing, there was a significant increase in the intensity and a decrease in the line widths of both the tungsten and the tungsten diselenide high-angle diffraction maxima, suggesting crystal growth perpendicular to the substrate. Figure 42 shows the diffraction pattern collected on this sample after being annealed at 750 °C for 23 h. The appearance of subsidiary peaks around the (002) diffraction peak of the tungsten diselenide near 13° and the (110) diffraction maxima of the tungsten near 40° clearly indicates the formation of a superlattice. The rocking curve data collected on these two Bragg diffraction peaks suggested that the alternating layers of tungsten and tungsten diselenide were crystallized preferentially: The (00l) Miller planes of the dichalcogenide layer and the (110) Miller plane of the tungsten layer are parallel to the interfaces [81].

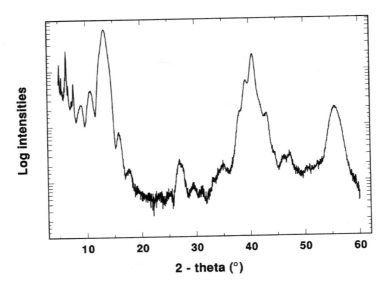

Fig. 42. The diffraction patterns of the $W_n[WSe_2]_8$ superlattice obtained by annealing a multilayer reactant. The subsidiary diffraction maxima resulting from the superlattice unit cell are apparent as shoulders of the 00l peak of the diselenide and the 110 peak of the tungsten.

5. Conclusions

5.1. Controlling Nucleation of Amorphous Intermediates Prepared Using Multilayer Reactants

The ability to design the structure of multilayer reactants provides access to amorphous intermediates by avoiding interfacial nucleation of binary compounds. Multilayer reactants below a critical layer thickness interdiffuse, forming metastable amorphous reaction intermediates. In ternary systems, the order of the deposited layers and the ability to control the diffusion distances in the initial reactant provide additional parameters to control the reaction pathway. The composition of the amorphous intermediate controls the relative activation energies required to nucleate different crystalline compounds.

A major challenge remains, however, in developing further techniques to control nucleation. The goal of these studies should be to develop control of the structure of the nucleated compound. Several techniques have been suggested as potential ways to achieve this control. The amorphous intermediate could be "seeded" with crystallites of the desired structure, templating the crystallization of the amorphous intermediate. The template for this seeding could be the substrate or a region within the initial multilayer having the composition of the desired seed compound.

A second goal is to develop techniques to control the nucleation density. Rapid thermal annealing of the wafer to high temperatures for a short period of time followed by low-temperature annealing might be an avenue to control nucleation density by the period of the high-temperature anneal. A second approach would be to prepare an isolated region on a wafer that is connected to the rest of the wafer by only a long and thin path. Nucleation would occur on one side of the path followed by crystal growth through the thin path. This would result in a single crystal region as only one crystallite would have the correct orientation to grow through the path.

A third area of exploration is the use of other deposition techniques as well as codeposition for preparing amorphous reaction intermediates. Codeposition may be one option for preparing amorphous alloys in systems that interfacially nucleate binary compounds easily. A difficulty in this area will be the characterization of the amorphous reactant produced.

Solid-state nuclear magnetic resonance (NMR) might be a powerful companion to X-ray diffraction and transmission electron microscopy for determining both the type and the frequency of local bonding arrangements within the amorphous precursors. Understanding the relationship between the structure of the amorphous intermediate and the activation energies for nucleating various compounds would add tremendous insight into the challenge of controlling nucleation.

5.2. Superlattice Formation from Multilayer Reactants

The results published on the formation of crystalline superlattices from multilayer reactants suggests that the reacting interfaces control the nucleation process and the identity of the nucleating compounds. The "first phase rule" developed by Walser and Bené is useful as a guide to predict the compounds that will nucleate. The initial structure of the superlattice reactant controls the subsequent kinetics of the solid-state reactions. In many respects, the use of the structure of the initial multilayer reactant to guide the subsequent evolution of products is analogous to the use of "protecting groups" as diffusion barriers in molecular chemistry. Although the results to date clearly demonstrate the ability to prepare crystalline superlattices, further experiments are necessary to demonstrate that the properties of the extended compounds formed can be controlled by manipulating the superlattice structure. Work toward this goal is underway.

References

1. F. J. DiSalvo, *Science* 247, 649 (1990).
2. H. Schmalzried, "Solid State Reactions," Vol. 12. Verlag Chemie, Deerfield Beach, FL, 1981.
3. J. D. Corbett, in "Solid State Chemistry Techniques" (A. K. Cheetham and P. Day, eds.), pp. 1–38. Clarendon Press, Oxford, UK, 1987.
4. H. Schäfer, *Angew. Chem., Int. Ed. Engl.* 10, 43 (1971).
5. G. M. Kanatzidis, *Curr. Opin. Solid State Mater. Sci.* 2, 139 (1996).
6. J. Rouxel, M. Tournmoux, and R. E. Brec, *Mater. Sci. Forum* 152–153 (1994).
7. R. Schollhorn, *Angew. Chem., Int. Ed. Engl.* 35, 2338 (1996).
8. J. DuMond and J. P. Youtz, *J. Appl. Phys.* 11, 357 (1940).
9. L. A. Greer, *Curr. Opin. Solid State Mater. Sci.* 2, 300 (1997).
10. A. L. Greer and F. Spaepen, in "Synthetic Modulated Structures" (L. C. Chang and B. C. Giessen, eds.), pp. 419–486. Academic Press, New York, 1985.
11. T. Novet, Ph.D. Thesis, University of Oregon, 1993.
12. T. Ishikawa, A. Iida, and T. Matsushita, *Nucl. Instrum. Methods Phys. Res., Sect. A* 246, 348 (1986).
13. D. L. Rosen, D. Brown, J. Gilfrich, and P. Burkhalter, *J. Appl. Crystallogr.* 21, 136 (1988).
14. S. K. Sinha, E. B. Sirota, S. Garoff, and H. B. Stanley, *Phys. Rev. B* 38, 2297 (1988).
15. S. R. Andrews and R. A. Cowley, *J. Phys. C* 18, 6427 (1985).
16. D. G. Stearns, *J. Appl. Phys.* 65, 491 (1989).
17. J. M. Elson, J. P. Rahn, and J. M. Bennett, *Appl. Opt.* 19, 669 (1980).
18. C. K. Carniglia, *Opt. Eng.* 18, 104 (1979).
19. D. E. Savage, J. Kleiner, N. Schimke, Y.-H. Phang, T. Jankowski, J. Jacobs, R. Kariotis, and M. G. Lagally, *J. Appl. Phys.* 69, 1411 (1991).
20. W. Weber and B. Lengeler, *Phys. Rev. B* 46, 7953 (1992).
21. F. Nava, P. A. Psaras, H. Takai, and K. N. Tu, *J. Appl. Phys.* 59, 2429 (1986).
22. P. Gas, F. M. d'Heurle, F. K. LeGoues, and S. J. La Placa, *J. Appl. Phys.* 59, 3458 (1986).
23. B. Coulman and H. Chen, *J. Appl. Phys.* 59, 3467 (1986).
24. C. Canali, F. Catellani, G. Ottaviani, and M. Prudenziati, *Appl. Phys. Lett.* 33, 187 (1978).
25. C. Canali, G. Majni, G. Ottaviani, and G. Celotti, *J. Appl. Phys.* 50, 255 (1979).
26. R. W. Bené, *Appl. Phys. Lett.* 41, 529 (1982).
27. R. M. Walser and R. W. Bené, *Appl. Phys. Lett.* 28, 624 (1976).
28. R. Sinclair and T. J. Konno, *J. Magn. Magn. Mater.* 126, 108 (1993).
29. K. Holloway and R. Sinclair, *J. Appl. Phys.* 61, 1359 (1987).
30. K. L. Holloway, Ph.D. Thesis, Stanford University, 1989.
31. K. Holloway, K. B. Do, and R. Sinclair, *J. Appl. Phys.* 65, 474 (1989).
32. R. Benedictus, A. Bottger, and E. J. Mittemeijer, *Phys. Rev. B* 54, 9109 (1996).

33. R. B. Schwarz and J. B. Rubin, *J. Alloys Compd.* 194, 189 (1993).
34. R. B. Schwarz and W. L. Johnson, *Phys. Rev. Lett.* 51, 415 (1983).
35. R. B. Schwarz, K. L. Wong, and W. L. Johnson, *J. Non-Cryst. Solids* 61/62, 129 (1984).
36. M. Van Rossum, M. A. Nicolet, and W. L. Johnson, *Phys. Rev. B* 29, 5498 (1984).
37. B. M. Clemens, R. B. Schwarz, and W. L. Johnson, *J. Non-Cryst. Solids* 61/62, 817 (1984).
38. B. M. Clemens, *Phys. Rev. B* 33, 7615 (1986).
39. H. Schroder, K. Samwer, and U. Koster, *Phys. Rev. Lett.* 54, 197 (1985).
40. R. W. Bene, *J. Appl. Phys.* 61, 1826 (1987).
41. B. M. Clemens and R. Sinclair, *Mater. Res. Soc. Bull.* 19 (1990).
42. E. J. Cotts, W. J. Meng, and W. L. Johnson, *Phys. Rev. Lett.* 57, 2295 (1986).
43. H. Beck and H.-J. Güntherodt, in "Glassy Metals I: Ionic Structure, Electronic Transport, and Crystallization" (H. Beck and H.-J. Güntherodt, eds.), Vol. 46, pp. 1–17. Springer-Verlag, New York, 1981.
44. J. H. Brophy, R. M. Rose, and J. Wulff, "The Structure and Properties of Materials," Vol. 2. 1964.
45. H. E. Kissinger, *Anal. Chem.* 29, 1702 (1957).
46. J. H. Brophy, R. M. Rose, and J. Wulff, "Thermodynamics of Structure," Vol. 2. Wiley, New York, 1964.
47. H. J. Highmore, A. L. Greer, J. A. Leake, and J. E. Evetts, *Mater. Lett.* 6, 40 (1988).
48. U. Gösele and K. N. Tu, *J. Appl. Phys.* 66, 2619 (1989).
49. W. J. Meng, C. W. Nieh, and W. L. Johnson, *Appl. Phys. Lett.* 51, 1693 (1987).
50. L. Fister and D. C. Johnson, *J. Am. Chem. Soc.* 114, 4639 (1992).
51. M. Fukuto, J. Anderson, M. D. Hornbostel, D. C. Johnson, H. Haung, and S. D. Kevan, *J. Alloys Compd.* 248, 59 (1997).
52. C. A. Grant and D. C. Johnson, *Chem. Mater.* 6, 1067 (1994).
53. T. Novet and D. C. Johnson, *J. Am. Chem. Soc.* 113, 3398 (1991).
54. O. Oyelaran, T. Novet, C. D. Johnson, and D. C. Johnson, *J. Am. Chem. Soc.* 118, 2422 (1996).
55. L. M. Fister, Ph.D. Thesis, University of Oregon, 1993.
56. K. W. Richter and H. Ipser, *J. Alloys Comp.* 247, 247 (1997).
57. M. D. Hornbostel, E. J. Hyer, J. Thiel, and D. C. Johnson, *J. Am. Chem. Soc.* 119, to appear.
58. L. Brewer, *J. Chem. Educ.* 35, 153 (1958).
59. R. Schneidmiller, M. D. Hornbostel, and D. C. Johnson, *Inorg. Chem.* 36, 5894 (1997).
60. L. Fister and D. C. Johnson, *J. Am. Chem. Soc.* 116, 629 (1993).
61. W. Jeitschko and D. Braun, *Acta Crystallogr. Sect. B* 33, 3401 (1977).
62. B. C. Sales, D. Mandrus, and R. K. Williams, *Science* 272, 1325 (1996).
63. G. S. Nolas, G. A. Slack, D. T. Morelli, T. M. Tritt, and A. C. Ehrlich, *J. Appl. Phys.* 79, 4002 (1996).
64. M. D. Hornbostel, E. J. Hyer, J. H. Edvalson, and D. C. Johnson, *Inorg. Chem.*, to appear.
65. H. Sellinschegg, S. L. Stuckmeyer, M. D. Hornbostel, and D. C. Johnson, *Chem. Mater.*, to appear.
66. A. Cho, ed., "Molecular Beam Epitaxy," Vol. 1. AIP Press, New York, 1994.
67. K. v. Klitzing, G. Dorda, and M. Pepper, *Phys. Rev. Lett.* 45, 494 (1980).
68. D. C. Tsui, H. L. Stormer, and A. C. Gossard, *Phys. Rev. Lett.* 48, 1559 (1982).
69. N. Sano, H. Kato, and S. Chiko, *Solid State Commun.* 49, 123 (1984).
70. F. Capasso, *Physica B* 129, 92 (1985).
71. I. K. Schuller, *Phys. Rev. Lett.* 44, 1597 (1980).
72. X. Li, T. Kawai, and S. Kawai, *Jpn. J. Appl. Phys.* 33, L18 (1994).
73. D. P. Norton, B. C. Chakoumakos, J. D. Budai, D. H. Lowndes, B. C. Sales, J. R. Thomson, and D. K. Christen, *Science* 265, 2074 (1994).
74. A. Koma and K. Yoshimura, *Surf. Sci.* 174, 556 (1986).
75. A. Koma, K. Saiki, and Y. Sato, *Appl. Surf. Sci.* 41/42, 451 (1989).
76. M. Noh, J. Thiel, and D. C. Johnson, *Science* 270, 1181 (1995).
77. J. A. Wilson and A. D. Yoffe, *Adv. Phys.* 18, 193 (1969).
78. M. Noh and D. C. Johnson, *J. Am. Chem. Soc.* 118, 9117 (1996).
79. M. Noh and D. C. Johnson, *Angew. Chem. Int. Ed. Engl.* 35, 2666 (1996).
80. M. Noh, H. J. Shin, K. Jeong, J. Spear, D. C. Johnson, S. D. Kevan, and T. Warwick, *J. Appl. Phys.* 81, 7787 (1997).
81. S. Moss, M. Noh, K. H. Jeong, D. H. Kim, and D. C. Johnson, *Chem. Mater.* 8, 1853 (1996).

Chapter 6

STRAINED-LAYER HETEROEPITAXY TO FABRICATE SELF-ASSEMBLED SEMICONDUCTOR ISLANDS

W. H. Weinberg, C. M. Reaves, B. Z. Nosho, R. I. Pelzel, S. P. DenBaars

Departments of Chemical Engineering and Materials, University of California, Santa Barbara, California, USA

Contents

Handbook of Nanostructured Materials and Nanotechnology, edited by H.S. Nalwa
Volume 1: Synthesis and Processing
Copyright © 2000 by Academic Press
All rights of reproduction in any form reserved.

ISBN 0-12-513761-3/$30.00

1. INTRODUCTION

1.1. Trends in Semiconductor Nanostructures: Smaller in All Dimensions

In the study of physical properties, nanostructures often provide the best or the only testing ground for phenomena in fields such as quantum mechanics and condensed-matter physics. In electronic devices, there is a trend to use smaller numbers of electrons to get a task done. With devices that emit light such as laser diodes, the emission wavelengths must be controlled. To improve both types of devices, structures that exploit quantum mechanical behavior are an option. This is achieved by reducing the size of the structure. If only one dimension is made small, the electron will only be partially confined; it will still behave as a free electron in the remaining two large dimensions. Quantum structures are, therefore, classified by how many dimensions provide confinement or, inversely, how many dimensions allow free-electron behavior. If a structure provides confinement in one dimension, it is called a quantum well. If a structure provides confinement in two dimensions, it is called a quantum wire. If a structure provides confinement in three dimensions, it is called a quantum box or quantum dot. Although the quantum mechanics are well established, the creation of quantum structures, in particular, quantum dots, is difficult.

This difficult work is pursued because quantum dots emulate a single atom. One atomic property is that an atom has discrete energy levels. If a nanostructure that confines electrons can be fabricated small enough, then discrete energy levels can be observed. This has been done with thin semiconductor structures for several decades [1]. However, thin structures only provide quantum confinement, and discrete energy states, in one dimension. The goal with quantum dots is to achieve quantum confinement in all dimensions. One motivation for this arises from the concept of density of states. The total energy of an electron has kinetic energy components resulting from motion (momentum) in three Cartesian directions. To account for a particular amount of energy, there is usually a number of combinations of momentum components that can be considered. Even in a quantum well or a quantum wire, the discrete energy levels only partially define the energy; momentum in the unconfined dimensions can lead to a range of allowed states. A quantum dot provides confinement in all dimensions. The allowed energy states are completely defined by the quantum confinement, and the resulting density of states is, therefore, a delta function [2–4].

Why is this well-defined density of states so desirable? One reason is the increased accuracy in the energy. With the widely used quantum well, the energy of an electron can be narrowed down to a *minimum* of the lowest energy state. There are several instances where this uncertainty can be a problem. Quantum structures are often used in physical measurements to understand better quantum mechanics and the properties of materials. What if an external field (e.g., magnetic, electric, or stress) is applied to a sample and the experimentalist is looking for a shift in the quantized energy level? A shift in the measured energy of the electron could possibly be due to redistribution among allowed energy states, not just a shift in the level. Quantum structures are often used in electronic and optoelectronic devices. In the case of lasers, a partially continuous density of states may lead to the transformation of electrical energy into light at an unwanted energy. For many applications, the distribution of electronic energy resulting from thermal considerations can be a problem, yet would be nearly eliminated if the density of states were a delta function [5].

Other chapters in this book discuss the fabrication and use of many types of nanostructures, many of which use semiconductor epitaxy. One approach to nanostructure fabrication is to take advantage of the unique aspects of epitaxy to form self-assembled nanostructures. Such structures have several advantages. One major advantage is that they are formed during the epitaxial growth with no processing.

1.2. Processing: The Good and the Bad

Regardless of the technique, there are two key requirements in fabricating nanostructures: (1) achieving the desired size, shape, density, and spatial distribution and (2) maintaining high material quality. These requirements can be difficult to achieve. Producing structures that are nanometer sized in one dimension is relatively simple. Epitaxy grows material on an atomic level of a control. Thin layers such as quantum wells have been readily fabricated for about two decades [1]. They can be found in a number of commercial microelectronic devices and have been used in a range of physical studies.

Thin semiconductor layers can be formed simply by epitaxy. What about nanostructures that are wires and boxes? These structures require control not only in the epitaxial growth direction but also laterally. There are several methods to achieve this. There are many successful approaches that involve common processing steps such as lithography and etching. Some of these techniques are illustrated in Figure 1. Other methods involve processing after growth. A semiconductor sample, often starting with a thin layer, is patterned by placing a patterned etch mask (often a photoresist or a dielectric layer) on the sample, as shown in Figure 1a. The sample is then etched with either solution-based wet techniques or reactive-ion dry techniques such that the thin layer is laterally defined. This technique has been used to create a quantum box laser [6].

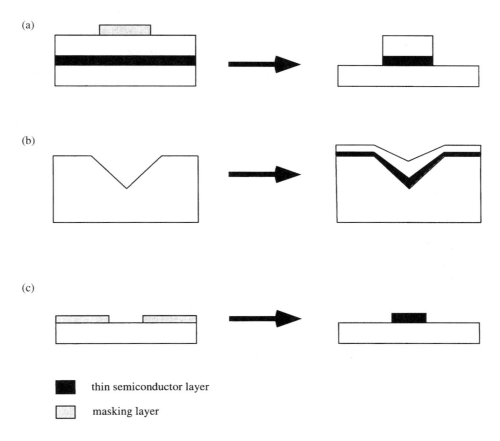

thin semiconductor layer

masking layer

Fig. 1. Typical methods of semiconductor nanostructure fabrication via processing. A thin layer sandwiched between other layers is etched (a) to reduce the lateral dimensions. Epitaxial growth can occur over a patterned surface (b) where variations in thickness will occur, leading to a wirelike region at the bottom of the groove. Epitaxial growth can also occur over a masked surface (c) such that the new material is only deposited in the exposed region.

A processed surface can also be used for semiconductor overgrowth to obtain laterally small nanostructures. For example, a V-shaped groove, shown schematically in Figure 1b, can be etched into a semiconductor by taking advantage of the etch selectivity of different crystallographic planes. These crystallographic planes will often exhibit different behavior during growth. Some work with this approach has formed material that is thicker at the bottom of the groove than on the side wall. The electronic behavior of such a structure has been used to make quantum wires [7]. Another approach that involves processing before growth consists of placing a patterned mask, often a dielectric, over the surface, as shown in Figure 1c. The growth can be done such that the new semiconductor material only deposits in the open area. The dielectric can later be removed to leave a laterally small semiconductor nanostructure.

There are many advantages of processing approaches to achieve semiconductor nanostructures. They can be very good in determining the shape, density, and spatial distributions of the structures. In some cases, these are major concerns. The size of the structure can also be controlled down to a lower limit. Lithographic techniques are evolving to smaller and smaller sizes and the 100–1000-Å range is readily achievable. In some cases, however, smaller nanostructures may be desired.

One disadvantage with processing techniques is that the commonly used etching processes often cause damage to the remaining material [2]. This damage may lower the material quality of the nanostructure. Another problem is exposing the sample to air between steps. For example, growth is usually done in one chamber, the sample removed, taken to another chamber for dielectric deposition, then removed, taken to a lithography system, patterned, then etched either in a beaker or in a chamber, removed, and so on. These multiple steps are often done in different environments. Changes in environments can introduce oxidation and contamination, also compromising material quality. One approach is to connect a processing chamber to a deposition chamber such that oxidation and contamination are reduced [8].

Many of the problems with processing routes to nanostructures are being addressed with ongoing research. A complementary approach is to find methods to fabricate semiconductor nanostructures without processing.

1.3. An Alternative: Self-Assembled Structures

Self-assembling approaches to quantum structures have the advantage that the structures are formed in the growth environment and no processing is needed either before or after the growth. There is no processing-related damage or contamination and the nanostructures can be smaller than lithographic dimensions, yet there is little direct control over the size, shape, density, and spatial distribution.

There have been several different types of self-assembled quantum structures. One class that can be fabricated *in situ* during growth are lateral superlattices. Two approaches have been demonstrated. One approach relies on the fact that some alloys undergo atomic ordering, leading to low band gap and high band gap regions [9]. This has been observed in the case of GaInP, which separates into gallium phosphide–rich and indium phosphide–rich regions that extend in one dimension within the sample. If the layer is thin, the resulting structures are quantum wires [10].

The other approach is to deposit fractional monolayers of one material alternately on a vicinal surface. During the step-flow growth mode [11, 12], new adatoms that adsorb on a terrace will attach to the up-step edge. Hence, if a half monolayer of material A is deposited on a vicinal surface followed by a half monolayer of material B and this process is repeated, a lateral superlattice can be formed consisting of vertical regions of different materials. This technique has been used to form tilted superlattices [13] and serpentine superlattices [14], which exhibited quantum wire behavior and have been used in laser structures [15].

There has also been considerable work in forming self-assembled quantum dots not using traditional epitaxy, but solution chemistry and other techniques to form small clusters [16, 17]. Although there have been a number of successes with this work, including precise control of cluster size by selection of template molecules, there are several disadvantages with producing quantum dots with these techniques. Passivation steps are vital to prevent a significant fraction of the cluster from oxidizing, and these clusters would also be difficult to integrate with traditional semiconductor structures.

The types of self-assembled quantum structures discussed in this chapter involve the formation of defect-free three-dimensional islands during strained-layer epitaxy. The previously mentioned techniques involving heteroepitaxy were demonstrated in lattice-matched materials systems.

1.4. Outline of the Chapter

Because the formation of these island nanostructures is highly dependent on the growth process, the basics of heteroepitaxy will be reviewed next (Section 2). Then comments will be made on the common experimental techniques used to fabricate and study these structures (Section 3). During the growth, there is an abrupt transition between two-dimensional growth and three-dimensional growth. One way to classify the self-assembling islands is to divide them into those that form before the transition (Section 4) and those that form after the transition (Section 5). Before summarizing, a brief discussion of the properties and applications of these islands will be given (Section 6).

2. BASICS OF HETEROEPITAXY

Heteroepitaxy is the process of depositing one crystalline material on a different material with an interface that is nearly perfect. The process is widely used, not only for research, but for manufacturing semiconductor devices such as lasers, light-emitting diodes, and transistors. With its well-established position in semiconductor research and manufacturing, using heteroepitaxy to fabricate nanostructures is a natural extension. Although epitaxy has been studied for many years, it is still not fully understood [18]. There have been several books, chapters, and reviews on the topic, a subset of which is listed here [19–24], and these can be consulted for more in-depth information. In this section, we review some basic concepts of heteroepitaxy that are important in understanding how self-assembled islands can be made. The basic surface processes will be reviewed and the most common growth modes will be introduced.

2.1. Fundamental Processes during Epitaxy

The key surface processes that occur during epitaxy are shown schematically in Figure 2 [22]. Regardless of the growth technique, atoms (and molecules) are delivered to the substrate surface, and a large fraction of these species adsorb on the surface. Once adsorbed, there are three things that can happen to the adatom. It can either form a strong chemical bond to the surface where it is trapped, diffuse on the surface to find an energetically preferred location prior to strong chemical bonding, or desorb. Once adsorbed chemically, the adatoms can also diffuse on the surface, and this diffusion can be highly anisotropic, depending on the symmetry and nature of the surface. These chemisorbed adatoms diffuse until they either (1) desorb from the surface, (2) find another adatom and nucleate into an island, (3) attach, or aggregate, into an existing island, (4) diffuse *into* the surface, or (5) react at defect sites. The last two effects are often considered relatively minor occurrences in epitaxy but are mentioned here for completeness. Diffusion into the surface, or interdiffusion, can be significant at times. The extent of interdiffusion can be thought of

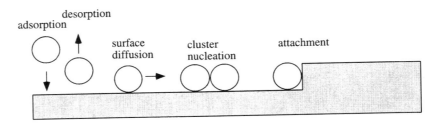

Fig. 2.   Basic processes during epitaxy.

as the solubility of one material into the other and clearly has a strong dependence on the material system. The segregation of atoms into other layers can be seen, for example, in the case of In segregation into surrounding barrier materials [25] and also in atomic diffusion from delta-doped layers [26]. The reactions at defect sites are often important. For example, reactions at step edges (a defect with respect to a perfect surface) are the foundations of step-flow growth.

The formation of clusters and the attachment of atoms to existing structures and clusters are important in the formation of self-assembled islands. When diffusing adatoms find each other, they can nucleate and form an island. Island growth continues either when other diffusing adatoms attach themselves or by direct impingement of gas phase atoms onto existing islands. Adatoms that directly impinge on an island can either incorporate into the island or lead to the next-layer growth, depending on the surface potential. Although the diffusion of adatoms attached to islands can be significantly reduced because of the local surface potential, it is still possible for adatoms to detach from the islands. Thus, islands have a "critical size" associated with them, at which they become "stable" with respect to "evaporation." Here, stable means that the islands are sufficiently large that the rate of attachment to the islands is the same or greater than the rate of detachment from the islands [27–31]. As the islands continue to grow further, and possibly migrate, they can find other islands and coalesce into one large island. The evolution of island formation can, therefore, be visualized as a progression through three different growth regimes. Initially, there is a high concentration of adatoms or monomers diffusing on the surface, resulting in a high probability of island nucleation. This is the nucleation regime, where the density of islands on the surface increases with coverage. The density continues to increase until the probability of a diffusing adatom finding an island is much higher than the probability of finding another adatom. The number of nucleation events is substantially reduced as the adatom diffusion length becomes large relative to the average island spacing, and, thus, the majority of events occurring are adatoms attaching to the existing islands, hence defining the aggregation regime. As further growth continues in the aggregation regime, the island density remains relatively constant while the islands continue to grow in size. Eventually, the islands will begin to merge with one another and enter into the coalescence regime, which is signified by a decrease in the island density with increasing coverage.

There has been a considerable amount of work in the literature in trying to describe analytically the atomic processes involved in thin-film growth through the use of kinetic rate equations. In these equations, each of the atomic processes can be represented by writing expressions for the time-dependent changes in the densities of single adatoms, clusters of a given size, and stable clusters. The specifics of these equations will be not be discussed in detail here as comprehensive reviews on this topic are widely available in the literature [22, 23, 27, 32, 33]. To illustrate the point, however, a possible equation for the time rate of change of the single adatom density is given by

$$\frac{dn_1}{dt} = U_{\text{dep}} + U_{\text{dis}} - U_{\text{evap}} - 2U_1 - U_{\text{cap}} \tag{1}$$

where similar expressions could be written for the rate of change of islands of a given size. The single adatom density can increase by the rate of deposition and dissociation (or detachment) from larger islands as embodied by U_{dep} and U_{dis}, respectively, or decrease by the rate of evaporation, the nucleation of two single adatoms, or the capture of a single adatom by a larger island as embodied by U_{evap}, $2U_1$, and U_{cap}, respectively. The factor of 2 in front of U_1 is to account for the two adatoms that nucleation requires. Clearly, additional terms could be added to represent diffusion into the surface, or reaction at defect sites, or any other surface process one could imagine.

To continue with this description, explicit expressions for the various elementary rates must be determined. The terms describing the deposition rate and the evaporation rate are fairly straightforward. The other terms involving the nucleation and aggregation of islands are functions of the diffusion coefficient, the densities of the single adatoms, the densities of islands of any given size, and a "capture number," which is a variable that takes into account the local distribution of adatoms around an island. To help discard some of the terms, certain regimes of the growth are studied to find terms that are minimal in that regime and, thus, reduce the rate equations. For example, in considering the aggregation regime of growth, an assumption could be made that the density of single adatoms on the surface is much smaller than the total density of islands, and the rate equations can be modified accordingly. By simplifying the rate equations sufficiently, the variables can be separated and then integrated to give general expressions for the densities. With the appropriate approximations, the equations describing the densities of the single adatoms and islands can be expressed as simple functions of the coverage and the ratio of flux to diffusion. This is the basis for the scaling relations derived for thin-film growth, and they have been used extensively in attempting to model epitaxial growth [28–30, 34–36].

2.2. Heteroepitaxial Growth Models

There are three general ways in which one material, say B, can grow epitaxially on a dissimilar material, say A [19]. These growth modes can be described by the equilibrium morphology, as determined from the surface free energies [18, 19, 21]. Following the notation used by Tsao [21], the three surface free energies considered are the energies associated with the substrate–vacuum, substrate–epilayer, and epilayer–vacuum interfaces and are denoted by $\gamma_{sub-vac}$, $\gamma_{sub-epi}$, and $\gamma_{epi-vac}$, respectively. The $\gamma_{sub-vac}$ term can be thought of as the initial energy term before the epilayer formation, and the remaining two terms, $\gamma_{epi-vac}$ and $\gamma_{sub-epi}$ are associated with the epilayer formation. Based on work by Bruinsma and Zangwill [37], a "spreading pressure" can be defined as

$$S = \gamma_{sub-vac} - \gamma_{sub-epi} - \gamma_{epi-vac} = \gamma_{sub-vac} - (\gamma_{sub-epi} + \gamma_{epi-vac})$$

It is evident that the relative contributions from these terms change as the epilayer evolves, resulting in a competition to determine the lowest-energy surface.

We will start the discussion with the two extremes. If $\gamma_{sub-epi} + \gamma_{epi-vac} < \gamma_{sub-vac}$, then material B will grow in a layer-by-layer fashion. This means that the B atoms will try to cover completely the surface of A because this growth minimizes the surface free energy. During the growth of the first layer of B, new atoms landing on the bare A surface will diffuse on the surface until they attach to existing clusters of B atoms. If a new B atom lands on top of an existing cluster of B atoms, it will diffuse to the edge of the structure and jump down, attaching to the edge of the cluster. Growth occurs in a two-dimensional fashion. Thus, each complete layer is thermodynamically stable as a two-dimensional layer. This is known as the Frank–van der Merwe growth mode and is illustrated in Figure 3a.

On the other hand, when $\gamma_{sub-epi} + \gamma_{epi-vac} > \gamma_{sub-vac}$, it is thermodynamically unfavorable for the epilayer to be flat and the B atoms will cluster and form islands to try to minimize the interface between A and B. Water beading on a waxed car is a result of a high interfacial energy. New atoms that land on top of an existing cluster of B atoms will

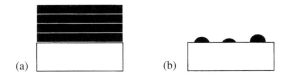

Fig. 3.   Comparison of Volmer–Weber (a) and Frank–Van der Merwe (b) growth modes.

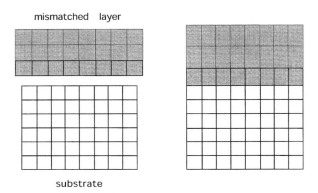

Fig. 4.   Heteroepitaxy of lattice-mismatched materials. Note the (tetragonal) deformation of the epitaxial layer.

remain on top instead of jumping down. Growth occurs in a three-dimensional fashion; new material will add to the height of the existing islands more than the lateral size. This is known as the Volmer–Weber growth mode and is illustrated in Figure 3b.

Before describing the third growth mode, we should discuss one important point that has been excluded from the previous discussion, namely, the concept of lattice mismatch. If the lattice structure or lattice constant of A and B is dissimilar, then elastic strain must be considered. The material in the layer being deposited must stretch or compress, as shown in Figure 4, to match the lattice of the underlying material. The fraction of lattice mismatch, f, is given by

$$f = \frac{a_f - a_s}{a_s}$$

where a_f is the lattice constant of the film and a_s is the lattice constant of the substrate. This definition is widely used [38] although other similar definitions do exist [39, 40].

There are two paths that lattice-mismatched films can take. For small lattice mismatches, approximately 2% or less, the growth will occur in a layer-by-layer fashion for many layers. At some point, the strain energy will build up and bonds within the sample, often at the heterointerface, will break. These patterns of broken bonds are known as dislocations. Dislocations in semiconductors and other materials are a widely studied field [40, 41]. Materials that contain dislocations and other crystal defects are often referred to as incoherent, in contrast to a coherent material with no defects.

For moderate to large lattice mismatches, approximately 3% and larger, the growth also initially occurs in a layer-by-layer fashion. In some cases, the growth of the first layer is heavily impacted by strain, a topic that is discussed further in Section 4. Returning to our discussion based on thermodynamics, consider a material system where the initial stages of growth resemble the Frank–van der Merwe growth mode. Recall that in this scenario $\gamma_{sub-epi} + \gamma_{epi-vac} < \gamma_{sub-vac}$. After the first layer is grown, we should now replace "substrate" terms with wetting layer terms, such that γ_{WL-epi}, $\gamma_{epi-vac}$, and γ_{WL-vac} are the relevant free energies resulting from the wetting layer–epilayer, epilayer–vacuum, and wetting layer–vacuum interfaces, respectively. These new terms take into account the strained wetting layer. As the film thickness increases and strain builds up, the contribution from the wetting layer–epilayer interface will begin to dominate such that $\gamma_{WL-epi} + \gamma_{epi-vac} > \gamma_{WL-vac}$. As in the Volmer–Weber growth mode, the surface will form three-dimensional islands to minimize the free energy and accommodate the strain. This is known as the Stranski–Krastanov growth mode [19], and is the most often observed growth mode in the lattice-mismatched heteroepitaxy of semiconductors.

The phenomenon of three-dimensional island formation during epitaxy has been documented for many decades [19]. However, there are two reasons why the islands so formed were not readily explored as potential quantum structures. First, there is a bias in semiconductor epitaxy toward flat (smooth) surfaces and interfaces. One reason behind this bias is that, with thin layers such as quantum wells and electron tunneling barriers, the thickness of the layer is critical to the performance of the device, such as the emission wavelength of a laser. If the interfaces are rough, leading to thickness variations of the layer, the emission energy will vary. Traditionally in semiconductor epitaxy, there has been an emphasis on developing and using flat surfaces and interfaces [1]. In more recent efforts, the formation of islands during strained-layer growth was seen as a problem, namely, a rough surface. In point of fact, considerable work has been done to suppress island formation by varying growth conditions [42] and by using surface treatments (surfactants) during growth [43, 44]. Second, the islands that were observed, often in metal epitaxy, contained dislocations and other defects. It was assumed that all such islands would be dislocated and, hence, not be suitable for quantum structures. However, many lattice-mismatched heteroepitaxial systems have been found to grow in a Stranski–Krastanov growth mode, where three-dimensional islands evolve after the formation of the two-dimensional wetting layer. Often, these systems possess a coverage range, or "window," in which defect formation is suppressed and three-dimensional coherent islands form. These coherent islands are the topic of Section 5. Eventually, the islands will dislocate; however, the coverage regime for coherent structures is easily obtained with current epitaxial growth techniques.

3. COMMON EXPERIMENTAL TECHNIQUES

Before discussing the details of the self-assembled islands, a few comments will be made on the growth and characterization techniques commonly used. These techniques will be codified briefly with appropriate references. The techniques to be discussed will be molecular beam epitaxy (MBE) and chemical vapor deposition (CVD) for synthesis, and reflection high-energy electron diffraction (RHEED), transmission electron microscopy (TEM), scanning tunneling microscopy (STM), and atomic force microscopy (AFM) for characterization.

3.1. Synthesis Techniques

3.1.1. Molecular Beam Epitaxy [21, 45–47]

Molecular beam epitaxy refers to the growth of a crystalline material in ultrahigh vacuum (UHV) using collimated gas phase reactants. The UHV environment facilitates the growth of extremely pure materials. The sources for growth can range from solids to gases and can be either elemental or compound. If the precursors are solid or liquid, they are heated in crucibles and their vapor is used to generate a molecular beam, or gaseous sources can be used directly. The geometry of the MBE system is such that there is a line of sight between the source and dopant beams and a temperature-controlled rotating substrate. Because growth occurs in UHV, the mean free path of the molecules is rather large, ensuring that the source molecules impinge onto the substrate directly. Growth is controlled by varying such parameters as substrate temperature, source flux, the sequence and duration of source beam(s) [i.e., alternating beam epitaxy and migration-enhanced epitaxy (MEE)], and group V overpressure (for III–V growth). Fluxes are controlled by modulating the molecular beams, usually through the use of high-speed mechanical shutters. These shutters provide the control necessary to deposit the desired quantity of material with better than 0.05 monolayer (ML) accuracy.

3.1.2. Chemical Vapor Deposition [48–50]

In chemical vapor deposition, growth occurs at a much higher pressure (1–760 torr) than in MBE. Often, the sources for growth are organometallic compounds (e.g., trimethyl gallium and arsine or tertiarybutylarsine for GaAs growth). These CVD techniques using organometallic sources go by a variety of names, two of the most common of which are organometallic vapor phase epitaxy (OMVPE) and metalorganic chemical vapor deposition (MOCVD) [48]. During growth, precursor compounds are flowed (using an inert carrier gas) over a substrate located on a heated susceptor. Flow rates are usually such that transport is governed by mass transport within a boundary layer that is present near the substrate surface. Growth usually occurs at relatively high temperatures (600–1000 °C) such that the metalorganic precursors are cracked in the boundary layer, facilitating the diffusion of the alkyl fragments through the boundary layer (away from the surface) into the free-stream flow of the carrier gas. Growth is controlled by varying such things as substrate temperature and reagent flow rates. Flow rates are controlled by metering the gaseous sources through flow controllers and fast switching valves.

3.2. Characterization Techniques

3.2.1. Reflection High-Energy Electron Diffraction [51–54]

One distinct advantage of MBE growth in comparison to CVD techniques is the ability to monitor MBE growth (*in situ*) using reflection high-energy electron diffraction. RHEED can be used to infer information about surface cleanliness, surface order and smoothness, and the growth rate. For RHEED, monoenergetic electrons (3–15 keV) are diffracted from the substrate (angle of incidence <1°) onto a fluorescent screen. The small angle of incidence used in RHEED corresponds to a relatively small penetration depth (a few monolayers), making RHEED a very surface-sensitive technique. In essence, RHEED is similar to X-ray diffraction for a surface. (The theoretical analysis for RHEED is identical to the formalism used to explain X-ray diffraction.) Thus, RHEED offers insight about the periodicity present at the substrate surface. Furthermore, by monitoring the intensity of the oscillations of a single RHEED spot, one is able to determine the growth rate because a single oscillation corresponds to the deposition of one monolayer.

3.2.2. Transmission Electron Microscopy [55–57]

The foundation of TEM is the wave behavior exhibited by electrons. In an experiment, periodic atomic planes within a thin crystalline sample ($\sim$1000 Å) diffract a monochromatic electron beam, typically accelerated by a 60–200-kV potential. Because TEM is sensitive to variations in the spacing of atomic planes [58], it is a useful technique for the study of islands formed during strained-layer epitaxy. Strain effects can be clearly seen as well as the presence of crystal defects [59, 60].

3.2.3. Scanning Tunneling Microscopy [61–66]

As the name suggests, STM relies on the quantum mechanical phenomenon of electron tunneling. In an STM experiment, a sharp metallic tip is brought sufficiently close to a conducting surface that electron tunneling between the sample and the tip will occur. A bias voltage (either positive or negative) is applied to the tip, and a tunneling current flows from the surface to the tip (for negative sample bias) or from the tip to the surface (for positive sample bias). In the constant-current mode, the tip is rastered in the plane of the surface and the tip–sample separation is altered, using a feedback circuit, such that the tunneling current is maintained at a constant value. The tip position in this rastering procedure follows a constant density of electronic states topograph of the surface. It is important to realize that an STM image is not a map of atomic position. Rather, it represents a constant density of electronic states contour of the surface for a given bias voltage. Even so, by varying the bias voltage, the electronic states of the surface involved in tunneling are changed. Furthermore, both negative and positive sample biases can be used to probe the filled and empty states. Thus, bias-dependent imaging can be used to infer details about the atomic structure and electronic nature of the surface. In practice, STM has been an invaluable technique used in the study of surface reconstructions and atomic scale features, providing information at resolutions unavailable by almost any other technique.

3.2.4. Atomic Force Microscopy [64, 67–69]

In an AFM experiment, a sharp tip mounted on the end of a flexible cantilever is brought into sufficiently close proximity with the sample that a detectable force is generated. Detection of the tip–sample force has been achieved in different ways. One common method involves the deflection of a laser beam reflected from the back of the cantilever. The tip is rastered above the surface and a feedback loop is used to keep the separation between the tip and the sample at a constant value through the actuation of a piezoelectric translator that moves the sample (in the z direction perpendicular to the sample).

As mentioned, the AFM relies on force detection. Depending on the tip–sample separation, the type and magnitude of the tip–sample force will vary. In the so-called contact mode, the separation distance between the tip and the sample is small, and the detected force is a result of core–core repulsion (the Pauli principle). Although contact mode AFM can provide atomic resolution, it requires a very rigid substrate. In the noncontact mode, the tip–sample separation is larger than for contact mode AFM, and the gradient of the van der Waals potential is the relevant force. Generally, the spatial resolution of noncontact mode AFM is inferior to the resolution achievable with contact mode AFM. Yet, noncontact mode AFM is less susceptible to imaging artifacts resulting from deformation of the sample by the tip. To image a surface nonintrusively and still achieve a high level of spatial resolution, tapping mode AFM was developed. In the tapping mode, the cantilever of the AFM is forced to oscillate at a certain distance from the sample while the probe is scanned laterally. When the tip encounters a surface feature, the resonant frequency of the cantilever changes. A change in the resonant frequency of the cantilever will result in a change in the amplitude of oscillation, which is detected by the laser reflecting from the back of the cantilever. When this happens, the feedback loop lowers the sample so that the original amplitude of cantilever oscillation is restored. An on-line computer records the voltages applied to the z-direction piezoelectric actuator for feedback control and converts them into a topographic map of the surface.

4. TWO-DIMENSIONAL GROWTH AND ISLAND FORMATION BEFORE TRANSITION TO THREE-DIMENSIONAL GROWTH

There has been much recent interest in creating reduced-dimensional structures with carrier confinement in two and three dimensions by self-assembling mechanisms in lattice-mismatched heteroepitaxy. It is hoped that the *in situ* formation of these low-dimensional structures can be utilized as a suitable alternative to damage-inducing *ex situ* processes such as etching and lithography [70–74]. Many lattice-mismatched heteroepitaxial systems follow the Stranski–Krastanov growth mode, where at least one monolayer grows in a layer-by-layer mode, forming a two-dimensional layer, sometimes known as the two-dimensional layer. The strain from the lattice mismatch is accommodated in elastic deformation and, thus, determines the critical thickness of the two-dimensional layer. In the layer-by-layer regime of growth, two-dimensional islands (islands that are one monolayer in height) can form as a mechanism to help minimize surface energy. As the film thickness increases beyond the critical thickness, three-dimensional islands form on the two-dimensional layer. Considering that high-quality materials preclude dislocation formation, a given heterostructure is limited by the amount of coherent material that can be deposited before dislocation formation becomes energetically favorable. Hence, it is of great importance to determine the mechanisms of strain relief for various technologically important semiconductor systems because a better understanding of the initial stages of strained-layer heteroepitaxy is crucial for the further development of this technology.

The importance of two-dimensional islanding should also be extended to the understanding of interfaces. Interfaces can play a critical role in the device performance of heteroepitaxial systems. As relevant length scales decrease, the importance of morphological issues at these interfaces becomes even more crucial. For example, resonant tunneling devices (RTDs) are currently of considerable interest because of their potential use in high-speed terahertz devices. RTDs are double-barrier heterostructures, where the barriers are typically on the order of 5 to 10 monolayers (ML) thick. A theoretical study examined the effects of islands at the interface on the performance of RTDs, as predicted by the peak-to-valley ratios determined from calculated I–V curves as a function of terrace sizes [75]. A modest amount of lattice mismatch can be tolerated because the barrier thickness can be substantially smaller than the critical thickness to three-dimensional growth. However, new issues arise when two compounds with different group-V components are grown together. As an example, consider the case of InAs and AlSb, which are used in RTDs. AlSb has a lattice constant slightly larger than InAs (1% mismatch) and also a significant band offset, making it an ideal candidate as a barrier material to InAs. The interface bonds for this system must be either In–Sb or Al–As. Most MBE growths occur under group-V overpressures, and the traditional approach is to "soak" the surface under the new group-V flux when changing materials at an interface. In either case, the InSb-like or AlAs-like material on the InAs is significantly lattice mismatched to the InAs substrate, thus making strain issues relevant.

In this section, we will review the issues relating to the initial stages of Stranski–Krastanov growth. In particular, we will first discuss the beginning stages of lattice-mismatched heteroepitaxial growth as the two-dimensional layer forms up through the transition from two-dimensional to three-dimensional growth. Furthermore, the effects of reconstruction and surface orientation on the formation and morphology of two-dimensional islands will be discussed. Clearly, the literature is far too rich to attempt to review every material system that has been studied throughout the years. Rather, we will discuss specific examples from the literature that we believe are representative of the concepts we bring forth. What we wish to emphasize are concepts, because the ideas are general for lattice-mismatched heteroepitaxy. We will focus this discussion on elemental or compound semiconductor systems with the diamond or zincblende structure.

4.1. Initial Stages of the Two-Dimensional Layer Formation

Insight into the mechanisms involved in the formation of the two-dimensional layer is needed to understand some of the fundamental aspects of lattice-mismatched heteroepitaxial growth. Figure 5 shows STM images of the initial stages of InAs growth on GaAs(001)–(2 × 4) at various submonolayer coverages. Parts a and b of Figure 5 show a coverage of 0.15 and 0.29 ML, respectively. The InAs forms two-dimensional islands, which are one monolayer in height above the GaAs surface. Higher-resolution STM images show that these islands also exhibit a c(4 × 4) reconstruction, which further distinguishes the islands

InAs   GaAs - (2 × 4)

(a)

[Ī10]

[110]

(b)

(c)

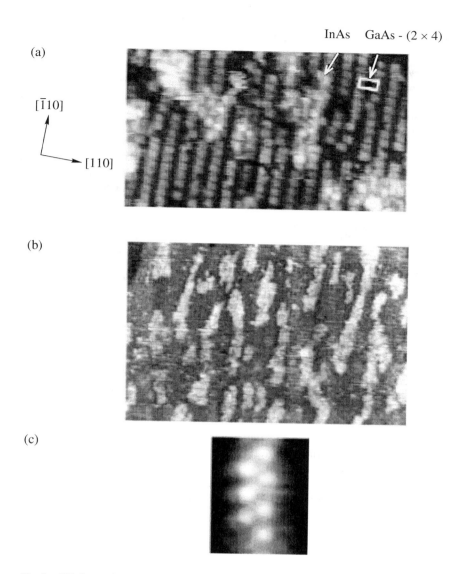

Fig. 5.   Filled-state STM images of InAs islands on GaAs(001)–(2 × 4): (a) 30-nm × 15-nm image of 0.15 ML of InAs deposition and (b) 100-nm × 70-nm image of 0.29 ML of InAs deposition. The islands in both images are elongated in the [Ī10] direction and exhibit a c(4 × 4) reconstruction and easily distinguishable from the (2 × 4) reconstruction still observed on the bare substrate. Reprinted with permission from V. Bressler-Hill et al., *Phys. Rev. B* 50, 8479 (© 1994 American Physical Society) and from J. Tersoff and R. M. Tromp, *Phys. Rev. Lett.* 70, 2782 (© 1993 American Physical Society).

from the initial (2 × 4) reconstructed surface (cf. Fig. 5c). In addition, the islands are observed to be anisotropic and elongated in the [Ī10] direction at all submonolayer coverages studied. Qualitatively, there appears to be a difference in the energetics of the two types of island edges. The edges that run along the [Ī10] direction are relatively straight compared to the edges that are parallel to the [110] direction, which are seen to have more kinks (cf. Fig. 5b). This indicates that rearrangement along the steps in the [Ī10] direction is relatively rapid. A statistical analysis on the island dimensions shows that the InAs islands maintain a most probable width of approximately 4 nm in the [110] direction at all coverages examined, whereas the islands appear to grow freely in the [Ī10] direction with increasing coverage. From this, it follows that the aspect ratio of these islands will significantly increase with coverage until the onset of coalescence. It should be mentioned that anisotropic island shapes are common in semiconductor systems, and are even seen in homoepitaxial systems. Several factors could lead to this observed anisotropy, including an anisotropy in the step energies, a preferential growth direction resulting from different step edge reactivities, anisotropic barriers to surface diffusion, or some combination of them all [76–78]. Strain, in addition to the other possible anisotropies mentioned, is clearly expected to influence the surface morphology for the case of lattice-mismatched heteroepitaxial growth.

One other noteworthy result seen in the aforementioned InAs/GaAs study was found when growing InAs islands on a 1° vicinal, B-type (anion-terminated steps running parallel to the [110] direction) GaAs(001)–(2 × 4) surface [79]. On this substrate, the steps limited the In diffusion in the [$\bar{1}$10] direction and, hence, the extent of growth possible in the [$\bar{1}$10] direction. Moreover, the B-type steps were found to be more reactive than the A-type steps (cation-terminated steps running parallel to the [$\bar{1}$10] direction) and to act as adatom sinks. Even with these growth constraints, the islands still showed a preferred width of around 4 nm in the [110] direction, indicating that the strain is likely being accommodated in that crystallographic direction. The observation of a preferred two-dimensional island size is consistent with work done by Massies and Grandjean [80], who considered the effects of nontetragonal elastic distortion at the edges of two-dimensional islands. A simplified version of the valence force field model was used in which the shape of a one-dimensional island on top of a completely rigid substrate could be calculated. For a surface coverage of 0.5 ML of $In_{0.49}Ga_{0.51}As$ islands on GaAs (lattice mismatch of 4%), the model predicted a preferred island size of 11 unit cells, or 4.4 nm. This preferred size would be expected to be smaller for the increased lattice mismatch of InAs on GaAs(001).

The observation of islands with a preferred size is also consistent with the theoretical predictions of Tersoff and Tromp [81], who proposed a shape transition for coherently strained islands. From their model, small islands initially evolve in a compact shape, but at a critical size, the islands take on a rectangular shape and become progressively elongated, where the width of these islands approaches a constant. The islands can then minimize their energy by keeping the optimal size in one direction, while continuing to grow in the other direction. The question then arises as to what determines which direction will elongate. On a surface terminated with group-V atoms, dimers form with the dimer bonds aligned in the [$\bar{1}$10] direction. Deposition of group-III atoms will form bonds between the group-III and the group-V atoms that will also be aligned in the [$\bar{1}$10] direction. If the atoms wish to move to help accommodate strain, they could modestly move perpendicular to the directions of their back bonds to the surface; that is, they could move in the [110] direction. Thus, strain energy could be partially relieved by the same nontetragonal deformation model of Massies and Grandjean [80] in the [110] direction.

4.2. Transition from the Two-Dimensional Layer to Three-Dimensional Islands

Following the Stranski–Krastanov growth mechanism, the two-dimensional layer will eventually give way to the formation of three-dimensional islands. The exact mechanism of this transition is presently not well understood and is a subject of much study [82–93]. It is reasonable to believe that the size and spatial distribution of the two-dimensional islands formed in the two-dimensional layer of these systems is somehow related to the distribution of three-dimensional islands formed after further material deposition. In other words, the two-dimensional islands, or platelets, are precursors to three-dimensional islands and, ultimately, grow into the larger structures. Priester and Lannoo [90] have proposed a model that follows this line of reasoning in an attempt to describe the formation of three-dimensional InAs islands on GaAs(001). In their model, a complete monolayer of InAs would first cover the surface forming the two-dimensional layer. At a total coverage of 1.4 ML, the model predicts the formation of large two-dimensional islands that are randomly distributed across the surface. These two-dimensional islands are uniform in size and are a single monolayer in height. This prediction of uniformly sized two-dimensional islands is reported by Chen and Washburn [84], who believe that the increase in strain at island edges affects the surface potential near the island edges and, subsequently, makes it increasingly difficult for adatoms to aggregate further into existing islands. Similarly, in kinetically driven models, such as Monte Carlo simulations, the detachment probabilities of adatoms from islands as a function of island size have been studied [30, 36]. As more material is deposited, the islands continue to grow and eventually evolve into uniformly sized three-dimensional islands.

In an attempt to study the two-dimensional-to-three-dimensional morphology transition, Heitz et al. [85] have examined this same InAs/GaAs system at InAs coverages ranging from 0.87 to 1.61 ML on the GaAs(001)–c(4 × 4) surface. Images from their study are shown in Figure 6. There is no evidence of the large platelets that were predicted by Priester and Lannoo, but rather a surprising result regarding the two-dimensional-to-three-dimensional transition. At a coverage of approximately 1.15 ML, features 2–4 ML high begin to appear on the surface. The density of these features increases until an approximate coverage of 1.35 ML, whereupon these three-dimensional features disappear. Upon further deposition to 1.45 ML, the three-dimensional islands reappear, but at a much higher density. This type of appearance, disappearance, and subsequent reappearance of three-dimensional islands at coverages below the critical film thickness has not been previously reported for systems evolving in the Stranski–Krastanov growth mode. To study this phenomenon further, samples were grown for photoluminescence (PL) measurements with InAs deposition ranging from 1 to 2 ML. Peaks attributed to three-dimensional islands are observed at 1.15 and 1.25 ML, where the peak at 1.25 ML is shifted toward lower energies because of the larger island sizes. However, at 1.45 ML, this peak disappears and does not reappear until about 1.55 ML of deposition, whereupon it evolves into the peak typically observed at the critical point.

Other examples of the transition from two-dimensional to three-dimensional growth can be found in the growth of antimony-based materials on GaAs. Thibado et al. [91] studied GaSb on GaAs(001)–c(4 × 4), and found that, after 1 ML of GaSb deposition, the surface was covered with two-dimensional islands, or platelets, which were approximately 10 nm in diameter, although slightly anisotropic in the [$\bar{1}$10] direction. An image in a related study of the same material system is shown in Figure 7, where the islanded two-dimensional layer is clearly observed. Furthermore, they found that adding a second monolayer of material primarily added onto the existing platelets, making them 2 ML high while approximately maintaining the same diameter. After 3 ML of deposition, they observed the formation of three-dimensional islands. Imaging the areas between the three-dimensional islands showed that the two-dimensional layer composed of the network of two-dimensional islands (2 ML in height) was still intact. Furthermore, a rough calculation based on the apparent island dimensions (neglecting convolution effects from the tip) showed that approximately 0.6 ML of material was incorporated into the three-dimensional islands, which is consistent with the two-dimensional layer remaining intact and not necessarily incorporating itself into the three-dimensional islands. Voigtländer and Kästner [92] have also tried to address the issues of three-dimensional island formation from two-dimensional growth by studying Ge growth on Si using *in vivo* STM during growth. They concluded that there did not appear to be any type of special morphology (e.g., step edges, large or high islands in the two-dimensional layer, domain boundaries, etc.) where the three-dimensional islands would nucleate and evolve, but rather it simply appeared to occur at random locations on the surface.

Based on these experimental observations, it appears that the large two-dimensional platelets are not the precursors to three-dimensional islands as predicted by equilibrium-based theoretical models [90]. In an earlier and related work to that of Heitz et al. [85], Ramachandran et al. [86] considered mass transfer and kinetics in the formation of

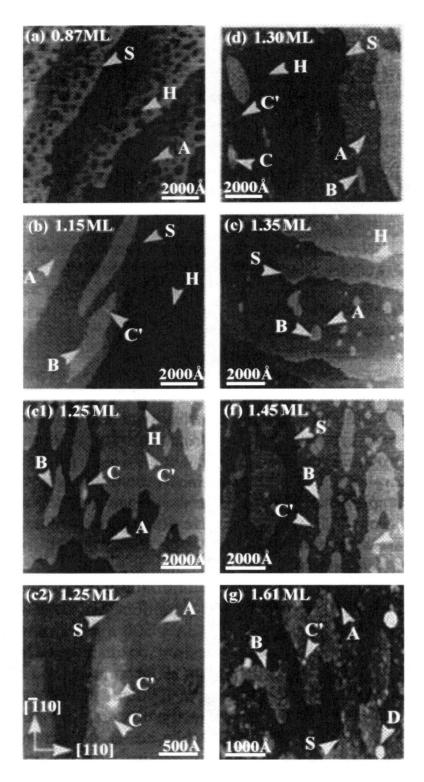

Fig. 6. Filled-state STM images showing the surface evolution for InAs deposition on GaAs(001)–c(4 × 4) at coverages of (a) 0.87, (b) 1.15, (c1, c2) 1.25, (d) 1.30, (e) 1.35, (f) 1.45, and (g) 1.61 ML. The labels represent small two-dimensional islands less than 20 nm in width (A), large two-dimensional islands greater than or equal to 50 nm (B), small multilayer (2–4 ML) clusters up to 20 nm wide (C′), larger multilayer clusters up to 50 nm wide (C), three-dimensional islands (D), atomic steps (S), and 1-ML-deep holes (H). Reprinted with permission from R. Heitz et al., *Phys. Rev. Lett.* 78, 4071 (© 1997 American Physical Society).

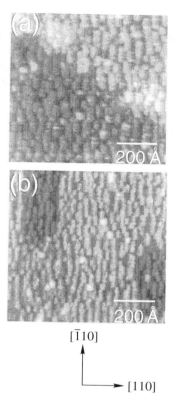

Fig. 7. Filled-state STM images of GaSb on GaAs(001)–c(4 × 4) at coverages of (a) 1.0 and (b) 3.5 ML. The image in (b) shows a region between three-dimensional islands that has already formed by that coverage. Reprinted from *J. Cryst. Growth*, 175/176, 888, B. R. Bennett et al. (© 1997), with kind permission of Elsevier Science–NL, Sara Borgerhartstraat 25, 1055 KV Amsterdam, The Netherlands.

three-dimensional InAs islands on GaAs. Again, no direct relationship between the large two-dimensional islands and the three-dimensional islands was observed. The major observation was that the highest observed density of the coherent three-dimensional islands was approximately an order of magnitude greater than the highest density of the two-dimensional islands. Some hints of a possible precursor to the three-dimensional islands were observed and denoted "quasi"-three-dimensional islands, or islands 2–4 ML in height (cf. Fig. 6, the features labeled C). These features are relatively high in density, nearly twice that of the coherent three-dimensional islands. These "quasi"-three-dimensional islands appear about 0.2 ML before the formation of the true three-dimensional islands, where upon they quickly disappear within about 0.2 ML prior to additional deposition. Thus, the changing densities of these different types of islands over very small changes in the total coverage indicate a mass transfer mechanism from the two-dimensional to the three-dimensional islands, which leads to a significant reorganization of mass on the surface. The authors conclude that to study and understand the formation and evolution of three-dimensional islands, the strain-dependent kinetics of the system must be considered [86]. Clearly, this is plausible as epitaxy by nature is a kinetically driven process. Furthermore, there may be island–island interactions that affect the surface morphology and that result from strain-related phenomena [86, 94].

4.3. Effects of Surface Reconstruction

Semiconductors often have a different symmetry at the surface compared to the bulk solid. This change of symmetry at the surface is a result of the displacement of atoms with respect to their bulk positions, thus creating a new unit cell on the surface. This rearrangement of surface atoms is referred to as a surface reconstruction, and significant effort has been devoted to understanding this phenomenon [95]. The motivation for the restructure itself is to minimize its free energy by forming bonds at the surface and maintaining charge neutrality. It should be noted that the reconstruction that is observed is not necessarily the lowest-energy surface, but rather the lowest-energy surface that is obtainable under the given set of kinetic conditions. Because the surface reconstruction represents a surface energy and symmetry (geometry), it is reasonable to believe that it will influence subsequent deposition of materials on the surface.

To illustrate the effects that surface reconstruction can have on the formation of islands, consider the study by Belk et al. [96] of InAs deposition at various coverages up to 1 ML on the GaAs(001)–c(4×4) surface at a substrate temperature of 420 °C. A number of STM images showing the progression of island formation from 0.1 to 1 ML are shown in Figure 8. At 0.1 ML of InAs, new domains of (1×3) reconstruction are observed at the step edges, and it appears that the (1×3) areas are on the same layer as the c(4×4) ones. As the coverage is increased to 0.3 ML (cf. Fig. 8b), two-dimensional islands be-

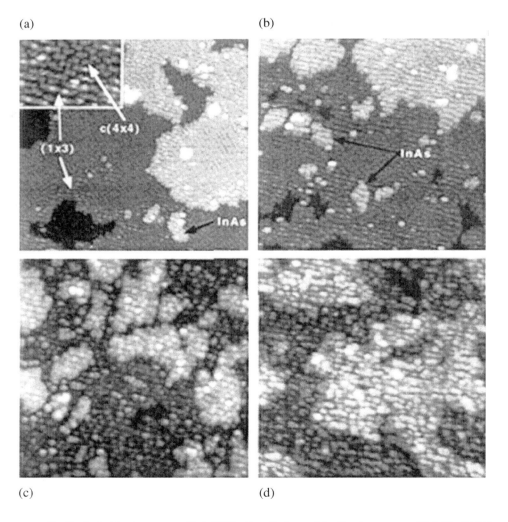

Fig. 8. InAs islands grown on GaAs(001)–c(4×4) shown at fractional coverages of (a) 0.1, (b) 0.3, (c) 0.6, and (d) 1.0 ML. The image sizes are (a) 50 nm × 50 nm and (b)–(d) 40 nm × 40 nm. Reprinted from *Surf. Sci.*, 365, 735, J. G. Belk et al. (© 1996), with kind permission of Elsevier Science-NL, Sara Burgerhartstraat 25, 1055 KV Amsterdam, The Netherlands.

gin to appear on the surface, which, at this point, is a mixture of (1×3) and $c(4 \times 4)$ domains. These islands are 1 ML in height and show the same (1×3) reconstruction. In contrast to the highly anisotropic, two-dimensional InAs islands formed on the (2×4) surface and shown in Figure 5, the islands formed on the $c(4 \times 4)$ surface have a different reconstruction and appear to be much more isotropic. At 0.6 ML (cf. Fig. 8c), the surface is entirely covered by the (1×3) structure. The RHEED pattern in the $[\bar{1}10]$ direction at 0.3 ML showed a mixture of 1/2- and 1/3-order spots, corresponding to the $c(4 \times 4)$ and the (1×3) reconstructions, respectively. By 0.6 ML, the 1/2-order spots were gone and the RHEED showed a clear (1×3). At 1.0 ML (cf. Fig. 8d), the surface is composed of large islands and terraces as some of the smaller islands have coalesced and the growth has proceeded in a reasonable, layer-by-layer fashion. Further study of the system revealed that the amount of InAs required to fill the surface completely with the (1×3) reconstruction varied with surface temperature. On this basis, the authors postulated that the (1×3) domains were actually composed of an $(In, Ga)As$ alloy. The exact mechanism for this reconstruction/transformation remains unknown.

Bennett et al. [82, 83] have studied the deposition of InSb on different reconstructions of GaAs(001). In particular, they deposited 1.5 ML of InSb on both a (2×4) and a $c(4 \times 4)$ reconstructed surface. Because of the significant lattice mismatch between InSb and GaAs (14.6%), this deposition is beyond the critical thickness and three-dimensional islands are present. The differences resulting from deposition on the different initial surface reconstructions can be seen in Figure 9. Figure 9a shows 1.5 ML of deposition on GaAs(001)–$c(4 \times 4)$, whereas Figure 9b shows the equivalent deposition on GaAs(001)–(2×4). The islands were observed to be anisotropic in both cases, with the elongation in the $[\bar{1}10]$ direction. The island separations, as determined from an autocorrelation analysis, were found to be 50 Å for growth on the $c(4 \times 4)$ surface and 40 Å for growth on the (2×4) surface.

The islands formed in the two-dimensional layers of these systems appear to be somewhat dependent on the reconstruction of the initial growth surface. Islands that are formed on the (2×4) surface are highly anisotropic (cf. Figs. 5 and 9b), whereas islands that are formed on the $c(4 \times 4)$ surface appear to be slightly more isotropic (cf. Figs. 8a–d and 9a). This dependence on the initial surface reconstruction could be due to several different factors. To consider how the reconstruction of the initial surface could play a part in subsequent island formation, it would be sensible to consider first differences in the atomic structure of the reconstructions. Although there have been several different proposals for the (2×4) reconstruction [97, 98], the commonly accepted structures of the (2×4) surface are shown in Figure 10a [95, 99]. The surface is composed of two As dimers and two missing dimers on the top level, with the exposed As atoms in the third layer forming dimers. There is a missing row of gallium atoms in the second layer, leaving 0.75 ML of Ga atoms in that layer. A proposed model for the $c(4 \times 4)$ reconstructed surface is shown in Figure 10b [100]. There is no missing row of gallium atoms in this case, but rather a complete layer of cations with 1.75 ML of arsenic on top of it. The top layer, composed of 0.75 ML of arsenic, forms the characteristic "brickwork" pattern of the $c(4 \times 4)$ reconstruction. Clearly, there are significant differences in the concentrations of group-III and group-V elements present at the surface for the different reconstructions.

There are a number of possibilities concerning how these differences could affect subsequent island morphology. Because these structures are typically grown in a group-V overpressure, it would seem that the concentration of group-III material could be the key variable. As previously noted, the (2×4) reconstruction has an incomplete layer of cations at the surface. Thus, significant variations in the energy barriers for diffusion on the surface may well result from the different atomic structures. These variations could lead to changes in the rate of diffusion of cations on the surface and also along step edges. Furthermore, the variations in the surface potential could easily produce sites of varying reactivity. This can be seen clearly in a study by Köhler et al. [88] in which Ge was deposited on the Si(111)–(7×7). When the deposition was at room temperature, Ge clusters with various

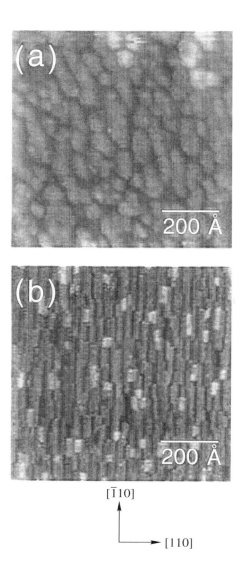

Fig. 9.   Filled-state STM images of 1.5 ML of InSb deposition on (a) GaAs(001)–c(4 × 4) and (b) GaAs (001)–(2 × 4). The islands in (b) are slightly more elongated in the [$\bar{1}$10] direction than in (a). Monolayer-high steps are also observed in both images. Reprinted from *J. Cryst. Growth*, 175/176, 888, B. R. Bennett et al. (© 1997), with kind permission of Elsevier Science-NL, Sara Burgerhartstraat 25, 1055 KV Amsterdam, The Netherlands.

sizes were observed to be randomly distributed on the surface. However, the clusters rarely occupied a corner hole in the (7 × 7) reconstruction, indicating a different reactivity at that site. Along this same line of reasoning, there may be differences in the structures that the surface will try to form to minimize the surface energy; that is, the equilibrium structures of the surface may be different.

4.4. Effects of Surface Orientation

Lattice-mismatched heteroepitaxy on a number of different (001) surfaces has been shown in many systems to follow the Stranski–Krastanov growth mode and eventually form coherently strained three-dimensional islands. Because the growth mode is dependent on the surface energy and reconstruction, another variable that can be used to affect the strained-layer growth of heteroepitaxial systems is the surface orientation. Obviously, the electronic structure and symmetry of the surface can lead to different reactivities and island shapes.

(a)   (b)

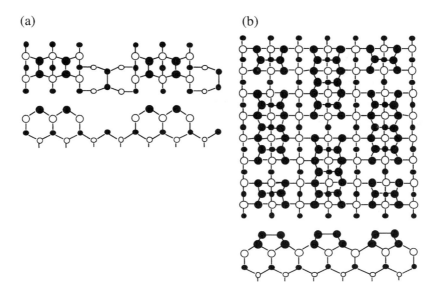

Fig. 10.   Ball-and-stick models of (a) (2×4) and (b) $c(4 \times 4)$ reconstructions.

For example, the Si(111) surface has threefold symmetry in the surface plane. Both Si homoepitaxy [88, 92] and Ge deposition [101] on the Si(111) surface result in triangularly shaped islands, reflecting the symmetry of the surface. On the other hand, both the (001) and the (110) surfaces are rectangular, with in-plane nearest neighbors in the $[\bar{1}10]$ and [110] directions. In GaAs homoepitaxy on the (001) surface, anisotropic islands elongated in the $[\bar{1}10]$ direction are observed. These islands are random in shape apart from their anisotropy [76, 102]. However, GaAs homoepitaxy on a (110) surface results in triangularly shaped islands pointing in the [110] direction with elongated sides running in the $\langle 113 \rangle$ and $\langle 115 \rangle$ directions. This island shape indicates a preference for adatoms to attach at the base of existing islands (edges that run in the $[\bar{1}10]$ direction) and illustrates the difficulty in predicting the shapes that islands will form on a given surface from surface symmetry alone [103].

In zincblende structures, certain crystal planes can actually have two types of surfaces because of the different ways in which the surface can be terminated. This is simply a result of the bilayer nature of zincblende structures. Diamond structures, such as silicon, do not exhibit this type of dependence because of their higher symmetry. To illustrate these differences, consider the {111} surfaces of a diamond and a zincblende structure. Figure 11a shows the unit cell of a zincblende structure. For the sake of discussion, let the white dots represent the anions and let the black dots represent the cations. The diamond structure, on the other hand, would have the same atomic positions, but, obviously, there would be no distinction between the types of atoms. The models for the (111)A and (111)B surfaces shown in parts b and c of Figure 11, respectively. For the unreconstructed compound surface, it can be seen that a surface will be either cation rich for the (111)A surface or anion rich for the (111)B surface. Clearly, different reactivities can be expected between the A- and B-type surfaces. For the diamond structure, the (111) surface would appear identical in either case because of its symmetry.

In some heteroepitaxial systems, growth on different surface orientations can dramatically change the growth mode. For example, in the case of InAs deposition on GaAs(001), which was discussed earlier, the growth was shown to proceed in a layer-by-layer manner, to the formation of coherent three-dimensional island and then, finally, to dislocated three-dimensional islands. When InAs is deposited on GaAs(110) and GaAs(111)A, however,

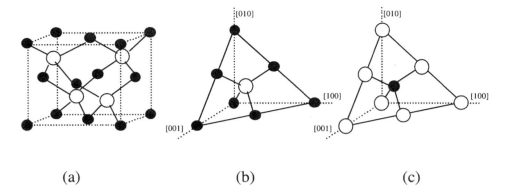

(a)   (b)   (c)

Fig. 11.   Ball-and-stick models of the (a) zincblende unit cell, (b) (111)A surface, and (c) (111)B surface. The white dots represent the cations and the black dots represent the anions. Two different (111) planes are possible because of the zincblende structure, with the difference being either cation [cf. (b)] or anion [cf. (c)] termination of the plane.

no coherent three-dimensional islands are observed [104–106]. On both surfaces, two-dimensional islands are observed in a layer-by-layer growth mode, but the formation of three-dimensional islands is replaced by the introduction of misfit dislocations as a strain-relieving mechanism. Interestingly, on the GaAs(111)A surface, Gonzalez-Borrero et al. [107] report quantum dot formation for InGaAs deposition. Altogether, these authors grew $In_{0.5}Ga_{0.5}As$ quantum dots on $(n11)A/B$ surfaces, where n was 1, 2, 3, 5, and 7, and compared the samples by PL measurements. In their work, they demonstrated that the $(n11)B$ surfaces had a higher integrated PL intensity compared to the $(n11)A$ surfaces, with the (311)B surface showing the most uniformity in island sizes as judged from PL. Nishi et al. [108] studied the $In_{0.5}Ga_{0.5}As$ quantum dot formation on the (311)B surface and found the dots to have an approximate diameter of 25 ± 2 nm and a height of 13.7 ± 2.2 nm, as determined from AFM. Luminescence peaks resulting from dots grown on the (311)B surface were shown to be much narrower than those measured from dots on a (100) surface.

Germanium deposition on different silicon surface orientations has been an area of recent interest because of its potential uses in optoelectronic devices. High-index silicon surfaces have been of particular interest because of their unusual density of states, which results in promising optical [109, 110] and transport properties [111, 112]. One drawback of high-index surfaces is that they tend to be high-energy surfaces and, consequently, form low-index facets upon annealing [113, 114]. One high-index surface that has been of particular interest is the (113) surface. One study by Gibson et al. [115] showed that annealing Si(110) produced various low-index planes, as well as (113) facets. Also, the Ge "hut clusters" that have been observed are faceted structures composed primarily of {113} planes [88, 89]. Thus, it appears that the Si(113) surface is a relatively low-energy surface and may be suitable as an MBE growth surface. One example of Ge growth on Si(113) was reported by Knall and Pethica [87]. In this work, a number of STM images were presented at various coverages ranging from clean Si(113) up through 5 ML, by which point three-dimensional islands had begun to evolve on the surface. Furthermore, they performed similar growths on Si(100) to contrast the differences resulting from the initial growth surface. The Ge growth on the (113) surface was predominantly two dimensional up to coverages of approximately 3 ML, whereas on the (100) surface second-layer growth of Ge islands began at submonolayer coverages. By 2.5 ML of deposition, at least four distinct levels could be observed on a given terrace. The differences in the growths were attributed to the preferred nucleation of islands at antiphase domain boundaries in the (100) surface reconstruction. Other studies have also shown that the nucleation of islands at antiphase domain boundaries can be a dominant growth mechanism [116, 117].

5. THREE-DIMENSIONAL ISLANDS

In the previous sections, the Stranski–Krastanov mode was introduced and the transition from layer-by-layer growth to three-dimensional island growth was discussed. In recent years, these islands have been utilized as nanostructures. These self-assembled islands will be the focus of this section.

5.1. Early Work

Although the Stranski–Krastanov growth mode has been known for many decades, it was not considered until relatively recently as a method to fabricate useful nanostructures. As previously mentioned, there is an emphasis in epitaxy on the creation of smooth surfaces and interfaces. Moreover, much of the early work in studying growth modes was done with metal-on-metal epitaxy (e.g., Ag/Mo) or metal-on-semiconductor epitaxy (e.g., Ag/Si) [19, 24]. These structures were not considered to be useful nanostructures because defects readily formed in these islands. With more work in strained-semiconductor epitaxy, evidence evolved that these islands were not defected.

One early report that showed evidence of island formation was presented by Goldstein et al. in 1985 [118]. They were studying InAs/GaAs superlattices and some samples exhibited three-dimensional nucleation. During growth, the RHEED pattern changed from a diffuse pattern to a spotty pattern, indicating three-dimensional growth for an InAs thickness greater than 2 ML. In TEM images, they observed localized strain fields that, in retrospect, were caused by the InAs islands that were present. Because the islands were surrounded by other material, direct observation of the islands was not possible; yet there was no evidence of dislocations.

In 1990, two important reports were published that peaked interest in the field of self-assembled semiconductor islands. In the first report, Eaglesham and Cerullo [59] examined Ge deposition onto Si(100). Their goal was to disprove the assumption that islands formed during Stranski–Krastanov growth are always dislocated. Using TEM, which is sensitive to dislocations and structural defects, they explored the islands. In plan-view images, they observed a strain contrast feature (cf. Fig. 12), consistent with a coherent particle within a lattice-mismatched matrix [58]. From their work, they developed the concept of the *coherent* Stranski–Krastanov growth mode. Similar results were reported later in this material system by Krishnamurthy and co-workers [119].

The second paper, by Guha and co-workers [60], presented coherent $Ga_{0.5}In_{0.5}As$ islands formed on GaAs, examined by cross-sectional TEM. Many of the islands, as shown

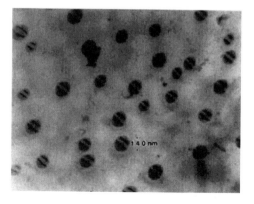

Fig. 12. Plan-view TEM image showing Ge islands surrounded by Si. The strain contrast features indicate that the islands are defect free. Reprinted with permission from D. J. Eaglesham and M. Cerullo, *Phys. Rev. Lett.* 64, 1943 (© 1990 American Physical Society).

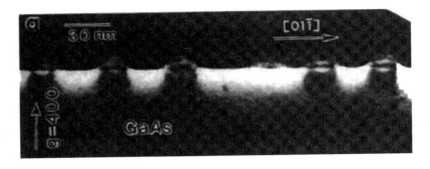

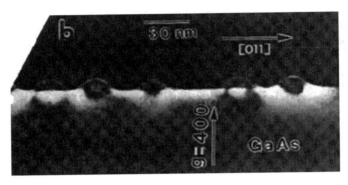

Fig. 13. Cross-sectional TEM image showing InAs on a GaAs surface. No defects are observed in these islands. Reprinted with permission from S. Guha et al., *Appl. Phys. Lett.* 57, 2110 (© 1990 American Institute of Physics).

in Figure 13, were found to be defect free. Those that were defected exhibited dislocations and stacking faults that were injected from the edge of the islands. These results not only support those seen for Ge on Si, but extend the phenomenology to results for compound semiconductor systems.

Semiconductor nanostructures are often considered candidates for quantum structures such as quantum dots. Such structures are discussed in detail elsewhere in this book as well as in the literature [16, 70]. Leonard et al. [120] first demonstrated the application of coherent Stranski–Krastanov islands as quantum dots in 1993. In this work, $Ga_{0.5}In_{0.5}As$ was deposited on GaAs. As in the case of the previous work, the GaInAs was deposited by MBE until the RHEED pattern indicated a three-dimensional growth. The quantized properties of the islands were determined from photoluminescence. This work was the precursor to additional studies of these self-assembled islands as quantum dots, which is ongoing today, because it is extremely promising for many applications. Additional comments concerning applications of such islands are discussed in the following section.

Many additional reports have more recently disclosed self-assembled semiconductor islands during multilayer strained heteroepitaxy. The remainder of this section will address some topics in this field, including how the islands provide strain relief, different types of three-dimensional islands, the impact of deposition conditions, and the formation of islands on different surface orientations. This section will finish with a few unique approaches to arranging the islands.

5.2. Strain Relief from the Islands

Islands form as a way for the system to relieve strain. The complete details of how this occurs are not known, but a few observations and theories exist. In the Stranski–Krastanov growth mode, the material will begin to grow in a layer-by-layer fashion. At some point, the strain energy can be initially relieved by surface roughening, as shown in Figure 14 [121].

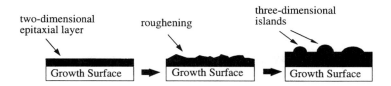

Fig. 14. Stranski–Krastanov growth mode. The growth is initially two dimensional. As the strain energy increases, the surface starts to roughen. With additional deposition, three-dimensional islands form to provide additional surface area to relieve strain. In many cases, the roughening cannot be clearly observed before island formation.

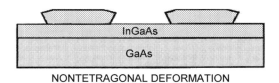

Fig. 15. Schematic diagram showing nontetragonal deformation of three-dimensional islands. The layer relaxes toward its natural lattice constant with distance from the heterointerface. Reprinted from *J. Cryst. Growth* 134, 51, N. Grandjean and J. Massies (© 1993), with kind permission of Elsevier Science–NL, Sara Burgerhart-straat 25, 1055 KV Amsterdam, The Netherlands.

A rough surface has a larger surface area than a smooth surface; free surfaces can provide strain relief [122]. This roughening has been observed to occur in some material systems. In other material systems, the growth goes directly to the formation of three-dimensional islands at a low coverage without a clear roughening stage.

Regardless of whether a layer undergoes roughening, the islands form to relieve strain. This relief results most likely from a combination of two effects. First, three-dimensional islands have a larger surface-to-volume ratio than a flat layer. Although this increased area has a related increase in the total surface energy, there is more freedom for the lattice within the island to relax. This relaxation is often referred to as nontetragonal deformation [80, 121]. In a strained two-dimensional layer, there is tetragonal deformation, as shown in Figure 4. The lateral or in-plane lattice constant is constant in each layer. In an island, however, the in-plane lattice constant may be different on the top of a unit cell than on the bottom; that is, the cell is trying to return to its natural lattice size, as shown in Figure 15. In a study of GaInAs islands on GaAs, Guha et al. [60] measured the lattice spacing within the islands and found that it increases with the distance from the base or interface.

Second, the material surrounding the islands can accommodate some of the strain [59]. The lattice constant of the material surrounding the island deviates from its natural value; the strain energy is distributed over a larger area. A schematic representation of this is shown in Figure 16, where the substrate lattice is locally deformed. This has been observed in several transmission electron microscopy studies [59, 123].

5.3. Different Types of Islands

Several stages of islanding have been observed in heteroepitaxial growth. These stages are often overlapping with different types or sizes of islands coexisting. The three types of islands are the following: (1) small islands, also known as precursor [90] or quasi-two-dimensional [86] islands; (2) medium-sized three-dimensional islands, which are still lattice-matched to the substrate (i.e., which are still strained); and (3) large, defected islands, which are no longer lattice matched to the substrate (i.e., which have relieved their strain by defect formation).

One example of these different types of islands was observed in the work by Reaves et al. [123] for InP islands formed on GaInP/GaAs(100) surfaces (cf. Fig. 17). The small

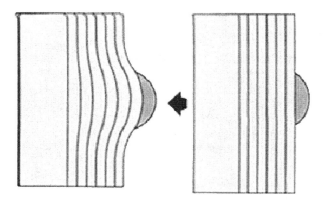

Fig. 16. Schematic diagram showing deformation of the substrate resulting from concentration of the strain by a three-dimensional coherent island. Reprinted with permission from D. J. Eaglesham and M. Cerullo, *Phys. Rev. Lett.* 64, 1943 (© 1990 American Physical Society).

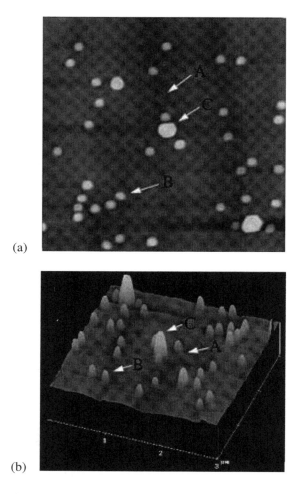

(a)

(b)

Fig. 17. Example of the three different types of islands. In this atomic force micrograph of InP growth on GaInP/GaAs(100), small islands are labeled A, medium-sized islands are labeled B, and large defected islands are labeled C. Reprinted from *Surf. Sci.*, 326, 209, C. M. Reaves et al. (© 1995), with kind permission of Elsevier Science–NL, Sara Burgerhartstraat 25, 1055 KV Amsterdam, The Netherlands.

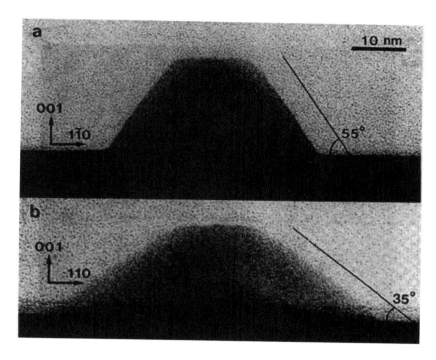

Fig. 18. Deduced shape of coherent InP islands. Reprinted with permission from K. Georgsson et al., *Appl. Phys. Lett.* 67, 2981 (© 1995 American Institute of Physics).

islands, an example of which is labeled A, are about 20 Å high and have a base width of 1200 Å. The medium-sized islands, an example of which is labeled B, are about 240 Å high and have a base width of 1200 Å. The defected islands, an example of which is labeled C, can vary in size, depending on the extent of deposition. The defected islands continue to grow with deposition, with each new dislocation in an island allowing a significant jump in the island size [124]. These islands become more obvious after they have grown to be much larger than the other islands that are present. Note that, for a given type, the islands appear to be similar in size. Size distribution studies for this and other material systems illustrate this [120, 125].

The exact shape of self-assembled islands is difficult to measure. One study discussing the shape of these islands was presented by Georgsson et al. [126] for InP islands. Many reports of islands observed by AFM or low-resolution electron microscopy conclude that the islands are cap-shaped and featureless. As shown in Figure 18, however, the islands are polyhedral. The tops are flat and the side walls consist of {110} and {111} planes. The shape varies according to the different crystallographic orientations that are present.

5.4. Impact of Deposition Conditions

In fabricating nanostructures for a particular application, the size is often important. What determines the size of self-assembled islands? Because they form as a way for a heteroepitaxial layer to relieve strain, one thought would be that only the lattice mismatch and the elastic properties of the layers should matter. This is not the case, however, because the island size can vary for the same material system deposited under different conditions. This implies that the formation of the islands is kinetically controlled under technologically relevant growth conditions. A few observations on these variations in island size and density will be discussed next.

One interesting observation is that the size of the islands for a given material system and set of deposition conditions will lock into a defined size. Only defected islands increase in

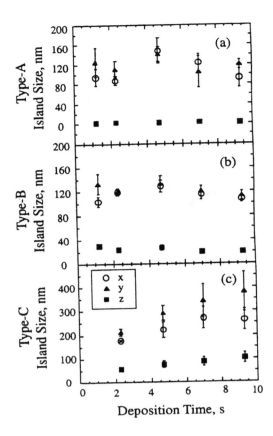

Fig. 19.   Island size vs deposition time for different types of InP islands. Note that the size of the small (type A) and medium-sized (type B) islands does not change with additional deposition. Reprinted from *Surf. Sci.*, 326, 209, C. M. Reaves et al. (© 1995), with kind permission of Elsevier Science–NL, Sara Burgerhartstraat 25, 1055 KV Amsterdam, The Netherlands.

size as additional material is deposited. This was quantified for the InP/GaInP/GaAs(100) system [123]. The height and base width of the small and medium-sized coherent islands are constant with additional deposition, as shown in Figure 19. Such observations indicate that coherent islands have a preferred size and remain at that size until they develop dislocations. As indicated by the error bars in Figure 19, the size distributions for the coherent islands, particularly for the medium-sized islands, are narrow.

If the islands are not getting bigger, then where is the newly deposited material going? Although some of this material goes into the growth of defected islands, the primary result is that the density of the islands increases. Reports in the literature show that the onset of three-dimensional islands, that is, the medium-sized islands, is abrupt. For the case of InAs/GaAs(100), the critical coverage for this transition is around 1.5 ML [86, 127]. Within 0.1 ML of additional coverage, the density of the medium-sized islands is already 10^9 cm^{-2}. This density continues to increase throughout the deposition [123, 128, 129].

Hence, the formation of these islands is governed by kinetics. With this in mind, the substrate temperature is expected to be a major influence on the size and density of the islands. Indeed, this has been observed to be the case. At low temperatures, island formation can be suppressed; diffusion is insufficiently rapid for islands to nucleate and grow [129]. The island size, in particular, the height, increases with increasing substrate temperature [125, 128, 130]. One explanation for this observation is that adatoms have a larger diffusion range at higher temperatures. They are more likely to find an existing island and add to that island rather than nucleate a new island. As a result, a lower density of larger islands is expected at higher temperatures. This is observed to be the case in reports where both island size and density are tracked as a function of substrate temperature [125, 128, 130].

Another deposition condition that can influence island size is the deposition rate or, equivalently, the arrival flux of the growth species. If the deposition rate is relatively high (relative to the rate of surface diffusion), the number of unattached adatoms on the surface is high and, therefore, the probability of nucleation of a new island is high. This would lead to a higher density of smaller islands. Surprisingly, there have been few studies of the impact of deposition rate with few consistent trends having been observed to date [128, 131].

5.5. Impact of Surface Orientation

As discussed in the previous section, the surface orientation will impact many of the processes occurring during epitaxy. A large fraction of semiconductor epitaxy is carried out on the (100) surface. The dominance of this surface extends into self-assembled islands formed during strained-layer epitaxy. Studies of growth on other surface orientations have, in general, shown some interesting results such as unique reconstructions [132] and different step-flow growth modes [133, 134]. Interesting results have also been observed in the case of self-assembled islands.

One of the more interesting results for island growth on higher-index surfaces has been demonstrated by Nötzel, Temmyo, and their co-workers [135–139]. For example, on a GaAs(n11)B substrate, with $n = 1$–5, GaInAs has been deposited on AlGaAs layers. What happens is that the GaInAs moves below the AlGaAs surface and forms coherent disks. Such structures, as shown in Figure 20, are uniform in size and often well aligned. The self-assembled nanostructures formed on (311)B surfaces have exhibited the best alignment. Similar disks have also been formed on InP(311)B surfaces. Several reasons have been suggested for these phenomena, based on the high surface energies of these surfaces [140]. These structures have been used for several optical studies as well as the active region of injection lasers [141].

Self-assembled islands, produced by the coherent Stranski–Krastanov growth mode, also form on high-index planes. The (311) surface, for example, has been used to form coherent islands in several systems [108, 142]. One of the more obvious differences with

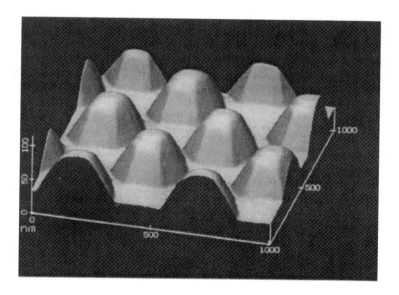

Fig. 20. Self-organized quantum disks formed by interlayer mixing of $Al_{0.5}Ga_{0.5}As$ on top of $In_{0.2}Ga_{0.8}As$ on a GaAs (311)B surface. Reprinted with permission from R. Nötzel, *Semicond. Sci. Technol.* 11, 1365 (© 1996 IOP Publishing).

respect to the (100) surface is that the islands are smaller and denser on the (311) surface. For example, with InP islands on GaInP/GaAs, the density of islands on the (311)A surface is 10^{10} cm^{-2}, whereas, on the (100) surface, it is 10^9 cm^{-2} [108, 142]. The height of these islands decreases from 240 Å on (100) to 60 Å on (311)A. Island growth has also been observed for InAs on GaAs(111)A. These islands appear, however, to be incoherent from the initial stages of growth [106].

As self-assembled islands are used more frequently for physical studies and device applications, the ability to adjust their size and density will become increasingly important. One of the disadvantages of self-assembled approaches with respect to lithographic avenues is the lack of direct control on sample morphology. The situation with self-assembled islands is not hopeless, however. There is clear tunability of island morphology, not just with deposition parameters, but also with surface orientation.

5.6. Controlling the Location of Self-Assembled Islands

There have been several interesting phenomena observed with respect to where self-assembled coherent islands are located on the growth surface. These phenomena can readily be traced back to basic epitaxial processes. One case is the vertical alignment of islands grown in sequential layers. If an array of islands is formed and then overgrown with buffer materials, the next layer of islands that is formed will be positioned directly above the first array of islands. This will continue to occur with additional layers of islands, as shown in Figure 21. This phenomenon has been observed by several groups [143–148]. Because

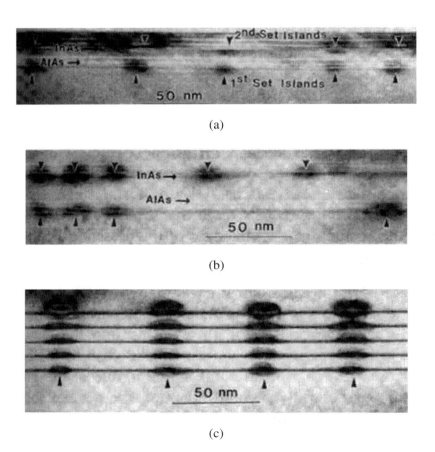

Fig. 21. Vertical alignment of InAs islands formed during sequential layers of island growth. Reprinted with permission from Q. Xie et al., *Phys. Rev. Lett.* 75, 2542 (© 1995 American Physical Society).

the strain fields from the islands extend into the surrounding materials, the thickness and modulus of the spacer layer are important. For the case of InAs islands separated by GaAs layers, spacers thinner than 150 Å lead to good vertical alignment [144]; for Ge islands separated by Si, spacers thinner than 1000 Å are needed [147]. This alignment arises from the impact of strain on surface diffusion [144]. The strain fields from an island can extend into the capping layer. One simple picture is that for a compressively strained island, the strain field in the spacer layer will make the local lattice constant larger. Adatoms for the next layer of islands will preferentially locate in these regions above previous islands. One result of this strain propagation is that the islands may grow larger because of less local lattice constant mismatch. This can be seen clearly in Figure 21.

Surface morphology will also impact where islands form. Vicinal surfaces with mono-layer steps are common as epitaxial growth substrates. Two reports for InAs islands on GaAs(100) have presented evidence that these steps are impacting island formation. Ikoma and Ohkouchi [149] found islands aligning along [1$\bar{1}$0] steps on a surface that had a 1° vicinal miscut towards [110]. A similar preferential nucleation of islands at step edges was observed by Leonard et al. [127] for a nominally flat surface. Moison et al. [150], how-ever, have observed that the majority of islands nucleate on the terraces away from the step edges. It is known that surface steps influence surface diffusion and attachment [151, 152]. These influences may provide local variations in adatom concentrations that increase the probability of island formation. Such variations in the basic epitaxial growth process would also arise from surface features larger than monolayer steps.

Under certain growth conditions, various surface features will form. These are com-monly identified as bunches of monolayer-high steps [133, 134, 153]. When these step bunches are present, they will act as nucleation sites for self-assembled islands. This has been observed for InAs/GaAs(100) [127, 154], for InGaAs/GaAs(100) [155], and for InP/GaInP/GaAs(100) [156]. In these reports, strings of several islands were found to be aligned along these surface features.

Large surface features can also be formed by patterning a substrate. Several groups have demonstrated various results. Mui et al. [157] performed InAs deposition on a GaAs(100) surface that had etched ridges. For [01$\bar{1}$] ridges, the islands only formed on the (100) planes at the top of and in the valleys between the ridges. The islands were observed to form along the side wall of [011] ridges with no islands formed on the (100) top and valley planes of the ridges. Saitoh et al. [158] etched tetrahedral pits into a GaAs(111)B surface and then formed InAs islands. Depending on the deposition temperature, the islands were observed to form either on the side wall of the pit or only in the bottom vertex.

6. PHYSICAL PROPERTIES AND APPLICATIONS OF SELF-ASSEMBLED ISLANDS

In this chapter, there has been a focus on the formation and structural properties of self-assembled semiconductor islands formed during heteroepitaxy. One motivation for work on this topic is the unique physical properties of the islands. These properties are discussed in detail elsewhere in this book and in other reviews [159, 160]. In this section, a few properties and uses of self-assembled islands will be discussed.

6.1. Physical Properties: Some Examples

As previously mentioned, semiconductor nanostructures are of interest for their often unique physical properties. These structures can be engineered such that they can con-fine electrons into small regions, leading to potentially useful effects. A few studies and applications of these effects will be discussed here.

Luminescence is the optical emission resulting from an electronic relaxation [161, 162]. The electrons can be initially excited with higher-energy photons as in photoluminescence or with higher-energy electrons as in cathodoluminescence. Such techniques are commonly used to study the properties of solids and engineered structures. They also provide some insight into how materials will behave in some applications such as lasers and light-emitting diodes. There has been some effort to study the luminescence from a single quantum dot. As previously discussed for self-assembled islands, the sizes have a narrow, yet finite, distribution. The quantization of energy levels in these nanostructures is a function of size. Even small variations in size will lead to a range of electron energy levels. The resulting luminescence from an ensemble of islands is, therefore, broadened. This convolution can hide the true properties of the nanostructures.

There were two early reports in the literature regarding luminescence from small ensembles of self-assembled islands. Marzin et al. [163] have studied self-assembled InAs quantum dots with photoluminescence. Using electron beam lithography, the authors were able to pattern a sample containing the nanostructures into separated mesas. Because each mesa had a limited number of islands, the luminescence spectrum, shown in Figure 22, exhibits isolated peaks. Grundmann et al. [164] have made similar studies with cathodoluminescence in which the size of the excitation electron beam could be decreased to a 50 nm diameter. Their results from a self-assembled InAs island sample, shown in Figure 23, also exhibit isolated peaks. Both studies conclude that the narrow peaks (<0.15 meV) arise from single quantum dots. The spacing between peaks has been linked to an equivalent island size difference of 16 InAs molecules [164].

Some interesting studies have also been done that consider the charge storage of self-assembled quantum dots. A traditional capacitor relies on the storage and discharge of electrons. By using capacitance spectroscopy to charge and discharge the dots in a controlled way, the energetics of electrons in InAs quantum dots formed on GaAs(100) substrates were studied [165]. The difference between the first energy level and the second energy level in these 200 Å islands was measured to be 41 meV. It was also observed that the Coulombic repulsion between the first and second electrons in an energy level (with opposite spins) was small. By using both n-type and p-type structures, electron and hole

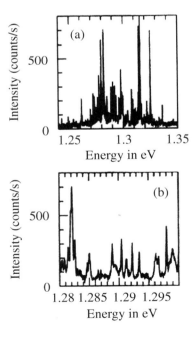

Fig. 22. Photoluminescence spectrum from self-assembled InAs quantum dots. The narrow lines are believed to arise from single dots. Reprinted with permission from J.-Y. Marzin et al., *Phys. Rev. Lett.* 73, 716 © 1994 American Physical Society).

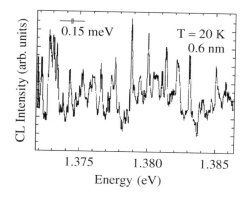

Fig. 23. Cathodoluminescence spectrum from self-assembled InAs quantum dots. The narrow lines are believed to arise from single dots. Reprinted with permission from M. Grundmann et al., *Phys. Rev. Lett.* 74, 4043 (© 1995 American Physical Society).

energy levels can be separately explored [166]. For example, InAs self-assembled islands may have two electron states, but only one hole state.

6.2. Self-Assembled Islands in Devices

One of the more exciting applications for quantum dots is the fabrication of an injection laser. Although lasers have been made from quantum wells and quantum wires, making one from a quantum dot has been difficult [167]. One problem with conventional lasers is that, as they heat up, because of either operation or ambient temperature, the energy distribution for carriers spreads. This spreading populates unwanted energy levels. The recombination from these levels does not add to the primary recombination, lowering efficiency. One way in which this is characterized is that the laser is tested at different temperatures, and the threshold current density for the start of stimulated emission is determined as a function of temperature. This behavior can be fitted by an exponential function of the form $\exp(T/T_0)$, where T_0 is a characteristic temperature [5, 168]. The higher this value, the less temperature sensitive is the laser. Because of the delta-function density of states in a quantum dot, thermal spreading of the carrier energy distribution is not allowed. Only well-defined energy transitions are allowed. The result is that the ideal quantum dot laser is insensitive to temperature variations, effectively leading to a T_0 of infinity. Along with the improved temperature performance, quantum dot lasers will have better gain properties; yet, one critical challenge is to obtain a very high density of dots to provide the necessary material [169]. Multiple layers of islands may be one approach to achieve this.

Some of the work using self-assembled islands to create lasers has exhibited large values of T_0 in an InAs-based device [170, 171]. Other work with lasers and optical structures based on islands has focused on devices that emit at 1.3 μm, a wavelength important for fiberoptic telecommunications. In these systems, either InAs or InGaAs alloys are used to form the islands [172]. Islands may also be used for integrating compound semiconductor devices with silicon. One report by Gérard et al. [173] used InAs islands formed on silicon substrates to obtain strong photoluminescence. When III–V semiconductor layers are grown on silicon, there is a large number of defects, and the resulting structures are often of low quality. This work found that, for InGaAs quantum wells grown on silicon, the photoluminescence was weak, whereas self-assembled quantum dot structures exhibited luminescence as strong as control samples deposited on GaAs. It is believed that self-assembled islands localize electrons and prevent them from finding defects where they may undergo nonradiative recombination.

Another potential application of quantum dots is in data storage. Most memory devices involve the storage of a charge: the larger that charge is, the more time that is needed to read and write to the memory. It has been observed that some self-assembled quantum dots only store one electron. If such devices can be used in the heart of a memory unit, then faster and higher-capacity memories can be created. Phenomena such as spin–spin interactions can be used for certain logic features currently not exploited [174]. For data storage, will each dot be solely responsible for a single bit of data? This would be the more economical way, but this also requires high fault tolerance for each quantum dot. This may be impractical with each dot storing such a small charge. Another approach is to have each memory cell based on hundreds or thousands of quantum dots [175]. If some dots fail, there is replication. This use of multiple dots will be limited when device density increases to the extreme point where feature sizes are 1–10 nm and each dot must be an individual, reliable device [176].

6.3. Use of Islands to Make Other Nanostructures

Although self-assembled islands are themselves high-quality nanostructures that can be used directly as quantum dots, they can also be used for other purposes. Some of these applications include making other nanostructures. One such application takes advantage of the fact that strain alone can modify the electronic band structure of a material [177]. This effect is realized by using the resulting strain field from a self-assembled island to modulate the band structure of the underlying material. In some cases, this is a quantum well located below the islands. The resulting potential in the quantum well resembles a parabolic quantum dot and exhibits good luminescence properties [178].

It has also been observed that, when material A is deposited on an islanded surface consisting of material B, and if these two materials are naturally lattice matched, then material A will initially grow exclusively on top of the islands of material B. The specific example for which this was demonstrated is that islands of InP were formed on a GaAs substrate. The alloy $Ga_{0.47}In_{0.53}As$ (which has the same lattice constant as bulk InP) was deposited on this surface and the new material was observed to grow only on top of the InP islands. This technique has been used to demonstrate quantum structures [179].

These two methods of using self-assembled islands arise from the different paths by which islands provide strain relief. The strain modulation is an example of substrate deformation, and the selective growth on the islands is likely a result of nontetragonal relaxation of the islands. As more is learned about self-assembled islands and strain relief, these and other applications will be further developed.

As mentioned earlier, one of the problems with etching to achieve quantum structures is the limitation of lithography. One alternate approach is to use self-assembled islands as an etch mask. This was done to create small GaAs structures by etching through a surface covered with InAs islands using chlorine gas [180]. These structures can then be overgrown and used as quantum dots.

7. SUMMARY

Although there are many methods for fabricating nanostructures discussed elsewhere in the literature and in this book, they have their limitations. New approaches are desired to complement these established methods. Self-assembled islands formed during strained-layer epitaxy constitute a unique and promising approach to the fabrication of semiconductor nanostructures. This chapter has both reviewed the basic concepts of strained-layer epitaxy needed to understand the formation of these self-assembled islands and described the techniques commonly used to fabricate and characterize these nanostructures.

During growth, the deposited material grows in a two-dimensional fashion until a transformation occurs and three-dimensional growth begins. The two-dimensional islands that grow during the early stages are often anisotropic and depend on the nature of the deposition surface. The transition is abrupt and the resulting islands are uniform in size. These islands have special properties and are being used more frequently for physical studies and device fabrication.

Regardless of whether the desired islands are being formed before or after the transition to three-dimensional growth, the fundamental aspects of epitaxy must be considered. Thermodynamic issues such as elastic strain and kinetic issues such as surface diffusion are part of the complex process of self-assembled island formation. This topic is still in its early stages of study. With time, a better understanding will evolve and an improved control of these nanostructures will almost certainly be possible.

Acknowledgment

The financial support of the NSF (Grant DMR-9504400) and QUEST, an NSF Science and Technology Center (Grant DMR-9120007), is very much appreciated.

References

1. C. Weisbuch and B. Vinter, "Quantum Semiconductor Structures: Fundamentals and Applications," Chap. 1. Academic Press, San Diego, 1991.
2. C. Weisbuch and B. Vinter, "Quantum Semiconductor Structures: Fundamentals and Applications," Chap. 6. Academic Press, San Diego, 1991.
3. Y. Arakawa and A. Yariv, *IEEE J. Quantum Electron.* QE-22, 1887 (1986).
4. G. Bastard, "Wave Mechanics Applied to Semiconductor Heterostructures." Halsted Press, New York, 1988.
5. Y. Arakawa and H. Sakaki, *Appl. Phys. Lett.* 40, 939 (1982).
6. H. Hirayama, K. Matsunaga, M. Asada, and Y. Suematsu, *Electron. Lett.* 30, 142 (1994).
7. E. Kapon, *Proc. IEEE* 80, 398 (1992).
8. T. A. Strand, B. J. Thibeault, D. S. L. Mui, L. A. Coldren, P. M. Petroff, and E. L. Hu, *Appl. Phys. Lett.* 66, 1966 (1995).
9. K. Y. Cheng, K. C. Hsieh, and J. N. Baillargeon, *Appl. Phys. Lett.* 60, 2892 (1992).
10. A. C. Chen, A. M. Moy, P. J. Pearah, K. C. Hsieh, and K. Y. Cheng, *Appl. Phys. Lett.* 62, 1359 (1993).
11. W. K. Burton, N. Cabrera, and F. C. Frank, *Philos. Trans.* 243, 299 (1951).
12. W. W. Mullins and J. P. Hirth, *J. Phys. Chem. Solids* 24, 1391 (1963).
13. M. Tsuchiya, P. M. Petroff, and L. A. Coldren, *Appl. Phys. Lett.* 54, 1690 (1989).
14. M. S. Miller, H. Weman, C. E. Pryor, M. Krishnamurthy, P. M. Petroff, H. Kroemer, and J. L. Merz, *Phys. Rev. Lett.* 68, 3464 (1992).
15. S. Y. Hu, J. C. Yi, M. S. Miller, D. Leonard, D. B. Young, A. C. Gossard, N. Dagli, P. M. Petroff, and L. A. Coldren, *IEEE J. Quantum Electron.* QE-31, 1380 (1995).
16. A. P. Alivisatos, *Science* 271, 933 (1996).
17. J. R. Heath, *Science* 270, 1315 (1995).
18. E. G. Bauer, B. W. Dodson, D. J. Ehrlich, L. C. Feldman, C. P. Flynn, M. W. Geis, J. P. Harbison, R. J. Matyi, P. S. Peercy, P. M. Petroff, J. M. Phillips, G. B. Stringfellow, and A. Zangwill, *J. Mater. Res.* 5, 852 (1990).
19. E. Bauer and H. Poppa, *Thin Solid Films* 12, 167 (1972).
20. W. A. Tiller, "The Science of Crystallization: Microscopic Interfacial Phenomena." Cambridge University Press, New York, 1991.
21. J. Y. Tsao, in "Materials Fundamentals of Molecular Beam Epitaxy." Academic Press, San Diego, 1993.
22. J. A. Venables and G. L. Price, in "Epitaxial Growth, Part B" (J. W. Matthews, ed.), pp. 381–436. Academic Press, San Francisco, 1975.
23. J. A. Venables, G. D. T. Spiller, and M. Hanbucken, *Rep. Prog. Phys.* 47, 399 (1984).
24. R. Kern, G. LeLay, and J. J. Metois, in "Current Topics in Materials Science" (E. Kaldis, ed.), Vol. 3, pp. 131–419. North-Holland, New York, 1979.
25. M. Pfister, M. B. Johnson, S. F. Alvarado, H. W. M. Salemink, U. Marti, D. Martin, F. Morier-Genoud, and F. K. Reinhart, *Appl. Surf. Sci.* 104/105, 516 (1996).
26. P. M. Koenraad, M. B. Johnson, H. W. M. Salemink, W. C. van der Vleuten, and others, *Mater. Sci. Eng., B* 35, 485 (1995).
27. J. A. Venables, *Philos. Mag.* 27, 697 (1973).
28. J. G. Amar and F. Family, *Phys. Rev. Lett.* 74, 2066 (1995).
29. M. Schroeder and D. E. Wolf, *Phys. Rev. Lett.* 74, 2062 (1995).
30. C. Ratsch, A. Zangwill, and P. Smilauer, *Surf. Sci.* 314, L937 (1994).
31. J. Tersoff, A. W. Denier van der Gon, and R. M. Tromp, *Phys. Rev. Lett.* 72, 266 (1994).
32. J. A. Venables, G. G. Hembree, J. Lui, and J. S. Drucker, in "Thirteenth International Congress on X-Ray Optics and Microanalysis," 1993, p. 415.

33. J. A. Venables, *Surf. Sci.* 299/300, 798 (1994).
34. G. S. Bales and D. C. Chrzan, *Phys. Rev. B* 50, 6057 (1994).
35. J. W. Evans and M. C. Bartelt, *Surf. Sci. Lett.* 284, L437 (1993).
36. C. Ratsch, P. Smilauer, A. Zangwill, and D. D. Vvedensky, *Surf. Sci.* 329, L599 (1995).
37. R. Bruinsma and A. Zangwill, *Europhys. Lett.* 4, 729 (1987).
38. V. Swaminathan and A. T. Macrander, "Materials Aspects of GaAs and InP Based Structures," Chap. 3. Prentice–Hall, Englewood Cliffs, NJ, 1991.
39. L. B. Freund, *Mater. Res. Soc. Bull.* 52 (1992).
40. J. W. Matthews, in "Epitaxial Growth: Part B" (J. W. Matthews, ed.), Chap. 8. Academic Press, San Francisco, 1975.
41. J. P. Hirth and J. Lothe, "Theory of Dislocations." Wiley, New York, 1982.
42. C. W. Snyder, B. G. Orr, and H. Munekata, *Appl. Phys. Lett.* 62, 46 (1993).
43. B. Voigtländer and A. Zinner, *Surf. Sci.* 351, L233 (1996).
44. C. W. Oh, E. Kim, and Y. H. Lee, *Phys. Rev. Lett.* 76, 776 (1996).
45. R. F. C. Farrow, in "Molecular Beam Epitaxy." Noyes Publications, Park Ridge, NJ, 1995.
46. M. A. Herman and H. Sitter, "Molecular Beam Epitaxy: Fundamentals and Current Status." Springer-Verlag, New York, 1989.
47. M. B. Panish and H. Temkin, "Gas Source Molecular Beam Epitaxy." Springer-Verlag, Berlin/Heidelberg, 1993.
48. G. B. Stringfellow, "Organometallic Vapor-Phase Epitaxy: Theory and Practice." Academic Press, San Diego, 1989.
49. H. O. Pierson, "Handbook of Chemical Vapor Deposition." Noyes Publications, Park Ridge, NJ, 1992.
50. T. P. Pearsall, "GaInAsP Alloys." Wiley, New York, 1982.
51. B. A. Joyce, J. H. Neave, P. J. Dobson, and P. K. Larsen, *Phys. Rev. B* 29, 814 (1984).
52. P. J. Dobson, B. A. Joyce, J. H. Neave, and J. Zhang, *J. Cryst. Growth* 81, 1 (1987).
53. T. Sakamoto, T. Kawamura, S. Nago, G. Hashiguchi, G. Sakamoto, and K. Kuniyoshi, *J. Cryst. Growth* 81, 59 (1987).
54. M. Prutton, "Surface Physics." Clarendon Press, Oxford, UK, 1983.
55. J. W. Edington, "Practical Electron Microscopy in Materials Science." TechBooks, Herndon, VA, 1976.
56. P. R. Busek, J. M. Cowley, and L. Eyring, "High-Resolution Transmission-Electron Microscopy." Oxford University, New York, 1988.
57. L. Reimer, "Transmission Electron Microscopy." Springer-Verlag, Berlin/Heidelberg, 1984.
58. P. Hirsch, A. Howie, R. Nicholson, D. W. Pashley, and M. J. Whelan, "Electron Microscopy of Thin Crystals." Krieger Publishing, Malabar, FL, 1977.
59. D. J. Eaglesham and M. Cerullo, *Phys. Rev. Lett.* 64, 1943 (1990).
60. S. Guha, A. Madhukar, and K. C. Rajkumar, *Appl. Phys. Lett.* 57, 2110 (1990).
61. G. Binnig and H. Rohrer, *Helv. Phys. Acta* 55, 726 (1982).
62. G. Binnig, H. Rohrer, C. Gerber, and E. Weibel, *Appl. Phys. Lett.* 40, 178 (1982).
63. J. A. Kubby and J. J. Boland, *Surf. Sci. Rep.* 26, 63 (1996).
64. S. N. Magonov, *Appl. Spectrosc. Rev.* 28, 1 (1993).
65. J. Tersoff and D. R. Hamann, *Phys. Rev. Lett.* 50, 1998 (1983).
66. J. Tersoff and D. R. Hamann, *Phys. Rev. Lett. B* 31, 805 (1985).
67. G. Binnig, C. F. Quate, and C. Gerber, *Phys. Rev. Lett.* 56, 930 (1986).
68. E. T. Yu, *Mater. Sci. Eng., R* 17, 147 (1996).
69. Q. Zhong, D. Inniss, K. Kjoller, and V. B. Elings, *Surf. Sci.* 290, L688 (1993).
70. C. Weisbuch and B. Vinter, "Quantum Semiconductor Structures: Fundamentals and Applications." Academic Press, San Diego, 1991.
71. U. A. Griesinger, S. Kronmuller, M. Geiger, D. Ottenwalder, F. Scholz, and H. Scheizer, *J. Vac. Sci. Technol., B* 14, 4058 (1996).
72. T. Koster, B. Hadam, J. Gondermann, B. Spangenberg, H. G. Roskos, H. Kurz, J. Brunner, and G. Abstreiter, *Microelectron. Eng.* 30, 341 (1996).
73. M. Illing, G. Bacher, T. Kummell, A. Forchel, D. Hommel, B. Jobst, and G. Landwehr, *J. Vac. Sci. Technol., B* 13, 2792 (1995).
74. M. Lopez, N. Tanaka, I. Matsuyama, and T. Ishikawa, *Solid-State Electon.* 40, 627 (1996).
75. P. Roblin, R. C. Potter, and A. Fathimulla, *J. Appl. Phys.* 1996, 2502 (1996).
76. E. J. Heller and M. G. Lagally, *Appl. Phys. Lett.* 60, 2675 (1992).
77. M. G. Lagally, Y.-W. Mo, R. Kariotis, B. S. Swartzentruber, and M. B. Webb, in "Kinetics of Ordering and Growth at Surfaces" (M. G. Lagally, ed.), pp. 145–168. Plenum, New York, 1990.
78. Y.-W. Mo, B. S. Swartzentruber, R. Kariotis, M. B. Webb, and M. G. Lagally, *Phys. Rev. Lett.* 63, 2393 (1989).
79. V. Bressler-Hill, A. Lorke, S. Varma, P. M. Petroff, K. Pond, and W. H. Weinberg, *Phys. Rev. B* 50, 8479 (1994).
80. J. Massies and N. Grandjean, *Phys. Rev. Lett.* 71, 1411 (1993).
81. J. Tersoff and R. M. Tromp, *Phys. Rev. Lett.* 70, 2782 (1993).

82. B. R. Bennett, P. M. Thibado, M. E. Twigg, E. R. Glaser, R. Magno, B. V. Shanabrook, and L. J. Whitman, *J. Vac. Sci. Technol.*, *B* 14, 2195 (1996).
83. B. R. Bennett, B. V. Shanabrook, P. M. Thibado, L. J. Whitman, and R. Magno, *J. Cryst. Growth* 175/176, 888 (1997).
84. Y. Chen and J. Washburn, *Phys. Rev. Lett.* 77, 4046 (1996).
85. R. Heitz, T. R. Ramachandran, A. Kalburge, Q. Xie, I. Mukhametzhanov, P. Chen, and A. Madhukar, *Phys. Rev. Lett.* 78, 4071 (1997).
86. T. R. Ramachandran, R. Heitz, P. Chen, and A. Madhukar, *Appl. Phys. Lett.* 640, 640 (1997).
87. J. Knall and J. B. Pethica, *Surf. Sci.* 265, 156 (1992).
88. U. Köhler, O. Jusko, G. Pietsch, B. Müller, and M. Henzler, *Surf. Sci.* 248, 321 (1991).
89. Y.-W. Mo, D. E. Savage, B. S. Swartzentruber, and M. G. Lagally, *Phys. Rev. Lett.* 65, 1020 (1990).
90. C. Priester and M. Lannoo, *Phys. Rev. Lett.* 75, 93 (1995).
91. P. M. Thibado, B. R. Bennett, M. E. Twigg, B. V. Shanabrook, and L. J. Whitman, *J. Vac. Sci. Technol.*, *B* 14, 885 (1996).
92. B. Voigtländer and M. Kästner, *Appl. Phys. A* 63, 577 (1996).
93. M. Berti, A. V. Drigo, G. Rossetto, and G. Torzo, *J. Vac. Sci. Technol.*, *B* 15, 1794 (1997).
94. N. P. Kobayashi, T. R. Ramachandran, P. Chen, and A. Madhukar, *Appl. Phys. Lett.* 68, 3299 (1996).
95. C. B. Duke, *Chem. Rev.* 96, 1237 (1996).
96. J. G. Belk, J. L. Sudijono, D. M. Holmes, C. F. McConville, T. S. Jones, and B. A. Joyce, *Surf. Sci.* 365, 735 (1996).
97. H. H. Farrell and C. J. Palmstrøm, *J. Vac. Sci. Technol.*, *B* 8, 903 (1990).
98. D. J. Chadi, *J. Vac. Sci. Technol.*, *A* 5, 834 (1987).
99. A. R. Avery, D. M. Holmes, J. Sudijono, T. S. Jones, and B. A. Joyce, *Surf. Sci.* 323, 91 (1995).
100. D. K. Biegelsen, R. D. Bringans, J. E. Northrup, and L.-E. Swartz, *Phys. Rev. B* 41, 5701 (1990).
101. B. Voigtländer and A. Zinner, *Appl. Phys. Lett.* 63, 3055 (1993).
102. J. L. Sudijono, M. D. Johnson, M. B. Elowitz, C. W. Snyder, and B. G. Orr, *Surf. Sci.* 280, 247 (1993).
103. D. M. Holmes, J. G. Belk, J. L. Sudijono, J. H. Neave, T. S. Jones, and B. A. Joyce, *Surf. Sci.* 341, 133 (1995).
104. J. G. Belk, J. L. Sudijono, H. Yamaguchi, X. M. Zhang, D. W. Pashley, C. F. McConville, T. S. Jones, and B. A. Joyce, *J. Vac. Sci. Technol.*, *A* 15, 915 (1997).
105. J. G. Belk, J. L. Sudijono, X. M. Zhang, J. H. Neave, T. S. Jones, and B. A. Joyce, *Phys. Rev. Lett.* 78, 475 (1997).
106. H. Yamaguchi, M. R. Fahy, and B. A. Joyce, *Appl. Phys. Lett.* 69, 776 (1996).
107. P. P. Gonzalez-Borrero, D. I. Lubyshev, E. Marega, Jr., E. Petitprez, and P. Basmaji, *J. Cryst. Growth* 169, 424 (1996).
108. K. Nishi, R. Mirin, D. Leonard, G. Medeiros-Ribeiro, P. M. Petroff, and A. C. Gossard, *J. Appl. Phys.* 80, 3466 (1996).
109. L. Sham, S. J. Allen, A. Kamgar, and D. C. Tsui, *Phys. Rev. Lett.* 40, 472 (1978).
110. D. Tsui and E. Gornik, *Appl. Phys. Lett.* 32, 365 (1978).
111. T. Cole, A. Lakhani, and P. J. Stiles, *Phys. Rev. Lett.* 38, 722 (1977).
112. T. J. Thornton, J. M. Fernandez, S. Kaya, P. W. Green, and K. Fobelets, *Appl. Phys. Lett.* 70, 1278 (1997).
113. B. Z. Olshanetsky and V. I. Mashanov, *Surf. Sci.* 111, 414 (1981).
114. T. Berghaus, A. Brodde, H. Neddermeyer, and S. Tosch, *J. Vac. Sci. Technol.*, *A* 6, 478 (1988).
115. J. M. Gibson, M. L. McDonald, and F. C. Unterwald, *Phys. Rev. Lett.* 55, 1765 (1985).
116. A. Oral and R. Ellialtioglu, *Surf. Sci.* 323, 295 (1995).
117. M. J. Bronikowski, Y. Wang, and R. J. Hamers, *Phys. Rev. B* 48, 12361 (1993).
118. L. Goldstein, G. Glas, J. Y. Marzin, M. N. Charasse, and G. L. Rous, *Appl. Phys. Lett.* 47, 1099 (1985).
119. M. Krishnamurthy, J. S. Drucker, and J. A. Venables, *J. Appl. Phys.* 69, 6461 (1991).
120. D. Leonard, M. Krishnamurthy, C. M. Reaves, S. P. DenBaars, and P. M. Petroff, *Appl. Phys. Lett.* 63, 3203 (1993).
121. N. Grandjean and J. Massies, *J. Cryst. Growth* 134, 51 (1993).
122. D. J. Srolovitz, *Acta Metall.* 37, 621 (1989).
123. C. M. Reaves, V. Bressler-Hill, S. Varma, W. H. Weinberg, and S. P. DenBaars, *Surf. Sci.* 326, 209 (1995).
124. F. K. LeGoues, M. C. Reuter, J. Tersoff, M. Hammar, and R. M. Tromp, *Phys. Rev. Lett.* 73, 300 (1994).
125. V. Bressler-Hill, C. M. Reaves, S. Varma, S. P. DenBaars, and W. H. Weinberg, *Surf. Sci.* 341, 29 (1995).
126. K. Georgsson, N. Carlsson, L. Samuelson, W. Seifert, and L. R. Wallenberg, *Appl. Phys. Lett.* 67, 2981 (1995).
127. D. Leonard, K. Pond, and P. M. Petroff, *Phys. Rev. B* 50, 11687 (1994).
128. M. Sopanen, H. Lipsanen, and J. Ahopelto, *Appl. Phys. Lett.* 67, 3768 (1995).
129. M. Taskinen, M. Sopanen, H. Lipsanen, J. Tulkki, T. Tuomi, and J. Ahopelto, *Surf. Sci.* 376, 60 (1997).
130. D. Leonard, M. Krishnamurthy, S. Fafard, J. L. Merz, and P. M. Petroff, *J. Vac. Sci. Technol.*, *B* 12, 1063 (1994).
131. C. M. Reaves, V. Bressler-Hill, W. H. Weinberg, and S. P. DenBaars, *J. Electron. Mater.* 24, 1603 (1995).
132. A. Zangwill, "Physics at Surfaces." Cambridge University Press, New York, 1988.

133. M. Krishnamurthy, A. Lorke, M. Wassermeier, D. R. M. Williams, and P. M. Petroff, *J. Vac. Sci. Technol.*, *B* 11, 1384 (1993).

134. K. Pond, A. Lorke, J. Ibbetson, V. Bressler-Hill, R. Maboudian, W. H. Weinberg, A. C. Gossard, and P. M. Petroff, *J. Vac. Sci. Technol.*, *B* 12, 2689 (1994).

135. R. Nötzel, J. Temmyo, and T. Tamamura, *J. Cryst. Growth* 145, 990 (1994).

136. R. Nötzel, J. Temmyo, and T. Tamamura, *Jpn. J. Appl. Phys.* 33, L275 (1994).

137. R. Nötzel, T. Fukui, H. Hasegawa, J. Temmyo, and T. Tamamura, *Appl. Phys. Lett.* 65, 2854 (1994).

138. J. Temmyo, A. Kozen, T. Tamamura, R. Nötzel, T. Fukui, and H. Hasegawa, in "Proceedings of the Indium Phosphide and Related Materials Conference," 1995, p. 766.

139. J. Temmyo, A. Kozen, T. Tamamura, R. Nötzel, T. Fukui, and H. Hasegawa, *J. Electron. Mater.* 25, 431 (1996).

140. R. Nötzel, *Semicond. Sci. Technol.* 11, 1365 (1996).

141. J. Temmyo, E. Kuramochi, M. Sugo, T. Nishiya, R. Nötzel, and T. Tamamura, *IEICE Trans. Electron.* E79, 1495 (1996).

142. C. M. Reaves, R. I. Pelzel, G. C. Hsueh, W. H. Weinberg, and S. P. DenBaars, *Appl. Phys. Lett.* 69, 3878 (1996).

143. G. S. Solomon, M. C. Larson, and J. J. S. Harris, *Appl. Phys. Lett.* 69, 11897 (1996).

144. Q. Xie, A. Madhukar, P. Chen, and N. P. Kobayashi, *Phys. Rev. Lett.* 75, 2542 (1995).

145. Q. Xie, N. P. Kobayashi, T. R. Ramachandran, A. Kalburge, P. Chen, and A. Madhukar, *J. Vac. Sci. Technol.*, *B* 14, 2203 (1996).

146. G. S. Solomon, J. A. Trezza, A. F. Marshall, and J. J. S. Harris, *J. Vac. Sci. Technol.*, *B* 14, 2208 (1996).

147. B. Rahmati, W. Jäger, H. Trinkaus, R. Loo, L. Vescan, and H. Lüth, *Appl. Phys. A* 62, 575 (1996).

148. N. N. Ledenstov, V. A. Shchukin, M. Grundmann, N. Kristaedter, J. Böhrer, O. Schmidt, D. Bimber, V. M. Ustinov, A. Y. Egorov, A. E. Zhukov, P. S. Kop'ev, S. V. Zaitsev, N. Y. Gorbeev, Z. I. Alferov, A. I. Borovkov, A. O. Kosogov, S. S. Ruvimov, P. Werner, U. Gösele, and J. Heydenreich, *Phys. Rev. B* 54, 8743 (1996).

149. N. Ikoma and S. Ohkouchi, *Jpn. J. Appl. Phys.* 34, L724 (1995).

150. J. M. Moison, L. Leprince, F. Barthe, F. Houzay, N. Lebouchè, J. M. Gèrard, and J. Y. Marzin, *Appl. Surf. Sci.* 92, 526 (1996).

151. R. L. Schwoebel and E. J. Shipsey, *J. Appl. Phys.* 37, 3682 (1966).

152. J. Ishizaki, S. Goto, M. Kishida, T. Fukui, and H. Hasegawa, *Jpn. J. Appl. Phys.* 33, 721 (1994).

153. M. Kitamura, M. Nishioka, J. Oshinowo, and Y. Arakawa, *Appl. Phys. Lett.* 66, 3663 (1995).

154. S. Jeppesen, M. S. Miller, D. Hessman, B. Kowalski, I. Maximov, and L. Samuelson, *Appl. Phys. Lett.* 68, 2228 (1996).

155. M. Kitamura, M. Nishioka, J. Oshinowo, and Y. Arakawa, *Appl. Phys. Lett.* 66, 3663 (1995).

156. S. Varma, C. M. Reaves, V. Bressler-Hill, S. P. DenBaars, and W. H. Weinberg, *Surf. Sci.* 393, 24 (1997).

157. D. S. L. Mui, D. Leonard, L. A. Coldren, and P. M. Petroff, *Appl. Phys. Lett.* 66, 1620 (1995).

158. T. Saitoh, A. Tanimura, and K. Yoh, *Jpn. J. Appl. Phys.* 35, 1370 (1996).

159. A. C. Gossard and S. Fafard, *Solid State Commun.* 92, 63 (1994).

160. P. M. Petroff and S. P. DenBaars, *Superlattices Microstruct.* 15, 15 (1994).

161. V. Swaminathan and A. T. Macrander, "Materials Aspects of GaAs and InP Based Structures," Chap. 5. Prentice–Hall, Englewood Cliffs, NJ, 1991.

162. J. I. Pankove, "Optical Processes in Semiconductors," Chap. 6. Dover, New York, 1971.

163. J.-Y. Marzin, J.-M. Gérard, A. Izraël, D. Barrier, and G. Bastard, *Phys. Rev. Lett.* 73, 716 (1994).

164. M. Grundmann, J. Christen, N. N. Ledentsov, J. Böhrer, D. Bimberg, S. S. Ruvimov, P. Werner, U. Richter, U. Gösele, J. Heyenreich, V. M. Ustinov, A. Y. Egorov, A. E. Zhukov, P. S. Kop'ev, and Z. I. Alferov, *Phys. Rev. Lett.* 74, 4043 (1995).

165. H. Drexler, D. Leonard, W. Hansen, J. P. Kotthaus, and P. M. Petroff, *Phys. Rev. Lett.* 73, 2252 (1994).

166. G. Medeiros-Ribeiro, D. Leonard, and P. M. Petroff, *Appl. Phys. Lett.* 66, 1767 (1995).

167. R. F. Service, *Science* 271, 920 (1996).

168. L. A. Coldren and S. W. Corzine, "Diode Lasers and Photonic Integrated Circuits," Chap. 2. Wiley, New York, 1995.

169. K. J. Vahala, *IEEE J. Quantum Electron.* QE-24, 523 (1988).

170. N. Kirstaedter, N. N. Ledentsov, M. Grundmann, D. Bimberg, V. M. Ustinov, S. S. Ruvimov, M. V. Maximov, P. S. Kop'ev, Z. I. Alferov, U. Richter, P. Werner, U. Gosele, and J. Heydenreich, *Electron. Lett.* 30, 1416 (1994).

171. N. Kirstaedter, O. G. Schmidt, N. N. Ledentsov, D. Bimberg, V. M. Ustinov, A. Y. Egorov, A. E. Zhukov, M. V. Maximov, P. S. Kop'ev, and Z. I. Alferov, *Appl. Phys. Lett.* 69, 1226 (1996).

172. A. Tackeuchi, Y. Nakata, S. Muto, Y. Sugiyama, T. Inata, and N. Yokoyama, *Jpn. J. Appl. Phys.* 34, L405 (1995).

173. J. M. Gérard, O. Cabrol, and B. Sermage, *Appl. Phys. Lett.* 68, 3123 (1996).

174. S. Bandyopadhyay and V. P. Roychowdhury, in "1995 International Conference on Solid State Devices and Materials," Osaka, Japan, p. 180.

175. R. S. Williams, personal communication.

176. N. Yokoyama, S. Muto, K. Imamura, M. Takatsu, T. Mori, Y. Sugiyama, Y. Sakuma, H. Nakao, and T. Adachihara, *Solid-State Electron.* 40, 505 (1996).
177. V. Swaminathan and A. T. Macrander, "Materials Aspects of GaAs and InP Based Structures." Prentice–Hall, Englewood Cliffs, NJ, 1991.
178. M. Sopanen, H. Lipsanen, and J. Ahopelto, *Appl. Phys. Lett.* 66, 2364 (1995).
179. H. Lipsanen, J. Ahopelto, T. Koljonen, and M. Sopanen, *J. Cryst. Growth* 145, 988 (1994).
180. G. Yusa, H. Noge, Y. Kadoya, T. Someya, T. Suga, P. Petroff, and H. Sakaki, *Jpn. J. Appl. Phys.* 34, L1198 (1995).

Chapter 7

NANOFABRICATION VIA ATOM OPTICS

Jabez J. McClelland

Electron Physics Group, National Institute of Standards and Technology, Gaithersburg, Maryland, USA

Contents

1. INTRODUCTION

Nanotechnology, because it is concerned with the construction of objects and devices a few nanometers in size, is dependent on the control of matter on the near-atomic scale. Remarkably, we have in recent decades developed tools for working in this regime despite the fact that it deals with objects more than seven orders of magnitude smaller than those encountered in everyday experience. Indeed, 30 years ago, it seemed almost inconceivable that we could engineer materials with such precision. Nevertheless, exciting developments such as scanning probe microscopy, high-resolution electron microscopy, and self-assembled fabrication have contributed to the beginnings of a fast-developing field.

Although we have seen a broad range of new nanoscale scientific studies and novel nanotechnologies emerge from the various techniques developed to date, there is still a great deal of progress to be made. Ultimately, it is desirable to have the ability to build rapidly

Handbook of Nanostructured Materials and Nanotechnology, edited by H.S. Nalwa
Volume 1: Synthesis and Processing
ISBN 0-12-513761-3/$30.00

any structure or array of structures with atomic precision using any material (i.e., any atoms) of choice. To reach this end, it is clear that simple refinement of existing techniques will not suffice. All of the tools currently in use, though they represent impressive advances over previous efforts, have fundamental limitations that prevent them from providing the ultimate in nanotechnology. To make further advances, it is important to examine continually completely different approaches to nanofabrication, with the hope that at least some of the fundamental obstacles will be circumvented with the introduction of new techniques.

It is in this spirit that nanofabrication with atom optics has become a subject of investigation in recent years. In this new technique, the motion of neutral atoms is controlled with nanoscale precision, allowing high-resolution structures to be constructed when the atoms are incident on a surface. The term atom optics is used because the ways in which the atomic motion is controlled have strong analogies to the ways that light rays are manipulated in light optics, or charged particle beams are steered and focused in electron (or ion) optics. In each case, optical elements, such as lenses, mirrors, beam splitters, diffraction gratings, and so forth, are constructed to transport a beam from an input (or object) region to an output (or image) region. Usually, there is some form of magnification or demagnification during the process, yielding a desired pattern at the output.

As a new approach to nanofabrication, atom optics offers the possibility of several advantages over existing techniques. For one thing, the fundamental diffraction limit imposed on resolution, present in any process where one attempts to focus particles (whether photons, charged particles, or neutral atoms), can be very small for atoms. This is because the De Broglie wavelength of a thermal atom is small—typically on the order of 10 pm—due to the relatively large mass of an atom. Also, there is no resolution limit resulting from Coulomb repulsion (as is found in charged-particle optics) because the atoms are charge neutral. Furthermore, atom optics can be used both in a direct deposition mode, where neutral atoms are focused by atom lenses into an extremely fine spot as they deposit onto a substrate, and also in a lithography mode, where focused atoms are used to expose a suitable resist material. In the direct deposition mode, nanostructures can be fabricated in a clean, resist-free environment, with little or no damage to the underlying substrate (because of the low kinetic energy of the atoms). This is important where issues of contamination and defect-free growth are critical. In the lithography mode, exposure of the resist is done with neutral atoms at thermal energies (the energy for resist exposure comes from internal atomic energy). Thus, the process can be very localized, with very little scattering and resist penetration. In either mode, parallelism, which is advantageous when issues of fabrication speed and/or long-range spatial coherence are important, can be achieved with very high dimensional accuracy over a large area of the substrate using laser focusing of atoms in a laser interference pattern.

With all these potential advantages, it has become apparent that nanofabrication with atom optics could provide some new avenues for manipulation of matter on the near-atomic scale. This chapter will present a review of the basic concepts that are used for atom-optical nanofabrication, as well as a discussion of the progress to date in realizations of the techniques. Because the field is relatively new, there is still a great deal to be learned, and many of the studies discussed represent the very first work in this field. As research continues, it is likely that more innovations will be forthcoming, and the full potential of the technique will be realized.

2. MANIPULATION OF ATOMS

In this section, we discuss the various ways in which neutral atoms are manipulated, as a prelude to discussing atom optics and its use in nanofabrication. First, we examine the production and characteristics of neutral atomic beams, as these are a fundamental ingredient of atom optics. We then proceed with a discussion of the manipulation of atoms with electrostatic and magnetostatic fields, and lasers. It is worth noting at the outset that atoms, because of their charge neutrality, are much more difficult to manipulate than ions or electrons. For many years, the only manipulation of atoms that could be conceived of consisted of using apertures or slits to collimate a beam and perhaps a mechanical shutter to turn it on and off. As shall see, however, recent developments have introduced new ways to manipulate atoms, especially with lasers. These new techniques have paved the way for the establishment of the new concept of atom optics.

2.1. Atomic Beams

Beams of neutral atoms or molecules, since their first implementation in the early decades of the 20th century, have become a mainstay of both atomic and molecular physics experiments and also of thin-film deposition technologies. A very detailed understanding of their behavior has been developed over the decades, and this will only be summarized here. Several texts cover the subject in detail [1], notably the classic work by Ramsey [2], which has an atomic and molecular orientation, or Maissel and Glang [3], which treats the subject from a thin-film point of view.

The most basic way to make a beam of atoms is by thermal evaporation in a vacuum system. Typically, a small cell or crucible of material is heated to the point where the vapor pressure is on the order of 100 Pa (i.e., around 1 torr) and the evaporated atoms effuse from a small orifice into a vacuum system with pressure low enough (typically less than 10^{-7} Pa, or 10^{-5} torr) so that the mean free path of the atoms is on the order of at least a meter or so (Fig. 1).

As the atoms emerge from the orifice, they fly in nominally straight lines across the vacuum system, eventually striking a substrate (or the vacuum chamber wall), where they stick or bounce, depending on the local temperature and their particular chemical nature. If the pressure behind the orifice is not too high, so that few collisions occur as the atoms leave the aperture, the intensity distribution of the beam follows a cosine distribution, falling off as the cosine of the angle relative to the axis of the aperture. At a distance l from the orifice, the total flux on the axis (in atoms per unit area per second) is given by

$$I = \frac{1}{4\pi} \frac{p \bar{v} a}{k_{\mathrm{B}} T l^2} \tag{1}$$

where p is the vapor pressure of atoms in the cell, $\bar{v}$ is the average velocity, a is the orifice area, k_{B} is Boltzmann's constant, and T is the cell temperature in kelvins. Although a wide

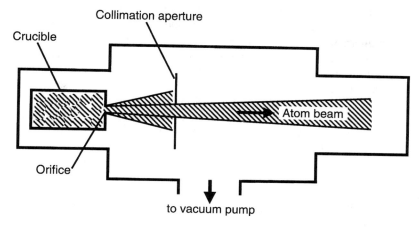

Fig. 1. Schematic of a generic atom beam apparatus. In a vacuum chamber, heating a crucible containing the desired material produces atomic vapor. Atoms effuse through an orifice and are collimated by an aperture.

range of fluxes is, in principle, possible because of the very steep dependence of vapor pressures on temperature, as a practical matter for many atomic species typical fluxes a few centimeters beyond the orifice tend to be in the range of 10^{19} atoms·m^{-2}·s^{-1}.

The velocities of the atoms in the beam follow the Maxwell–Boltzmann thermal distribution based on the temperature of the cell, given by

$$F(v)\, dv = \frac{1}{2}\left(\frac{m}{k_\mathrm{B}T}\right)^2 v^3 \exp\left(\frac{-mv^2}{2k_\mathrm{B}T}\right) dv \tag{2}$$

where $F(v)$ is the flux distribution, v is the velocity, and m is the atomic mass [4]. The most probable velocity for this distribution is $(3k_\mathrm{B}T/m)^{1/2}$, which works out to be in the range 200 to 1000 m/s for most atomic species. The spread in velocities, as given by the root mean square of this distribution, is $2(k_\mathrm{B}T/m)^{1/2}$.

Although thermal evaporation provides a good source for atomic beams of many atomic species, there are a few materials that present some difficulty because of their particularly low vapor pressures. Refractory metals such as W, Mo, and Ta, for example, do not achieve significant vapor pressures until they reach temperatures greater than 3000 K. If beams of these materials are desired, other methods such as sputtering [5] or laser ablation [6] can be used. In sputtering, a beam of energetic ions is directed at a solid target of the desired material and atoms are dislodged collisionally. Laser ablation also uses a solid target, but a high energy, pulsed laser beam is focused onto the target, locally heating the material to a very high temperature and generating a plume of the desired atoms. Both of these methods have the advantage that very high fluxes can be obtained; indeed, they are often employed even for nonrefractory materials if a very high flux is desired. One disadvantage, however, is that they tend to produce velocity distributions with most probable velocities corresponding to several electronvolts of kinetic energy, which is much higher than the fraction of an electronvolt typically seen in thermal sources. The widths of these distributions tend to be in the range of several electronvolts as well, with long tails extending to high velocity.

Because atomic beams tend, in general, to have very broad velocity distributions, the implementation of any atom-optical system is potentially complicated by what amounts to chromatic aberration—that is, atoms with different velocities behave differently in the optical system. To minimize the effects of chromatic aberration, it is often desirable to narrow the velocity spread in the atomic beam. One way of achieving this is to use a supersonic expansion [7]. This is done by increasing the vapor pressure of atoms in the crucible and making the orifice very small, so that a large number of collisions occur in the beam as it expands into the vacuum. These collisions, in combination with the rapid expansion of the gas of atoms, lower the effective temperature of the beam. The longitudinal velocity distribution is narrowed accordingly, in some cases to a width of about 10% of the mean velocity. The expansion can be done either using the vapor pressure of the atoms being evaporated or, alternatively, using a carrier gas, typically a light noble gas such as helium or neon. The use of a carrier gas has proven particularly useful in combination with laser ablation [8] to make nearly monoenergetic beams of a wide range of atomic species.

Another way of narrowing the velocity distribution is by velocity selection. This is typically done by passing the atoms through a series of rotating slotted disks [9]. The slots are offset from each other so that, for a given disk rotation speed, only one band of atomic velocities can pass through. With refinement of this technique, velocity monochromization (i.e., reduction of the velocity spread) in the range of a few percent is attainable. However, a corresponding loss of flux is encountered, and the beam is, by necessity, pulsed.

2.2. Manipulating Atoms with Static Electric and Magnetic Fields

The interaction between a ground-state neutral atom and an electrostatic field is extremely small and, in most cases, can be neglected completely. Whatever force does exist arises from an induced electric dipole moment in the atom and the presence of a gradient in the electric field. The induced dipole moment is $\mathbf{p} = \alpha \mathbf{E}$, where α is the polarizability of the atom and $\mathbf{E}$ is the electric field. The electrostatic energy of this dipole in the electric field is $W = -\frac{1}{2}\mathbf{p} \cdot \mathbf{E} = -\frac{1}{2}\alpha E^2$. Thus, the force on the atom is given by $\frac{1}{2}\nabla(\alpha E^2)$. To get an estimate of the size of this interaction, we note that typical atomic polarizabilities lie in the range from 2×10^{-41} farad·m^2 for helium to 7×10^{-39} farad·m^2 for cesium. Considering a cesium atom in the presence of a typical laboratory electric field of about 1000 V/m with a gradient of about 10^6 V/m^2, the acceleration comes out to be about 3×10^{-5} m/s^2—a very small value.

Whereas the interaction with electrostatic fields requires an induced dipole moment, magnetostatic interactions can take advantage of the permanent magnetic dipole moment present on many atoms. The force arises from the energy shift $-\boldsymbol{\mu} \cdot \mathbf{B}$ felt by an atom with a net magnetic dipole moment $\boldsymbol{\mu}$ in the presence of a magnetic field $\mathbf{B}$. An atomic magnetic dipole moment occurs when the atom has nonzero angular momentum—either orbital, spin, or both. For example, a ground-state spin-1/2 a atom has a magnetic moment with magnitude μ_B, the Bohr magnetron (9.274×10^{-24} J/T). Atoms with more complex angular momentum configurations have magnetic dipole moments that depend on the details of the angular momentum coupling [11], but, in most cases, the values are usually within a factor of 3 of μ_B.

As with electrostatic fields, the force on the atom arises from a gradient in the magnetic field, that is, $\mathbf{F} = \nabla(\boldsymbol{\mu} \cdot \mathbf{B})$. Considering again a cesium atom in its ground state and taking a typical maximal laboratory gradient of 100 T/m, we obtain an acceleration of 4400 m/s^2. Because the geometry of creating such a field gradient allows it to be applied over a fair amount of time, we see that a reasonable (though not huge) deflection of an atomic trajectory can be realized with a magnetostatic field.

2.3. Manipulating Atoms with Laser Light

Whereas static electric and magnetic fields exert relatively weak forces on atoms, laser light, on the other hand, can be used to alter dramatically their trajectories. For example, light forces can be used to slow a thermal atom beam, compressing its velocity distribution, or even bring it to a complete stop [12]. For light forces to have a strong effect, however, a narrow-band laser must be used, and it must be tuned near an atomic resonance. If this can be done, two types of forces arise, one from the effects of spontaneous emission and the other from induced dipole effects [13].

Because of the wide range of applications for laser-manipulated atoms, a number of comprehensive reviews have been compiled on this subject. For more detailed information, the reader is referred to the reviews in [14–21]. In what follows, we provide only a basic discussion of the origins of the effects.

2.3.1. Spontaneous Force

The spontaneous force arises from the scattering of photons in spontaneous emission. When an atom is exposed to near-resonant laser light, it will absorb photons and make transitions from its ground state to an excited state. In this absorption process, the momentum carried by the photons, which points in the direction of the laser beam, is transferred to the atom (see Fig. 2). As the atom makes transitions back to the ground state, it spontaneously gives off photons with momenta pointing in all directions. These momenta average to zero, so there is no recoil (on average) during the spontaneous decay. The result is that there is a net transfer of momentum to the atom in the direction of the laser beam. The magnitude of this momentum transfer depends on the number of photons scattered, each photon carrying a unit of momentum equal to $\hbar k$, where $k = 2\pi/\lambda$ is the wavenumber of the light. The rate of momentum transfer, or the force, depends on how

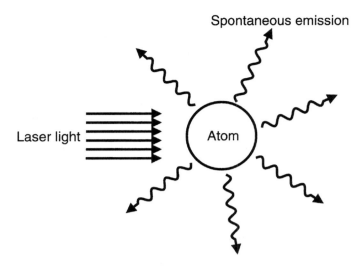

Fig. 2. The spontaneous force exerted by laser light on an atom. Photons incident from one direction only are absorbed and reemitted by spontaneous emission in all directions. The result, on average, is a transfer of momentum to the atom.

frequently the atoms can give off a spontaneous photon, as governed by the atomic lifetime τ and the probability that the atom is in the excited state f_{ex}. Thus, the force is given by

$$\langle F \rangle = \frac{\hbar k f_{\mathrm{ex}}}{\tau} \tag{3}$$

We note that the force in Eq. (3) is an *average* force; the spontaneous emission process is, of course, random, and also leads to a diffusive component in the force, which can play an important role in some circumstances. The excited state fraction is given by

$$f_{\mathrm{ex}} = \frac{\Omega^2}{2\Omega^2 + 4\Delta^2 + \Gamma^2} \tag{4}$$

where, in this expression, $\Gamma = 1/\tau$ is the atomic transition probability, Δ is the detuning of the laser from resonance, and Ω is the Rabi frequency, which contains information about the laser intensity I and the strength of the laser–atom interaction via $\Omega = \Gamma(I/2I_{\mathrm{s}})^{1/2}$, I_{s} being the saturation intensity associated with the atomic resonance [22].

Considering again the example of a cesium atom, we have a $6S_{1/2} \rightarrow 6P_{3/2}$ resonance at $\lambda = 852$ nm, so $\hbar k = 7.8 \times 10^{-28}$ kg·m/s. The spontaneous lifetime of this resonance is 32 ns, so the force can be as high as 1.2×10^{-20} N if 50% of the atoms are in the excited state. This corresponds to an acceleration of 5×10^4 m/s²—much larger than what is observed with electrostatic or magnetostatic fields.

As long as the atom and laser stay in resonance, the spontaneous force can be applied continuously, resulting in very significant changes in its motion. The resonance condition, however, can be affected by the Doppler shift, present when the atom has a velocity component toward or away from the laser light source. The Doppler shift is included in the force expression of Eq. (4), by replacing the detuning Δ by a modified detuning $\Delta' = \Delta - \mathbf{k} \cdot \mathbf{v}$, where $\mathbf{v}$ is the atomic velocity. By evaluating $\mathbf{k} \cdot \mathbf{v}$ for typical parameters, it readily becomes clear that at thermal velocities the Doppler shift can be very significant relative to the natural linewidth of the atomic transition or the Rabi frequency, unless the angle between the atomic motion and the laser beam is very close to 90°.

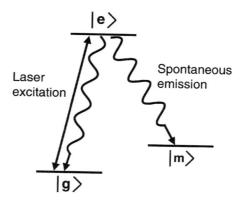

Fig. 3.   Optical pumping. A laser is used to excite an atom from the ground state $|g\rangle$ to the excited state $|e\rangle$. While the atom is in the excited state, there is some probability that it will fall into the metastable state $|m\rangle$ by spontaneous emission. Once in $|m\rangle$, the atom can no longer interact with the laser.

The Doppler shift is especially important in experiments where atoms are being slowed by a counterpropagating laser beam. As the atoms slow, their velocity component along the laser beam changes by enough to shift them completely out of resonance very quickly. To counteract this effect, a number of approaches can be taken. For example, a spatially varying magnetic field can be used to keep the atoms in resonance by the Zeeman shift [23], the laser frequency can be varied in time ("chirped") to slow repeatedly one group of atoms after another [24], or very broad band laser light can be used [25].

So far, our discussion of the spontaneous force has tacitly assumed only two atomic levels—the ground state and the excited state. If there are other states in the atom, a possible pitfall arises as a result of what is referred to as optical pumping (see Fig. 3). Optical pumping occurs if somewhere below the excited state there is a metastable state that has even a small probability of receiving an atom by spontaneous decay. Over many excitation–decay cycles (required if the spontaneous force is to have a significant effect on the atomic motion), a sizable fraction of the atoms can get trapped in the metastable state. As this happens, these atoms will stop participating in the spontaneous force process and become a potentially troublesome background unaffected by the laser light.

For example, in the case of sodium, the $3^2S_{1/2}$ ground state is split into two hyperfine levels separated by 1772 MHz. If the spontaneous force is exerted by tuning a laser from one of these levels to a hyperfine level of the excited $3^2P_{3/2}$ state, there is a finite probability that some atoms will decay into the other ground-state hyperfine level. If nothing is done to prevent this, eventually all the atoms will be optically pumped into an off-resonance state and the spontaneous force will cease acting. Another example is chromium, where the spontaneous force can be applied by tuning 425-nm laser light from the 4^7S_3 ground state to the 4^7P_4 state located at 23,500 cm^{-1} ($\sim$2.9 eV). About 8000 cm^{-1} ($\sim$1 eV) above the ground state lie the 4^5D metastable levels, which are weakly coupled to the excited state with a transition rate of about 6000 s^{-1}. Here, too, if the atoms are exposed to resonant radiation too long they will be lost and the spontaneous force will cease to have an effect.

In many cases, optical-pumping population traps can be remedied by the addition of laser frequencies to pump the lost atoms back into the ground-to-excited-state loop. For sodium, an acoustooptic or electrooptic modulator can be used to put sidebands on the main frequency, and, for chromium, laser beams in the 660-nm range can be introduced. In some cases, however, the problem can become quite difficult if, for example, there are no lasers available at the necessary wavelengths or if there are too many metastable states. This latter situation occurs, for instance, if an attempt is made to apply the spontaneous force to molecules. Because of the manifolds of vibrational and rotational levels that exist in even the simplest molecular spectra, it is very unlikely for a given molecule to return to the original ground state and absorb more than one photon from the laser.

2.3.2. Dipole Force

The other significant form of interaction between an atom and laser light is the dipole force. In this case, the force arises from a shift in the energy of the atom induced by the presence of the light field. If there is a spatial dependence in this shift, that is, a gradient in the energy, there will be an associated force. This force can be many times larger than the spontaneous force because it does not rely on the rate of spontaneous emission, but rather on how strong the laser–atom coupling is and how steep a gradient in light intensity can be achieved. As will be seen later, it is the interaction of choice in a majority of the atom-optical implementations involving laser light.

To get some sense of the origin of the dipole force, a classical picture is helpful as a starting point. Consider the atom to be a charged harmonic oscillator (electron on a spring) with resonant frequency ω_0. One can then ask what happens when this atom is placed in a near-resonant oscillating electric field, that is, a laser field, with frequency ω. The effect of this field is to induce an oscillating electric dipole moment on the atom with magnitude that depends on the atomic polarizability and how close ω is to ω_0. The phase of the oscillating dipole relative to the phase of the electric field will vary from $0°$ (in phase) to $180°$ (out of phase) as ω goes from below the atomic resonance to above. Just as in the case for an atom in an electrostatic field, there will be an energy $-\frac{1}{2}\mathbf{p} \cdot \mathbf{E}$ associated with the induced dipole in the presence of the external field. This energy will, however, be positive for detuning above resonance (positive detuning) and negative for detuning below resonance (negative detuning). Thus, if the electric field has a gradient, the atom will feel a force away from high field strength for positive detuning or toward high field strength for negative detuning.

Although the classical description gives a good physical picture for the qualitative behavior of the dipole force, to model the interaction correctly, a fully quantum treatment must be implemented. This can be done fairly easily via the dressed-state formalism for a two-level atom interacting with a monochromatic light field [26]. To determine the force on the atom in this approach, the energy shift as a function of laser intensity is derived, and then a spatial derivative can be taken. The energy shift in the atom is obtained by forming the two dressed-atom wave functions $|1\rangle$ and $|2\rangle$, each a linear combination of the ground state and excited state. For positive detuning, the state $|1\rangle$ consists of mostly the ground state, with an increasing admixture of the excited state as the laser intensity increases, and the state $|2\rangle$ is mostly the excited state with an increasing admixture of the ground state. The energies of states $|1\rangle$ and $|2\rangle$ are given by

$$E_1 = \frac{\hbar}{2}\left([\Omega^2 + \Delta^2]^{1/2} - \Delta\right) \tag{5}$$

$$E_2 = \frac{\hbar}{2}\left([\Omega^2 + \Delta^2]^{1/2} + \Delta\right) \tag{6}$$

where $\Omega = \Gamma(I/2I_s)^{1/2}$ is the Rabi frequency and $\Delta = \omega_0 - \omega$ is the detuning of the laser light from resonance. E_1 and E_2 are often referred to as light shifts of the atomic energy levels.

In situations where the laser is tuned relatively far from resonance (i.e., when $\Delta \gg \Omega$), nearly all the population is in the state $|1\rangle$, and this state is nearly identical to the ground state. At this limit, the energy of the atoms in the field is approximately equal to $\hbar\Omega^2/(4\Delta)$. If the detuning is relatively small, however, the situation is a little more complex. Ignoring spontaneous emission, one simply has atomic populations in two states, with possible coherence between them, and the motion of the atoms is governed by the two distinct potentials [27]. Taking spontaneous emission into account, one can consider the limit in which the atom stays at rest while many spontaneous photons are emitted. In this case, one

can speak of a mean potential felt by the atoms, weighted by their relative populations in
$|1\rangle$ and $|2\rangle$. This potential is given by [26]:

$$U = \frac{\hbar\Delta}{2} \ln\left(1 + \frac{2\Omega^2}{\Gamma^2 + 4\Delta^2}\right) \qquad (7)$$

We note that in the limit $\Delta \gg \Omega$, the expression for U approaches the same limit as Eq. (5),
that is, $U \cong \hbar\Omega^2/(4\Delta)$.

Although the potentials given in Eqs. (5) and (7) provide a simple basis for calculat-
ing the effects of a light field on the motion of an atom, often the real situation is more
complicated. Most atoms are not two-level atoms, because, at a minimum, they will have
some magnetic sublevels in the ground or excited state (or usually both). Considering that
the laser light will have some definite polarization state, one must take into account which
transitions between magnetic sublevels are allowed by optical selection rules and what
their relative strengths are, as governed by the Clebsch–Gordan coefficients. The dressing
of such a multilevel atom is possible, and it leads to an array of potentials, each associated
with a state that is a linear combination of the various magnetic sublevels of the undressed
atom. The motion of the atom on these potentials can be calculated, but, in practice, this
must be done numerically because there are too many level populations and potentials to
keep track of analytically.

The situation is further complicated when the atoms move across the potentials too
quickly for the population to settle into the dressed levels, thereby inducing nonadiabatic
transitions between the dressed levels. More complexity is introduced when spontaneous
emission is taken into account, as this introduces random transitions between the dressed
levels. To account fully for all these effects and, hence, calculate exactly the motion of
actual atoms in a light field, quantum Monte Carlo calculations are performed [28]. These
consider the time evolution of the density matrix of the atoms, tracing many wave packets
and allowing random spontaneous emission events to occur. After accumulating a large
number of wave packets, the resulting distribution of atoms can be determined with fairly
high accuracy. In the context of atom optics, these calculations find their most utility in
providing confirmation of approximate models and modeling subtle experimental effects.

2.3.3. Laser Cooling

One of the most dramatic forms that laser manipulation of atoms can take is the cooling
of an ensemble of atoms—that is, the reduction of the width of its velocity distribution.
Laser cooling can be remarkably efficient, reducing the effective temperature of a cloud
of atoms to as low as 200 nK or, in some cases, even lower. In addition to being a very
useful tool for atom optics, laser cooling has a number of other applications, ranging from
the development of very high precision atomic clocks [29] to the generation of a Bose–
Einstein condensate [30].

The simplest form of laser cooling is referred to as Doppler cooling [31]. In this version,
counterpropagating laser beams tuned below resonance are directed at the ensemble of
atoms (see Fig. 4). If an atom in the ensemble has a velocity component toward any of
the incoming laser beams, it will see the laser frequency of that beam as being shifted
higher because of the Doppler effect. Thus, the incoming laser will appear to be closer to
resonance and the atom will feel a stronger spontaneous force from this beam. The force
will be greater the closer the frequency is shifted toward resonance, or, equivalently, the
larger the velocity component toward the incoming laser beam is. The result is a velocity-
dependent force in a region of space sometimes referred to as "optical molasses" [32]. In
such a region, the atoms move exactly as if they are in a viscous medium that dissipates
their kinetic energy and results in a narrower velocity spread; in other words, they become
cooled.

The limits of Doppler cooling are set by a balance between heating caused by continued
spontaneous emission, which adds random momentum kicks to the atomic velocities, and

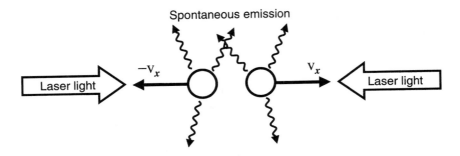

Fig. 4. Doppler cooling. Counterpropagating laser beams, tuned below the atomic resonance, interact with a population of atoms with random velocities. Those atoms with velocity component v_x toward one of the laser beams will be Doppler-shifted closer to resonance and, hence, will feel a stronger spontaneous force from that laser. Thus, atoms feel a velocity-dependent force, which reduces the velocity spread of the population.

cooling from the optical molasses. The minimum achievable temperature is given by [33]:

$$k_B T_{min} = \frac{\hbar \Gamma}{2} \qquad \text{(Doppler cooling)} \qquad (8)$$

where k_B is Boltzmann's constant and Γ is the spontaneous decay rate. For most atoms, this value is in the range of a few hundred microkelvins.

Since the discovery of Doppler cooling, a number of new mechanisms that produce even colder atoms have been uncovered. Polarization–gradient cooling, in particular, has been shown to cool atoms below the Doppler limit [34, 35]. In this form of laser cooling, use is made of the potential hills created by the light shifts induced by the laser light. Atoms repeatedly climb these hills only to find themselves optically pumped to the bottom again— hence, the term "Sisyphus cooling" is often applied, after the character in Greek mythology who was forced to push a stone continually up a hill only to see it roll down again. The necessary configuration of potential hills is created by giving the counterpropagating laser beams different polarizations; for example, they may be linearly polarized perpendicular to each other (lin ⊥ lin configuration), or one could be $\sigma+$ while the other is $\sigma-$. The other necessary ingredient, besides a laser tuned below resonance, is that the atom must have some magnetic sublevel structure in the form of at least two Zeeman levels in the ground state. When such an atom is placed in a lin ⊥ lin field, it can be viewed as moving on two light-shift-induced potentials, one for each Zeeman level. Each potential is sinusoidal in shape, but they are shifted by a half period relative to each other (see Fig. 5). As an atom in a given Zeeman level moves along the light-shift potential starting at the bottom of a hill, it must go up the potential and give up a corresponding amount of kinetic energy. At the peak of the hill, it is closest to resonance with the laser and, hence, has the greatest chance to be optically pumped to the other Zeeman level. If this pumping takes place, the atom finds itself at the bottom of the hill again because it is now on the other potential. As this process is repeated over and over, the atoms gradually lose energy and the ensemble can become cooled well below the Doppler limit. The minimum temperature obtainable in polarization–gradient cooling is conveniently expressed in terms of the recoil energy $E_R = \hbar^2 k^2/m$, which is the kinetic energy associated with the absorption and emission of a single photon. The smallest values seen experimentally are in the range of $10 E_R$ to $15 E_R$, and these are reasonably well explained by detailed theoretical calculations [36].

Although it may seem that the recoil limit would represent an absolute minimum for any laser cooling process, recent research has shown that even this limit can be surpassed. Two schemes of interest that have demonstrated subrecoil velocity spreads have been velocity-selective coherent population trapping (VSCPT) [37, 38] and stimulated Raman cooling [39]. Both these processes are not, strictly speaking, cooling processes, but rely instead on creating a situation in which atoms can fall into a state that for a very narrow

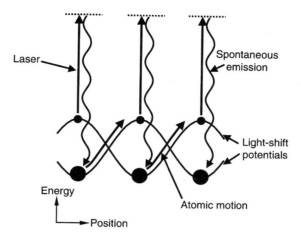

Fig. 5. Polarization–gradient cooling. In a lin ⊥ lin laser field, an atom with two Zeeman levels in the ground state experiences two sinusoidal light shift potentials offset by one half period. As the atoms move along these potentials, optical pumping from one potential to the other occurs more readily at the peaks because the laser is red-detuned. Thus, atoms are forced to travel "uphill" more frequently, resulting in a net loss of kinetic energy.

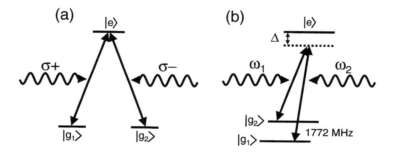

Fig. 6. (a) Velocity-selective coherent population trapping (VSCPT). In a Λ configuration, an atom with two degenerate ground states $|g_1\rangle$ and $|g_2\rangle$ interacts with counterpropagating $\sigma+$ and $\sigma-$ laser beams via a single excited state $|e\rangle$. If the atom has momentum $\pm\hbar k$, a coherent superposition state results that cannot absorb photons. Over time, atoms with momenta in very narrow ranges around $\pm\hbar k$ accumulate in the superposition state, resulting in cooled populations. (b) Stimulated Raman cooling. In a sodium atom, counterpropagating laser pulses with frequencies differing by the hyperfine splitting of the ground state (1772 MHz) generate Raman transitions with a detuning Δ from the excited state $|e\rangle$. By adjusting the frequency width, detuning, and propagation direction of the pulses, selective population transfer for only atoms with near-zero velocity can be achieved. The result is an accumulation of cold atoms in one of the hyperfine states.

velocity range does not interact with the laser. By repeatedly giving atoms a chance to fall into this state, population can be accumulated with a very narrow velocity spread.

In VSCPT, use is made of a coherent superposition of two degenerate ground states coupled through an excited state in a "Λ configuration" (see Fig. 6a). Such a configuration is realizable, for instance, with the metastable helium 2^3S_1–2^3P_1 transition, where the $M = +1$ and $M = -1$ sublevels of the 2^3S_1 lower state can be coupled to each other through the $M = 0$ sublevel of the 2^3P_1 excited state [37]. The coherent coupling between the two M sublevels is created by counterpropagating lasers of opposite circular polarization, which results in a superposition state that cannot absorb photons (i.e., is "dark") if the atom has translational momentum $\pm\hbar k$. Thus, if atoms fall into states with momentum $\pm\hbar k$ during a random-walk process, they will remain there without being heated by scattering photons. The result is an accumulation of atoms in two very narrow velocity bands around the velocities $+\hbar k/m$ or $-\hbar k/m$. Using this scheme, temperatures (referring now to the width of the velocity distributions around $\pm\hbar k/m$) well below the recoil limit have been observed.

In the stimulated Raman process, cooling below recoil is achieved by making use of the "recoil-less" nature of stimulated Raman transitions. This type of cooling has been demonstrated in sodium, where Raman transitions are induced between the $F = 1$ and $F = 2$ hyperfine levels by pulses of two counterpropagating laser beams differing in frequency by the spacing of the hyperfine levels (1772 MHz) (see Fig. 6b). By varying the frequency width, detuning, and propagation direction of the Raman pulses, all atoms except those in a very narrow velocity band around zero are transferred from the $F = 1$ to the $F = 2$ hyperfine level. From there, they are optically pumped back to the $F = 1$ level, in the process randomizing their velocity and giving them a chance to have zero velocity again. After several Raman pulse-optical pumping cycles, a population of very cold atoms is accumulated, and this has been shown in one dimension to have a velocity spread as small as 1/10th the recoil limit [39].

2.4. Atom Trapping

Another important development involving the manipulation of atoms is atom trapping. Motivated in part by the opportunities for extremely high resolution spectroscopy, and also the study of collective effects such as Bose–Einstein condensation, there has been a great deal of research recently into ways to generate potential wells that will trap neutral atoms. Although, to date, the trapping of atoms has not been employed in any form of nanofabrication, the degree of control over atomic motion that it affords suggests that applications might be forthcoming in the near future.

In principle, all the interactions discussed in the earlier part of this chapter can be put to use to trap atoms, with varying degrees of success. Generally speaking, because any interactions with neutral atoms tend to be weak, atom traps tend to be quite shallow. For this reason, the study of atom traps has historically been intimately connected with the study of atom cooling. An atomic population must be made very cold before it will be confined by the types of potentials available for trapping.

Whereas electrostatic fields are generally too weak to trap atoms, magnetostatic traps have been used with considerable success. Considering the energy $-\boldsymbol{\mu}\cdot\mathbf{B}$ of an atom with a magnetic dipole moment $\boldsymbol{\mu}$ in a magnetic field $\mathbf{B}$, one can see that atoms whose moment is aligned along the field will have a minimum energy at a minimum in the magnetic field strength. Such a local magnetic field minimum can be produced in three dimensions in a number of ways. For example, a quadrupole trap for sodium atoms has been demonstrated using a pair of coils in a Helmholtz geometry, but with current flowing in opposite directions in the two coils (anti-Helmholtz configuration) [40]. This type of trap generates a magnetic field that increases linearly in all directions from a value of zero at the center and, hence, can have a relatively narrow confinement. It has a disadvantage, however, in that very cold atoms can escape in a very small region around the zero of magnetic field at the center of the trap by flipping their spins (i.e., they undergo Majorana transitions). Another scheme, demonstrated with spin-polarized hydrogen atoms [41], employs a "pinch"-type trap made from a superconducting quadrupole magnet for radial confinement and two auxiliary solenoids for axial confinement. Still another scheme uses six permanent magnets oriented along three mutually orthogonal axes around a region in such a way as to create a local minimum at the center and a quadratic dependence of the field in the radial direction. This arrangement has been used to trap lithium atoms with good efficiency [42].

Although magnetostatic traps are simple in concept and are usable when no optical means are available (such as with hydrogen), by far the most popular atom trap has proven to be the magnetooptical trap (MOT) [43]. This trap makes use of the spontaneous force from resonant laser light to confine atoms. It relies on there being some magnetic sublevel

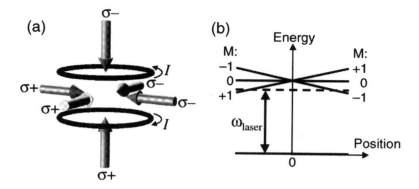

Fig. 7.   Magnetooptical trap (MOT). (a) A magnetic field that increases linearly from zero in all directions is produced by two coils with current I flowing in opposite directions (anti-Helmholtz configuration), and three pairs of oppositely, circularly, polarized laser beams counterpropagate through the center. (b) Energy of a $J = 0 \rightarrow J = 1$ atom in the presence of the magnetic field of an MOT. The magnetic sublevels $M = -1, 0, 1$ are shifted in opposite directions on opposite sides of the center. When the laser frequency ω_{laser} is tuned below resonance, atoms at negative positions are closer to resonance with the $\sigma+$ laser beam, while atoms at positive positions are closer to resonance with the $\sigma-$ laser beam. Thus, all atoms feel a net spontaneous force toward the center.

structure in the atom, and also on the fact that $\sigma+$ light excites only $\Delta M = +1$ transitions, whereas $\sigma-$ light excites only $\Delta M = -1$ transitions. The atoms are placed in a quadrupole magnetic field generated by a pair of coils in the anti-Helmoltz configuration, and irradiated with three pairs of counterpropagating laser beams along three mutually orthogonal axes (see Fig. 7). The restoring force necessary to keep the atoms in the center of the trap is generated by a combination of (a) a laser tuned below resonance, (b) opposite circular polarization in the counterpropagating laser beams, and (c) a radially increasing Zeeman shift of the atomic energy levels resulting from the magnetic field. Referring to Figure 7b, which shows the energy levels of an idealized $J = 0 \rightarrow J = 1$ atom along one dimension of the trap, we see that with a laser tuned below resonance the $M = +1$ state is Zeeman-shifted closer to resonance for negative positions, whereas the $M = -1$ state is Zeeman-shifted closer to resonance for positive positions. Thus, for negative positions, the atoms will interact most strongly with the $\sigma+$ light, which is incident from negative to positive, and vice versa for positive positions. The result is a restoring force that keeps atoms trapped at the minimum of the magnetic field.

Magnetooptic traps owe their popularity to their relative simplicity of construction and their relative robustness of operation. Typically, up to 10^8 atoms can be confined with peak densities up to approximately 10^{11} atoms/cm^3. An added advantage of the traps is that the negative detuning of the laser contributes some velocity damping to the force, and, hence, the atoms are cooled as well as confined while in the trap. Temperatures around 1 mK can be readily achieved, and with some care even sub-Doppler cooling is possible [44].

Atoms can also be confined with laser light alone. Making use of the dipole potential (see Section 2.3.2), a trap for sodium atoms has been demonstrated by tightly focusing a single red-detuned laser beam into a region of optical molasses [45]. The tuning below resonance of the trapping beam creates a dipole potential with a minimum at the highest laser intensity. Because the laser is a focused Gaussian beam, an ellipsoidal potential well is formed with its long axis along, and short axis transverse to, the laser beam. This concept has been further developed by making use of the fact that, for large detuning, the potential depth is proportional to I/Δ and the excited state fraction is proportional to I/Δ^2. Thus, a reasonable trap depth can be had with a very large detuning by using a very high laser intensity, all the while keeping the excited state population, and, hence, spontaneous emission and the associated heating, to a minimum. Such a trap has been demonstrated for rubidium atoms with a detuning up to 65 nm below the D$_1$ resonance at 794 nm [46].

An intriguing example of dipole force atom trapping is the transverse confinement of atoms inside a hollow optical fiber. In the first demonstration of this [47], rubidium atoms were guided down the bore of a hollow optical fiber in which red-detuned laser light also propagated. The laser light in the fiber had a maximum along the axis, so the atoms felt a radially inward dipole force that prevented them from sticking to the fiber walls. Successful guiding was seen through a 31-mm length of fiber with a hollow-core diameter of 40 μm. Recently, a similar guiding has been accomplished using blue-detuned laser light coupled into the shell of the hollow fiber [48]. This allowed the atoms to be confined in a low-intensity region, thereby reducing the effects of spontaneous emission.

2.5. Bose–Einstein Condensation

One of the ultimate goals of atom cooling and trapping research has been the formation of a Bose–Einstein condensate. For many years, it has been theoretically predicted [49] that a gas of atoms with the correct nuclear spin, if cold and dense enough, would undergo a phase transition, coalescing into a macroscopic occupation of a single quantum state with unique properties. Recently, this phenomenon has been demonstrated for three different atoms: Rb [30], Li [50], and Na [51]. In each case, a population of trapped atoms is cooled and compressed to the point where the predicted phase transition occurs, as evidenced by measurements on the spatial and velocity distributions of the atoms. In the case of Rb, the atoms were first trapped and cooled in an MOT. Then the MOT was shut off and the atoms were retrapped in a quadrupole magnetic trap that had an additional transverse rotating magnetic field component. The time orbiting potential (TOP) created by this configuration prevented the atoms from undergoing Majorana transitions at the trap center. The trapped atoms were then subjected to evaporative cooling by turning on a radio frequency (rf) field, which selectively allowed hotter atoms to escape the trap. It was this evaporative cooling step that provided enough reduction in temperature and increase of phase space density for condensation to occur. The experiments with Li were similar, except the trap was purely magnetostatic, formed by six permanent magnets arranged to produce a magnetic field minimum at the center and a quadratic radial dependence. The Na experiments used a quadrupole trap as in the Rb experiments, but the leak at the zero field point was sealed by focusing a far-off-resonant blue-detuned laser beam into the center of the trap.

Bose–Einstein condensation is of interest to atom-optical methods for nanostructure fabrication mainly because of the type of atomic source it represents. As will be discussed in more detail later, thermal beams of atoms present some serious restrictions on what kind of atomic focusing can be achieved because of their spatial incoherence and broad velocity distributions. A Bose–Einstein condensate, on the other hand, represents an extremely coherent group of atoms that could, in principle, be focused with much higher precision, or even diffracted to generate complex patterns. Just as the laser, which in a way represents a Bose–Einstein condensate of photons, has introduced a wide range of new optical applications, we can imagine that a Bose–Einstein condensate of atoms could open many new possibilities for atom optics. Although these possibilities may be far in the future, progress is presently encouraging, as evidenced by the very recent demonstration of an "atom laser" produced by coupling Na atoms out of a Bose–Einstein condensate [52].

3. Atom Optics

We now turn to a more specific discussion of the types of atomic manipulation that can generally be grouped under the concept of atom optics. As the name implies, atom optics is concerned with producing "optical" elements for beams of neutral atoms. These optical elements include, for example, lenses, mirrors, or gratings that manipulate atoms in ways analogous to the ways photons or charged particles are manipulated by similarly named

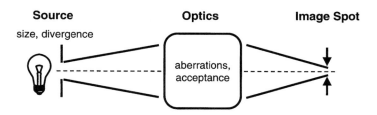

Fig. 8. Components of an optical system, illustrating the separation into source and optics characteristics.

objects in other forms of optics. We give here a summary of some of the various types of atom-optical elements that have been discussed in the literature. A number of reviews of this subject have also been published; in particular, references [17] and [53] are quite useful.

The analogy between atom optics and ordinary optics, which has both classical and quantum mechanical aspects, is a very useful concept. On the classical trajectory level, the analogy arises from the fact that the motion of any particles traveling predominantly in one direction and affected relatively weakly by a conservative potential can be treated with a paraxial approach. This allows the separation of longitudinal motion from transverse motion and makes the concept of lenses useful. On the quantum level, there is a fundamental similarity between the time-independent Schrödinger equation for a particle traveling in a conservative potential and the Helmholtz equation for an electromagnetic wave traveling in a dielectric medium [17]. Because these two take the same functional form, most of the results of scalar diffraction theory developed for light optics can be applied directly to atom optics. Thus, many insights can be had into the behavior of atom optics just by considering the light or charged-particle analog.

Another advantage to using the concept of atom optics is that the analysis of atom beam manipulation can be separated into the roles played by the object (i.e., atom source) properties and the optical system properties (see Fig. 8). One can then concentrate on two separate problems: (1) developing the best optical system, assuming the source to be, for example, a perfect plane wave; and (2) developing the best possible source. This simplifies the analysis and often points to where the weakness of a system is. Having separated the problem in this way, one can then go a step further and see if there is a way to modify the optical system to accommodate the atom source, such as is done, for example, in light optics with achromatic lenses.

3.1. Atom Lenses

Because nanofabrication with atom optics is naturally concerned with concentrating atoms into nanoscale dimensions, atom lenses are of central importance for this field. Quite a few types of atom lenses have been discussed or demonstrated, utilizing a wide variety of atom manipulation methods. Although, so far, the application to nanofabrication has only been done with a limited subset of the types of lenses available, it is nevertheless useful to consider what possibilities exist, because future developments may broaden the field.

To construct an atom lens, the most important requirement is a force that is exerted radially toward the axis of the optical system with a magnitude proportional to the distance from the axis, that is, $F = -kr$, where k is a constant (see Fig. 9). This is the necessary condition for Gaussian optics to hold, and it is the situation in which pure imaging takes place according to the elementary laws of optics, for example, the Gaussian lens law [54]:

$$\frac{1}{s_1} + \frac{1}{s_2} = \frac{1}{f} \tag{9}$$

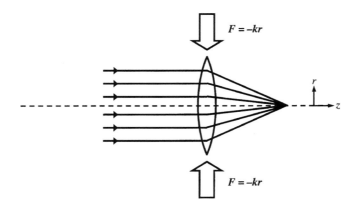

Fig. 9. The essential property of a Gaussian lens: A transverse force F must be exerted that is proportional to the distance r from the axis, so that all initially parallel rays cross the axis at the same point.

where s_1 is the distance from the object to the lens, s_2 is the distance from the lens to the image, and f is the focal length of the lens.

In general, it is only necessary for the linear radial force dependence to hold in the vicinity of the axis of the optical system. In fact, nearly all optical systems deviate from linear dependence away from the axis, but as long as there is linearity near the axis these deviations can be treated as aberrations. If the radial force acts only over a short axial distance (compared to the focal length), the additional approximation of a thin lens can be made. If this is not the case, though, formalisms exist for treating the lens as a thick and possibly an immersion lens without undue complications.

The construction of an atom lens then reduces to the production of a linear radial force dependence in the vicinity of an axis. Such a force can be achieved for neutral atoms using basically the same interactions that are used in atom traps—that is, either magnetostatic or optical forces [55]. Many arrangements of laser or magnetic fields that form atom lenses have been discussed in the literature; we discuss a few of them here to illustrate the variety of possibilities available.

3.1.1. Magnetic Hexapole Lens

One of the earliest demonstrations of an atom lens utilized a magnetic hexapole field [56]. The radial dependence of such a field is quadratic near the center of the lens, resulting in the necessary linear dependence of the force on a spin-polarized atom. A recent demonstration of this type of lens [57] used NdFeB permanent magnet pole pieces arranged as shown in Figure 10. Using a laser-slowed atom beam, this experiment showed imaging of a pattern of holes drilled in a screen placed at the object plane of the lens. The focal length of the lens is governed by the velocity v of the atoms and the second derivative of the magnetic field B at the center of the lens, and is given in the thin-lens approximation by

$$f = \frac{mv^2}{2\mu_B \int (\partial^2 B/\partial z^2)\, dz} \tag{10}$$

where m is the mass of the atom. Typical focal lengths of 40–50 mm were obtained with Cs atoms slowed to 60–70 m/s and a magnetic field second derivative of 2.66×10^4 T/m^2.

3.1.2. Coaxial Laser Lens

The first demonstration of the use of laser light to focus atoms was done using a Gaussian, red-detuned laser beam copropagating with a thermal sodium atom beam [58]. Because

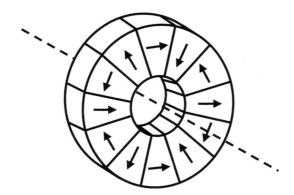

Fig. 10.   Pole configuration for the magnetic hexapole lens discussed in [57]. Arrows indicate the direction of magnetization.

of the red detuning, the atoms felt a dipole force toward higher laser intensity and were, therefore, attracted toward the center of the laser beam. Concentration of the atoms was observed by comparing the transverse atom beam profiles with the laser on and off. Using a 200-μm laser beam diameter, focusing of the atom beam to a spot size of 28 μm was achieved [59], demonstrating for the first time the concentration of atoms by laser light.

3.1.3. "Doughnut"-Mode Laser Lens

A major limitation on the spot size for atoms focused by a copropagating Gaussian laser beam is the diffusion of the atom trajectories caused by spontaneous emission. An alternative approach is to use a "doughnut"-mode, or TEM_{01}^{*} laser beam, which has a hollow center [60–62]. In this case, the laser is blue-detuned so the atoms are concentrated in the lower intensity regions of the laser beam and, hence, experience less spontaneous emission. Calculations of the behavior of such a lens have shown that, if the laser beam is brought to a diffraction-limited focus of approximately 1 μm, and if the atoms are constrained to travel through the center of this focus, focal spot sizes of 1 nm or less are, in principle, possible (see Fig. 11). An intriguing aspect of this "doughnut"-mode atom lens is that the axial dependence of the potential is such that the first-order (paraxial) equation of motion takes on exactly the same mathematical form as the equation of motion of an electron in a magnetostatic lens in the Glaser bell model [63]. This model, which allows an analytic solution to the equation of motion, has been analyzed in detail in the context of electron optics, so results can be transferred directly to the atom-optical case. The result provides an opportunity to analyze an atom-optical lens in great detail, examining all the common aberrations such as spherical aberration, chromatic aberration, and diffraction, as well as some unique ones such as spontaneous emission and dipole force fluctuations [62].

3.1.4. Spontaneous Force Lens

Although the dipole force seems a natural choice for high-resolution focusing, it is also possible to focus atoms with the spontaneous force. Such a lens has been demonstrated using four diverging near-resonant laser beams aimed transversely at a sodium atomic beam from four sides (Fig. 12) [64]. The approximately linear force dependence in this case comes from the fact that the laser beams are diverging as they propagate toward the atom beam. Atoms traveling through this light field experience a higher laser intensity the farther away from the axis they are, and so the spontaneous force is greater (as long as the atomic transition is not saturated). With this lens, it was possible to create an easily discernible image of a two-aperture atomic source, demonstrating the imaging capability of

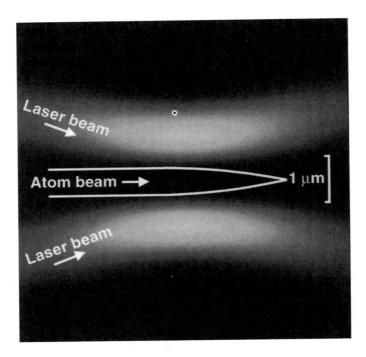

Fig. 11.   Schematic of atom focusing in a "doughnut"-mode (TEM_{01}^*) laser beam. Atoms traveling coaxially through the focus of the laser beam feel a dipole force toward the axis, focusing them into a very small spot. Analysis of aberrations indicates that focal spots in the few-nanometer regime are possible.

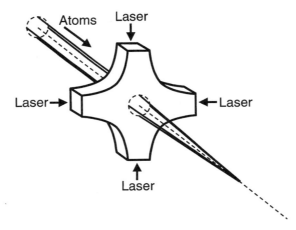

Fig. 12.   Spontaneous force lens. Four diverging resonant laser beams propagate transversely to the atom beam. Because the laser light becomes more intense as a function of distance from the axis, atoms feel a radially increasing spontaneous force, resulting in first-order focusing.

the technique. The two oven apertures were 0.5 mm in diameter and the resulting image spot sizes were 1.3 mm in diameter. The spot size was found to be limited by chromatic and spherical aberrations, as well as the random component of the spontaneous force.

3.1.5. Large-Period Standing-Wave Lens

Another lensing technique demonstrated recently involved sending metastable helium atoms through a large-period standing wave [65]. The large-period standing wave was

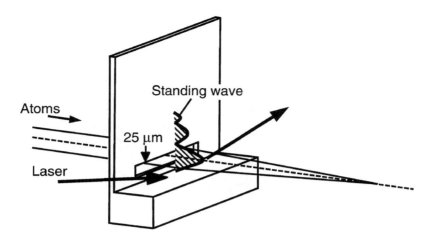

Fig. 13. Large-period standing-wave lens. A below-resonance laser beam reflects at grazing incidence from a substrate, creating a standing wave with a 45-μm-wide antinode. Atoms, apertured by a 25-μm slit aligned with the peak of the antinode, feel a dipole force toward the highest intensity, resulting in focusing.

formed by reflecting a laser beam, tuned just below the $2^3S_1 \rightarrow 2^3P_2$ transition at 1083 nm, at grazing incidence from a substrate placed transversely to the atom beam (see Fig. 13). The atom beam was apertured to 25 μm, so that it filled only a portion of a single antinode of the standing wave, which was 45 μm wide. Clear imaging, at unity magnification, of a 2-μm slit and also a grating with 8-μm periodicity was observed with this cylindrical lens. An image spot size of 6 μm was observed under optimal focusing conditions. The major contribution to this spot size was considered to be diffraction, arising from the long focal length (28 cm) and small lens aperture (25 μm). Chromatic aberrations were held to a minimum because the atomic beam in this case was produced in a supersonic expansion. An additional interesting feature of this lens is that it was formed under conditions of relatively high intensity and small detuning. Ordinarily, spontaneous emission would be a major effect under these conditions, but, in this case, the transit time through the lens was too short for any significant amount to occur. Thus, the atomic motion in the lens was governed by the two potentials given in Eqs. (5) and (6), with a fair fraction (15%) of the atoms in the state that feels a repulsive potential.

3.1.6. Standing-Wave Lens Array

An atom-focusing technique that has seen a great deal of attention recently is the focusing of atoms in an array of lenses created by a laser standing wave. This technique has been used successfully for nanostructure fabrication [66–69], and will be discussed in detail later on in this chapter. The principle of the approach is to make use of each node of a near-resonant, blue-detuned laser standing wave as an individual lens, so that the entire standing wave acts as a large lens array (see Fig. 14). Near the center of the nodes of the standing wave, the intensity increases quadratically as a function of distance from the node center. This intensity variation leads to a quadratically varying light-shift potential (as long as the excited-state fraction is low), and, hence, the force on the atom is linear and conditions are consistent with first-order focusing. Because of the high intensity gradient inside the node (the intensity goes from zero to full value in a fourth of an optical wavelength), it is relatively easy to get quite short focal lengths (on the order of a few tens of micrometers) with a standing-wave lens and, hence, small spot sizes, reaching into the nanometer regime.

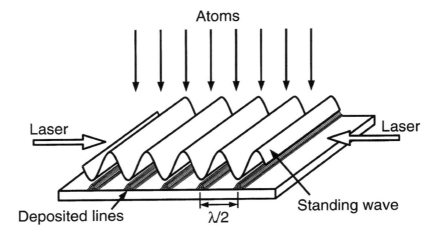

Fig. 14. Standing-wave lens array. An above-resonance laser standing wave propagates parallel to and as close as possible to a surface. Collimated atoms, incident perpendicular to the surface, are focused in each of the nodes of the standing wave by the dipole force. Nanometer-scaled focusing has been demonstrated with this lens (see Section 4.1).

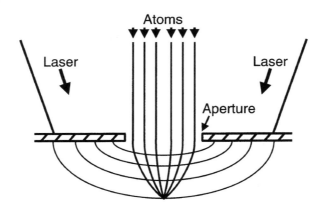

Fig. 15. Near-field lens. Below-resonance laser light propagates through a subwavelength-sized aperture. The longitudinally and transversely decaying transmitted laser light produces a light-shift potential that can focus atoms on the nanoscale.

3.1.7. Near-Field Lens

Another recently proposed way to achieve nanometer-scale spot sizes makes use of the intensity pattern found in the vicinity of a small aperture irradiated by near-resonant red-detuned laser light [70]. In this scheme, atoms are passed through an aperture that is illuminated with light copropagating with the atoms (see Fig. 15). The aperture is typically made smaller than the optical wavelength, so the intensity pattern of the light on the far side of the aperture is dominated by near-field effects. Close to the aperture, the intensity falls off rapidly in both the radial and the axial directions. The radial dependence of the intensity approaches a quadratic form near the axis, so, again, the correct spatial variation of the light-shift potential for focusing is obtained. Because of the small size of the lens, short focal lengths can be obtained, and calculations involving the standard aberrations result in predicted spot sizes of 4–7 nm.

3.1.8. Channeling Standing-Wave Lens

Although a laser standing wave can be used to construct an array of lenses for nanoscale focusing as discussed previously, it is also worth noting that it can be used in a macroscopic sense as well. A recent demonstration has shown that a diverging sodium atom beam passing through a standing wave can be concentrated by making use of the channeling that occurs in the nodes of the standing wave [71]. In this arrangement, the laser intensity is high enough to cause the atoms to be confined by the dipole potential and oscillate within a node as they traverse the standing wave. As they emerge from the standing wave, their trajectories are concentrated into groups traveling either toward the axis or away from the axis. Those atoms approaching the axis can be considered to be focused.

3.1.9. Fresnel Lens

Although the bulk of atom lenses make use of magnetostatic or light forces, there is another type of focusing that has also been represented in atom optics. Fresnel lenses create focusing conditions by relying on a diffraction phenomenon. Typically, a mask is fabricated that transmits incident radiation or particles in a pattern of concentric rings, the radii of which increase as the square root of the ring number, counting from the center out. Diffraction from this pattern of rings creates a spherical wavefront that is convergent on a spot beyond the lens, resulting in focusing (see Fig. 16). The focal length is given by $f = r_1^2/\lambda_{dB}$, where r_1 is the radius of the innermost ring and λ_{dB} is the De Broglie wavelength of the atoms. Such a lens has been demonstrated for atoms [72] using a freestanding Fresnel zone plate 210 μm in diameter microfabricated from gold. The plate had 128 zones and a first zone diameter of 18.76 μm. Focusing of metastable He atoms in the 2^1S_0 and 2^3S_1 states was observed, as a result of diffraction caused by the atomic De Broglie wavelength. The atoms were produced in a cooled supersonic expansion, so the velocity spread was narrow and the mean velocity was variable (by varying the source temperature). The De Broglie wavelength of the atoms was, therefore, well defined, and was variable from 0.055 to 0.26 nm. Clear images of a single and a double slit were observed with approximately 1:1 imaging and a focal length of 0.45 m. The observed images of the 10-μm slit were 18 μm wide, in agreement with numerical calculations of the expected diffraction limit. Whereas the advantages of a Fresnel lens include no requirement for near-resonant laser light and, hence, no restriction on the atomic species that can be focused, the disadvantages include multiple focal lengths arising from multiple diffraction orders and spot sizes limited to no smaller than the smallest feature that can be fabricated in the zone plate.

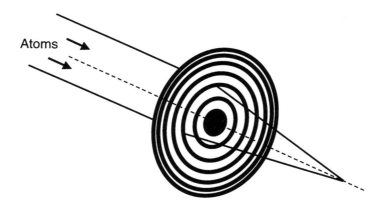

Fig. 16. Fresnel lens. A transmission mask diffracts atoms to form a converging spherical wavefront, thereby focusing them. The mask consists of concentric rings that increase in radius as the square root of the ring number, in accordance with the Fresnel zone formula.

3.1.10. Atom-Optical Calculations

Whatever particular geometry of laser or magnetic fields is chosen to make an atom-optical lens, it is usually of interest to perform some calculations of the behavior of atoms in the lens to find out what focal lengths and resolutions might be expected. With the exception of the Fresnel lens, which must be treated with diffraction theory, most atom lenses can be treated quite successfully with a particle optics approach. Diffraction comes into play only in determining a limit on focal spot size. As long as the dimensions of the lens are large compared to the De Broglie wavelength, the spot size is well approximated by the diffraction limit formula used in conventional optics:

$$d = \frac{0.61 \lambda_{dB}}{\alpha} \tag{11}$$

where d is the full width at half maximum of the spot, λ_{dB} is the De Broglie wavelength, and α is the convergence half-angle of the beam at the focus.

To trace the trajectories of atoms in a lens, the starting point is with the basic equations of motion derived from classical mechanics. In a cylindrically symmetric potential, these reduce to

$$\frac{d^2 r}{dt^2} + \frac{1}{m} \frac{\partial U(r, z)}{\partial r} = 0 \tag{12}$$

$$\frac{d^2 z}{dt^2} + \frac{1}{m} \frac{\partial U(r, z)}{\partial z} = 0 \tag{13}$$

where r is the radial coordinate, z is the axial coordinate, m is the mass of the particle, and $U(r, z)$ is the potential. We note that Eqs. (12) and (13) are also applicable in a one-dimensional focusing geometry, such as is found in a one-dimensional laser standing wave, with the substitution of the coordinate x for r. Thus, the following discussion also applies for this geometry.

One approach to analyzing an atom-optical lens is simply to integrate Eqs. (12) and (13) numerically. This approach certainly gives useful information [73], but for motion that is generally axial it is often useful to eliminate time in these equations and write them as a single equation for r as a function of z. This is done by using the conservation of energy to reduce Eqs. (12) and (13) to

$$\frac{d}{dz} \left[\left(1 - \frac{U(r, z)}{E_0} \right)^{1/2} \left(1 + r'^2 \right)^{-1/2} r' \right]$$
$$+ \frac{1}{2E_0} \left(1 - \frac{U(r, z)}{E_0} \right)^{-1/2} \left(1 + r'^2 \right)^{1/2} \frac{\partial U(r, z)}{\partial r} = 0 \tag{14}$$

where prime denotes differentiation with respect to z and E_0 is the initial kinetic energy of the atom.

To simplify Eq. (14), it is very useful to make the paraxial approximation. This concentrates on trajectories that are not affected too greatly by the potential, that is, those that are near the axis, and is made by taking the limit $r' \ll 1$ and $U(r, z)/E_0 \ll 1$. In the paraxial limit, Eq. (14) reduces to

$$r'' + \frac{1}{2E_0} \frac{\partial U(r, z)}{\partial r} = 0 \tag{15}$$

Equation (15) provides a very simple equation that can often be solved analytically, or at least with minimal numerical assistance. This allows first-order lens properties to be derived, such as focal lengths and principal plane locations if the lens is thick. Such an analysis is invaluable in determining the basic behavior of the lens in terms of the external parameters, such as magnetic field strength or laser intensity and detuning [62, 74].

After the paraxial approximation is made, it is then possible to determine the spot size limitations introduced by lens aberrations. Aberrations originate from the higher-order terms in the expansion of Eq. (14), as well as from any spread that may be present in the velocities of the atoms entering the lens (chromatic aberration). The effects of aberrations can be analyzed by taking the next-order terms in the expansion of Eq. (14), as is done in conventional aberration theory [62]. Alternatively, it may be more straightforward to solve Eq. (14) numerically. This can be done by introducing the slope of the trajectory $\alpha \equiv dr/dz$ as an independent variable and separating the equation into two first-order equations:

$$r' = \alpha \qquad (16)$$

$$\alpha' = \frac{1 + \alpha^2}{2(E_0 - U)} \left(\alpha \frac{\partial U}{\partial z} - \frac{\partial U}{\partial r} \right) \qquad (17)$$

Equations (16) and (17) can be solved readily with conventional numerical integration techniques.

Without going any further into the details of aberrations, we note only that, in the types of atom lenses that have been analyzed so far, spherical aberration (which results from higher-order terms in the expansion of the potential about the axis) tends to be relatively minor. Thus, in the absence of other aberrations, a diffraction-limited spot size can often be achieved. On the other hand, chromatic aberration, arising from the velocity spread of the incident atoms, tends to be rather significant. One way to see this is to solve the paraxial equation of motion for a particular lens and derive the velocity dependence of the focal length. For an immersion lens, the focal length is proportional to the velocity, and, for a thin lens, it is proportional to v^2 [74]. Because atom beams tend to have relatively broad velocity spreads, this velocity dependence can lead to large chromatic aberration effects. For this reason, research has been carried out on the possibility of an achromatic lens for atoms [75].

3.1.11. Focusing vs Concentrating

Before moving on to discuss other atom-optical elements, we mention briefly one more aspect of atom lenses. Because forces on atoms are generally weak, in practice, atom lenses are often generated by fields that extend over some distance in the axial direction (i.e., along z). In this situation, lenses are more often thick than not, and, in many instances, multiple crossovers can occur within an atom lens. In such a situation (the coaxial laser lens of Section 3.1.2 is a good example of this), the concept of focal length is less useful. The atoms are essentially "channeled" by the focusing field, and the lens acts more as a concentrator than a true lens, in the sense that a small spot can be generated but no image could ever be formed at the focus. If the purpose of a lens is to produce a very small spot of atoms, however, this is not necessarily a disadvantage and, in many cases, it can be an advantage. If the atoms are channeled in a lens, one is not relying on focusing at a particular focal point, but rather on the average effect of many oscillations within the lens. Thus, the effects of velocity spread in the atom beam are drastically reduced and the tolerance for focal location is greatly increased. The final spot size is, of course, not as small as can be achieved with true focusing, but, nevertheless, this technique can be used to focus to nanometer-scale sizes [66–69].

3.2. Atom Mirrors

Although atom lenses are the main optical elements of interest for nanofabrication with atom optics, there are other ways to manipulate atoms that parallel the methods of light optics. For example, atom mirrors have been the subject of some research recently. Here, the object is to reflect a beam of atoms coherently by generating a surface from which the atoms bounce specularly. Uses for such a mirror include reflective focusing elements or cavities for storing atoms.

To make an atom mirror, efforts have concentrated on using the same sorts of forces as those employed for trapping and focusing. Using a real physical surface is not, in general, practical because the interaction with the surface will typically not be at all elastic or specular because of the complex processes that are involved, such as chemisorption, physisorption, or phonon excitation [76]. Resorting to magnetostatic or laser forces, one is faced with the same limitations experienced with trapping or focusing; that is, the types of interactions available are generally rather weak. Thus, it is unreasonable to contemplate redirecting the high-velocity atoms in a thermal beam through a large angle because their kinetic energy is much higher than any potential that could be generated. To realize an atom mirror, very slow atoms (such as those falling from an atom trap) must be used or else the reflection must be at grazing incidence.

Atom mirrors have been demonstrated with both magnetostatic and laser fields. In the case of magnetostatic fields, use has been made of the strong gradients found near the surface of a magnetic material with a periodic array of alternating magnetizations. The magnetic field in the vicinity of such a surface falls off exponentially with distance from the surface, creating a repulsive potential for atoms with spins parallel to the magnetic field direction.

Utilizing this approach, reflection of rubidium atoms has been demonstrated from a section of magnetic recording media on which a sine wave was recorded [77]. The frequency of the sine wave was such that the magnetization reversed with a periodicity of 9.5 μm. The Rb atoms were first captured in a magnetooptic trap (MOT) and then released, so their velocity was only that attained by falling in the earth's gravitational field, that is, 0.7 m/s in this case. The potential barrier created by the magnetic field was sufficient to reflect suitably polarized atoms with $(94 \pm 8)\%$ probability in this experiment.

Another approach to this same type of atom mirror employed a stack of alternatingly magnetized sheets of the high-coercive-strength material NdFeB [78]. Using 18 1.04-mm-thick sheets, a mirror was constructed that could reflect Cs atoms dropped from an MOT with $(100 \pm 2)\%$ probability. Multiple bounces were observed, and the specular nature of the reflection was verified by observing the dependence of reflection angle on incidence angle.

The reflection of atoms from a light-shift potential was first proposed in 1982 [79]. In this scheme, the reflective light field is produced by total internal reflection of a near-resonant laser beam from within a prism. The light field above the surface of the prism does not propagate, but has an intensity that decays exponentially with a characteristic length $\lambda/2\pi$, which is on the order of 100 nm. Because of the exponential decay over such a short distance, there is a large gradient in the light-shift potential and, hence, a fairly strong repulsive force if the laser is detuned below resonance.

Demonstration of the reflection of atoms from such an evanescent light wave has been carried out either with atoms dropped from an MOT or with atoms incident at a grazing angle. The bouncing of dropped atoms has been observed with sodium [80], cesium [81], and rubidium [82]. Multiple bounces have generally been observed, although there are usually some losses with each bounce because of the transverse velocity components of the atoms. In addition, there has been some indication that diffuse scattering arising from surface roughness plays a significant role in this type of mirror [82].

The reflection of thermal atoms at grazing incidence from an evanescent wave has been demonstrated with sodium atoms at incident angles of up to 7 mrad [83], and with metastable argon atoms at angles up to 6.4 mrad [84]. In the latter case, improved reflection was obtained by using a surface plasmon–enhanced evanescent wave, which was obtained by applying a multilayer planar optical waveguide to the surface of the prism.

3.3. Diffraction of Atoms

In addition to focusing and reflecting atoms, a third atom-optical process that has been studied is diffraction. Diffraction of neutral atoms arises because of the basic quantum mechanical principle that any particle is described by a wave function that obeys Schrödinger's equation. The existence of this wave function means that the particle has a De Broglie wavelength $\lambda_{dB} = h/p$, where h is Planck's constant and p is the particle's momentum. The result is that diffractive wave phenomena exist for particles in analogy to the common diffraction effects observed with light. This analogy is made rigorous by the fact that, in the presence of a conservative potential, the time-independent Schrödinger equation takes on a form identical to the form of the equation for the propagation of an electromagnetic wave in a dielectric medium [17]. Because the equation is the same, all wave phenomena that are predicted for light must occur as well for particles in an equivalent situation.

An important consideration in discussing diffraction effects with neutral atoms is the size scale set by the De Broglie wavelength. Large, easily observable diffraction effects will, in general, only be seen when the diffracting body has feature sizes in the same range as the wavelength, which is on the order of 10 pm for thermal atoms. Such a situation occurs, for example, in scattering atoms from the regular atomic array found on a single-crystal surface, where atom diffraction has actually been put to use as a diagnostic tool to observe surface structure [85]. Most atom-optical versions of diffraction, however, are done with larger structures. Thus, the diffraction effects are generally small, and observation is not easy because of the difficulty in obtaining a source of atoms with enough longitudinal coherence (i.e., narrow velocity spread) and transverse coherence (i.e., collimation) to see an effect.

The motivation for demonstrating atom diffraction has, for the most part, been the production of coherent beam splitters for the construction of atom interferometers [86, 87]. Focusing of atoms with a Fresnel lens has been observed (see Section 3.1.9), and holography via atom interference has been demonstrated (see the following discussion), but the bulk of the research so far has concentrated on making use of the fact that diffraction is one of the few ways to separate a beam of atoms coherently into two or more paths. Direct application of atom diffraction to nanostructure fabrication has so far not been demonstrated; nevertheless, we discuss a few of the implementations here, because these techniques constitute an integral part of atom-optical research and could possibly find application to nanostructure fabrication in the future.

The diffraction of atoms from physical structures has been demonstrated in several forms in addition to the Fresnel zone plate discussed previously [72], where metastable He atoms were diffracted and focused. In the first demonstration, sodium atoms were diffracted on passing through a microfabricated grating with a 200-nm spatial period fabricated in a freestanding gold membrane [88]. A clear diffraction pattern was observed with first-order peaks at 85 μrad, using a highly collimated (10 μrad) atom beam produced in a supersonic expansion with $\Delta v/v = 12\%$. The pattern was observed by detecting atoms with a scanning hot wire detector located 1.5 m beyond the grating. A similar observation was made using a supersonic beam of metastable helium atoms with De Broglie wavelength 0.1 nm diffracting from a gold grating with a period of 500 nm [89]. More recently, interesting near- and intermediate-field diffraction effects have been observed in diffracting potassium atoms [90] and sodium atoms [91] from a series of gratings.

Perhaps the most dramatic demonstration of atom-optical diffraction has been the generation of a recognizable pattern by diffracting atoms through a computer-generated, microfabricated hologram [92, 93]. In this demonstration, metastable neon atoms were first trapped in an MOT and then released by a push from a near-resonant laser beam. The atoms fell through a screen that had an array of 500-nm-scale holes cut into it by microlithography, arranged in a pattern calculated to be the hologram of a desired image. After diffracting from the screen, the atoms were detected by a microchannel plate in the far field, where a clear image was observed (see Fig. 17). The transverse and longitudinal coherences of the atoms were kept high in this case because the atoms were launched from a trap, where their transverse and longitudinal velocity spreads were highly reduced. This experiment shows that the possibility of holographically imaging atoms is real, and opens up the possibility of fabrication of an arbitrary nanoscale pattern in the future.

—— 1 mm ——

Fig. 17. Image formed by diffraction of metastable Ne atoms passing through a computer-generated hologram fabricated as an array of 500-nm-scale holes in a 100-nm-thick silicon nitride film. The image was detected with a microchannel plate. Reprinted with permission from M. Morinaga et al., *Phys. Rev. Lett.* 77, 802 (© 1996 American Physical Society).

There have also been a number of demonstrations of atom diffraction from gratings created by light force potentials. In this case, there is not a particularly good match between the De Broglie wavelength (around 10 pm) and the grating period (typically half an optical wavelength, or about 300 nm), but the regularity of the period, the strength of the interaction, and the ease of producing the grating with an optical standing wave have led to quite a few measurements.

Diffraction has been observed from a standing wave in free space and also from a standing evanescent wave. The first demonstration of diffraction from a standing wave in free space was performed with sodium atoms produced in a supersonically cooled beam collimated with two 10-μm slits separated by 0.9 m [94]. Using a red-detuned, narrow-waist standing wave with detunings ranging from 93 to 1116 MHz, and a variety of laser powers, clear diffraction peaks were observable. This work was further extended into the Bragg scattering regime, where the standing-wave grating was made with a broader laser beam waist [95]. Here the diffraction was restricted to angles that satisfy the Bragg condition, and good agreement was seen between the predictions and experiment.

Subsequently, a number of examples of atom diffraction from a standing wave have been demonstrated. With emphasis on producing a large diffraction angle, metastable helium atoms have been diffracted from an optical standing wave in the presence of a magnetic field [96]. Here the shape of the light force potentials is modified by the magnetic field such that the grating is effectively "blazed" to emphasize higher-order diffraction. This sort of blazing of a light force grating for metastable helium has also been demonstrated without a magnetic field, making use of a pair of differentially detuned standing waves offset in spatial phase by $\pi/2$ [97].

Further demonstrations of atom diffraction from a standing-wave light field have been performed with specific emphasis on creating an atom interferometer. Both metastable argon [98] and metastable neon [99] have been used with success, showing good diffraction effects in experimental arrangements suitable for measuring small angles.

To show that atoms can also be diffracted from a standing wave in the reflection mode, several experiments have been done using total internal reflection of a light wave from within a prism, much as was done for the atomic mirrors discussed in Section 3.2. In this case, two counterpropagating laser beams were superimposed to create a standing-wave grating along the surface that decayed exponentially in the direction perpendicular to the surface. The first demonstration of this effect was done with sodium atoms [100], with further work done using metastable neon atoms [101, 102].

3.4. Collimation and Velocity Compression of Atoms

The parallels that exist between atom optics and particle optics provide an extremely useful framework for understanding the behavior of atom-optical elements. However, there is one aspect of atom-optics that has no parallel in other forms of optics. This aspect is the availability, through laser cooling, of dissipative forces that can be put to use in the collimation and velocity compression of atom beams. These nonconservative forces permit circumvention of a number of restrictions that apply to both particle and light optics and, hence, provide additional flexibility for atom optics.

In particle and light optics, a fundamental conservation law, referred to variously as the law of sines, the Abbé sine law, or the Helmholtz–Lagrange law, imposes limitations on the flux, or brightness, of beams in an imaging system [103]. This law can be derived from thermodynamic principles and, ultimately, stems from Liouville's theorem on the conservation of phase space volume in the presence of a conservative potential. For any bundle of rays in an optical system, the law states that the product $AV \sin^2 \theta$ is conserved at the object plane and at any image plane in the system. Here A is the area of the beam, θ is the divergence angle of the bundle of rays, and V represents either the local kinetic energy of the particles in the particle optics version or n^2, the square of the index of refraction, in the light optics version.

The impact of this law becomes evident when one wishes to increase the flux of a beam by focusing it down to a small spot. The law implies that, to reduce the area of the beam, it is necessary to increase θ and/or V. Because there are usually external constraints on V, the practical result is that it is not possible to brighten a beam and simultaneously collimate it (i.e., reduce θ) using only conservative optical elements.

Another limitation that occurs when optics are formed with conservative potentials is the lack of freedom in altering the velocity distribution. If an ensemble of particles with a certain velocity distribution and associated mean kinetic energy passes through a conservative optical system, the velocity distribution on exiting the system is constrained by the conservation of energy to have the same mean kinetic energy spread as before. Thus, particle beams cannot be monochromized by conservative optics alone. Of course, conservative, dispersive elements can be used in conjunction with apertures to eliminate unwanted velocities, but there is always an associated loss of flux with this approach.

The availability of nonconservative forces in atom optics allows the circumvention of both of these limitations. The law of sines is no longer a restriction because the conservation of phase space volume dictated by Liouville's theorem no longer holds. Also, the restrictions on altering the velocity distribution of a beam because of energy conservation are no longer in place. Thus, it is possible, in principle, to both collimate and brighten a beam of atoms to any degree desirable, as well as to compress the velocity distribution in atom beams without any loss of flux. These unique capabilities make atom optics especially interesting from an optical point of view, and also lead to practical advantages that can, in principle, be exploited.

The collimation of an atom beam by laser cooling has been demonstrated in quite a few experiments. A number of geometries have been used, including transversely counterpropagating laser light in one and two dimensions [104], laser light focused by an axicon (conical lens) [105], laser light enhanced by a one-dimensional buildup cavity [106], and laser light in the form of a spherical standing wave [107, 108]. Different cooling schemes have been utilized, including Doppler cooling [104], stimulated cooling with blue-detuned laser light [106, 109, 110], and polarization–gradient cooling [111]. More extensive beam brightening in conjunction with collimation has also been accomplished, making use of not only laser cooling, but also magnetooptical forces. This has been done with what is referred to as an "atom funnel," in which slowed atoms are sent into a two-dimensional magnetooptic trap where they are compressed and cooled radially [112, 113]. Alternatively, use can be made of an extended beam line consisting of first capturing and collimating atoms with laser cooling, then focusing them in a magnetooptic lens, and, finally, recollimating them with another stage of laser cooling [114]. In all these experiments, collimation was demonstrated on one of the alkali atoms Na, Rb, or Cs, one of the metastable rare gases He* or Ne*, or Cr.

Velocity compression of an atom beam, as well, has been demonstrated in a number of experiments. In the process of slowing a sodium beam with counterpropagating light, compression has been seen with either no compensation for Doppler shifts [115], compensation with a tapered magnetic field [116], or compensation by chirping of the laser frequency [117]. Also, velocity compression of a supersonic metastable argon beam has been demonstrated with two symmetrically detuned, counterpropagating laser beams traveling along the atom beam axis, cooling the atoms in a reference frame moving with the mean velocity of the beam [118].

4. NANOFABRICATION WITH ATOM OPTICS

We now turn to a discussion of some specific examples of the use of atom optics to create nanostructures on a surface. Although the discussion up to now has shown that there is a rich variety of ways to manipulate atomic beams, it is perhaps surprising that until very recently little of this knowledge base has been turned toward controlling atoms on the nanoscale to make features on a surface. One possible reason for this is the mismatch that exists between the atomic species generally studied in atom optics on the one hand, and the materials traditionally used in nanofabrication on the other. Another could be that the research methods involved in the two fields are sufficiently different that the combination is not readily accomplished. This situation has begun to change in recent years with the introduction of a number of key developments that have brought these two fields together and stimulated significant interest in what might now be possible.

Two basic approaches to nanofabrication with atom optics have evolved from the preliminary research in this area. In one method, atoms are manipulated during direct deposition, growing nanostructures by adding atoms to the surface in a specific pattern. In the other, atoms are manipulated as they impact and expose a resist, in what is referred to as neutral atom lithography. In the next two sections, we discuss some of the results that have been obtained with these two approaches.

4.1. Direct Deposition Techniques

To date, nanofabrication by direct deposition of laser-focused atoms has been demonstrated using only one configuration—the standing-wave lens array, as discussed in Section 3.1.6. This configuration has the advantage that it is relatively simple to set up, and it also represents one route toward fabrication of a large, coherent array of nanostructures in parallel.

The basic experimental arrangement for atom focusing in a standing wave is illustrated in Figure 18. An effusive oven produces a beam of atoms that is first collimated by a pair of slits or apertures, and then further collimated by transverse laser cooling. This second stage of collimation is required because the angular divergence of the atom beam must typically be reduced to substantially less than a milliradian. Such a high degree of collimation is necessary because the standing-wave potential wells typically have a depth of only approximately 1 μeV. Considering that the longitudinal kinetic energy is generally

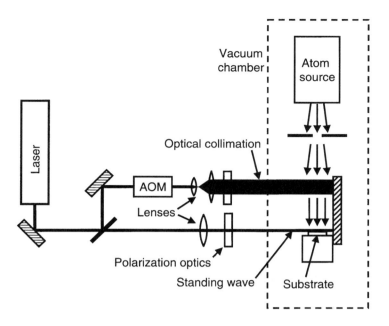

Fig. 18. Generic experimental setup for atom focusing in a standing wave. Near-resonant laser light is produced in a single-frequency, tunable laser source, such as a dye laser, a diode laser, or a Ti:sapphire laser. The laser, typically tuned several hundred megahertz above resonance, passes into the vacuum chamber and propagates very closely above the surface of a substrate, after which it is retroreflected to create a standing wave. A portion of the beam is split off and frequency-shifted in an acoustooptic modulator (AOM) to just below the atomic resonance to provide laser cooling for optical collimation of the atom beam. Both the standing-wave and optical collimation beams are carefully shaped and polarized with appropriate optics. Within the vacuum system, an atom beam is produced in a heated oven, is precollimated by an aperture, and propagates transversely through the cooling and standing-wave laser beams. As the atoms deposit onto the substrate, they are focused in the nodes of the standing wave, creating an array of nanostructures.

about 100 meV, even a small transverse component of the velocity can result in a transverse kinetic energy that is larger than the potential well depth. Also, even if the transverse kinetic energy is less than the depth of the well, the ultimate spot size obtainable is very sensitive to the degree of collimation (see Section 4.1.4).

It is worth noting that, although the collimation of the atomic beam could, in principle, be done with very small apertures at a large separation, laser cooling has a number of advantages. For one thing, it produces the required collimation with essentially no loss of flux, in contrast with the huge flux loss associated with using small apertures. Furthermore, the incident atoms can be automatically aligned perpendicular to the standing wave if the cooling laser beam is retroreflected from the same mirror that produces the standing-wave beam. Finally, it is generally convenient to take advantage of the near-resonant laser light that is available anyway, so the method does not pose too much additional complication for the experiment.

The laser light for generating the standing wave, and also the optical collimation beam, is, by necessity, produced in a single-frequency, tunable source such as a dye laser, a diode laser, or a Ti:sapphire laser. The frequency must be tightly controlled because it is important to be able to tune the laser very close to the atomic resonance, especially for the laser cooling. Both the standing wave and the laser cooling beam can be obtained from the same laser source with the aid of an acoustooptic (or electrooptic) frequency shifter. Typically, the laser cooling frequency must be about one atomic linewidth (e.g., 5 MHz for Cr) below the atomic resonance, whereas the standing-wave frequency is set several hundred megahertz above the atomic resonance to minimize spontaneous emission and ensure concentration of atoms into the low-intensity regions of the standing wave. The laser frequency must be locked relative to the atomic resonance to eliminate drifts during deposition, and this is accomplished either by using a saturable absorption cell or by imaging the fluorescence from a transverse probe laser beam crossing a diverging part of the atom beam onto a split photodiode [119].

The polarization of the laser beams must be carefully controlled. The laser cooling light is typically circularly polarized for Doppler cooling or put in a lin ⊥ lin configuration for polarization–gradient cooling. The latter is achieved by placing a quarter-wave plate in front of the retroreflection mirror. The standing-wave polarization can, in fact, be either circular or linear, depending on the magnetic sublevel distribution present in the atoms as they enter the standing wave. If Doppler cooling is used in the collimation, the atoms will typically be fully optically pumped into a pure $|M| = J$ state (stretched state). Then circularly polarized light is most appropriate, as it will have the strongest interaction with the atom. If polarization–gradient cooling is used, the distribution of the magnetic sublevel population tends to be spread over several levels, and linear polarization is more appropriate as it interacts more evenly with the different levels.

In mounting the sample and the standing-wave mirror, it is important to ensure that their relative position is stabilized at the nanometer level so that drift or vibrations do not smear out the deposited structures. This can be done either passively, by building a rigid mount, or actively, by sensing the relative position with a Fabry–Perot interferometer and correcting it with a feedback loop connected to a piezoelectric transducer. Although the mirror's positional stability is critical, it is interesting to note that the laser beam's positional stability is not nearly so important. This is because the locations of the nodes of the standing wave are determined to first order only by the mirror position and the laser wavelength. The slight shifting of the nodes that results from a variation in the angle of the laser beam is only a second-order effect because it depends on a cosine, which deviates from unity only quadratically for small angles.

4.1.1. Sodium

The first demonstration of nanofabrication by laser focusing of atoms was done with sodium [66], using an experimental setup containing the essential elements of Figure 18. Building on earlier work in which a focused Gaussian laser beam was used as a stencil to create a millimeter-scaled pattern in a deposited sodium beam [120], this work showed that a periodic nanoscale pattern could be deposited directly onto a silicon wafer. The sodium pattern was detected by observing the diffraction of a laser beam of shorter wavelength from the surface after deposition. Clear diffraction peaks were observed at the angles corresponding to the expected periodicity of the pattern, that is, half the sodium resonance wavelength, or 294.5 nm. In subsequent work, scanning tunneling microscopy (STM) images verified the existence of the sodium features [121–124]. This was done by transferring the samples to an STM operating within the vacuum system, because the sodium features reacted on exposure to air.

To achieve focusing of the sodium atoms, the laser beam, produced with a dye laser, was tuned near the $3^2S_{1/2}(F = 2)$ to the $3^2P_{3/2}(F = 3)$ D$_2$ line at 589 nm (this transition has a natural linewidth of 10 MHz and a saturation intensity of 6.3 mW/cm^2). The atom beam originated in an effusive source at 420 °C and was collimated in the first experiments with Doppler cooling utilizing circularly polarized laser light. In subsequent work, polarization–gradient cooling was also used. The standing wave, with detunings ranging from 70 MHz to 1.7 GHz and laser powers in the few milliwatts regime, was positioned (in different experiments) at a number of distances from the substrate, ranging from directly on the surface to 200 μm above it. Optimum focusing of the sodium atoms was found with the standing wave propagating directly along the surface, such that its profile was cut in half and the maximum intensity was at the surface. For this optimum case, the detuning was 1.7 GHz, the standing-wave beam waist diameter was 29 μm, and the traveling-wave power was 8 mW [125].

4.1.2. Chromium

Shortly after the first demonstration using sodium atoms, laser control of chromium atoms was introduced, allowing the first permanent nanostructures to be made with atom optics. Chromium has the advantage that on exposure to air a very thin ($\sim$1 nm) passivating oxide layer is formed, so samples can be prepared in vacuum and then removed and inspected in air with an atomic force microscope (AFM). Using the generic setup for laser focusing in a standing wave depicted in Figure 18, a number of demonstrations of chromium nanofabrication have been realized [67, 126–129].

For nanofabrication with chromium, the laser light must have a wavelength of 425.43 nm in air (425.55 nm in vacuum), which matches the 4^7S_3-to-4^7P_4 transition. This is provided either by a dye laser operating with stilbene-3 laser dye and pumped with an ultraviolet (UV) argon ion laser [67] or by a Ti:sapphire laser doubled in an external ring cavity [129]. The natural linewidth of this transition is 5 MHz, and the saturation intensity is 8.5 mW/cm^2. Typically, polarization–gradient cooling is used to collimate the atoms, resulting in a collimation angle as small as 0.16 mrad [111]. The atom beam is produced in an effusive cell operating at 1550–1650 °C.

Figure 19 shows two AFM images of chromium lines obtained with laser-focused atomic deposition in a standing wave. Figure 19a shows a three-dimensional rendering of a 1-μm-square image taken from a deposition on silicon dioxide, and Figure 19b shows a plan view of an 8-μm-square image showing lines deposited on sapphire in a somewhat thicker deposition. These nanostructures were produced with the Gaussian standing wave (single beam power, 33 mW; $1/e^2$ beam diameter, 0.13 mm) propagating so that its profile was cut in half by the substrate. Figure 19a indicates how narrow the lines can be made— 38 nm full width at half maximum (FWHM) in this particular case. The 8-μm-square image in Figure 19b gives some idea of the uniformity of the lines. The peak-to-valley height of the lines, as governed by the contrast of the deposition, as well as the flux of Cr and the deposition time, is 8 nm in Figure 19a and 16 nm in Figure 19b. The area covered extends for approximately 1 mm in the direction transverse to the lines and 0.15 mm along the lines.

The samples shown in Figure 19, and, in fact, all samples of laser-focused Cr nanostructures made so far, have some amount of background deposition in the regions between the lines. This background arises in part from the 16% of the atom beam that consists of isotopes other than ^{52}Cr, which do not interact with the laser. Other contributions possibly include the high-velocity tail of the longitudinal velocity distribution, wings in the transverse velocity distribution, or spherical aberration in the lens. Experimentally, the background level is found to vary somewhat with deposition conditions, with a minimum value of about one third the peak-to-valley height.

In addition to making arrays of lines, laser-focused atomic deposition of Cr has also been used to fabricate two-dimensional arrays of peaks. For example, a square array has been fabricated by superimposing a second standing wave at 90° over the first [130]. An AFM image of this pattern is shown in Figure 20. Although this orthogonal geometry is the simplest conceptually, it is necessary to be aware of the possible problems that can arise as a result of the relative temporal phase between the two standing waves. In general, the two standing waves will interfere with each other to create additional nodes, the pattern of which depends on the relative temporal phase. Because this phase is determined by the difference in path lengths taken by the two laser beams making up the two standing waves, the nodal pattern can vary if the path difference varies, as it would if influenced by acoustical vibrations. To circumvent this problem, the path difference would ordinarily have to be

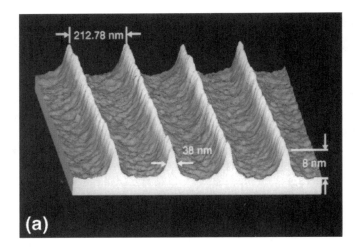

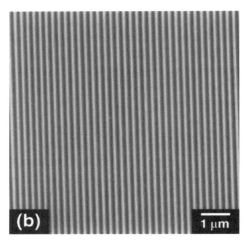

Fig. 19. Atomic force microscope (AFM) images of nanostructures formed by laser-focused atomic deposition of chromium. (a) 1-μm-square image of a relatively thin deposition on SiO_2, showing 212.78-nm pitch, 38-nm linewidth (full width at half maximum), and 8-nm peak-to-valley distance. [Source: McClelland et al., *Aust. J. Phys.* 49, 555 (1996).] (b) 8-μm-square image in plan view of a thicker deposition on sapphire, illustrating the long-range uniformity of the lines.

actively stabilized. Alternatively, as described by Gupta et al. [130], the standing waves can be given orthogonal linear polarizations, with one polarization parallel to the substrate and the other perpendicular. This eliminates any interference between the standing waves and leads to a stable pattern. Another approach to the temporal phase problem is to use the interference pattern generated by three laser beams crossing at mutual angles of 120° [131]. In this case, a two-dimensional pattern with hexagonal symmetry is formed, with no dependence on relative phase. Figure 21 shows an AFM image of a pattern formed in such a configuration.

In the one- and two-dimensional patterns shown in Figures 19 and 20, the periodicity is fixed at half the laser wavelength, or 212.78 nm. Recently, it has been demonstrated with chromium that the line spacing can be reduced to $\lambda/8$, or 53.2 nm [132]. This is accomplished by changing the polarization in the standing wave to the lin $\perp$ lin configuration, in which two traveling waves with orthogonal linear polarizations counterpropagate. With this polarization configuration, the optical potential generated by the standing wave can no longer be considered a simple sine wave. Instead, the complex interactions of all

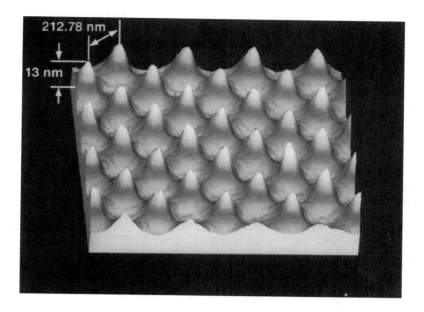

Fig. 20. Atomic force microscope (AFM) image of a two-dimensional array formed by laser-focused atomic deposition of chromium using two orthogonal standing waves. Shown is a 2-μm-square image. [Source: Gupta et al., *Appl. Phys. Lett.* 67, 1378 (1995).]

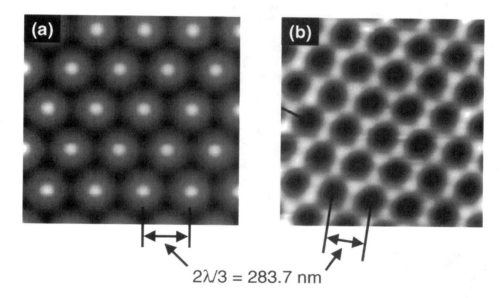

Fig. 21. Atomic force microscope (AFM) images of two-dimensional arrays formed by laser-focused atomic deposition of chromium using three laser beams crossing at mutual angles of 120°. (a) Using red detuning, an array of dots is created by focusing atoms into the antinodes of the light field; and (b) using blue detuning, an array of rings is formed. Reprinted with permission from U. Drodofsky et al., *Appl. Phys. B* 65, 755 (© 1997 Springer).

the magnetic sublevels in the Cr ground state with the light polarization that varies across the wavelength must be considered. Taking proper account of these interactions, one finds that the motion of the atoms is governed by an array of seven adiabatic potentials, some of which have minima at even multiples of $\lambda/8$, and some of which have minima at odd multiples of $\lambda/8$. The result is an array of lines with four times the periodicity of the lines produced with ordinary polarization. An example of this type of pattern is seen in the AFM image of Figure 22, which shows an array of lines with 53.2-nm spacing. We note that, in this type of deposition, the background level tends to be higher, with the modulation depth of the Cr surface reaching only about 50% of the average film thickness.

In discussing nanofabrication by laser focusing of chromium, it is worth mentioning a number of extensions that have been realized making use of the Cr lines to construct nanostructures out of more diverse materials. One example of this is the use of the Cr lines as a template for replica molding [133]. In this process, a mold of liquid prepolymer poly(dimethylsiloxane) (PDMS) is cast against the original Cr lines. The resulting elastomeric mold is peeled from the substrate and used as a mold for photochemically curable polyurethane. The result is a rigid replica of the original lines with excellent fidelity. An AFM image of such a mold is shown in Figure 23.

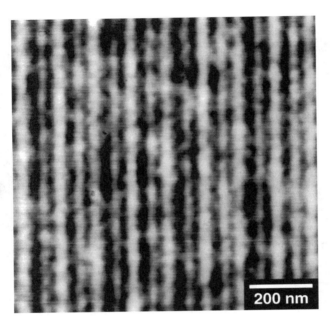

Fig. 22. Atomic force microscope (AFM) image of chromium lines with $\lambda/8$ spacing, produced by focusing atoms in a lin $\perp$ lin standing wave. The average pitch is 53.2 nm. [Source: Gupta et al., *Phys. Rev. Lett.* 76, 4689 (1996).]

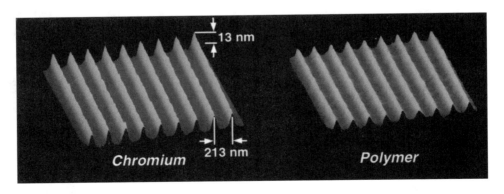

Fig. 23. Replication of chromium features made by laser-focused atomic deposition. Shown are AFM images of the original chromium lines and a polyurethane cast of them made by forming an intermediate mold of poly(dimethylsiloxane). [Source: Xia et al., *Adv. Mater.* 9, 147 (1997).]

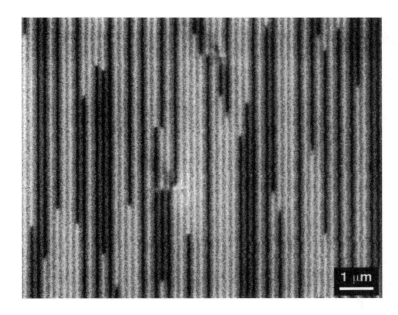

Fig. 24. Magnetic nanowires formed by evaporation of iron at grazing incidence onto lines made by laser-focused atomic deposition of chromium. The image is taken with scanning electron microscopy with polarization analysis (SEMPA), which is sensitive to the magnetization of the surface of the specimen. In this image, black indicates magnetization in one direction along the lines and white indicates the other. The gray regions between the lines indicate the lack of magnetization of the chromium substrate exposed between the iron wires. [Source: McClelland et al., *SPIE Proc.* 2995, 90 (1997).]

Another example is the fabrication of nanoscale magnetic materials. In a demonstration of this, Fe was evaporated at a grazing angle of $10°$ onto the sides of the Cr lines, resulting in an array of magnetic lines with width less than approximately 100 nm and length of about 150 μm [134]. Figure 24 shows an image of such an array of lines, taken with a scanning electron microscope with polarization analysis (SEMPA) [135]. In this image, magnetization in one direction along the lines shows up as white, and magnetization in the other direction shows up as black, with unmagnetized Cr showing up as gray between the lines. Clear black and white domains are seen in the magnetic nanolines, indicating that shape anisotropy has forced the magnetization to be oriented only along the long axes of the lines, in one direction or the other.

In addition to replicating from the Cr lines and depositing on them, it is also possible to use reactive-ion etching to modify their shape and/or transfer the pattern to the substrate. A recent demonstration of this has shown that a number of distinct forms can be made [136]. The various forms are created as a result of the interplay between slow sputtering of the Cr lines and rapid etching of the Si substrate. Initially, the Cr film is sputtered uniformly until the regions between the lines are cleared of Cr. Then reactive-ion etching takes over on the substrate, forming rapidly deepening trenches while the Cr lines are gradually thinned and narrowed by sputtering. After a fixed etching time, the result will be either narrow trenches in the silicon substrate, an array of well-separated Cr ribbons on silicon pedestals, or an array of narrow ($\sim$70 nm) Cr wires (see Fig. 25). Which form occurs depends on the initial contrast of the Cr deposition: Lower contrast leads to narrow trenches, and higher contrast leads to increasingly narrow Cr lines and deeper trenches.

4.1.3. Aluminum

Besides sodium and chromium, the only other element that has been successfully focused and deposited on the nanoscale using atom optics is aluminum [68]. As with the other two

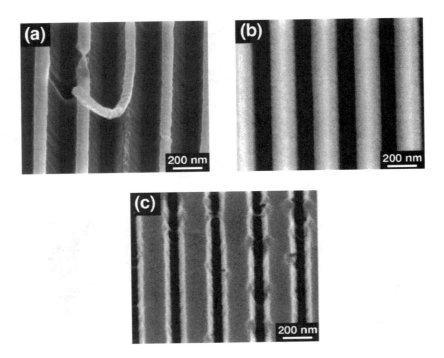

Fig. 25. Scanning electron microscope (SEM) images showing reactive-ion etching of chromium nanostructures formed by laser-focused atomic deposition. (a) Region of relatively high contrast in the original deposition, showing narrow (68 nm wide) chromium wires atop sharp silicon ridges; (b) region of intermediate original contrast, showing the formation of well-separated chromium ribbons; and (c) region of low original contrast, showing narrow trenches cut into the silicon. [Source: McClelland et al., *Appl. Phys. B* 66, 95 (1998)].

elements, the configuration employed was the standing-wave lens array, and a setup similar to that shown in Figure 18 was employed.

Although good reasons exist for making nanostructures from aluminum, such as its superconducting properties and its use as an interconnect material in integrated circuits, it also poses some special challenges for laser-focused atomic deposition. The optical transition in the aluminum atom most appropriate for optical manipulation is between the $3^2P_{3/2}(F = 4)$ ground-state level and the $3^2D_{5/2}(F = 5)$ level. This transition has a wavelength of 309.4 nm, a natural linewidth of 13 MHz, and a saturation intensity of 57 mW/cm^2 [137]. Because this transition is in the ultraviolet, the simplest sources of tunable laser light are not useful. Nevertheless, a dye laser can be frequency doubled to provide the necessary light for both collimating and focusing the atoms. In the work described in [68], a single-frequency ring dye laser operating at 618 nm was doubled in an astigmatically compensated external ring buildup cavity using a LiIO$_3$ nonlinear crystal. Up to 40 mW of tunable laser light at 309.4 nm was obtained, which was sufficient for both the transverse cooling and laser focusing of the atoms.

Unlike sodium and chromium, where essentially all the atoms are in (or can be put into) a single ground state that interacts with the laser, aluminum has its ground-state population spread out over a number of levels that are thermally populated. Not only are the other hyperfine levels of the $3^2P_{3/2}$ state statistically populated, but so also are the levels of the $3^2P_{1/2}$ state, which lies 14 meV below the $3^2P_{3/2}$ state. As a result, only 25% of the atoms in a thermally produced beam are accessible to a single laser frequency (in principle, of course, additional lasers could be used to pump optically more atoms into the resonant level if desired).

Despite these experimental complications, clear nanostructures of aluminum have been fabricated. For the experiments described in [68], the standing-wave laser beam had a

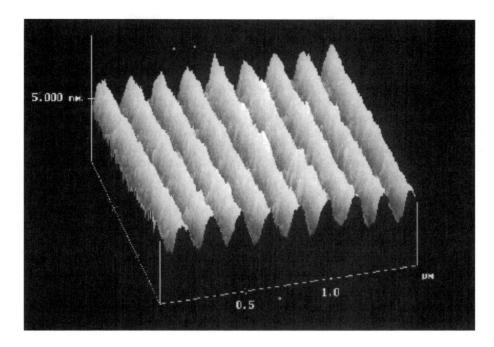

Fig. 26. Atomic force microscope (AFM) image of aluminum lines formed by laser-focused atomic deposition. These lines have a periodicity of 155 nm, a full width at half maximum (FWHM) of approximately 80 nm, and a peak-to-valley height of 3 nm. Reprinted with permission from R. W. McGowan et al., *Opt. Lett.* 20, 2535 (© 1995 Optical Society of America).

single-beam power of 16.7 mW, a waist radius of 0.11 mm, and was linearly polarized. The laser cooling was performed with 7.6 mW of single-beam laser power in a lin $\perp$ lin configuration for polarization–gradient cooling. The detuning was 210 MHz below resonance for the standing wave and 10 MHz below resonance for the laser cooling.

Like chromium, the aluminum nanostructures are passivated by exposure to air, and can be removed from vacuum for examination with an AFM. Figure 26 shows an AFM image of the aluminum lines, where the expected 155-nm periodicity is clearly visible. These lines have an average height of 3 nm, an FWHM of approximately 80 nm, and an average background level of about 30 nm.

4.1.4. Modeling of a Standing-Wave Lens

Along with the experimental investigations of focusing atoms in a standing-wave lens, several calculational tools for analyzing the lens have also been developed. These tools cover three levels of increasing complexity, and these can be put to use according to the degree of detailed information desired.

The most basic approach is to consider the motion of the atoms through the standing wave as governed by the paraxial approximation to the trajectory equation (see Section 3.1.10, Eq. (15), and [74]). From this equation, the first-order focusing properties of the lens can be obtained, from which gross estimates of the behavior can be derived. Defining z as the coordinate along the direction in which the atoms predominantly travel (i.e., perpendicular to the surface) and x as transverse to z, along the laser propagation direction, the paraxial equation becomes [74]:

$$\frac{d^2x}{dz^2} + q^2 g(z)x = 0 \qquad (18)$$

where

$$q^2 = k^2 \frac{\hbar \Delta}{2E_0} \frac{I_0}{I_s} \frac{\Gamma^2}{\Gamma^2 + 4\Delta^2} \tag{19}$$

in which $k = 2\pi/\lambda$ is the laser wavenumber, Δ is the detuning, E_0 is the kinetic energy of the atoms, I_0 is the peak intensity of the standing wave, I_s is the saturation intensity of the atom, and Γ is the atomic transition probability. The function $g(z)$ describes the z dependence of the laser intensity, that is, its profile as encountered by the atoms on the approach to the surface.

Although $g(z)$ could, in principle, take on a range of forms, in practice, it has almost exclusively been very close to a Gaussian profile with $1/e^2$ width σ, that is, $g(z) = \exp(-2z^2/\sigma^2)$. With this profile, the focal properties of the lens can be derived in either the thin-lens condition, when the atoms focus well beyond the standing wave, or the thick-lens condition, when the atoms focus within the standing wave. In the thin-lens case, the focal length is given by [74]:

$$f = \left(\frac{2}{\pi}\right)^{1/2} \frac{1}{\sigma q^2} \tag{20}$$

Considering Eq. (20) together with Eq. (19) for q^2, we see that the focal length is inversely proportional to the laser intensity and directly proportional to the kinetic energy of the atoms or, equivalently, the square of their velocity. These simple facts provide useful insight into the basic behavior of the lens.

The thick, immersion lens limit, where the atoms can focus within the lens, is the most appropriate for the experimental situation when the standing wave propagates as close as possible to the substrate surface. In this case, it is not sufficient to consider only the focal length of the lens; one must also take into account a principal plane whose position relative to the center of the Gaussian beam envelope varies with lens strength. Although an analytic solution of the paraxial equation in this limit, yielding the principal plane location and focal length, is not possible, it is, nevertheless, feasible to carry out a simple numerical integration with scalable parameters that gives the behavior of the lens. This has been done in [74], where more details can be found of the exact behavior.

To gain some insight into the general behavior of the lens in the thick, immersion regime, it is instructive to consider the constant-intensity limit, that is, the case when $g(z) = 1$. The trajectories are particularly simple in this case, as the solution to Eq. (18) is a linear combination of $\sin qz$ and $\cos qz$. From this solution, the principal plane location z_p can be derived to be

$$z_p = \frac{\pi - 2}{2q} \tag{21}$$

and the focal length takes on the simple form

$$f = q^{-1} \tag{22}$$

Interestingly, we note that the focal length in this regime is inversely proportional only to the first power of the velocity, instead of to the square of the velocity as it is in the thin-lens case.

Given the focal length of the lens, some basic properties of the focusing can be estimated. For instance, the diffraction limit to the deposited linewidth can be estimated using Eq. (11). Also, the linewidth limitation d resulting from an atom beam collimation angle θ can be estimated by the formula $d = f\theta$, which follows from the Gaussian lens law Eq. (9) in the limit of large object distance s_1 [74].

Although the paraxial equation is very useful for determining the general operating parameters of a standing-wave lens, as well as seeing the true lenslike behavior in terms of focal lengths and principal planes, more is needed if a realistic estimate of deposition

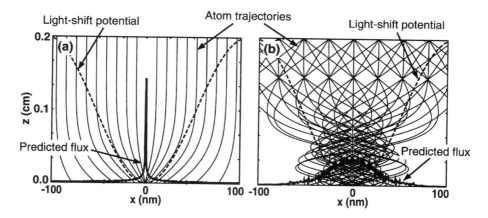

Fig. 27. Calculations of atomic trajectories during focusing induced by the light shift potential in a laser standing wave. Shown is a single period of the standing wave. The light is assumed to have a Gaussian profile with maximum at the surface ($z = 0$). (a) Calculation for monochromatic atoms with zero angular divergence. Note the predicted flux at the surface is extremely narrow (on the order of 1 nm). (b) Calculation for atoms with a thermal velocity spread and angular spread corresponding to typical collimation by laser cooling. Note the atoms are "channeled," rather than focused, yet still maintain a relatively well concentrated flux distribution at the surface. The laser intensity and detuning in (a) are chosen so that the lens is "in focus" in the paraxial approximation; for (b), the intensity is many times higher to achieve effective channeling.

linewidths is to be approached. This is especially true if there are significant velocity and angular spreads in the incident beam of atoms. The next level of analysis that has proven useful for analyzing the focusing of atoms in a standing wave is a classical trajectory approach that numerically solves the exact trajectory equations Eqs. (16) and (17) for a large number of trajectories of different starting conditions. Starting with an assumption (or measurement) of the velocity and angle distributions in the incident atom beam, the calculated trajectories can be used to derive a flux distribution at any point along the focusing by adding up the contributions from the various trajectories. The result is a calculation of an atom spatial distribution that provides the necessary information on the focusing of the lens.

Two examples of such a calculation are shown in Figure 27. In Figure 27a, the calculation is carried out assuming a monochromatic, parallel beam of Cr atoms, and the laser beam parameters are chosen for optimum focusing. The resulting flux distribution at the focal plane is surprisingly narrow, considering that the paraxial condition does not hold for many of the trajectories. The small width and pedestal of the central peak that are visible are, in fact, caused by trajectories that do not quite make it to the focus because they enter the lens too far from the axis. In the language of lens aberrations, these features can be attributed to spherical aberration. This calculation shows how important (or unimportant) deviations from the paraxial equation are in a given lensing situation and, hence, gives information on their contribution to the deposition linewidth.

Figure 27b shows a calculation for incident atom beam conditions that are closer to what is often found in an experiment—a thermal velocity spread and an angular spread as might come out of polarization–gradient transverse laser cooling. We see that, in this case, the trajectories do not all converge to a well-defined focus, but are, in fact, "channeled" into a small region (see Section 3.1.11). This channeling now becomes the object of attention in the calculation, because it determines what sort of linewidth can be expected. Clearly, the paraxial approximation trajectories and the lens focal properties derived from them are less relevant in this case. The type of ray-tracing calculations displayed in Figure 27, however, can be very helpful in understanding the behavior of realistic attempts to focus atoms in a standing wave, and can provide an invaluable guide for adjusting laser beam parameters.

Although classical ray-tracing calculations provide good insight into the behavior of a standing-wave lens and are relatively easy to implement, they leave out certain details that could prove important in some situations. Two potentially significant aspects of the laser–atom interaction that are neglected are (1) the wave nature of the atomic motion and (2) the true light shift potential in terms of the dressed atomic energy levels, in particular, the subtle effects that arise from spontaneous emission and the magnetic sublevels of the atom. To approximate the effects of the wave nature of the atomic motion, use can be made of the powerful analogies that exist between light optics and particle optics. These allow one simply to infer many diffraction phenomena, for example, the diffraction limit of a focal spot that is estimated with Eq. (11). However, it must be recognized that, in some cases, for example, in the channeling regime, optical diffraction analogies do not exist and a more complete theory must be employed. The same holds true when the true light shift potentials must be taken into account, for instance, when multiple light-shifted potentials play a role or spontaneous emission makes the potential no longer conservative.

In such situations, resort can be made to fully quantum calculations that treat all possible interactions exactly and quantize both atomic internal and external degrees of freedom. Although analogies between the atomic motion and common optical phenomena are lost with this approach, what is gained is the possibility of an exact prediction of the sort of atomic distribution that can be expected during deposition. Such calculations have been performed to analyze the quantized motion of atoms in a standing wave [28, 138], in laser cooling [36, 139], and in the deposition of a $\lambda/8$-period pattern in Cr as discussed in Section 4.1.2 [132].

4.2. Neutral-Atom Lithography

As discussed in the previous section, a number of demonstrations of nanofabrication with atom optics have been carried out using the direct deposition of atoms. An alternative approach to this is to employ neutral-atom lithography. In this process, a resist-coated surface is exposed to a patterned neutral-atom beam, and then etched with an agent that differentially penetrates the damaged resist and transfers the pattern to the surface. Thus, neutral atoms replace the conventional exposure agents—photons, electrons, or ions—in a process akin to conventional lithography. The advantages gained by using neutral atoms lie in the new opportunities afforded by atom optics, including the possibility of exposing large areas with nanometer-scaled resolution, the reduction of substrate damage, and the elimination of diffraction and space charge as resolution limitations.

Because neutral-atom beams typically have very small kinetic energies, being thermally produced, exposure of a resist requires going beyond the simple transfer of kinetic energy that is the main exposure mechanism with electron and ion beam lithography. Two separate processes have been identified as being useful for pattern transfer in this respect. One process involves chemically altering the surface, and the other involves transferring internal energy from the atom to the surface.

Exposing a resist by chemical means can be achieved by using a reactive species of neutral atom. The alkali atoms are ideal for this purpose because they are easily manipulated by lasers and they are also quite chemically active. Li, Na, K, Rb, and Cs have all been the subject of extensive studies of atom manipulation, and so are natural candidates for neutral atom lithography.

Internal energy transfer is readily done with metastable rare-gas atoms. These atoms, which are all susceptible to laser manipulation via optical transitions from the metastable state to a higher-lying state, carry a significant amount of energy, ranging from 20 eV for He* to 10 eV for Kr*. This energy cannot escape from the atom unless a perturbation occurs, such as when contact with a surface is made. Thus, the metastable atoms can travel through a vacuum system and only lose their energy when they strike the surface of the resist, causing damage, or exposure, in the process.

One special property of the metastable rare gases Ne*, Ar*, and Kr* that makes them particularly interesting from an atom optics point of view is their ability to be quenched by laser light—that is, to be radiatively transferred out of the metastable state to the ground state. This opens the possibility of patterning the atom beam by selectively quenching regions of the beam instead of modifying the atom trajectories. For example, a laser standing wave of sufficient intensity can be used to impose on an atomic beam a pattern of lines with widths as small as $\lambda/25$, as atoms are quenched everywhere except in narrow regions near the nodes [140].

Quenching in a metastable atom is achieved by tuning a laser from the metastable state to a higher-lying state that has a large transition probability to the ground state. This opens a pathway to the otherwise inaccessible ground state, quickly removing the energy content of the atom and eliminating its resist-damaging capability. Quenching transitions for the metastable species Ne*, Ar*, and Kr* generally lie in the visible to near-infrared region, so they can be readily accessed with available lasers.

Whether chemical exposure or energy deposition is used to transfer the pattern to the resist, care must be used in selection of a suitable resist material. Conventional resist materials, such as polymethylmethacrylate (PMMA), are generally not appropriate for neutral-atom lithography because they are too thick—the damage mechanisms involved with neutral atoms are restricted to only the very surface of the resist material, so insufficient exposure would occur. A class of chemicals that do work well as resists for atom lithography are those that form self-assembled monolayers (SAMs) on a surface. Examples of these include alkanethiolates, which self-assemble on gold surfaces, and siloxanes, which form on silicon substrates. With SAMs, the monomolecular layer that forms is tightly bound to the surface and generally protects it from etching. When damaged, however, the films allow etchants to penetrate, providing the mechanism for pattern transfer.

In addition to SAMs, another resist system has been shown to be effective. This involves using what is referred to as "contamination lithography." Here, an unprotected surface is placed in a vacuum system with some background pressure of hydrocarbons and/or silicone pump oil. During exposure to the atoms, the background material that has accumulated on the surface is polymerized and becomes hardened. This turns out to be an excellent etch mask, and some very interesting features have been fabricated using this process.

Demonstrations of neutral-atom lithography have until very recently been restricted to showing that the basic process works, without particular attention to combining it with atom optics. The first demonstration was done using Ar* atoms to expose an SAM of dodecanethiolate (DDT) on a gold film evaporated onto a silicon substrate [141]. In this work, Ar* was produced in a flowing-afterglow direct-current (dc) discharge source, collimated, and directed at the SAM-covered substrate through a mask made from a copper grid with 5-μm-wide lines. After exposure to a dose of about 10^{16} metastable atoms per square centimeter, the substrate was removed from the vacuum system and etched in a ferricyanide solution. On examination by eye, and in a scanning electron microscope (SEM), a clear shadow of the grid was seen etched into the gold (see Fig. 28). To verify that the exposure of the SAMs was caused only by the metastable atoms, and not by possible photons or fast neutrals in the beam, a control exposure was carried out with the Ar* atoms quenched by a laser tuned to the $4s^3P_2 \rightarrow 4p^1D_2$ transition at 764 nm. Significantly less damage was observed with the metastables quenched, indicating that the effect was real. Building on this initial demonstration, a number of additional experiments have yielded significant results. Under similar experimental conditions in two separate laboratories, He* has also been shown to expose DDT SAMs, and with somewhat higher efficiency [142, 143]. In addition, metastable atoms have been found to be very effective at producing contamination resist. This has been demonstrated using Ar* to expose Si, SiO_2, or gold through a SiN membrane mask. After wet etching, clear features as small as 50 nm were seen [144]. Further refinement of this process has been achieved by switching the exposing atom to Ne*,

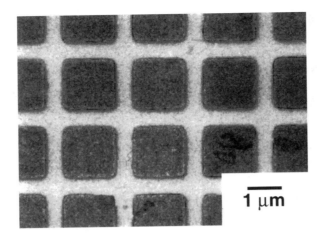

Fig. 28. Neutral-atom lithography with metastable argon atoms. A gold film evaporated onto a silicon substrate is coated with a dodecanetiolate self-assembled monolayer (SAM). A beam of metastable argon atoms, shadowed by a mesh, strikes the surface. On placement in a ferricyanide solution, the regions of the SAM damaged by metastable argon impact allow the underlying gold to be etched. Shown is a scanning electron microscope image of the grid shadow etched into the gold surface. [Source: Berggren et al., *Science* 269, 1255 (1995).]

the substrate to GaAs, and using chemically assisted ion beam etching [145]. The result, an example of which is shown in Figure 29, is a remarkably well formed pattern.

Atom lithography through chemical damage with alkali atoms has also seen demonstration via a number of routes. Alkanethiolate SAMs on gold have successfully been used to transfer a pattern formed in a Cs atom beam. Figure 30 shows the result of exposing an SAM of nonanethiolate to thermal Cs atoms passing though a grid, followed by etching in a ferricyanide solution [146]. Maximal contrast was achieved in this work with a dose of only 1.6–2.5×10^{15} Cs atoms/cm^2 (3–5 Cs atoms per SAM molecule). Similar results were found using a SiN membrane mask to create 50-nm structures [147], as well as using an octylsiloxane SAM on SiO$_2$ instead of the alkanethiolate SAM on a gold resist system [148].

Very recently, the work with Cs atoms has been extended to include the first atom-optical implementation of neutral-atom lithography [149]. To demonstrate an atom-optical process, Cs atoms were focused using an optical standing wave in an experimental setup similar to that used with direct deposition (see Section 4.1). Using a nonanethiolate SAM on gold and etching with ferricyanide after exposure, an array of nanoscale lines was formed in the gold. Figure 31 contains an AFM image of this pattern, which clearly shows the expected periodicity of half the Cs resonance wavelength, or 426 nm.

5. FUTURE PROSPECTS

As can be seen by the discussion in the previous sections, the field of nanofabrication via atom optics is just beginning to emerge as an area of study leading to new ways to fabricate structures on the nanoscale. A strong foundation has been laid by a wealth of research on fundamental atom optics, covering many different ways to manipulate free, neutral atoms. However, this research is only just now starting to be applied to the control of atomic motion during the approach toward a surface, with the goal of creating nanostructures. Based on the initial results, it seems that a number of possibilities have been opened; it remains now to explore further to find out what extensions and limitations will be forthcoming. Though it is difficult to predict the evolution of any field of research, a few areas in which it appears that progress can be made do present themselves.

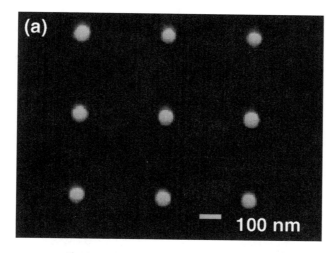

Fig. 29. Contamination lithography with metastable neon atoms. A beam of metastable neon atoms passes through a mask and strikes a surface. Surface contaminants, such as hydrocarbons and/or silicone diffusion pump oil, form a hard resist when impacted by a metastable atoms. On etching, the areas of the substrate not impacted by metastable atoms are removed. Shown are scanning electron microscope (SEM) images of (a) 50-nm pillars and (b) 500-nm square posts, formed in GaAs by exposure through a SiN mask, followed by chemically assisted ion beam etching. Reprinted with permission from S. J. Rehse et al., *Appl. Phys. Lett.* 71, 1427 (© 1997 American Institute of Physics).

5.1. Resolution Limits

One of the most fundamental issues that needs to be investigated is the ultimate practical resolution that can be expected with atom-optical techniques. In discussing resolution, the factors that play a role can be divided into two main categories—those that arise from the atom optics (including source properties) and those that arise from the atom interaction with the surface.

From the atom-optical perspective, resolution can be addressed by analyzing the aberrations and spot size limitations expected for a variety of atom lenses. These considerations show that atom-optical elements are, in fact, capable of extremely high resolution, but the attainment of this is hampered by the optical quality of atom sources. In all the lensing scenarios discussed in Section 3.1, chromatic aberration and source collimation (or effective source brightness) play a major role in determining the ultimate spot size. Yet atom sources are typically very far from the ideal in this respect, having broad, thermal velocity

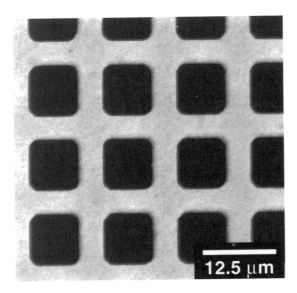

Fig. 30.   Neutral-atom lithography with cesium atoms. A beam of cesium atoms passes through a mesh and strikes a gold film covered with a nonanethiolate self-assembled monolayer. On etching in ferricyanide solution, the gold film is removed in the regions exposed to the cesium atoms. Shown is a scanning electron microscope (SEM) image of the etched gold film. Reprinted with permission from M. Kreis et al., *Appl. Phys. B* 63, 649 (© 1996 Springer).

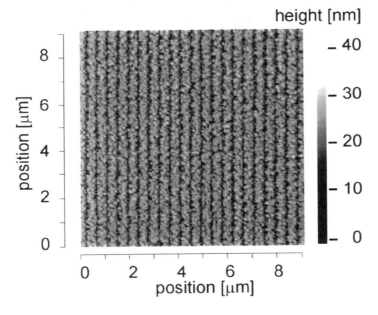

Fig. 31.   Nanoscale neutral-atom lithography with cesium atoms. In an experimental set up similar to that shown in Figure 18, cesium atoms are focused in a standing-wave lens array as they are incident on a nonanethiolate-coated gold film. On etching in ferricyanide solution, the nanoscale pattern of focused atoms is transferred to the gold film. Reprinted with permission from F. Lison et al., *Appl. Phys. B* 65, 419 (© 1997 Springer).

distributions and wide angular spreads. Improvements in atom sources, ranging all the way from the use of laser manipulation to brighten and monochromize a beam, to the production of a practical Bose–Einstein condensate, will have a significant impact on the attainable focal spot size.

Even if atom-optical methods could focus atoms into an infinitely narrow spot, however, the behavior of the atoms during and after surface impact could play a crucial role in determining the size of the resulting feature. This is true in either the direct deposition or the atom lithography case. In direct deposition, surface diffusion and grain growth could play significant roles in determining where the deposited atoms finally come to equilibrium. This could seriously impact, or even completely determine, the size of the structures deposited, regardless of the resolution of the atom-optical focusing. To complicate matters, these effects would be dependent on the particular atom–substrate system under investigation, as well as the contamination level. In atom lithography, where resists are used, the energy deposition or chemical process that exposes the resist will have a certain spatial range over which it has an effect. This spatial range is presently unknown and could conceivably be very small. On the other hand, it could be larger than the focal spot created by atom focusing, becoming the limitation on the resolution of the process.

5.2. More Complex Patterns

Another avenue of future investigation with potentially fruitful outcome concerns the extension of the basic atom-optical processes demonstrated to date to encompass the fabrication of more complex patterns. This could proceed on a number of levels. Utilizing the basic standing-wave concept, with which large arrays of lines and dots have been made, a fairly straightforward extension would be to make an array of dots and then scan the substrate during deposition. Scanning within the range of a unit cell of the array, for example, with a piezoelectrically actuated stage, the atom lenses could be used like an array of "atom pens," writing many identical complex patterns with a periodicity governed by the standing wave.

Going beyond this, the standing-wave field could be generalized into a more complex interference pattern. Using a number of laser beams incident from a range of angles, one could conceive of creating a very complex pattern. The challenge here would be to start with a given, desired pattern and determine how many laser beams incident from which angles and with what relative phase will be required to generate this pattern. This would have to be done working within the laws of diffraction, as well as with the limitation that the laser wavelength is the same for all beams because of the atomic resonance requirement. Although it appears that a broad range of patterns could be generated in this way, it also seems that limitations will be encountered. It remains for future research to find out just what can be accomplished.

Another approach is to work on development of an atom-optical projection system. Here, an atom beam patterned by a mask is imaged by an atom-optical system with some demagnification, creating an arbitrary pattern on a surface with potentially nanoscale feature size. With this approach, advantage is taken of the true imaging characteristics of atom-optical lenses, which, in principle, can be very good because the diffraction limit can be very small. Along with these good imaging characteristics, however, come the technical challenges of atomic source optical quality, lens aberrations, field of view, off-axis aberrations, and so on, which become much more critical in a true imaging situation.

5.3. Other Atoms

So far, nanofabrication with atom optics has been demonstrated with few atomic species and materials. Direct deposition has been implemented with sodium, chromium, and aluminum, and atom lithography has been shown to be effective with metastable rare gases and alkalis on self-assembled monolayer resists. To realize the potential of this technique fully, it is worth considering the issues that govern what other atoms could be used.

The first consideration that must be taken into account in evaluating the feasibility of manipulating a particular atom is the possibility of making a monoatomic beam of sufficient flux. Any element will evaporate if heated to a high enough temperature, but making atomic beams can be tricky, sometimes involving complex high-temperature materials issues. Furthermore, some atom beams tend to have a large fraction of dimers or other molecular forms, which can add a background level if not separated out somehow (e.g., by laser deflection of the desired atoms). There is, however, a fairly well established technology of making a wide range of beams, so, in many cases, methods have already been established [150].

The next consideration depends on what sort of atom optics will be implemented. For purely magnetic focusing, any atom with a ground-state magnetic dipole moment is, in principle, manipulable. This encompasses quite a few atoms, because any atom with an unfilled shell will, in general, have some angular momentum and, hence, a magnetic dipole moment. If, however, the additional capability of laser-based atom optics is desirable, the range of possible atoms is restricted to those with appropriate optical transitions. In fact, many atoms do have the right sort of transition; the issue becomes whether a sufficiently powerful laser source exists that can be tuned near enough to the resonance to have a strong enough effect. Many atoms have their strongest transitions in the deep ultraviolet, making this requirement technologically more challenging. A further complication arises if an atom has a number of naturally occurring isotopes. These will typically have shifted resonance frequencies resulting from the hyperfine structure and isotope shifts, causing them to interact differently with the laser light and possibly resulting in unwanted background effects.

If the laser-based atom optics to be used relies on creating an excited-state population that persists for more than a short time, such as if laser cooling is to be implemented, an even further restriction is imposed. Now there must be no population sinks—that is, intermediate metastable levels into which an atomic population can decay from the excited state, remaining there without interacting with the laser. This last requirement, being quite restrictive, is the major reason why the study of atom manipulation has so far been restricted to only a handful of atoms. It should be noted, however, that, in many cases where population sinks exist, additional laser light can be introduced to repump the atoms back into the ground state, allowing them to interact with the manipulating laser again.

Given these requirements, it appears that the generalization of laser-based atom optics to other atoms is not trivial. Nevertheless, there are many atoms in the periodic table, and each of them has a very large number of optical transitions. The feasibility of manipulating any particular atom cannot be dismissed without carefully considering all the possible optical transitions to see if some combination of laser frequencies and transitions, perhaps involving a metastable state, could be used. An invaluable resource for such a study is a compilation of atomic energy levels and transition probabilities, such as is found in [151].

5.4. Applications

Because nanofabrication with atom optics is in the very early stages of development, it is perhaps premature to discuss specific applications for which it will be useful. Nevertheless, it is already possible to see generic directions in which the process has natural capabilities, as well as a few direct applications for the technology as it exist today.

One of the clear advantages that laser-focused atomic deposition has over most other nanofabrication techniques is the inherently resist-free nature of the process. By directly depositing atoms on a surface, it is possible to avoid many of the disadvantages of using a resist, such as the extra steps involved in applying and stripping the resist and the possibility of contamination associated with these steps. Thus, in contamination-sensitive nanofabrication processes, it could be very useful to develop atom-optical alternatives to current methods.

Another general area in which atom optics has an inherent advantage is the fabrication of any structure that requires a large array of identical substructures. This type of application can take advantage of the standing-wave lens array and its natural periodicity to pattern conveniently a large area in parallel. A good example of such an application is the fabrication of photonic materials. Here, arrays are often required that not only need to be large, but also need to have a very high degree of long-range spatial coherence. A standing-wave lens array is perfectly suited to this because it has a very long coherence length (up to 30 km) resulting from the narrow-frequency width of the laser required for atom-optical focusing. Another example is the construction of a sensor array. Recent studies have found that nanostructures, just by the nature of their size and shape, can greatly enhance certain signals (an example of this is surface-enhanced Raman scattering [152]). Covering a large area with an array of identical nanostructures is, thus, a good way to make a very efficient sensor.

As a final example, we mention the possibility of using an array of dots or lines made by laser focusing in a standing wave as a nanoscale length standard. Because the standing-wave laser beam must be tuned to an atomic resonance for the pattern to be formed, the wavelength of the laser is determined with spectroscopic accuracy—that is, to better than a part per million, in the case of chromium. Thus, the standing-wave periodicity, which is simply half the wavelength with some small corrections for wavefront curvature, is known with essentially the same accuracy. In the basic laser-focused deposition geometry (see Section 4.1), the standing wave propagates parallel to the surface of the substrate, so its periodicity transfers directly to the deposited pattern, with some small corrections for angular deviations. The result is a deposited pattern that has an extremely accurate periodicity, or pitch, the major uncertainty of which is governed by the stability (thermal or other) of the substrate on which it is deposited.

These few examples concern applications of nanofabrication via atom optics assuming essentially a status quo of development, that is, assuming that only relatively incremental improvements will be added to the techniques already demonstrated. Taking a broader view, it is clear that there are many aspects of atomic manipulation that have not yet been brought to bear on nanofabrication. As research progresses, it is very likely that new, innovative ways will evolve applying some of the more recent atom manipulation techniques to the control of atoms as they impact a surface. With these new developments, it is quite possible that the ultimate capabilities of atom optics will be realized, focusing atoms with atomic resolution for unprecedented nanostructure fabrication.

References

1. See, for example, F. B. Dunning and R. G. Hulet, eds., "Atomic, Molecular and Optical Physics: Atoms and Molecules, Experimental Methods in the Physics Sciences," Vol. 29B. Academic Press, San Diego, 1996.
2. N. Ramsey, "Molecular Beams." Oxford University Press, London, 1956.
3. L. I. Maissel and R. Glang, eds., "Handbook of Thin Film Technology." McGraw–Hill, New York, 1970.
4. The presence of a v^3 factor in Eq. (1) is a result of this being a *flux* distribution rather than a simple probability distribution. If the number density of atoms within a given velocity range is desired, the factor should be v^2. Further discussion of this is found in [2], and also in F. Reif, "Fundamentals of Statistical and Thermal Physics," pp. 273–277. McGraw–Hill, New York, 1965. An interesting alternative treatment is found in C. C. Leiby, Jr., and A. L. Besse, *Am. J. Phys.* 47, 791 (1979).
5. L. I. Maissel and R. Glang, eds., "Handbook of Thin Film Technology," Chap. 3. McGraw–Hill, New York, 1970.
6. G. K. Hubler, ed., "Pulsed Laser Deposition," *Mater. Res. Soc. Bull.* 17, 26 (1992).
7. M. D. Morse, in "Atomic, Molecular and Optical Physics: Atoms and Molecules, Experimental Methods in the Physics Sciences" (F. B. Dunning and R. G. Hulet, eds.), Vol. 29B, p. 21. Academic Press, San Diego, 1996.
8. T. G. Dietz, M. A. Duncan, D. E. Powers, and R. E. Smalley, *J. Chem. Phys.* 74, 6511 (1981).
9. H. U. Hostettler and R. B. Bernstein, *Rev. Sci. Instrum.* 31, 872 (1960).
10. W. M. Feist, in "Electron Beam and Laser Beam Technology" (L. Marton and A. B. El-Kareh, eds.), p. 1. Academic Press, New York, 1968.

11. See, for example, E. E. Anderson, "Modern Physics and Quantum Mechanics," p. 317. W. B. Saunders, Philadelphia, 1971.
12. J. Prodan, A. Migdall, W. D. Phillips, L. So, H. Metcalf, and J. Dalibard, *Phys. Rev. Lett.* 54, 992 (1985).
13. J. P. Gordon and A. Ashkin, *Phys. Rev. A* 21, 1606 (1980).
14. P. Meystre and S. Stenholm, eds., "The Mechanical Effects of Light," *J. Opt. Soc. Am. B* 2, 1706 (1985).
15. S. Chu and C. Wieman, eds., "Laser Cooling and Trapping of Atoms," *J. Opt. Soc. Am. B* 6, 2019 (1989).
16. J. Mlynek, V. Balykin, and P. Meystre, eds., "Optics and Interferometry with Atoms," *Appl. Phys. B* 54, 319 (1992).
17. C. S. Adams, M. Sigel, and J. Mlynek, *Phys. Rep.* 240, 143 (1994).
18. A. Aspect, R. Kaiser, N. Vansteenkiste, and C. I. Westbrook, *Phys. Scr.* T58, 69 (1995).
19. Y.-Z. Wang and L. Liu, *Aust. J. Phys.* 48, 267 (1995).
20. C. C. Bradley and R. G. Hulet, in "Atomic, Molecular and Optical Physics: Atoms and Molecules, Experimental Methods in the Physics Sciences" (F. B. Dunning and R. G. Hulet, eds.), Vol. 29B, p. 129. Academic Press, San Diego, 1996.
21. C. S. Adams and E. Riis, *Prog. Quantum Electron.* 21, 1 (1997).
22. R. Loudon, "The Quantum Theory of Light," pp. 22–37. Oxford University Press, Oxford, 1973.
23. W. D. Phillips and H. Metcalf, *Phys. Rev. Lett.* 48, 596 (1982).
24. V. I. Balykin, V. S. Letokhov, and V. I. Mushin, *Pis'ma Zh. Eksp. Teor. Fiz.* 29, 614 (1979); W. Ertmer, R. Blatt, J. L. Hall, and M. Zhu, *Phys. Rev. Lett.* 54, 996 (1985).
25. M. Zhu, C. W. Oates, and J. L. Hall, *Phys. Rev. Lett.* 67, 46 (1991).
26. J. Dalibard and C. Cohen-Tannoudji, *J. Opt. Soc. Am. B* 2, 1706 (1985).
27. This phenomenon has been put to use to demonstrate an "optical Stern–Gerlach" effect. See T. Sleator, T. Pfau, V. Balykin, O. Carnal, and J. Mlynek, *Phys. Rev. Lett.* 68, 1996 (1992).
28. R. Dum, P. Zoller, and H. Ritsch, *Phys. Rev. A* 45, 4879 (1992).
29. J. Vanier and C. Audoin, "Quantum Physics of Atomic Frequency Standards." Hilger, Bristol, 1989.
30. M. H. Anderson, J. R. Ensher, M. R. Matthews, C. E. Wieman, and E. A. Cornell, *Science* 269, 198 (1995).
31. T. Hänsch and A. Schawlow, *Opt. Commun.* 13, 68 (1975); D. J. Wineland and W. M. Itano, *Phys. Rev. A* 20, 1521 (1979).
32. S. Chu, L. Hollberg, J. E. Bjorkholm, A. Cable, and A. Ashkin, *Phys. Rev. Lett.* 55, 48 (1985).
33. P. D. Lett, W. D. Phillips, S. L. Rolston, C. E. Tanner, R. N. Watts, and C. I. Westbrook, *J. Opt. Soc. Am. B* 6, 2084 (1989).
34. P. D. Lett, R. N. Watts, C. I. Westbrook, W. D. Phillips, P. L. Gould, and H. J. Metcalf, *Phys. Rev. Lett.* 61, 169 (1988).
35. J. Dalibard and C. Cohen-Tannoudji, *J. Opt. Soc. Am. B* 6, 2023 (1989).
36. P. Marte, R. Dum, R. Taïeb, P. D. Lett, and P. Zoller, *Phys. Rev. Lett.* 71, 1335 (1993).
37. A. Aspect, E. Arimondo, R. Kaiser, N. Vansteenkiste, and C. Cohen-Tannoudji, *Phys. Rev. Lett.* 61, 826 (1988).
38. A. Aspect, E. Arimondo, R. Kaiser, N. Vansteenkiste, and C. Cohen-Tannoudji, *J. Opt. Soc. Am. B* 6, 2112 (1989).
39. M. Kasevitch and S. Chu, *Phys. Rev. Lett.* 69, 1741 (1992).
40. A. L. Migdall, J. V. Prodan, W. D. Phillips, T. H. Bergeman, and H. J. Metcalf, *Phys. Rev. Lett.* 54, 2596 (1985).
41. H. F. Hess, G. P. Kochanski, J. M. Doyle, N. Masuhara, D. Kleppner, and T. J. Greytak, Phys. Rev. Lett. 59, 672 (1987).
42. J. J. Tollett, C. C. Bradley, C. A. Sackett, and R. G. Hulet, *Phys. Rev. A* 51, R22 (1995).
43. E. L. Raab, M. Prentiss, A. Cable, S. Chu, and D. E. Pritchard, *Phys. Rev. Lett.* 59, 2631 (1987).
44. A. M. Steane, M. Chowdhury, and C. J. Foot, *J. Opt. Soc. Am. B* 9, 2142 (1992).
45. S. Chu, J. Bjorkholm, A. Ashkin, and A. Cable, *Phys. Rev. Lett.* 57, 314 (1986).
46. J. D. Miller, R. A. Cline, and D. J. Heinzen, *Phys. Rev. A* 47, R4567 (1993).
47. M. J. Renn, D. Montgomery, O. Vdovin, D. Z. Anderson, C. E. Wieman, and E. A. Cornell, *Phys. Rev. Lett.* 75, 3253 (1995).
48. H. Ito, K. Sakaki, W. Jhe, and M. Ohtsu, *Opt. Commun.* 141, 43 (1997).
49. S. N. Bose, *Z. Phys.* 26, 178 (1924); A. Einstein, *Sitz. Kgl. Preuss. Acad. Wiss.* XVIII–XXV, 3 (1925).
50. C. C. Bradley, C. A. Sackett, J. J. Tollet, and R. G. Hulet, *Phys. Rev. Lett.* 75, 1687 (1995). ERRATUM: *Phys. Rev. Lett.* 79, 1170 (1997).
51. K. B. Davis, M.-O. Mewes, M. R. Andrews, N. J. van Druten, D. S. Durfee, D. M. Kurn, and W. Ketterle, *Phys. Rev. Lett.* 75, 3969 (1995).
52. M.-O. Mewes, M. R. Andrews, D. M. Kurn, D. S. Durfee, C. G. Townsend, and W. Ketterle, *Phys. Rev. Lett.* 78, 582 (1997).
53. V. I. Balykin and V. S. Letokhov, *Physics Today* 42, 23 (1989).
54. See, for example, E. Hecht, "Optics," 2nd ed., Chap. 5. Addison–Wesley, Reading, MA, 1987.
55. Electrostatic forces are typically too weak to form an atom lens because atoms do not have a permanent electric dipole moment. An electrostatic lens has been demonstrated for ammonia molecules, which do have a dipole moment. See J. P. Gordon, *Phys. Rev.* 99, 1253 (1955).

56. H. Friedburg, Z. Phys. 130, 493 (1951).
57. W. G. Kaenders, F. Lison, A. Richter, R. Wynands, and D. Meschede, Nature 375, 214 (1995).
58. J. E. Bjorkholm, R. R. Freeman, A. Ashkin, and D. B. Pearson, Phys. Rev. Lett. 41, 1361 (1978).
59. J. E. Bjorkholm, R. R. Freeman, A. Ashkin, and D. B. Pearson, Opt. Lett. 5, 111 (1980).
60. V. I. Balykin and V. S. Letokhov, Opt. Commun. 64, 151 (1987).
61. G. M. Gallatin and P. L. Gould, J. Opt. Soc. Am. B 8, 502 (1991).
62. J. J. McClelland and M. R. Scheinfein, J. Opt. Soc. Am. B 8, 1974 (1991).
63. W. Glaser, Z. Phys. 117, 285 (1941).
64. V. I. Balykin, V. S. Letokhov, Yu. B. Ovchinnikov, and A. I. Sidorov, J. Mod. Opt. 35, 17 (1988).
65. T. Sleator, T. Pfau, V. Balykin, and J. Mlynek, Appl. Phys. B 54, 375 (1992).
66. G. Timp, R. E. Behringer, D. M. Tennant, J. E. Cunningham, M. Prentiss, and K. K. Berggren, Phys. Rev. Lett. 69, 1636 (1992).
67. J. J. McClelland, R. E. Scholten, E. C. Palm, and R. J. Celotta, Science 262, 877 (1993).
68. R. W. McGowan, D. M. Giltner, and S. A. Lee, Opt. Lett. 20, 2535 (1995).
69. U. Drodofsky, J. Stuhler, B. Brezger, T. Schulze, M. Drewsen, T. Pfau, and J. Mlynek, Microelectron. Eng. 35, 285 (1997).
70. V. I. Balykin, V. V. Klimov, and V. S. Letokhov, J. Phys. II 4, 1981 (1994).
71. Qiming Li, K. G. H. Baldwin, H.-A. Bachor, and D. E. McClelland, J. Opt. Soc. Am. B 13, 257 (1996).
72. O. Carnal, M. Sigel, T. Sleator, H. Takuma, and J. Mlynek, Phys. Rev. Lett. 67, 3231 (1991).
73. K. K. Berggren, M. Prentiss, G. L. Timp, and R. E. Behringer, J. Opt. Soc. Am. B 11, 1166 (1994).
74. J. J. McClelland, J. Opt. Soc. Am. B 12, 1761 (1995).
75. M. Drewsen, R. J. C. Spreeuw, and J. Mlynek, Opt. Commun. 125, 77 (1996).
76. For extremely slow atoms, whose De Broglie wavelength is much larger than the distance over which the surface interaction potential changes, the sticking probability will go to zero because of quantum reflection. See, for example, J. Boheim, W. Brenig, and J. Stutzki, Z. Phys. B 48, 43 (1982).
77. T. Roach, H. Abele, M. G. Boshier, H. L. Grossman, K. P. Zetie, and E. A. Hinds, Phys. Rev. Lett. 75, 629 (1995).
78. A. I. Sidorov, R. J. McLean, W. J. Rowlands, D. C. Lau, J. E. Murphy, M. Walkiewicz, G. I. Opat, and P. Hannaford, Quantum Semiclassical Opt. 8, 713 (1996).
79. R. J. Cook and R. K. Hill, Opt. Commun. 43, 258 (1982).
80. M. A. Kasevich, D. S. Weiss, and S. Chu, Opt. Lett. 15, 607 (1990).
81. C. G. Aminoff, A. M. Steane, P. Bouyer, P. Desbiolles, J. Dalibard, and C. Cohen-Tannoudji, Phys. Rev. Lett. 71, 3083 (1993).
82. A. Landragin, G. Labeyrie, C. Henkel, R. Kaiser, N. Vansteenkiste, C. I. Westbrook, and A. Aspect, Opt. Lett. 21, 1591 (1996).
83. V. I. Balykin, V. S. Letokhov, Yu. B. Ovchinnikov, and A. I. Sidorov, Phys. Rev. Lett. 60, 2137 (1988).
84. W. Seifert, C. S. Adams, V. I. Balykin, C. Heine, Yu. Ovchinnikov, and J. Mlynek, Phys. Rev. A 49, 3814 (1994).
85. D. R. Frankl, Prog. Surf. Sci. 13, 285 (1983).
86. J. H. Muller, D. Bettermann, V. Rieger, F. Ruschewitz, K. Sengstock, U. Sterr, M. Christ, M. Schiffer A. Scholtz, and W. Ertmer, AIP Conf. Proc. 323, 240 (1996).
87. See, for example, P. Berman, ed., "Atom Interferometry." Academic Press, San Diego, 1997.
88. D. W. Keith, M. L. Schattenberg, H. I. Smith, and D. E. Pritchard, Phys. Rev. Lett. 61, 1580 (1988).
89. O. Carnal, A. Faulstich, and J. Mlynek, Appl. Phys. B 53, 88 (1991).
90. J. F. Clauser and Shifang Li, Phys. Rev. A 49, R2213 (1994).
91. M. S. Chapman, C. R. Ekstrom, T. D. Hammond, J. Schmiedmayer, B. E. Tannian, S. Wehinge, and D. E. Pritchard, Phys. Rev. A 51, R14 (1995).
92. J. Fujita, M. Morinaga, T. Kishimoto, M. Yasuda, S. Matsui, and F. Shimizu, Nature 380, 691 (1996).
93. M. Morinaga, M. Yasuda, T. Kishimoto, F. Shimizu, J. Fujita, and S. Matsui, Phys. Rev. Lett. 77, 802 (1996).
94. P. L. Gould, G. A. Ruff, and D. E. Pritchard, Phys. Rev. Lett. 56, 827 (1986).
95. P. J. Martin, B. G. Oldaker, A. H. Miklich, and D. E. Pritchard, Phys. Rev. Lett. 60, 515 (1988).
96. T. Pfau, Ch. Kurtsiefer, C. S. Adams, M. Sigel, and J. Mlynek, Phys. Rev. Lett. 71, 3427 (1993).
97. K. S. Johnson, A. Chu, T. W. Lynn, K. K. Berggren, M. S. Shahriar, and M. Prentiss, Opt. Commun. 20, 1310 (1995).
98. E. M. Rasel, M. K. Oberthaler, H. Batelaan, J. Schmiedmayer, and A. Zeilinger, Phys. Rev. Lett. 75, 2633 (1995).
99. D. M. Giltner, R. W. McGowan, and S. A. Lee, Phys. Rev. A 52, 3966 (1995).
100. J. V. Hajnal and G. I. Opat, Opt. Commun. 71, 119 (1989).
101. M. Christ, A. Scholz, M. Schiffer, R. Deutschmann, and W. Ertmer, Opt. Commun. 107, 211 (1994).
102. R. Brouri, R. Asimov, M. Gorlicki, S. Feron, J. Reinhardt, V. Lorent, and H. Haberland, Opt. Commun. 124, 448 (1996).
103. For an electron-optical discussion, see O. Klemperer, "Electron Optics," 2nd ed., pp. 12–15. Cambridge University Press, London, 1953; for a light-optical discussion, see M. Born and E. Wolf, "Principles of Optics," 3rd ed., pp. 166–169. Pergamon Press, Oxford, UK, 1965.

104. B. Sheehy, S.-Q. Shang, R. Watts, S. Hatamian, and H. Metcalf, *J. Opt. Soc. Am. B* 6, 2165 (1989).

105. V. I. Balykin, V. S. Letokhov, and A. I. Sidorov, *Pis'ma Zh. Eksp. Teor. Fiz.* 40, 251 (1984).

106. C. E. Tanner, B. P. Masterson, and C. E. Wieman, *Opt. Lett.* 13, 357 (1988).

107. V. I. Balykin, V. S. Letokhov, Yu. B. Ovchinnikov, and S. V. Shul'ga, *Opt. Commun.* 77, 152 (1990).

108. W. Rooijakkers, W. Hogervorst, and W. Vassen, *Opt. Commun.* 123, 321 (1996).

109. A. Aspect, J. Dalibard, A. Heidmann, C. Salomon, and C. Cohen-Tannoudji, *Phys. Rev. Lett.* 57, 1688 (1986).

110. Y. Wang, Y. Cheng, and W. Cai, *Opt. Commun.* 70, 462 (1989).

111. R. E. Scholten, R. Gupta, J. J. McClelland, R. J. Celotta, M. S. Levenson, and M. G. Vangel, *Phys. Rev. A* 55, 1331 (1997).

112. E. Riis, D. S. Weiss, K. A. Moler, and S. Chu, *Phys. Rev. Lett.* 64, 1658 (1990).

113. J. Yu, J. Djemaa, P. Nosbaum, and P. Pillet, *Opt. Commun.* 112, 136 (1994).

114. M. D. Hoogerland, J. P. J. Driessen, E. J. D. Vredenbregt, H. J. L. Megens, M. P. Schuwer, H. C. W. Beijer-inck, and K. A. H. van Leeuwen, *Appl. Phys. B* 62, 323 (1996).

115. S. V. Andreev, V. I. Balykin, V. S. Letokhov, and V. G. Minogin, *Pis'ma Zh. Eksp. Teor. Fiz.* 34, 463 (1981).

116. W. D. Phillips, J. V. Prodan, and H. J. Metcalf, *J. Opt. Soc. Am. B* 2, 1751 (1985).

117. J. Nellsen, J. H. Müller, K. Sengstock, and W. Ertmer, *J. Opt. Soc. Am. B* 6, 2149 (1989).

118. A. Faulstich, A. Schnetz, M. Sigel, T. Sleator, O. Carnal, V. Balykin, H. Takuma, and J. Mlynek, *Europhys. Lett.* 17, 393 (1992).

119. J. J. McClelland and M. H. Kelley, *Phys. Rev. A* 31, 3704 (1985).

120. M. Prentiss, G. Timp, N. Bigelow, R. E. Behringer, and J. E. Cunningham, *Appl. Phys. Lett.* 60, 1027 (1992).

121. V. Natarajan, R. E. Behringer, D. M. Tennant, and G. Timp, *J. Vac. Sci. Technol., B* 13, 2823 (1995).

122. R. E. Behringer, V. Natarajan, and G. Timp, *Appl. Phys. Lett.* 68, 1034 (1996).

123. R. E. Behringer, V. Natarajan, and G. Timp, *Appl. Surf. Sci.* 104/105, 291 (1996).

124. V. Natarajan, R. E. Behringer, and G. Timp, *Phys. Rev. A* 53, 4381 (1996).

125. R. E. Behringer, V. Natarajan, G. Timp, and D. M. Tennant, *J. Vac. Sci. Technol., B* 14, 4072 (1996).

126. J. J. McClelland, R. Gupta, Z. J. Jabbour, and R. J. Celotta, *Aust. J. Phys.* 49, 555 (1996).

127. R. E. Scholten, J. J. McClelland, E. C. Palm, A. Gavrin, and R. J. Celotta, *J. Vac. Sci. Technol., B* 12, 1847 (1994).

128. R. J. Celotta, R. Gupta, R. E. Scholten, and J. J. McClelland, *J. Appl. Phys.* 79, 6079 (1996).

129. U. Drodofsky, J. Stuhler, B. Brezger, Th. Schulze, M. Drewsen, T. Pfau, and J. Mlynek, *Microelectron. Eng.* 35, 285 (1997).

130. R. Gupta, J. J. McClelland, Z. J. Jabbour, and R. J. Celotta, *Appl. Phys. Lett.* 67, 1378 (1995).

131. U. Drodofsky, J. Stuhler, Th. Schulze, M. Drewsen, B. Brezger, T. Pfau, and J. Mlynek, *Appl. Phys. B* 65, 755 (1997).

132. R. Gupta, J. J. McClelland, P. Marte, and R. J. Celotta, *Phys. Rev. Lett.* 76, 4689 (1996).

133. Y. Xia, J. J. McClelland, R. Gupta, D. Qin, X.-M. Zhao, L. L. Sohn, R. J. Celotta, and G. M. Whitesides, *Adv. Mater.* 9, 147 (1997).

134. D. A. Tulchinsky, M. H. Kelley, J. J. McClelland, R. Gupta, and R. J. Celotta, *J. Vac. Sci. Technol., A* 16, 1817 (1998).

135. M. R. Scheinfein, J. Unguris, M. H. Kelley, D. T. Pierce, and R. J. Celotta, *Rev. Sci. Instrum.* 60, 1 (1989).

136. J. J. McClelland, R. Gupta, R. J. Celotta, and G. A. Porkolab, *Appl. Phys. B* 66, 95 (1998).

137. The value of 57 mW/cm^2 is calculated assuming a two-level atom. In McGowan et al., *Opt. Lett.* 20, 2535 (1995), the value quoted is 140 mW/cm^2, which is an effective saturation intensity taking into account linearly polarized excitation of several magnetic sublevels having Clebsch–Gordan coefficients less than unity.

138. J. Chen, J. G. Story, and R. G. Hulet, *Phys. Rev. A* 47, 2128 (1993).

139. T. Bergeman, *Phys. Rev. A* 48, R3425 (1993).

140. A. P. Chu, K. K. Berggren, K. S. Johnson, and M. G. Prentiss, *Quantum Semiclassical Opt.* 8, 521 (1996).

141. K. K. Berggren, A. Bard, J. L. Wilbur, J. D. Gillaspy, A. G. Helg, J. J. McClelland, S. L. Rolston, W. D. Phillips, M. Prentiss, and G. M. Whitesides, *Science* 269, 1255 (1995).

142. S. Nowak, T. Pfau, and J. Mlynek, *Appl. Phys. B* 63, 203 (1996).

143. A. Bard, K. K. Berggren, J. L. Wilbur, J. D. Gillaspy, S. L. Rolston, J. J. McClelland, W. D. Phillips, M. Prentiss, and G. M. Whitesides, *J. Vac. Sci. Technol., B* 15, 1805 (1997).

144. K. S. Johnson, K. K. Berggren, A. Black, C. T. Black, A. P. Chu, N. H. Dekker, D. C. Ralph, J. H. Thywissen, R. Younkin, M. Tinkham, M. Prentiss, and G. M. Whitesides, *Appl. Phys. Lett.* 69, 2773 (1996).

145. S. J. Rehse, A. D. Glueck, S. A. Lee, A. B. Goulakov, C. S. Menoni, D. C. Ralph, K. S. Johnson, and M. Prentiss, *Appl. Phys. Lett.* 71, 1427 (1997).

146. M. Kreis, F. Lison, D. Haubrich, D. Meschede, S. Nowak, T. Pfau, and J. Mlynek, *Appl. Phys. B* 63, 649 (1996).

147. K. K. Berggren, R. Younkin, E. Cheung, M. Prentiss, A. J. Black, G. M. Whitesides, D. C. Ralph, C. T. Black, and M. Tinkham, *Adv. Mater.* 9, 52 (1997).

148. R. Younkin, K. K. Berggren, K. S. Johnson, M. Prentiss, D. C. Ralph, and G. M. Whitesides, *Appl. Phys. Lett.* 71, 1261 (1997).

149. F. Lison, H.-J. Adams, D. Haubrich, M. Kreis, S. Nowak, and D. Meschede, *Appl. Phys. B* 65, 419 (1997).

150. K. J. Ross and B. Sonntag, *Rev. Sci. Instrum.* 66, 4409 (1995).

151. W. L. Weise, M. W. Smith, and B. M. Glennon, *NSRDS–NBS* 4 (1966); W. L. Weise, M. W. Smith, and B. M. Miles, *NSRDS–NBS* 22 (1969); G. A. Martin, J. R. Fuhr, and W. L. Weise, *J. Phys. Chem. Ref. Data* 17, Suppl. 3 (1988); J. R. Fuhr, G. A. Martin, and W. L. Weise, *J. Phys. Chem. Ref. Data* 17, Suppl. 4 (1988); see also the Web site http://aeldata.phy.nist.gov/nist_atomic_spectra.html.

152. A. Otto, I. Mrozek, H. Grabhorn, and W. Akemann, *J. Phys.: Conden. Matter.* 4, 1143 (1992).

Chapter 8

NANOCOMPOSITES PREPARED BY SOL–GEL METHODS: SYNTHESIS AND CHARACTERIZATION

Krzysztof C. Kwiatkowski, Charles M. Lukehart

Department of Chemistry, Vanderbilt University, Nashville, Tennessee, USA

Contents

1. INTRODUCTION

In typical sol–gel syntheses, metal or main-group element compounds undergo hydrolysis and condensation reactions giving gel materials with extended three-dimensional structures [1–10]. As shown in the following equation for silicon, the addition of an acid or base catalyst to a solution of an alkoxysilane reagent, such as tetramethoxysilane (TMOS), water,

Handbook of Nanostructured Materials and Nanotechnology, edited by H.S. Nalwa
Volume 1: Synthesis and Processing
ISBN 0-12-513761-3/$30.00

and methanol leads to hydrolysis of the Si–OMe bonds to form Si–OH functional groups:

$$Si(OMe)_4 \; + \; 4H_2O \xrightarrow[\text{catal.}]{\text{MeOH}} \; \text{``Si(OH)}_4\text{''} \; + \; 4MeOH \qquad (1)$$

$$\Big\downarrow {\scriptstyle -2H_2O}$$

$$SiO_2 \text{ gel}$$

Subsequent elimination of water from two such Si–OH groups gives eventually an extended silica gel matrix (known as a xerogel when dry). Because hydrolysis and condensation reactions occur concurrently, monomeric silanols proceed to xerogel through oligomeric and polymeric intermediates. As these reactions progress, the viscosity of sol–gel solutions increases and can reach a "spinable" stage at which point thin films or fibers can be produced. Otherwise, xerogel products are obtained as porous powders or as monoliths that assume the shape of their container. If desired, residual Si–OH groups remaining in the xerogel product can be removed at elevated temperature to give fully densified SiO_2.

A wide variety of elements exhibit sol–gel chemistry. Alkoxy, or related, compounds of any one such element undergo sol–gel conversions to produce a gel containing that element, as shown for the silica gel synthesis in Eq. (1). However, mixtures of precursor compounds for two or more elements can undergo heterocondensation to form mixed-element gels. Similarly, mixtures of different precursor compounds, such like a TMOS, and other reactants containing reactive groups, such as $Si(OR)_3$ substituents, likewise undergo sol–gel conversion to give gels of complex composition.

Sol–gel methods are commonly used to prepare nanocomposite materials because these conversions occur readily with a wide variety of precursors and can be conducted at or near room temperature. In addition, the gel products frequently have properties ideal for desired applications. Porous ceramic xerogels of high surface area can serve as supports for chemical catalysts [11–13], and thin-film deposition is useful for materials possessing desired optical or magnetic properties [14].

Reports appearing since January 1990 on the synthesis and characterization of nanocomposite materials prepared by sol–gel methods are reviewed. Nanocomposite materials are defined as materials consisting of a host matrix along with an identifiable, discrete nanoparticulate guest substance having a maximum dimension of less than 1000 nm. Either the host matrix, the guest substance, or both must be derived from sol–gel methods. Discussion of the nonlinear optical and magnetic properties of such nanocomposite materials is provided elsewhere in this book.

Important sol–gel materials not explicitly covered in this review include organic/inorganic hybrid gels lacking a nanoparticulate guest phase [15–21]; gels containing entrapped polymers [22–32]; organic [33–39] or inorganic [40, 41] molecules as gel dopants; gels doped with metal ions [42–46]; pure noncomposite materials, such as thin films or powders, prepared by sol–gel processing [47–68]; or mixed metal oxides lacking nanoparticulate phases [69–109].

2. NANOCOMPOSITES CONTAINING ELEMENTAL NANOPARTICULATES

Elemental nanoclusters are the least complex nanoparticulates by composition. Available review articles for nanocomposites of this type include (1) nano-Co/Mo, Cu, Fe, Ni, Pd, Pt, or Ru particles in Al_2O_3, SiO_2, TiO_2, or ZrO_2 gels [110]; (2) nano-C, Cu/Ni, Pd/Ni, or Pt in silica gel [111]; (3) nano-Ag, Ge, or Os in silica xerogel [112]; (4) nano-Ag, Cu, Fe, Mo, Os, Pd, Pt, Re, Ru, or PtSn in silica xerogel [113]; and (5) nano-Ge or Si composites [114].

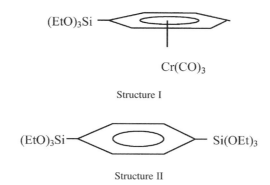

Structure I

(EtO)$_3$Si ⬡ Si(OEt)$_3$

Structure II

2.1. Group VI Metal Nanocomposites

A 2:98 molar mixture of (alkoxy)silanes (Structures I and II) undergoes sol–gel conversion in aqueous tetrahydrofuran (THF) solution in the presence of an acid catalyst to give a poly(1,4-phenylene)-bridged silsesquioxane condensation product. The host matrix material is an organic/inorganic hybrid of typical sol–gel ceramic matrices in that bifunctional phenylene groups serve as spacers between extended silicate networks. The compound shown in Structure I is an η^6-arene complex of chromium. This complex serves as both an internal dopant and a precursor source of chromium. When the condensation product is thermally treated at 120 °C under high vacuum, decomposition of the Cr(CO)$_3$ complex results in the formation of irregularly shaped nanoclusters of cubic Cr metal approximately 10 nm in diameter [115, 116]. When using a low-valent Cr complex as a precursor, Cr nanoparticle formation occurs directly upon precursor decomposition without proceeding through a metal oxide intermediate.

2.2. Group VIII Metal Nanocomposites

Iron/silica xerogel nanocomposites are formed by dissolving Fe(III) nitrate or chloride salts in a conventional silica sol–gel formulation using tetraethylorthosilicate (TEOS) as a source of silica xerogel [117–125]. Ferric ions within the gel are reduced to metallic iron upon thermal treatment under hydrogen gas at 450–700 °C for up to 4 h. Iron particles with average diameters of 3–30 nm are produced depending on the weight percentage of Fe ion present in the xerogel and the temperature and period of thermal treatment. Higher Fe ion loading and more robust heating conditions yield larger average particle sizes. Fe/silica xerogel nanocomposites are usually pressed into pellets or prepared as thin films on glass slides for magnetic or electrical conductivity measurements. Experimental observations include the following: (1) Crystalline iron nanoclusters are usually formed; (2) FeO, Fe$_3$O$_4$, and Fe$_2$O$_3$ phases can be present also; (3) Fe$_2$SiO$_4$ is formed at elevated temperatures (700 °C) if the Fe ion content exceeds 16 wt%; (4) electrical resistivity decreases with increasing Fe nanocluster size, presumably because a greater number of charge percolation chains involve Fe particles; and (5) the effective magnetic anisotropy of Fe/silica gel nanocomposites can be greater than that of bulk iron.

Iron nanocomposites have been produced in an alumina matrix using Al(Osec-Bu)$_3$ as a sol–gel precursor [126]. Using (alkoxy)aluminum compounds in combination with TMOS affords a mullite matrix (3Al$_2$O$_3$–2SiO$_2$). Iron ion is introduced into these matrices using immersion infiltration of ferric nitrate solutions. Reduction under hydrogen at 600 °C produces crystalline Fe nanoclusters poorly dispersed throughout the solid matrix for Fe doping levels of 25 wt%. The Fe/alumina nanocomposite exhibits only weak saturation magnetization. Reaction of the Fe/mullite nanocomposite in air at 650 °C produces Fe$_3$O$_4$ particulates, whereas heating to 800 °C gives Fe$_2$O$_3$ (maghemite) particles.

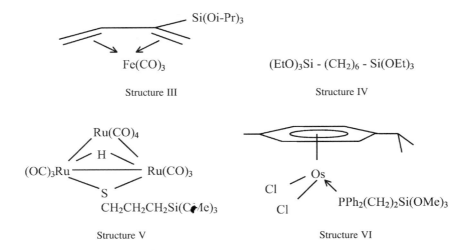

Structure III

(EtO)$_3$Si - (CH$_2$)$_6$ - Si(OEt)$_3$

Structure IV

Structure V

Structure VI

Sol–gel condensation in a 2:98 molar ratio of the Fe(CO)$_3$ complex shown in Structure III with either Structure II or IV gives inorganic/organic hybrid polysilsesquioxane gels containing Fe(III) as an internal dopant [127]. Irradiation of the resulting doped gel with ultraviolet (UV) light under vacuum leads to loss of CO and formation of Fe nanoclusters with an average diameter of 4 (0.5) nm. By using a low-valent iron precursor, Fe particles are formed without requiring calcination and subsequent reduction. The electrical conductivity of this poly(1,4-phenylene)-bridged silsesquioxane nanocomposite is in the semiconductor range.

Acid hydrolysis of ferric nitrate/cupric nitrate solutions affords inorganic gels that, upon thermal reduction, give nanocomposites consisting of Fe nanoclusters dispersed throughout a Cu metal matrix [124, 128]. Reduction by hydrogen at 425 °C gives Fe nanoclusters with an average diameter of 10 nm at a doping level of 29.6 wt% Fe. Mössbauer spectra confirm the presence of α-Fe particles, and magnetic measurements indicate that each Fe particle is a single magnetic domain.

Ru/silica xerogel nanocomposites have been prepared by conventional metal ion doping or by more unconventional covalent internal doping of silica xerogels. The addition of Ru(III) chloride to sol–gel solutions containing TEOS at the levels of 0.5, 1.0, and 2.0 wt% Ru leads to silica xerogels doped with Ru(III) ions. Thermal treatments up to 600 °C give Ru/silica xerogel nanocomposites that serve as hydrogenation catalysts [129]. Ru average particle diameters are less than 3 nm.

The addition of the triruthenium cluster complex, Ru$_3$(CO)$_{10}$[μ-S(CH$_2$)$_3$Si(OMe)$_3$][μ-H] (Structure V), which contains a bifunctional thiolate ligand, to a TMOS silica sol–gel formulation gives covalent, internal incorporation of this Ru precursor complex into the silica xerogel matrix as it is being formed. Thermal treatment of this molecularly doped xerogel under hydrogen at 700 °C gives a Ru/silica xerogel nanocomposite [130]. The Ru nanoclusters exhibit near hexagonal shape and have average diameters of 3.1 nm [by transmission electron microscopy (TEM)] or 7.5 nm [by X-ray diffraction (XRD)]. Use of a low-valent Ru precursor complex gives Ru metal particles directly without proceeding through calcination thermal treatments that would produce highly toxic and volatile ruthenium oxides.

Using a similar strategy, covalent incorporation of the (η^6-arene)Os complex (Structure VI) into a silica xerogel using TMOS as a sol–gel precursor occurs through hydrolysis and condensation reactions involving the Si(OMe)$_3$ functional group of the bifunctional phosphine ligand. Thermal treatment of the resulting molecularly doped xerogel under hydrogen at 900 °C gives an Os/silica xerogel nanocomposite directly without calcination, thereby avoiding formation of the very volatile and toxic OsO_4 [131]. The measured average diameters of the Os nanoclusters are 2.7 nm (by TEM) or 5.5 nm (by XRD).

2.3. Group IX Metal Nanocomposites

Co/silica xerogels have been prepared by metal ion doping of conventional sol–gel formulations followed by reductive thermal treatment [118]. Similar Co nanocomposites have been prepared using ZrO_2-modified silica aerogels. Aerogels are formed using conventional sol–gel chemistry but are usually dried under supercritical conditions. Such composites have large surface areas, and these Co/aerogel nanocomposites are active Fischer–Tropsch catalysts [132].

In situ coordination of $Co(OAc)_2$ by the bifunctional diamine (Structure VII) in a sol–gel formulation based on TEOS leads to apparent covalent incorporation of Co ion into a silica xerogel matrix as it is being formed [133, 134]. Calcination of this molecularly doped xerogel in air at 500 °C affords CoO nanoclusters with an average diameter of 19.5 nm. Thermal reduction under hydrogen at 500 °C produces a Co/silica xerogel nanocomposite having crystalline Co nanoclusters with an average diameter of 17.4 nm. The particle size range for these Co nanoclusters is 11.0–24.9 nm.

Similarly, lysine (Structure VIII) reacts with $Ti(OEt)_4$ or $Zr(OPr)_4$ to form bifunctional complexes of the type $[(RO)_3M(lysinate)]_n$, in which the butylamine functional group of lysine is chemically available to coordinate to metal ions. The addition of cobalt(II) acetate, $Co(OAc)_2$, to solutions of these precursor complexes leads to *in situ* complexation of Co ion. Subsequent sol–gel condensation gives gels containing Co ion as an internal dopant [135]. Calcination in air at 500 °C affords $CoTiO_3$/titania or CoO/zirconia nanocomposites. Thermal reduction under hydrogen at 500 °C gives the corresponding Co/titania or Co/zirconia gel nanocomposites. With a molar doping level of 20% Co, the average diameter of the Co nanoclusters is 19 nm (by XRD).

Rh/silica or Rh/zirconia-modified silica gel nanocomposites have been prepared by conventional internal metal ion doping of sol–gel formulations, as well as by using metal ion impregnation techniques [136]. The Rh nanoclusters produced by external doping are more active as hydrogenolysis catalysts, presumably because these metal nanoclusters have more exposed surface area.

2.4. Group X Metal Nanocomposites

Nanoscale electrodeposition of nickel in thin films of silica xerogel has been reported [137]. Thin-film nanocomposites of Ni in silica xerogel have also been prepared by doping TEOS sol–gel formulations with nickel nitrate followed by reduction under hydrogen at 650 °C [121]. In the latter films, the average diameter of the Ni particles increases from 5.2 to 10.5 nm as the period of reduction is increased from 15 to 35 min. The electrical conductivity of these films has been studied.

Bulk Ni/silica xerogel nanocomposites are prepared by doping silica sol–gel formulations with nickel salts, such as nickel(II) chloride or nitrate, followed by reduction under hydrogen at 500–650 °C [120, 138–140]. The average Ni particle diameters range from 5 to 13.4 nm, depending on the severity of thermal treatment. Bulk powder Ni/silica xerogels

$$H_2NCH_2CH_2CH_2CH_2 - \underset{\underset{NH_2}{|}}{CH} - CO_2H$$

$$H_2NCH_2CH_2NH(CH_2)_3Si(OEt)_3$$

Structure VII   Structure VIII

containing 7 wt% Ni and an average particle size of 5 nm show good catalytic activity in hydrogenation reactions [138]. Compacted Ni/silica xerogels [140] having Ni particles with diameters of 5–7 nm possess semiconductor electrical conductivity at 235–340 K [139], whereas similar compacts having Ni particles 9–13.4 nm in size are ferromagnetic [120].

In situ complexation of Ni(OAc)$_2$ by the bifunctional diamine shown in Structure VII gives molecularly doped silica xerogels following sol–gel processing [133, 134, 141]. The addition of TEOS to these sol formulations dilutes the doping level of nickel salt. Calcination in air at 500–525 °C and reduction under hydrogen at 500 °C or higher gives Ni/silica xerogel nanocomposites. Further sintering of these composites leads to growth in Ni nanocluster particle size. Sintering for 2 h at 500, 100, or 200 °C affords Ni nanoclusters of 30-, 100-, or 200-nm average diameter, respectively, as determined by XRD [141].

In situ complexation of Ni(OAc)$_2$ by bifunctional Ti or Zr complexes of the type [(RO)$_3$M(lysinate)]$_n$, where R is Et or Pr, respectively, leads to Ni/titania or Ni/zirconia nanocomposites after sol–gel condensation and reduction under hydrogen [135]. At a 20-mol% loading of Ni, the average diameters of Ni nanoclusters in the titania or zirconia gel matrices are 11 and 7 nm, respectively. Ni/Y$_2$O$_3$–ZrO$_2$ gel nanocomposites are formed by *in situ* complexation of Ni(OAc)$_2$ to bifunctional amine ligands, such as glycine or triethanolamine, in the presence of Zr(OPr)$_4$ and yttrium nitrate [142, 143]. Calcination in air and reduction under hydrogen affords Ni nanoclusters having average sizes of 5–10 nm or 30–40 nm, depending on the conditions of thermal treatment and the doping level of Ni. These nanocomposites become electrically conductive upon hot pressing the powder when the Ni loading exceeds 30 mol% [143].

Ni/Al$_2$O$_3$ gel nanocomposites are formed by sol–gel condensation of aluminum alkoxides, such as Al(OR)$_3$, where R is isopropyl or *sec*-butyl, in the presence of nickel nitrate or formate salts followed by calcination and reduction [126, 144, 145]. Nickel loadings of 5–50 vol% give Ni nanoclusters with average diameters of 5–60 nm. The mechanical [144], magnetic [126], and catalytic properties [145] of these nanocomposites have been reported.

Nickel nanocomposites in BaTiO$_3$ or Pb(Zr$_{0.58}$Ti$_{0.42}$)O$_3$ gel matrices are prepared by sol–gel condensation of a mixture of appropriate metal alkoxides along with added nickel salts [146, 147]. The ceramic matrix is formed upon calcination, and Ni nanoclusters with diameters of 2.0–7.6 nm are formed during subsequent thermal reduction under hydrogen. Such ferroelectric materials are of interest for capacitor fabrication. Compacted Ni/BaTiO$_3$ nanocomposite gels have a dielectric constant of 400–900 at 1 kHz [146], whereas the dielectric constant of thin films of the Ni/Pb(Zr$_{0.58}$Ti$_{0.42}$)O$_3$ gel nanocomposites is 220–410 at 1 kHz at room temperature [147].

Nanocomposites of Pd and Pt are of interest primarily for use as chemical catalysts or as materials exhibiting interesting nonlinear optical (NLO) properties [148]. A variety of sol–gel preparative methods are available for fabricating such nanocomposites as either bulk powders or as thin films.

Palladium or Pt nanocomposites in γ-Al$_2$O$_3$ [149–152], TiO$_2$ [153–156], or Y$_2$O$_3$-stabilized ZrO$_2$ [157] gels are formed by doping conventional sol–gel formulations with metal salts or complexes. Sols are formed from metal alkoxides, such as TEOS, Ti(Oi-Pr)$_4$, or M(OBu)$_4$, where M is Ti or Zr, and Pd or Pt is introduced as palladium nitrate, PdCl$_2$, PdCl$_4^{2-}$, H$_2$PtCl$_6$, or as the acetylacetonate (acac) complexes. Ethylenediaminetetraacetic acid (EDTA) can also be added to the sol to effect *in situ* complex formation [149–151]. Thermal calcination and reduction of gels doped with Pd or Pt ions gives metal nanoclusters, although direct thermal reduction of these more noble metals can occur without using an explicit reducing agent. Palladium or Pt nanocluster average sizes typically range from 4 to 115 nm with narrow particle size distributions obtained through careful control of metal doping levels and the severity of thermal treatment.

Pd or Pt nanocomposites are also formed using organically modified gels or gel precursors, as shown in the following equation [158–163]:

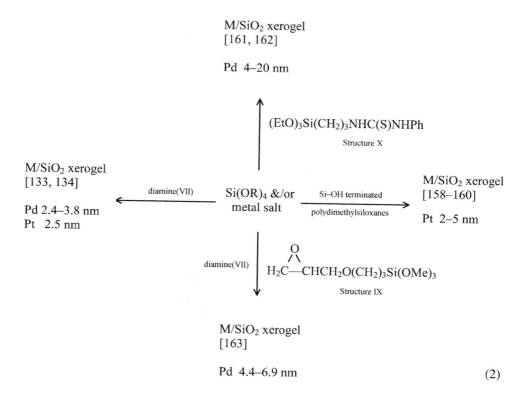

M/SiO$_2$ xerogel
[161, 162]

Pd 4–20 nm

(EtO)$_3$Si(CH$_2$)$_3$NHC(S)NHPh

Structure X

M/SiO$_2$ xerogel
[133, 134]

Pd 2.4–3.8 nm
Pt   2.5 nm

diamine(VII)

Si(OR)$_4$ &/or
metal salt

Si–OH terminated

polydimethylsiloxanes

M/SiO$_2$ xerogel
[158–160]

Pt   2–5 nm

diamine(VII)

H$_2$C——CHCH$_2$O(CH$_2$)$_3$Si(OMe)$_3$

Structure IX

M/SiO$_2$ xerogel
[163]

Pd 4.4–6.9 nm

(2)

The bifunctional diamine (Structure VII) gives composites having average metal particle diameters of 2.4–3.8 nm [133, 134]. Pd particle sizes are essentially independent of the precursor, Pd(acac)$_2$, loading level. For the Pt nanocomposite, the Pt particle size distribution broadens and shifts to larger average diameters when treated thermally at higher temperatures (750–900 °C) [164]. Platinum silicide formation, such as Pt$_3$Si and Pt$_{12}$Si$_5$, also occurs at these conditions [165]. Use of the bifunctional diamine (Structure VII) in combination with the bifunctional epoxide (Structure IX) affords both *in situ* complexation of Pd ion and binding of the silica xerogel matrix to glass surfaces for casting and dip-coating applications [163]. *In situ* complexation of Pd salts to silica xerogel derivatized with the bifunctional benzoylthiourea (Structure X) gives a Pd/silica xerogel that acts as a hydrogenation catalyst [161, 162]. Either bulk or thin films of Pt/silica xerogel nanocomposites containing up to 5 wt% Pt are obtained from conventional sol–gel formulations using silanol-terminated polydimethylsiloxanes as additives [158–160]. Reduction of Pt ion using H$_2$PtCl$_6$ as a metal dopant is effected either thermally or photochemically. Other reports of Pd/silica xerogel nanocomposite formation using organic binders or modified sol–gel methods are available [166–168].

Addition of the bifunctional Pt complex, PtCl$_2$[PPh$_2$(CH$_2$)$_2$Si(OMe)$_3$]$_2$, to TMOS at a Si/Pt molar ratio of 100:1 under sol–gel conditions gives a silica xerogel containing this complex as a covalent, internal dopant. Subsequent thermal calcination in air and reduction under hydrogen at 700 °C gives a Pt/silica xerogel nanocomposite with Pt nanoclusters having an average diameter of 3.8 nm [131].

2.5. Group XI Metal Nanocomposites

A vast literature exists on nanocomposite materials containing nanoclusters of Cu, Ag, and Au. Principal interest in these materials is derived from their interesting optical properties. All three metals have surface plasmon resonances in the visible region for nanoclusters of typical size.

Cu/silica xerogel nanocomposites are formed as bulk powders [139, 140, 169] or as thin films [121, 122, 170, 171] by adding Cu(II) salts to conventional TEOS sol–gel formulations. Metal ion doped gels are usually calcined at 500–600 °C, and reduction to metal occurs upon heating from 400 to 700 °C in the presence of hydrogen. Average Cu particle sizes varying from 3 to 28 nm are obtained, depending on the level of metal doping and thermal conditions. Reduction at 400 °C for a gel containing 20 wt% Cu ion gives Cu nanoclusters of average diameter 6.2 nm with a monomodal size distribution. Thermal treatment at 700 °C gives a bimodal Cu particle size distribution with maxima at 8.5 and 28.0 nm [169].

Another route to Cu/silica gel nanocomposites is *in situ* complexation of Cu ion to organically modified gels followed by calcination and thermal reduction under hydrogen [158]. Use of the bifunctional diamine (Structure VII) permits a high doping level of Cu ion in a silica sol–gel formulation. The average diameter of the resulting Cu nanoclusters is 3.9 nm [133, 134]. A similar synthesis using a much lower concentration of Cu ion and thermal treatment at 900 °C gives a Cu/silica xerogel nanocomposite having Cu particles with an average diameter of 15 nm [172]. Silica sol–gel formulations containing the bifunctional diamine shown in Structure VII, 3-glycidoxypropyl triethoxysilane, TEOS, and Cu(II) ion give gels that coat glass through dip coating. Thermal reduction of these coatings gives a film of Cu/silica xerogel containing Cu nanoclusters 5–30 nm in diameter [173a]. Sol–gel condensation of the bifunctional titanium lysinate complex, $[Ti(OEt)_3(lysinate)]_2$, in the presence of $Cu(OAc)_2$ forms a TiO_2-based xerogel. Upon calcination and reduction, a bulk Cu/TiO_2 gel nanocomposite having Cu particles 20 nm in average diameter is formed [135]. A related preparation of a Cu/TiO_2 nanocomposite has appeared recently [173b].

Cocondensation of TMOS and the bifunctional complex, $CuCl[PPh_2(CH_2)_2Si(OEt)_3]_3$, in a 10:1 Si/Cu molar ratio under sol–gel conditions gives a silica xerogel covalently doped with this Cu precursor complex. Direct thermal reduction at 600 °C under hydrogen gives a Cu/silica xerogel nanocomposite containing Cu nanoclusters with an average diameter of 24 nm [131].

Nanocomposites of Ag and Au have received much interest primarily as materials exhibiting interesting optical properties. Because these metals are easily reduced, a variety of sol–gel preparative methods are available for fabricating Au or Ag nanocomposites as either bulk powders or thin films. Thin-film methods are particularly useful for optical studies to maintain adequate optical transmission.

Silver or Au nanocomposites in γ-Al_2O_3 [174, 175], SiO_2 [176–196], SiO_2/B_2O_3 [197], SiO_2/TiO_2 [198], TiO_2 [199–201], $BaTiO_3$ or PZT [202–205] gels, or micellular [206, 207] matrices are formed by direct metal ion doping of conventional sol–gel formulations with metal salts or complexes. Sols are formed from $AlCl_3$ or metal alkoxides, such as $Al(Osec$-$Bu)_3$, TEOS, H_3BO_3, $Ti(Oi$-$Pr)_4$, $M(OBu)_4$, where M is Ti or Zr, and Ba or Pb acetates. The most common source of Ag is silver nitrate; gold is usually introduced as $HAuCl_4$ or $NaAuCl_4$ or as acetylacetonate complexes [177, 178]. Thermal calcination and hydrogen reduction of metal ion doped gels affords metal nanoclusters, although thermal reduction of these more noble metals, particularly Au, occurs without using an explicit reducing agent. Hydrazine hydrate has been used to reduce gold ion at reduced temperatures [175], and both Ag and Au ions can be reduced to metal by light [177–179] or by γ-irradiation [195, 196, 200].

Silver or Au nanocluster sizes typically fall in a range of 4 to 40 nm with narrow particle size distributions obtained by careful control of metal doping levels and the severity of thermal treatment. The growth of Ag particles as a function of annealing temperature and gaseous environment has been studied [167, 170, 194]. Ag/silica xerogel nanocomposites heated in air show some loss of silver at temperatures as low as 650 °C. Annealing such samples in inert or N_2/H_2 atmospheres from 200 to 600 °C leads to particle growth by Ostwald ripening growth processes, whereas particle growth at more elevated temperatures gives bimodal particle size distributions. Annealing further to 850 °C leads to a decrease in average particle size and to a sharpening of Ag particle size distributions as larger Ag clusters undergo fragmentation. The thermal bleaching effect observed for Ag/silica xerogel nanocomposites is attributed to thermal fragmentation of the larger Ag nanoclusters within the composite. The addition of Li_2O completely inhibits this bleaching effect [193].

Silver and Au nanocomposites exhibit properties useful for many applications including (1) resonant and nonlinear optical effects [171, 174–184, 193, 197, 199, 201, 202], (2) use as high-dielectric-strength media for capacitors [146, 187–189, 203], (3) enhanced electrical conductivity of ceramic media [204, 205], (4) pattern etching using HF [186], (5) augmentation of electrode response [185], and (5) as magnetic spin glasses for Ag nanocomposites [191].

Ag or Au/silica nanocomposites have been prepared using organically modified gels or gel precursors, as shown in the following equation:

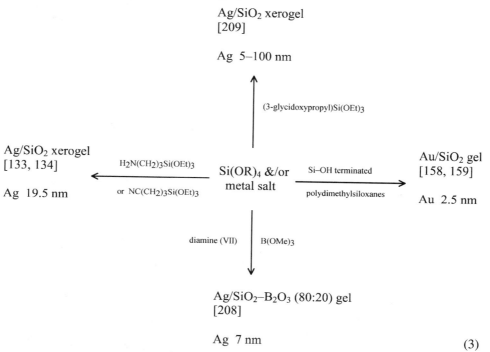

$$\tag{3}$$

Bulk or thin-film products are obtained. The sources of the metals are $AgNO_3$, AgOAc, or $HAuCl_4$. Reduction to metal is accomplished either thermally [158, 159], by reaction with hydrogen at elevated temperature [133, 134, 209], or by photolysis [158, 159, 208]. The particle sizes shown in Eq. (3) represent either average particle size or a range of average particle sizes produced under controlled reduction conditions. Control over average metal particle size can be partially achieved by variation of reduction conditions. Peak absorption for the surface plasmon resonance of Ag nanoclusters in borosilicate gels ranges from 394 to 405 nm, depending on the metal particle size [208].

Sol–gel condensation of $Zr(Oi\text{-}Pr)_4$ in the presence of $HAuCl_4$, $CH_2 = C(Me)CO_2(CH_2)_3Si(OMe)_3$, $CH_2 = C(Me)CO_2H$, and diamine (Structure VII) leads to a zirconia gel doped with Au(III) ions. Photolysis initiates alkene polymerization and reduction of Au ion to form Au nanoclusters. Controlled irradiation gives nanocomposites having Au nanoclusters with average diameters of 10–40 nm [210].

Hydrogen reduction of $HAuCl_4$ by photolysis in the presence of poly(N-vinyl-2-pyrrolidone) (PVP) forms polymer-stabilized Au nanoclusters. The addition of this dispersion to $Ti(Oi\text{-}Pr)_4$ along with acetylacetone and water leads to sol–gel condensation and formation of a hybrid gel containing Au nanoclusters 2.8 nm in diameter [160]. The size variation of the Au particulates is accomplished by controlling the reducing conditions. Maxima of the surface plasmon resonance of the Au nanoclusters range typically from 530 to 620 nm, depending on the metal particle size. Analogous silica hybrid gels could not be produced by this procedure because of unfavorable precipitation upon mixing PVP with TEOS.

The addition of the bifunctional Ag mixed-ligand complex, $[AgS(CH_2)_2Si(OMe)_3]_7$ $[AgCl]_3$, to a silica sol–gel formulation based on TMOS at a Si/Ag molar ratio of 100:1 gives a silica xerogel containing this complex as a covalent, internal dopant. Subsequent thermal calcination and reduction under hydrogen at 600 °C gives a Ag/silica xerogel nanocomposite containing Ag nanoclusters with an average diameter of 5.3 nm [131].

Au nanoclusters have also been prepared in silica materials using sol–gel conversions within inverse micelles as microreactors [211]. Poly(diethoxysilane) and $AuCl_3$ are dissolved in the presence of an inverse micelle. Reduction to Au metal is effected using $LiBH_4$ affording Au nanoclusters with diameters of 6.0 (1.5) nm. Control of metal particle size is complicated by the effects of gel precursor and sol–gel condensation products on the stability of the inverse micelle.

Silver or gold nanoclusters have also been prepared in silica/zirconia matrices using γ-radiation to initiate reduction [212].

2.6. Metal Alloy Nanocomposites

Four Cu-based alloy/silica xerogel nanocomposites have been prepared using sol–gel chemistry, as shown in the following equation:

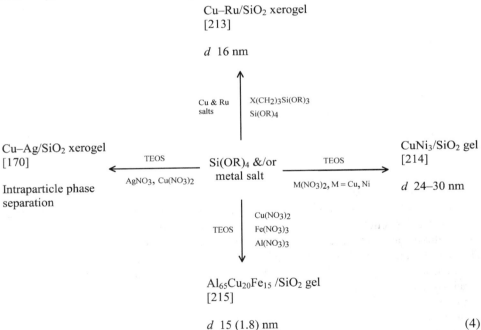

$$\tag{4}$$

In situ complexation of Cu and Ru salts by bifunctional ligands gives a silica gel containing coordination complexes of these metals [213]. Subsequent calcination in air and thermal reduction under hydrogen affords a Cu/Ru nanocomposite containing particles with an average diameter of 16 nm. Alcohol solutions of copper and silver nitrate salts containing TEOS form, upon sol–gel conversion, a gel matrix doped with these metal ions [170]. Thermal treatment of the resulting mixed-metal-ion-doped xerogels under hydrogen gives Cu/Ag alloy particulates; however, phase separation of the metals occurs within each particle. Compaction at elevated temperatures of a physical mixture of silica xerogels containing either Cu or Ni metal nanoclusters at a Cu/Ni atomic ratio of 1:3 gives a CuNi$_3$ nanocomposite [214]. XRD data confirm a single-phase composition for these alloy particulates. The average diameter of the alloy particles ranges from 27 to 30 nm as the temperature of compaction increases from 450 to 650 °C. Sol–gel processing of solutions of Cu(II), Fe(III), and Al(III) nitrates with TEOS affords a mixed-metal-ion-doped silica gel [215]. Particles of the ternary alloy, Al$_{65}$Cu$_{20}$Fe$_{15}$, with an average diameters of 15 nm are formed in the gel matrix along with nanoparticulate Fe upon thermal treatment with hydrogen. XRD data indicate that the alloy particles are quasicrystalline, and Mössbauer spectra are consistent with these particle compositions.

Two nanocomposites containing binary alloys of Pt have been prepared using sol–gel methods. Gel formation using Al(Osec-Bu)$_3$ and acetic acid in the presence of Pt(acac)$_2$ and Re$_2$(CO)$_{10}$ gives a Pt–Re/Al$_2$O$_3$ composite upon calcination at 400 °C [152]. At loadings of 1 wt% Pt and Re, nanoparticles with an average size of 3.5 nm are obtained. Although the presence of Re appears to inhibit Pt metal aggregation, direct evidence for the formation of alloy nanoclusters is lacking.

Sol–gel condensation of mixtures of TMOS and the bifunctional complex, Pt(PPh$_3$)$_2$(Ph) [SnPh$_2$S(CH$_2$)$_3$Si(OMe)$_3$], leads to covalent incorporation of this binuclear complex into the resulting silica xerogel matrix [130, 131]. Thermal treatment of this molecularly doped xerogel under hydrogen at 700 °C gives a nanocomposite containing crystalline nanoclusters of niggliite, PtSn, with an average diameter of 6 nm. No other Pt/Sn phases are detected by either XRD or electron diffraction. The 1:1 Pt/Sn stoichiometry of the precursor complex apparently gives selective formation of niggliite nanocrystals.

An Fe$_{20}$Ni$_{80}$/silica xerogel nanocomposite has been prepared by sol–gel processing [119]. The alloy particles have an average diameter of 8(1) nm and face-centered cubic (fcc) structure. Magnetic measurements reveal an effective magnetic anisotropy for the composite that is larger than that of the bulk alloy.

2.7. Group XIV Nanocomposites

Fullerene nanocomposites have been prepared by (1) entrapment of C$_{60}$/C$_{70}$ mixtures in silica aerogels through the addition of fullerenes to a conventional sol–gel mixture [216–220]; (2) absorption of C$_{60}$ solutions into silica gel formed from TEOS, followed by impregnation and polymerization of an organic polymer giving a multiphasic inorganic/organic composite [221, 222]; or (3) derivatization of C$_{60}$ with a bifunctional substituent, H$_x$C$_{60}$[NH(CH$_2$)$_3$Si(OEt)$_3$]$_x$, where $x = 1$–3, followed by sol–gel conversion using TEOS [223]. These aerogel composites are produced by supercritical fluid drying. Photoluminescence is observed at 2.26 eV when using laser excitation at 488 nm. Fullerene molecules are presumably entrapped in nanopores within the areogel matrix. For the multiphasic composites, the organic polymeric phase is formed from polymerization of methyl methacrylate or bisbenzothiazole 3,4-didecyloxythiophene or mixtures of the two. These composites are optically transparent and might be useful as materials having tailored optical responses. Composites prepared from covalently derivatized C$_{60}$ absorb light at various wavelengths, depending on the C$_{60}$ loading. Femtosecond excited-state absorption dynamics studies of C$_{60}$ in inorganic sol–gel matrices reveal a crucial role of C$_{60}$ in relaxation dynamics. This property may find application in optical-limiting devices [224, 225].

Pyrolysis of gaseous organic compounds in sol–gel aerogels forms carbon nanotubual composites [226]. The presence of carbon nanotubes reduces the infrared (IR) transmission of the aerogel and improves its thermal performance relative to that of the undoped aerogel.

Nanosilicon/silica xerogel composites can be formed by ultrasonic fracturing of porous silicon. The nanoscale Si fragments produced are mixed into conventional silica sol–gel formulations giving an n-Si/xerogel composite [227, 228]. These materials exhibit red (680 nm) and blue (415–460 nm) electroluminescence with decay times different from those observed from porous silicon.

Elemental Ge nanocomposites are accessible using conventional sol–gel methods. Reduction of intermediate germanium oxides under hydrogen and subsequent Ge particle growth occur at temperatures below 1000 °C. Ge/silica gel composites result from cocondensation of TEOS and GeCl$_4$ or Ge(OEt)$_4$ followed by thermal treatment [229–232]. Annealing temperatures above 800 °C usually give crystalline Ge nanoclusters of approximately 5-nm average diameter, although larger particle sizes are observed upon prolonged heating at elevated temperatures. Ge particles of 5 nm show an optical absorption edge onset near 2.2 eV and a broad photoluminescence peak at the same energy. Larger Ge particles do not exhibit photoluminescence. A decrease in Ge average particle size leads to an intensity increase in the observed photoluminescence peak along with a blue shift in peak position.

The addition of the bifunctional organogermane, Me$_3$GeS(CH$_2$)$_3$Si(OMe)$_3$, to a conventional TMOS sol–gel formulation leads to covalent incorporation of this molecular precursor into the silica xerogel matrix as it is being formed [131, 233]. Subsequent calcination in air at 600 °C followed by reduction in hydrogen at 600–900 °C gives a Ge/silica xerogel nanocomposite containing Ge particles with an average diameter of 6.8 nm. Optical absorption spectra of this composite are consistent with expected quantum confinement effects.

A Ge/alumina gel nanocomposite has also been prepared [234]. The host matrix is formed by sol–gel processing of AlCl$_3$.

3. NANOCOMPOSITES CONTAINING NANOPARTICULATE SUBSTANCES

In nanocomposites of chemical substances, two or more elements are present in the nanocluster, and, ideally, a single stoichiometry is desired. Although many such compositions are expected to show catalytic activity, such as supported metal oxides, there is much interest in preparing nanocomposites of semiconductor materials for property measurements or for device fabrication. The NLO and magnetic properties of such materials are reviewed elsewhere in this book.

3.1. Metal Carbide Nanocomposites

A mixed-metal gel is formed when Ti(Oi-Pr)$_4$ is added in a 70:30 molar ratio to a prehydrolyzed sol derived from SiMe$_2$(OEt)$_2$ [235]. Subsequent pyrolysis of this gel under Ar at 400–1600 °C gives eventual formation of cristobalite and crystalline TiC phases. XRD data indicate the appearance and growth of TiC particles between 1000 and 1600 °C, although particle size estimates have not been determined. Formation of a TiC/SiO$_2$–TiO$_2$ nanocomposite presumably occurs at intermediate temperatures.

The addition of the intensely purple tricobalt carbido cluster complex (Structure XI) to a TMOS sol–gel formulation leads to covalent incorporation of the complex to the silica xerogel matrix as it is being formed [130, 131, 236]:

Structure XI   (5)

Heterocondensation of the $Si(OH)_3$ functional group of the complex shown in Structure XI with hydrolysis products of TMOS assures a covalent molecular doping of the resulting silica xerogel, as confirmed by IR spectroscopy and by the inability to remove the complex (Structure XI) from the matrix by solvent extraction. A control reaction using a derivative of the complex shown in Structure XI in which the $Si(OH)_3$ group is replaced by a hydrogen atom substituent leads to a physically doped silica xerogel. Washing this latter xerogel with organic solvents leads to complete removal of the tricobalt cluster from the matrix. Thermal degradation of the covalently doped gel under hydrogen at $500\,°C$ leads to a Co_3C/silica xerogel. Selected-area electron diffraction (SAED) and both powder and single-crystal XRD indicate that crystalline particles of Co_3C are formed. These nanoparticles have an average diameter of 25 nm and a stoichiometry identical to that of the precursor core. Phase separation into crystalline particles of other Co/C stoichiometries is not observed. Use of a more complex molecular precursor apparently provides control over the elemental stoichiometry of the resulting cobalt carbide nanoclusters.

Nanocomposites containing SiC particulates of either 30- or 600-nm average diameter have been prepared in an alumina host matrix by ultrasonic dispersion of preformed SiC nanoparticles in a sol of pseudo-boehmite, aluminum acetate, and acetic acid [237]. Gel formation is initiated by the addition of ammonia. Calcination in air at $900\,°C$ and thermal treatment in vacuum at $1300\,°C$ converts the host matrix to α-Al_2O_3. Equilibration of the resulting gel with solutions of ^{220}Ra and ^{224}Ra leads to a radioactive doping of the gel in the top 80 nm of the surface. Emanation thermal analysis and differential thermal analysis (DTA) studies indicate that the onset of structural change and grain growth of the alumina matrix shifts to a higher temperature when SiC nanoparticulates are present. This effect is greatest for smaller SiC particles. A related study has also been reported [238].

3.2. Metal Pnictide Nanocomposites

Hydrolysis and heterocondensation of TEOS and phenyl(trimethoxy)silane gives an organic functionalized silica gel. Thermal reaction of this gel with nitrogen gas at $1500\,°C$ forms a nano-Si_3N_4/SiC composite. Changes in the composition of the resulting gel as a function of thermal treatment conditions have been studied [239].

Covalent incorporation of low-valent transition metal complexes containing bifunctional phosphine ligands into a silica gel matrix as it is being formed gives molecularly doped gels. Subsequent thermal treatment under hydrogen at $600–900\,°C$ provides a convenient route to silica xerogel composites containing crystalline nanoclusters of the transition metal phosphides, Fe_2P, RuP, Co_2P, Rh_2P, Ni_2P, Pd_5P_2, or PtP_2 [130, 131, 240–243]:

$$
\begin{array}{ll}
Fe(CO)_4[PPh_2CH_2CH_2Si(OMe)_3] & Fe_2P(4.7\ nm)/SiO_2\ xerogel \\[4pt]
RuCl_2(\eta^6\text{-cymene})[PPh_2(CH_2)_2Si(OEt)_3] & RuP(4.7\ nm)/SiO_2\ xerogel \\[4pt]
Co_2(CO)_6[PPh_2CH_2CH_2Si(OEt)_3]_2 & Co_2P(5.0\ nm)/SiO_2\ xerogel \\[4pt]
RhCl[PEt_2CH_2CH_2Si(OEt)_3]_3 & Rh_2P(2.0\ nm)/SiO_2\ xerogel \\[4pt]
Ni[PPh_2CH_2CH_2Si(OEt)_3]_4 & Ni_2P(2.6\ nm)/SiO_2\ xerogel \\[4pt]
NiCl_2[PPh_2CH_2CH_2Si(OEt)_3]_2 & Ni_2P(35.5\ nm)/SiO_2\ xerogel \\[4pt]
PdBr_2[PEt_2CH_2CH_2Si(OEt)_3]_2 & Pd_5P_2(11.3\ nm)/SiO_2\ xerogel \\[4pt]
PtCl_2[PPh_2CH_2CH_2Si(OMe)_3]_2 & PtP_2(4.0\ nm)/SiO_2\ xerogel
\end{array}
\tag{2}
$$

with center reaction conditions: (1) TMOS sol–gel (2) H_2, $>600\,°C$

Each molecular complex serves as a single-source precursor for the specific metal phosphide phase shown. For the RuP and PtP_2 composites, the metal phosphide stoichiometry is the same as that of the molecular precursor. Selective formation of the other metal phosphides requires a relative loss of phosphorus from the molecular precursor during thermal treatment. In each case, the metal phosphide stoichiometry of the nanocluster is that of a congruently melting phase or that of a particularly stable composition. Equivalent precursor loadings of the Ni(0) and Ni(II) precursor complexes gives Ni_2P nanocrystals differing in average size by nearly 14-fold (by TEM). The origin of this effect is not known.

Methanolysis of the cadmium phosphide compound, $Cd[P(SiPh_3)_2]_2$, in a pyridine solution gives nanoclusters of Cd_3P_2 having diameters of 2–7 nm [244]. These particles are assumed to be surface passivated with OMe and $SiPh_3$ groups. This synthetic route is a variation of typical sol–gel procedures in which solvent evaporation from the reaction solution provides an initial composite of nano-Cd_3P_2 along with byproducts, such as $HP(SiPh)_2$ and $Si(OMe)Ph_3$.

Several processing techniques, including sol–gel methods, have been used to introduce 4-nm GaAs nanoclusters into amorphous matrices [245]. Sol–gel synthesis of GaAs/silica gels is one such technique. These bulk materials show quantum confinement effects as revealed by optical absorption spectroscopy.

3.3. Metal Oxide Nanocomposites

3.3.1. Al_2O_3 Gel Matrix

Alumina gel nanocomposites containing nanoclusters of CoO, FeO, TiO_2, or ZrO_2 have been prepared by sol–gel methods [246–250]. Common sources of the alumina matrix include aluminum trialkoxides or hydrated alumina powder. Titania and zirconia particles are usually derived from tetraalkoxide reagents, although metal chlorides are also used. The CoO and FeO composites test well as catalysts for selective reduction of NO by unsaturated hydrocarbons [246]. Individual TiO_2 particles are observed when the Ti/Al molar ratio is less than 17.5% and can be imaged by TEM when the particles have diameters greater than 1.2 nm [248]. Alumina-stabilized zirconia films are of interest for high-temperature protection of turbine blades [249].

3.3.2. SiO_2 Gel Matrix

Nanoparticulates of numerous metal oxides have been formed in silica gel matrices, as shown in Table I. Synthetic methods include *in situ* coordination of metal ions by a bifunctional ligand, such as $X(CH_2)_3Si(OEt)_3$, where X is NH_2, CN, or $NH(CH_2)_3NH_2$, in conventional sol–gel formulations, Method A; dissolution of metal salts in conventional sol–gel formulations, Method B; heterocondensation of metal alkoxide precursors and tetraalkoxide silicates, Method C; or direct sol–gel condensation of zirconium oxychloride octahydrate onto a silica–glass surface followed by thermal annealing, Method D. Drying-control chemical additives, such as formamide, are occasionally used in Methods B and C. Triethoxysilane [258], H_2SiO_3 [258], or organically modified silicates (Ormosils), such as silanol-terminated polydimethyl siloxane [259], can be used as sources of the host matrix. Ormosil composites are inorganic/organic hybrid matrices. The chemical composition of the resulting gel depends on the weight fraction of polymeric silanol used in the sol formulation and the severity of the resulting thermal treatment.

Metal oxide particle sizes are usually determined by electron microscopy. In some cases, however, metal oxide nanoparticulates are intermediates in the conversion of metal ions to metal nanoclusters, and their sizes are inferred from the sizes of the resulting metal nanoclusters if direct observation is not possible [133].

The properties of interest for metal oxide/silica gel nanocomposites include optical absorption in the visible region [271] and NLO effects (particularly for nanoparticulate ferroelectric materials, such as metal titanates and niobates). Nanoclusters of magnetic materials, such as maghemite (γ-Fe_2O_3), are stabilized through entrapment in silica gel matrices over a considerable temperature range [258]. These composites are superparamagnetic materials.

Table I. Metal Oxide/Silica Gel Nanocomposites

M_xO_y	Synthetic method (see text)	Particle size (nm)	Properties	Reference(s)
Ag_2O	A			[133]
CdO	A			[133]
CoO	B		Alternating-current electrical conductivity	[251]
Co_3O_4	B	8.9–9.7	Effect on optical transmittance	[181, 182]
Cr_2O_3	B	12		[181]
CuO	A			[133]
	B	5–8, 17–21	NLO	[181, 252]
FeO	B		Color variation	[253, 254]
Fe_2O_3	B	2–10, <300	Effect on optical transmittance Mössbauer, magnetism	[64, 181, 182, 254–258]
Mn_2O_3	B	8	IR filters	[181, 184]
NiO	A	1.4, 34.8		[133]
	B	7.5–8.1		[181, 260]
PdO	A			[133]
	B	10.7–27.1		[181]
PtO	A			[133]
TiO_2	C	20–50	NLO	[260, 261]
V_2O_5	C		Photoinitiated polymerization	[262]
$(Ni, Zn)Fe_2O_4$		13–35	Superparamagnetism	[263]
$(Li, K)NbO_3$	C		NLO	[259]
$PbTiO_3$	C	20–40	NLO	[264–266]
$Pb(Zr, Ti)O_3$	C	8–10	NLO	[259, 267]
$(In, Sn)O_2$ (ITO)	B	1–8.5	NLO	[268]
$LiMn_2O_4$	B	Submicrometer		[269]
ZrO_2	D	<15	Oxygen-deficient ZrO_2	[270]

3.3.3. MO_2 Gel Matrices, where M Is Ti, Zr, or Sn

Nanoclusters of several metal oxides have been prepared in TiO_2, ZrO_2, or SnO_2 gels, as shown in Table II. Synthetic methods include sol–gel conversion of metal halide or coordination complexes with alkoxide compounds of the second metal, Method A; hydrolysis and heterocondensation of homoleptic alkoxide compounds of both metals, Method B; sol–gel condensation of metal acetate salts with either $M(OBu)_4$ (Method C) or $M(OEt)_3$(lysinate) compounds, where M is Ti or Zr (Method D); sol–gel preparation followed by supercritical fluid drying, Method E; or hydrolysis and condensation of mixtures of metal halide or acetate salts, Method F. Metal oxide particle sizes are obtained from electron microscopy, XRD, or fluorescence measurements.

Table II. Metal Oxide/TiO_2, ZrO_2, or SnO_2 Nanocomposites

M_xO_y	Synthetic method (see text)	Particle size (nm)	Properties	Reference(s)
		TiO_2 Matrix		
Al_2O_3	A	<100	Membrane structure	[272]
CeO_2	A, B	1–5	Transparent, colorless films	[273–276]
CuO	C	8		[135]
	D	61		[135]
NiO				[277]
ZnO				[277]
$CoTiO_3$	C	28		[135]
	D	47		[135]
$NiTiO_3$	C	7		[135]
	D	22		[135]
		ZrO_2 Matrix		
CoO	C, D			[135]
CuO	C, D, E		Aerogel or xerogel	[135, 278, 279]
NiO	C, D			[135]
		SnO_2 Matrix		
CuO	F	20–23, 25–37		[280, 281]
$Sn(Eu)O_2$	F	2–3	Time-resolved fluorescence	[282]

3.3.4. Al or $MgAl_2O_4$ Matrices

Nanocomposites containing nanocrystals of mullite or ZrO_2 in an aluminum metal matrix have been reported; the metal oxide particles are formed using sol–gel processing [283]. The Vickers hardness values of these composites are substantially higher than that of aluminum.

Dispersion of commercial spinel powders in a sol solution of $Zr(OPr)_4$, acetylacetone, n-propanol, and water gives, after gelation and thermal treatment, nanocrystals of ZrO_2 in $MgAl_2O_4$ as a host matrix [284]. Intergranular ZrO_2 particles are 250 nm in size, whereas intragranular particles have diameters of only 50–100 nm. Orthorhombic ZrO_2 particles are observed.

3.3.5. Polymer Host Matrices

Metal oxide/organic polymer composites have been prepared through sol–gel processing under mild conditions. Polyacrylate composites containing nanoparticulate SiO_2 [285, 286], TiO_2 [287], ZrO_2 [288, 289], or $BaTiO_3$ [290] are formed by *in situ* sol–gel conversion of metal alkoxide precursors within a preformed polymer matrix or by cocondensation of metal alkoxides with derivatized monomers followed by initiated polymerization.

Hydrolysis and condensation of $Ti(Oi\text{-}Pr)_4$ in the presence of polyvinylacetate gives a polymer composite containing TiO_2 particles with an average diameter of 6 nm. Strong covalent cross-linking occurs between the organic and inorganic phases [287].

Prehydrolysis of TEOS affords colloidal silica gel particles 10.5 nm in size [285]. Nuclear magnetic resonance (NMR) data indicate that these particles possess ethoxy and hydroxyl groups at surface sites. Subsequent reaction of this colloid with the methacrylate monomer, $(MeO)_3Si(CH_2)_3O_2CC(Me)=CH_2$, leads to sol–gel derivatization of the surface hydroxy groups and to formation of a 2-nm layer of methacrylate monomer on the silica gel particles. Copolymerization of these derivatized particles with excess methyl methacrylate under photochemical initiation gives the final inorganic/organic composite. At least 85% of the surface-bound methacrylate groups cross-link to the bulk polymer matrix.

Prehydrolysis of $Zr(OPr)_4$ in the presence of methacrylic acid and subsequent polymerization gives ZrO_2 particles with a diameter of 4 nm within a polymer matrix [288]. The presence of ZrO_2 alters the refractive index of the polymer matrix, and patterning techniques afford strip waveguides and other microoptical devices made from this material. Similar polymers containing ZrO_2 particles and an azobenzene chromophore formed under electrical poling exhibit second-harmonic NLO properties [289].

Sol–gel processing of methacryl(tri-isopropoxytitanium), barium alkoxide, and styrene forms $BaTiO_3$ particles of approximately 3 nm in diameter within a polymer matrix [290]. The dielectric properties of these inorganic/organic hybrid materials have been measured.

Several types of SiO_2 or SiO_2/TiO_2 nanocomposites in perfluorosulfonate ionomer matrices, such as Nafion, have been prepared [291–294]. Diffusion of TEOS into ionomer membranes occurs in aqueous alcohol solutions, and subsequent sol–gel hydrolysis and condensation affords particles of silica gel dispersed within the ionomer matrix. Particulate features have a dimension of approximately 5 nm. Postreaction with titanium alkoxides forms a TiO_2 gel layer as a shell on the silica particles and leads to interparticle binding. This percolative growth process occurs in the near-surface region of the membrane and controls the mechanical properties of the resulting inorganic/ionomer polymer.

Nanoparticulate SiO_2 or TiO_2 gels have been prepared in polymer or polymer composites of styrene-maleic anhydride copolymer [295, 296], poly(vinyl alcohol) [297], silicones [298,299], polyimides [300, 301], polyurethanes [302], or poly(hydroxypropylcellulose) [303]. The inorganic gel particulates are usually formed by *in situ* sol–gel processing of metal alkoxide precursors, such as TEOS. Gel particles are typically found in polymer pore structures with sizes smaller than 100 nm, although direct determination of ceramic particle sizes is lacking in many cases. These inorganic/organic composites are of interest because the structure and composition of the inorganic phase controls selected physical properties of the organic polymer. For example, a silica gel/poly(dimethylsilicone) nanocomposite has a butane/methane selectivity double that of the undoped polymer based on permeability measurements. Optical absorption spectra of films of a titania gel/poly(hydroxypropylcellulose) nanocomposite demonstrate that this composite is transparent in the visible region but acts as an efficient block for UV radiation.

3.4. Metal Chalcogenide (S, Se, or Te) Nanocomposites

3.4.1. CdS/Silica Gel Composites

Cadmium sulfide/silica gel nanocomposites are formed by a variety of synthetic methods, as shown in Figure 1. The most common procedure involves dissolving cadmium nitrate [148, 304–311] or cadmium acetate [312–319] in conventional silica sol–gel formulations, using TEOS, followed by reaction with H_2S (Method A). CdS particles with average diameters of 1.6–10 nm are produced, depending on the weight percentage of CdS present in the xerogel and the conditions of thermal annealing. The addition of particle stabilizers, such as poly(vinyl alcohol) or sodium hexametaphosphate (Method B) [320], or the bifunctional ligand, $H_2N(CH_2)_3Si(OEt)_3$, to solutions of a cadmium(II) salt (Method C) [321, 322] gives CdS particle sizes in the range of 2.3–4.2 nm. Sol–gel condensation of bifunctional $(EtO)_3SiX$ compounds, where X is NH_2, SH, or carboxylate, and cadmium

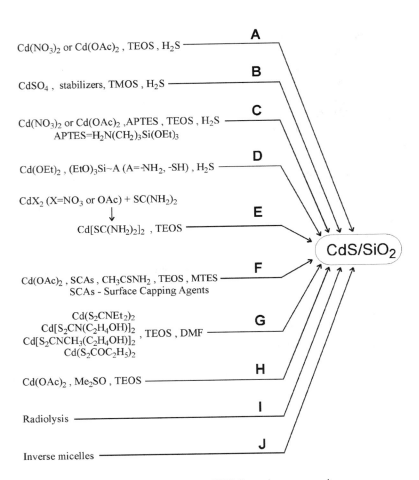

Fig. 1. Preparative routes to CdS/silica gel nanocomposites.

ethoxide gives a silica gel presumably doped with Cd(II) complexes [323]. Subsequent reaction with H_2S affords a nanocomposite containing CdS particles of less than 5 nm in diameter (Method D).

CdS/silica gel composites are also obtained by sol–gel condensation of TEOS, Cd(II) salts, and sources of sulfur other than H_2S. Using thiourea complexes of cadmium(II) nitrate [324–327] or acetate [328] together with TEOS gives CdS particles with average diameters of 2.1–8 nm (Method E). Thin-film CdS/silica gel nanocomposites have been prepared by mixing preformed CdS colloids with silica sols (Method F) [329]. CdS particles are also formed by the reaction of cadmium acetate with thioacetamide. Control of CdS particle size is achieved by using surface-capping agents as additives. Hydrolysis of TEOS along with dithiocarbamatocadmium complexes using dimethylformamide (DMF) as a cosolvent gives CdS particulates having average diameters of 2.7–7.2 nm (Method G) [330, 331]. Dimethylsulfoxide (DMSO) can serve as a source of sulfur and as solvent for the sol–gel conversion of TEOS and cadmium acetate to form a CdS/silica gel nanocomposite (Method H) [332]. The observed average CdS particle size is 2.7 nm.

Cadmium sulfide/silica gel composites can also be prepared using γ-radiolysis (Method I) [196, 333] and inverse micelle techniques (Method J) [196]. Other reports of CdS/silica gel nanocomposite formation are available [334–345].

Precipitation of CdS nanoparticles occurs in gel matrices at low temperatures, particularly when H_2S is the source of sulfur. More unreactive sources of sulfur, such as dithiocarbamate or thioacetamide, require mild heating for sulfur generation (usually as H_2S in an aquated sol–gel matrix). CdS/silica gel composites, once prepared, are frequently annealed at temperatures up to 1000 °C under nonoxidizing conditions. CdS particles are obtained as a distribution of sizes, although there is much interest in controlling this polydispersity [313, 317].

The spectral properties of CdS nanocomposites, in general, have been extensively studied because CdS nanoclusters behave as semiconductor quantum dots, and optical confinement effects are expected. CdS/silica gel nanocomposites have been studied by Raman spectroscopy [310, 311, 318, 335, 341, 343] and optical absorption spectroscopy [310, 312, 314, 316, 317, 319, 328, 337, 342, 346–348] including spectral hole burning [334]. Of particular interest are photoluminescence [307, 309, 323, 330, 331, 336, 340, 349] and NLO [148, 304–306, 314, 338, 339, 341, 342, 344, 345, 350] properties, as discussed elsewhere. Charge carrier dynamics has also been studied in CdS nanoparticles [320, 351].

Raman data reveal CdS vibrations at 300 and 600 cm^{-1} consistent with the bulk material. Exciton formation occurs by absorption of light with specific quantum confinement effects being observed depending on the relative size of the nanoparticle and the exciton radius. The silica xerogel represents a dielectric matrix surrounding the CdS particles, and significant third-order NLO effects are observed. Nanoparticulate CdS shows optical absorption that is blue-shifted relative to the bulk band gap of 512 nm at room temperature, and photoluminescence across the CdS band gap has been observed near 515 and 525 nm.

3.4.2. CdS/Sodium Borosilicate Gel Composites

Cadmium sulfide nanocomposites are formed in $Na_2O/B_2O_3/SiO_2$ matrices using sol–gel routes like those shown in Figure 2. Most commonly, a mixture of TMOS, $B(OEt)_3$, and aqueous $NaOAc/Cd(OAc)_2$ solution serves as a source of both the matrix and the CdS guest phases (Method A) [352–358]. In this process, $Cd(OAc)_2$ microcrystallites precipitate as a dopant in the wet gel. Thermal treatment in air converts the $Cd(OAc)_2$ to CdO, and subsequent exposure of the doped gel to H_2S transforms the CdO particles into CdS nanocrystals. The CdS-doped gel can then be densified further through drying procedures. The doping level and size distribution of the CdS nanoparticles, which affect the nonlinear optical properties of these composites, depends on specific experimental conditions. Typically, CdS average particle sizes range from 3.5 to 10 nm.

Nanosized CdS quantum dots with narrow size distribution can be fabricated by the addition of the bifunctional ligand, $H_2N(CH_2)_3Si(OEt)_3$, to a mixture of TEOS, triethylborate, NaOAc, and $Cd(OAc)_2$ (Method B) [359]. The amine ligand presumably binds to the Cd(II) ion through *in situ* coordination and anchors these ions homogeneously to the matrix through hydrolysis and condensation reactions of the triethoxysilyl group.

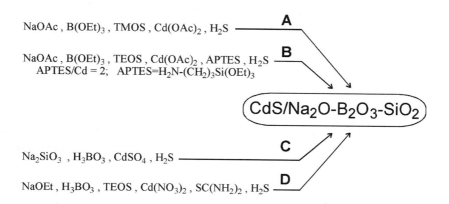

Fig. 2. Preparative routes to CdS/sodium borosilicate gel nanocomposites.

Sol–gel processing of sodium metasilicate, Na_2SiO_3, boric acid, H_3BO_3, with $CdSO_4$ as an additive gives a gel doped with Cd(II) ions. Following thermal treatment in air, these gels are heated to different temperatures while being exposed to H_2S gas (Method C) [360–363]. Careful control over the conditions of H_2S exposure affords nanocomposites containing CdS nanocrystals having average diameters of 2.15–9.2 nm.

Sodium borosilicate gels containing $Cd(NO_3)_2$ and $SC(NH_2)_2$ are formed using these compounds as additives and TEOS, boric acid, and sodium alkoxide as precursors of the host matrix. After heat treatment, the doped gels are exposed to H_2S gas to convert the formed CdO particles to CdS, giving average particle sizes of 8–10 nm (Method D) [364]. Other reports on the synthesis and properties of CdS/sodium borosilicate composites are available [365–367].

Because of the strong dependence of optical properties on semiconductor particle size, there is great interest in controlling the average size and particle size distributions in CdS nanocomposite materials. Regarding CdS/sodium borosilicate nanocomposites, studies of CdS particle size distributions on annealing conditions or exposure to H_2S have been reported [356, 360]. In general, CdS particle size increases with the temperature of reaction or annealing but is essentially independent of the period of exposure to H_2S. Narrow CdS particle size distributions [2.8 (0.9) nm in diameter] are obtained when $H_2N(CH_2)_3Si(OEt)_3$ is used as an additive [358, 359]. These materials have been successfully fabricated into channel waveguides.

Spectroscopic studies reported on CdS/sodium borosilicate gel nanocomposites include Raman spectroscopy [366], linear optical absorption spectroscopy [352, 354, 357, 359, 360, 362, 363], optical gain spectroscopy [354, 357], NLO properties [352, 353, 355, 367], and photoluminescence [358, 364]. The presence of CdS quantum dots is confirmed, and band edge luminescence near 495 nm is blue-shifted from the CdS bulk band gap energy of 512 nm with decreasing CdS particle size.

3.4.3. CdS/SiO₂–TiO₂ Gel Composites

Silica–titania ($30TiO_2$–$70SiO_2$) optical waveguides doped with CdS quantum dots have been prepared from TEOS, $Ti(OBu)_4$, and cadmium acetate [368]. Sulfide is obtained from thiourea, which is present as an additive. A more stable sol is achieved by adding acetylacetone to the cadmium acetate solution. The average CdS particle sizes obtained by this method range from 3.5 to 4.8 nm. Silica–titania (70:30) thin films containing CdS quantum dots form upon mixing CdS colloidal sols with a prehydrolyzed mixture of TEOS, methyltriethoxysilane (MTES), and $Ti(OBu)_4$ [329]. Cadmium sulfide particles are also obtained by the reaction of cadmium acetate with thioacetamide, and control of CdS particle size is achieved by using surface-capping agents (SCAs), such as acetylacetone, 3-aminopropyltriethoxysilane, 3-aminopropyltrimethoxysilane, or 3-mercaptopropyltrimethoxysilane. This procedure gives good control of CdS particle sizes in the 2- and 20-nm range.

In another approach, a silica–titania host layer 2.2 μm thick is deposited onto a Si substrate using sol–gel processing, and this layer is covered with a pure SiO_2 buffer layer 7.5 μm thick [369]. CdS crystallites precipitate in the porous matrix when this bilayer is soaked in CdF_2 solution and is subsequently treated with H_2S gas. The absorption spectrum of the resulting composite indicates that CdS crystallites having a maximum radius of approximately 2.6 nm are present.

A large blue shift (62 nm) of the CdS absorption edge is observed for SiO_2-TiO_2 films containing inclusions of CdS microcrystals [370]. This shift is attributed to quantum confinement effects in very small CdS quantum dots.

3.4.4. CdS Composites in Other Sol–Gel Matrices

Cadmium sulfide/alumina composites form when a mixture of cadmium acetate and $AlCl_3$ (which has been treated with NH_3 and peptized with acetic acid) is heated in a stream of

H_2S gas at temperatures between 100 and 600 °C [234, 371]. The NLO properties of these composites have been measured.

Stable confinement of CdS quantum dots in their lowest excitonic state has been achieved in a transparent ZrO_2 gel film. Optical absorption and fluorescent measurements of these films have been reported [372]. The addition of TiO_2 to the ZrO_2 gel matrix attenuates fluorescence emanating from impurities in the composite.

Polysilsequioxanes doped with CdS particles have been prepared by sol–gel condensation of either 1,4-bis(triethoxysilyl)benzene (Structure II) or 1,6-bis(triethoxysilyl)hexane (Structure IV) to give inorganic/organic gels. Soaking these xerogels for 2 days in aqueous $CdCl_2$ solution followed by treatment with a solution of Na_2S leads to precipitation of CdS [373, 374]. Because the host xerogels produced from the compound shown in Structure II or IV have different average pore sizes, the resulting CdS nanocomposites have average diameters of 5.8 and 9 nm, respectively.

Inorganic/organic composites containing nanoclusters of CdS and Co metal result when a Co nanocomposite prepared from 1,4-bis(triethoxysilyl)benzene (Structure II) and chromium tricarbonyl(triethoxysilyl)benzene (Structure I) is soaked with a solution of cadmium chloride. Sulfide is introduced by soaking the cadmium-doped gel with a solution of sodium sulfide, and CdS precipitation is induced by a subsequent heat treatment [116]. A mixture of phase-separated chromium and CdS nanocrystallites is obtained.

Optically transparent ZnO films on glass substrates (prepared by sol–gel dip processing) serve as a host matrix for CdS quantum dots [375]. In these CdS/ZnO composites, a rapid electron transfer from CdS particles to the ZnO matrix takes place rapidly following photoinduced electron–hole pair generation in the CdS nanoclusters. The stored electrons cause a blue shift of the absorption threshold of ZnO, an effect attributed to exciton polarization within the created excess electric field.

Silica glasses doped with $Cd_xZn_{1-x}S$ nanoclusters are formed by hydrolysis and polymerization of TEOS, cadmium acetate, and zinc acetate in DMSO solution [376]. Optical absorption measurements of the resulting composite indicate a greater shift of the optical absorption edge toward the blue than observed with related CdS/SiO_2 composites. The magnitude of this shift has been attributed to the possible formation of a solid solution in the glass matrix affording particularly small CdS clusters.

Sandwiched CdS–PbS composites are synthesized by the addition of a THF solution of lead methoxy ethanolate to a THF solution of $(CH_3)_3Si–S–Si(CH_3)_3$, cadmium ethoxide, and a bifunctional diamine that stabilizes the quantum dots [377]. PbS formation completely quenches the intense room temperature photoluminescence of the CdS clusters, indicating that PbS growth occurs at the surface of the CdS nanoparticles. PbS–CdS composites having cluster sizes less than 5 nm also exhibit third-order nonlinear optical properties.

CdS/Mn particles 1.2–2.4 nm in diameter have been prepared in silica xerogels using inverse micelle synthetic methods [336, 378]. Reagents include cadmium nitrate, manganese nitrate, sodium sulfide, pyridine (as a surface-capping agent), and $SiMe(OEt)_3$. Extensive studies (extended X-ray absorption fine structure, electron spin resonance, and electron–nuclear double resonance) reveal that incorporation of Mn(II) ions as a dopant in metal sulfide nanocomposites induces a bright photoluminescence (quantum yield of 7%) resulting from the $^4T_1 \rightarrow {}^6A_1$ transition centered on the Mn(II) ion.

3.4.5. CdS_xSe_{1-x}/Silica Gel Composites

Because of the difference in bulk band gaps between CdS (2.4 eV) and CdSe (1.7 eV), cadmium sulfoselenide/glass composites are commercially produced as color filters. Variation in the relative amounts of sulfur and selenium provides control over the transmission window of the glass filter. A desire to prepare color filters having cadmium chalcogenide particles of more uniform size has inspired preparations of CdS_xSe_{1-x}/silica gel composites [379–381]. Synthetic routes include mixing soluble sources of Cd(II) ion, sulfur, and selenium into conventional sol–gel formulations [379]; treatment of silica gels doped with

Cd(II) ions and selenide ions with gaseous H_2S [380]; successive diffusion of Cd(II) and chalcogenide solutions into preformed silica gels [381]; diffusion of chalcogenide ions into silica gels doped with Cd(II) ions [381]; or solid-couple diffusion of two separate gels, one containing Cd(II) ion and the other containing a source of chalcogenide ion [381]. $CdSeO_4$, NH_4SCN, K_2SeO_4, and elemental Se dissolved in nitric acid are common sources of chalcogenide ion. Ammonium thiocyanate is converted to thiourea in aqueous solution. Subsequent thermal treatment is the last step in these syntheses.

Variation of the S/Se atomic ratio is controlled through reactant stoichiometry. For those composites prepared by heating doped silica gels in an atmosphere of H_2S, the sulfide content increases with increasing temperature. XRD data confirm a single hexagonal nanoparticulate phase consistent with formation of a Cd/S/Se solid solution. The average sizes of the nanoparticles are usually less than 10 nm. Raman peaks at 300 and 200 cm^{-1} are assigned to the expected Cd–S and Cd–Se phonon bands, respectively. Optical absorption spectra reveal the expected blue shift in absorption onset.

3.4.6. CdSe/Gel Composites

CdSe nanocomposites have been prepared in silica gel [325, 338, 382–385], organically modified silica gel [386], sodium borosilicate gel [362, 363, 365, 387], or related matrices [337, 339, 346]. Synthetic methods parallel those used to prepare CdS/gel composites by sol–gel processing. Sources of selenide ion include KSeCN, H_2SeO_3, or Se dissolved in nitric acid. However, molecular precursors of Se, such as $Cd(SePh)_2$ or salts of $Cd[Se(NH_2)_2]_2^{2+}$, have also been used [325, 385]. The initially formed composites are usually treated thermally under nonoxidizing conditions to form CdSe quantum dots.

CdSe nanoparticle average sizes usually are 10 nm or less. Impregnation of silica gels doped with Cd(II) ions by solutions of KSeCN leads to bimodal CdSe particle size distributions [382]. Some control of CdSe average particle size is obtained through careful heat treatment [384]. As found with CdS nanocomposites, CdSe quantum dots provide both linear [337, 338, 384, 385, 387] and nonlinear [338, 339, 346] optical properties of interest.

3.4.7. CdTe/Gel Composites

CdTe nanocomposites in silica gel [388, 389], sodium borosilicate [141, 390, 391], or related matrices [346] are known. Sources of telluride ion include $CdTeO_4$, H_2TeO_4, or Te dissolved in nitric acid. The most common synthetic strategy involves adding soluble sources of both Cd(II) ion and tellurium to a conventional sol–gel formulation followed by thermal treatment. The CdTe average particle size is typically below 10 nm. Under controlled heat treatment, CdTe/silica gel composites having average CdTe particle sizes in the range of 4 to 9 nm have been prepared. Although the optical absorption edge is blue-shifted as CdTe nanocluster size decreases [388, 389], this shift is not always attributed to quantum confinement effects [390]. A mixed-metal telluride composite, $Cd_xHg_{1-x}Te$/silica gel, has been prepared by sol–gel methods using $Hg(OAc)_2$ as a soluble source of Hg [337, 338, 392]. The optical absorption spectrum and NLO properties of this composite have been measured. A blue shift in absorption edge is observed with decreasing nanocluster size [392].

3.4.8. ZnS/Gel Composites

Zinc sulfide is a semiconductor substance with an optical band gap of 3.6 eV (344 nm). As expected, quantum dots of ZnS have electronic structures between those of bulk and molecular materials, and quantum confinement effects should be evident by a blue shift of the absorption edge with decreasing particle size. As with cadmium chalcogenide quantum dots, sol–gel preparations of ZnS nanocomposites provide an entry into the fabrication of devices having tunable optical properties, including nonlinear optical effects.

ZnS nanocomposites have been prepared by sol–gel processing in silica gel [325, 326, 337–339, 393–396], aluminoborosilicate gel [397, 398], and related glassy matrices [333]. Synthetic routes include (1) the addition of zinc and sulfide sources, such as $Zn(OAc)_2$ and thiourea or a zinc thiourea complex, to a conventional sol–gel formulation [325, 326, 339]; (2) a two-step process in which Zn(II) ion is incorporated directly into a sol–gel matrix using zinc additives followed by the introduction of sulfide using H_2S thermal treatment [337, 338, 393–396]; (3) the preparation of a sol–gel matrix of well-defined pore structure followed by successive incorporation of Zn(II) by solution absorption and sulfide ion by H_2S thermal treatment [397, 398]; or (4) γ-radiolysis or membrane methods [333]. Thermal aging or annealing of the initially prepared nanocomposite induces ZnS particle growth.

ZnS average particle sizes are typically less than 10 nm. Synthetic control of ZnS particle size distribution by careful thermal treatment and regulation of doping level leads to size distributions of approximately 2–3 nm in breadth. Size distributions of approximately 6 nm in breadth are otherwise more common. ZnS particle size increases with increasing dopant level and annealing temperature. Careful preparations of ZnS/aluminoborosilicate composites indicate that the pore size of the host matrix constrains the size of ZnS particulates. Average pore sizes of 2.0–3.5 nm give ZnS average particle sizes of 2.3–4.6 nm. The larger ZnS particles are formed at higher annealing temperatures where pore size restraint begins to break down [397, 398].

ZnS quantum dots crystallize with the sphalerite structure [396], and Raman spectra reveal Zn–S vibrations at 283 and 544 cm^{-1} [395]. The optical absorption edges of ZnS quantum dots of average sizes 2.3, 4.0, or 4.6 nm appear at 280, 310, or 315 nm, respectively, showing the expected blue shift of absorption onset with decreasing dot size [398]. NLO properties of the ZnS nanocomposites have also been observed [338, 339, 393–395], including photoluminescence [394].

3.4.9. ZnSe or ZnTe/Gel Composites

Zinc chalcogenide/silica gel nanocomposites of the heavier chalcogens, Se and Te, have been prepared using synthetic routes analogous to those used for the preparation of ZnS composites [325, 338, 339]. Ascorbic acid can be used as an antioxidant additive [325]. Optical absorption spectra reveal band gaps of the ZnX quantum dots greater than that of the bulk substances, and NLO measurements have been reported [338, 339].

3.4.10. PbS/Gel Composites

Lead sulfide is a semiconductor with an optical band gap near 0.41 eV (3020 nm). Composite materials containing PbS quantum dots are of interest as possible optical devices based on expected quantum confinement effects.

PbS nanocomposites have been prepared via sol–gel processing in silica gel [329, 393, 399–402], aluminoborosilicate gel [397, 403], silica–titania gels [329, 368], and related glassy matrices using methods similar to those used to prepare ZnS nanocomposites [333, 339, 404]. Synthetic routes include (1) the addition of lead and sulfide sources, such as lead acetate or nitrate salts or lead thiourea complexes and thiourea or thioacetamide, to conventional sol–gel formulations [329, 368, 401]; (2) a two-step process in which Pb(II) ion is incorporated directly into a sol–gel matrix using lead additives followed by the introduction of sulfide using H_2S thermal treatment [339, 393, 399]; (3) the preparation of a sol–gel matrix of well-defined pore structure followed by successive incorporation of Pb(II) by solution absorption and sulfide ion by H_2S thermal treatment [397]; (4) mixing colloidal PbS dispersions into conventional sols [400]; or (5) γ-radiolysis and membrane methods [333, 404]. Thermal annealing of the formed nanocomposites is a common practice.

Average PbS particle sizes are typically less than 10 nm, although PbS particles with average diameters of 25–31 nm have also been reported [368]. Synthetic control of PbS particle size distribution by careful thermal treatment and regulation of doping level leads to distributions of approximately 0.8–3.5 nm in breadth. Size distributions of approximately 2–6 nm in breadth are otherwise more commonly observed. The PbS average particle size increases with increasing dopant level and annealing temperature [401]. Careful preparations of PbS/aluminoborosilicate composites reveal that the pore size of the host matrix constrains PbS particle size formation, but to a lesser degree than observed for ZnS particulate formation [397].

The optical absorption edge of bulk PbS near 3020 nm can be blue-shifted to near 650 nm through quantum confinement effects within PbS quantum dots 2–3 nm in size [401]. Intermediate absorption edge onsets can be tailored by control of the PbS average particle size. NLO properties of the PbS nanocomposites have also been observed [339, 400].

3.4.11. PbSe or PbTe/Gel Composites

PbSe/porous glass composites are obtained by soaking gels doped with Pb(II) ions with solutions containing selenide ion [339]. The resulting glassy composites show quantum confinement effects and third-order harmonic generation.

PbTe nanocomposites in silica or borosilicate gels are formed by conventional sol–gel methods using lead nitrate and H_6TeO_6 as additives [405]. The initial gel is treated in air at 450 °C followed by further annealing in nitrogen/hydrogen atmosphere. Annealing temperatures of 400–500 °C for periods of 10–120 min give crystalline PbTe particulates 2–15 nm in size. PbTe particle growth is more rapid in a borosilicate matrix. Although the band gap of bulk PbTe is in the infrared (0.31 eV, 4000 nm), the absorption edge onset is shifted into the near-visible region for these PbTe nanocomposites.

3.4.12. Other Metal Chalcogenide/Gel Nanocomposites

The addition of bismuth nitrate and thiourea to a conventional silica sol–gel formulation followed by thermal treatments of 350 °C in air and 400–500 °C in nitrogen gives a Bi_2S_3/silica gel nanocomposite [148, 406–408]. The Bi_2S_3 average particle size is less than 10 nm. Complex optical absorption spectra show absorption valleys ranging from 510 to 660 nm with a red shift being observed for gels subjected to longer heating. Thermal treatment for 2 h at 400 °C affords Bi_2S_3 particles with an average diameter of 2 nm. This composite emits luminescence ranging from 345 to 540 nm upon photoexcitation.

Sol–gel processing using the techniques mentioned previously has been used to prepare several metal chalcogenide nanocomposites for designed applications or property enhancements. Materials of these types include (1) WSe_3 nanocomposites as patternable silicate–copolymer films and monoliths for optical applications [386], (2) submicrometer silica gel particles suspended in styrene–butadiene latex rubber for strength reinforcement [409], (3) silica gel particulates on silicon for integrated optical devices [316], (4) SnO_2 deposition in the pores of porous silicon for gas sensor applications [410], (5) ZnO membrane formation from sol–gel-derived ZnO colloids as luminescent chemical sensors [411], (6) nanoparticles of Cu or CuO as coatings for coloration [412], or (7) boehmite coatings of Si_3N_4 particles to provide fracture-free green-state deformation of solid compacts [413].

3.5. Metal Halide Nanocomposites

3.5.1. Silver Halide/Gel Nanocomposites

Colloidal and surface-passivated submicrometer particles of a Ag halide have been prepared as a dispersion in a gel matrix using freeze–dry and sol–gel methods [414]. This nanocomposite exhibits large NLO properties because of quantum size effects.

3.5.2. Copper Halide/Gel Nanocomposites

Suspensions of Cu_2O or CuX (X = Cl or Br) in acetonitrile have been used in the preparation of glasses doped with quantum dots of CuCl and CuBr, as shown in Figure 3 [148, 304, 415, 416]. Complete dissolution of the copper reagent occurs in the sol.

Thin-film glass composites of CuCl or CuBr having waveguide properties are obtained by the addition of a Cu_2O/acetonitrile suspension to partially hydrolyzed TEOS [304]. Sol–gel processing is initiated by the addition of water, methanol, and either HCl or HBr. Copper halide average particle size is controlled (48.2–51.9 nm) through regulation of the Cu_2O concentration and by using montmorillonite as an entrapping support (23.9 nm). These nanocomposites exhibit NLO properties.

Silica composites containing cubic CuCl nanocrystals 3–6 nm in size are formed by sol–gel processing of solutions of TEOS and CuCl using HCl as a catalyst [415]. Optical absorption measurements indicate strong quantum confinement effects, and peaks observed at approximately 370 and 380 nm for samples heated above 700 °C are attributed, respectively, to excitations of the confined $Z_{1,2}$ and Z_3 excitons in CuCl microcrystals. As the size of the CuCl crystals decreases, exciton energies show a blue shift. A resonant third-order nonlinear susceptibility of 1.1×10^{-8} esu is observed at 77 K for this nanocomposite.

CuBr doped silica–alumina composites are obtained through sol–gel processing of solutions containing TEOS, $Al(OC_4H_9)_3$, and CuBr [416]. Heating the resulting gels to 900 °C in N_2 results in the formation of cubic CuBr nanocrystals 5–8 nm in diameter. Optical absorption spectra of samples heated above 700 °C show bands near 3.05 and 3.20 eV at 77 K. These peaks are attributed to the $Z_{1,2}$ and Z_3 excitons, respectively, in CuBr nanocrystals and show a blue shift as the size of the CuBr crystals decreases.

3.5.3. Magnesium Halide/Gel Nanocomposites

MgF_2/silica gel composites, containing MgF_2 particles with an average size of 5–10 nm, depending on the conditions of thermal treatment, are formed when a MgF_2 sol is mixed with silicate sols. The MgF_2 sol is formed by reacting HF with a sol prepared from methanolic H_2O_2 and $Mg(OCH_3)_2$, whereas the silicate sols are prepared from acid-catalyzed hydrolysis and condensation of TMOS [417].

3.5.4. Lead Halide/Gel Nanocomposites

PbI_2/silica gel nanocomposite films containing PbI_2 particles with an average diameter of 6 nm have been prepared by sol–gel methods [148, 418–420]. The Pb(II) ions are introduced into the sol as an additive, while iodide ions are added at a later stage. X-ray

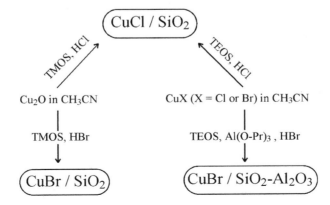

Fig. 3. Preparative routes to copper halide/gel nanocomposites.

diffraction from the PbI_2 particles is observed only when the average particle size is larger than 8 nm [418]. Low-temperature luminescence from the PbI_2 nanoparticles consists of a series of exciton lines near 2.5 eV and additional donor-acceptor combination luminescent bands at lower energies, centered at 2.44 eV (green) and 2.07 eV (red), respectively. A correlation between donor–acceptor recombination emission processes and relaxation by lattice imperfection was examined utilizing optically detected magnetic resonance (ODMR) spectroscopy.

4. SUMMARY

Sol–gel processing of nanocomposite materials has attracted much interest because of the generality of this synthetic method. A wide variety of host matrices can be formed under very mild conditions with relatively inexpensive reagents using sol–gel chemistry. Conversion of a sol to a gel using hydrolysis and intermolecular condensation reactions occurs, under proper conditions, even for complex reactant mixtures affording convenient syntheses of host matrices having complex compositions. This flexibility enables the tuning of the pore size, surface area, density, dielectric constant, refractive index, and chemical composition of a host matrix. Inorganic/organic matrices are also accessible, and sol–gel chemistry even occurs within inverse micelles.

Formation of the nanoparticulate phase is achieved most commonly through reduction processes for elemental particles or precipitation reactions for chemical substances. A variety of methods are available to initiate particle formation. Nanoparticle growth is usually promoted by thermal annealing, and, for many substances, thermal energy is required for single-crystal formation. This growth process typically affords nanoclusters of the guest phase having a distribution of particle sizes. Synthetic control of nanoparticle size has been, and remains, a challenge in sol–gel processing. Although control of host matrix pore size gives some degree of control of nanoparticle size, convenient routes to monodisperse nanocomposites are still needed. Although a high degree of monodispersity might not be required for some applications, it is highly desired for most optical applications.

Classical homogeneous nucleation theories, as applied to the growth of CdS or CdSe quantum dots in a glass matrix, predict (1) a nucleation stage in which nanocluster size distribution is Gaussian; (2) a normal growth stage where the average radius of the nanoclusters is proportional to the square root of the annealing time; and (3) a coalescence stage in which a supersaturated glass solid solution is established, and the average radius of the nanoclusters is proportional to the cube root of the annealing time [421]. Coalescence-like particle growth is observed for CdSe particles of average radius 2.5 (1.4) nm grown in a silicate-based glass matrix. The growths of Ag nanoclusters in a sputtered SiO_2 matrix and of ZnS nanoclusters in a sol–gel silica matrix also follow a cube root dependence on annealing time [396, 422]. Given the dependence of average nanoparticle size and particle size distribution on the time and temperature of thermal treatment, less energy intensive synthetic strategies are needed to achieve highly monodisperse nanocomposites.

Acknowledgments

C.M.L. thanks the U.S. Army Research Office under Grant DAAH04-95-1-0146 for support. We also thank Mr. Stan Griffin for assistance in preparing the manuscript.

References

1. C. J. Brinker and G. W. Scherer, "Sol–Gel Science." Academic Press, Boston, 1990.
2. A. K. Cheetham, C. J. Brinker, M. L. Mecartney, and C. Sanchez, eds., "Better Ceramics through Chemistry VI," Vol. 346. Materials Research Society, Pittsburgh, PA, 1994.

3. M. J. Hampden-Smith, W. G. Klemperer, and C. J. Brinker, eds., "Better Ceramics through Chemistry V," Vol. 271. Materials Research Society, Pittsburgh, PA, 1992.
4. L. L. Hench and J. K. West, *Chem. Rev.* 90, 33 (1990).
5. R. J. P. Corriu and D. Leclercq, *Angew. Chem., Int. Ed. Engl.* 35, 1420 (1996).
6. J. Zarzycki, *J. Sol.–Gel Sci. Technol.* 8, 17 (1997).
7. S. Sakka, Proc. SPIE-Int. *Soc. Opt. Eng.* 1758, 2 (1992).
8. A. M. Klonkiowski, *Wiad. Chem.* 47, 497 (1993).
9. J. D. Mackenzie and D. R. Ulrich, *Proc. SPIE-Int. Soc. Opt. Eng.* 1328, 2 (1990).
10. (a) R. M. Laine, *Proc. SPIE-Int. Soc. Opt. Eng.* 1328, 16 (1990). (b) J. Y. Ying, ed., "Sol–Gel Derived Materials," *Chem. Mater.* 9 (1997).
11. M. A. Cauqui and J. M. Rodriguez-Izquierdo, *J. Non-Cryst. Solids* 147–148, 724 (1992).
12. G. M. Pajonk, *Appl. Catal.* 72, 217 (1991).
13. U. Schubert, *New J. Chem.* 18, 1049 (1994).
14. D. R. Uhlmann, S. Motakef, T. Suratwala, J. Young, J. M. Boulton, B. J. J. Zelinski, Z. Gardlund, G. Teowee, and J. Cronin, *Proc. SPIE* 2288, 2 (1994).
15. B. K. Coltrain, C. Sanchez, D. W. Schaefer, and G. L. Wilkes, eds., "Better Ceramics through Chemistry VII," Vol. 435. Materials Research Society, Pittsburgh, PA, 1996.
16. U. Schubert, N. Husing, and A. Lorenz, *Chem. Mater.* 7, 2010 (1995).
17. B. Lebeau, S. Brasselet, J. Zyss, and C. Sanchez, *Chem. Mater.* 9, 1012 (1997).
18. L. Ukrainczyk, R. A. Bellman, and A. B. Anderson, *J. Phys. Chem. B* 101, 531 (1997).
19. J. C. Schrotter, M. Smaihi, and C. Guizard, *Mater. Res. Soc. Symp. Proc.* 435, 199 (1996).
20. G. M. Jamison, D. A. Loy, K. A. Opperman, J. V. Beach, and R. M. Waymouth, *Polym. Prepr.* 37, 297 (1996).
21. S. K. Young, Q. Deng, and K. A. Mauritz, *Polym. Mater. Sci. Eng.* 74, 309 (1996).
22. C. L. Beaudry and L. C. Klein, *ACS Symp. Ser.* 622, 382 (1996).
23. G. W. Jang, C. Chen, R. W. Gumbs, Y. Wei, and J. M. Yeh, *J. Electrochem. Soc.* 143, 2591 (1996).
24. F. Leroux, B. E. Koene, and L. F. Nazar, *J. Electrochem. Soc.* 143, L181 (1996).
25. C.G. Wu, D. C. DeGroot, H. O. Marcy, J. L. Schindler, C. R. Kannewurf, Y.-J. Liu, W. Hirpo, and M. G. Kanatzidis, *Chem. Mater.* 8, 1992 (1996).
26. Y.J. Liu, J. L. Schindler, D. C. DeGroot, C. R. Kannewurf, W. Hirpo, and M. G. Kanatzidis, *Chem. Mater.* 8, 525 (1996).
27. P. R. McDaniel and T. L. St Clair, *Mater. Res. Soc. Symp. Proc.* 405, 535 (1996).
28. M. A. Harmer, W. E. Farneth, and Q. Sun, *J. Am. Chem. Soc.* 118, 7708 (1996).
29. G. M. Kloster, J. A. Thomas, P. W. Brazis, C. R. Kannewurf, and D. F. Shriver, *Chem. Mater.* 8, 2418 (1996).
30. M. A. Harmer and Q. Sun, PCT Int. Appl. WO 9619288 A1 960627, 59 pp.
31. H. Krug, N. Merl, and H. Schmidt, *Teubner-Texte Phys.* 27, 192 (1993).
32. C. Sanchez, M. In, P. Toledano, and P. Griesmar, *Mater. Res. Soc. Symp. Proc.* 271, 669 (1992).
33. M. A. Meneses-Nava, S. Chavez-Cerda, and J. J. Sanchez-Mondragon, *Proc. SPIE-Int. Soc. Opt. Eng.* 2778, 395 (1996).
34. P. N. Prasad, R. Gvishi, G. S. He, J. D. Bhawalkar, N. D. Kumar, G. Ruland, C. F. Zhao, and B. A. Reinhardt, *Mater. Res. Soc. Symp. Proc.* 413, 203 (1996).
35. E. J. A. Pope, *Mater. Res. Soc. Symp. Proc.* 345, 331 (1994).
36. D. L'Esperance and E. L. Chronister, in "Act. Mater. Adapt. Struct., Proc. ADPA/AIAA/ASME/SPIE Conference" (J. G. Knowles, ed.), pp. 875–878. IOP Publishing, Bristol, UK 1992.
37. Y. Wei, D. Jin, and T. Ding, *J. Phys. Chem. B* 101, 3318 (1997).
38. S. Yano and M. Kodomari, *Nihon Reoroji Gakkaishi* 24, 15 (1996).
39. F. Bentivegna, M. Canva, A. Brun, F. Chaput, and J. P. Boilot, *J. Appl. Phys.* 80, 4655 (1996).
40. G. Qian and M. Wang, *J. Phys. Chem. Solids* 58, 375 (1997).
41. R. L. Matthews and E. T. Knobbe, *Chem. Mater.* 5, 1697 (1993).
42. Y. J. Lin and C. J. Wu, *Surf. Coat. Technol.* 88, 238 (1997).
43. R. Reisfeld, V. Chernyak, M. Eyal, and C. K. Jorgensen, in "Int. Sch. Excited States Transition Elem." (W. Strek, ed.), 2nd ed., Vol. 247. World Scientific, Singapore, 1992.
44. T. N. Vasilevskaya and R. I. Zakharchenya, *Fiz. Tverd. Tela* 38, 3129 (1996).
45. G. E. S. Brito, S. J. L. Ribeiro, V. Briois, J. Dexpert-Ghys, C. V. Santilli, and S. H. Pulcinelli, *J. Sol.–Gel Sci. Technol.* 8, 261 (1997).
46. S. P. Feofilov, A. A. Kaplyanskii, A. B. Kutsenko, T. N. Vasilevskaya, and R. I. Zakharchenya, *Mater. Sci. Forum* 239–241, 687 (1997).
47. K. Yao, K. Kong, L. Kong, L. Zhang, and X. Yao, *Xi'an Jiaotong Daxue Xuebao* 30, 40, 59 (1996).
48. M. C. Gust, L. A. Momoda, and M. L. Mecartney, *Ceram. Trans.* 55, 85 (1995).
49. Z. Xu, H. K. Chae, M. H. Frey, and D. A. Payne, *Mater. Res. Soc. Symp. Proc.* 271, 339 (1992).
50. Z. Zhong, Z. Hu, O. Yan, X. Fu, and H. Peng, *Hunan Shifan Daxue Ziran Kexue Xuebao* 19, 48 (1996).
51. Y. Wu, T. Yamaguchi, and C. Xianghao, *Nippon Kagaku Kaishi* 12, 1376 (1993).
52. S. Zhao, H. Cai, Z. Peng, C. Zhao, B. Xu, M. Zhao, and J. Xing, *Wuji Huaxue Xuebao* 10, 409 (1994).

53. Z. Zhong, L. Chen, Q. Yan, X. Fu, and J. Hong, *Stud. Surf. Sci. Catal.* 91, 647 (1995).

54. W. H. Shih and Q. Lu, *Ferroelectrics* 154, 1379 (1994).

55. W. L. Zhong, B. Jiang, P. L. Zhang, J. M. Ma, H. M. Cheng, Z. H. Yang, and L. X. Li, *J. Phys.: Condens. Matter.* 5, 2619 (1993).

56. S. Lu, L. Zhang, and X. Yao, *Proc. IEEE Int. Symp. Appl. Ferroelectr.* 8, 385 (1992).

57. Q. Zhang, Z. Zhang, D. Shen, W. Xue, H. Wang, and M. Zhao, *Proc. IEEE Int. Symp. Appl. Ferroelectr.* 8, 551 (1992).

58. Q. F. Zhou, Y. Han, L. Y. Zhang, and X. Yao, *Proc. IEEE Int. Symp. Appl. Ferroelectr.* 8, 621 (1992).

59. Q. Zhou, Y. Li, L. Zhang, and X. Yao, *Chin. Sci. Bull.* 37, 1402 (1992).

60. T. R. Narayanan Kutty and P. Padmini, *J. Mater. Chem.* 7, 521 (1997).

61. K. No, D. S. Yoon, and J. M. Kim, *J. Mater. Res.* 8, 245 (1993).

62. Y. A. Attia, D. K. Sengupta, and H. A. Hamza, in "Sol.–Gel Process. Appl." (Y. A. Attia, ed.), pp. 159–168. Plenum, New York, 1994.

63. A. R. Modak and S. K. Dey, *Integr. Ferroelectr.* 5, 321 (1994).

64. J. F. Meng, B. K. Rai, R. S. Katiyar, and G. T. Zou, *Phys. Lett. A* 229, 254 (1997).

65. A. K.Bhattacharya, K. K. Mallick, A. Hartridge, and J. L. Woodhead, *Mater. Lett.* 18, 247 (1994).

66. J. Mira, J. Rivas, D. Fiorani, R. Caciuffo, D. Rinaldi, C. Vazquez Vazquez, J. Mahia, M. A. Lopez Quintela, and S. B. Oseroff, *J. Magn. Magn. Mater.* 164, 241 (1996).

67. Y. K. Sun, I. H. Oh, and S. A. Hong, *J. Mater. Sci.* 31, 3617 (1996).

68. Y. Fan and D. Chen, *Wuji Huaxue Xuebao* 8, 194 (1992).

69. B. K. Kim, N. Mizuno, and I. Yasui, *Kongop Hwahak* 7, 1034 (1996).

70. R. M. Almeida, M. I. DeBarros Marques, and X. Orignac, *J. Sol.–Gel Sci. Technol.* 8, 293 (1997).

71. S. M. Koo, D. H. Lee, C. S. Ryu, and Y. E. Lee, *Kongop Hwahak* 8, 301 (1997).

72. G. Dagan, S. Sampath, and O. Lev, *Chem. Mater.* 7, 446 (1995).

73. I. M. Miranda Salvado, F. M. A.Margaca, and J. Teixeira, *J. Mol. Struct.* 383, 271 (1996).

74. A. Hanprasopwattana, S. Srinivasan, A. G. Sault, and A. K. Datye, *Langmuir* 12, 3173 (1996).

75. S. M. Melpolder, A. W. West, and C. L. Bauer, *Mater. Res. Soc. Symp. Proc.* 239, 371 (1992).

76. G. W. Koebrugge, L. Winnubst, and A. J. Burggraaf, *J. Mater. Chem.* 3, 1095 (1993).

77. N. Ozer, S. De Souza, and C. M. Lampert, *Proc. SPIE-Int. Soc. Opt. Eng.* 2531, 143 (1995).

78. K. Kameyama, S. Shohji, S. Onoue, K. Nishimura, K. Yahikozawa, and Y. Takasu, *J. Electrochem. Soc.* 140, 1034 (1993).

79. S. Onoue, K. Kameyama, K. Yahikozawa, and Y. Takasu, *Proc. Electrochem. Soc.* 93-14, 158 (1993).

80. K. E. Swider, C. I. Merzbacher, P. L. Hagans, and D. R. Rolison, *Chem. Mater.* 9, 1248 (1997).

81. J. Y. Jeong, H. M. Lee, and H. L. Lee, *Yoop Hakhoechi* 33, 1138 (1996).

82. H. L. Lee, J. Y. Jeong, and H. M. Lee, *Yoop Hakhoechi* 33, 1147 (1996).

83. F. Lu, Q. Xu, Y. Lu, R. Ye, and B. Song, *J. Univ. Sci. Technol. Beijing* 3, 1 (1996).

84. D. L. Bourell and K. W. Parimal, *J. Am. Ceram. Soc.* 76, 705 (1993).

85. G. B. Prabhu and D. L. Bourell, *Nanostruct. Mater.* 5, 727 (1995).

86. V. M. Sglavo, R. Dal. Maschio, G. D. Soraru, and G. Carturan, *J. Eur. Ceram. Soc.* 11, 439 (1993).

87. Y. Murakami, H. Ohkawauchi, M. Ito, K. Yahikozawa, and Y. Takasu, *Electrochim. Acta* 39, 2551 (1994).

88. J. Yuan, Z. Liu, Q. Fang, and X. Xu, *Huadong Ligong Daxue Xuebao* 22, 316 (1996).

89. M. Sando, A. Towata, and A. Tsuge, *Funtai Kogaku Kaishi* 29, 755 (1992).

90. L. Armelao, R. Bertoncello, L. Crociani, G. Depaoli, G. Granozzi, E. Tondello, and M. Bettinelli, *J. Mater. Chem.* 5, 79 (1995).

91. C. Suerig, K. A. Hempel, and C. Sauer, *J. Magn. Magn. Mater.* 157/158, 268 (1996).

92. A. K. Bhattacharya, A. Hartridge, K. K. Mallick, C. K. Majumdar, D. Das, and S. N. Chintalapudi, *J. Mater. Sci.* 32, 557 (1997).

93. T. Ishiwaki, H. Inoue, and A. Makishima, *J. Non-Cryst. Solids* 203, 43 (1996).

94. C. Liu, Y. Shen, and Q. Xi, *Yingyong Huaxue* 13, 72 (1996).

95. P. G. Harrison, *Catal. Today* 17, 483 (1993).

96. H. Wang and W. J. Thomson, *AIChE J.* 41, 1790 (1995).

97. M. Nofz, R. Stoesser, B. Unger, and W. Herrmann, *J. Non-Cryst. Solids* 149, 62 (1992).

98. M. Zhou, S. Zhou, and Y. Wang, *Huadong Huagong Xueyuan Xuebao* 18, 74 (1992).

99. Z. Liu, Y. Sun, X. Du, and J. Cheng, *Huadong Huagong Xueyuan Xuebao* 18, 1 (1992).

100. N. Satyanarayana, R. Patcheammalle, P. Muralidharan, and M. Venkateswarlu, *Bull. Electrochem.* 12, 658 (1996).

101. N. Satyanarayana, P. Muralidharan, R. Patcheammalle, M. Venkateswarlu, and G. V. Rama Rao, *Solid State Ionics* 86–88, 543 (1996).

102. S. Komarneni and L. Rani, *Proc. SPIE-Int. Soc. Opt. Eng.* 1622, 147 (1992).

103. M. Ishida, H. Jin, and T. Okamoto, *Energy Fuels* 10, 958 (1996).

104. N. N. Ghosh and P. Pramanik, *Br. Ceram. Trans.* 95, 209 (1996).

105. X. Li, H. Zhang, F. Chi, S. Li, B. Xu, and M. Zhao, *Mater. Sci. Eng., B* 18, 209 (1993).

106. W. Nie, G. Boulon, C. Mai, C. Esnouf, R. Xu, and J. Zarzycki, *J. Non-Cryst. Solids* 121, 282 (1990).

107. W. Nie, G. Boulon, C. Mai, C. Esnouf, R. Xu, and J. Zarzycki, *Chem. Mater.* 4, 216 (1992).

108. W. Nie, G. Boulon, C. Mai, C. Esnouf, R. Xu, and J. Zarzycki, *Mater. Chem. Phys.* 25, 105 (1990).

109. F. Keller-Besrest and S. Benazeth, *J. Phys. IV* 4, C9/117 (1994).

110. G. M. Pajonk, *Heterog. Chem. Rev.* 2, 129 (1995).

111. U. Schubert, F. Schwertfeger, and C. Doersmann, *ACS Symp. Ser.* 622, 366 (1996).

112. C. M. Lukehart, J. P. Carpenter, S. B. Milne, and K. J. Burnam, *Chem. Tech.* 23, 29 (1993).

113. J. P. Carpenter, C. M. Lukehart, S. B. Milne, S. R. Stock, J. E. Wittig, B. D. Jones, R. Glosser, D. O. Henderson, and R. Mu, *Int. SAMPE Tech. Conf.* 27, 549 (1995).

114. W. D. King, D. L. Boxall, and C. M. Lukehart, *J. Cluster Sci.* 8, 267 (1997).

115. K. M. Choi and K. J. Shea, *Mater. Res. Soc. Symp. Proc.* 346, 763 (1994).

116. K. M. Choi and K. J. Shea, *J. Am. Chem. Soc.* 116, 9052 (1994).

117. J. P. Wang and H. L. Luo, *J. Magn. Magn. Mater.* 131, 54 (1994).

118. A. Basumallick, K. Biswas, S. Mukherjee, and G. C. Das, *Mater. Lett.* 30, 363 (1997).

119. J. P. Wang, H. Han, H. L. Luo, B. Liu, S. B. Hu, and T. S. Low, *IEEE Trans. Magn.* 32, 4496 (1996).

120. S. Roy, D. Das, D. Chakravorty, and D. C. Agrawal, *J. Appl. Phys.* 74, 4746 (1993).

121. A. Chatterjee and D. Chakravorty, *J. Phys. D: Appl. Phys.* 23, 1097 (1990).

122. A. Chatterjee and D. Chakravorty, *J. Mater. Sci.* 27, 4115 (1992).

123. S. Roy and D. Chakravorty, *Jpn. J. Appl. Phys., Part 1* 32, 3515 (1993).

124. J. P. Wang, D. Han, H. L. Luo, Q. X. Lu, and Y. W. Sun, *Appl. Phys. A* 61, 407 (1995).

125. J. P. Wang, D. Han, H. L. Luo, Y. W. Sun, and Q. X. Lu, *Wuli Xuebao* 44, 963 (1995).

126. V. Vendange and P. Colomban, *Mater. Sci. Eng., A* 168, 199 (1993).

127. K. M. Choi and K. J. Shea, *J. Sol.–Gel Sci. Technol.* 5, 143 (1995).

128. A. Chatterjee, A. Datta, A. K. Giri, and D. Chakravorty, *J. Appl. Phys.* 72, 3832 (1992).

129. T. Lopez, R. Gomez, O. Novaro, A. R. Solis, E. S. Mora, S. Castillo, E. Poulain, and J. M. Magadan, *J. Catal.* 141, 114 (1993).

130. J. P. Carpenter, C. M. Lukehart, S. B. Milne, S. R. Stock, and J. E. Wittig, *Inorg. Chim. Acta* 251, 151 (1996).

131. K. J. Burnam, J. P. Carpenter, C. M. Lukehart, S. B. Milne, S. R Stock., B. D. Jones, R. Glosser, and J. E. Wittig, *Nanostruct. Mater.* 5, 155 (1995).

132. Y. Zhang, Q. Wang, and B. Zhong, *Ranliao Huaxue Xuebao* 24, 517 (1996).

133. B. Breitscheidel, J. Zieder, and U. Schubert, *Chem. Mater.* 3, 559 (1991).

134. C. Goersmann, B. Breitscheidel, and U. Schubert, in "Organosilicon Chemistry" (N. Auner and J. Weis, eds.), p. 319. VCH Publisher, Weinheim, Germany, 1994.

135. U. Schubert, S. Tewinkel, and R. Lamber, *Chem. Mater.* 8, 2047 (1996).

136. P. Reyes, M. Morales, and G. Pechhi, *Bol. Soc. Chil. Quim.* 41, 221 (1996).

137. T. J. Lee, K. G. Sheppard, A. Gamburg, and L. Klein, *Proc. Electrochem. Soc.* 94, 231 (1994).

138. R. Monaci, A. Musinu, G. Piccaluga, and G. Pinna, *Mater. Sci. Forum* 195, 1 (1995).

139. S. Roy and D. Chakravorty, *J. Phys.: Condens. Matter* 6, 8599 (1994).

140. S. Roy, A. Chatterjee, and D. Chakravorty, *J. Mater. Res.* 8, 689 (1993).

141. J. Petrullat, S. Ray, U. Schubert, G. Guldner, C. Egger, and B. Breitscheidel, *J. Non-Cryst. Solids* 147–148, 594 (1992).

142. D. Spron, J. Grossmann, A. Kaiser, R. Jahn, and A. Berger, *Nanostruct. Mater.* 6, 328 (1995).

143. J. Grossmann, K. Rose, and D. Sporn, *Ceram. Trans.* 51, 713 (1995).

144. E. D. Rodeghiero, O. K. Tse, B. S. Wolkenberg, and E. P. Giannelis, *Process. Fabr. Adv. Mater. IV, Proc. Symp., 4th*, 775 (1995). CA: 126:10481.

145. T. Okubo, M. Watanabe, K. Kusakabe, and S. Morooka, *Key Eng. Mater.* 61–62, 71 (1991).

146. T. K. Kundu and D. Chakravorty, *J. Phys. Soc. Jpn.* 65, 3357 (1996).

147. T. K. Kundu and D. Chakravorty, *Appl. Phys. Lett.* 66, 3576 (1995).

148. R. Reisfeld, *Struct. Bonding* 85, 99 (1996).

149. A. W. Li, G. X. Xiong, H. B. Zhao, J. H. Gu, S. S. Sheng, W. Cui, and L. B. Zheng, *Chin. Chem. Lett.* 7, 193 (1996).

150. C. He and R. Wang, *Cuihua Xuebao* 18, 93 (1997).

151. H. Zhao, A. Li, J. Gu, and G. Xiong, *Cuihua Xuebao* 17, 263 (1996).

152. S. Rezgui, R. Jentoft, and B. C. Gates, *J. Catal.* 163, 496 (1996).

153. S. Sakka, H. Kozuka, and G. Zhao, *Proc. SPIE-Int. Soc. Opt. Eng.* 2288, 108 (1994).

154. E. Sanchez, T. Lopez, R. Gomez, I. L. Bokhimi, A. Morales, and O. Novaro, *J. Solid State Chem.* 122, 309 (1996).

155. M. Boutonnet Kizling, C. Bigey, and R. Touroude, *Appl. Catal., A* 135, L13 (1996).

156. J. L. Bokhimi, J. L. Boldu, E. Munoz, O. Novaro, T. Lopez, and R. Gomez, *Mater. Res. Soc. Symp. Proc.* 405, 523 (1996).

157. H. Shiga, T. Okubo, and M. Sadakata, *Ind. Eng. Chem. Res.* 35, 4498 (1996).

158. C. Y. Li, J. Y. Tseng, C. Lechner, and J. D. Mackenzie, *Mater. Res. Soc. Symp. Proc.* 272, 133 (1992).

159. J. Y. Tseng, C. Y. Li, T. Takada, C. Lechner, and J. D. Mackenzie, *Proc. SPIE-Int. Soc. Opt. Eng.* 1758, 612 (1992).

160. M. Ohtaki, Y. Oshima, K. Eguchi, and H. Arai, *Chem. Lett.* 11, 2201 (1992).

161. C. Ferrari, G. Predieri, A. Tiripicchio, and M. Costa, *Chem. Mater.* 4, 243 (1992).
162. D. Cauzzi, G. Marzolini, G. Predieri, A. Tiripicchio, M. Costa, G. Salviati, A. Armigliato, L. Basini, and R. Zanoni, *J. Mater. Chem.* 5, 1375 (1995).
163. T. Burkhart, M. Mennig, H. Schmidt, and A. Licciulli, *Mater. Res. Soc. Symp. Proc.* 346, 779 (1994).
164. C. Goersmann, U. Schubert, J. Leyrer, and E. Lox, *Mater. Res. Soc. Symp. Proc.* 435, 625 (1996).
165. C. Hippe, R. Lamber, G. Schulz-Ekloff, and U. Schubert, *Catal. Lett.* 43, 195 (1997).
166. M. Catalano, E. Carlino, M. A. Tagliente, A. Licciulli, and L. Tapfer, *Microsc. Microanal. Microstruct.* 6, 611 (1995).
167. (a) M. Catalano, G. De, A. Licciulli, and L. Tapfer, *Mater. Sci. Forum* 195, 87 (1995). (b) U. Werner, M. Schmitt, and H. Schmidt, *Mater. Res. Soc. Symp. Proc.* 435, 637 (1996).
168. A. Julbe, C. Balzer, A. Larbot, C. Guizard, and L. Cot, *Recent Prog. Genie Proc.* 6, 127 (1992).
169. S. Szu, C. Y. Lin, and C. H. Lin, *J. Sol.–Gel Sci. Technol.* 2, 881 (1994).
170. G. De, M. Gusso, L. Tapfer, M. Catalano, F. Gonella, G. Mattei, P. Mazzoldi, and G. Battaglin, *J. Appl. Phys.* 80, 6734 (1996).
171. G. De, L. Tapfer, M. Catalano, G. Battaglin, F. Caccavale, F. Gonella, P. Mazzoldi, and R. F. Haglund, Jr., *Appl. Phys. Lett.* 68, 3820 (1996).
172. G. De, M. Epifani, and A. Licciulli, *J. Non-Cryst. Solids* 201, 250 (1996).
173. (a) M. Mennig, M. Schmitt, B. Kutsch, and H. Schmidt, *Proc. SPIE-Int. Soc. Opt. Eng.* 2288, 120 (1994). (b) X. Bokhimi, A. Morales, O. Novaro, T. Lopez, O. Chimal, M. Asomoza, and R. Gomez, *Chem. Mater.* 9, 2616 (1997).
174. Y. Hosoya, T. Suga, T. Yanagawa, and Y. Kurokawa, *J. Appl. Phys.* 81, 1475 (1997).
175. H. Kozuka, M. Okuno, and T. Yoko, *J. Ceram. Soc. Jpn.* 103, 1305 (1995).
176. M. Lee, T. S. Kim, and Y. S. Choi, *J. Non-Cryst. Solids* 211, 143 (1997).
177. F. Akbarian, B. S. Dunn, and J. I. Zink, *Proc. SPIE-Int. Soc. Opt. Eng.* 2288, 140 (1994).
178. F. Akbarian, B. S. Dunn, and J. I. Zink, *J. Raman Spectrosc.* 27, 775 (1996).
179. I. Tanahashi and T. Tohda, *J. Am. Ceram. Soc.* 79, 796 (1996).
180. H. Kozuka, T. Hashimoto, T. Uchino, and T. Yoko, *ICR Ann. Rep.* 2, 22 (1996).
181. K. Yasumoto, N. Koshizaki, and K. Suga, *Proc. SPIE-Int. Soc. Opt. Eng.* 1758, 604 (1992).
182. K. Yasumoto and N. Koshizaki, *Busshitsu Kogaku Kogyo Gijutsu Kenkyusho Hokoku* 4, 121 (1996).
183. A. Maegawa, *Rep. Ind. Res. Cent. Shiga Prefect.* 8, 7 (1994).
184. T. Miura, H. Sakai, and Y. Asahara, Jpn. Kokai Tokkyo Koho JP 02225342 A2 900907 Heisei, 11 pp.
185. L. Zhang, C. Y. Fan, and L. Jiang, *Chin. Chem. Lett.* 7, 77 (1996).
186. B. E. Baker, M. J. Natan, H. Zhu, and T. P. Beebe, Jr., *Supramol. Sci.* 4, 147 (1997).
187. G. C. Vezzoli and M. F. Chen, *Mater. Res. Soc. Symp. Proc.* 286, 271 (1993).
188. G. C. Vezzoli, M. F. Chen, and J. Caslavsky, *Nanostruct. Mater.* 4, 985 (1994).
189. G. C. Vezzoli, M. F. Chen, and J. Caslavsky, *Ceram. Int.* 23, 105 (1997).
190. G. De, A. Licciulli, C. Massaro, L. Tapfer, M. Catalano, G. Battaglin, C. Meneghini, P. Mazzoldi, *J. Non-Cryst. Solids* 194, 225 (1996).
191. S. K. Ma and J. T. Lue, *Solid State Commun.* 97, 979 (1996).
192. P. Innocenzi and H. Kozuka, *J. Sol.–Gel Sci. Technol.* 3, 229 (1994).
193. B. Ritzer, M. A. Villegas, and J. M. Fernandez Navarro, *J. Sol.–Gel Sci. Technol.* 8, 917 (1997).
194. A. Licciulli, G. De, P. Mazzoldi, M. Catalano, L. Mirenghi, and L. Tapfer, *Mater. Sci. Forum* 203, 59 (1996).
195. Y. Zhu, Y. Qian, M. Zhang, Z. Chen, and G. Zhou, *J. Mater. Chem.* 4, 1619 (1994).
196. T. Gacoin, F. Chaput, J. P. Boilot, M. Mostafavi, and M. O. Delcourt, *Eur. Mater. Res. Soc. Mon.* 5, 159 (1992).
197. W. Xiang, C. Wang, Z. Wang, Q. Yang, W. Zhao, Z. Ding, M. Wang, and J. Wang, *Guangxue Xuebao* 16, 967 (1996).
198. P. Innocenzi, G. Brusatin, A. Martucci, and K. Urabe, *Thin Solid Films* 279, 23 (1996).
199. (a) G. Zhao, H. Kozuka, and T. Yoko, *J. Ceram. Soc. Jpn.* 104, 164 (1996). (b) G. Zhao, H. Kozuka, and T. Yoko, *Thin Solid Films* 277, 147 (1996).
200. Y. Zhu, Y. Qian, H. Huang, M. Zhang, and S. Liu, *Mater. Lett.* 28, 259 (1996).
201. H. Kozuka, G. Zhao, and S. Sakka, *Bull. Inst. Chem. Res. Kyoto Univ.* 72, 209 (1994).
202. T. Kineri, E. Matano, and T. Tsuchiya, *Proc. SPIE* 2288, 145 (1994).
203. T. K. Kundu and D. Chakravorty, *Appl. Phys. Lett.* 67, 2732 (1995).
204. T. K. Kundu and D. Chakravorty, *J. Mater. Res.* 11, 200 (1996).
205. D. Chakravorty, *Ferroelectrics* 102, 33 (1990).
206. A. N. Krasovskii and A. I. Andreeva, *Zh. Nauchn. Prikl. Fotogr.* 41, 7 (1996).
207. A. N. Krasovskii and A. I. Andreeva, *Zh. Prikl. Khim.* 69, 834 (1996).
208. M. Mennig, J. Spanhel, H. Schmidt, and S. Betzholz, *J. Non-Cryst. Solids* 147–148, 326 (1992).
209. M. Mennig, M. Schmitt, and H. Schmidt, *J. Sol.–Gel Sci. Technol.* 8, 1035 (1997).
210. M. Mennig, U. Becker, M. Schmitt, and H. Schmit, in "Advanced Materials in Optics, Electro-Optics and Communication Technologies" (P. Vincenzini, ed.), p. 39. Techna Sri., 1995.
211. A. Martino, S. A. Yamanaka, J. S. Kawola, and D. A. Loy, *Chem. Mater.* 9, 423 (1997).
212. T. Gacoin, F. Chaput, J. P. Boilot, and G. Jaskierowicz, *Chem. Mater.* 5, 1150 (1993).

213. C. Goersmann, B. Breitscheidel, and U. Schubert, in "Organosilicon Chemistry" (N. Auner and J. Weis, eds.), p. 319. VCH Publishers, Weinheim, Germany, 1994.
214. S. K. Pradhan, A. Datta, M. Pal, and D. Chakravorty, *Metall. Mater. Trans. A* 27, 4213 (1996).
215. A. Chatterjee, A. K. Giri, D. Das, and D. Chakravorty, *J. Mater. Sci. Lett.* 11, 518 (1992).
216. L. Zhu, Y. Li, J. Wang, and J. Shen, *J. Appl. Phys.* 77, 2801 (1995).
217. L. Zhu, Y. Li, J. Wang, and J. Shen, *Chem. Phys. Lett.* 239, 393 (1995).
218. J. Shen, J. Wang, W. Jiang, B. Zhou, B. Zhang, and L. Zhu, *Gongneng Cailiao* 26, 168 (1995).
219. J. Shen, L. Zhu, J. Wang, Y. Li, and X. Wu, *Chin. Phys. Lett.* 12, 693 (1995).
220. J. Shen, L. Zhu, J. Wang, X. Wu, and Y. Li, *Wuji Cailiao Xuebao* 11, 371 (1996).
221. R. Gvishi, J. D. Bhawalkar, N. D. Kumar, G. Ruland, U. Narang, P. N. Prasad, and B. A. Reinhardt, *Chem. Mater.* 7, 2199 (1995).
222. P. N. Prasad, R. Gvishi, G. Ruland, N. D. Kumar, J. D. Bhawalkar, U. Narang, and B. A. Reinhardt, *Proc. SPIE-Int. Soc. Opt. Eng.* 2530, 128 (1995).
223. H. Peng, S. M. Leung, C. F. Au, X. Wu, N. T. Yu, and B. Z. Tang, *Polym. Mater. Sci. Eng.* 75, 247 (1996).
224. D. McBranch, V. Klimov, L. Smilowitz, M. Grigorova, J. M. Robinson, A. Koskelo, B. R. Mattes, H. Wang, and F. Wudl, *Proc. SPIE-Int. Soc. Opt. Eng.* 2854, 140 (1996).
225. D. McBranch, V. Klimov, L. Smilowitz, M. Grigorova, and B. R. Mattes, *Proc. Electrochem. Soc.* 96-10, 384 (1996).
226. W. Cao, X. Y. Song, and A. J. Hunt, *Mater. Res. Soc. Symp. Proc.* 349, 87 (1994).
227. G. R. Delgado, H. W. H. Lee, and K. Pakbaz, *Mater. Res. Soc. Symp. Proc.* 452, 669 (1997).
228. H. W. H. Lee, J. E. Davis, M. L. Olsen, S. M. Kauzlarich, R. A. Bley, S. H. Risbud, and D. J. Duval, *Mater. Res. Soc. Symp. Proc.* 351, 129 (1994).
229. M. Nogami and Y. Abe, *J. Sol.–Gel Sci. Technol.* 9, 139 (1997).
230. M. Nogami and Y. Abe, *Appl. Phys. Lett.* 65, 2545 (1994).
231. M. Nogami, T. Kasuga, and Y. Abe, *Proc. SPIE-Int. Soc. Opt. Eng.* 2288, 193 (1994).
232. Y. Maeda, Jpn. *J. Appl. Phys., Part 1* 34, 254 (1995).
233. J. P. Carpenter, C. M. Lukehart, D. O. Henderson, R. Mu, B. D. Jones, R. Glosser, S. R. Stock, J. E. Wittig, and J. G. Zhu, *Chem. Mater.* 8, 1268 (1996).
234. Y. Kurokawa and Y. Kobayashi, *Zairyo Gijutsu* 11, 197 (1993).
235. S. Dire and F. Babonneau, *J. Sol.–Gel Sci. Technol.* 2, 139 (1994).
236. J. P. Carpenter, C. M. Lukehart, S. R. Stock, and J. E. Wittig, *Chem. Mater.* 7, 201 (1995).
237. V. Balek, E. Klosova, M. Murat, and N. A. Camargo, *Am. Ceram. Soc. Bull.* 75, 73 (1996).
238. Y. Xu, A. Nakahira, and K. Niihara, *Ceram. Trans.* 44, 275 (1994).
239. J. Y. Choi, C. H. Kim, and D. K. Kim, *Ceram. Trans.* 74, 153 (1996).
240. C. M. Lukehart, S. B. Milne, S. R. Stock, R. D. Shull, and J. E. Wittig, *ACS Symp. Ser.* 622, 195 (1996).
241. C. M. Lukehart, S. B Milne, S. R. Stock, R. D. Shull, and J. E. Wittig, *Mater. Sci. Eng., A* 204, 176 (1995).
242. S. B. Milne, 176 pp. Available from Univ. Microfilms Int., Order No.: DA9541802. From: *Diss. Abstr. Int., B 1995* 56, 4309 (1995).
243. C. M. Lukehart, S. B. Milne, S. R. Stock, R. D. Shull, and J. E. Wittig, *Polym. Mater. Sci. Eng.* 73, 436 (1995).
244. M. A. Matchett, A. M. Viano, N. L. Adolphi, R. D. Stoddard, W. E. Buhro, M. S. Conradi, and P. C. Gibbons, *Chem. Mater.* 4, 508 (1992).
245. S. H. Risbud and H. B. Underwood, *J. Mater. Synth. Process.* 1, 225 (1993).
246. H. Hamada, Y. Kintaichi, M. Inaba, M. Tabata, T. Yoshinari, and H. Tsuchida, *Catal. Today* 29, 53 (1996).
247. A. Towata, M. Awano, M. Sando, and A. Tsuge, *Kagaku Kogaku Ronbunshu* 18, 315 (1992).
248. G. Pacheco-Malagon, A. Garcia-Borquez, D. Coster, A. Sklyarov, S. Petit, and J. J. Fripiat, *J. Mater. Res.* 10, 1264 (1995).
249. A. Nazeri and S. B. Qadri, *Surf. Coat. Technol.* 86–87, 166 (1996).
250. J. Kuo and D. L. Bourell, in "Synth. Process. Nanocryst. Powder, Proc. Symp." (D. L. Bourell, ed.), pp. 237–248. Minerals, Metals and Materials Society, Warrendale, PA, 1996.
251. S. Hazra, A. Ghosh, and D. Chakravorty, *J. Phys.: Condens. Matter* 8, 10279 (1996).
252. Q. Zhou, Q. Zhang, L. Zhang, and X. Yao, *Chin. Sci. Bull.* 41, 555 (1996).
253. M. G. F. Da Silva and J. M. F. Navarro, *J. Sol.–Gel Sci. Technol.* 6, 169 (1996).
254. M.G. F. Da Silva and C. B. Koch, *J. Sol.–Gel Sci. Technol.* 8, 311 (1997).
255. D. Niznansky, N. Viart, and J. L. Rehspringer, *J. Sol.–Gel Sci. Technol.* 8, 615 (1997).
256. T. P. Sinha, M. Mukherjee, D. Chakravorty, and M. Bhattacharya, *Indian J. Phys., A* 70, 741 (1996).
257. F. del Monte, M. P. Morales, D. Levy, A. Fernandez, M. Ocana, A. Roig, E. Molins, K. O'Grady, and C. J. Serna, *Langmuir* 13, 3627 (1997).
258. C. Chaneac, E. Tronc, and J. P. Jolivet, *J. Mater. Chem.* 6, 1905 (1996).
259. C. H. Cheng, Y. Xu, J. D. Mackenzie, J. Chee, and J. M. Liu, *Proc. SPIE-Int. Soc. Opt. Eng.* 1758, 485 (1992).
260. K. Takeuchi, T. Isobe, and M. Senna, *J. Non-Cryst. Solids* 194, 58 (1996).
261. Q. F. Zhou, Q. Q. Zhang, J. X. Zhang, L. Y. Zhang, and X. Yao, *Mater. Lett.* 31, 39 (1997).

262. A. E. Stiegman, H. Eckert, G. Plett, S. S. Kim, M. Anderson, and A. Yavrouian, *Chem. Mater.* 5, 1591 (1993).

263. A. Chatterjee, D. Das, S. K. Pradhan, and D. Chakravorty, *J. Magn. Magn. Mater.* 127, 214 (1993).

264. Q. Zhou, S. Wang, Y. Han, L. Zhang, and X. Yao, *Ferroelectrics* 154, 1433 (1994).

265. Q. Zhou, J. Zhang, L. Zhang, and X. Yao, *Cailiao Yanjiu Xuebao* 9, 71 (1995).

266. K. Yao, L. Zhang, and X. Yao, *Ferroelectrics* 19, 113 (1995).

267. K. Yao, L. Zhang, and X. Yao, *Guisuanyan Xuebao* 24, 355 (1996).

268. T. Sei, H. Takeda, T. Tsuchiya, and T. Kineri, *Ann. Chim.* 18, 329 (1993).

269. B. Ammundsen, G. R. Burns, A. Amran, and S. E. Friberg, *J. Sol.–Gel Sci. Technol.* 2, 341 (1994).

270. S. Jana and P. K. Biswas, *Mater. Lett.* 30, 53 (1997).

271. H. Inoe, Jpn. Kokai Tokkyo Koho JP 08175823 A2 960709 Heisei, 5 pp.

272. S. Luo and L. Gui, *Mater. Res. Soc. Symp. Proc.* 371, 303 (1995).

273. D. Keomany, C. Poinsignon, and D. Deroo, *Sol. Energy Mater. Sol. Cells* 33, 429 (1994).

274. D. Keomany, J. P. Petit, and D. Deroo, *Proc. SPIE-Int. Soc. Opt. Eng.* 2255, 363 (1994).

275. D. Keomany, J. P. Petit, and D. Deroo, *Sol. Energy Mater. Sol. Cells* 36, 397 (1995).

276. D. Camino, D. Deroo, J. Salardenne, and N. Treuil, *Sol. Energy Mater. Sol. Cells* 39, 349 (1995).

277. G. Zhou, S. Liu, and D. Peng, *Huaxue Wuli Xuebao* 9, 54 (1996).

278. J. Shi, J. Liu, Q. Zhu, L. Zhang, and J. Zhang, *Cuihua Xuebao* 17, 277 (1996).

279. J. Shi, J. Liu, Q. Zhu, L. Zhang, and J. Zhang, *Cuihua Xuebao* 17, 323 (1996).

280. G. Fang, Z. Liu, Z. Zhang, Y. Hu, I. A. Ashur, and K. L. Yao, *Phys. Status Solidi A* 156, 15 (1996).

281. G. Fang, Z. Liu, Y. Hu, and K. Yao, *Wuji Cailiao Xuebao* 11, 537 (1996).

282. S. J. L. Ribeiro, S. H. Pulcinelli, and C. V. Santilli, *Chem. Phys. Lett.* 190, 64 (1992).

283. S. K. Pradhan, A. Datta, A. Chatterjee, M. De, and D. Chakravorty, *Bull. Mater. Sci.* 17, 849 (1994).

284. R. Guinebretiere, Z. Oudjedi, and A. Dauger, *Scr. Mater.* 34, 1039 (1996).

285. R. Joseph, S. Zhang, and W. T. Ford, *Macromolecules* 29, 1305 (1996).

286. M. Motomatsu, T. Takahashi, H. Y. Nie, W. Mizutani, and H. Tokumoto, *Polymer* 38, 177 (1997).

287. B. Lantelme, M. Dumon, C. Mai, and J. P. Pascault, *J. Non-Cryst. Solids* 194, 63 (1996).

288. H. Krug and H. Schmidt, *New J. Chem.* 18, 1125 (1994).

289. L. Kador, R. Fischer, D. Haarer, R. Kasemann, S. Brueck, H. Schmidt, and H. Duerr, *Adv. Mater.* 5, 270 (1993).

290. S. Hirano, T. Yogo, K. Kikuta, and S. Yamada, *Ceram. Trans.* 68, 131 (1996).

291. P. L. Shao, K. A. Mauritz, and R. B. Moore, *J. Polym. Sci., Part B: Polym. Phys.* 34, 873 (1996).

292. Q. Deng, K. M. Cable, R. B. Moore, and K. A. Mauritz, *J. Polym. Sci., Part B: Polym. Phys.* 34, 1917 (1996).

293. M. A. F. Robertson and K. A. Mauritz, *Polym. Prepr.* 37, 248 (1996).

294. P. L. Shao, K. A. Mauritz, and R. B. Moore, *Polym. Mater. Sci. Eng.* 73, 427 (1995).

295. Z. Zhao, Z. Gao, Y. Ou, Z. Qi, and F. Wang, *Gaofenzi Xuebao* 2, 228 (1996).

296. Z. Gao, Z. Zhao, Y. Ou, Z. Qi, and F. Wang, *Polym. Int.* 40, 187 (1996).

297. S. Yano, T. Furukawa, and M. Kodomari, *Kimio Kurita* 53, 218 (1996).

298. J. E. Mark, *Int. SAMPE Tech. Conf.* 27, 539 (1995).

299. S. P. Nunes, J. Schultz, and K. V. Peinemann, *J. Mater. Sci. Lett.* 15, 1139 (1996).

300. N. D. Kumar, G. Ruland, M. Yoshida, M. Lal, J. Bhawalkar, G. S. He, and P. N. Prasad, *Mater. Res. Soc. Symp. Proc.* 435, 535 (1996).

301. Y. Chen, X. Wang, Z. Gao, X. Zhu, Z. Qi, and C. L. Choy, *Gaofenzi Xuebao* 1, 73 (1997).

302. H. Kaddami, S. Cuney, J. P. Pascault, and J. F. Gerard, *AIP Conf. Proc.* 354, 522 (1996).

303. V. J. Nagpal, R. M. Davis, and S. B. Desu, *J. Mater. Res.* 10, 3068 (1995).

304. R. Reisfeld and H. Minti, *J. Sol.–Gel Sci. Technol.* 2, 641 (1994).

305. R. Reisfeld, *Proc. SPIE-Int. Soc. Opt. Eng.* 1758, 546 (1992).

306. H. Minti, M. Eyal, R. Reisfeld, and G. Berkovic, *Chem. Phys. Lett.* 183, 277 (1991).

307. R. Reisfeld, H. Minti, and M. Eyal, *Proc. SPIE-Int. Soc. Opt. Eng.* 1513, 360 (1991).

308. E. Blanco, R. Litran, M. Ramirez-del-Solar, N. De La Rosa-Fox, and L. Esquivias, *J. Mater. Res.* 9, 2873 (1994).

309. T. Fujii, Y. Hisakawa, E. J. Winder, and A. B. Ellis, *Bull. Chem. Soc. Jpn.* 68, 1559 (1995).

310. M. Pinero, R. Litran, C. Fernandez-Lorenzo, E. Blanco, M. Ramirez-Del-Solar, N. De La Rosa-Fox, L. Esquivias, A. Craievich, and J. Zarzycki, *J. Sol.–Gel Sci. Technol.* 2, 689 (1994).

311. R. Litran, R. Alcantara, E. Blanco, and M. Ramirez-Del-Solar, *J. Sol.–Gel Sci. Technol.* 8, 275 (1997).

312. M. Nogami, K. Nagasaka, and M. Takata, *J. Non-Cryst. Solids* 122, 101 (1990).

313. A. Othmani, J. C. Plenet, E. Bernstein, F. Paille, C. Bovier, J. Dumas, and C. Mai, *J. Mater. Sci.* 30, 2425 (1995).

314. A. Othmani, J. C. Plenet, E. Bernstein, C. Bovier, J. Dumas, P. Riblet, P. Gilliot, R. Levy, and J. B. Grun, *J. Cryst. Growth* 144, 141 (1994).

315. A. Othmani, C. Bovier, J. Dumas, and B. Champagnon, *J. Phys. IV* 2, C2-275 (1992).

316. E. M. Yeatman, M. Green, E. J. C. Dawnay, M. A. Fardad, and F. Horowitz, *J. Sol.–Gel Sci. Technol.* 2, 711 (1994).

317. M. Nogami, K. Nagaska, and E. Kato, *J. Am. Ceram. Soc.* 73, 2097 (1990).
318. J. Huang, Q. Dai, G. Guo, and L. Gui, *Wuli Huaxue Xuebao* 12, 621 (1996).
319. G. Tu, Z. Sui, and C. Wang, *J. Less-Common Met.* 175, 205 (1991).
320. T. Rajh, O. I. Micic, D. Lawless, and N. Serpone, *J. Phys. Chem.* 96, 4633 (1992).
321. Y. L. Chia, Y. Kao, K. Hayashi, T. Takada, J. D. Mackenzie, K. Kang, S. Lee, N. Peyghambarian, and M. Yamane, *Proc. SPIE-Int. Soc. Opt. Eng.* 2288, 151 (1994).
322. T. Takada, C. Li, J. Y. Tseng, and J. D. Mackenzie, *J. Sol.–Gel Sci. Technol.* 1, 123 (1994).
323. L. Spanhel, E. Arpac, and H. Schmidt, *J. Non-Cryst. Solids* 147–148, 657 (1992).
324. N. Tohge, M. Asuka, and T. Minami, *Proc. SPIE-Int. Soc. Opt. Eng.* 1328, 125 (1990).
325. N. Tohge and T. Minami, *Proc. SPIE-Int. Soc. Opt. Eng.* 1758, 587 (1992).
326. N. Tohge, M. Asuka, and T. Minami, *J. Non-Cryst. Solids* 147–148, 652 (1992).
327. T. A. King, D. West, D. L. Williams, C. Moussu, and M. Bradford, *Adv. Sci. Technol.* 11, 21 (1995).
328. N. Tohge, M. Asuka, and T. Minami, *Chem. Express* 5, 521 (1990).
329. M. Guglielmi, A. Martucci, E. Menegazzo, G. C. Righini, S. Pelli, J. Fick, and G. Vitrant, *J. Sol.–Gel Sci. Technol.* 8, 1017 (1997).
330. M. Tsutomu, I. Tomoaki, T. Kiyoharu, and T. Masahiro, *Proc. SPIE-Int. Soc. Opt. Eng.* 2288, 183 (1994).
331. T. Iwami, K. Tadanaga, M. Tatsumisago, T. Minami, and N. Tohge, *J. Am. Ceram. Soc.* 78, 1668 (1995).
332. E. Cordoncillo, P. Escribano, G. Monros, M. A. Tena, V. Orera, and J. Carda, *J. Solid State Chem.* 118, 1 (1995).
333. T. Gacoin, J. P. Boilot, F. Chaput, and A. Lecomte, *Mater. Res. Soc. Symp. Proc.* 272, 21 (1992).
334. K. Kang, A. D. Kepner, Y. Z. Hu, S. W. Koch, N. Peyghambarian, C. Y. Li, T. Takada, Y. Kao, and J. D. Mackenzie, *Appl. Phys. Lett.* 64, 1487 (1994).
335. A. Othmani, C. Bovier, J. C. Plenet, J. Dumas, B. Champagnon, and C. Mai, *Mater. Sci. Eng., A* 168, 263 (1993).
336. G. Counio, S. Esnouf, T. Gacoin, P. Barboux, A. Hofstaetter, and J.-P. Boilot, *Mater. Res. Soc. Symp. Proc.* 452, 311 (1997).
337. H. L. Liu, S. S. Wang, L. Y. Zhang, and X. Yao, *Ferroelectrics* 196, 245 (1997).
338. H. Liu, S. Wang, L. Zhang, and X. Yao, *Xi'an Jiaotong Daxue Xuebao* 29, 13 (1995).
339. N. Toge, T. Minami, I. Tanahashi, and T. Mitsuyu, Jpn. Kokai Tokkyo Koho JP 04270131 A2 920925 Heisei, 5 pp.
340. S. Lu, Q. Xie, L. Zhang, and X. Yao, *Chin. Sci. Bull.* 39, 96 (1994).
341. B. Honerlage, P. Gilliot, and R. Levy, *Nuovo Cimento Soc. Ital. Fis., D* 17, 1247 (1995).
342. J. C. Plenet, A. Othmani, F. Paille, J. Mugnier, E. Bernstein, and J. Dumas, *Opt. Mater.* 7, 129 (1997).
343. G. Yu, K. Wang, J. Huan, and B. Zhao, *Wuli Huaxue Xuebao* 13, 230 (1997).
344. S. Lu, H. Yu, L. Zhang, and X. Yao, *Guisuanyan Xuebao* 22, 71 (1994).
345. C. S. Mun, J. B. Kang, and K. M. Kim, *Yoop Hakhoechi* 33, 1353 (1996).
346. N. Herron, *ACS Symp. Ser.* 455, 582 (1991).
347. C. Y. Li, M. Wilson, N. Haegel, J. D. Mackenzie, E. T. Knobbe, C. Porter, and R. Reeves, *Mater. Res. Soc. Symp. Proc.* 272, 41 (1992).
348. H. Ye, D. He, and Z. Jiang, *Guisuanyan Xuebao* 22, 434 (1994).
349. J. Zhao, K. Dou, Y. Chen, C. Jin, L. Sun, S. Huang, J. Yu, W. Xiang, and Z. Ding, *J. Lumin.* 66-67, 332 (1995).
350. E. J. C. Dawnay, J. Fick, M. Green, M. Guglielmi, A. Martucci, S. Pelli, G. C. Righini, G. Vitrant, and E. M. Yeatman, *Adv. Sci. Technol.* 11, 15 (1995).
351. E. Vanagas, J. Moniatte, M. Mazilu, P. Riblet, B. Honerlage, S. Juodkazis, F. Paille, J. C. Plenet, J. G. Dumas, M. Petrauskas, and J. Vaitkus, *J. Appl. Phys.* 81, 3586 (1997).
352. T. Takada, T. Yano, A. Yasumori, M. Yamane, and J. D. Mackenzie, *J. Non-Cryst. Solids* 147–148, 631 (1992).
353. T. Takada, J. D. Mackenzie, M. Yamane, K. Kang, N. Peyghambarian, R. J. Reeves, E. T. Knobbe, and R. C. Powell, *J. Mater. Sci.* 31, 423 (1996).
354. J. Butty, N. Peyghambarian, Y. H. Kao, and J. D. Mackenzie, *Appl. Phys. Lett.* 69, 3224 (1996).
355. J. D. Mackenzie, *J. Ceram. Soc. Jpn.* 101, 1 (1993).
356. M. Yamane, T. Takada, J. D. Mackenzie, and C. Y. Li, *Proc. SPIE-Int. Soc. Opt. Eng.* 1758, 577 (1992).
357. J. Butty, Y. Z. Hu, N. Peyghambarian, Y. H. Kao, and J. D. Mackenzie, *Appl. Phys. Lett.* 67, 2672 (1995).
358. C. Li, Y. Kao, K. Hayashi, T. Takada, J. D. Mackenzie, K. Kang, S. Lee, N. Peyghambarian, and M. Yamane, *Proc. SPIE-Int. Soc. Opt. Eng.* 2288, 151 (1994).
359. Y. Kao, K. Hayashi, L. Yu, M. Yamane, and J. D. Mackenzie, *Proc. SPIE-Int. Soc. Opt. Eng.* 2288, 752 (1994).
360. H. Mathieu, T. Richard, J. Allegre, P. Lefebvre, G. Arnaud, W. Granier, L. Boudes, J. L. Marc, A. Pradel, and M. Ribes, *J. Appl. Phys.* 77, 287 (1995).
361. W. Granier, L. Boudes, A. Pradel, M. Ribes, J. Allegre, G. Arnaud, P. Lefebvre, and H. Mathieu, *J. Sol.–Gel Sci. Technol.* 2, 765 (1994).

362. T. Richard, J. Allegre, P. Lefebvre, H. Mathieu, C. Jouanin, L. Boudes, J. L. Marc, W. Granier, A. Pradel, and M. Ribes, in "Int. Conf. Phys. Semicond." (D. J. Lockwood, ed.), Vol. 3, p. 2003. World Scientific, Singapore, 1995.

363. P. Lefebvre, T. Richard, J. Allegre, H. Mathieu, A. Pradel, J. L. Marc, L. Boudes, W. Granier, and M. Ribes, *Proc. SPIE-Int. Soc. Opt. Eng.* 2288, 163 (1994).

364. J. Zhao, K. Dou, S. Lu, Y. Chen, S. Huang, J. Yu, W. Xiang, and Z. Ding, *J. Mater. Sci. Lett.* 15, 702 (1996).

365. P. Lefebvre, T. Richard, J. Allegre, H. Mathieu, A. Pradel, J. L. Marc, L. Boudes, W. Granier, and M. Ribes, *Superlattices Microstruct.* 15, 447 (1994).

366. W. Granier, L. Boudes, A. Pradel, M. Ribes, J. Allegre, G. Arnaud, P. Lefebvre, and H. Mathieu, *Mater. Sci. Forum* 152–153, 351 (1994).

367. J. D. Mackenzie and Y. Kao, *Proc. SPIE-Int. Soc. Opt. Eng.* 2145, 90 (1994).

368. M. Guglielmi, A. Martucci, G. C. Righini, and S. Pelli, *Proc. SPIE-Int. Soc. Opt. Eng.* 2288, 174 (1994).

369. M. A. Fardad, E. M. Yeatman, E. J. C. Dawnay, M. Green, J. Fick, M. Guntau, and G. Vitrant, *IEE Proc.: Optoelectron.* 143, 298 (1996).

370. J. Fick, G. Vitrant, A. Martucci, M. Guglielmi, S. Pelli, and G. C. Righini, *MCLC S&T, Sect. B: Nonlinear Opt.* 12, 203 (1995).

371. H. Kawaguchi, T. Miyakawa, N. Tanno, Y. Kobayashi, and Y. Kurokawa, *Jpn. J. Appl. Phys., Part 2* 30, L280 (1991).

372. A. K. Atta, P. K. Biswas, and D. Ganguli, *Polym. Other Adv. Mater.* 3, 645 (1995).

373. K. M. Choi and K. J. Shea, *J. Phys. Chem.* 98, 3207 (1994).

374. K. M. Choi and K. J. Shea, *Chem. Mater.* 5, 1067 (1993).

375. L. Spanhel, *Eur. Mater. Res. Soc. Monogr.* 5, 407 (1992).

376. E. Cordoncillo, J. B. Carda, M. A. Tena, G. Monros, and P. Escribano, *J. Sol.–Gel Sci. Technol.* 8, 1043 (1997).

377. L. Spanhel, H. Schmidt, A. Uhrig, and C. Klingshirn, *Mater. Res. Soc. Symp. Proc.* 272, 53 (1992).

378. G. Counio, S. Esnouf, T. Gacoin, and J. P. Boilot, *J. Phys. Chem.* 100, 20021 (1996).

379. M. Nogami and A. Kato, *J. Sol.–Gel Sci. Technol.* 2, 751 (1994).

380. M. Nogami, A. Kato, and Y. Tanaka, *J. Mater. Sci.* 28, 4129 (1993).

381. C. M. Bagnall and J. Karzycki, *Proc. SPIE-Int. Soc. Opt. Eng.* 1328, 108 (1990).

382. D. C. Hummel, I. L. Torriani, A. Y. Ramos, A. F. Craievich, N. Fox, and L. Esquivias, *Mater. Res. Soc. Symp. Proc.* 346, 673 (1994).

383. D. A. Hummel, I. L. Torriani, A. F. Craievich, N. De La Rosa Fox, A. Y. Ramos, O. Lyon, *J. Sol.–Gel Sci. Technol.* 8, 285 (1997).

384. M. Nogami, S. Suzuki, and K. Nagasaka, *J. Non-Cryst. Solids* 135, 182 (1991).

385. J. L. Coffer, G. Beauchamp, and T. W. Zerda, *J. Non-Cryst. Solids* 142, 208 (1992).

386. L. Spanhel, M. Popall, and G. Mueller, *Proc. Indian Acad. Sci., Chem. Sci.* 107, 637 (1995).

387. J. L. Marc, W. Granier, A. Pradel, M. Ribes, T. Richard, J. Allegre, and P. Lefebvre, *Mater. Res. Soc. Symp. Proc.* 346, 901 (1994).

388. M. Nogami, K. Nagasaka, and T. Suzuki, *J. Am. Ceram. Soc.* 75, 220 (1992).

389. M. Nogami, I. Kojima, and K. Nagasaka, *Proc. SPIE-Int. Soc. Opt. Eng.* 1758, 557 (1992).

390. P. Lefebvre, T. Richard, J. Allegre, H. Mathieu, A. Combette-Roos, and W. Granier, *Phys. Rev. B: Condens. Matter* 53, 15440 (1996).

391. P. Lefebvre, T. Richard, J. Allegre, H. Mathieu, A. Pradel, J. L. Marc, L. Boudes, W. Granier, and M. Ribes, *Superlattices Microstruct.* 15, 447 (1994).

392. H. Liu, S. Wang, L. Zhang, and X. Yao, *Ferroelectrics* 19, 83 (1995).

393. M. Nogami, M. Watabe, and K. Nagasaka, *Proc. SPIE-Int. Soc. Opt. Eng.* 1328, 119 (1990).

394. S. Lu, H. Liu, L. Zhang, and X. Yao, in "Proc. Int. Conf. Electron. Compon. Mater., Sens. Actuators," p. 230. International Academic Publishers, Beijing, 1995.

395. S. Lu, H. Liu, L. Zhang, and X. Yao, *Chin. Sci. Bull.* 41, 1923 (1996).

396. S. Lu, H. Liu, L. Zhang, and X. Yao, *Chin. Sci. Bull.* 41, 83 (1996).

397. Y. Zhang, J. K. Bailey, C. J. Brinker, N. Raman, R. M. Crooks, and C. S. Ashley, *Proc. SPIE-Int. Soc. Opt. Eng.* 1758, 596 (1992).

398. Y. Zhang, N. Raman, J. K. Bailey, C. J. Brinker, and R. M. Crooks, *J. Phys. Chem.* 96, 9098 (1992).

399. M. Nogami, K. Nagasaka, and K. Kotani, *J. Non-Cryst. Solids* 126, 87 (1990).

400. N. Pellegri, R. Trbojevich, O. De Sanctis, and K. J. Kadono, *J. Sol.–Gel Sci. Technol.* 8, 1023 (1997).

401. N. Narvathy, G. M. Pajonk, and A. V. Rao, *Mater. Res. Bull.* 32, 397 (1997).

402. T. K. Singh, S. K. Sharma, and C. L. Jain, *Def. Sci. J.* 46, 105 (1996).

403. T. Takada, T. Yano, A. Yasumori, and M. Yamane, *J. Ceram. Soc. Jpn.* 101, 73 (1993).

404. T. Gacoin, J. P. Boilot, M. Gandias, C. Ricolleau, and M. Chamarro, *Mater. Res. Soc. Symp. Proc.* 358, 247 (1995).

405. G. Li and M. Nogami, *J. Sol.–Gel Sci. Technol.* 1, 79 (1993).

406. Z. Jiang and Y. Hui, *Proc. SPIE-Int. Soc. Opt. Eng.* 2288, 200 (1994).

407. H. Ye, D. He, Z. Jiang, and Y. Ding, *J. Sol.–Gel Sci. Technol.* 3, 235 (1994).

408. H. Ye, D. He, and Z. Jiang, *Guangxue Xuebao* 14, 518 (1994).

409. K. Yoshikai, M. Yamaguchi, and K. Nishimura, *Nippon Gomu Kyokaishi* 69, 485 (1996).

410. C. Cobianu, C. Savaniu, O. Buiu, D. Dascalu, M. Zabarescu, C. Parlog, A. van den Berg, and B. Pecz, in "CAS 96 Proc., Int. Semicond. Conf.," Vol. 2, p. 633. IEEE, New York, 1996. CA: 126:53304.

411. S. Sakohara, L. D. Tickanen, and M. A. Anderson, *J. Phys. Chem.* 96, 11086 (1992).

412. N. Maliavski, O. Dushkin, E. Tchekounova, P. Innocenzi, G. Scarinci, and M. Guglielmi, *Chim. Chron.* 23, 181 (1994).

413. W. H. Shih, D. J. Farrell, and W. Y. Shih, *Ceram. Trans.* 62, 233 (1996).

414. T. Oozeki and S. Urabe, Jpn. Kokai Tokkyo Koho JP 05127200 A2 930525 Heisei, 34 pp.

415. M. Nogami, Y. Q. Zhu, Y. Tohyama, K. Nagasaka, T. Tokizaki, and A. Nakamura, *J. Am. Ceram. Soc.* 74, 238 (1991).

416. M. Nogami, Y. Q. Zhu, and K. Nagasaka, *J. Non-Cryst. Solids* 134, 71 (1991).

417. A. A. Rywak and J. M. Burlitch, *Chem. Mater.* 8, 60 (1996).

418. E. Lifshitz, M. Yassen, L. Bykov, I. Dag, and R. Chaim, *NATO ASI Ser., Ser. E* 260, 503 (1994).

419. E. Lifshitz, *Mater. Res. Soc. Symp. Proc.* 452, 377 (1997).

420. E. Lifshitz, L. Bykov, and M. Yassen, *Proc. Electrochem. Soc.* 95-25, 477 (1996).

421. L. C. Liu and S. H. Risbud, *J. Appl. Phys.* 68, 28 (1990).

422. I. Tanahashi, M. Yoshida, Y. Manabe, and T. Tohda, *J. Mater. Res.* 10, 361 (1995).

Chapter 9

CHEMICAL PREPARATION AND CHARACTERIZATION OF NANOCRYSTALLINE MATERIALS

Qian Yitai

Department of Chemistry, University of Science and Technology of China, Hefei, Anhui, People's Republic of China

Contents

1. INTRODUCTION

In this chapter, four chemical preparation methods are introduced: the solvothermal technique and γ-irradiation for nanocrystalline powders and hydrothermal deposition and spray pyrolysis for nanocrystalline thin films.

By substitution of nonaqueous solvents such as benzene and toluene for water, solvothermal methods have been developed to prepare nanocrystalline materials especially for nonoxide compounds. The benzene thermal synthesis method was used to prepare nanometer GaN with a particle size of 32 nm. High-resolution electron microscopy (HREM) microphotographs indicate that three phases coexist: wurtzite, zincblende, and rock salt cubic, the latter being apparently a kinetic product previously observed only at very high pressure. By using this method, InP (12 nm), InAs (15 nm), and CdS nanorods (36 nm $\times$ 720 nm) were synthesized.

Handbook of Nanostructured Materials and Nanotechnology, edited by H.S. Nalwa
Volume 1: Synthesis and Processing
Copyright © 2000 by Academic Press

ISBN 0-12-513761-3/$30.00

The γ-irradiation method with a relatively high yield of product and low cost can be carried out under normal pressure at room temperature. This method was successfully used to prepare nanocrystalline metals, nonmetals, and oxides such as Cd (20 nm), Ru (12 nm), Ag–Cu alloy (25 nm), Se (70 nm), Cr_2O_3 (6 nm), and nanocomposite Ag (11 nm)/PAm (polyacrylamide) in which the metal salt and organic monomer are mixed homogeneously at the molecular level.

Two kinds of chemical techniques to fabricate nanocrystalline film have been developed, resulting in particle sizes less then 100 nm. In the hydrothermal method, which is usually carried out at 100–150 °C, metal sulfide and metal oxide nanometer thin films, such as ZnS (60 nm × 10 nm), Fe_3O_4 (50 nm), and highly oriented TiO_2 epitaxy thin film (10 nm), have been prepared. The spray pyrolysis method has been used to prepare Cr_2O_3 (100 nm × 20 nm) and Cr-doped Al_2O_3 (15 nm), the size of which is less than that of pure Al_2O_3 (40 nm) film.

2. SOLVOTHERMAL SYNTHETIC ROUTE TO NANOCRYSTALLINE MATERIALS

The engineering of materials and devices on the nanometer scale is of considerable interest in electronics [1], optics [2], catalysis [3], ceramics [4], and magnetic storage [5]. Nanocrystalline materials display optical, electronic, and structural properties that often are not present in either isolated molecules or macroscopic solids. Well-defined ordered solids prepared from tailored nanocrystalline building blocks provide opportunities for optimizing the properties of materials and offer possibilities for observing interesting and potentially useful new collective physical phenomena. For example, the theoretical prediction [6, 7] that semiconductor band gaps should increase as the dimension of the materials grows smaller has been investigated in luminescence [8, 9] and other spectroscopic studies [9–11].

The hydrothermal technique has been widely used to prepare oxide nanocrystallines. Under high temperatures (>80 °C) and pressures (>1 atm), a number of compounds, which are practically insoluble in water under ordinary conditions, turn out to be soluble in aqueous solution. However the hydrothermal synthetic method is confined, some reactants will decompose or some reactions will not occur in the presence of water. So it is difficult to obtain many kinds of powders, such as carbonides, nitrides, phosphides, silicides, and so on. Thus, by substitution of nonaqueous solvents such as C_2H_5OH and C_6H_6 for water, solvothermal methods have been developed to prepare nanocrystalline compounds whose precursors are sensitive to water.

In a solvothermal method, one or several kinds of precursors are dissolved in nonaqueous solvents. The reactions occur in liquid phase or under supercritical conditions, where the reactants are dispersed in solution and become more active. The products are slowly produced in liquid phase or under supercritical conditions. Such a process is relatively simple and easy to control, and the sealed system can effectively prevent the contamination of the toxic and air-sensitive precursors. In addition, the phase formation and the particle size and shape can also be controlled.

In this section, the solvothermal synthetic route to nanocrystalline III–V compounds and sulfides is discussed.

2.1. III–V Group Compounds

There are many reports on the preparation and characterization of nanocrystalline III–V group semiconductors [12–21], which contain dehalosilylation reaction, alkali metal halide elimination, thermolysis and thermolysis–ammonolysis, dihydrogen elimination, and alcoholysis [22]. All these methods need either toxic organometallic precursors or high reaction temperature or posttreatment. Recently, we developed a new, low-temperature solvothermal synthetic route to nanocrystalline III-V compounds.

2.1.1. Gallium Nitride [23, 24]

GaN usually exists as the wurtzite structure. There are two other metastable cubic structures of GaN: zincblende (space group: T_d^2) and rock salt (space group: O_h^5). Zincblende GaN was only formed during the film preparation process [25], whose substrate is of zincblende phase such as (100). Rock salt GaN can be transferred from the wurtzite phase under ultrahigh pressure of at least 37 GPa and was observed with high-pressure energy-dispersive X-ray diffraction by using diamond-anvil techniques [26]. The rock salt phase would be transferred back to the wurtzite phase as the pressure is released [26]. In previous work, no rock salt–phase GaN was observed under normal pressure.

A new benzene thermal process was developed to synthesize nanocrystalline GaN based on the following liquid–solid reaction:

$$GaCl_3 + Li_3N \xrightarrow{\text{benzene}} GaN + 3LiCl$$

An appropriate amount of $GaCl_3$ benzene solution and Li_3N powder was added to stainless-steel autoclave with a silver liner, filling the autoclave with benzene up to 90% volume. The autoclave was kept at about 280 °C for 6–12 h and then cooled to room temperature. A dark-gray product was collected for characterization.

The X-ray powder diffraction (XRD) pattern of as-prepared GaN is shown in Figure 1b. Reflections, denoted by solid circles, can be indexed to the wurtzite phase of GaN with lattice parameters $a = 3.188$ Å and $c = 5.176$ Å, which are consistent with the reported values [27]. However, an unusually strong (002) peak in the pattern indicates a preferential orientation of [001] in nanocrystalline GaN. Other small reflections, which are denoted by open circles, can be indexed to rock salt–phase GaN with $a = 4.100$ Å, which is greater than the reported value of 4.006 Å at 50 GPa [26]. In Figure 1a, the reflection peaks, denoted by asterisks, correspond to the zincblende phase with $a = 4.400$ Å.

Transmission electron microscopy (TEM) images show that as-prepared GaN has an average size of 32 nm and displays a uniform shape. High-resolution transmission electron microscopy (HRTEM) images of GaN particles are shown in Figure 2. There is a very interesting area, in which region A belongs to zincblende GaN with $a = 4.42$ Å, as

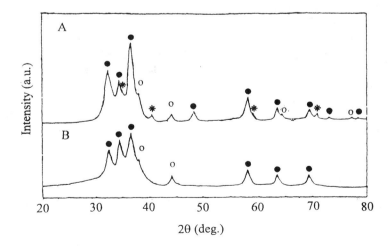

Fig. 1. XRD patterns of GaN prepared at (a) 300 °C and (b) 280 °C: • wurtzite phase, ○ rock-salt phase, ∗ zincblende phase. (Source: Reprinted with permission from [24]. © 1996 American Institute of Physics.)

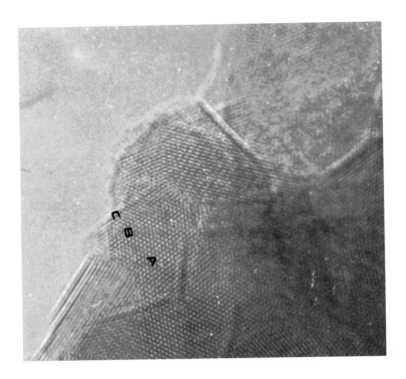

Fig. 2.   HRTEM image of GaN.

calculated from the (111) plane intervals of 2.56 Å and (100) plane intervals of 2.21 Å, which is in good agreement with that from XRD. Region C belongs to rock salt GaN with $a = 4.08$ Å, as calculated from the (111) plane intervals of 2.38 Å and (100) plane intervals of 2.04 Å, which is also consistent with XRD results. From Figure 2, there are two dislocation strips between zincblende GaN and rock salt GaN, respectively. Region B between the two dislocation strips also shows a [110] oriented, face-centered cubic lattice structure. The (111) plane intervals in region B were measured to be 2.46 Å, which falls between the 2.56 Å of zincblende GaN and the 2.38 Å of rock salt GaN; and the (100) plane intervals were calculated to be 2.13 Å, which falls between the 2.21 Å of zincblende GaN and the 2.04 Å of rock salt GaN. Thus, region B can be considered as a transitional region between zincblende GaN and rock salt GaN.

The samples are also characterized by X-ray photoelectron spectra (XPS) and photoluminescence (PL) analysis. XPS gave the Ga 3d peaks at 20.0 eV and Ga 2p at 1118.5 eV with N 1s peaks at 397.5 eV, which is close to that of bulk GaN [28]. The content of Ga and N was quantified by Ga 3d and N 1s peak areas and an average composition of $Ga_{0.95}N$ was given. The PL spectrum consists of one broad emission feature at 370 nm, which is in agreement with that of the bulk GaN [29]. This result indicates that as-prepared GaN is too large for quantum confinement and, in fact, the Bohr exciton radius for GaN is 11 nm, which could be calculated according to the literature [30].

2.1.2. Indium Phosphide [31, 32]

In a solvothermal process, InP can also be obtained according to the following reaction:

$$Na_3P + InCl_3 \rightarrow InP + 3NaCl$$

An appropriate amount of sodium phosphide and the 1,2-dimethoxyethane (DME) solution of indium chloride was added to the autoclave. The autoclave was kept at 180 °C for 12 h and then cooled to room temperature.

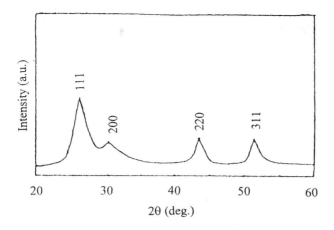

Fig. 3.   XRD pattern of InP nanocrystalline. (Source: Adapted from Xie et al. [32].)

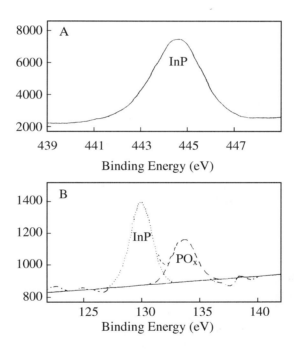

Fig. 4.   XPS spectra of InP.

The XRD pattern of as-prepared InP (Fig. 3) shows the reflections at d spacings of 3.39, 2.94, 2.08, and 1.77 Å, corresponding to the 111, 200, 220, and 311 reflections of InP, respectively. After refinement, the cell constant, $a = 5.87$ Å, is close to that reported in the literature. The crystalline size of the sample, which is calculated from the half-width of diffraction peaks using the Scherrer equation, is about 10 nm. The TEM image shows that the InP particles consist of uniform spherical crystallites with an average size of 12 nm, which is consistent with that obtained from the XRD pattern.

Figure 4 is an XPS analysis performed on the nanocrystalline InP samples. In Figure 4a, curve-fitting analysis of the asymmetrical In $3d_{5/2}$ peaks shows the existence of two signals caused by indium containing two species: one from indium (at 444.60 eV) in InP and the other from indium (at 442.70 eV) in In_2O_3. There is very little In_2O_3 in the sample

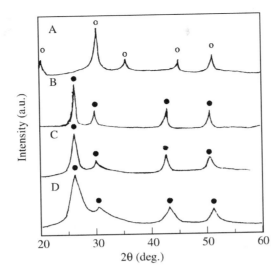

Fig. 5. XRD patterns of InP prepared with different solvents (• InP, ○ In$_2$O$_3$): (a) 3-methyl-2-pentanone, (b) benzene, (c) 1,4-dioxane, and (d) 1,2-dimethoxyelenthane.

(less than 5%), and it cannot be detected by XRD. As seen in the P2p spectrum in Figure 4b, clearly two phosphorus signals were observed. Curve-fitting analysis of the signals indicates two distinct phosphorus environments: one from InP (at 129.9 eV) and the other from phosphorus containing oxidized species (at 133.60 eV). The XPS analysis of InP also shows the presence of oxygen impurities; however, in contrast to the III–V materials produced by the solid-state method, halogen or alkali metal impurities were not detected. The surface oxidation of the nanocrystalline InP is a result of the absorbed oxygen, which is common to all powder samples exposed to atmosphere and is more intense for high-surface-area samples such as nanocrystallines.

Figure 5 shows the XRD patterns of InP prepared in different solvents. The crystalline sizes in benzene and 1,4-dioxane are about 25 nm and 40 nm, respectively, which indicates that the solvent is key to the formation of InP in nanocrystalline-sized semiconductors. In general, group III halides exist as dimers because of the Lewis acidity of the group III metal atoms. Glyme solvents can break up the dimeric structures of these halides and form ionic coordination complexes by expanding the coordination sphere of the metal center [33]. DME is an open-chain polyether and can act as a bidentate ligand, which could facilitate the formation of chelate complexes with group III halides. Also, it causes dissociation of indium halides, thereby forming ionic coordination complexes. It is suggested that such dissociation and the formation of ionic complexes may play an important role in the mechanism of chelation. Experiments showed that merely adduct formation was not sufficient in nanocrystalline InP formation when using other coordinating solvents such as dioxane. Furthermore, the coordinating solvents must contain no active hydrogen; otherwise, it would react with Na$_3$P and no InP would be obtained.

Ultraviolet–visible (UV–vis) spectra of the solution of InP clusters prepared at 160 °C for different times are shown in Figure 6. All display absorption peaks at 285 nm, indicating the existence of InP clusters [34]. There is a red shift of the absorption edge from 330 to 375 nm, showing the growth of clusters with time, which is similar to the InP clusters from methanolysis of Cp*(Cl)In[μ-P(SiMe$_3$)$_2$] in methanol solution [35].

The cluster size is evaluated from the equation [36]:

$$E = E_g + \left(\frac{h}{2\pi}\right)^2 \frac{\pi^2}{2R^2}\left(\frac{1}{m_e} + \frac{1}{m_h}\right) - \frac{1.8e^2}{4\pi\varepsilon_0\varepsilon R}$$

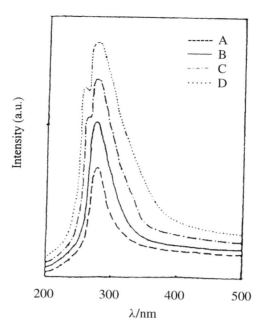

Fig. 6. UV–vis absorption spectra of the solution after various periods of thermal treatment: (a) 0.5 h, (b) 3 h, (c) 6 h, and (d) 12. (Source: Adapted from Xie et al. [32].)

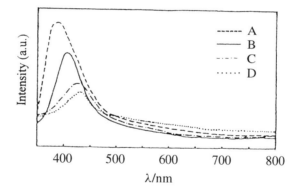

Fig. 7. PL spectra of the solution after various periods of thermal treatment: (a) 0.5 h, (b) 3 h, (c) 6 h, and (d) 12. (Source: Adapted from Xie et al. [32].)

where m_e and m_h are the electron and hole effective masses, respectively, R is the cluster radius, and ε is the static dielectric constant. According to the data in [37], the InP clusters range in size from 3.0 to 3.6 nm. The Bohr exciton radius in InP is 28 nm, which could be calculated according to the equation [30]:

$$R = \frac{\hbar^2 \varepsilon}{e^2} \left(\frac{1}{m_e} + \frac{1}{m_h} \right)$$

Thus, the clusters are much smaller than the Bohr exciton radius, exhibiting the quantum size effects.

Figure 7 shows the PL spectra of InP clusters in the solution taken at various periods of time. From the PL spectra of the solution (lines a–d in Fig. 7), the spectra have emission peaks ranging from 385 to 430 nm, showing the growth of the clusters. The PL spectra also show a continuous decrease of the emission intensity with time, indicating that the concentration of clusters decreases with time as the precipitation of nanocrystalline InP increases. After 12 h, that the PL intensities of solution remain constant indicates a possible equilibrium between the clusters and the nanocrystallines established.

From the UV–vis and PL spectra, a possible mechanism in the preparation of nanocrystalline InP is that the clusters act as the intermediate and combine to form amorphous and finally nanocrystalline InP particles during the solvothermal process.

2.1.3. Indium Arsenide [38]

A solvothermal coreduction route was employed to prepare nanocrystalline InP. The reaction was carried out in a xylene solution and involved the simultaneous reduction of $InCl_3$ and $AsCl_3$ by zinc powder:

$$InCl_3 + AsCl_3 + 3Zn \rightarrow InAs + 3ZnC_2$$

Zinc powder was added to the xylene solution of analytically pure $AsCl_3$ and $InCl_3$ in the autoclave. The autoclave was kept at 150 °C for 48 h and then cooled to room temperature. A dark-gray product was obtained.

Figure 8 shows the XRD pattern of the InAs sample. All the reflections can be indexed to be a pure zincblende phase InAs with $a = 6.058$ Å, which is equal to the value reported in the literature. There is a small noncrystalline peak in the figure, which can be attributed to the amorphous precursor. The average particle size of the product was determined by the Scherrer equation to be about 12 nm.

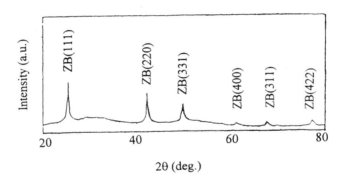

Fig. 8.   XRD pattern of nanocrystalline InAs. (Source: Reprinted with permission from [38]. © 1997 American Chemical Society.)

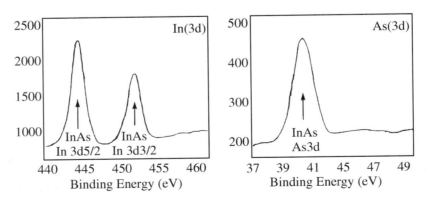

Fig. 9.   XRS spectra of InAs. (Source: Reprinted with permission from [38]. © 1997 American Chemical Society.)

Further evidence for the formation of InAs can be obtained from the XPS of the product (Fig. 9). The two strong peaks at 40.8 and 444.6 eV correspond to As 3d and In 3d in InAs, respectively. In the pattern, no obvious peaks resulting from In_2O_3 (444.9 eV), $In(OH)_3$ (445.0 eV), As_2O_3 (44.9 eV), and As_2O_5 (46.2 eV) were observed. The quantification of peaks gives a ratio of As to In of 51:49.

The TEM image revealed that the average size of the particles in the sample is about 15 nm. This is a bit larger than that obtained from the XRD patterns. The difference may be caused by the coexistence of the amorphous precursor. The size of the particles is hard to determine exactly.

Within the reducing environment caused by metallic zinc powder, no extreme conditions such as an absolutely nonaqueous, nonoxygen environment are required. The oxygen in the reaction system can be reduced to a negligible amount. A trace amount of water in the system proves not to be a serious problem. On the contrary, a small amount of water in the system may aid the electron transfer process of the reaction:

$$3Zn + 6H_2O \rightarrow 3Zn(OH)_2 + 6H$$

$$InCl_3 + AsCl_3 + 6H \rightarrow InAs + 6HCl$$

$$3Zn(OH)_2 + 6HCl \rightarrow 3ZnCl_2 + 6H_2O$$

Atomic hydrogen generated by the reaction between zinc powder and water can act as a charge carrier during the coreduction procedure. Experiments demonstrate that less than 1% (volume ratio) water in the reaction solution has no deleterious effect on the product. However, when the amount of water is increased to approximately 3%, an indium hydroxide [$In(OH)_3$] obviously exists in the sample. It may be caused by the hydrolysis of $InCl_3$.

With the coreduction of $AsCl_3$ and $InCl_3$, the newly reduced arsenic and indium are evenly dispersed in the solution. The reaction between them can result in two possible products, amorphous InAs and nanocrystalline InAs. In a conventional solution reaction, amorphous InAs is the dominant product. In a solvothermal system, the higher solvent temperature causes nanocrystalline InAs to become the major product. By increasing the processing temperature and solvent pressure further, a better crystallized product can be obtained. When the processing temperature is increased to 300 °C, a perfectly crystallized product is obtained.

2.2. Sulfides

2.2.1. Tin Sulfide [39]

Tin sulfide (SnS_2) has interesting optical and electrical properties [40–42] and has been widely used as a semiconductor and photoconductor [43]. The traditional preparation method of SnS_2 is carried out in sealed tubes with temperatures ranging from 400 to 700 °C [44–48]. Chianlli and Dines [49] synthesized a number of transition metal dichalcogenides in nonaqueous solution at room temperature by the reaction between anhydrous transition metal chloride and either lithium sulfide or ammonium hydrogen sulfide. The products had high surface areas, but were poorly crystalline, or amorphous.

Recently, we prepared nanocrystalline β-SnS_2 through a solvothermal process by the reaction of tin chloride and sodium sulfide; toluene was used as a solvent instead of water because tin chloride is very susceptible to water.

In a typical reaction, stoichiometric amounts of $SnCl_4$ and anhydrous Na_2S were added to an autoclave filled with toluene. There was no observable reaction of $SnCl_4$ with Na_2S at room temperature. The autoclave was maintained at 150 °C for 6–8 h and then cooled to room temperature naturally. A yellow powder was collected.

The typical XRD pattern of the sample is shown in Figure 10. All the peaks could be indexed as the hexagonal β-SnS_2 phase with cell constants $a = 3.65$ Å and $c = 5.90$ Å, which are consistent with the literature [50]. The crystalline size of the sample was about 12 nm, as estimated by the Scherrer equation.

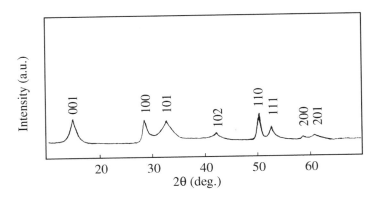

Fig. 10. XRD pattern of SnS₂. (Source: Reprinted from [39] with permission of Elsevier Science.)

The TEM micrograph of β-SnS$_2$ particles shows the particles consisted of uniform spherical crystallites with an average size of about 12 nm, which is consistent with that deduced from the XRD pattern.

In the preparation of nanocrystalline β-SnS$_2$ through the solvothermal procedure, several factors such as temperature, reaction time, and solvent were considered. Toluene was chosen because it is a weakly polar organic solvent that could prevent the immediate reaction of Na$_2$S with SnCl$_4$ at room temperature, which is beneficial to controlling the reaction and forming crystalline tin sulfide. In the solvothermal process, the optimum condition for preparing β-SnS$_2$ was 150 °C for 6–8 h. If the temperature was lower than 100 °C or the time was shorter than 4 h the yield of β-SnS$_2$ was lower and as-prepared β-SnS$_2$ was of low quality. In the reaction process, toluene was especially rigorous in the absence of oxygen and water, which prevented the oxidation of SnCl$_4$ and Na$_2$S. The infrared (IR) spectra of the final product also indicated that there were no tin oxides, because no Sn–O vibrations were detected in the range 500 to 720 cm^{-1} [51]. The byproduct NaCl could be removed by washing with absolute ethanol. Tetrahydrofuran (THF) was also used as a solvent. As Na$_2$S was slowly added the THF solution of SnCl$_4$ at room temperature in a dry box, a reaction immediately occurred, which yielded a yellow solid. XRD indicated the product was amorphous. Even if the product was treated under similar conditions to the solvothermal process, SnS$_2$ was still poorly crystalline. Because THF is a polar solvent, SnCl$_4$ and Na$_2$S dissolve in it and react with each other to form amorphous product immediately, which is not beneficial to forming crystalline SnS$_2$.

2.2.2. Cobalt Sulfide [52]

CoS$_x$ ($x > 0.89$) is of importance as a hydrodesulfurization catalyst and for its magnetic properties [53, 54]. Conventionally, it is synthesized through the direct reaction of elements [53, 55] (>500 °C). Some low-temperature routes are known. Reactions between anhydrous hexamine cobalt(II) and hydrogen sulfide at room temperature produce poorly crystalline CoS$_2$ [56]. When anhydrous cobalt sulfate salt is exposed to a mixture of hydrogen and hydrogen sulfide at 525 °C, Co$_9$S$_8$ is formed [57].

We successfully prepared nanocrystalline Co$_9$S$_8$ and its serial compounds by the toluene thermal reaction between Na$_2$S$_3$ and CoCl$_2$·6H$_2$O in the temperature range of 120 to 170 °C. The phase formation can be controlled under appropriate reaction conditions.

In a typical reaction, appropriate amounts of CoCl$_2$·6H$_2$O and Na$_2$S$_3$ (excess 50%) are added to an autoclave that is filled with toluene and maintained at 120–170 °C for 12–24 h and then cooled to room temperature naturally. The different reaction conditions are shown in detail in Table I.

Table I.   Reaction Conditions of Cobalt Sulfides

Reagent	Solvent	Temperature (°C)	Time (h)	Product
$Na_2S_3 + CoCl_2 \cdot 6H_2O$	Toluene	110–130	24	Co_9S_8
$Na_2S_3 + CoCl_2 \cdot 6H_2O$	Toluene	150–170	12	$Co_9S_8 + Co_3S_4 + CoS_2$
$Na_2S_3 + CoCl_2 \cdot 6H_2O + Zn$	Toluene	170	12	Co_9S_8
$Co_9S_8 + I_2$ [a]	Toluene	170	12	$CoS_2 + Co_3S_4$
$Na_2S_3 +$ anhydrous $CoCl_2$	Toluene	140–170	12	CoS_2

(Source: Data from X. F. Qian et al., *Inorg. Chem.*, 1999.)

[a] Co_9S_8 is first formed by the toluene thermal process at 120 °C for 24 h, without any posttreatment, I_2 is added, and then the autoclave is continually maintained at 170 °C for 12 h to form CoS_2 and Co_3S_4.

In the preparation process, toluene is chosen because of its appropriate boiling point (110.6 °C) and pressure (about 5 atm at 178 °C), which is much lower than that of water (about 10 atm at the same temperature). So the toluene thermal process is much safer in comparison with the hydrothermal process. Furthermore, toluene is a poor polar organic solvent; it can avoid the immediate reaction of Na_2S_3 and $CoCl_2 \cdot 6H_2O$ at room temperature, which is beneficial to controlling the rate of the reaction and forming nanocrystalline materials.

The toluene thermal process may be a liquid–solid reaction according to the theory of Gouw and Jentoft [59]. Because of the coexistence of S_3^{2-} ions, which have a weak reduction property, with water at high temperature [60], there is a redox reaction. The equation can be written as follows:

$$9CoCl_2 \cdot 6H_2O + 9Na_2S_3 \rightarrow Co_9S_8 + 18NaCl + 19S + 54H_2O$$

Figure 11a shows the XRD pattern of a typical sample prepared by the toluene thermal process. All the peaks can be indexed to the single phase of Co_9S_8 with lattice parameter $a = 9.92$ Å, which is close to the reported data. The crystalline size of Co_9S_8 estimated by the Scherrers equation is about 20 nm. The TEM micrograph of the as-prepared Co_9S_8 particles shows the Co_9S_8 particles consisting of uniform spherical crystallites. The average size is about 20 nm, which is consistent with the result from the XRD pattern.

In the preparation process of Co_9S_8, the water of crystallization in the precursor $CoCl_2 \cdot 6H_2O$ is very important. If we use anhydrous $CoCl_2$ as the precursor, we only obtained the single phase of CoS_2 (Fig. 11c) and no Co_9S_8 occurred. The average size of the particles is about 30 nm. In the anhydrous atmosphere, ionic sulfur may be present mainly as S_3^{2-} and it is difficult to disproportionate into S and S^{2-}. The reaction may be written as

$$CoCl_2 + Na_2S_3 \rightarrow CoS_2 + 2NaCl + S$$

The reaction temperature and time have an important effect on the preparation processes of cobalt sulfides. The optimum condition for preparing Co_9S_8 is about 110–130 °C for 24 h. If the temperature is lower than 100 °C or the time is shorter than 12 h, the reaction is very slow and incomplete. If the reaction temperature is higher, 150–170 °C, a mixed phase of Co_9S_8, CoS_2, and Co_3S_4 results (Fig. 11b), which may be caused by the high reactivity of Co_9S_8 and an enriched sulfur environment in the autoclave. The process may be written as follows:

$$Na_2S_3 + 2H_2O \rightarrow H_2S + 2NaOH + 2S$$
$$Co_9S_8 + (9x - 8)H_2S \leftrightarrow 9CoS_x + (9x - 8)H_2 \qquad x > 1.06$$

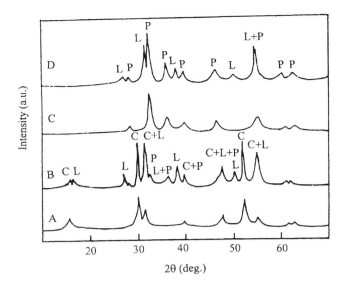

Fig. 11. XRD pattern of nanocrystalline cobalt sulfide powders. (C = Co$_9$S$_8$, L = Co$_3$S$_4$, P = CoS$_2$.) (a) Co$_9$S$_8$ prepared by the reaction of CoCl$_2$·6H$_2$O and Na$_2$S$_3$ at 120 °C for 24 h. (b) A mixed phase of Co$_9$S$_8$, CoS$_2$, and Co$_3$S$_4$ prepared by the reaction of CoCl$_2$·6H$_2$O and Na$_2$S$_3$ in the range of 150 to 170 °C for 12 h. (c) CoS$_2$ prepared by the reaction of anhydrous CoCl$_2$ and Na$_2$S$_3$ at 140 °C for 12 h. (d) A mixed phase of Co$_3$S$_4$ and CoS$_2$ prepared by the reaction of Co$_9$S$_8$ and I$_2$. (Source: Reprinted with permission from [52]. © 1999 American Chemical Society.)

The process can be demonstrated by adding Zn and I$_2$, respectively. As we add Zn to the autoclave with CoCl$_2$·6H$_2$O and Na$_2$S$_3$, we only obtain a single phase of Co$_9$S$_8$, even though the temperature reached 170 °C for 24 h. This is due to the H$_2$, which results from the reaction of the zinc and the water of crystallization and which shifts the equilibrium toward Co$_9$S$_8$. At the same time, if we add I$_2$ to the Co$_9$S$_8$ (which is prepared according to the preceding procedure with no posttreatments) and react the solution at 170 °C for 12 h, we obtain a mixed-phase Co$_3$S$_4$ and CoS$_2$ (Fig. 11d). No Co$_9$S$_8$ phase occurs. This may be due to the formation of HI, which makes the equilibrium transfer to CoS$_x$ ($x > 1.06$) in accordance with the literature [61].

The sulfur contents of the as-prepared single phases of Co$_9$S$_8$ and CoS$_2$ are determined by heating the samples to constant weights in a stream of oxygen [53]. The sulfides are converted to CoO at a temperature that does not exceed 900 °C in order to prevent overoxidation of CoO. Products prepared by this procedure correspond to Co$_9$S$_{7.93}$ and CoS$_{1.97}$, respectively.

2.2.3. Cadmium Sulfide [62]

CdS is an important material in nonlinear optics [63], quantum size effect semiconductors [64], electroluminescent devices [65], and other interesting physical and chemical technological applications [66–68]. Many methods have been developed to synthesize solid nanocrystalline CdS [63–69], but they all produce solid semiconductor particles or small clusters with morphologies.

CdS nanorods were first prepared in ethylenediamine nonaqueous solvent systems with the reaction of cadmium metal powder and sulfur under pressure in an autoclave:

$$Cd + S \rightarrow CdS$$

An appropriate amount of sulfur in ethylenediamine and cadmium powder were added to an autoclave filled with ethyleneamine. The autoclave was maintained at 120–190 °C for 3–6 h and cooled to room temperature. A bright-yellow product was obtained.

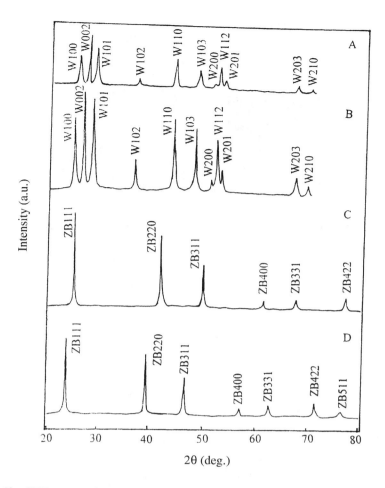

Fig. 12. XRD patterns of obtained samples: (a) CdS (prepared with Cd and S), (b) CdS (prepared with CdCl$_2$ and S), (c) CdSe, and (d) CdTe (W, wurtzite structure, ZB, zincblende structure). (Source: Reprinted with permission from [62]. © 1998 American Chemical Society.)

In the XRD pattern (Fig. 12), all the peaks can be indexed by the hexagonal cell of CdS (wurtzite structure) with lattice parameters $a = 4.141$ Å and $c = 6.72$ Å. The unusually strong (002) peak in the pattern indicates a preferential orientation of [001] in CdS crystallines.

Elemental analysis of the sample confirmed the element composition of CdS: Cd:S in 50:50 atomic ratios. This was also confirmed by energy-dispersive spectrometry (EDS) analysis.

TEM images showed that the CdS crystallites synthesized in ethylenediamine solvent appear to display rodlike monomorphology with lengths of 300–2500 nm and widths of 25–75 nm (Fig. 13). Figure 13 shows that the CdS nanorod is a single crystal. The nanorod axis (growth direction) was [002], which is consistent with the XRD pattern (Fig. 12).

In this synthetic system, the result of CdS nanorod formation indicated that the nucleation and growth were well controlled. The ethylenediamine plays an important role in controlling the nucleation and growth of the CdS nanorod. It was supposed that, owing to the ethylenediamine intermolecular interactions including hydrogen bond, van der Waals force, and electrostatic interaction, a supermolecular structure [70] was formed that provided a template for inorganic atom or ion self-assembling. Thus, the nanorod growth mechanism can be viewed as a templating mechanism. Using a similar process, CdSe and CdTe nanorods can also be prepared.

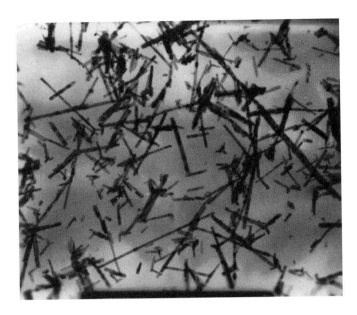

Fig. 13. TEM image of CdS nanorods. (Source: Reprinted with permission from [62]. © 1998 American Chemical Society.)

2.2.4. Bismuth Sulfide [71]

$A_2^V B_3^{VI}$ (A = Sb, Bi, As and B = S, Se, Te) group semiconductors have potential applications in television cameras with photoconducting targets, thermoelectric devices, electronic and optoelectronic devices, and in IR spectroscopy [72]. Bismuth sulfide (Bi_2S_3) has a direct band gap of 1.3 eV [73] and is useful for photodiode arrays or photovoltaics [74, 75].

The preparation methods of bismuth sulfide usually include direct reaction, chemical deposition, and thermal decomposition. Direct reaction and thermal decomposition need high reaction temperature. The products obtained by chemical deposition are amorphous or poorly crystallized. Here, we prepared Bi_2S_3 through a solvothermal reaction of $BiCl_3$ and thiourea in ethanol.

In a typical synthesis, thiourea was dissolved in the ethanol in the autoclave, and then an appropriate amount of $BiCl_3$ was added with stirring. The autoclave was sealed and maintained at 140 °C for 12 h and then cooled to room temperature. The dark-brown product was washed and dried in vacuum.

The XRD pattern of the product is shown in Figure 14. All the peaks can be indexed with orthorhombic cells with $a = 11.128$, $b = 11.264$, and $c = 3.978$ Å, which are close to the reported data. A TEM micrograph (Fig. 15) shows that the particles have a size of 500 nm × 30 nm on average, which indicates that the morphology of Bi_2S_3 is rodlike. In comparison, nanocrystalline Bi_2S_3 prepared by the hydrothermal process also displays a rodlike shape with a particle size of only about 150 nm × 40 nm. This means that the solvothermal method can control the growth of nanorods, which has been observed in II–VI group semiconductor nanorods.

The formation of the bismuth sulfide in ethanol may be through two steps: (i) the formation of Bi–thiourea complex in ethanol and (ii) the thermal decomposition of Bi–thiourea complex in ethanol at appropriate temperature and the formation of bismuth sulfide.

The influence of thermal treatment temperature and time on the formation of Bi_2S_3 in ethanol was investigated. The appropriate temperature for the preparation is 140 °C. If the temperature was lower than 100 °C, the product was poorly crystallized and some unidentified phases were detected. If the time was shorter than 6 h, there was some amorphous phase in the sample.

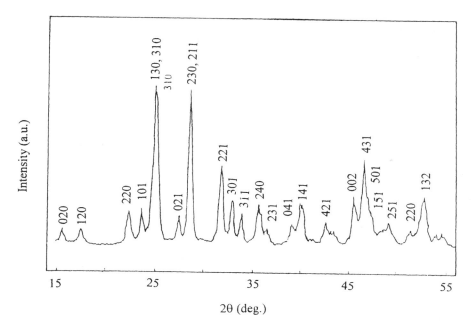

Fig. 14. XRD pattern of Bi_2S_3. (Source: Reprinted from [71] with permission of Elsevier Science.)

Fig. 15. TEM image of Bi_2S_3. (Source: Reprinted from [71] with permission of Elsevier Science.)

The IR spectrum of Bi_2S_3 showed no peaks of Bi_2O_3 and no evidence of organic impurities in the compound. The absorption peaks ranging from 3600 to 3000 cm^{-1} correspond to the $-OH$ group of H_2O absorbed in the sample. The absorption peaks centered at approximately 1630 cm^{-1} correspond to CO_2 absorbed on the surface of the particle. In fact, the absorbed water and CO_2 are common to all powder samples that have been exposed to atmosphere and were more pronounced for high-surface-area particles.

3. γ-IRRADIATION SYNTHESIS AND CHARACTERIZATION OF NANOMETER MATERIALS

Fujita et al. [76] began the synthesis of metal aggregates by the radiolytic reduction of metal cations in solution. In recent decades, this method has been developed further [77–80]. The technique of pulse radiolysis has been used to study the yield of short-lived clusters and their optical spectra in dilute aqueous solutions of about 10^{-4} M metallic ions [81–83]. By this method, silver cluster and colloidal silver could form [84, 85]. The formation of other colloidal metals from correspondent ions such as Cu^{2+} [86–88], Ti^{4+} [89, 90], Pb^{2+} [91, 92], and Pd^{2+} [93] has also been studied. By high-energy electron irradiation of 1.7×10^4 M NaAu(CN)$_2$ solution, Mosseri et al. [94] prepared ultrafine Au particles. Marignier et at. [95] prepared ultrafine Ni and Cu–Pd alloy particles by separate γ-irradiation of NiSO$_4$ solution and CuSO$_4$–PdCl$_2$ solution.

However, the solutions used in the aforementioned studies are about 10^{-4} M metal ions, and the ultrafine particles produced are in the colloidal state. We have developed a new method—γ-irradiation of solutions containing 10^{-2}–10^{-3} M metal ions—to prepare nanometer-sized powders of not only metals, but also nonmetals, alloys, metal oxides, metal sulfides, and nanocomposites. This method with relatively high yield of product and low cost can process under normal pressure at room temperature.

During γ-irradiation, the formation of hydrated electrons can be shown as follows [96, 97]:

$$H_2O \rightarrow e_{aq}^-, H_3O^+, H, H_2, OH, H_2O_2$$

Then the radiation reduction of metal ions by hydrated electrons leads to the formation of metal nanoparticles:

$$e_{aq}^- + M^{m+} \rightarrow M^{(m-1)+}$$
$$e_{aq}^- + M^+ \rightarrow M^0$$
$$nM^0 \rightarrow M_2 \rightarrow M_n \rightarrow M_{agg}$$

where n is a number of aggregation of a few units and M_{agg} is the aggregate in the final stable state [87].

3.1. Nanocrystalline Metals

3.1.1. Nanocrystalline Silver [98]

Highly pure silver powders of fine, narrowly distributed, and uniform particles are very important in many fields of technology. For example, ultrafine silver powders constitute the active part of conductive ink pastes and adhesives used in the manufacture of various electronic parts [99]. Conductive silver pastes and inks form the bases for thick-film technology, for producing electronic components such as hybrid microcircuits [100], and for the internal electrodes of multilayer ceramic capacitors [101].

The stability of Ag_4^{2+} and colloidal silver has been investigated [102]. After 5 min of γ-irradiation of a solution containing 5×10^{-2} M AgNO$_3$, 1.0 M (CH$_3$)$_2$CHOH, and 1.0×10^{-2} M C$_{12}$H$_{25}$NaSO$_4$ at a dose of 18.5 Gy/min (sample 1), the result is as shown in Figure 16. From this figure, it is obvious that the absorption at about 290 nm (due to the Ag_4^{2+} cluster [103]) decays very slowly. The Ag_4^{2+} cluster can exist for more than 1 month in γ-irradiated concentrated AgNO$_3$ solution in air. This is much longer than that in dilute Ag^+ ion solutions previously reported [103–109]. On the other hand, the absorption band of colloidal silver (around 400–410 nm [110]) broadens and shifts toward longer wavelengths at longer aging time, indicating a slow increase in the size of the colloidal silver.

Because the colloid was stable, the hydrothermal treatment method was used to aggregate the colloid to metal powders. Solutions were prepared with silver nitrate of analytical grade in distilled water. Sodium dodecyl sulfate, poly(vinyl alcohol), and sodium polyphosphate were chosen as the surfactants, and isopropanol was used as a scavenger for hydroxyl radicals. Both the surfactant and the scavenger were added to solutions in various concentrations. All solutions were deaerated by bubbling with nitrogen for 1 h and then irradiated in the field of a 2.59×10^{15}-Bq ^{60}Co γ-ray source with different doses. Then the irradiated solutions were put into autoclaves with Teflon inners and heated in an oven at different temperatures ranging from 105 to 200 °C for different periods of time. After cooling to

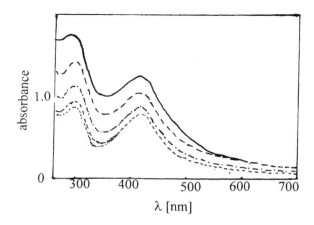

Fig. 16. Aging effect in open air on the solution containing 5×10^{-2} M $AgNO_3$, 1.0×10^{-2} M $C_{12}H_{25}NaSO_4$, 1.0 M $(CH_3)_2CHOH$ with 5 minutes of γ-irradiation at a close rate of 18.5 Gy/min. Aging time after irradiation: ($\cdots$) just after irradiation, (- - - -) 38 h, (-·-·-) 73 h, (– –) 144 h, (—) 240 h. (Source: Adapted from Zhu et al. [102].)

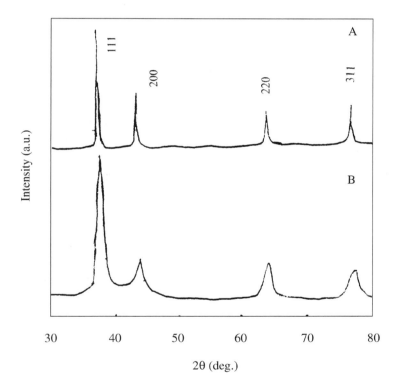

Fig. 17. X-ray diffraction pattern of the product prepared by the γ-irradiation–hydrothermal treatment combined method: (a) sample 1 and (b) sample 4. (Source: Reprinted from [98] with permission of Elsevier Science.)

room temperature, the products were separated and washed with distilled water and 25% ammonia aqueous solution. The final product was a black powder.

XRD patterns show the products are metallic silver (Fig. 17). The average crystallite sizes calculated from the peak broadening of XRD patterns by the Scherrer equation [111] are listed in Table II, from which one can see that the smallest average particle size of silver powder is 8 nm for sample 4, which was prepared by irradiating a mixed solution

Table II. Experimental Conditions and Silver Particle Diameters [98]

Sample number	Solution	Irradiation dose ($\times 10^4$ Gy)	Temperature (°C) and time (h) of hydrothermal treatment	Silver particle size (nm)
1	0.1 M $AgNO_3$ + 4.0 M $(CH_3)_2CHOH$	1.8	105, 3	47
2	0.1 M $AgNO_3$ + 4.0 M $(CH_3)_2CHOH$ + 0.1 M $C_{12}H_{25}NaSO_4$	4.3	105, 2	13
3	0.05 M $AgNO_3$ + 6.0 M $(CH_3)_2CHOH$ + 0.1 M $C_{12}H_{25}NaSO_4$	3.0	105, 1	15
4	0.05 M $AgNO_3$ + 6.0 M $(CH_3)_2CHOH$ + 0.1 M $C_{12}H_{25}NaSO_4$	2.4	105, 1	8
5	0.05 M $AgNO_3$ + 6.0 M $(CH_3)_2CHOH$ + 0.1 M $C_{12}H_{25}NaSO_4$	1.1	105, 1	9
6	0.05 M $AgNO_3$ + 6.0 M $(CH_3)_2CHOH$ + 0.1 M $C_{12}H_{25}NaSO_4$	0.77	105, 1	10
7	0.05 M $AgNO_3$ + 6.0 M $(CH_3)_2CHOH$ + 0.1 M $C_{12}H_{25}NaSO_4$	1.1	Precipitation	23
8	0.05 M $AgNO_3$ + 6.0 M $(CH_3)_2CHOH$ + 0.1 M $C_{12}H_{25}NaSO_4$	1.1	105, 1	31
9	0.05 M $AgNO_3$ + 6.0 M $(CH_3)_2CHOH$ + 0.1 M $C_{12}H_{25}NaSO_4$	1.1	105, 26	35
10	0.05 M $AgNO_3$ + 6.0 M $(CH_3)_2CHOH$ + 0.1 M $C_{12}H_{25}NaSO_4$	1.1	210, 1	43
11	0.05 M $AgNO_3$ + 6.0 M $(CH_3)_2CHOH$ + 0.2 M $C_{12}H_{25}NaSO_4$	3.0	105, 1	13

(Source: Data from Y. Zhu et al., *Mater. Lett.*, 1993.)

Fig. 18. TEM microphotographs of silver particles produced by the γ-irradiation–hydrothermal treatment combined method: irradiation solution of sample 5. (Source: Reprinted from [98] with permission of Elsevier Science.)

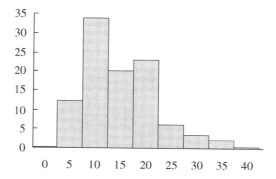

Fig. 19. Particle size distribution of sample 2. (Source: Reprinted from [98] with permission of Elsevier Science.)

of 0.05 M AgNO$_3$, 6.0 M (CH$_3$)$_2$CHOH, and 0.1 M C$_{12}$H$_{25}$NaSO$_4$ with a dose of 2.4 $\times$ 10^4 Gy and hydrothermally treated at 105 °C for 1 h. The TEM image (Fig. 18) shows that the silver powders consist of quasispherical crystallites. The particle size distribution was determined by the photographic image microstructure densitometry analysis method. The distribution of particle sizes of sample 2 is shown in Figure 19, from which we can see silver particle sizes ranging from 6 to 40 nm, and the largest percentage is about 35% in the size range 10 to 15 nm. The average particle size is 16 nm calculated from this figure. The experiments reveal that the surfactant, the γ-irradiation dose, the hydrothermal temperature, and time influence the silver particle size, as shown in Table II and Figure 20. The shape of the silver particles produced by γ-irradiation depends on the surfactant used. For example, when sodium dodecyl sulfate is used as a surfactant, the silver particles are quasispherical. However, when poly(vinyl alcohol) is used, the silver particles have various shapes (Fig. 21).

The yield and radiation chemistry yield (G value) [112] were studied and the results are given in Table III. Table III shows that the yield increased on increasing the irradiation dose when the concentration of AgNO$_3$ was fixed. On the other hand, the G value had little

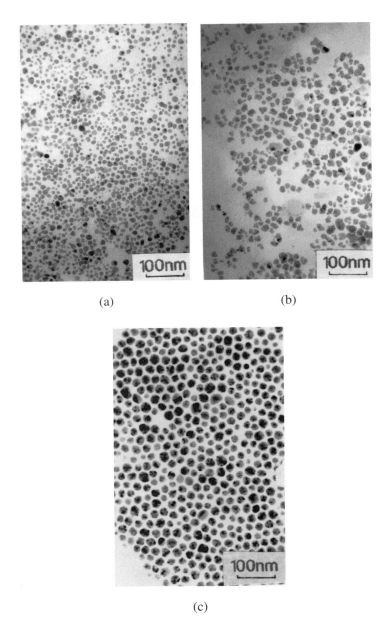

Fig. 20. TEM images of silver particles produced by the γ-irradiation method. Solution: (a) 0.01 M AgNO$_3$, 0.01 M C$_{12}$H$_{25}$NaSO$_4$, and 1.0 M (CH$_3$)$_2$CHOH; (b) the same solution as in (a); and (c) 0.05 M AgNO$_3$, 0.01 M C$_{12}$H$_{25}$NaSO$_4$, and 1.0 M (CH$_3$)$_2$CHOH. Dose rate (Gy·min^{-1}) and radiation time: (a) 18.4, 5 min; (b) 59.3, 30 min; and (c) 72, 5.5 h. (Source: Reprinted from [98] with permission of Elsevier Science.)

relation to the irradiation dose when the AgNO$_3$ concentration was fixed. However, the G value increased rapidly from 2.6 to 9.65 when the AgNO$_3$ concentration increased from 0.01 to 0.05 M. With further increase in the AgNO$_3$ concentration, the G value increased slowly. This implies that the mechanism of radiolytic reactions in the solutions containing 10^{-2}–10^{-3} M metal ions is different from the mechanism in the dilute solutions (about 10^{-4} M metal ions) that contain only primary free-radical reactions [104–108]. In concentrated Ag$^+$ solutions, Ag$^+$ ions could go into the spurs in which they reacted with primary radicals produced during radiolysis, and this led to the higher G value.

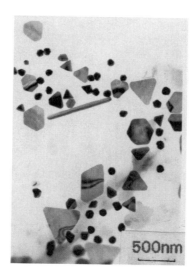

Fig. 21. TEM image of silver particles produced by the γ-irradiation method. Solution: 0.1 M AgNO$_3$, saturated poly(vinyl alcohol), and 1.0 M (CH$_3$)$_2$CHOH. Dose rate: 76.3 Gy·min^{-1}; radiation time: 12 h. (Source: Reprinted from [98] with permission of Elsevier Science.)

Table III. Effect of Experimental Parameters on the Yield and G value of Silver Produced from AgNO$_3$ Solution

Solution	Irradiation dose (10^4 Gy)	Yield (%)	G value
0.01 M AgNO$_3$ + 2.0 M (CH$_3$)$_2$CHOH	3.09	95.1	2.66
0.01 M AgNO$_3$ + 2.0 M (CH$_3$)$_2$CHOH	2.50	84.5	2.91
0.01 M AgNO$_3$ + 2.0 M (CH$_3$)$_2$CHOH	1.34	40.9	2.64
0.05 M AgNO$_3$ + 2.0 M (CH$_3$)$_2$CHOH	3.09	69.3	9.62
0.1 M AgNO$_3$ + 2.0 M (CH$_3$)$_2$CHOH	3.09	37.4	10.3

3.1.2. Nanocrystalline Copper [113]

Figure 22 shows XRD patterns of the products prepared from solution containing sodium dodecyl sulfate as the surfactant. One can see that the products only washed with distilled water contain the impurity copper(I) oxide. The impurity could be removed by washing with 25% ammonia aqueous solution. Because of the formation of Cu(NH$_3$)$_2^+$, pure copper powders could be obtained.

The process of reducing the copper ions in the solution may be understood as follows: Cu^{2+} ions are rapidly reduced to Cu$^+$ ions by hydrated electrons and organic radicals produced by γ-ray radiation. Cu$^+$ ions can be further reduced in reaction with e$_{aq}^-$, giving atomic copper. The copper atoms are generators of small clusters (Cu$_2^+$ and others), which then form aggregates (Cu$_n$). The clusters also may act as nuclei on which the dismutation of monovalent copper ions takes place:

$$\text{Cu}_n + 2\text{Cu}^+ = \text{Cu}_{n+1} + \text{Cu}^{2+}$$

This results from the extremely negative reduction potential for the copper atoms [114]. The two simultaneous processes of the growth of the metallic aggregates through

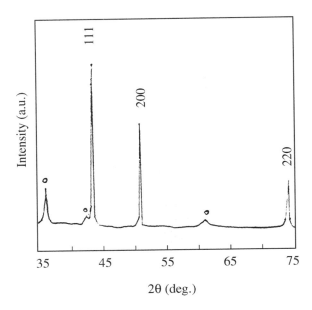

Fig. 22. XRD pattern of the product prepared by γ-irradiation combined with hydrothermal treatment. Solution: 0.01 M Cu(NO$_3$)$_2$, 0.1 M C$_{12}$H$_{25}$NaSO$_4$, and 3.0 M (CH$_3$)$_2$CHOH; dose: 8.6 × 10^4 Gy; ○ copper(I) oxide. (Source: Reprinted from [113] with permission of Elsevier Science.)

dismutation and coalescence result in the formation of colloidal copper. There is a competitive reaction with the reduction and dismutation of monovalent copper ions, that is, the formation of poorly soluble copper(I) hydroxide (CuOH). As a result of the competition of the preceding reactions, the products consist of both copper and copper(I) oxide.

When the pH value is greater than 4, CuOH is very unstable in solution and decomposes rapidly to copper(I) oxide (Cu$_2$O). To obtain pure copper powders, the pH value of solution should be smaller. Although poly(vinyl alcohol) is used as the surfactant instead of sodium dodecyl sulfate, copper(I) oxide could not form in the solution for the lower pH value.

The experimental result shows that copper powders produced by γ-irradiating copper salt solutions without ethylenediaminetetraacetic acid (EDTA) as a complex agent consist of relatively large particles. When EDTA was added to the solution, the final product consists of much smaller copper particles. Thus, the complexation of copper ions with EDTA is favorable for preparing nanocrystalline copper powders. This may be due to a dramatic decrease in the reduction reaction rate caused by the complexation of copper ions with the EDTA ligand. Also, the ligand on the copper ion may act as a bridge for electron transfer from the solvent to the copper ions. Copper powder prepared from a solution of 0.01 M CuSO$_4$, 0.01 M EDTA, 0.1 M C$_{12}$H$_{25}$NaSO$_4$, and 3.0 M (CH$_3$)$_2$CHOH with a radiation dose of 3.6 × 10^4 Gy consists of quasispherical particles with an average particle size of 16 nm (Fig. 23a). In addition to the quasispherical shape, acicular copper particles (Fig. 23d) were observed in the sample prepared from a solution of 0.01 Me Cu$_2$SO$_4$, 0.01 M EDTA, 0.02 M C$_{12}$H$_{25}$NaSO$_4$, and 3.0 M (CH$_3$)$_2$CHOH. Therefore, by controlling the conditions of the experiment, different-shaped copper particles can be obtained.

The growth process of nanocrystalline copper to single crystals induced by electron irradiation was observed by TEM, as shown in Figure 23a–c. Figure 23a shows the original nanocrystalline copper particles with an average particle size of 16 nm. After electron irradiation under the electron microscope, the particles became larger. Finally, after 12 s of electron irradiation, the particles became a large spheroid about 1.5 μm in diameter (Fig. 23b). The electron diffraction result is shown in Figure 23c, from which one can see that it is a single crystal of copper. Because of the thickness of the single crystal, Kikuchi

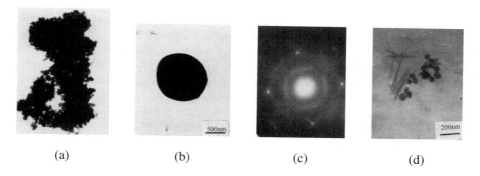

(a)   (b)   (c)   (d)

Fig. 23. TEM image of the sample prepared by γ-irradiation combined with hydrothermal treatment: (a) solution: 0.01 M $CuSO_4$, 0.01 M $C_{12}H_{25}NaSO_4$, and 3.0 M $(CH_3)_2CHOH$; dose: 3.6×10^4 Gy; (b) after 12 s of electron irradiation under an electron microscope; (c) electron diffraction of the sample in (b) in the [001] direction; and (d) solution: 0.01 M $CuSO_4$, 0.01 M EDTA, 0.02 M $C_{12}H_{25}NaSO_4$, and 3.0 M $(CH_3)_2CHOH$; dose: 3.6×10^4 Gy. (Source: Reprinted from [113] with permission of Elsevier Science.)

lines can be observed in its electron diffraction microphotograph (Fig. 23c), where the beam of electrons was incident in the [001] direction. From the occurrence of the Kikuchi lines, we may conclude that the thickness of the single crystal is more than 100 nm.

3.1.3. Nanocrystalline Ruthenium [115]

Ultrafine powders of metal could also be obtained from acid group ions. Nanocrystalline Ru(12) has been prepared successfully from RuO_4^{2-}. The mechanism of the reduction of ruthenate ions during γ-irradiation may be explained as follows. The hydrated electrons react with RuO_4^{2-} to form Ru^V:

$$RuO_4^{2-} + e_{aq}^- \rightarrow Ru^V$$

The dismutation of pentavalent ruthenium occurs [116]:

$$Ru^V + Ru^V \rightarrow Ru^{IV} + Ru^{VI}$$

The tetravalent ruthenium reacts further with hydrated electrons to form ruthenium atoms.

3.1.4. Nanocrystalline Nickel [117,118] and Cobalt

In nickel salt solutions, Ni^{2+} ions are rapidly reduced to Ni^+ ions by hydrated electrons (e_{aq}^-) and organic radicals produced by γ-irradiation. Ni^+ ions can further react with e_{aq}^- and organic radicals [119]:

$$e_{aq}^- + Ni^+ \rightarrow Ni^0$$
$$Ni^+ + C(CH_3)_2OH \rightarrow NiC(CH_3)_2OH^+$$

$NiC(CH_3)_2OH^+$ reacts in the following ways:

$$NiC(CH_3)_2OH^+ + H^+ \rightarrow Ni^{2+} + (CH_3)_2CHOH$$
$$NiC(CH_3)_2OH^+ + Ni^+ \rightarrow Ni^{2+} + Ni^0 + (CH_3)_2CHOH$$
$$NiC(CH_3)_2OH^+ + (CH_3)_2COH \rightarrow Ni^0 + (CH_3)_2CO + (CH_3)_2COH + H^+$$

The metal atoms are the generators of small clusters, which may act as nuclei on which aggregates form, resulting in the formation of colloidal nickel.

However, the zero valence state of nickel, which resulted from the two-electron transfer, could also undergo oxidation:

$$Ni_n + H_3O^+ \rightarrow Ni_{n-1} + Ni^+ + \tfrac{1}{2}H_2 + H_2O$$

In the presence of $NH_3 \cdot H_2O$, which acts as an alkalizing agent, the pH of the solution is kept in the range of 10 to 11, so the reoxidation of atoms or aggregates of nickel in solution is greatly suppressed. The product produced by γ-irradiating a solution containing 0.01 M $NiSO_4$, 0.1 M $NH_3 \cdot H_2O$, 0.1 M $C_{12}H_{25}NaSO_4$, and 2.0 M $(CH_3)_2CHOH$ with a dose of 6.0×10^4 Gy is a single phase of nickel with an average particle size of 8 nm.

Figure 24 shows the differential thermal analysis (DTA) curve of the sample, from which it can be seen that two exothermic peaks appeared in the temperature range of 40 to $1040\,°C$, because of the oxidation of metallic nickel, which began at approximately $220\,°C$ in air. The maximum of the first exothermic peak, located at approximately $340\,°C$, corresponds to the formation of NiO in the oxidation of metallic nickel. The second exothermic peak, located at approximately $385\,°C$, corresponds to the formation of Ni_2O_3 because of the oxidation of NiO. The irradiated solutions were black and stable in air. When $Ni(NO_3)_2$ was used instead of $NiSO_4$, no nickel powder product was formed. This may be due to the reaction of NiO_3^- ions with hydrated electrons during γ-irradiation.

The yields and G values are given in Table IV. From Table IV, it can be seen that the weight and G value of the product increased on increasing the concentration of $NiSO_4$. However, the yield of nickel powder decreased as the $NiSO_4$ concentration increased. This was because more Ni^{2+} ions were left unreduced in the solution with larger concentration when the irradiation dose was fixed. The G value increased rapidly from 2.17 to 5.91 when the $NiSO_4$ concentration increased from 0.01 to 0.03 M, but with further increase in the $NiSO_4$ concentration, the G value increased slowly. This implies that the mechanism of

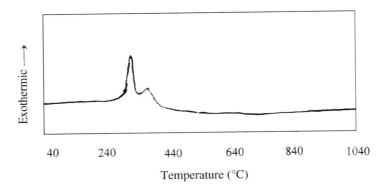

Fig. 24. The DTA curve of the sample prepared by the γ-irradiation method. Solution: 0.01 M $NiSO_4$, 0.1 M $NH_3 \cdot H_2O$, 0.01 M $C_{12}H_{25}NaSO_4$, and 2.0 M $(CH_3)_2CHOH$; dose: 6.0×10^4 $J \cdot kg^{-1}$. (Source: Reprinted from [117] with kind permission from Kluwer Academic Publishers.)

Table IV. Effect of Experimental Parameters on the Yield and G Value of Nickel Produced from $NiSO_4$ Solution (Dose: 2.78×10^4 Gy) [118]

Solution (in molar concentration)				Weight of product in 200 mL solution/g	Yield (%)	G value
$NiSO_4$	$NH_3 \cdot H_2O$	$(NH_4)_2SO_4$	$(CH_3)_2CHOH$			
0.01	0.16	0.1	3.0	0.088	75.2	2.17
0.03	0.28	0.1	3.0	0.241	68.7	5.91
0.05	0.28	0.1	3.0	0.261	44.6	6.36
0.1	0.7	0.1	3.0	0.280	23.9	7.72
0.01	0.16	0.1	0	0	0	0

(Source: Data from Y. Zhu et al., *Chin. Sci. Bull.*, 1997.)

Table V. Effect of the Concentration on the Yield and G Value of Nickel Produced from $Ni(CH_3COO)_2$ Solution (Dose: 2.71×10^4 Gy)

Solution (in molar concentration)				Weight of product in 200 mL solution/g	Yield (%)	G value
$Ni(CH_3COO)_2$	$NH_3 \cdot H_2O$	CH_3COONH_4	$(CH_3)_2CHOH$			
0.01	0.16	0.2	3.0	0.106	90.6	2.68
0.03	0.28	0.2	3.0	0.117	33.4	2.94
0.05	0.4	0.2	3.0	0.174	29.8	4.35
0.1	0.7	0.2	3.0	0.175	15.0	4.31

(Source: Data from Y. Zhu et al., *Chin. Sci. Bull.*, 1997.)

radiolytic reaction was different in dilute and concentrated solutions. In concentrated solutions, Ni^{2+} ions could go into the spurs where they reacted with primary radicals produced during radiolysis, and this led to the higher G value. The maximum G value in these experiments is as large as 7.72, much larger than that previously reported 0.03 [87]. On the other hand, in the absence of isopropanol, nickel powder failed to form.

Table V shows the trends of the yield and G value with the change of $Ni(CH_3COO)_2$ concentration. The G value increased from 2.68 to 4.35 corresponding to a $Ni(CH_3COO)_2$ concentration increase from 0.01 to 0.05 M. However, further increase in the concentration did not lead to an increase in the G value, indicating the maximum G value was reached (Fig. 25) at a dose of 2.71×10^4 Gy.

The G value and yield of nickel decreased on decreasing the dose rate at a given irradiation time (Table VI). This may be due to a positive effect of the nucleation of crystallites caused by an increase in the reduction reaction rate at a larger dose rate. The anion in the salt had an influence on the yield and G value of nickel produced (Table VII). Our experiments showed that the optimum G value of nickel produced was reached by using nickel sulfate.

In this way, nanocrystalline cobalt with an average particle size of 22 nm was prepared by γ-irradiation of a solution containing $CoCl_2$ instead of $NiSO_4$.

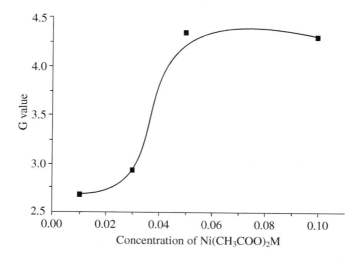

Fig. 25. The effect of the concentration on the G value of nickel. Solution: $Ni(CH_3COO)_2$; dose: 2.71×10^4 Gy. (Source: Adapted from Zhu et al. [118].)

Table VI.   Effect of Dose Rate of γ-Irradiation on the Yield and G Value of Nickel Produced from the Solution Containing 0.05 M Ni(CH$_3$COO)$_2$, 0.4 M NH$_3$·H$_2$O, 0.5 M CH$_3$COONH$_4$, and 1.0 M (CH$_3$)$_2$CHOH (Irradiation time: 6.5 h)

Dose rate/Gy·min^{-1}	Weight of product in 300 mL solution/g	Yield (%)	G value
80	0.200	22.3	3.12
45	0.065	7.4	1.83
30	0.038	4.4	1.67

(Source: Data from Y. Zhu et al., *Chin. Sci. Bull.*, 1997.)

Table VII.   Yields and G Values of Nickel Produced from Solutions of Different Nickel Salts (Dose: 3.02×10^4 Gy)

Solution (in molar concentration)			Weight of product in 300 mL solution/g	Yield (%)	G value
Salt	NH$_3$·H$_2$O	(CH$_3$)$_2$CHOH			
0.05 M NiSO$_4$	4.0	2.5	0.277	31.5	3.88
0.05 M NiCl$_2$	4.0	2.5	0.235	26.7	3.29
0.05 M Ni(CH$_3$COO)$_2$	4.0	2.5	0.224	25.5	3.14
0.05 M Ni(NO$_3$)$_2$	4.0	2.5	0	0	0

(Source: Data from Y. Zhu et al., *Chin. Sci. Bull.*, 1997.)

3.1.5. Nanocrystalline Cadmium [120], Tin [121], Indium [122], Antimony [123], and Lead

Nanocrystalline metals with low melting points are difficult to prepare because of their lower melting points (Table VIII). Some methods such as electrolysis and reduction with more active metals have been developed for the manufacture of these fine particles. However, the particle sizes of the powders prepared by these methods are relatively larger [124–126]. By using the γ-irradiation method, we have successfully prepared nanocrystalline Cd (20 nm), Sn (28 nm), In (32 nm), Sb (8 nm), and Pb (45 nm) under ambient pressure at room temperature.

The mechanism of Cd^{2+} reduction during γ-irradiation may be understood as follows: The hydrated electrons form in the aqueous solution during γ-irradiation. Then the hydrated electrons react rapidly with Cd^{2+} to form Cd$^+$ [127,128]:

$$e_{aq}^- + Cd^{2+} \rightarrow Cd^+ \tag{1}$$

The OH radicals react with isopropanol to yield organic radicals:

$$OH^- + (CH_3)_2CHOH \rightarrow H_2O + (CH_3)_2COH \tag{2}$$

Subsequently, these intermediates react among each other:

$$Cd^+ + Cd^+ \rightarrow Cd_2^{2+} \rightarrow Cd^0 + Cd^{2+} \tag{3}$$

$$Cd^+ + (CH_3)_2CHOH + H_2O \rightarrow Cd^{2+} + (CH_3)_2CH_2OH + OH^- \tag{4}$$

Reaction (3) competes with reaction (4), and reaction (3) is much faster than reaction (4). Colloidal cadmium is finally formed via dismutation and association reactions of Cd^{2+}. By γ-irradiating a solution containing 0.01 M CdSO$_4$, 0.01 M (NH$_4$)$_2$SO$_4$, 1 M NH$_3$·H$_2$O, 0.01 M C$_{12}$H$_{25}$NaSO$_4$, and 6.0 M (CH$_3$)$_2$CHOH with a dose of 1.6×10^4 Gy,

Table VIII.   Melting Points of Metals

Metal	Cd	Sn	Pb	In	Sb
Melting point (°C)	302.9	231.9	327.5	156.3	630.0

nanocrystalline Cd with an average particle size of 20 nm is obtained. Similarly, nanocrystalline Sn, In, Sb, and Pb can also be obtained.

3.2. Alloys [129]

Ultrafine powders of alloys are important in many applications such as coatings, conductor pastes, and parts requiring good electrical and thermal conductivity. Various methods are used to prepare the Ag–Cu alloy [130–135]. The gas-condensation method is commonly adopted, but a high temperature is needed and the product yield is relatively low. We have prepared the Ag–Cu alloy by the γ-irradiation method. We dissolved analytically pure $Cu(NO_3)_2 \cdot 3H_2O$ and $AgNO_3$ in distilled water and added $NH_3 \cdot H_2O$ as a complex agent. A surfactant, scavenger, and bubbling were also necessary.

To obtain a Ag–Cu alloy, a NH_3 ligand was used to adjust the condition of the solution. In the absence of the NH_3 ligand, the product is a mixture of metallic silver and copper. When the solution contains the NH_3 ligand, the single phase of the Ag–Cu alloy is obtained. This may be caused by the change in the rate of reduction reactions of Ag^+ and Cu^+ ions during γ-irradiation resulting from the complexation of metal ions with the NH_3. On the other hand, the NH_3 ligand on metal ions may act as a bridge for electron transfer from the solution to the metal ions.

The product prepared by γ-irradiating a solution containing 0.01 M $AgNO_3$, 0.05 M $Cu(NO_3)_2$, 0.3 M $NH_3 \cdot H_2O$, and 2.0 M $(CH_3)_2CHOH$ at a dose of 2.3×10^4 Gy does not contain the phase of metallic copper, and all the diffraction peaks shift toward larger diffraction angles compared with those of metallic silver, indicating the formation of the Ag–Cu alloy (25 nm). From the XRD data, the cell parameter a of the product is calculated to be 4.0395 Å, which is smaller than that of metallic silver ($a = 4.0862$ Å). Figure 26 is the TEM micrograph of the sample. The composition of the Ag–Cu alloy was analyzed using X-ray photoelectron spectroscopy (XPS), and the results are listed in Table IX.

3.3. Nanocrystalline Nonmetals

We extended the γ-irradiation method to the preparation of ultrafine powders of nonmetallic elements.

3.3.1. Tellurium [136]

The XRD pattern of the product prepared by γ-irradiating a solution containing 0.0063 M TeO_2, 0.05 M $C_{12}H_{25}NaSO_4$, 0.7 M HCl, and 1.6 M $(CH_3)_2CHOH$ with a dose of 2.32×10^4 Gy indicates that the product is a single phase of hexagonal tellurium. Figure 27 shows the TEM micrograph of the sample. It shows that the tellurium powder consisted of aciculate particles of size ranging from 10 nm × 80 nm to 40 nm × 300 nm.

3.3.2. Selenium [137]

It is difficult to prepare nanometer-sized selenium powders because of its low melting point (217 °C). We have successfully prepared nanometer-sized powders of both amorphous and crystalline selenium at room temperature by γ-irradiation. We dissolved analytically pure SeO_2 in hydrochloric acid, or in distilled water, or in NaOH solution, and added a surfactant ($C_{12}H_{25}NaSO_4$) and a scavenger (C_2H_5OH) for hydroxyl radicals. After irradiation,

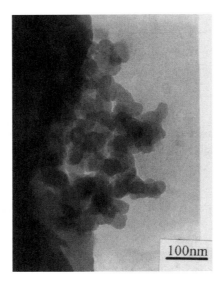

Fig. 26. The TEM micrograph of the Ag–Cu alloy prepared by the γ-irradiation method. (Source: Reprinted from [129] with permission of Elsevier Science.)

Table IX. Composition of the Ag–Cu Alloy Prepared by the γ-Irradiation Method

Element	at%	wt%
Ag	84.97	90.56
Cu	15.03	9.44

(Source: Data from Y. Zhu et al., *J. Alloys Comp.*, 1995.)

Fig. 27. TEM micrograph of the sample prepared by the γ-irradiation method. Solution: 0.0063 M TeO_2, 0.05 M $C_{12}H_{25}NaSO_4$, 0.7 M HCl, and 1.6 M $(CH_3)_2CHOH$; radiation dose: 2.32×10^4 Gy. (Source: Reprinted from [136] with kind permission from Kluwer Academic Publishers.)

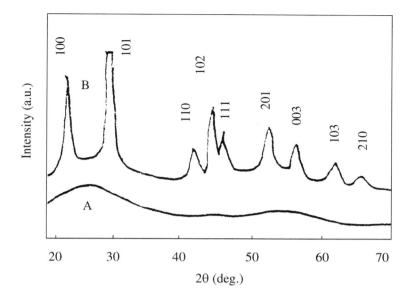

Fig. 28. XRD pattern of the products prepared by γ-irradiation (dose: 3.32×10^4 Gy) of the solution containing 0.01 M SeO_2, 0.46 M HCl, 0.01 M $C_{12}H_{25}NaSO_4$, and 2.6 M C_2H_5OH. (a) The product dried at room temperature, and (b) the product dried at 80 °C. (Source: Reprinted from [137] with permission of Elsevier Science.)

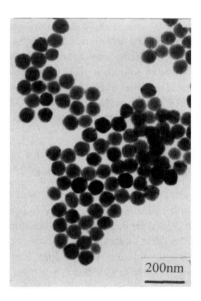

Fig. 29. TEM micrograph of the same sample as in Figure 28a. (Source: Reprinted from [137] with permission of Elsevier Science.)

the powders were dried at room temperature or at 80 °C. The powders obtained from the irradiated solution containing hydrochloric acid and dried at room temperature were amorphous selenium (Fig. 28a) with a uniform particle size of 70 nm (Fig. 29), whereas the product dried at 80 °C was crystalline hexagonal selenium with an average particle size of 8 nm, as calculated using the Scherrer equation (Fig. 28b). On the other hand, the product prepared in an irradiated solution containing NaOH and dried at room temperature was nanocrystalline hexagonal selenium with an average particle size of 17 nm.

Table X. Yields and G Values of Selenium Powders Prepared by γ-Irradiation

Sample number	Solution (in molar concentration)					Dose $(10^4$ Gy)	Yield (%)	G value (atoms/100 eV)
	SeO_2	HCl	C_2H_5OH	$C_{12}H_{25}NaSO_4$	NaOH			
1	0.01	0.46	2.6	0.01	0	1.31	39.6	2.25
2	0.01	0.46	2.6	0.01	0	1.74	47.8	2.71
3	0.01	0	2.6	0.01	0	1.74	31.0	1.78
4	0.01	0	2.6	0.01	0.05	1.74	14.3	0.82
5	0.1	0.46	2.6	0.01	0	1.74	5.60	3.14
6	0.01	0.46	2.6	0.01	0	3.32	84.3	4.79
7	0.01	0.46	0	0.01	0	3.32	0	0

(Source: Data from Y. Zhu et al., *Mater. Lett.*, 1996.)

Table X shows that the yields and G values of the selenium powders prepared in hydrochloric acid (sample 2) are much larger than those obtained from solutions of water (sample 3) and sodium hydroxide (sample 4). This is due to the different mechanisms involved in radiation reduction of selenium(IV) by the hydrated electrons produced during γ-irradiation in acidic and alkaline solutions. Selenium(IV) exists as Se^{4+} ions and SeO_3^{2-} ions in hydrochloric acid and NaOH solutions, respectively. The reduction of SeO_3^{2-} ions was more difficult than that of Se^{4+} ions. This is due to the stability of SeO_3^{2-} ions caused by the strong covalent bonding between selenium and oxygen.

The G value increased with increasing concentration of SeO_2 when other conditions were fixed. The yield and G value of selenium prepared increased with increasing irradiation dose. From Table X, it can also be seen that, when there was no ethanol as a scavenger for hydroxyl radicals in the solution, no selenium powder was obtained. In the absence of ethanol, the zero valent state of selenium resulting from the four-electron transfer could undergo oxidation by the hydroxyl radicals.

3.3.3. Nanometer-Sized Amorphous Powders of Arsenic [138]

After γ-irradiation of the hydrochloric acid solution of analytically pure arsenic(III) oxide in the absence of sodium dodecyl sulfate, a deeply red precipitate of arsenic formed immediately. The product produced from a solution containing 0.05 M CH_3CH_2OH and 1.0 M HCl with a dose of 3.32×10^4 Gy and dried at $60\,°C$ in air was amorphous. Electron diffraction also confirmed the amorphism of the product. However, the product treated at $350\,°C$ in N_2 consisted of single-phase crystalline arsenic particles (average size: 20 nm) with a hexagonal structure. The average particle size increased with heat-treated temperature and time. For example, when the sample was heated at $500\,°C$ for 48 h, the average particle size increased to 25 nm. Figure 30 shows the TEM micrographs of the samples. The particle size of arsenic powder prepared by γ-irradiation from the alkaline solution is 10 nm on an average (Fig. 30c), which is much smaller than that from the acidic solution, 30 nm on an average (Fig. 30a), or the water solution, 40 nm on an average (Fig. 30d).

3.4. Nanometer-Sized Metal Oxides [139–142]

Metal oxides can be prepared by the reaction of solvated electrons and high-valence multivalent metal ions. In a previous section, copper was obtained from a $CuSO_4$ solution after γ-irradiation. Now, by controlling the conditions, a nanocrystalline powder of cuprous oxide can also be obtained after γ-irradiation with a dose of 2.4×10^4 Gy of a 0.01 M

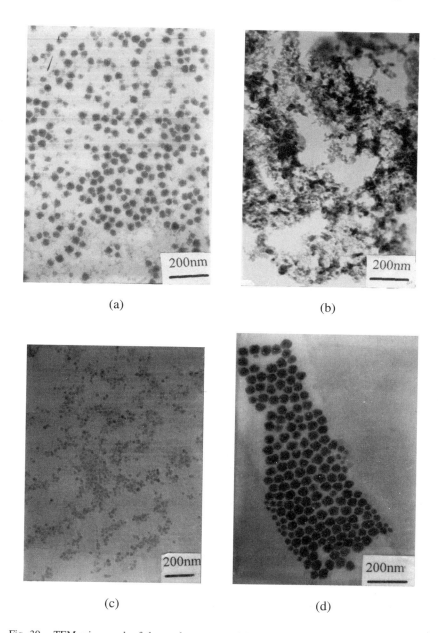

Fig. 30. TEM micrograph of the products prepared by γ-irradiating the solution containing 0.05 M As_2O_3, 0.5 M $C_{12}H_{25}NaSO_4$, 0.5 M C_2H_5OH, and 1.0 M HCl. (a) The product dried in air at 60 °C; (b) the product heat-treated at 350 °C in N_2 for 15 h; (c) the sample prepared from an alkaline solution; and (d) the sample prepared from a solution of water. (Source: Reprinted from [138] with permission of Elsevier Science.)

Cu_2SO_4 solution containing 0.01 M $C_{12}H_{25}NaSO_4$, 2.0 M $(CH_3)_2CHOH$, and a 0.02 M CH_3COOH/0.03 M CH_3COONa buffer. Without the CH_3COOH/CH_3COONa buffer pair, the pH of the solution is about 3.0–3.5 and the final product is a mixture of copper and cuprous oxide. However, when the solution contains a CH_3COOH/CH_3COONa buffer pair that keeps the pH in the range of 4.0 to 4.5, the final product is pure cuprous oxide. In this case, the reduction and dismutation of cuprous ions are completely suppressed. Because cuprous hydroxide is very unstable in a solution of pH > 4.0, it decomposes rapidly to cuprous oxide immediately after its formation. On the other hand, the precipitate of cupric hydroxide forming in the solution should be controlled in the range of 4.0 to 5.0 in the preparation of cuprous oxide. This can be achieved by using the CH_3COOH/CH_3COONa

Table XI. Experimental Conditions, Products, and Particle Sizes of Cuprous Oxide

Sample number	Solution	Irradiation dose ($\times 10^4$ Gy)	Product	Particle
1	0.01 M CuSO$_4$ + 0.01 M C$_{12}$H$_{25}$NaSO$_4$ + 0.02 M CH$_3$COOH + 0.03 M CH$_3$COONa + 2.0 M (CH$_3$)$_2$CHOH	2.4	Cu$_2$O	14
2	0.01 M CuSO$_4$ + 0.01 M C$_{12}$H$_{25}$NaSO$_4$ + 0.02 M CH$_3$COOH + 0.075 M CH$_3$COONa + 2.0 M (CH$_3$)$_2$CHOH	2.4	Cu$_2$O	20
3	0.01 M CuSO$_4$ + 0.05 M C$_{12}$H$_{25}$NaSO$_4$ + 0.05 M CH$_3$COOH + 0.075 M CH$_3$COONa + 2.0 M (CH$_3$)$_2$CHOH	2.4	Cu$_2$O	19
4	0.01 M CuSO$_4$ + 0.1 M C$_{12}$H$_{25}$NaSO$_4$ + 0.05 M CH$_3$COOH + 0.075 M CH$_3$COONa + 2.0 M (CH$_3$)$_2$CHOH	2.4	Cu$_2$O	16
5	0.01 M CuSO$_4$ + 0.05 M CH$_3$COOH + 0.075 M CH$_3$COONa + 2.0 M (CH$_3$)$_2$CHOH	2.4	Cu$_2$O	50
6	0.1 M CuSO$_4$ + 0.1 M C$_{12}$H$_{25}$NaSO$_4$ + 0.05 M CH$_3$COOH + 0.075 M CH$_3$COONa + 2.0 M (CH$_3$)$_2$CHOH	2.4	Cu$_2$O	28
7	0.05 M Cu(CH$_3$COO)$_2$ + 0.05 M C$_{12}$H$_{25}$NaSO$_4$ + 2.0 M (CH$_3$)$_2$CHOH	2.4	Cu$_2$O	36
8	0.05 M CuSO$_4$ + 0.05 M C$_{12}$H$_{25}$NaSO$_4$ + 0.05 M CH$_3$COOH + 0.075 M CH$_3$COONa + 0.05 M EDTA + 2.0 M (CH$_3$)$_2$CHOH	2.4	Cu + 12% Cu$_2$O	
9	0.01 M CuSO$_4$ + 1.6% poly(vinyl alcohol) + 0.02 M CH$_3$COOH + 0.03 M CH$_3$COONa + 0.2 M (CH$_3$)$_2$CHOH	2.4	Cu$_2$O	28
10	0.01 M CuSO$_4$ + 0.05 M C$_{12}$H$_{25}$NaSO$_4$ + 0.05 M CH$_3$COOH + 0.075 M CH$_3$COONa + 2.0 M (CH$_3$)$_2$CHOH	6.2	Cu$_2$O	24
11	0.01 M CuSO$_4$ + 0.1 M C$_{12}$H$_{25}$NaSO$_4$ + 2.0 M (CH$_3$)$_2$CHOH	3.6	Cu + 30% Cu$_2$O	

(Source: Data from Y. Zhu et al., *Mater. Res. Bull.*, 1994.)

buffer solution. The experiment shows that other conditions also influence the particle size (Table XI): (1) The particle size of cuprous oxide increases as the concentrations of acetic acid and sodium acetate increase. (2) The particle size of cuprous oxide decreases as the concentration of sodium dodecyl sulfate increases. (3) In the concentration range of 0.01 to 0.1 M of cupric ions, the particle size increases with the concentration of cupric ions. (4) In the case of using a higher dose, the product consists of relatively larger particles.

Similarly, we can obtain nanometer-sized Cr$_2$O$_3$ powders from a solution containing 0.05 M K$_2$Cr$_2$O$_7$, 0.05 M C$_{12}$H$_{25}$NaSO$_4$, and 3.0 M (CH$_3$)$_2$CHOH at a dose of 1.0×10^4 Gy. After heat treatment at 500 °C, the amorphous Cr$_2$O$_3$ turned into a single phase of crystalline Cr$_2$O$_3$, and the particle size increased from 6 to 15 nm. Cr$_2$O$_3$ powders were not produced from γ-irradiated solutions containing Cr$_2$O$_4^{2-}$ ions. This may be due to the difference in the structure of these two ions. In the case of the CrO$_4^{2-}$ ion, there exists a tetrahedral arrangement in which a chromium atom is located at the center. However, in the dichromate ion, there are two tetrahedral units linked together by one oxygen atom. The Cr–O distance for the bridging oxygen is greater than that for the other oxygen atoms. The $\sigma-\pi$ donor properties of the bridging oxygen atom are much less compared with the terminal oxygen atoms. Thus, the Cr–O bond for the bridging oxygen in the Cr$_2$O$_7^{2-}$ ion could break down to form a CrO$_3$ radical in the process of γ-irradiation. The CrO$_3$ radical is unstable and is reduced rapidly by hydrated electrons to form Cr$_2$O$_3$.

MoO$_2$ and Mn$_2$O$_3$ have been successfully prepared by γ-irradiation of an aqueous solution of [(NH$_4$)$_6$Mo$_7$O$_{24}$·4H$_2$O] and KMnO$_4$ [141, 142], respectively.

3.5. Nanocomposites

3.5.1. Oxide/Metal Nanocomposites

Nanocomposite materials are very important because of their interesting electrical and optical properties, their possible commercial exploitation, and their importance in improving the stability of nanometals and providing models for understanding the physics of nanocrystalline particles. We have developed a new method—sol–gel γ-irradiation—to prepare titania–silver and silica–silver nanocomposites.

3.5.1.1. SiO$_2$/Ag [143]

First, a solution of colloidal silver was obtained by γ-irradiation. Then the sol–gel method was used to prepare silica–silver nanocomposites. Tetraethoxysilane [(C$_2$H$_5$O)$_4$Si, 5 mL] was dissolved in isopropyl alcohol (10 mL) with water (5 mL), and dilute nitric acid (2N) was wadded to keep the pH close to 2, after continuous stirring for 1.5 h. For gelation, the pH of the solution was increased to 8 by the addition of aqueous ammonia under mild stirring. The hydrogels obtained were dried overnight in air. Figure 31 gives the XRD pattern of a typical sample containing metallic silver particles prepared by γ-irradiation of a solution containing 0.01 M AgNO$_3$, 0.01 M C$_{12}$H$_{25}$NaSO$_4$, and 2.0 M (CH$_3$)$_2$CHOH with a dose of 8.1×10^3 Gy. This shows that the sample consists of two phases, namely, metallic silver (6 nm) and noncrystalline silica. The amount of silver present as a metallic species in the composite glass is 1.24%, as measured by atomic absorption spectroscopy. A TEM micrograph of the sample is shown in Figure 32. The silica glass contains a dispersion of fine metallic silver grains that are quasispherical and well separated.

3.5.1.2. TiO$_2$/Ag [144]

In the preparation of titania-based nanocomposites, the formation of particulate materials occurs because of the vigorous hydrolysis of titanium alkoxides with water. This can

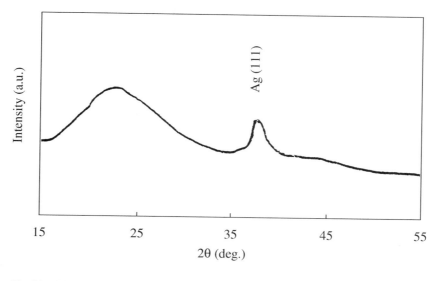

Fig. 31. XRD pattern of the sample prepared by the γ-irradiation sol–gel method. Solution: 0.01 M AgNO$_3$, 0.01 M C$_{12}$H$_{25}$NaSO$_4$, and 2.0 M (CH$_3$)$_2$CHOH; dose: 8.1×10^3 Gy. (Source: Reprinted with permission from [143]. © 1994 Royal Chemical Society.)

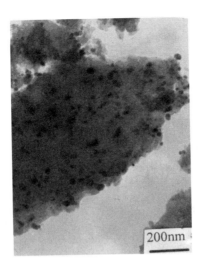

200nm

Fig. 32. TEM micrograph of the sample in Figure 31. (Source: Reprinted with permission from [143]. © 1994 Royal Chemical Society.)

be prevented by controlling the amount of water and by the addition of acids [145]. The hydrogels were prepared by the sol–gel method. Solutions were prepared by dissolving tetrabutyl titanate [$(C_4H_9O)_4Ti$] in isopropyl alcohol with water and adding a few drops of concentrated nitric acid to keep the pH at 2.5. The solutions were stirred continuously for 1 h before adding $AgNO_3$ solution and then adding nitric acid to adjust the pH to 2.5. The two solutions were mixed together under continuous stirring for 0.5 h. The gelation was reached after aging for 1 h, and colorless transparent hydrogels were obtained. After irradiation, brown transparent hydrogels were produced. The irradiated hydrogels were dried at 60 °C in air overnight. The XRD pattern of a typical sample containing metallic silver particles shows that the sample consisted of two phases, namely, metallic silver and crystalline titania, which could be indexed as the anatase structure. The titania matrix contained a dispersion of fine particles that are quasispherical and well separated (Fig. 33). The selected-area diffraction pattern confirms that these particles were metallic silver. The diameters of the silver particles ranged from 4 to 30 nm, and the average particle size was 8 nm.

3.5.2. Polymer/Metal Nanocomposites [146]

3.5.2.1. Preparation in Aqueous Solution

More recently, organic–inorganic nanocomposites have aroused much interest and attention [147, 148]. Only a few methods have been used to prepare polymer–metal nanocomposites [149, 150]. In the most common methods, the polymerization of organic monomers and the formation of nanocrystalline metal particles are performed separately, and the polymer matrix and metal nanocomposites are hybridized physically to form polymer–metal nanocomposites. Thus, metal nanoparticles are not well dispersed homogeneously in the polymer matrix. Furthermore, thermal treatment or pressure is necessary in these methods. So we developed a new method to prepare polymer–metal nanocomposites by γ-irradiation at room temperature and under atmospheric pressure. In this method, the metal salt and organic monomer are mixed homogeneously at the molecular level in the solution, and the formation of nanocrystalline metal particles and the polymerization of monomers are performed simultaneously in solution, leading to a homogeneous dispersion of nanocrystalline metal particles in the polymer matrix. Figure 34 shows the XRD pattern of a typical sample of a polyacrylamide–silver nanocomposite prepared by γ-irradiation of a solution

Fig. 33. TEM micrograph of the titania–silver nanocomposite prepared by the sol–gel γ-irradiation method. Solution: 0.1 M AgNO$_3$, 0.5 M (C$_4$H$_9$O)$_4$Ti, and 4.0 M (CH$_3$)$_2$CHOH; radiation dose: 3.86×10^4 Gy. (Source: Reprinted from [144] with permission of Elsevier Science.)

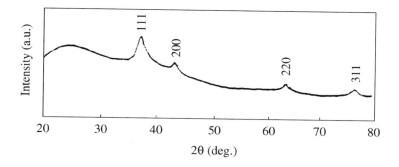

Fig. 34. XRD pattern of the sample prepared by γ-irradiation. Solution: 0.05 M AgNO$_3$ and 3.52 M acrylamide; radiation dose: 2.7×10^4 Gy. (Source: Reprinted with permission from [146]. © 1997 Royal Chemical Society.)

containing 0.05 M AgNO$_3$ and 3.52 M acrylamide with a radiation dose of 2.7×10^4 Gy. This shows that the sample consisted of two phases, namely, metallic silver, as indicated by diffraction peaks (111), (200), (220), and (311), and noncrystalline polyacrylamide. The average particle size of the silver was 11 nm, as estimated by the Scherrer equation. The amount of silver present as a metallic species in the nanocomposite was 2.09 wt%, as measured by precipitation titration.

Figure 35 shows the TEM micrograph of the sample. We can see that the nanocomposite contained fine quasispherical metallic silver particles homogeneously dispersed and well separated in the polyacrylamide matrix.

3.5.2.2. Preparation in Nonaqueous Solution

To date, little work has been published regarding the preparation of polymer–metal nanocomposites at ambient conditions in a nonaqueous solution. We have developed a new method—γ-irradiation in a nonaqueous solution—to prepare these kinds of materials from nonaqueous solutions. By γ-irradiating an ethanol solution containing 0.05 M AgNO$_3$ and 2.82 M acrylamide with a dose of 3.0×10^4 Gy, we obtained a polyacrylamide–silver

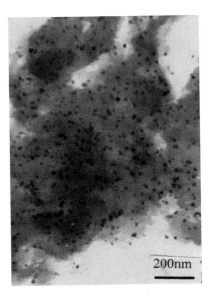

Fig. 35. TEM microphotograph of the sample in Figure 34. (Source: Reprinted with permission from [146]. © 1997 Royal Chemical Society.)

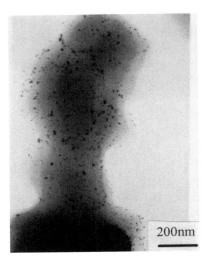

Fig. 36. TEM micrograph of the sample prepared by γ-irradiation. Solution: 0.05 M $AgNO_3$ and 2.82 M acrylamide. Absolute ethanol was used as the solvent. Radiation dose: 3.0×10^4 Gy.

nanocomposite in which the average Ag particle size was 5 nm (ranging from 1 to 10 nm) and the amount of silver present as a metallic species was approximately 2.5 wt%. A TEM micrograph of the sample is shown in Figure 36. Compared with the sample prepared from an aqueous solution (Fig. 35), this sample contained smaller metallic silver nanoparticles with a more narrow size distribution and a more homogeneous dispersion in the polyacrylamide matrix.

3.6. Nanometer-Sized Metal Sulfides

Transition metal sulfides are important in many applications such as photoluminescence [151], electroluminescence [151], photoconductive materials [152], photoacoustic modulation materials [153], and catalysts [154, 155]. Various methods are used to synthesize metal sulfides [156–159], but high temperatures are needed or the product will be larger.

During γ-irradiation, H_2S and HS^- can be prepared from an aqueous solution containing sulfur alcohol [160]. Hayes et al. [161] reported the formation of colloid CdS from a Cd^{2+} dilute aqueous solution (10^{-4} M) containing sulfur alcohol.

Recently, we prepared nanometer-sized CdS (6 nm) and ZnS (5 nm) by γ-irradiating the solution containing 0.01 M $CdSO_4$ or 0.01 M $ZnSO_4$, 0.1 M $Na_2S_2O_3$, and 3.0 M $(CH_3)_2CHOH$ at a dose of 3.7×10^4 Gy. The progress of the reaction that has been reported could be shown as follows:

$$H_2O \rightarrow e_{aq}^-, H_3O^+, H, H_2, OH, H_2O_2$$
$$S_2O_3^{2-} + \tfrac{1}{2}O_2 + 2e_{aq}^- \rightarrow SO_4^{2-} + S^{2-}$$
$$2M^{n+} + nS^{2-} \rightarrow M_2S_n$$
$$nM_2S_n \rightarrow (M_2S_n)_n$$

From the reaction, it is obvious that the $S_2O_3^{2-}$ ions act as the source of S^{2-} ions.

In this way, we also obtained NiS powders.

4. PREPARATION OF NANOCRYSTALLINE THIN FILMS BY CHEMICAL SOLUTION PROCESS

Because of the important physical and chemical properties of thin films, they have been developed for ever-increasing applications in various fields. Thus, as a young branch of the solid physics and solid chemistry fields, the research on the preparation of thin films, especially nanocrystalline thin films, is now receiving more attention.

Usually, the chemical preparation methods of thin films can be divided into the chemical solution method and the chemical vapor deposition (CVD) method. The chemical solution method mainly includes the sol–gel method and the electrochemistry method. The sol–gel method is mainly used for the preparation of metal oxide thin films. The sol–gel method requires high substrate temperatures followed by annealing for crystallization. The heating process, however, would bring about the reaction of the film with the substrate and/or atmosphere, resulting in cracking and/or peeling of the films because of shrinkage caused by the crystallization of deposited amorphous films. Furthermore, in the sol–gel process, if an organometallic precursor is used, it is difficult to form dense films, because the decomposition of an organometallic precursor would give porous ceramics containing fine grains. The electrochemistry method is only used to prepare metal thin films.

Chemical solution routes to single-crystal oxide thin films are relatively new and have been explored for possible device applications as an alternative to the more costly vapor phase routes [162, 163].

The hydrothermal method of preparation has been used extensively in the past to synthesize ceramic powders [164], but recently it is being applied to prepare $BaTiO_3$ films [165, 166]. As one of the chemical solution routes, the newly introduced hydrothermal epitaxy route offers processing temperatures that approach ambient conditions, in which the single-crystal thin films are directly synthesized on a substrate in water at a temperature lower than 150 °C [167]. This would be very advantageous for sequential processing of devices where interdiffusion of previously processed components must be avoided.

In the various vapor phase routes, single-crystal substrates are used as a supporting structure and a template for the oriented overgrowth of the film material—a process known as epitaxy. The chemical vapor deposition method mainly includes chemical molecular beam epitaxy and metallic organic chemical vapor deposition. The advantages for this method include great development of surface morphology, quaternary alloy, more precise composition, rapid speed of formation, easy control of the composition of films, and so on.

In the spray pyrolysis method, an organic metallic is used as a precursor. As the solution evaporates, the process that follows is quite similar to that of the CVD method. Compared with CVD, however, the spray pyrolysis method is conducted under ambient pressure. Therefore, the equipment costs less and is easily applied in industry. This technique is widely used to prepare thin films.

In the following sections, we mainly introduce the fabrication of thin films by the hydrothermal technique and the spray pyrolysis method based on the work in our laboratory.

4.1. Hydrothermal Preparation of Thin Films

Recently, Chien et al. [166] reported that $BaTiO_3$ heteroepitaxial single-crystal thin films and $Pb(Zr_xTi_{1-x})O_3$ heteroepitaxial thin films [168] were produced at 90–150 °C by the hydrothermal method, and Shi et al. [165] also reported the preparation of $BaTiO_3$ thin films by the hydrothermal method. The preparative process involved, first, putting a metal B substrate into an ion A aqueous solution. After hydrothermal reaction at a temperature above 100 °C and about 1 bar pressure for a suitable time, crystalline thin films of ABO_3 can be obtained without postannealing. Usually, metal B is needed to sputter onto the Si substrate. The advantage of this method is that the atoms of the thin films have a chance to arrive at their balance position when they are deposited on the substrate. Therefore, the films have less defect and are relatively dense.

In recent years, we have developed a hydrothermal deposition method and successfully prepared metal sulfide and metal oxide nanocrystalline thin films such as ZnS [30 nm on Si(100) and 60 nm × 10 nm on α-Al_2O_3] [169], Cu-doped ZnS (60 nm × 10 nm on α-Al_2O_3) [170], highly oriented TiO_2 [150 nm thickness on Si(100)] [167], ZnO (thickness of 65 nm) [171], SnO_2 [3.5 nm on Si (100)] [172], Fe_3O_4 [150 nm on α-Al_2O_3 and 50 nm on Si(100)] [173], and α-Fe_2O_3 [10 nm on Si(111) and 30 nm × 5 nm on Si(100)] [174] thin films.

The preparation procedure includes two steps: (1) preparation of precursors and (2) hydrothermal treatment of the precursors to grow thin films. The precursors generally used are sols or solutions. The substrate was ultrasonically treated in solutions of $H_2O:NH_3H_2O:H_2O_2$ and $H_2O:HCl:H_2O_2$ in the volume ratio 6:1:1, followed by ultrasonic cleaning in deionized water prior to deposition. The substrate was then attached to the bottom of a Teflon vessel. The precursor was added to a Teflon vessel until 70% of its volume was filled. Then the vessel was placed into a stainless-steel tank to perform the hydrothermal treatment. The tank was kept at 60–200 °C for 6–24 h.

The properties of the as-prepared thin films were characterized by the XRD, scanning electron microscopy (SEM), IR, Raman, and Mössbauer techniques. The detailed preparation procedures and properties of each thin film will be discussed next.

4.1.1. Zinc Sulfide Thin Films and Doped Zinc Sulfide Thin Films

4.1.1.1. Zinc Sulfide Thin Films [169]

Various techniques have been used to prepare ZnS thin films, such as radio frequency (rf) sputtering [175], molecular beam epitaxy (MBE) [176], and chemical vapor deposition (CVD) [177]. All of these techniques must be carried out in high vacuum or in an inertgas system without oxygen to avoid oxidation of the ZnS, making the preparation procedure complicated.

We successfully prepared thin films of zinc sulfide on Si(100) and polycrystalline α-Al_2O_3 substrates by hydrothermal deposition. An appropriate amount of analytical zinc acetate [$Zn(Ac)_2$·$2H_2O$] was dissolved in distilled water to form a 1-M solution. Then an appropriate amount of 2 M Na_2S solution was added to form a gel-like precipitate. After that 4 M of nitric acid solution was used to adjust the pH value of the system and

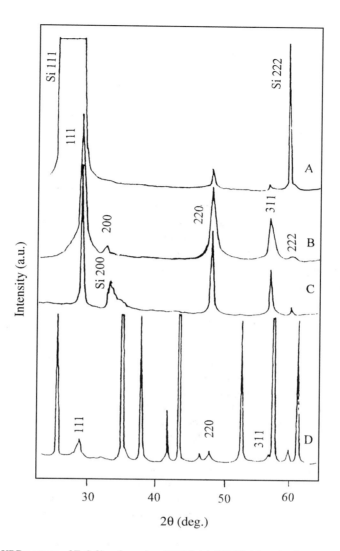

Fig. 37.   XRD patterns of ZnS films formed on Si(111) (a), Si(100) (c), and polycrystalline α-Al$_2$O$_3$ (d) substrates, and powder formed in the same system (b). (Source: Reprinted from [169] with permission of Elsevier Science.)

a semitransparent sol (pH = 5–7) formed. The substrates used were treated by the standard procedure mentioned previously. The hydrothermal treatment was carried out at 100–200 °C for 6–24 h. After cooling, the samples were washed with distilled water and dried under infrared light.

XRD patterns of the films prepared at 140 °C for 12 h (parts a, c, and d of Fig. 37) and the powder formed in the same system (Fig. 37b) show that cubic zinc sulfide is the only crystalline phase formed. Although the intense peaks of the cubic phase coexist with that of the hexagonal phase at the same 2θ, the unique peaks for the cubic phase are weak. However, there were no unique peaks, such as (100) at 26.9° and (101) at 30.5°, for the hexagonal phase either in the film or in the residual powder diffraction pattern, indicating the absence of a hexagonal phase.

The X-ray pattern for the film of ZnS on Si(111) is shown in Figure 37a. The intense peak at $2\theta = 28.6°$ ($d = 3.12$ Å) corresponding to the cubic (111) planes of ZnS is concealed by the Si(111) peak. However, the appearance of peaks at $2\theta = 47.8°$ and 56.6°, especially the peak unique to cubic ZnS at 59.3°, can confirm that the film on Si(111)

461

consists of cubic ZnS. The absence of a (200) peak at 33.1°, which is more intense than that of (222) in the polycrystalline powder diffraction pattern (Fig. 37b), indicates that the ZnS film on Si(111) has a 111 preferred orientation. In contrast to this result, the film formed on Si(100) demonstrates a typical polycrystalline cubic nature (Fig. 37c). Its diffraction pattern is consistent with that of the coexisting powder (Fig. 37b).

Consistent with the results, the polycrystalline cubic phase of zinc sulfide was also formed on a polycrystalline α-Al$_2$O$_3$ substrate in the same system (Fig. 37d). The peaks at 28.6°, 47.8°, and 56.6° can be assigned to the cubic (111), (220), and (311), respectively. All the other peaks in the pattern arose from the diffraction of the polycrystalline α-Al$_2$O$_3$ substrate.

The surface appearance of the zinc sulfide films deposited on Si(100) and the α-Al$_2$O$_3$ substrate at 140 °C for 6 h are dense, smooth, and homogeneous without visible pores. The grain size for the films on Si(100) and α-Al$_2$O$_3$ are 30 nm (spherical) and 60 nm × 10 nm (sticklike), respectively. This reveals that the substrates can affect the nucleation and growth of ZnS film to some degree. The cross section shows that a layer of 200-nm film formed on the polycrystalline α-Al$_2$O$_3$ substrate. The resulting films display high transparency in the IR region.

4.1.1.2. Doped ZnS Ultrafine Thin Films [170]

Luminescent thin films of doped zinc sulfide have been studied extensively in the past [178–180]. They are usually prepared at high substrate temperatures or with subsequent heat treatments, which are necessary for the chemical reaction of the precursors, the thermal diffusion of optically active luminescence centers (i.e., Cu) from a salt or carbonate, and the increasing combination forces between the film and the substrate. However, high substrate temperatures or subsequent thermal treatment would increase the grain size of the thin film. Therefore, it is difficult to obtain doped ultrafine thin films of luminescent material.

Hydrothermal preparation and doping of the ZnS thin film was carried out at 140 °C. The formation of ZnS film and doping are accomplished at the same time in an equilibrium way. The semitransparent sol of ZnS is prepared as described previously. The polycrystalline α-Al$_2$O$_3$ substrates used were ultrasonically cleaned using the process described in the literature [180]. Then the prepared sol together with an appropriate amount of 0.05 M CuCl$_2$·2H$_2$O solution was added in a Teflon vessel until 70% to 80% of its volume was filled. The vessel was placed into a stainless-steel tank and the hydrothermal treatment was carried out at 140 °C for 24 h. After cooling, the sample was rinsed in distilled water and dried under infrared light.

As shown in Figure 38, peaks at 28.6°, 47.8°, and 56.6° can be assigned to the cubic ZnS 111, 220, and 311, respectively. All the peaks denoted by solid circles in the pattern arise from the diffraction of the polycrystalline α-Al$_2$O$_3$ substrate. No characteristic peaks of CuS, Cu$_2$S, ZnO, and CuO appear in XRD patterns, indicating that the impurity phases are not significant. The refined cell constant of cubic ZnS:Cu film on an α-Al$_2$O$_3$ substrate is $a = 5.4124$ Å, which is consistent with that of ZnS bulk material [181]. SEM images show that the film is uniform without visible pores and defects and consists of 60 nm × 10 nm sticklike particles, similar to that of the pure ZnS film [169]. The cross section shows that the thickness of the prepared film is about 200 nm.

Parts a–c of Figure 39 are the photoluminescence spectra for the Cu dopant at 0.2×10^{-3}, 1.4×10^{-3}, and 5×10^{-3} g/g, respectively. Figure 39 shows that the intensity of the luminescence bands (both the blue and green bands) increases with an increase in the Cu dopant concentration. It also reveals that the blue band of the photoluminescence spectra in the ZnS:Cu thin film is not evenly patterned as generally described for the bulk sample but shifts to a higher energy (from 445 to 430 nm), and a possible new blue band with a peak at 415 nm appears as the Cu-doped concentration reaches 5×10^{-3} g/g. The intensity

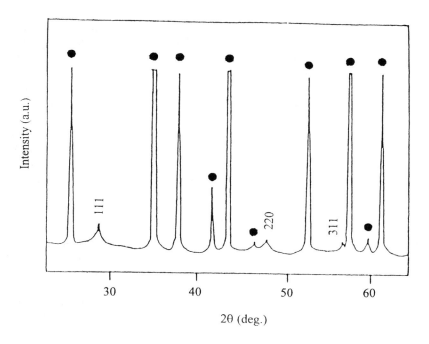

Fig. 38. The XRD pattern of hydrothermally deposited ZnS:Cu thin film. Peaks denoted by solid circles arise from the polycrystalline α-Al$_2$O$_3$ substrate. (Source: Reprinted with permission from [170]. © 1995 American Institute of Physics.)

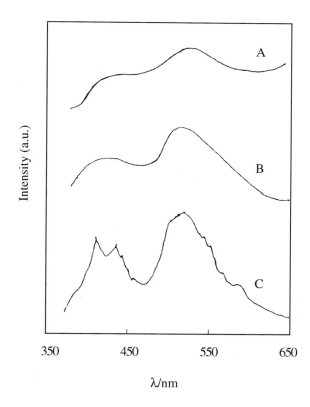

Fig. 39. The photoluminescence spectra of ZnS:Cu thin films with a Cu dopant concentration of 0.2 × 10^{-3} g/g (a), and 1.4 × 10^{-3} g/g (b), and 5 × 10^{-3} g/g (c). (Source: Reprinted with permission from [170]. © 1995 American Institute of Physics.)

of the 415-nm band decreases with the annealing temperature and an increase in time. For example, when the prepared film was annealed at 950 °C for 8 h, only a 430-nm blue band and a 520-nm green band existed in the luminescence spectra. The disappearance of the 415-nm blue band may be the result of grain growth caused by thermal treatment. This indicates that the blue-band peaking at 415 nm is a new luminescence band related to the nanometer-sized grains, and the 430-nm band is the normal blue band that also exists in the bulk ZnS:Cu material.

The possible new blue band (Fig. 39c) peaks at 415 nm, corresponding to an energy value of about 3.0 eV, which is smaller than the host energy band gap (3.66 eV), and indicates that the luminescence is a phenomenon in the band gap. It is certain that this is not the result of an unknown impurity because no other impurities besides copper were detected by X-ray fluorescence measurements. The sharp line of the α-Al_2O_3 substrate has been deducted from the spectra pattern. Combined with the fact that the new blue band has particle size dependence, we tend to think that it is caused by the combination of dangling sulfur bonds on the interface of the fine grains with copper ions, which leads to a new localized energy level in the energy band gap. At a relatively low concentration of Cu dopant, the new blue band is indistinct (parts a and b of Fig. 39), indicating that the localized energy level in the s–p electron energy band gap of ultrafine ZnS film arising from the combination of copper ions with sulfur dangling bonds appears at higher Cu-doping concentration.

4.1.2. Highly Oriented Thin Films

4.1.2.1. TiO₂ Thin Films [167]

TiO_2 thin films can be prepared by various methods such as chemical vapor deposition (CVD) [182, 183] and the sol–gel technique [184]. These methods require a high substrate temperature followed by annealing for crystallization. The heating process, however, would bring about the reaction of the films with the substrates, resulting in cracking or peeling of the films because of the shrinkage due to the crystallization of amorphous films. Furthermore, in organometallic processes it is difficult to dense films, because the decomposition of organometallic amorphous precursors would give porous ceramics containing fine grains [185].

We successfully prepared the preferred oriented anatase TiO_2 films on the Si(100) substrate by hydrothermal epitaxy. In this method, metallic Ti is oxidized in ammonia solution using H_2O_2 to form a TiO_4^{2-} transparent solution. An appropriate concentration solution of TiO_4^{2-} was added to the vessel until 70% of its volume was filled. Then the Teflon vessel was placed into a stainless-steel tank to undergo hydrothermal treatment. The tank was kept at a temperature between 60 and 100 °C for 6 h and then at a temperature between 100 and 200 °C for another 6 h.

XRD patterns of the film prepared (Fig. 40a) and powder formed in the system (Fig. 40b) show that anatase is the only crystalline phase formed in the process. Pretreatment at a lower temperature (60–100 °C) was found to be necessary for the titania nuclei on the Si(100) substrate; without the pretreatment, no films were obtained. Furthermore, the pretreatment time together with the concentration of the TiO_4^{2-} solution determines the thickness of the as-prepared film. An amorphous film formed on the substrate after pretreatment at 60 °C for 6 h. The optimum pH value was 6–7 when the Si substrate was used. The minimum time for subsequent high-temperature (above 100 °C) treatment for crystallization was 6 h; a longer time would not change the thickness of the film. It was possible to decrease the time for crystallization by increasing the temperature of hydrothermal treatment. However, higher temperatures tend to form discontinuous films.

SEM images show that these films are dense, smooth, and homogeneous without visible pores and defects. The cross section shows that a layer of 150-nm film formed on the substrate.

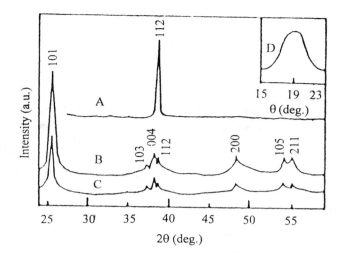

Fig. 40. XRD patterns of the as-prepared film (a), powder formed in the system (b), film derived by the sol–gel technique after treatment at 600 °C for 4 h (c), and rocking curve of (112) reflection (d). (Source: Reprinted with permission from [167]. © 1995 American Institute of Physics.)

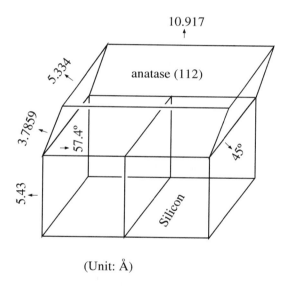

Fig. 41. Scheme of match between (112) lattice plane of anatase and Si(100) substrate. (Source: Reprinted with permission from [167]. © 1995 American Institute of Physics.)

Only the 112 and 224 reflections of anatase phase were present in the XRD pattern, whereas the strongest diffraction peak 101 of anatase polycrystalline was not detected (Fig. 40a). The disappearance of other TiO_2 diffraction peaks suggests that the as-prepared films have preferentially grown along the (112) direction. Silicon is cubic with $a = 5.43$ Å; anatase titania is tetragonal with $a = 3.7859$ Å and $c = 9.514$ Å. Taking lattice match into consideration, the size of the (112) lattice plane is 5.354×10.917 as shown in Figure 41, which is close to two times of Si(100) (5.43×10.86). This indicates that the preparation condition determines the orientation of the film obtained. The SEM micrograph reveals that the film is composed of square platelike particles (average size 10 nm), which are deposited on the substrate with their 112 plane positioned parallel to the substrate surface, consequently resulting in (112) orientation of the film. The extent of the film orientation can be estimated in terms of the Lotgering orientation factor (f) from the XRD peak intensities (I). The factor f [25] is defined as where p or $p_0 = \sum I(112)/\sum I(hkl)$ over a certain range of 2θ values.

4.1.2.2. Polycrystalline ZnO Thin Films [171]

Highly oriented ZnO thin films on Si(100) are also prepared by the hydrothermal method. The zincite sol was made by passing NH_3 into a zinc acetate solution. The hydrothermal treatment is first conducted at 60–100 °C for 6 h and then at 100–200 °C for 6–12 h.

The extent of the grain orientation was determined by examining the XRD diffraction intensities of the films and ZnO powder obtained from the same system. As shown in Figure 42a, the (101) peak of hexagonal ZnO is more intense than that of the hydrothermally prepared powder. However, for the ZnO film, which was heated at 60 °C for 6 h and then at 160 °C for 4 h (Fig. 42b), the (100) reflection is the most intense. The shift in intensity is a result of the preferred grain orientation in the ZnO film on Si(100). The extent of the orientation can also be estimated in terms of the Lotgering orientation factor (f) [186] from the XRD peak intensities (I). The factor f is defined as $f = (P - P_0)/(1 - p_0)$, where p or $p_0 = \sum I(101)/\sum I(hkl)$ over a certain range of 2θ values, the values of all the (hkl) intensities for the ZnO films and the powder formed from material in the same system, respectively. The orientation factor f for films varied with the heating time of the higher temperature, as shown in Table XII. After hydrothermal treatment at 60 °C for 6 h and then at 160 °C for 4 h, a film showing the (10$\bar{1}$0) preferred orientation with $f = 0.398$ formed. When the time was increased to 12 h, a highly (1010) oriented ZnO film (Fig. 43a) with $f = 1$ was obtained. The orientation can be further confirmed by the rocking curve (Fig. 43b) for the (100) reflection of the as-prepared film (Fig. 43a) at a Bragg angle of 31.827°.

Silicon is cubic with $a = 5.43$ Å; ZnO is hexagonal with $a = 3.249$ Å and $c = 5.205$ Å. Taking lattice match into consideration, the c axis for ZnO is close to the a axis for silicon, while five times 3.249 Å gives 16.245 Å for ZnO and three times 5.43 Å gives 16.29 Å for silicon. Thus, Si(100) is a good match for the (10$\bar{1}$0) orientation; the (10$\bar{1}$0) film appears to consist of hexagonal platelets oriented with the platelet surfaces perpendicular to the film

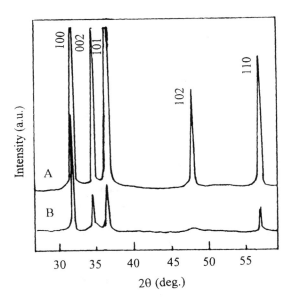

Fig. 42. XRD patterns of ZnO powder (a) and film (b) obtained by hydrothermal treatment of zincite sol at 60 °C for 6 h and then increased to 160 °C for 4 h. (Source: Reprinted from [171] with permission of Elsevier Science.)

Table XII. Lotgering Orientation Factor, f, for ZnO Films Grown on Si(100) at 60 °C for 6 h and Then Increased to 170 °C for the Time Noted

Time (h)	f
4	0.398
6	0.603
8	0.796
10	0.962
12	1.0
14	1.0

(Source: Data from Q. W. Chen, *Mater. Lett.*, 1995.)

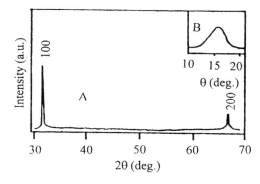

Fig. 43. XRD patterns of ZnO film prepared by hydrothermal treatment of zincite sol at 60 °C for 6 h and then increased to 160 °C for 12 h (a) and rocking curve of (100) reflection in (a) at a Bragg angle of 32.827 °C (b). (Source: Reprinted from [171] with permission of Elsevier Science.)

surface. Heating at a higher temperature (200 °C) for a longer time (>12 h) would not change the orientation of the film. As we can see in Figure 7a, only (100) and (200) peaks appear in the XRD diffraction pattern of the as-prepared film, which indicates that the film exhibits single-crystal features to some degree.

Treatment at the lower temperature (60–100 °C) was found to be indispensable for the formation of ZnO nuclei on the substrate. After the preceding treatment, XRD patterns showed that the formed layer was amorphous, and the higher-temperature (100–200 °C) treatment resulted in the formation of polycrystalline hexagonal ZnO films. No films were obtained without the first relatively low temperature pretreatment. The films obtained by hydrothermal treatment of the zincite sol (pH = 7) exhibited a $(10\bar{1}0)$ preferred orientation.

SEM images indicate that the films are dense, smooth, and homogeneous without visible pores and defects. The cross section shows that an extremely homogeneous layer of 65-nm film formed on the Si(100) substrate.

4.1.3. Tin Oxide (SnO₂) Thin Films [172]

Transparent conductive SnO_2 thin films are widely used in many optoelectronic devices, such as displays and solar cells [187–189], transparent electrodes in electronic devices [188], and gas sensors [190]. These SnO_2 films are prepared by various methods, including chemical vapor deposition [191, 192], spray pyrolysis [193–195], vacuum evaporation [196], direct-current (dc) glow discharge [197], reactive sputtering [198], and sol–gel procedures [199, 200]. The sol–gel technique has succeeded in lowering the thin-film processing temperature. However, the subsequent heat treatment, necessary for the chemical reaction of the precursors, brings about unwanted substrate–film reactions and cracking or peeling of the film because of thermal reactions. Furthermore, the annealing treatment tends to result in large particles, which are less sensitive to reducing gases and increase the operating temperature [201, 202].

We prepared SnO_2 thin films on a Si(100) substrate by hydrothermal deposition from a tin hydroxide sol. The typical procedure is as follows: An appropriate amount of $SnCl_4 \cdot 5H_2O$ was dissolved in water and then NH_3 gas was passed into the solution. The precipitate formed was centrifugally separated from the solution and washed with deionized water several times until no chlorine ion was detected by $AgNO_3$ solution. After adding an appropriate amount of concentrated nitric acid to dissolve the precipitate, a semi-transparent sol (pH = 5–7) was formed. The weight ratio of Sn:H_2O in the sol was maintained at nearly 8 wt%. In the second step, the Si(100) substrate was cleaned and fixed in the autoclave.

XRD patterns of the film prepared (Fig. 44a) and powder formed in the same system (Fig. 44b) show that SnO_2 is the only crystalline phase formed in the reaction. Pretreatment at low temperature (60–100 °C) was found to be necessary for the formation of SnO_2 nuclei on the Si(100) substrate. After the pretreatment, an amorphous layer was formed on the substrate, and the subsequent higher-temperature (100–200 °C) hydrothermal treatment led to the crystallization of the SnO_2 film. Without the pretreatment, no films were deposited on the substrate.

The optimum pH value is 5–7 when a Si substrate is used. A basic solution tends to result in a reaction between the substrate and the solution, whereas a lower acidic solution decreases the deposition ratio. Figure 44a shows the XRD pattern of the as-prepared film, with distinct broadening of the peaks. The grain size of the films was calculated using the Scherrer equation. The calculated grain sizes for reflections (110), (101), and (211) from XRD spectra (Fig. 44a) were all 3.5 nm. For the powder formed in the same system (Fig. 44b), on the other hand, the particle sizes (3 nm) were smaller than those of the as-prepared film. This indicates that the Si substrate affects the growth of the particles. Such fine grains tend to form a smooth, homogeneous film. The SEM micrograph shows that the film consisted of squarelike particles with an average particle size of 4 nm, which is consistent with that calculated from the Scherrer equation. A 200-nm layer of film formed on the substrate. The gas sensitivity is defined as the resistance ratio R_{air}/R_{gas}. The measurement

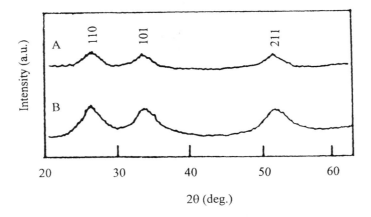

Fig. 44. XRD patterns of (a) the prepared thin film and (b) the powder formed in the same system. (Source: Reprinted from [172] with permission of Elsevier Science.)

results of transients on the prepared films with a thickness of 200 nm show that the maximum sensitivity of the film to 120 ppm of ethanol vapor (86 at 180 °C) is larger than that to gasoline gas (2.8 at 290 °C). Hence, the film has good selectivity for detecting C_2H_5OH in the presence of gasoline. Park et al. [200] have reported that the highest sensitivity of pure SnO_2 film to 1000 ppm of ethanol was 12 at 300 °C.

4.1.4. Iron Oxide Thin Films

Iron oxide is known to have interesting properties as a photoanode [203–205]. Moreover, it has been reported that thin films of iron have higher photoenergy conversion efficiencies than those of the electrodes normally used [206].

Usually, the preparation of α-Fe_2O_3 requires subsequent heat treatment, which is necessary for the chemical reaction of the precursors and which would bring about unwanted substrate–film reactions and cracking or peeling of the films because of thermal contraction [207].

Various techniques have been employed for the preparation of iron oxide films [208–213]. A direct deposition of Fe_3O_4 from an iron target in an Ar/O_2 gas environment followed by oxidation in air to γ-Fe_2O_3 is of major interest because of its short procedure and good reproducibility, as well as its reasonable magnetic properties. However, the formation of Fe_3O_4 is carried out in a nonequilibrium manner, which would affect the non-stoichiometry ratio, chemical stability, crystallinity, and cation order of the film and then, undoubtedly, affect the magnetic properties. Furthermore, contamination of α-Fe_2O_3 and $Fe_{1-x}O$ is often encountered [209, 210].

We prepared thin films of Fe_3O_4 and α-Fe_2O_3 by the hydrothermal method.

4.1.4.1. Magnetic (Fe_3O_4) Thin Films [173]

Fe_3O_4 thin films have been prepared on silicon (100) and α-Al_2O_3 substrates by the hydrothermal treatment of an iron (III) nitrate solution in the presence of reduced iron powder and urea at 140 °C for 2–6 h, which was carried out using a 0.5-L autoclave. The substrates used were treated as mentioned before. The vessel was filled to 70% of its volume. After treatment, the films were washed with distilled water and dried under infrared light.

XRD patterns of the synthesized films (parts a and b of Fig. 45) and powder (Fig. 45c) formed in the system show that magnetite is the only crystalline phase formed.

The formation of magnetite is caused by the reaction of $Fe(OH)_3$ with $Fe(OH)_2$, which comes from the hydrothermal oxidation of iron powder. Hence, the mole proportion of $Fe(OH)_2$ to $Fe(OH)_3$ should be at least 1:2 to form magnetite and to inhibit the deposition of $Fe(OH)_3$ during the formation of α-Fe_2O_3. The optimum condition is Fe^{3+}/Fe/urea = 1:6:3 and time 2–6 h. Urea was used as a pH-adjusting agent; the optimum pH value for the deposition of magnetite is 6–7. A higher hydrothermal temperature can decrease the deposition rate. The optimum temperature in the present case is 140–200 °C.

The Mössbauer spectra analysis demonstrates two sextuplets, one for Fe^{3+} on a tetrahedral site, which is characteristic of Fe_3O_4. To confirm further that Fe_3O_4 is the grown phase, the Fe^{3+}/Fe^{2+} ratios of the coexisting powders were determined by titrating the ferrous and total iron content with a $K_2Cr_2O_7$ solution. The result shows that the mole ratio of Fe^{3+}/Fe^{2+} is 2:0.98, which is consistent with the stoichiometric ratio of the Fe_3O_4 phase. Combining the XRD results, Mössbauer spectra analysis, and black appearance of the sample, it can clearly be established that the as-prepared films are Fe_3O_4.

SEM indicates that the films on polycrystalline α-Al_2O_3 and Si(100) consist of 150- and 54-nm grains, respectively. This indicates that the α-Al_2O_3 substrate promotes the growth of Fe_3O_4 grains. The film on Si(100) is more homogeneous, smooth, and dense than that on α-Al_2O_3 because of the finer grains.

The results of vibrating sample magnetometer measurements show that the M_s's of the as-prepared films on Si(100) and α-Al_2O_3 are 456 and 448 emu/cm^3, respectively, which

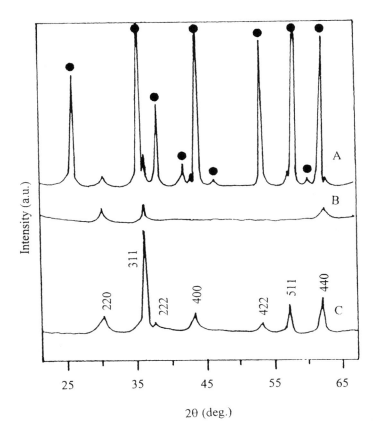

Fig. 45. XRD patterns of the films deposited on (a) α-Al$_2$O$_3$ and (b) Si(100) substrates and (c) the powder formed in the same system. (Source: Reprinted from [173] with permission of Elsevier Science.)

is a little lower than the bulk value (480 emu/cm^3) and larger than that of reactively sputtered Fe$_3$O$_4$ thin films [214]. This was thought to be the result of the high quality of the films without visible pores, and especially of the preparation process carried out under equilibrium, resulting in less cation disorder and precise stoichiometry ratio. The coercivities found for the Fe$_3$O$_4$ films on Si(100) and α-Al$_2$O$_3$ are 470 and 450 Oe, respectively, which is also a little larger than the reported value [214]. Relatively fewer defects in the film because of the hydrothermal process may be responsible for the increase of H_c. The higher M_s and coercivity of the film formed on Si(100) than those of the film on α-Al$_2$O$_3$ were also thought to be the result of fewer pores present in the film on Si(100) because of the finer grains.

4.1.4.2. Iron(III) Oxide (α-Fe$_2$O$_3$) Thin Films [174]

The experimental procedure consisted of placing substrates into a Teflon beaker with a solution of 1 M iron nitrate and 1.5 M urea. Then the substrates were treated as mentioned before. The hydrothermal treatment was carried out in the temperature range 100–200 °C for 4–24 h. After treatment, the films were washed with distilled water and dried under infrared light.

XRD patterns of the films (parts a and b of Fig. 46) and powder (Fig. 46c) coexisting in the system show that α-Fe$_2$O$_3$ (hematite) is the only ciystalline phase formed in the hydrothermal process, and the refined cell constants of α-Fe$_2$O$_3$ film on Si(100) are $a = 5.0343$ Å and $c = 13.74$ Å, which are consistent with the reported data ($a = 5.0340$ Å, $c = 13.752$ Å) [215].

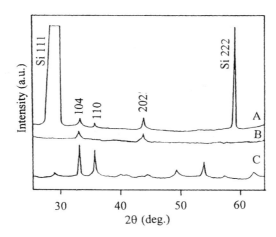

Fig. 46.   XRD patterns of the films prepared on Si(100) (a), Si(100) (b) substrates, and coexisting powder (c). (Source: Adapted from Chen et al. [174].)

Urea was used as the homogeneous precipitate. As it decomposed at a temperature above 70 °C, the pH value of the solution increased. Although a high pH value can increase the deposition rate, it tends to cause large particle size and inhomogeneities in the film. The optimum pH value for the nucleation and growth of α-Fe_2O_3 films is 5–6. Also, the mole proportion of Fe^{3+} to urea is another factor related to the quality of thin films of α-Fe_2O_3. The optimum mole ratio of Fe^{3+} to urea for preparing homogeneous films is 2:3.

The films deposited on Si(100) and Si(111) have a (101) preferred orientation. As can be seen from Figure 46c, the 202 peak is weak for the powder formed in the system. However, for the α-Fe_2O_3 films deposited on Si(111) and Si(100) (parts a and b of Fig. 46), the 202 peak is the most intense peak. This indicates that both films on Si(111) and Si(100) have a (101) preferred orientation.

SEM images show that the films deposited on the two substrates are all homogeneous and uniform. The films consist of spherical grains for Si(111) and columnar grains for Si(100) with a typical grain size of 10 nm and 30 nm × 5 nm, respectively. These indicate the crystalline direction of the nucleation and growth of α-Fe_2O_3 particles on the substrate. The cross section from SEM shows that a layer of 120-nm film formed on the Si(100) substrate.

4.2. Spray Pyrolysis Preparation of Nanocrystalline Oxide Films

The spray pyrolysis process is a traditional deposition process. It is simple, low cost, and easy to apply in industry. Recently, a novel spray pyrolysis process has been developed for the preparation of thin oxide films of high quality [216, 217]. Deposition of metal oxide thin films by the spray pyrolysis process was performed by ultrasonically nebulizing, spraying, and thermally decomposing the metal acetylacetonate solution in oxygen at 430–450 °C in a reactor. The reactor was heated by a two-zone mirror furnace (Trans-Temp, Chelsea, MA). The solution was nebulized by a commercial ultrasonic humidifier (Holms Air), and the resulting mist was swept into the reactor by a stream of oxygen. The nebulized solution was delivered to the substrate in 5-s pulses with 30-s intervals between pulses. The substrates consist of 1-cm squares of n-type (100) silicon with 0.01 Ω·cm resistivity. They were cleaned and etched immediately prior to the deposition according to the procedure described by Fournier et al. [218].

In this section, we introduce recent studies on the preparation and properties of several metal oxide nanometer thin films such as α-Fe_2O_3 [211], α-Cr_2O_3 [219], Al_2O_3 [220], Cr-doped Al_2O_3 [221], and NiO [222] by the spray pyrolysis technique.

4.2.1. Corundum-Type Oxide Thin Films and Doped Thin Films

4.2.1.1. Iron(III) Oxide Films and IR Spectroscopy [211]

Various techniques have been employed for the preparation of iron oxide films including pulsed laser evaporation [223], sputtering [224], sol–gel [225], and chemical vapor deposition [226].

IR spectra of polycrystalline α-Fe_2O_3 have been studied extensively. However, the position, width, and relative intensity of the observed absorption bands lack consistency and depart from theoretical expectations. This discrepancy is caused by the diversity in the size and shape of the sample particles [227, 228]. Preparative conditions, as well as the matrix in which the powder is embedded, play important roles in determining the nature of the product. Most of the spectral measurements described in the literature were made on powder samples using KBr as a matrix. Because of the difficulty in controlling the parameters involved in preparing such pellets, this technique generally does not yield reproducible spectra.

Uniform, mirror-like oxide films with a submicrometer texture can be easily deposited on a silicon substrate. Such films afford a unique opportunity to obtain IR spectra of oxides directly without the use of KBr as a matrix. This study concerns the preparation and properties of α-Fe_2O_3 thin films deposited on silicon substrates by spray pyrolysis. An iron acetylacetonate solution was ultrasonically nebulized, sprayed onto the substrates, and thermally decomposed.

Iron acetylacetonate was used as the precursor. A solution of 0.01 M iron acetylacetonate in a 50% methanol/water mixture was ultrasonically nebulized, sprayed onto the substrates, and thermally decomposed at 500 °C in the reactor described previously. For measurement of the optical band gap, silica substrates were used. These were cleaned with hydrochloric acid, distilled water, and semiconductor-grade acetone prior to deposition.

Films of iron oxide approximately 250 nm thick have been deposited on silicon by an ultrasonic nebulization and pyrolysis technique. The films had good adherence to both substrates. They appeared uniform and shiny. The thickness of the films on the silicon substrate was measured at different positions on the film and was found to be uniform to within 1% on a substrate 1 cm $\times$ 1 cm. The bright and uniform color also revealed that the films were of uniform thickness.

Usually, phase identification of the thin film, which is deposited on some substrate, is carried out by X-ray diffraction [229]. If there is a small amount of impurity, which only appears in the range near the substrate, X-ray diffraction is unable to detect it. We successfully used IR transmission spectra to study the phase information of metal oxide thin films. The IR spectrum of α-Fe_2O_3 obtained by subtracting the spectra of the bare substrates from the spectra of the coated substrates is shown in Figure 47.

The results of X-ray diffraction and IR showed that the composition of the annealed films was α-Fe_2O_3. Moreover, annealing improves the optical and gap and current–voltage characteristics of the films. The transmission spectra of iron oxide films on silica ranged from 350 to 1000 nm. The optical absorption from such data was used to generate plots of $(\alpha h\nu)^{1/2}$ vs $h\nu$, which gave a band gap energy of 1.97 eV for the as-deposited film and a band gap energy of 2.2 eV for the annealed film. The reported band gap for pure α-Fe_2O_3 is 2.2 eV [230], which is in agreement with the value obtained for the annealed films. The dc current–voltage measurements were made on several films approximately 250 nm thick. A significant change of current was observed at 3.5 and 5.5 V of applied potential for the as-deposited and annealed films, respectively. Differences in the band gap and the current–voltage behavior can be attributed to the existence of Fe(II) in the films as prepared. In a previous study [230], it was shown that pure α-Fe_2O_3 is not a photoconductor and a necessary condition for the observation of any appreciable conductivity is the existence of Fe(II). Postannealing of the films in oxygen is essential for the production of high-quality films.

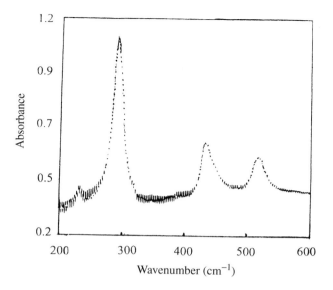

Fig. 47. IR spectrum of α-Fe$_2$O$_3$ film of 250 nm thickness. (Source: Reprinted with permission from [211]. © 1991 Academic Press.)

4.2.1.2. Chromium(III) Oxide Thin Films [219]

Chromium(III) oxide α-Cr$_2$O$_3$ is a good candidate for thin insulting layers in electronic devices [231]. These films would be useful for numerous applications such as protective coatings [232], as well as for isolation masks on the alumina output windows of high-power klystron tubes [233] to prevent damage arising from the high yield of secondary electrons by alumina.

Previous efforts to prepare films of α-Cr$_2$O$_3$ on various substrates have used evaporation [231], sputtering [231,232], magnetron sputtering of metallic chromium followed by oxidation [233], and MOCVD [234]. Other low-temperature methods for the preparation of films of α-Cr$_2$O$_3$ need to be investigated. Among the alternative methods, spray pyrolysis processing has been shown to be relatively inexpensive and effective for the preparation of oxide films [235].

Chromium acetylacetonate Cr(acac)$_3$ was prepared as described in the literature [236]. Three different chromium acetylacetonate solutions were used for spray pyrolysis in this investigation. All the films appear to be homogeneous and free of pinholes, but there are some differences. The film prepared from an acetic acid solution is composed of densely packed small particles of about 40 nm diameter. Their distinct shape indicates appreciable crystallinity. The film prepared from a 30% ethanol/water solution consists of less densely packed, acicular particles approximately 100 nm × 20 nm. These also show appreciable crystallinity. The films prepared from a pure ethanol solution again consist of densely packed small particles of about 40 nm diameter, but their less distinct shape indicates a less crystalline character. High-quality, uniform films free of pinholes, composed of small particles, and with high dc breakdown voltages were obtained from all three solutions.

4.2.1.3. Alumina and Chromium-Doped Alumina Thin Films [220, 221]

Aluminum oxide is an attractive candidate for thin insulating layers in electronic devices because of its high dielectric constant, low sensitivity to Na$^+$ diffusion, and high radiation resistance [237].

Earlier efforts to prepare homogeneous films of Al$_2$O$_3$ on an InP support by a sol–gel method indicated that such films were porous and resulted in low-breakdown potentials [238]. Several investigations on the preparation of Al$_2$O$_3$ films on various semiconductors by the thermal decomposition of aluminum isopropoxide have been reported [218, 239–241]. Other low-temperature preparative methods of thin-films insulators need to be investigated and the resulting films compared as to uniformity, homogeneity, and electrical properties. Among the alternative methods, spray pyrolysis processing has been shown to be a relatively inexpensive method for the preparation of oxide films [242].

The nebulized aluminum acetylacetonate solution was delivered to the substrate in 5-s pulses with 20-s intervals between pulses. In addition, the oxygen carrier gas was allowed to flow continuously in order to minimize temperature fluctuations during spray pulsing. The n-type (100) silicon served as the substrate.

The thickness of a number of films was measured at different positions on the substrate with an ellipsometer. It was found that the thickness was uniform to within 1% except near the very edge. From the IR data, there was no absorption band from 2700–3700 cm^{-1}, which is characteristic of an OH group, thus indicating no residual hydration in the films. Electron micrograph indicates that the film is composed of densely packed small particles, with an average size of 30–40 nm.

Direct-current current–voltage measurements were made on several films about 200 nm thick. There was no observed voltage breakdown up to 10 V of applied potential. Films prepared by thermal decomposition of aluminum isopropoxide [218] had voltage breakdown of 4 V even when grown at a higher temperature (500 °C).

Although either aluminum(III) oxide or chromium(III) oxide films themselves are promising insulating films that are useful for numerous applications as protective coatings, it was reported that thin films of chromium-doped aluminum oxide ($Al_{2-x}Cr_xO_3$) that were applied as insulating and corrosion protective layers exhibited a good corrosion resistance in 50% HCl solution and H_2SO_4 solution.

Previous efforts to prepare films of $Al_{2-x}Cr_xO_3$ on various substrates have used flame spray, plasma spray, plasma-activated CVD and so on. Also the novel spray pyrolysis process developed for the deposition of thin oxide films of high quality has been shown to be a relatively inexpensive method and an effective alternative for the preparation of thin oxide films.

We prepared amorphous $Al_{2-x}Cr_xO_3$ films ($x = 0.072$) with the spray pyrolysis technique. The film was high performance in electrical insulating properties. A characteristic of the film that merits special attention is its excellent chemical corrosion resistance even in severe environments such as 40% HF and 10% $KMnO_4$ in H_2SO_4.

Chromium acetylacetonate $Cr(acac)_3$ and aluminum acetylacetonate $Al(acac)_3$ were prepared as described in the literature [236]. Specifically, 0.156 g of $Cr(acac)_3$ and 0.669 g of $Al(acac)_3$ were slowly dissolved in 75 mL of ethanol in a 254-mL volumetric flask, which was then filled to volume with distilled water.

The carrier gas was oxygen with a flow rate of 3.5 L/min, which was allowed to flow continuously in order to minimize temperature fluctuations during spray pulsing. The distance of the nozzle to the substrates was about 8 cm. The solution was delivered to the substrates in 5-s pulses with 15-s intervals between pulses. The substrates consisted of 1-cm squares of n-type (100) silicon with 0.01 Ω·cm resistivity.

The X-ray diffraction pattern indicates that the $Al_{2-x}Cr_xO_3$ film is amorphous. XPS measurements show that the ratio of Al/Cr (at%) in the $Al_{2-x}Cr_xO_3$ film is 96.42:3.58 by determining the normalized integral area of the Al_{2p} peak and the Cr_{2p} peak. The x value in $Al_{2-x}Cr_xO_3$ is calculated to be 0.072.

The SEM image indicates that the film is homogeneous and free of pinholes. The $Al_{2-x}Cr_xO_3$ film is composed of densely packed ultrafine particles with size less than 20 nm, whereas the particles in films of Al_2O_3 are 40 nm [220].

Direct-current current–voltage measurement of the film shows that no voltage breakdown is observed for potentials up to 20 V applied to any of the evaporated gold electrodes, whereas films prepared by thermal decomposition of aluminum isopropoxide have voltage breakdown of 4 V. This result is consistent with the electron micrograph, which indicated the formation of a relatively pinhole-free, dense film on silicon substrates.

The $Al_{2-x}Cr_xO_3$ film deposited by the spray pyrolysis method is superior in corrosion resistance to films of Al_2O_3 and Cr_2O_3. The $Al_{2-x}Cr_xO_3$ film is undissolved after immersion for 5 h either in 40% HF or in 10% $KMnO_4$ in H_2SO_4, whereas the Al_2O_3 film is easily dissolved in 40% HF because of the formation of AlF_6^{3-}, and the Cr_2O_3 film dissolves in a $KMnO_4$ solution of H_2SO_4 because of the strong oxidation of MnO_4^-. This property of chromium-doped aluminum oxide film might also be related to the densely packed ultrafine particles in the $Al_{2-x}Cr_xO_3$ film. The mechanism needs to be investigated further.

4.2.2. NiO Thin Films [222]

The interest in nickel oxide thin films is growing fast because of their importance in many applications in science and technology. Recently, NiO thin films have been investigated for use as active materials for electrochromic devices [243]. NiO thin films are usually deposited by magnetron sputtering [243, 244] and electrodeposition [245].

We successfully deposited thin films of nickel oxide with a thickness of 200 nm onto Si(111) and fleshly cleaved NaCl(100) substrates by the spray pyrolysis method. The ethanol solution of nickel acetylacetonate were ultrasonically nebulized, sprayed onto the substrates, and thermally decomposed at a low temperature. The deposited NiO films have a preferred (110) orientation.

Nickel acetylacetonate $Ni(acac)_2$ was chosen as the precursor for the preparation of NiO films. The process of preparation of $Ni(acac)_2$ was similar to that of $Al(acac)_3$.

The Si(111) substrate was ultrasonically treated as described before. The NaCl(100) substrate of 0.1 cm thickness was cleaved to 0.7 cm × 0.7 cm just prior to deposition.

An ethanol solution of 0.01 M $Ni(acac)_2$ was ultrasonically nebulized, sprayed on the substrates, and thermally decomposed in the temperature range 400 to 430 °C in the reactor. The flow rate of carrier gas was 3.5 L/min. Experiments showed that the distance of the nozzle to the substrates was between 8.0 and 8.5 cm.

The resulting films were identified as NiO films with a poor preferred (110) orientation as shown by X-ray powder diffraction. Uniformity of the films was confirmed by SEM and the bright uniform color of the films. SEM and TEM investigations showed that the NiO film on Si(111) substrates consisted of particles less than 20 nm, whereas the films on NaCl(100) consisted of spherical particles with sizes of 100 nm and these particles were, in fact, composed of densely packed nanocrystallines less than 10 nm.

Acknowledgment

This work was supported by the Chinese National Foundation of Natural Science Research and the National Nanometer Materials Climbing Program.

References

1. M. A. Kastner, *Phys. Today* 46, 24 (1993).
2. L. E. Brus, *Appl. Phys. A* 53, 465 (1991).
3. L. N. Lews, *Chem. Rev.* 93, 2693 (1993).
4. R. Freer, "Nanoceramics." Institute of Materials, London, 1993.
5. D. D. Awsonalom and D. P. DiVincenzo, *Phys. Today* 48, 43 (1995); J. F. Smyth, *Science* 258, 414 (1992).
6. L. E. Brus, *J. Chem. Phys.* 80, 4403 (1984).
7. M. V. Rama Krishna and R. A. Friesner, *J. Chem. Phys.* 95, 8309 (1991).
8. M. Grundmann, J. Christen, N. N. Ledentsov, J. Buhrer, D. Bimberg, S. S. Ruvimov, P. Werner, U. Richter, U. Gusele, J. Heydenreich, V. M. Ustinov, A. Yu. Egorov, A. E. Zhukov, P. S. Kopev, and Zh. I. Alferov, *Phys. Rev. Lett.* 74, 4043 (1995).
9. K. Kash, *J. Lumin.* 46, 6982 (1990).

10. A. P. Alivisatos, *Science* 271, 933 (1996).
11. R. J. D. Miller, G. L. McLendon, A. J. Nozik, W. Schmickler, and F. Willig, "Surface Electron Transfer Process," Chap. 6 and references therein. VCH Publishers, New York, 1995.
12. R. L. Wells, C. G. Pitt, A. T. Mcphail, A. P. Purdy, S. Shafieezad, and R. B. Hallock, *Chem. Mater.* 1, 4 (1989).
13. M. A. Olshavsky, A. N. Goldstein, and A. P. Alivisatos, *J. Am. Chem. Soc.* 112, 9438 (1990).
14. S. S. Kher and R. L. Wells, *Chem. Mater.* 6, 2056 (1994).
15. S. S. Kher and R. L. Wells, *Mater. Res. Soc. Symp. Proc.* 351, 293 (1994).
16. K. E. Gonsalves, G. Carlson, S. P. Rangarajan, M. Bemaissa, and M. Jose-Yacaman, *J. Mater. Chem.* 6, 1451 (1996).
17. J. F. Janik and R. L. Wells, *Chem. Mater.* 8, 2708 (1996).
18. J.-W. Hwang, S. A. Hansen, D. Britton, J. F. Evans, K. F. Jensen, and W. L. Gladfelter, *Chem. Mater.* 7, 517 (1990).
19. J.-W. Hwang, J. P. Campbell, L. Kozubowski, S. A. Hansen, J. F. Evans, and W. L. Gladfelter, *Chem. Mater.* 7, 517 (1995).
20. E. K. Byrne, L. Parkanyi, and K. H. Theopold, *Science* 241, 332 (1988).
21. T. J. Trentler, K. M. Hicjman, S. C. Goel, A. M. Viano, P. C. Gibbons, and W. E. Buhro, *Science* 270, 1791 (1995).
22. R. L. Wells and W. L. Gladfelter, *J. Cluster Sci.* 8, 217 (1997).
23. Y. Xie, Y. Qian, W. Wang, S. Zhang, and Y. Zhang, *Science* 272, 1926 (1996).
24. Y. Xie, Y. Qian, S. Zhang, W. Wang, X. Liu, and Y. Zhang, *Appl. Phys. Lett.* 69, 334 (1996).
25. N. Kuwano, Y. Nagatomo, and K. Kobayshi, *Jpn. J. Appl. Phys.* 33, 18 (1994); S. Logothetidis, J. Petalas, M. Cardona, and T. D. Moustakas, *Mater. Sci. Technol.*, B 29, 65 (1995).
26. H. Xia, Q. Xia, and A. L. Ruoff, *Phys. Rev. B* 47, 12925 (1993).
27. P. Perlin, C. J. Canilon, J. P. Itie, and A. S. Miguel, *Phys. Rev. B* 45, 83 (1992).
28. C. D. Wagner, "Handbook of X-ray Photoelectron Spectroscopy." Perkin-Elmer Corporation, Minnesota, 1979.
29. H. P. Maruska and J. J. Tietjent, *Appl. Phys. Lett.* 15, 327 (1969).
30. B. K. Ridley, "Quantum Process in Semiconductors," p. 62. Clarendon, Oxford, UK, 1982.
31. Y. Xie, W. Wang, X. Qian, and X. Liu, *Chin. Sci. Bull.* 41, 997 (1996).
32. Y. Xie, Y. Qian, W. Wang, and G. Zhou, *Chin. J. Chem. Phys.* 10, 39 (1997).
33. T. Douglas and K. H. Theopold, *Inorg. Chem.* 30, 594 (1991); O. I. Micic, C. J. Curtis, K. M. Jones, et al., *J. Phys. Chem.* 98, 4966 (1994).
34. G. Wilkinson, "Comprehensive Coordination Chemistry," Vol. 3, Chaps. 24 and 25. Pergamon Press, Oxford, UK, 1987.
35. R. L. Wells, S. R. Aubuchon, and S. S. Kher, *Chem. Mater.* 7, 793 (1995).
36. L. E. Brus, *J. Chem. Phys.* 79, 5566 (1983).
37. J. Q. Fang and D. Lu, eds., "Solid State Physics," Vol. 2, p. 2. Shanghai Science and Technology Press, 1981.
38. Y. Li, X. Duan, Y. Qian, L. Yang, M. Ji, and C. Li, *J. Am. Chem. Soc.* 119, 7869 (1997).
39. X. F. Qian, X. M. Zhang, C. Wang, W. Z. Wang, Y. Xie, and Y. T. Qian, *J. Phys. Chem. Solid* 60, 415 (1999).
40. C. R. Whitehouse and J. Robin, *Solid State Commun.* 24, 363 (1970).
41. B. Palosz, W. Palosz, and S. Gierlotka, *Bull. Mineral.* 109, 143 (1986).
42. J. C. Bailar, H. J. Emeleus, R. S. Nyholm, and A. F. Trotman Dickenson, "Comprehensive Inorganic Chemistry," Vol. 2. Pergamon Press, Oxford, UK, 1973.
43. C. Raisin, Y. Bertrand, and J. Robin, *Solid State Commun.* 24, 353 (1977).
44. R. W. Parry, "Inorganic Synthesis," Vol. 12. McGraw–Hill, New York, 1955.
45. K. Kourtakis, J. Dicarlo, R. Kershaw, K. Dwight, and A. Wold, *J. Solid State Chem.* 76, 186 (1988).
46. S. K. Arora, D. H. Patel, and M. K. Agarwal, *J. Cryst. Growth* 131, 268 (1993).
47. K. Chihiro, S. Yoshio, and F. Kazuo, *J. Cryst. Growth* 94, 967 (1989).
48. I. P. Parkin and A. T. Rowley, *Polyhedron* 12, 2961 (1993).
49. R. R. Chianlli and M. B. Dines, *Inorg. Chem.* 31, 2127 (1992).
50. *Nat. Bur. Stand. Mon.* 25, Sect. 9.
51. Sadtler Commercial Spectra, "IR Grating Inorganics," Vol. 1, $Y_{157}O_2Sn$.
52. X. F. Qian, X. M. Zhang, C. Wang, Y. Xie, and Y. Q. Qian, *Inorg. Chem.* 38, 2621 (1999).
53. B. Morris, V. Johnson, and A. Wold, *J. Phys. Chem. Solids* 28, 1565 (1967).
54. T. A. Pecoraro and R. R. Chianelli, *J. Catal.* 67, 430 (1981).
55. J. Covino, D. Pasquariello, K. Kim, K. Dwight, A. Wold, and R. R. Chianelli, *Mater. Res. Bull.* 17, 1191 (1982).
56. J. D. Passaretti, K. Dwight, A. Wold, W. J. Croft, and R. R. Chianelli, *Inorg. Chem.* 20, 2631 (1981).
57. D. M. Pasquariello, R. Kershaw, J. D. Passaretti, K. Dwight, and A. Wold, *Inorg. Chem.* 23, 872 (1984).
58. C. W. Robert, "CRC Handbook of Chemistry and Physics," 63rd ed., D-219. CRC Press, Boca Raton, FL.
59. T. H. Gouw and R. E. Jentoft, *Adv. Chrematogr.* 13 (1972).
60. F. Y. Yao, D. W. Guo, and M. D. Gui, *Inorg. Chem. Ser.* 5, 174 (1990).

61. G. Krabbes, H. Oppermann, and J. Henke, *Z. Anorg. Allg. Chem.* 442, 79 (1978).
62. Y. Li, H. Liao, and Y. Qian, *Chem. Mater.* 10, 2301 (1998).
63. L. E. Brus, *Appl. Phys. A* 53, 465 (1991).
64. M. L. Steigerwald and L. E. Brus, *Ann. Rev. Mater. Sci.* 19, 471 (1988).
65. V. L. Colvin, M. C. Kagan, and A. P. Alivisatos, *Nature* 370, 354 (1994).
66. P. V. Braum, P. Osenar, and S. I. Stupp, *Nature* 380, 325 (1996).
67. C. B. Murray, D. J. Norris, and M. G. Bawendi, *J. Am. Chem. Soc.* 115, 8706 (1993).
68. S. Mann, *Nature* 322, 119 (1988).
69. C. T. Dameron, *Nature* 338, 596 (1989).
70. M. R. Ghadiri, J. R. Granja, and L. K. Buehler, *Nature* 369, 301 (1994).
71. S. H. Yu, Y. T. Qian, L. Shu, L. Yang, and C. S. Wang, *Mater. Lett.* 35, 169 (1998).
72. D. Arivuoli, F. D. Gnanam, and P. Ramasamy, *J. Mater. Sci. Lett.* 7, 711 (1988).
73. J. Black, E.M. Conwell, L. Seigle, and C. W. Spencer, *J. Phys. Chem. Solids* 2, 240 (1957).
74. B. B. Nayak, H. N. Acharya, G. B. Mitra, and B. K. Mathur, *Thin Solid Films* 105, 17 (1983).
75. S. H. Pawar, P. N. Bhosale, M. D. Uplane, and S. Tanhankar, *Thin Solid Films* 110, 165 (1983).
76. H. Fujita, M. Izawa, and H. Yamazaki, *Nature* 196, 666 (1962).
77. J. Belloni, M. O. Delcourt, and C. Leclere, *Nouv. J. Chim.* 6, 507 (1982).
78. M. O. Delcourt, N. Keghouche, and J. Belloni, *Nouv. J. Chim.* 7, 131 (1983).
79. M. O. Delcourt, J. Belloni, J. L. Marignier, C. Mory, and C. Collex, *Radiat. Phys. Chem.* 23, 485 (1984).
80. Z. Chen, B. Chen, Y. Qian, M. Zhang, L. Yang, and C. Fan, *Acta Metall. Sinica B* 5, 407 (1992).
81. R. Tausch-Trekml, A. Henglein, and J. Lilie, *Ber. Bunsen-Ges. Phys. Chem.* 82, 1335 (1978).
82. B. G. Ershov and N. L. Sukhov, *Radiat. Phys. Chem.* 36, 93 (1990).
83. J. H. Baxendale, E. M. Fielden, and J. P. Keene, *Proc. Roy. Soc. London, Ser. A* 286, 320 (1965).
84. A. Henglein and R. Tausch-Trekml, *Colloid J. Interface Sci.* 80, 84 (1981).
85. M. Mostafavi, J. L. Marignier, J. Amblard, and J. Belloni, *Radiat. Phys. Chem.* 34, 605 (1989).
86. Y. Ilan, Y. A. Ilan, and G. Czapski, *Biochem. Biophys. Acta* 503, 339 (1978).
87. N. L. Sukhov, M. A. Akinshin, and B. G. Ershov, *Khim. Vysokikh Energii* 20, 392 (1986).
88. B. G. Ershov, E. Janata, M. Michaelis, and A. Henglein, *J. Phys. Chem.* 95, 8996 (1991).
89. B. Cercek, M. Ebert, and A. J. Swallow, *J. Chem. Soc. A* 612 (1966).
90. J. Butler and A. Henglein, *Radiat. Phys. Chem.* 15, 603 (1980).
91. M. Breitenkamp, A. Henglein, and J. Lilie, *Ber. Bunsen-Ges. Phys. Chem.* 80, 973 (1976).
92. A. Henglein, E. Janata, and A. Fojtik, *J. Phys. Chem.* 96, 4734 (1992).
93. M. Michaelis and A. Henglein, *J. Phys. Chem.* 96, 4719 (1992).
94. S. Mosseri, A. Henglein, and E. Janata, *J. Phys. Chem.* 93, 6791 (1989).
95. J. L. Marignier, J. Belloni, M. O. Delcourt, and J. P. Chevalier, *Nature* 317, 344 (1985).
96. M. Anbar, M. Bambeneck, and A. B. Ross, *Natl. Bur. Stand. Ref. Data Set.* 43 (1973).
97. A. B. Ross, *Natl. Bur. Stand. Ref. Data Set.* 43, Suppl. (1975).
98. Y. Zhu, Y. Qian, M. Zhang, and Z. Chen, *Mater. Lett.* 17, 314 (1993).
99. H. D. Glickman, "Metals Handbook," 9th ed., Vol. 7, p. 147. American Society for Metals, Metals Park, OH, 1984.
100. J. R. Larry, R. M. Rosenberg, and R. G. Uhler, *IEEE Trans. Compon. Hybrids Manuf. Technol.* CHMT-3, 211 (1980).
101. G. Fisher, *Ceram. Ind. (Chicago)* 120, 80 (1983).
102. Y. Zhu, Y. Qian, M. Zhang, Z. Chen, and C. Fan, *Chin. J. Chem. Phys.* 8, 435 (1995).
103. P. Mulvaney and A. Henglein, *J. Phys. Chem.* 94, 4182 (1990).
104. A. Henglein, *Chem. Phys. Lett.* 154, 473 (1989).
105. A. Henglein, *Chem. Rev.* 89, 1861 (1989).
106. M. Mostafavi, N. Keghouche, M. O. Delcourt, and J. Belloni, *Chem. Phys. Lett.* 167, 193 (1990).
107. A. Henglein, T. Linnert, and P. Mulvaney, *Ber. Bunsen-Ges. Phys. Chem.* 94, 1449 (1990).
108. T. Linnert, P. Mulvaney, A. Henglein, and H. Weller, *J. Am. Chem. Soc.* 112, 4657 (1990).
109. B. G. Ershov, J. Janata, and A. Henglein, *J. Phys. Chem.* 97, 339 (1993).
110. E. Wiegel, *Z. Phys.* 136, 642 (1954).
111. C. N. J. Wagner and E. N. Aqua, *Adv. X-ray Anal.* 7, 46 (1964).
112. J. W. T. Spinks and R. J. Woods, "An Introduction to Radiation Chemistry," 2nd ed. Wiley-Interscience, New York, 1976.
113. Y. Zhu, Y. Qian, M. Zhang, and Z. Chen, *Mater. Sci. Eng., B* 23, 116 (1994).
114. A. Henglein, *Ber. Bunsen-Ges. Phys. Chem.* 556, 81 (1977).
115. Y. Zhu, Y. Qian, H. Huang, M. Zhang, and L. Yang, *J. Mater. Sci. Lett.* 15, 1346 (1996).
116. M. Haissonsky and J. C. Dran, *J. Chem. Phys.* 5, 321 (1968).
117. Y. Zhu, Y. Qian, M. Zhang, Z. Chen, and G. Zhou, *J. Mater. Sci. Lett.* 13, 1243 (1994).
118. Y. Zhu, Y. Qian, and M. Zhang, *Chin. Sci. Bull.* 42, 644 (1997).
119. M. Kelm, J. Lilie, A. Henglein, and E. Janata, *J. Phys. Chem.* 882, 78 (1974).
120. Y. Zhu, Y. Qian, M. Zhang, Y. Li, W. Wang, and Z. Chen, *Mater. Trans. JIM* 36, 80 (1995).
121. Y. Zhu, Y. Qian, Y. Li, W. Wang, M. Zhang, Z. Chen, and S. Tan, *Nanostruct. Mater.* 4, 915 (1994).

122. Y. Liu, Y. Qian, M. Zhang, Z. Chen, and C. Wang, *Mater. Lett.* 26, 81 (1996).

123. Y. Liu, Y. Qian, C. Wang, and Y. Zhang, *Mater. Res. Bull.* 31, 973 (1996).

124. E. R. Burkhardt and R. D. Ricke, *J. Org. Chem.* 50, 416 (1985).

125. A. K. Al-Nimr and M. M. Martynyuk, *Poroshk. Metall. (Kiev)* 3, 21 (1988).

126. V. P. Artamonov and T. V. Shchetinina, *Poroshk. Metall. (Kiev)* 4, 1 (1990).

127. M. Kelm, J. Lilie, and A. Henglein, *J. Chem. Soc., Faraday Trans. 1* 71, 1132 (1975).

128. A. Henglein, M. Gutierrez, E. Janata, and B. G. Ershov, *J. Phys. Chem.* 96, 4598 (1992).

129. Y. Zhu, Y. Qian, M. Zhang, Y. Li, Z. Chen, and G. Zhou, *J. Alloys Comp.* 221, 4 (1995).

130. B. Kemp, Jr. and A. Johnson, U.S. Patent 4,711,611, 1987.

131. P. S. Gilman and W. E. Mattson, U.S. Patent 4,627,959, 1986.

132. J. Januseviciene, R. Pranculyte, and D. Poskute, *Issled. Obl. Osazhdeniya Met., Resp. Konf. Elektrokhim. Lit. SSR* 19, 218 (1983).

133. M. Ya. Gen, I.V. Plate, N. I. Soenko, C. B. Storozhev, and E. A. Fedorova, *Fizikokhim. Ultradispersnykh Sist.* 151 (1987).

134. S. M. Ryabykh and Yu. Yu. Sidorin, *Fizikokhim. Ultradispersnykh Sist.* 127 (1987).

135. M. Uda, *Nanostruct. Mater.* 11, 101 (1992).

136. Y. Zhu, Y. Qian, H. Huang, and M. Zhang, *J. Mater. Sci. Lett.* 15, 1700 (1996).

137. Y. Zhu, Y. Qian, H. Huang, and M. Zhang, *Mater. Lett.* 28, 119 (1996).

138. Y. Zhu and Y. Qian, *Mater. Sci. Eng., B* 47, 184 (1997).

139. Y. Zhu, Y. Qian, M. Zhang, Z. Chen, D. Xu, L. Yang, and G. Zhou, *Mater. Res. Bull.* 29, 377 (1994).

140. Y. Zhu, Y. Qian, and M. Zhang, *Mater. Sci. Eng., B* 4, 294 (1996).

141. Y. Liu, Y. Qian, M. Zhang, Z. Chen, and C. Wang, *Mater. Res. Bull.* 31, 1029 (1996).

142. Y. Liu, Y. Qian, Y. Zhang, M. Zhang, Z. Chen, L. Yang, C. Wang, and Z. Chen, *Mater. Lett.* 28, 357 (1996).

143. Y. Zhu, Y. Qian, M. Zhang, Z. Chen, and G. Zhou, *J. Mater. Chem.* 4, 1619 (1994).

144. Y. Zhu, Y. Qian, H. Huang, M. Zhang, and S. Liu, *Mater. Lett.* 28, 259 (1996).

145. B. E. Yoldas, *J. Mater. Sci.* 21, 1087 (1986).

146. Y. Zhu, Y. Qian, X. Li, and M. Zhang, *Chem. Commun.* 1081 (1997).

147. P. Judeinstein, P. W. Oliveire, H. Krug, and H. Schmidt, *Chem. Phys. Lett.* 220, 35 (1994).

148. R. Flitton, J. Johal, S. Maeda, and S. P. Armes, *J. Colloid Interface Sci.* 173, 135 (1995).

149. J. J. Watkins and T. J. McCarthy, *Polym. Mater. Sci. Eng.* 73, 158 (1995).

150. L. M. Bronstein, E. Sh. Mirzoeva, M. V. Seregina, P. M. Valetsky, S. P. Solodovnikov, and R. A. Register, *ACS Symp. Ser.* 622, 102 (1996).

151. R. N. Bhargava et al., *Phys. Rev. Lett.* 72, 416 (1994).

152. Jpn. Kokai 83-193551 (1983), 83-193552 (1983), 83-194716 (1983), 83-194739 (1983), (Cannon K. K.), Jpn. Kokai 85-210526 (1985), Suzki, K. (Cannon K. K.).

153. T. Toyoda and H. Shin-Yasu, *J. Phys. IV* 4, 365 (1994).

154. R. R. Chianelli et al., *Adv. Catal.* 40, 177 (1994).

155. S. Yogintomi et al., *Nippon Enerugi Gakkaishi* 74, 147 (1995).

156. A. H. Thompson et al., *Mater. Res. Bull.* 10, 915 (1975).

157. D. Coucouvanis, *Prog. Inorg. Chem.* 26, 301 (1979).

158. T. Vossmeyer et al., *J. Phys. Chem.* 98, 7665 (1994).

159. S. Modes and P. Lianos, *J. Phys. Chem.* 93, 5854 (1989).

160. R. A. Hoxson et al., *J. Chem. Soc., Chem. Commun.* 823 (1994).

161. D. Hayes et al., *J. Phys. Chem.* 93, 4603 (1989).

162. F. F. Lange, *Science* 273, 903 (1990).

163. E. G. Bauer et al., *J. Mater. Res.* 5, 852 (1990).

164. W. J. Dawson, *Am. Ceram. Soc. Bull.* 67, 1673 (1988).

165. E.-W. Shi and W. Z. Zhong, *Chin. J. Synth. Cryst.* 23, 207 (1994).

166. A. T. Chien, J. S. Speck, F. F. Lange, A. C. Daykin, and C. G. Levi, *J. Mater. Res.* 10, 1784 (1995).

167. Q. W. Chen, Y. T. Qian, Z. Y. Chen, W. B. Wu, Z. W. Chen, G. E. Zhou, and Y. H. Zhang, *Appl. Phys. Lett.* 66, 1 (1995).

168. A. T. Chien, J. S. Speck, and F. F. Lange, *J. Mater. Res.* 12, 1176 (1997).

169. Q. W. Chen, Y. T. Qian, Z. Y. Chen, L. Shi, X. G. Li, G. E. Zhou, and Y. H. Zhang, *Thin Solid Films* 272, 1 (1996).

170. Q. W. Chen, X. G. Li, Y. T. Qian, J. S. Zhu, G. E. Zhou, W. P. Zhang, and Y. H. Zhang, *Appl. Phys. Lett.* 68, 1 (1995).

171. Q. W. Chen, Y. T. Qian, Z. Y. Chen, G. E. Zhou, and Y. H. Zhang, *Mater. Lett.* 22, 93 (1995).

172. Q. W. Chen, Y. T. Qian, et al., *Thin Solid Films* 264, 25 (1995).

173. Q. W. Chen, Y. T. Qian, Z. Y. Chen, Y. Xie, G. E. Zhou, and Y. H. Zhang, *Mater. Lett.* 24, 85 (1995).

174. Q. W. Chen, Y. T. Qian, et al., *Mater. Res. Bull.* 30, 443 (1995).

175. B. R. Critchley and P. R. C. Steevens, *J. Phys. D* 11, 491 (1978).

176. R. C. F. Farrow and G. M. Williams, *Thin Solid Films* 55, 303 (1978).

177. T. Matsumoto, T. Morita, and T. Tshida, *J Cryst. Growth* 53, 225 1981.

178. W. A. Thornton, *J. Electrochem. Soc.* 107, 895 (1960).

179. N. A. Vlasenko and S. A. Zynio, *Phys. Status Solidi* 20, 311 (1967).

180. X. M. Xu, J. Q. Yu, and G. Z. Zhang, *J. Lumin.* 101, 36 (1986).

181. JCPDS Card Files, No. 5-566.

182. K. S. Yeung and Y. W. Law, *Thin Solid Films* 109, 169 (1983).

183. H. J. Hoval, *J. Electrochem. Soc.* 125, 983 (1978).

184. U. Selvaraj, A. V. Prasadarao, S. Komarneni, and R. Roy, *J. Am. Ceram. Soc.* 75, 1167 (1992).

185. M. Yoshimura, S. E. Yoo, M. Hagashi, and N. Ishizowa, *Jpn. Appl. Phys.* 28, L2008 (1989).

186. F. K. Lotgering, *J. Inorg. Nucl. Chem.* 9, 1113 (1959).

187. A. L. Dawar and J. C. Joshi, *J. Mater. Sci.* 19,1 (1984).

188. K. L. Chopra, S. Major, and D. K. Pandya, *Thin Solid Films* 102, 1 (1983).

189. J. L. Vosson, *Phys. Thin Films* 9, 1 (1977).

190. R. Lalauze, P. Breuil, and C. Pijolat, *Sens. Actuators, B* 3, 175 (1991).

191. J. Kane, H. P. Scheweize, and W. Kern, *J. Electrochem. Soc.* 123, 270 (1976).

192. B. J. Baliga and S. K. Gandhi, *J. Electrochem. Soc.* 123, 941 (1976).

193. A. Rohatgi, T. R. Viverio, and L. H. Slack, *J. Am. Ceram. Soc.* 57, 278 (1974).

194. H. Kim and H. A. Laitinen, *J. Am. Ceram. Soc.* 58, 23 (1975).

195. M. Miki-Yoshida and E. Andrade, *Thin Solid Films* 224, 87 (1993).

196. H. Watanabe, *Jpn. J. Appl. Phys.* 9, 1551 (1970).

197. D. E. Carlson, *J. Electrochem. Soc.* 122, 1334 (1975).

198. H. W. Lahmann and R. Widmei, *Thin Solid Films* 27, 359 (1993).

199. S. J. L. Ribeiro, S. H. Pulcinelli, and C. V. Santilli, *Chem. Phys. Lett.* 190, 64 (1992).

200. S. Park, H. Zhang, and J. D. Mackenzie, *Mater. Lett.* 17, 346 (1993).

201. H. Ogawa, A. Abe, N. Nishikana, and S. Hayakawa, *J. Electrochem. Soc.* 128, 2020 (1981).

202. T. Pisarkiewiez and T. Stapinski, *Thin Solid Films* 174, 277 (1989).

203. Y. T. Qian, Y. Xie, C. He, and Z. Y. Chen, *Mater. Res. Bull.* 29, 953 (1994).

204. R. K. Quinm, R. D. Nasby, and R. J. Baughman, *Mater. Res. Bull.* 11, 1011 (1976).

205. K. L. Hardee and A. J. Bard, *J. Electrochem. Soc.* 123, 1024 (1976).

206. K. Itoh and J. O. M. Bockris, *J. Electrochem. Soc.* 131, 1266 (1984).

207. M. E. Pilleux, C. R. Grahmann, and V. M. Fuenzalida, *Appl. Surf. Sci.* 65, 283 (1993).

208. J. K. Lin, J. M. Sivertsen, and J. H. Judy, *IEEE Trans. Magn.* MAG-22 (1986).

209. S. Joshi, R. Nawathey, V. N. Koinkar, V. P. Godbole, S. M. Chaudhari, and S. B. Ogale, *J. Appl. Phys.* 64, 5647 (1988).

210. M. Langiet, M. Labeau, and J. C. Joubert, *IEEE Trans. Magn.* MAG-22, 600 (1986).

211. Y. T. Qian, C. M. Niu, C. Hannigan, S. Yang, K. Dwight, and A. Wold, *J. Solid State Chem.* 92, 208 (1991).

212. G. Kordas and R. A. Weeks, *J. Appl. Phys.* 57, 3812 (1985).

213. M. Abe, Y. Tamaura, M. Oishi, T. Saitoh, T. Itoh, and M. Gomi, *IEEE Trans. Magn.* MAG-23, 3432 (1987).

214. Y. K. Kim and M. Oliveria, *J. Appl. Phys.* 75, 431 (1994).

215. JCPDS Card Files, No. 13-534.

216. W. Desisto, M. Ocana, F. Smith, J. Deluca, R. Kershaw, K. Dwight, and A. Wold, *Mater. Res. Bull.* 24, 753 (1989).

217. P. Wu, Y.-M. Gao, J. Baglio, K. Dwight, and A. Wold, *Mater. Res. Bull.* 24, 905 (1989).

218. J. Fournier, W. Desisto, R. Brusasco, M. Sosnowski, R. Kershaw, K. Dwight, and A. Wold, *Mater. Res. Bull.* 23, 31 (1988).

219. Y. T. Qian, R. Kershaw, K. Dwight, and A. Wold, *Mater. Res. Bull.* 25, 1243 (1990).

220. W. J. Desisto, Y. T. Qian, C. Hannigan, J. O. Edward, R. Kershaw, K. Dwight, and A. Wold, *Mater. Res. Bull.* 25, 183 (1990).

221. Y. T. Qian, Y. Xie, Z. Chen, J. Lu, and J. Zhu, *Chin. J. Chem. Phys.* 8, 549 (1995).

222. Y. Xie, W. Z. Wang, Y. T. Qian, L. Yang, and Z. Chen, *J. Cryst. Growth* 167, 656 (1996).

223. S. Joshi, R. Nawathey, V. N. Koinkar, V. P. Godbole, S. M. Chaudhari, and S. B. Ogale, *J. Appl. Phys.* 64, 5647 (1988).

224. J. K. Lin, J. M. Sivertsen, and J. H. Judy, *IEEE Trans. Magn.* MAG-22, 50 (1986).

225. G. Kordas and R. A. Weeks, *J. Appl. Phys.* 7, 3812 (1985).

226. M. Langlet, M. Labeau, and J. C. Joubert, *IEEE Trans. Magn.* MAG-22, 600 (1986).

227. C. J. Serna, J. L. Rendon, and J. E. Iglesias, *Spectrochim. Acta, Part A* 38, 797 (1982).

228. J. E. Iglesias, M. Ocana, and C. J. Serna, *Appl. Spectrosc.* 44, 418 (1990).

229. S.Yoon, H. Lee, and H. Kim, *Thin Solid Films* 171, 251 (1989).

230. P. Merchant, R. Kershaw, K. Dwight, and A. Wold, *J. Solid State Chem.* 27, 307 (1979).

231. R. D. Evans and C. M. Cooke, "1977 Annual Report, Conference on Electronical Insulating and Dielectrics Phenomena," p. 142. Committee on Dielectrics Assembly of Engineering National Research Council, National Academy of Science, Washington, DC, 1979.

232. R. C. Ku and W. L. Winterbottom, *Thin Solid Films* 127, 241 (1985).

233. R. E. Kirby, E. L. Garwin, F. K. King, and A. R. Nyaiesh, *J. Appl. Phys.* 62, 1400 (1987).

234. M. Seo, K. Takemasa, and N. Sato, *Appl. Surf. Sci.* 33/34, 120 (1988).

235. D. Albin and S. H. Risbud, *Adv. Ceram. Mater.* 2, 243 (1987).

236. W. C. Fernelius and J. E. Blanch, in "Inorganic Synthesis" (T. Moeller, ed.), Vol. 5, p. 130. McGraw–Hill, New York, 1957.
237. M. Balog, M. Schrieber, S. Patai, and M. Michman, *J. Cryst. Growth* 17, 291 (1972).
238. R. Busasco, R. Kershaw, J. Baglio, K. Dwight, and A. Wold, *Mater. Res. Bull.* 21, 301 (1986).
239. J. A. Aboat, *J. Electrochem. Soc.* 114, 948 (1967).
240. M. T. Duffy and W. Kern, *RCA Rev.* 754 (December 1970).
241. J. Saraie, J. Kwon, and Y. Yodogawa, *J. Electrochem. Soc.* 132, 890 (1985).
242. D. Albin and S. H. Risbud, *Adv. Ceram. Mater.* 2, 243 (1987).
243. D. A. Wruck et al., *J. Vac. Sci. Technol., A* 9, 5927 (1991).
244. J. S. E. M. Svensson and C. G. Grangvist, *Appl. Phys. Lett.* 49, 1566 (1986).
245. S. I. Cordoba-Torrsi et al., *J. Electrochem. Soc.* 138, 1548 (1991).

Chapter 10

SEMICONDUCTOR QUANTUM DOTS: PROGRESS IN PROCESSING

David J. Duval, Subhash H. Risbud

Department of Chemical Engineering and Materials Science, University of California, Davis, California, USA

Contents

1. INTRODUCTION

Electronic and optical materials based on quantum dots are at the very forefront of materials-processing research and development and present a rich variety of preparation and processing opportunities. Ever since the discovery of the transistor, electronic devices have continued a trend toward smaller and smaller sizes with the attendant challenges of nano-size materials processing. The processing science of semiconductors of a size small enough (nanometers) to achieve quantum confinement effects, preferably in all three spatial dimensions (quantum dots), has made rapid progress in the last 15 years and represents perhaps the best integration of materials synthesis, quantum physics, and state-of-the-art electronic and photonic technologies. The device applications of quantum wells, superlattices, and quantum wires and dots have spurred a revolution in the speed of signal processing and miniaturization of circuitry. Parallel with rapid exploration of the applications of quantum dot–based devices and structures, the last decade has seen major strides in cultivating the understanding of the underlying physics and science of quantum size materials. From a materials chemistry point of view, many synthesis and processing issues still remain difficult to solve such as, for example, how to form reproducible, organized arrays of quantum dots of any semiconductor on any surface or in a bulk matrix of glass, polymer, or ceramic. Because this nanoprocessing science is yet to be fully developed, it represents a frontier area of materials research. The rewards for making controlled quantum size nanostructures with

Handbook of Nanostructured Materials and Nanotechnology, edited by H.S. Nalwa
Volume 1: Synthesis and Processing

ISBN 0-12-513761-3/$30.00

monodispersed semiconductor features on the nanoscale is almost certain to have a major impact on many new technologies including flat panel displays, light-emitting materials, and nanoelectronics. For example, the development of light-emitting diodes (LEDs) and semiconductor lasers is critically dependent on materials capable of efficient light emission in the visible and ultraviolet (UV) regimes.

From the publication of the first reported data in the literature demonstrating quantum confinement effects for microcrystalline semiconductors [1] to the present, there has been a need to control the size and distribution of quantum dots in a stable matrix. This is particularly true for Type I quantum confinement devices, where the confined material's band gap lies inside that of the matrix. Type II devices generally exhibit confined electrons and delocalized holes because the valance band of the confined material is below the valence band of the matrix. These latter devices are generally used for infrared applications rather than for visible light applications. We will limit our discussion to Type I quantum confinement materials.

The size distribution and morphology of quantum dots in their matrices are associated with the energy of the optical absorption edge, control and efficiency of electric field modulated optical absorption (quantum confined Stark effect), and higher-order absorption phenomena. Many techniques have been developed to enhance uniformity, control morphology, and determine the spatial distribution of quantum dots in bulk materials, as well as thick and thin films. Techniques include sol–gel and solid-state precipitation in glass or crystalline matrices [2, 3], molecular beam epitaxy [4], chemical vapor deposition techniques [5], and lithographic mechanisms [6]. Advances in these processing methods, and others, will be discussed in this chapter.

2. QUANTUM DOT PROCESSING TECHNOLOGIES

2.1. High-Temperature Formation of Quantum Dot Materials

It has been known for quite some time that coloration in many commercial filter glasses is due, in large part, to the precipitation of semiconductor microcrystallites in the glasses [7]. Although commercially available glasses are not generally designed for optoelectronic applications, Cruz et al. [8] have observed the pure light-induced ultrafast optical Stark shift in such a semiconductor-doped glass. Use of these commercial glasses for fast optical switching devices would ensure a relatively inexpensive supply of homogeneous glasses with predictable lot-to-lot behavior. Recent efforts to control the structure and distribution of CuCl quantum dots (QDs) in host glasses for improved quantum confinement behavior have been reported in the glass system $10Na_2O–43B_2O_3–8Al_2O_3–39SiO_2$ [9]. As reported with precipitates in other systems [10], QD growth was proportional to the 1/3 power of the heat treatment time. Blue shifts in the absorption edge were reported for decreasing dot size. It is noteworthy that precipitation kinetics could be controlled in this very refractory glass, implying this system could be a candidate for robust quantum device applications. However, very strict control of precipitate size, morphology, and composition is necessary for the realization of efficient quantum devices.

Figure 1 shows the effect of heat treatment times and temperatures for a variety of semiconductors in a borosilicate glass matrix [11]. Color changes are associated with homogeneous nucleation and growth of QDs in the glass.

Improving the size distribution of QDs in glasses is possible by double annealing the glass into the (1) nucleation and then (2) growth regimes of precipitation [12]. Double annealing of CdTe QDs in a borosilicate glass resulted in a 3% improvement in size dispersion of quantum dots (2 nm average radius). Single annealing resulted in an actual dispersion of 10.5% for this system. Figure 2 is a high-resolution transmission electron micrograph of a CdTe microcrystallite in a borosilicate glass matrix, which was formed from a single anneal [13].

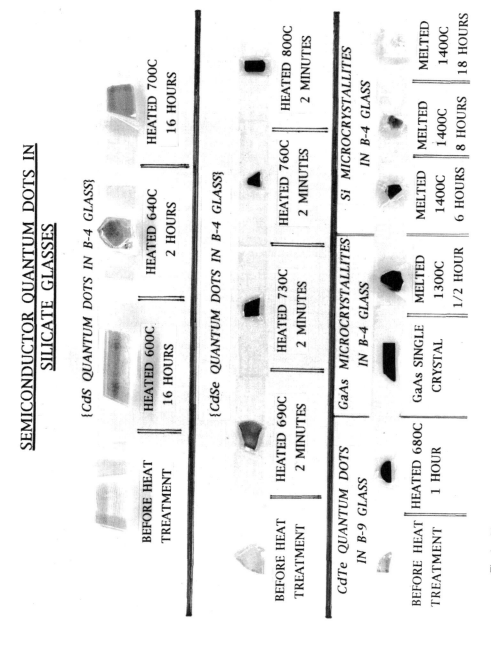

Fig. 1. Photographs of various semiconductor quantum dots in glasses synthesized in our laboratory during the last 8 years.

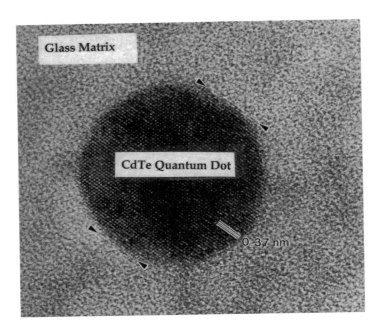

Fig. 2.   High-resolution electron micrograph of a cadmium telluride quantum dot in a glass matrix.

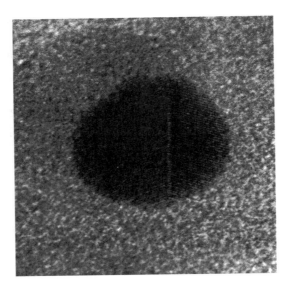

Fig. 3.   High-resolution transmission electron microscopy image of a twinned glass-sequestered ZnSe nanocrystal showing lattice fringes and the surrounding glass matrix. The majority of particles were single crystal, but a few larger internally twinned particles found.

Semiconductor concentration plays a large role in size dispersion of QDs in glass. Lower dopant concentration leads to larger quantum dot sizes with less dispersion when double annealed. It is the arithmetic averaging of concurrently coalescing and nucleating QDs in glasses with high semiconductor dopant levels that causes a broader dot size distribution [14]. Figure 3 is a micrograph of a twinned ZnSe precipitate in a borosilicate glass matrix, and Figure 4 is a micrograph of the same specimen with an electron diffraction pattern indicating that the quantum dots have a cubic structure [15]. Along with twinning, other evidence of coalescence was seen, and the dot size distribution was broad.

Fig. 4. TEM image of as-cast glass-sequestered ZnSe nanocrystals with an average particle diameter of 5.5 ± 1.7 nm. (The average particle diameter of the quenched sample is 3.7 ± 1.1 nm.) Inset is a selected-area electron diffraction (SAED) pattern for as-cast sample with diffraction ring spacings corresponding to cubic ZnSe. Visible rings are for (111), (220), and (311) lattice planes.

An alternative to precipitation of quantum dots in glasses is dissolution of crystallites into an unsaturated host matrix. The dissolution of silicon powders into a borosilicate matrix has been performed by our group [16], where silicon particles have been etched by oxygen permeation into the glass at high temperatures. This resulted in the formation of SiO_2 at the Si/glass interface. Because the solubility of SiO_2 in the glass is infinite, quantum dots of any size can be produced by stopping the high-temperature etch at prescribed times, depending on the initial crystallite size. Figure 1 shows a series of silicon-loaded glasses where thermal etching had been performed at different times. The increasing transparency with increasing time indicates that silicon is being digested, resulting in smaller quantum dots. The micrograph in Figure 5 is of a silicon nanocrystal in the borosilicate glass matrix.

Complex geometries of quantum dot composites and quantum dot core/shell morphologies have been reported. These systems may lead to novel quantum optoelectronic devices including hybrid Type I and Type II QD devices. Ko et al. [17] have used the melt/quench process to create silver-coated CdS QDs or CdS-coated silver nanocrystals in crown glass. The composition of the shell of this core/shell system is controlled by processing conditions. The advantages of these morphologies include the possibility that local electric field effects on the QD may be increased by exciton/plasmon resonance, resulting in enhanced intrinsic nonlinear optical behavior. Liu et al. [18] used the melt/quench process to produce a core/shell quantum dot of CdTeS in a modified borosilicate matrix. The shell, having more Te than the core, thus has a narrower band gap. This results in what those authors call a structural quantum dot. They suggest that this process may be used to grow a three-dimensional quantum well structure. The authors of both works on complex QD morphologies concede their materials have broad dot size distributions, but the possibility of using these methods to produce optoelectronic devices with novel applications is indicated.

It has been suggested that quantum dots precipitated in a crystalline matrix would be three-dimensionally aligned [19], resulting in more predictable bulk behavior than quantum

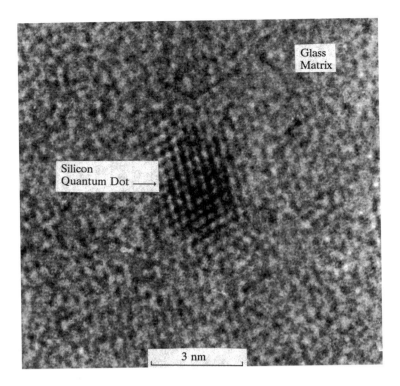

Fig. 5. Atomic resolution micrograph of a silicon nanocrystal in a glass matrix.

dot materials in a glassy matrix. Recent examples of QDs precipitated in crystals include CuCl quantum dots in a NaCl matrix [3] and Pb^{2+} aggregates in a CsCl crystal [20]. In the case of the former system, the presence of biexcitons has been determined. Although a broad envelope associated with a distribution of dot sizes was observed, the authors report that biexciton luminescent energy changed little with change in quantum dot nanocrystal radius. For optical applications, a device using this system may prove easier to engineer because of relaxed specifications on otherwise strict requirements for uniform dot size. Luminescence studies of the $CsCl:Pb^{2+}$ system showed evidence of a $CsPbCl_3$-like nanophase in the single crystal. A luminescence band at 420 nm was observed with picosecond decay kinetics. This appears to be the first investigation into Pb^{2+} ions aggregated in a body-centered alkali halide crystal. Precipitation kinetics are expected to be quite different from those of the NaCl structure, and these studies may prove useful for controlling the precipitate size distribution for other quantum dot materials in various alkali halide matrices.

2.2. Low-Temperature Formation of Quantum Dot Materials

2.2.1. Precipitation from Solution

Organometallics have been used as precursors mainly for the production of lead and cadmium selenides and sulfides in sol–gel matrices [2, 21, 22]. There are three advantages of the sol–gel method over the melt/quench method for fabricating QD materials:

1. Lower temperatures are used to consolidate the gels, thus decreasing dissolution, sublimation, or oxidation of the semiconductor.
2. Much higher concentrations of semiconductor can be incorporated into the matrix because complete solubility of the semiconductor in a melt is no longer necessary to obtain a homogeneous distribution of quantum dots.

3. Photodarkening from exposure to high-intensity light sources for long periods is lower in sol–gel-derived glasses than in conventionally melted glasses.

To produce a quantum dot material by precipitation from a solution, often a metal acetate crystal is precipitated in a matrix. The crystal is then converted to the metal oxide by heating in oxygen and, subsequently, converted either to the sulfide by exposure to H_2S [21] or to the selenide by exposure to a selenosulphate solution [22].

On the other hand, the matrix can be formed from a borosilicate sol where sodium and metal acetates are added after hydrolysis of the sol, but before condensation. The gel subsequently formed upon condensation contains microchannels of various sizes in which the acetate crystals precipitate on drying. In these cases, the matrix plays a large role in limiting crystallite size. The resulting gel can then be consolidated into a monolith below approximately 600 °C [21, 23]. The size dispersion of QDs formed by this method is, in large part, determined by the variation in microchannel size of the porous gel. Precipitation in controlled volumes is discussed in more detail in a later section.

Mackenzie and Kao [23] have enhanced the uniformity of cadmium distribution in the gel by complexing the metal with bulky ligands (3-aminopropyltriethoxysilane). The ligands uniformly anchor the metal ion to the silica gel matrix, which results in a narrower size distribution ($\leqslant\sim$5%) for quantum dots in the consolidated monolith [21].

With regard to CdSe and CdS quantum dots, Gorer et al. [22] found that these nanocrystals were protected against conversion from the cubic to the hexagonal form by being incorporated into a sol–gel matrix. Quantum dots of CdSe in an organic solvent started converting to the hexagonal phase below 300 °C. Those incorporated into either TiO_2 or ZrO_2 gels started conversion above 400 °C.

An alternative chemical method for the production of freestanding monodisperse QDs has been employed to produce "capped" CdS quantum dots with a well-defined tetrahedral morphology [24]. Thioglycerine was used to coat (cap) the nanocrystals formed when thiourea was used instead of hydrogen sulfide as the sulfur source in solution. The reaction was carried out in a dimethylformamide solvent. Interestingly, these polymer-capped QDs could be caused to orient themselves on a quartz slide into fibers. This resulted in an optically transparent and anisotropic quantum dot thick film.

A solution/growth technique has been used to grow CdS directly on glass [25]. Controllable chemical reactions among cadmium acetate, thiourea, triethanolamine, and aqueous ammonia allow for predictable growth of CdS nanocrystals. In this case, grains in the hexagonal structure with sizes ranging from 7.4 to 26.5 nm were obtained. Grain size and uniformity were optimal at the lowest reactant concentrations, demonstrating the major role of kinetics in this reaction.

Gindele et al. [26], using cadmium ethoxy acetates as the cadmium source and bis (trimethylsilyl)selenium for the selenide source, created closely packed CdSe quantum dot thick films on glass using a solution/growth technique. The cubic 3.2 ± 0.2-nm crystallites were immobilized in an aminopropyltriethoxysilane (APTS)-bridged array. APTS stabilized the particles against photodecomposition with its low ionic or atomic oxygen permeability. Optical studies on this highly concentrated system indicated that carrier (and energy) exchange between adjacent QDs is seen, meaning these materials behave differently from quantum dot materials previously discussed containing isolated QDs. Those authors suggest these materials behave more like quantum well materials.

With the observation that it is possible to attach many functional groups to semiconductor nanoparticles, Alivisatos et al. [27] and Mirkin et al. [28] have suggested ways of ordering QDs into well-defined arrays using deoxyribonucleic acid (DNA) technology. In their studies, both groups used nanocrystalline gold as their model quantum dot material. Whereas the Alivisatos group demonstrated the formation of well-defined "dimer" quantum dot molecules, the Mirkin group has shown it is possible to construct macroscopic materials with less symmetry using a similar methodology. In terms of molecular ordering,

Destri et al. [29] have succeeded in producing fully conjugated copolymers that act as one-dimensional arrays of quantum dots. Nonlinear quantum dot optical behavior in the visible spectrum was displayed for a film of these completely organic polymers. No attempt was made to correlate their results with some sort of size distribution parameter.

An interesting method for producing quantum dot materials with narrow size distributions has been suggested by the work of Gaponenko et al. [30]. This group produced CdS crystallites by bubbling H_2S through a cadmium acetate solution. The particles were stabilized for dispersion on a glass slide with an inert organic polymer. In this material, the authors reported permanent changes in the absorption spectra of the total quantum dot ensemble by size-selective photodecomposition. This technique could aid in the study of homogeneous and nonhomogeneous line broadening of CdS optical bands. It also may prove to be a technique capable of enhancing the uniformity of quantum dot size dispersion in a device by size-specific quantum dot dissolution.

As with high-temperature precipitation, composite geometries have been obtained for semiconductor nanoparticles formed at low temperatures. For example, nanocrystals of CdS precipitated from an aqueous solution of cadmium nitrate and sodium sulfide have been successfully coated with PbS [31]. The CdS nanocrystals of 6.0–8.0 nm were initially stabilized with an organic. Pb^{2+} was then added in the form of an aqueous nitrate solution. Transmission electron microscopy (TEM) analysis proved the specimens had a core/shell morphology, where Pb^{2+} had replaced Cd^{2+} at the surface of the crystallites. The specimens showed a red shift in the absorption spectrum with time, indicating that increasing shell thickness caused by PbS formation (and cadmium dissolution) was occurring. As the quantum dots in this matrix are not stable with time, they are of limited interest in device applications.

2.2.2. Precipitation in Controlled Volumes

Heath [32] has suggested it may be possible to obtain uniformly sized quantum dot materials by controlling nucleation and growth within a lithographically or electrochemically designed template or by using weak interactions to cause size-dependent phase separation into opal-like structures. One could easily imagine using naturally occurring or synthetic templates with precisely known dimensions for a similar purpose. Mu et al. [33] used commercially obtained Gelsil porous substrates with 2.5-, 5-, 10-, and 20-nm pore diameters as templates for melt/infusion of the semiconductor PbI_2. Their results indicated this material possessed a narrower quantum dot size distribution than that produced by nucleation and growth in solution.

Precious opals are the result of face-centered cubic (fcc) packing of identical spherical silica balls with less than 5% size variation within a specimen. Romanov et al. [34] used opals as templates for melt/infusion of the semiconductor InSb. Their results showed that the semiconductor completely occupied all 26% free volume in the structure, with a bimodal quantum dot size distribution. This distribution is in agreement with the expected octahedral and tetrahedral void distribution characteristic of the fcc structure.

Although the aforementioned methods involved some heat treatment for semiconductor incorporation, it is important to realize that these techniques are fundamentally different from high-temperature techniques. Here, it is necessary that temperatures are consistently low enough so that the pore structure of the host material remains unchanged during all stages of processing. Other minerals have been used as templates for CdSe quantum structures, such as magnesium silicates in the form of asbestos and sepiolite [35]. As is the case with refractory semiconductors, this system did not lend itself to a melt/infusion technique. Instead, a solution of cadmium acetate was forced into the pores at high pressures. Subsequently, the structure was dried and treated with H_2Se to precipitate CdSe within the nanochannels. However, the volume fraction of the specimens occupied by CdSe was only on the order of 10^{-3}, and spectroscopic data indicated that quantum wires, rather than dots, were probably formed.

Organic templates also have been used for the synthesis of CdS, ZnS, and PbS QD structures. Wang et al. [36] used multilayers of a channel protein found in mammalian eye lenses as their host for CdS and ZnS quantum dot structures. Films of the protein were deposited on glass slides, which were subsequently dipped into either zinc or cadmium chloride solutions. After washing the films to remove excess metal ions, the films were exposed to hydrogen sulfide. Channel size in the protein is controlled by pH, allowing the control of the quantum dot size. The existence of quantum dots in the final structure was deduced by a blue shift in the absorbance spectra.

Yang et al. [37] used a bicontinuous cubic phase of a lipid protein as a template for precipitation of PbS nanocrystals from aqueous solutions. The lipid phase consisted of a sodium sulfide solution mixed with sodium dioctylsulfosuccinate. It should be mentioned that this system is structurally different from a collection of inverse micelles in that this cubic phase consists of aqueous pores interconnected by narrow channels, rather than of isolated water pools. Hence, ions can diffuse between pores without having to pass through a hydrophobic membrane. This lipid phase was then exposed to a lead nitrate solution, where homogeneous nucleation of PbS in the pores was observed. The protein phase was then dissolved, and PbS nanocrystals were capped with n-dodecanethiol to inhibit coarsening or further growth of the crystallites. This material was then dispersed in chloroform for spectroscopic analysis. It was determined that particle size could be controlled by Na_2S concentration. Although absorption spectra showed a blue shift, no excitonic peaks were measured.

2.3. Thin-Film Techniques

Although expensive and capital intensive, thin-film techniques for producing quantum dot materials show advantages over the usual precipitation and growth methods. Increased control over stoichiometric composition, quantum dot size, and spatial distribution, as well as control over the design of certain quantum dot device geometries, will be discussed in the following sections.

2.3.1. Codeposition of Quantum Dots with Thin Films

Quantum dots produced by solution chemistry should form unagglomerated particles with good crystallinity. This would facilitate their introduction into reaction vessels for processing and deposition. Because the surface energy of quantum dots is high, exothermic "wetting" on a substrate may lower temperatures usually associated with bonding these materials to a substrate.

One technique for producing quantum dot–containing thin films involves incorporating solution-precipitated, organically capped, and size-specific quantum dots into metal–organic chemical vapor deposited (MOCVD) thin films [5]. Capped CdSe quantum dots were prepared by size-selective precipitation, subsequent dispersion in pyridine, and precipitation with hexane to produce the pyridine cap. A suspension of the nanocrystallites in a pyridine–acetonitrile solution was then injected into the MOCVD reactor via an electrospray capillary. Appropriate organometallic precursors for forming a ZnSe thin film using a hydrogen carrier gas were injected into the reactor farther down the flow path. Solvents and reactants were dissociated, and a thin film of ZnSe incorporating CdSe quantum dots was grown on externally heated glass substrates. The films were 3–5 μm thick. Although QD sizes were initially narrow, significant agglomeration reportedly occurred during processing.

An alternative method for depositing uniformly sized quantum dots on various substrates as thin films has been employed for GaAs, where the dots are formed and size-selected inside the chemical vapor deposition (CVD) reactor [38]. Gallium metal was evaporated in a tube furnace with hydrogen as a carrier gas and passed through a cool region, where partial precipitation of gallium droplets occurred. Particles were then exposed to α radiation. The charged particles passed through a differential mobility analyzer (DMA) for size separation. The size-selected particles were then reacted with arsine gas in the CVD reactor. GaAs particles were then exposed to α radiation and analyzed by a second DMA. The narrow-sized and charged GaAs nanocrystals were then deposited on various substrates by a perpendicular electric field. The authors suggest that the fine particles act as catalysts for cracking of arsine, resulting in self-limiting GaAs formation even at room temperature in the Ga-poor environment. The size of the resulting particles was measured to be 8.0 ± 1.1 nm by TEM.

Nanocrystallites are formed during laser ablation or evaporation of a polycrystalline target. The evaporated material cools and coalesces in an inert atmosphere, and is then deposited on a substrate. Laser evaporation could be used to produce a reactant of controlled size, such as the source of gallium droplets described in the previous paragraph. Figure 6 depicts a bright-field image of GaN quantum dots deposited on a cellulose nitrate membrane [39]. The quantum dots were formed by laser ablation of a gallium metal target and reaction of the resulting gallium clusters with nitrogen. The samples were subsequently annealed at 800 °C for 1 h under ammonia to enhance crystallinity. Figure 7 shows the resulting size distribution taken from a TEM dark-field image. Laser ablation can be used as a mechanism to produce nanocrystallites of robust compound semiconductors such as CdTe from polycrystalline targets. Often, layers of quantum dots produced by this method are deposited on a substrate, which is then coated with a high-band-gap material to act as an electrical insulator or diffusion barrier for further processing. Multiple layers of evaporated semiconductor and coating would result in thicker films useful for optical and/or electronic applications.

One method for producing CdTe quantum dots embedded in glass films has used laser evaporation to produce CdTe nanocrystals in conjunction with plasma-enhanced CVD (PECVD) to coat the crystallites with SiO_2 [40]. First, a rotating polycrystalline CdTe

Fig. 6. A representative bright-field TEM image of an annealed nanocrystalline GaN sample showing individual nanocrystals 2–20 nm in size. The larger features (about 50–100 nm in size) are aggregates of nanocrystals as confirmed by dark-field TEM. The SAED pattern (inset) confirms the hexagonal GaN structure.

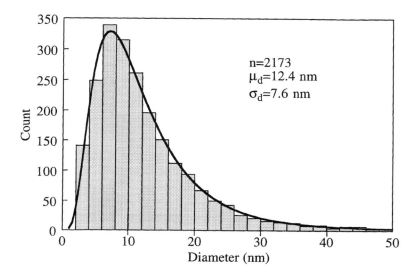

Fig. 7. A particle size histogram taken from representative TEM dark-field images of an annealed sample. The solid line represents a fit of the data to a log-normal distribution: μ_d is the mean particle diameter and σ_d is the standard deviation determined from the fit.

target was ablated by laser. A glass substrate was placed at an angle to the target shield aperture—the source of the evaporated nanocrystals—in order to optimize the deposited nanocrystallites' size distribution. Nanocrystallite size is controlled by inert gas pressure and laser power during ablation. Subsequently, a glass film produced by PECVD of tetramethoxysilane (TMOS) coated the QDs. The advantage of oxide films produced in this manner is their smoothness and purity as compared with other fabrication techniques (e.g., laser evaporation). The deposition/coating cycle was repeated 40–50 times to create films about 1 μm thick. These films showed lower scattering loss and more pronounced spectral features than specimens produced by laser evaporation alone.

Magnetron sputtering is also used to produce QDs in glass films. Argon ions [produced by radio frequency (rf) irradiation] sputter-ablate a target. The evaporated species are collected on a substrate. Films with various sizes of nanocrystalline silicon have been produced using this technique [41]. In this case, a target of SiO_2 glass containing silicon chips was used as a target to produce silicon QDs surrounded by SiO_2 deposited as a thin film on either silicon or glass substrates. By *in situ* substrate heating between 300 and 800 °C with a halogen lamp, and by varying other reactor conditions, a series of specimens with silicon QD sizes ranging from 2 to 5 nm were produced. The normalized standard deviation of the particle sizes was reported to be 0.3. From optical spectra, the authors deduced that complete surface passivation did not occur for quantum dots with diameters less than 3.0 nm. Other investigators varied the ratio of silicon to SiO_2 in the target to modify the $Si:SiO_2$ ratio at the substrate [42]. In this study, *ex situ* annealing under N_2 was performed to control the nanocrystallite size. Their extended X-ray absorption fine structure (EXAFS) studies suggest that QDs less than 4.5 nm in size have crystalline cores, but there is considerable disorder surrounding the cores.

Silicon quantum dots have been formed by very high frequency (VHF) plasma reaction [43]. These authors note that conventional magnetron radiation (MR) is not as favorable for efficient plasma generation. MR occurs at radio frequencies lower by an order of magnitude than those in the VHF range. A SiH_4 plasma is generated in a plasma cell attached to the CVD reactor. Si nanocrystals are formed in the gas phase by coalescence of the plasma's radicals. Deposited crystallites were measured by TEM to be 8.0 nm in diameter with a standard deviation of about 12.5%.

2.3.2. Chemically Designing Quantum Dots on Surfaces

QD fabrication techniques rely heavily on molecular and organometallic precursor chemistry. This is particularly true for chemical vapor deposition production methods. Understanding precursor chemistry and deposition conditions is important in understanding substrate mismatch, interdiffusion, control of nanocrystallite growth, and the occurrence and control of metastable phases.

Buhro [44] has divided precursors for semiconductor materials into two categories. "Conventional" precursors are organometallics that replace an element or simple salt in the production of simple or compound semiconductors. The other category is "single-source" precursors, where all of the precursors for compound semiconductors are replaced by a single type of precursor molecule. The major driving forces for using organometallics as sources are (1) decreasing impurity incorporation in the product, (2) increasing precursor purity, and (3) eliminating the use of toxic gases as precursors. Amines and alkylated metals are generally the bases of organometallics used to make QDs. The following sections will address the formation of quantum dots by molecular beam epitaxy (MBE) and other specialized vapor deposition techniques using organometallic precursors. Specifically, processing involving lithography, controlled growth on slightly off-axis (vicinal) and patterned surfaces, spontaneous surface segregation (islanding), and vertical alignment of quantum dots will be discussed.

2.3.2.1. Lithographic Techniques

Lithography entails shining radiation through a template onto a flat surface coated with a photoresist so that specific areas of the surface can be chemically treated. Photolithography uses photons to react with the photoresist, causing it to polymerize and adhere to the surface. Unexposed photoresist can be removed from the surface, and chemical treatment will only occur in the uncovered areas. This technique allows spatially selective chemical processing. The smaller the wavelength of the radiation, the greater the spatial resolution that can be obtained during chemical processing. Electron beam lithography (EBL) dispenses with the need of a template by allowing a rastered electron beam to expose the photoresist directly. Furthermore, because the wavelength of an electron beam is so small, greater spatial resolution for chemical processing is realized.

Sotomayor Torres et al. [6] used EBL to pattern ZnTe quantum dots onto GaAs substrates. ZnTe layers were grown on GaAs substrates by MOCVD to a thickness of 0.8 μm. The specimens were cleaned and coated with a poly(methyl methacrylate) (PMMA)-based photoresist. EBL patterned the resist to expose spots the size of the electron beam. A 50-nm film of titanium was deposited on the specimen, and the resist with Ti was removed by dipping in a warm acetone solution. The specimen was then exposed to a 1/8 mixture of methane to hydrogen at low pressures in a reactive ion etch (RIE) machine to remove ZnTe from areas uncovered by titanium. Eventually, etching started to undercut the Ti cap, and the size of the ZnTe pillars was reduced to 30 nm in diameter. Subsequent annealing eliminated surface traps and nonstoichiometry. Thin dots showed fewer dislocations at the dot/wafer interface, giving better photoluminiscence (PL) intensity.

Arrays of GaAs quantum dots with 57 ± 3.0-nm diameters have been produced by EBL [45]. A 500-nm buffer layer of GaAs was grown on a GaAs wafer, followed by the deposition of a 20-nm barrier layer of AlGaAs. A quantum well of GaAs was created by subsequently depositing a 5-nm layer of GaAs topped with a 50-nm barrier layer of AlGaAs on the previous layers. The assembly was then capped with a 5.0-nm layer of GaAs. EBL followed by $SiCl_4$ and Ar plasma etching down to the buffer produced the array. The dots' lateral dimensions being smaller than the beam diameter is attributed to the uniform etching by $SiCl_4$.

Many times, annealing of the dots produced by these dry etching techniques is necessary to remove surface damage caused by exposure to plasma. However, annealing may alter the stoichiometry of the dots or diminish their optical responses. Wet etching instead of annealing may circumvent this problem. Gourgon et al. [46] used this approach to produce CdTe quantum dots on a CdTe substrate. Quantum dots were produced by molecular beam epitaxy (MBE) deposition of a strain-free 2-μm-thick CdZnTe buffer layer on a CdTe substrate, followed by a 5-nm epilayer of CdTe under a 90-nm cap of CdZnTe. Normal EBL methods and an Ar$^+$ ion beam etch (IBE) were used to produce CdTe pillars, but a subsequent wet anodic etch in a KOH solution was used to remove IBE-induced defects and reduce the dots' widths via oxidation. In this process, white light irradiating the specimen was used to produce the holes necessary for oxidation to form TeO$_2$ and subsequently dissolve Cd. The oxide was electrochemically grown at a given voltage, resulting in a particular oxide thickness, and kept at that voltage until there was zero current. The rate of oxidation was determined by the electric field between the surface and the electrolyte. Constant voltage reduced localized oxide thickness fluctuations. Oxide was then removed by a wash in HF. Repeated wet etching and washing resulted in further size reduction and 70-nm QDs were fabricated. Oxidation appeared to remove IBE-induced damage and to decrease traps sites.

Quantum dots of CdZnSe/ZnSe were similarly produced by a preferential etch technique [47]. A buffer layer of GaAs 200 nm thick was MBE-deposited on a GaAs wafer. The CdZnSe dots were formed from a structure composed of a ZnSe (70 nm) buffer layer, a CdZnSe (28 nm) quantum well structure, and, finally, a ZnSe (20 nm) cap on the structure. EBL followed by IBE produced pillar structures. Wet etching using K$_2$Cr$_2$O$_7$:HBr:H$_2$O caused the pillars to become pyramids. The resulting dots were 28 nm at the base and 4 nm high.

Silicon quantum structures were formed using a wet-etch process as well [48]. A thermally grown layer of SiO$_2$ on a silicon wafer supported a poly-Si layer grown by low-pressure CVD (LPCVD). A mask of chloromethylpolystyrene was used for EBL processing. Following RIE using HBr and He, a wet etch with NH$_4$:H$_2$O$_2$:H$_2$O slowly reduced the size of the pillars to 70 nm high with 10×10 nm bases isolated on SiO$_2$. No spectroscopic results were reported, however.

In all these cases, some sort of mask and resist were used for lithography. Ezaki et al. [49] developed a technique combining photochemical etching with laser interferometry to etch a wafer lithographically at ambient pressure without the need of a mask. Quantum dots of GaAs (87 nm diameter) and InP (80 nm diameter) have been produced by this method. The direct photochemical etching was accomplished by light from a Nd:YAG laser impinging upon a quantum well heterostructure of InGaAs/InP or InGaAs/GaAs design formed by MBE in a CH$_3$Br:He atmosphere. Etching only occurs when the surface is activated by laser light. If two laser beams interfere at a surface, a holographic grating (interference pattern) is created. The spacing between lines of positive and negative interference is determined by the two beams' relative angle of incidence to the wafer. However, in the case of a single laser beam impinging on a wafer, surface electromagnetic waves (SEWs) of period determined by the dielectric constant of the material are created perpendicular to the direction of polarization. SEWs are created by a nonlinear process caused by interference between the incident laser beam and the resulting Stokes scattering. Appropriate geometrical and optical arrangements result in the creation of an optical grid on the wafer surface. The grid can act as a "mask" for photochemical etching of quantum dots in periodic arrangements equal to the wavelength of the incident light. Frequency doublers and quadruplers were used to bring the light into the 200-nm range, resulting in an array of dots spaced approximately 200 nm apart.

2.3.2.2. *Heterogeneous Deposition on Surfaces*

Direct lithographic techniques have a lower spatial resolution limit of approximately 0.1 μm, but resolutions of 0.01 μm are required to achieve true control of quantum dot design. Hence, indirect methods that rely on surface diffusion of deposited atoms and species are often employed. For example, if a groove is patterned into a surface and adsorbed atoms (adatoms) cluster at the bottom or ridge of the groove, then lithography has been used only indirectly to form quantum structures. Another way to form nanometer-sized features without the use of lithography is to polish a wafer (e.g., GaAs) at a 2° angle from (100). This results in terrace formation with 8-nm periodicity. To achieve room temperature lasing, the energy difference between excited states must be greater than kT, or about 25 meV as a lower limit [4]. Large interlevel spacings of this type are usually observed for quantum structures confined to sizes on the order of tens of nanometers. Calculations show that the smaller a quantum feature's size, the greater is the effect of size distribution on spectral bandwidth. Thus, tight control on the dimensions of submicrometer features is necessary to attain useful quantum confinement.

Gondermann et al. [50] formed SiGe quantum dots on Si using MBE and a "micro-shadow mask" technique. An oxide layer 1 μm thick was thermally grown on a (100) silicon wafer, upon which a thin layer of Si_3N_4 was deposited by LPCVD. A polymer was spin-coated onto the Si_3N_4 and Ti was sputtered over this. A PMMA-type photoresist was spin-coated onto the wafer. EBL was used to make 0.2-μm islands in the PMMA, and the rest of the resist was removed. An anisotropic BCl_3 RIE step followed by an isotropic O_2 RIE step converted the exposed Ti into TiO_2. The wafer was then etched with a CHF_3–O_2 RIE step to remove the remaining resist and make holes in the Si_3N_4. This is a negative etch where TiO_2 is inert and Ti and Si_3N_4 are etched. The polymer under the TiO_2 was removed by an O_2 RIE. The SiO_2 layer under the Si_3N_4 was etched by dipping in HF. The purpose of this process was to form a large undercut "cavern" where a small opening in the Si_3N_4 was located quite far away from the surface of the silicon wafer. The small opening in the Si_3N_4 located 1 μm above the wafer was the shadow mask for MBE deposition of SiGe mesas onto the wafer. Quantum dots with lateral dimensions of 200 nm were fabricated. After epitaxy, the entire mask was removed by washing in HF, and the surface was passivated with Si deposition.

2.3.2.3. Self-Organization and Islanding

Strain and interface energy result in spontaneous island formation for mismatched epilayer deposition. This is known as Stranski–Krastanov (S–K) growth. Multiple layers cause increased strain, where uniform coverage would have higher energy than surface energy from isolated islands (islanding) even with wetting. Under certain conditions, monosized islands are predicted. Critical deposition layer thicknesses for islanding are dependent on substrate temperature and deposition rate. Unfortunately, dots formed in this manner tend to have a squat, cylindrical shape. This means that, although confined in three dimensions, the exciton's envelope function is not isotropic: The photoemission from the "diameter" of smaller cylinders becomes coincident with the photoemission from the "height" of larger cylinders with similar aspect ratios. This may cause multiple peaks and notable PL broadening. Nevertheless, this technique shows great promise in the field of self-organizing nanotechnology, and recently there has been considerable interest in self-assembled nanostructures.

A combination of lithography and islanding has been used to create SiGe quantum wires and dots on the same structure [51]. A buffer layer of Si was grown on a Si (100) wafer, upon which a SiO_2 layer was deposited by LPCVD. Lithography was used to etch stripes into the oxide layer. An anisotropic KOH solution etched V-shaped grooves into the exposed silicon surface. LPCVD was used to deposit SiGe monolayers (MLs) epitaxially onto the walls of the silicon. There was a strong tendency for the deposited material to cluster preferentially at the bottom of the grooves to form quantum wires; however, SiGe coating the walls of the groove tended to form S–K islands at sufficient coverages. The entire structure was capped with a silicon epilayer to passivate the quantum structures.

Relying solely on islanding to form quantum dots, Apetz et al. [52] examined the electroluminescence (EL) of SiGe quantum dots deposited by LPCVD on silicon. Islands of

$Si_{0.7}Ge_{0.3}$ were grown on a 1-μm-thick p$^+$ silicon buffer layer deposited on a (100) silicon wafer. The structure was capped with about 300 nm of n$^+$ silicon for passivation to produce p$^+$/n$^-$/n$^+$ diodes. The islands were 10.0 nm tall and about 150 nm in diameter, with a broad distribution in diameters. A nonphonon radiant peak was observed at 940 meV with a full width at half-maximum (FWHM) of 26 meV. It appeared that inward diffusion of Ge resulted in a red shift of the peak from 1000 meV, and the peak breadth was credibly attributed to the size distribution of the dots. Nonradiative recombination was reduced when compared with similar specimens with a quantum well geometry.

Applications for islanding in the fabrication of InAs/GaAs lasers have been addressed by Moison et al. [53]. Quantum dot lasers produced from these compound semiconductors must consist of nanostructures less than 10 nm in diameter with less than 10% size variation, and the density of the dots must be about 10^{11}/cm^2. This means that the dots must be about 30 nm apart. In this study, InAs was deposited by MBE at 500 °C on a 1-μm-thick GaAs buffer layer. A strong correlation between the height and the volume of the quantum dots showed that dots with the same volume had the same shape at all coverage levels up to coalescence. Furthermore, atomic force microscopy (AFM) studies showed that flat surfaces were more conducive to spatial ordering of the dots than rough or vicinal surfaces. This was attributed to the isotropic stress fields generated at the dot/wafer interface as opposed to anisotropic stress fields associated with wafer irregularities. *In situ* PL imaging showed that dots form in less than 1 s using MBE, so size fluctuations cannot be controlled by MBE growth rates. Hence, uniformly sized dots would result when MBE growth rates are very high and their structural evolution is quenched by immediate GaAs capping. Dots in equilibrium at 500 °C had a height of 3.0 nm with a base 12.0 nm wide. Dots would have a separation distance of 60 nm at equilibrium, and dots closer together would tend to be in a metastable state with about 30% less volume.

An alternative to island formation by S–K growth known as faceting has been achieved for InAs quantum dots grown on GaAs wafers [54]. Elemental source MBE was used to deposit a series of superlattice structures composed of GaAs, AlGaAs, and InAs to ensure appropriate lattice mismatch strains between the InAs dot and the GaAs wafer. The InAs deposited on the superlattice structure formed regular pyramids aligned in rows along $\langle 100 \rangle$. Dots were measured to be 12 ± 1.0 nm at the base with 4–6-nm heights. The dots were capped with a similar series of superlattices. Photoluminescence from the dot array was intense because of the high dot density, approximately 10^{11}/cm^2.

Bimberg et al. [55] discussed whether faceting may result from morphological instabilities on bare surfaces, known as Volmer–Weber (V–W) growth, or on wetted surfaces by the S–K growth. In both cases, coherent islands are formed, albeit with different growth behavior predicted. These authors argued that neither theory adequately accounted for the quasiperiodic lateral periodicity. They included terms associated with elastic relaxation of surface stress and volume strain relaxation resulting from facet formation (pyramids) into the equations. To test this idea, alternate layers of 0.5-nm InAs and 1.5-nm GaAs were grown on an MBE-buffered GaAs wafer for three cycles. This resulted in vertically arranged pyramids, which the authors referred to as "split pyramids." They argued that strain fields around the InAs pyramid cause Ga adatoms to migrate away from the pyramids, resulting in local nonstoichiometry, which limits growth. Subsequent GaAs deposition restores equilibrium, and new InAs pyramids grow above the old pyramids to minimize surface stresses. These vertically coupled quantum dots showed narrower PL peaks than isolated dots, and injection lasing was observed at room temperature.

Other methods for nanoscopic self-arrangement have been discussed briefly by Nötzel [56]. When InGaAs was grown on an AlGaAs buffer layer deposited on a high-index [i.e., GaAs (311)B wafer] substrate, the InGaAs arranged into nanocrystal islands that were spontaneously buried below AlGaAs when deposition was interrupted. This was attributed to surface transport from the buffer layer and was only observed to occur with epilayers grown on high-index wafers. Notzel notes that this phenomenon has been observed in the GaInAs/AlInAs and GaInAs/InP structures grown on InP (311)B wafers by MBE. Other methods of self-organization he discusses are droplet epitaxy of GaAs on sulfur-terminated GaAs surfaces and precipitation of silicon in heavily doped GaAs grown by MBE.

Interesting temperature-dependent lasing characteristics have been observed for InGaAs quantum dot devices constructed on a GaAs (311)B substrate [57]. For these devices, lasing degraded below 100 K and was quite efficient at room temperature. The device was fabricated by doubly stacking InGaAs quantum disks in the AlGaAs buffer layer. The disks were 6 nm high with a base of 60–70 nm. Metal–organic vapor phase epitaxy (MOVPE) was used to deposit the semiconductor layers. The waveguide and cladding were composed of AlGaAs, and conventional patterning, dry etching by RIE, and cleaving were used to produce a transverse-mode stabilized ridge waveguide laser. The lasing efficiency decrease with temperature was associated with increasing exciton concentration in the cladding layers as temperature decreased. This would decrease carrier delivery to the quantum dots and increase radiative recombination in the "bulk" AlGaAs material.

The controversy over the appropriate model to explain self-organizing behavior for MOCVD-grown quantum dots has been addressed by Heinrichsdorff et al. [58] using models developed for MBE-grown quantum dots. The variables examined included growth interruption time, elemental delivery flux ratios, and growth temperature. Specimens were prepared by MOCVD using trimethylgallium, trimethylindium, and AsH_3 as source materials. Growth was performed on (001) GaAs substrates after deposition of a 200-nm GaAs buffer grown at 640 °C. The existence of a wetting layer was confirmed, indicating that islanding was most likely controlled by S–K growth. Equilibrium took place after several minutes, as opposed to the seconds associated with MBE. Quantum dot densities were not sensitive to AsH_3 concentration, but dot density was reduced with increased indium concentration, which is in agreement with Moison et al. [53] and MBE observations. The growth of high-quality InAs quantum dots was hindered by long-range adatom surface diffusion. Dots of InGaAs had square bases aligned along ⟨100⟩ and arranged in that direction, but at high dot densities the dots aligned in chains along ⟨110⟩. This was not observed for MBE-derived specimens. It was concluded that dot shapes are governed by energy considerations, whereas arrangement is controlled by delivery and diffusion kinetics.

Lee et al. [59] confirmed that there is a temperature dependence in the critical layer thickness for islanding at the boundary between migration-enhanced and continuous growth modes for InAs/GaAs quantum dots formed by MBE. The controlling mechanism is indium adatom desorption. When temperatures are high and desorption is significant, the islanding transition is reversible. Furthermore, the size distribution becomes bimodal when annealing under As at high temperatures. This latter observation is probably due to In desorption occurring at lower rates from quantum dots located on steps than dots located on terraces. This means that the critical wetting layer thickness would be higher on the steps, and larger islands accommodated. The size and density of quantum dots of InAs on GaAs is diffusion controlled at low temperatures, but indium desorption dominates in determining size distributions at high temperatures.

Malik et al. [60] prepared islands of InAs on GaAs by solid-source MBE for two cycles and capped them with GaAs in order to study the effect of annealing on quantum dot size distribution. Rather than annealing under As_4 or arsine, the anneal was carried out after a SiO_2 cap had been deposited on the structure. This ensured no out-diffusion of the III–V semiconductor. Annealing appeared to change the quantum dot composition, which lowered the confining potential and decreased the dot size distribution. This significantly narrowed PL spectral features.

An alternative method for producing InGaAs/GaAs quantum dots that does not involve islanding is atomic layer epitaxy (ALE). This technique differs from MOCVD or MBE in that source gases are supplied in a fast, pulsed stream. Delivery cycles are separated by a hydrogen purge. Dots formed by this technique on an InGaAs barrier layer [61] were measured to be 20 nm in diameter with a Gaussian distribution ($\sigma = 2.9$ nm). Dots were 10 nm high, separated by about 50 nm and connected with a thicker InGaAs quantum well structure than the one or two MLs associated with the other deposition techniques. A gradual change in stoichiometry around the dot is characteristic of ALE, as opposed to an abrupt change associated with islanding. Mukai et al. [61] report superior lasing characteristics for these devices.

The most comprehensive study in the literature to date attempting to explain self-organizing models with thermodynamics has come from Zhu et al. [62]. II–VI quantum dots of CdSe in ZnSe, ZnSe quantum dots in ZnS, and ZnO quantum pyramids on a ZnO barrier layer were fabricated by MBE. The role of interfacial energy γ_{12} on island formation in the S–K mode versus self-organized island formation dependent on epilayer surface energy σ_2 and substrate surface energy σ_1 is examined. Self-organized islands form if $\sigma_2 + \gamma_{12} > \sigma_1$. One result of this thermodynamic approach to understanding islanding is that the growth of faceted islands leads to self-limiting growth and uniform QD sizes. Alternatively, using substrates with a high index lowers the substrate surface energy, thus making islanding energetically favorable. A more spherical quantum dot is predicted because there is low lattice strain and organization is accomplished by high interfacial energy. This appeared to be the case when CdSe was deposited by ALE on a ZnSe barrier layer. Changes in Cd and Se concentration affected the dot density, but not the dot size or shape. In this study, the ZnSe buffer layer was grown on an atomically smooth GaAs (111)A substrate at 300 °C. In another study, ZnSe quantum dots were grown by ALE on a barrier layer of ZnS over a GaP (100) substrate, and Zn followed by Se was deposited on the ZnS at 300 °C. These specimens exhibited normal S–K growth modes. Further evidence supporting this thermodynamic approach to understanding self-organizing behavior came from a study of ZnO pyramids grown on a ZnO buffer layer formed by oxygen plasma-enhanced MBE. AFM showed triangular pyramids of uniform size measuring 140 ± 20 nm at the base. The epilayers were grown on a (0001) sapphire wafer: The high lattice strain encouraged faceted growth.

Lee and Liu [63] used the phenomenon of island formation to create a mask for spatially selective thermal evaporation of a quantum well material to make quantum dots. MBE was used to form a buffer layer of GaAs on a GaAs wafer, upon which an InGaAs quantum well structure was designed. The slow growth and thinness of the heterostructure minimized the tendency to form islands. Above the quantum well structure, MBE was used to deposit a thin layer of refractory AlAs. Here, islands of uniform size were formed at the InGaAs interface. Subsequent heating of the wafer caused semiconductor not covered by an AlAs island to evaporate. After thermal processing, the structures were capped and passivated with a second growth of GaAs, resulting in InGaAs quantum dots embedded in a GaAs matrix. Dot sizes were estimated by PL spectra to be 10 ± 1 nm in diameter. A blue shift in the PL features was observed with decreasing dot size.

2.3.2.4. Vertical Self-Alignment

A number of researchers have complemented and elaborated upon the work done by the Bimberg group [55] on vertical alignment of quantum dots. Ishikawa et al. [64] created horizontally aligned silicon quantum dot structures in SiO_2 deposited on a silicon wafer. The specimens were formed by low-energy oxygen ion implantation during MBE growth of Si on silicon substrates. Alternate MBE depositions of Si and oxygen ion implantation were carried out for five cycles. The specimens were then annealed at 1280 °C for 2 h in N_2 to form quantum dots of Si suspended in well-ordered layers inside a silica film. Although regularly faceted nanocrystallites were formed, sizes ranged from about 16 to 25 nm in diameter.

Electric coupling of vertically aligned GaAs/InAs quantum dots has been demonstrated by Solomon et al. [65] in the fabrication of light-emitting diodes. Up to 10 periods of InAs island dots over a sub-10-nm spacer region were grown by MBE at 457 °C on a bottom cladding layer of n-type GaAs 0.75 μm thick. A top cladding layer of 0.29-μm-thick p-type GaAs was then deposited over the structures. The diode intrinsic region was 0.11 μm thick. The InAs layers were the equivalent of 3 MLs of epitaxy. The spacing between the aligned dots was approximately 1.5 nm, and the dots were 4 nm high $\times$ 18 nm in diameter. Spectral bands were reported to be broad.

InAs quantum dot islands with the equivalent of 1.8 MLs of epitaxy and a 15-nm GaAs spacer were grown in a stacked manner, and were demonstrated to maintain vertical alignment up to about nine layers [66]. Beyond this layer, significant In segregation caused by increasingly high strain fields suppressed further alignment. This study also showed that island size decreases with increasing island density above 510 °C, but below this temperature island size and density both increase linearly with coverage. This phenomenon is attributed to In reevaporation above about 510 °C, and is in agreement with the work by other groups [53, 58, 59].

Arakawa et al. [67] succeeded in using the self-alignment of InGaAs quantum dots to fabricate a vertical microcavity InGaAs/GaAs quantum dot laser. S–K growth accounted for the dot sizes produced by MOCVD. The substrate was a GaAs (001) wafer with 2° misalignment. Trimethylgallium, trimethylindium, and arsine were the source gases. InGaAs layers were deposited at 500 °C. These InGaAs quantum dots grew almost exclusively on terraces with a high degree of two-dimensional misorientation. Quantum dots of about 20 nm resulted, where the quantum dots grown on terraces acted as the template for multilayer deposition of aligned quantum dots.

3. SUMMARY

This brief overview has summarized major progress in selected areas of quantum dot preparation and processing. Many new processes for making quantum dots through self-assembly, organization in polymer sphere templates, and aerosol-based techniques continue to receive attention. It is clear that exciting advances in the synthesis and optical/electronic properties of semiconducting quantum dots will remain a fascinating multidisciplinary challenge for materials scientists, physicists, and chemists for many years.

References

1. A. I. Ekimov and A. A. Onushchenko, *Fiz. Tekh. Poluprovodn.* 16, 1215 (1982) [*Sov. Phys. Semicond.* 16, 775 (1982)].
2. T. Takada, T. Yano, A. Yasumori, and M. Yamane, *J. Ceram. Soc. Jpn.* 101, 73 (1993).
3. Y. Masumoto, S. Katayanagi, and T. Mishina, *Phys. Rev. B* 49, 10782 (1994).
4. A. C. Gossard and S. Fafard, *Solid State Commun.* 92, 63 (1994).
5. M. Danek, K. F. Jensen, C. B. Murray, and M. G. Bawendi, *Appl. Phys. Lett.* 65, 2795 (1994).
6. C. M. Sotomayor Torres, A. P. Smart, M. Watt, M. A. Foad, K. Tsutsui, and C. D. W. Wilkinson, *J. Electron. Mater.* 23, 289 (1994).
7. J. Warnock and D. D. Awschalom, *Phys. Rev. B* 32, 5529 (1985).
8. C. H. B. Cruz, C. L. Cesar, L. C. Barbosa, A. M. de Paula, and S. Tsuda, *Appl. Surf. Sci.* 109–110, 30 (1997).
9. S.-T. Park, G.-S. Jeen, H.-Y. Park, and H.-K. Kim, *Ungyong Mulli* 9, 490 (1996).
10. P. G. Shewmon, "Diffusion in Solids," 2nd ed., p. 29. Minerals, Metals and Materials Society, Warrendale, PA, 1989.
11. L.-C. J. Liu, Ph.D. Thesis, University of California at Davis, 1993.
12. Y. Liu, V. C. S. Reynoso, L. C. Barbosa, C. H. Brito Cruz, C. L. Cesar, H. L. Fragnito, and O. L. Alves, *J. Mater. Sci. Lett.* 15, 142 (1996).
13. L.-C. Liu, M. J. Kim, S. H. Risbud, and R. W. Carpenter, *Philos. Mag. B* 63, 769 (1991).
14. V. C. S. Reynoso, Y. Liu, R. F. C. Rojas, N. Aranha, C. L. Cesar, L. C. Barbosa, and O. L. Alves, *J. Mater. Sci. Lett.* 15, 1037 (1996).

15. V. J. Leppert, S. H. Risbud, and M. J. Fendorf, *Philos. Mag. Lett.* 75, 29 (1997).
16. S. H. Risbud, L.-C. Liu, and J. F. Shackelford, *Appl. Phys. Lett.* 63, 1648 (1993).
17. M.-J. Ko, J. Plawsky, and M. Birnboim, *J. Non-Cryst. Solids* 203, 211 (1996).
18. Y. Liu, V. C. S. Reynoso, R. F. C. Royas, C. H. B. Cruz, C. L. Cesar, H. L. Fragnito, and L. C. Barbosa, *J. Mater. Sci. Lett.* 15, 980 (1996).
19. C. W. White, J. D. Budai, J. G. Zhu, and S. P. Withrow, in "Ion–Solid Interactions for Materials Modification and Processing Symposium" (D. B. Poker, D. Ila, Y.-T. Cheng, L. R. Harriott, et al., eds.), pp. 377–384. Materials Research Society, Pittsburgh, PA, 1996.
20. M. Nikl, K. Nitsch, K. Polak, G. P. Paazi, P. Fabeni, D. S. Citrin, and M. Gurioli, *Phys. Rev. B* 51, 5192 (1995).
21. T. Takada, J. D. Mackenzie, M. Yamane, K. Kang, N. Peyghambarian, R. J. Reeves, E. T. Knobbe, and R. C. Powell, *J. Mater. Sci.* 31, 423 (1996).
22. S. Gorer, G. Hodes, Y. Sorek, and R. Reisfeld, *Mater. Lett.* 31, 209 (1997).
23. J. D. Mackenzie and Y.-H. Kao, *Proc. SPIE* 2145, 90 (1994).
24. A. Chemseddine, *Chem. Phys. Lett.* 216, 265 (1993).
25. J. Lee and T. Tsakalakos, *Nanostruct. Mater.* 8, 381 (1997).
26. F. Gindele, R. Westphaling, U. Woggon, L. Spanhel, and V. Ptatschek, *Appl. Phys. Lett.* 71, 2181 (1997).
27. A. P. Alivisatos, K. P. Johnsson, X. Peng, T. E. Wilson, C. J. Loweth, M. P. Bruchez, Jr., and P. G. Schultz, *Nature* 382, 609 (1996).
28. C. A. Mirkin, R. L. Letsinger, R. C. Mucic, and J. J. Storhoff, *Nature* 382, 607 (1996).
29. S. Destri, W. Porzio, and Y. Dubitsky, *Synth. Met.* 75, 25 (1995).
30. S. V. Gaponenko, I. N. Germanenko, A. M. Kapitonov, and M. V. Artemyev, *J. Appl. Phys.* 79, 7139 (1996).
31. H. S. Zhou, I. Honma, H. Komiyama, and J. W. Haus, *J. Physics Chem.* 97, 895 (1993).
32. J. R. Heath, in "Proceedings of the Third International Symposium on Quantum Confinement: Physics and Applications" (M. Cahay, S. Bandyopadhyay, J. P. Leburton, and M. Razeghi, eds.), p. 244. Electrochemical Society, Pennington, NJ, 1996.
33. R. Mu, Y. Xue, Y. S. Tung, and D. O. Henderson, in "Proceedings of the Third International Symposium on Quantum Confinement: Physics and Applications" (M. Cahay, S. Bandyopadhyay, J. P. Leburton, and M. Razeghi, eds.), pp. 186–193. Electrochemical Society, Pennington, NJ, 1996.
34. S. G. Romanov, A. V. Fokin, D. K. Maude, and C. Portal, *Appl. Phys. Lett.* 69, 2897 (1996).
35. V. V. Poborchii, V. I. Al'perovich, Y. Nozue, N. Ohnishi, A. Kasuya, and O. Terasaki, *J. Phys.: Condens. Matter* 9, 5687 (1997).
36. J. Y. Wang, R. A. Uphaus, S. Ameenuddin, and D. A. Rintoul, *Thin Solid Films* 242, 127 (1994).
37. J. P. Yang, S. B. Qadri, and B. R. Ratna, *J. Phys. Chem.* 100, 17255 (1996).
38. K. Deppert, J.-O. Bovin, J.-O. Malm, and L. Samuelson, *J. Cryst. Growth* 169, 13 (1996).
39. T. J. Goodwin, V. J. Leppert, S. H. Risbud, I. M. Kennedy, and H. W. H. Lee, *Appl. Phys. Lett.* 70, 3122 (1997).
40. K. Tsunetomo, S. Ohtsuka, T. Koyama, and S. Tanaka, *Opt. Mater.* 6, 233 (1996).
41. K. Kohno, Y. Osaka, F. Toyomura, and H. Katayama, *Jpn. J. Appl. Phys., Part 1* 33, 6616 (1994).
42. Q. Zhang, S. C. Bayliss, and R. G. Pritchard, in "Proceedings of the Third International Symposium on Quantum Confinement: Physics and Applications" (M. Cahay, S. Bandyopadhyay, J. P. Leburton, and M. Razeghi, eds.), pp. 178–185. Electrochemical Society, Pennington, NJ, 1996.
43. T. Ifuku, M. Otobe, A. Itoh, and S. Oda, *Jpn. J. Appl. Phys., Part 1* 36, 4031 (1997).
44. W. E. Buhro, *Adv. Mater. Opt. Electron.* 6, 175 (1996).
45. T. D. Bestwick, M. D. Dawson, A. H. Kean, and G. Duggan, *Appl. Phys. Lett.* 66, 1382 (1995).
46. C. Gourgon, L. S. Dang, H. Mariette, C. Vieu, and F. Muller, *Appl. Phys. Lett.* 66, 1635 (1995).
47. M. Illing, G. Bacher, T. Kummell, A. Forchel, D. Hommel, B. Jobst, and G. Landwehr, *J. Vac. Sci. Technol., B* 13, 2792 (1995).
48. A. Nakajima, H. Aoyama, and K. Kawamura, *Jpn. J. Appl. Phys., Part 2* 33, L1796 (1994).
49. M. Ezaki, H. Kumagai, K. Toyoda, and M. Obara, *IEEE J. Select. Top. Quantum Electron.* 1, 841 (1995).
50. J. Gondermann, B. Spangenberg, T. Koster, B. Hadam, H. G. Roskos, H. Kurz, J. Brunner, P. Schittenhelm, G. Abstreiter, H. Gossner, and I. Eisele, *Mater. Sci. Technol.* 11, 407 (1995).
51. A. Hartmann, L. Vescan, C. Dieker, and H. Luth, *Mater. Sci. Technol.* 11, 410 (1995).
52. R. Apetz, L. Vescan, A. Hartmann, R. Loo, C. Dieker, R. Carius, and H. Luth, *Mater. Sci. Technol.* 11, 425 (1995).
53. J. M. Moison, L. Leprince, F. Barthe, F. Houzay, N. Lebouche, J. M. Gerard, and J. Y. Marzin, *Appl. Surf. Sci.* 92, 526 (1996).
54. D. Bimberg, M. Grundmann, N. N. Ledentsov, S. S. Ruvimov, P. Werner, U. Richter, J. Heydenreich, V. M. Ustinov, P. S. Kop'ev, and Zh. I. Alferov, *Thin Solid Films* 267, 32 (1995).
55. D. Bimberg, N. N. Ledentsov, M. Grundmann, N. Kirstaedter, O. G. Schmidt, M. H. Mao, V. M. Ustinov, A. Yu. Egorov, A. E. Zhukov, P. S. Kop'ev, Zh. I. Alferov, S. S. Ruvimov, U. Gosele, and J. Heydenreich, *Phys. Status Solidi B* 194, 159 (1996).
56. R. Nötzel, *Semicond. Sci. Technol.* 11, 1365 (1996).

57. E. Kuramochi, M. Sugo, M. Notomi, H. Kamada, T. Nishiya, J. Temmyo, R. Nötzel, and T. Tamamura, in "Proceedings of the Third International Symposium on Quantum Confinement: Physics and Applications" (M. Cahay, S. Bandyopadhyay, J. P. Leburton, and M. Razeghi, eds.), pp. 37–55. Electrochemical Society, Pennington, NJ, 1996.

58. F. Heinrichsdorff, A. Krost, M. Grundmann, D. Bimberg, F. Bertram, J. Christen, A. Kosogov, and P. Werner, *J. Cryst. Growth* 170, 568 (1997).

59. H. Lee, R. R. Lowe-Webb, W. Yang, and P. C. Sercel, *Appl. Phys. Lett.* 71, 2325 (1997).

60. S. Malik, C. Roberts, R. Murray, and M. Pate, *Appl. Phys. Lett.* 71, 1987 (1997).

61. K. Mukai, N. Ohtsuka, H. Shoji, and M. Sugawara, *Appl. Surf. Sci.* 112, 102 (1997).

62. Z. Zhu, E. Kurtz, K. Arai, Y. F. Chen, D. M. Bagnall, P. Tomashini, F. Lu, T. Sekiguchi, T. Yao, T. Yasuda, and Y. Segawa, *Phys. Status Solidi B* 202, 827 (1997).

63. C.-P. Lee and D.-C. Liu, *Appl. Surf. Sci.* 92, 519 (1996).

64. Y. Ishikawa, N. Shibata, and S. Fukatsu, *Jpn. J. Appl. Phys., Part 1* 36, 4035 (1997).

65. G. S. Solomon, M. C. Larson, and J. S. Harris, Jr., *Appl. Phys. Lett.* 69, 1897 (1996).

66. Y. Sugiyama, Y. Nakata, K. Imamura, S. Muto, and N. Yokoyama, *Jpn. J. Appl. Phys., Part 1* 35, 1320 (1996).

67. Y. Arakawa, M. Nishioka, H. Nakayama, and M. Kitamura, *IEEE Trans. Electron.* E79-C, 1487 (1996).

Chapter 11

RAPID SOLIDIFICATION PROCESSING OF NANOCRYSTALLINE METALLIC ALLOYS

I. T. H. Chang

School of Metallurgy and Materials, University of Birmingham, Edgbaston, Birmingham, United Kingdom

Contents

1. INTRODUCTION

Research on the processing and characterization of nanocrystalline materials has expanded enormously over the last decade. According to the report [1]: "Our increasing ability to artificially assemble and engineer condensed matter at length scales below 100 nm, where properties are often changed dramatically, is heralding a revolution in the synthesis and processing of high performance materials for a wide variety of applications." Nanocrystalline materials represent a new kind of material that is made up of crystallites with sizes less than 100 nm, thereby giving a high fraction of atoms at the surface of the crystallite. These materials exhibit unusual properties that are different from both fully amorphous (i.e., noncrystalline) and ordinary coarse-grained materials. The nanocrystalline structure contributes to novel chemical, physical, magnetic, and electronic behaviors. They form a new class of advanced materials with increased strength, reduced core loss and local magnetic anisotropy fluctuation, rapid atomic diffusion, and high activities in the grain boundaries [1–4].

Handbook of Nanostructured Materials and Nanotechnology, edited by H.S. Nalwa
Volume 1: Synthesis and Processing
Copyright © 2000 by Academic Press
All rights of reproduction in any form reserved.

ISBN 0-12-513761-3/$30.00

This chapter gives a review on the processing, properties, and mechanisms of the formation of nanocrystalline metallic alloys produced by rapid solidification processing (RSP) combined with thermal treatments.

1.1. What Are Nanocrystalline Materials?

Nanocrystalline materials generally consist of crystallites that are less than 100 nm. This covers a wide range of microstructures from single to multiphase systems. There are various forms of nanocrystalline structures: (1) a granular microstructure composed of nano-sized crystalline and amorphous phases, (2) an equiaxed microstructure composed of a mixture of different crystalline phases or just a single phase with an average grain size in the nanometer scale, and (3) a multilayer microstructure composed of ultrathin layers of different crystalline/amorphous phases. Rapid solidification processing of metals and alloys commonly produces a nanoscale mixed structure of phases (also known as nanocomposites) and rarely gives a nanoscale homogeneous single-phase structure except after polymorphous crystallization of the amorphous precursor. However, multilayer microstructures are readily prepared by vapor deposition and they will be discussed in other chapters of this book.

Nanocrystalline materials with a considerable volume fraction of interfaces provide an opportunity to study the nature of interfaces in solids. In common with most new categories of materials, nanocrystalline materials are in states far from equilibrium. This makes them of especial thermodynamic interest [5] for understanding the contributions to the excess free energy in these materials.

2. RAPID SOLIDIFICATION PROCESSING

Nanocrystalline metallic alloys have been produced by rapid solidification processing (RSP). In some cases, formation of the nanocrystalline microstructure requires a subsequent controlled heat treatment after rapid solidification processing.

Rapid solidification involves the manipulation of a stream of molten metal into a thin layer or fine droplets that are in good thermal contact with a cooling medium (e.g., solid substrate or fluid jet). This allows the thermal energy (e.g., latent heat) from the molten material to be conducted rapidly to the cooling medium.

2.1. Heat Flow Analysis of Rapid Solidification Processing

Jones [6] and Cantor [7] have reviewed the heat flow during rapid solidification processing. The heat flow analysis was based on a thin alloy layer of thickness X at an initial temperature T_i brought into contact with a cooling medium at a temperature T_s, which could be described by the one-dimensional thermal diffusion equation:

$$\frac{\partial T}{\partial t} = \frac{k}{\rho C} \frac{\partial^2 T}{\partial x^2} \tag{1}$$

where k, ρ, and C are the thermal conductivity, density, and specific heat capacity, respectively. Using Newton's law of cooling, the heat transferred to the medium is equal to the total heat associated with cooling and solidification of the alloy layer. This gives

$$h(T_x - T_s) = -\rho C X \frac{dT}{dt} + L X \frac{df}{dt} \tag{2}$$

where h is the heat transfer coefficient, T_x is the temperature at the interface between the layer and the cooling medium, L is the latent heat of fusion, and f is the fraction solidified.

The preceding heat balance equation can be simplified further by using an effective specific heat C' to give

$$h(T_x - T_s) = -\rho C' X \frac{dT}{dt} \tag{3}$$

where $C' = C - (L/\Delta T_o)$ and ΔT_o is the effective freezing range. For Newtonian cooling conditions with negligible thermal gradient in the alloy layer (i.e., Biot number, Bi $\ll$ 1), the solution to this differential equation becomes

$$\frac{T - T_s}{T_i - T_s} = \exp\left(-\text{Bi} \frac{k}{C'\rho} \frac{t}{X^2}\right) = \exp\left(-\frac{ht}{\rho C' X}\right) \tag{4}$$

However, for ideal (non-Newtonian) cooling conditions with perfect alloy–substrate thermal contact (i.e., Biot number, Bi $\gg$ 1), the analytical solution becomes

$$\frac{T - T_s}{T_i - T_s} = \sum_1^n \frac{2\cos\left(\frac{B_n x}{X}\right)\exp\left(-\frac{B_n^2 k}{C'\rho X^2}\right)}{B_n^2} \tag{5}$$

where $B_n = (n + 0.5)\pi$. Figure 1 shows the calculated average cooling rates ε as a function of thickness X and heat transfer h for the solidification of Al in contact with a Cu chill. The cooling rate increases with increasing heat transfer coefficient and decreasing thickness. The measured cooling rates of 10^4–10^6 K/s during rapid solidification are associated with heat transfer coefficients of 10^4–10^5 W/m^2 K, which correspond to cooling conditions intermediate between Newtonian and ideal cases. Typically, in rapid solidification processing, the cooling rate can reach in excess of 10^4 K/s for a layer thickness less than 100 μm. This implies that the solidification is completed within a few milliseconds.

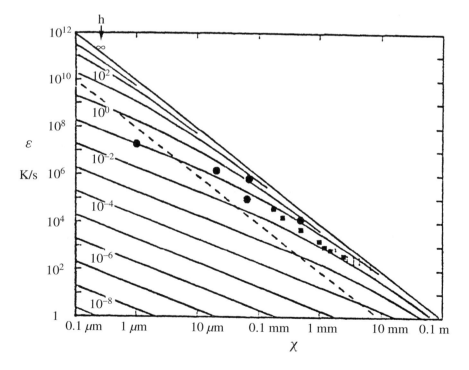

Fig. 1.   Calculated average cooling rate ε as a function of thickness X and heat transfer coefficient h (W/mm^2 K) for Al in contact with a Cu chill, together with experimental data points. (Source: Reprinted with permission from [6]. © 1982 Institute of Materials.)

The average solidification rate R can be approximated by equating the latent heat released to the heat removed to the cooling medium [6]. This gives the following expression:

$$R = \frac{dX}{dt} = B \frac{A_o}{A_f} h \tag{6}$$

but

$$B = \frac{T_f - T_s}{L\rho}$$

$$\frac{A_o}{A_f} = 1 \qquad \text{for a slab-like geometry}$$

$$\frac{A_o}{A_f} = \frac{Z}{Z - X} \qquad \text{for a cylinder}$$

$$\frac{A_o}{A_f} = \frac{Z^2}{(Z - X)^2} \qquad \text{for a sphere}$$

where T_f is the liquidus (freezing) temperature and Z is the radius of a cylinder or sphere.

2.2. Rapidly Solidified Microstructures

The rapid extraction of thermal energy associated with RSP permits a large deviation from equilibrium, as evidenced by the extension in solid solubility limits; the reduction or elimination of the detrimental effects of segregation; the development of new nonequilibrium crystalline, quasicrystalline, or noncrystalline (amorphous) phases; and the sharp reduction of grain size to the micrometer or nanometer scale.

Jones [6], Liebermann [8], Cahn [9], and Suryanarayana [10] have reviewed the rapid solidification processing technology. A range of rapid solidification processes have been developed to produce the metastable microstructure in materials. There are three main types of rapid solidification processes: (1) chilled, (2) spray, and (3) weld methods. The most common RSP methods used in the manufacture of nanocrystalline metallic alloys and composites are chill–block melt spinning and gas atomization.

2.3. Chill–Block Melt Spinning

In chill–block melt spinning [11], the molten material is forced through a nozzle to form a liquid stream, which is then spread continuously across the surface of a rotating wheel (or drum) under an inert atmosphere, to manufacture strip and ribbon products, as shown in Figure 2. The thickness of the ribbon varies from 10 to 100 μm. The heat flow from the liquid stream to the cold substrate under certain conditions can be treated as a Newtonian cooling mechanism [12], which gives an expression for the thickness of the ribbon, t, as

$$t = \left(\frac{\alpha_T}{V_R} \cdot \frac{\Delta T}{\Delta H} \right) \cdot l \tag{7}$$

where α_T is the empirical heat transfer coefficient, V_R is the linear velocity of the substrate (i.e., the speed of the rotating wheel), ΔT is the temperature difference between the opposite sides of the ribbon and ΔH is the latent heat per unit volume of the liquid metal, and l is the length of the contact zone between the liquid metal and substrate. The thickness of the ribbon is proportional to the cooling conditions and they depend on the substrate material (α_T). All process parameters for a given chill–block melt-spinning apparatus are adjusted to preserve the stability of the liquid metal. These parameters are nozzle size, nozzle-to-substrate distance, melt ejection pressure, and substrate speed, all of which, in concert, control the puddle length of the molten metal. This length limits the time available for the solidification of the ribbon and, therefore, governs the ribbon thickness. The width of the ribbon varies from 1 to 3 mm and is governed by the size of the nozzle.

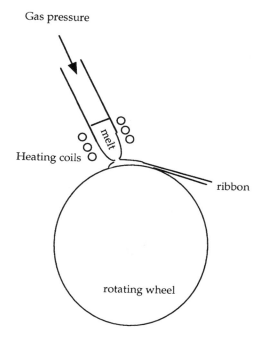

Fig. 2.   Chill–block melt-spinning apparatus.

Chill–block melt spinning is a simple technique and it can produce fully dense samples directly. However, it is predominantly used for metals because these materials can easily be melted by induction heating. There have been some cases where melt spinning has been employed in the rapid solidification of semiconductors [13].

2.4. Gas Atomization

In gas atomization, a fine dispersion of droplets is formed when molten metal is impacted by a high-energy fluid (i.e., inert gas), as shown in Figure 3. Atomization occurs as a result of the transfer of kinetic energy from the atomizing fluid to the molten metal. In general, the droplets are formed from Rayleigh instabilities and they grow on the surface of torn molten ligaments [14]. Further breakdown of the droplet may occur as a result of interactions with the atomizing gas if the dynamic pressure resulting from the gas stream velocity exceeds the restoring force resulting from the surface tension of the droplet [15]. This is followed immediately by the spheroidization of the individual droplets [16], as shown in Figure 4. The spherical or near-spherical droplets continue to travel down the atomization vessel, rapidly losing heat as a result of convection to the atomizing fluid. The disintegration of a molten metal by high-energy gas jets has been reported to obey a simple correlation to give the mass mean droplet diameter [17] (i.e., the opening of a screening mesh that lets through 50% of the mass of the powder resulting from an atomization) d_{50} as

$$d_{50} = K_d \left[\left(\mu_m d_0 \sigma_m / \mu_g V_{ge}^2 \rho_m \right) \left(1 + J_{melt} / J_{gas} \right) \right]^{1/2} \tag{8}$$

where K_d is an empirically determined constant with a value between 40 and 400; μ_m, σ_m, ρ_m, and J_{melt} are the viscosity, surface tension, density, and mass flow rate of the melt, respectively; m_g, V_{ge}, and J_{gas} are the viscosity, velocity, and mass flow rate of the atomizing gas, respectively; and d_0 is the diameter of the metal delivery nozzle. During flight, heterogeneous solidification occurs in all but the smallest droplets because of the following: (a) bulk heterogeneous nucleation within the droplet, (b) surface oxidation processes, or (c) interparticle collisions. A high cooling rate is readily achieved during atomization

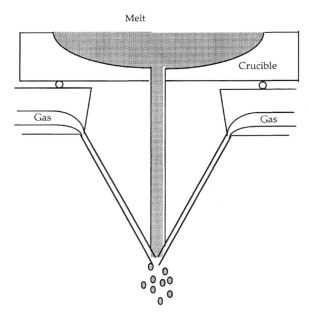

Fig. 3.   V- or cone-jet gas atomization apparatus.

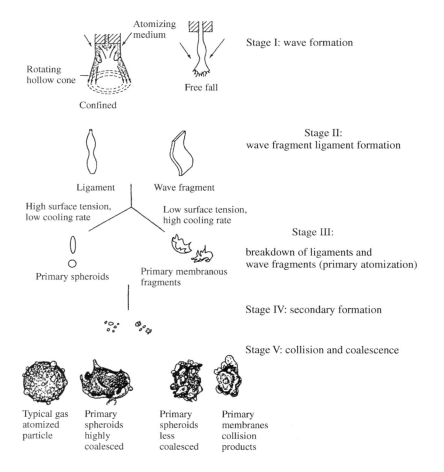

Fig. 4.   Schematic diagram showing particle formation during atomization. (Source: Reprinted with permission from [14]. © 1983 ASM International.)

where fast interfacial growth velocities (0.5–3.5 m/s) and, therefore, microstructural refinement are maintained as a result of (a) particles with large surface/volume ratios and (b) an efficient convective heat flux to the surrounding atomizing gas [18, 19].

Gas atomization is an ideal method to produce a large quantity of nanocrystalline alloy powder for subsequent hot consolidation to form bulk samples. However, it is limited to metals because they can easily be molten.

3. DEVELOPMENT OF NANOCRYSTALLINE METALLIC ALLOYS

3.1. Formation of Nanocrystalline Microstructure

Recently, the development of nanocrystalline metallic alloys using rapid solidification processing has focused on four main categories of nanoscale microstructures consisting of (1) crystalline plus amorphous phases, (2) quasicrystalline plus crystalline phases, (3) multiple crystalline phases, and (4) single crystalline phases. Chill–block melt spinning is a simple technique and it can produce (a) porosity-free samples and (b) samples of different grain size by controlling the processing parameters. Furthermore, because no artificial consolidation process is involved, the interfaces are clean and the product is dense. However, gas atomization has been employed to produce either fully amorphous alloy powders or partially crystallized alloy powders with nanocrystalline phase embedded in an amorphous matrix.

The size of the crystal in the rapidly solidified microstructure is controlled by the nucleation rate I and growth rate G during manufacture [20, 21]. The growth rate G and nucleation rate I both depend on the degree of undercooling ΔT of the melt prior to solidification according to the following expressions:

$$G = a\nu \exp\left(-\frac{Q}{kT}\right)\left(1 - \exp\left(\frac{\Delta G_{\mathrm{v}}}{kT}\right)\right) \approx \frac{D}{a}(1 - \exp(-A\,\Delta T)) \tag{9}$$

$$I = n\nu \exp\left(-\frac{Q}{kT}\right)\exp\left(-\frac{16\pi\sigma^3 f(\theta)}{3\,\Delta G_{\mathrm{v}}^2 kT}\right) \approx \frac{nD}{a^2}\exp\left(-\frac{B}{\Delta T^2}\right) \tag{10}$$

but

$$\Delta G = \frac{L\,\Delta T}{T_{\mathrm{m}}}$$

$$f(\theta) = \frac{1}{4}\left(2 - 3\cos\theta + \cos^3\theta\right)$$

where a is the interatomic spacing, ν is the atomic vibration frequency, Q is the activation energy required for an atom to transfer across the solid–liquid interface, ΔG_{v} is the driving force for solidification, T_{m} is the melting temperature, D is the diffusion coefficient in the liquid, θ is the contact angle for a solid nucleus on the substrate surface, n is the nucleation site density, σ is the solid–liquid surface energy, and A and B are constants. Cantor [20] has derived the following relationship between crystal size d, nucleation rate N, and growth rate G based on columnar solidification through the melt spun ribbons:

$$d = \sqrt[3]{\left(\frac{8G}{\pi N}\right)} \tag{11}$$

This gives the following relationship between the crystal size d and the undercooling ΔT:

$$d^3 = \frac{8a(1 - \exp(-A\,\Delta T))}{\pi n \exp(-B/\Delta T^2)} \tag{12}$$

Figure 5 shows the variation of crystal size with undercooling. Crystals cannot nucleate above the nucleation onset temperature T_{n} and they cannot grow below the glass transition temperature T_{g}. The crystal size reaches a minimum value with increasing undercooling.

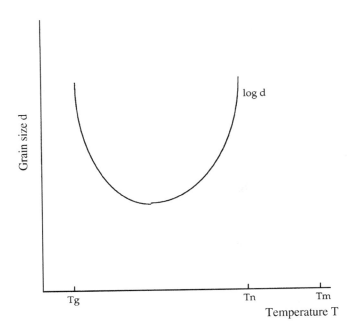

Fig. 5. Plot of grain size versus temperature. (Source: Reprinted with permission from [20]. © 1997 Cambridge University Press.)

Therefore, nanocrystalline grain structures can be obtained by rapid quenching to a high undercooling during solidification of a liquid alloy.

3.2. Nanoscale Mixed Structure of Crystalline and Amorphous Phases

3.2.1. Nanocrystalline Light Metals

The need for high-strength and lightweight materials has led to the development of Al–TM–Ln (TM = Ni, Cu, Ag, Co, Fe, Zr, Ti; Ln = Ce, La, Y, Mm, Nd) alloys [22–29] prepared by melt spinning and gas atomization processes. These materials are based on a composition of about 85–94 at% Al and exhibit a tensile strength ($\sigma_f > 1200$ MPa) greater than conventional high-strength Al alloys. The typical microstructure of a melt-spun Al–Y–Ni–Fe alloy is shown in Figure 6. It consisted of 10–30-nm-sized defect-free α-Al particles embedded in an amorphous matrix [22a]. The volume fraction of the α-Al particles varies from 0.1 to 0.3. These microstructures can be produced either directly from melt spinning at low rotating speed or by subsequent annealing of the fully amorphous structure produced by melt spinning at high rotating speed. A similar nanoscale mixed structure has also been found in gas-atomized Al–Ni–Mm–Zr [30] alloys.

It has been reported [31] that the Al-rich amorphous alloys with low concentration of solute elements has a two-stage crystallization process involving: (1) Am → α-Al and (2) Am′ (remaining amorphous phase) → intermetallic compounds, as evidenced by the two exothermic peaks found in the differential scanning calorimetry (DSC) trace in Figure 7. This type of crystallization is known as primary crystallization in which the amorphous phase decomposes into a crystalline phase with different composition. This provides a two-stage continuous cooling transformation behavior where an α-Al phase field is located at the lower-temperature side, as shown in Figure 8. The control of the cooling rate during RSP or annealing to primary crystallization for this type of Al-based alloy is expected to cause the production of a nanoscale mixed structure of α-Al particles embedded in an amorphous phase. This can be illustrated by the continuous cooling transformation

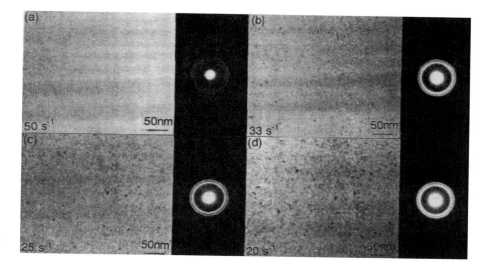

Fig. 6.   Bright-field TEM micrographs and selected-area diffraction patterns showing the change in microstructure of melt-spun $Al_{88}Ni_9Ce_2Fe_1$ ribbons produced at different rotation speeds. (Source: Reprinted with permission from [22b]. © 1992 Japan Institute of Metals.)

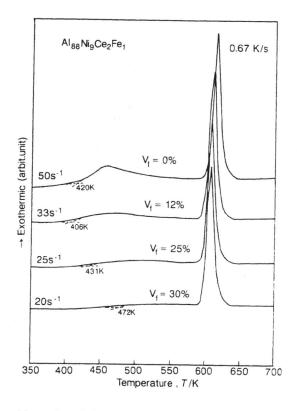

Fig. 7.   Differential scanning calorimetry (DSC) traces of melt-spun $Al_{88}Ni_9Ce_2Fe_1$ ribbons produced at different rotation speeds. (Source: Reprinted with permission from [22b]. © 1992 Japan Institute of Metals.)

(CCT) behavior of amorphous Al alloys, as shown in Figure 8. Because the nose of a CCT curve corresponds to the minimum time for the onset of crystallization from the melt, at high quench rates, the cooling curve misses the noses of both C curves of the

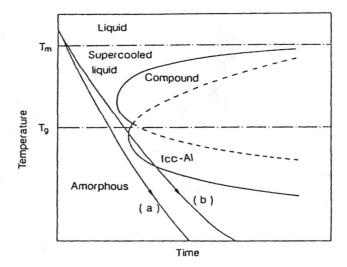

Fig. 8.   Schematic diagram of two-stage continuous cooling transformation (CCT) behavior for Al-rich amorphous alloys and two kinds of cooling curves. (Source: Reprinted with permission from [31]. © 1995 Japan Institute of Metals.)

crystalline α-Al and compound phases. This implies that crystallization is prevented during the quenching from the melt and a fully amorphous structure is developed. On the other hand, at lower quench rates, the cooling curve cuts the C curve of the α-Al phase and leads to the formation of α-Al particles. However, the transformation is incomplete and the remaining liquid is quenched into the amorphous structure. Alternatively, when an amorphous alloy is heated to sufficiently high temperature, thermal motion becomes sufficient for the nucleation and growth of the crystalline phase. To develop the nanocrystalline microstructure, a high nucleation rate together with a slow growth is required. In practice, annealing of the fully amorphous structure is commonly used because the volume fraction and size of the α-Al particles can be controlled more readily.

The primary crystallization of amorphous Al alloys involves transient heterogeneous nucleation, which is influenced by the quench rate. This provides a fine dispersion of quenched-in nucleation sites, giving a population of α-Al particles on the order of 10^{23} m^{-3} for Al$_{90}$Ni$_6$Nd$_4$ alloys and 10^{21}–10^{22} m^{-3} for Al$_{85}$Ni$_5$Y$_{10}$ alloys [32]. The growth behavior of α-Al particles in the primary crystallization is very unusual. For the Al$_{88}$Ni$_4$Y$_8$ alloy [33], the growth shows a sharp transition from a high coarsening rate in the first few minutes of annealing at temperatures between 190 and 220 °C to a much slower coarsening rate at longer annealing times, as shown in Figure 9. A similar growth behavior has also been observed in the primary crystallization of the amorphous Al$_{90}$Ni$_6$Nd$_4$ alloy [32, 34]. The particle size did not agree with the square-root dependence on the annealing time. This implies that the growth kinetics is not a simple diffusion-controlled growth of an isolated particle. There appears to be impingement of diffusion fields around the particle (i.e., soft impingement) during the growth process. The reduction in the growth rate at long annealing times may be due to the presence of a diffusion barrier between the α-Al particles and the matrix.

Field ion microscopy (FIM) has been employed to study the local composition of the nanocrystalline microstructure produced after primary crystallization of the amorphous Al alloys. Hono et al. [31] have shown through FIM measurements that the Ln element diffuses more slowly than the TM element in Al–Ni–Ce alloy. In the partially crystallized Al$_{87}$Ni$_{10}$Ce$_3$ alloy, the Ni and Ce atoms are rejected from the α-Al particles and the concentration of α-Al is approximately 98% Al. The rejected Ni and Ce atoms are partitioned into the amorphous matrix phase and its composition is approximately 25% Ni and 3% Ce,

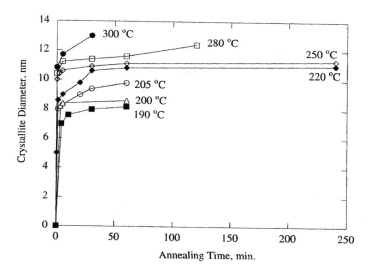

Fig. 9. Plot of α-Al particle diameter versus time at different annealing temperatures in melt-spun $Al_{88}Y_8Ni_4$ ribbons. (Source: Reprinted with permission from [33]. © 1997 Minerals, Metals, & Materials Society.)

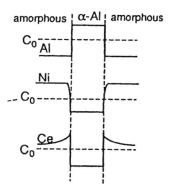

Fig. 10. Schematic diagram of the concentration profiles of Al, Ni, and Ce across the interface between the α-Al and the amorphous phases for an amorphous $Al_{87}Ni_{10}Ce_3$ alloy annealed for 180 s at 553 °C. (Source: Reprinted with permission from [31]. © 1995 Japan Institute of Metals.)

respectively, as shown in Figure 10. The Ce atoms are enriched within a distance of less than 3 nm at the α-Al/amorphous interface. Although both Ni and Ce are rejected from the α-Al phase, only Ce is enriched at the interface because of the lower diffusivity of large Ce atoms. Hence, during the growth of the α-Al particles, the rejected Ce atoms are enriched at the interface and the particle has to drag Ce for further growth. This effectively controls the grain growth. Therefore, the small interparticle spacing together with the gradual pile-up of the slow-diffusing solute species around the α-Al particle quickly arrests the growth of the particles.

The resistance against crystallization of the surrounding residual amorphous matrix is due to a combination of increasing TM contents [23] and the presence of a sharp concentration gradient produced by the pile-up of the Ln atoms [35]. This results in the suppression or reduction of the thermodynamic driving force for the nucleation of intermetallic compounds ahead of the growing particles, thereby stabilizing the residual amorphous matrix. Eventually, a metastable equilibrium state between the primary α-Al phase and the residual amorphous phase is reached. The metastable equilibrium composition of the amorphous

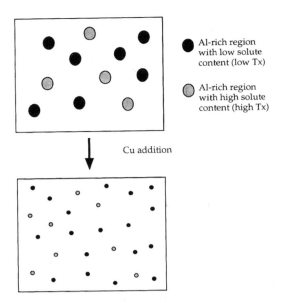

Fig. 11. Schematic diagram showing the refinement of α-Al particles by the addition of Cu. (Source: Reprinted with permission from [28]. © 1994 Japan Institute of Metals.)

matrix can be affected by the curvature of the crystallite/matrix when the crystal size becomes so small that the Gibbs–Thomson effect [36] is significant.

The addition of Cu, Ag, Ga, and Au to the Al alloys increases the α-Al particle density, leading to finer particles [28]. It was proposed that the addition of these soluble elements into Al changes the amorphous structure to one that contains a high number of small Al-rich regions distributed homogeneously in the disordered structure, as shown in Figure 11. These Al-rich regions are the preexisting nuclei of α-Al. However, recent X-ray absorption fine structure (XAFS) measurements have shown that the addition of Cu to the amorphous Al alloy induces the formation of Cu-rich regions and increases the inhomogeneity of the amorphous matrix [29]. Furthermore, the diffusivity of the Al element in the α-Al particles doped with solute atoms becomes more difficult because of the necessity of the solute redistribution. This suppresses further grain growth of the α-Al particles. This proposed mechanism is supported by the result that the addition of insoluble elements such as Fe does not have any effect on the refinement of the α-Al particles. Experimental evidence indicates that the amount of precipitation from the amorphous matrix is strongly influenced by the concentration of solute elements and the size of the particles decreases gradually with decreasing solute concentration.

A similar microstructure of 30–50-nm-sized α-Al particles surrounded by a 10-nm-thick amorphous phase, as shown in Figure 12a, has been achieved directly from the melt spinning of the $Al_{97}Ti_5Fe_2$ alloy. The solidification involves the nucleation of the α-Al phase and the solidification of the remaining liquid to the amorphous phase. The formation of the amorphous phase is believed to be caused by the low diffusivity of the Ti element [25].

A new type of nanoscale mixed structure of nanocrystalline and amorphous phases has been found in melt-spun $Al_{94}V_4Fe_2$. The typical melt-spun microstructure consisted of homogeneously mixed 20-nm-sized granular amorphous and 7-nm-sized α-Al phases, as shown in Figure 12b [25]. It has been proposed that the solidification takes place through the primary formation of an amorphous phase, followed by the nucleation of α-Al from the remaining liquid. Usually, the addition of V to Al alloys tends to promote the formation of a quasicrystalline phase (i.e., icosahedral) during solidification. In this case, the supercooling in rapid solidification processing may suppress the formation of a long-range icosahedral phase and lead to the development of a nanoscale granular amorphous phase.

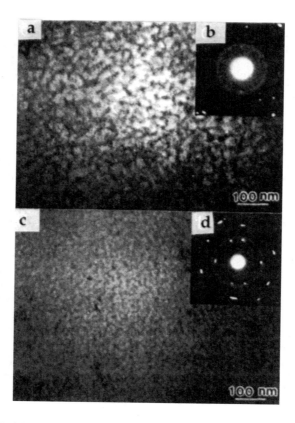

Fig. 12.   Bright-field electron micrographs and selected-area diffraction patterns of nanocrystalline melt-spun $Al_{93}Ti_5Fe_2$ (a, b) showing a mixed microstructure of 30–50-nm-sized α-Al particles embedded in an amorphous matrix and nanocrystalline melt-spun $Al_{94}V_4Fe_2$ (c, d) showing a mixed microstructure of 20-nm-sized granular amorphous and 7-nm-sized α-Al phases. (Source: Reprinted with permission from [25]. © 1997 Minerals, Metals & Materials Society.)

Another nanoscale mixed structure consisting of Mg particles embedded in an amorphous matrix has also been found in the annealing of melt-spun amorphous Mg–Zn–Ln alloys [37, 38]. The Mg particles have a hexagonal closed-pack (hcp) crystal structure. These annealed Mg alloys exhibit a tensile strength of approximately 920 MPa (i.e., three times that of commercial Mg-based alloys [19]).

It has also been reported that primary crystallization of amorphous $Ti_{60}Ni_{30}Si_{10}$ or $Ti_{56}Ni_{28}Si_{16}$ alloys produced 20–30-nm spherical Ti_2Ni or lozenge-shaped Ti_5Si_3 particles embedded in an amorphous matrix, respectively [39]. The Ti_2Ni phase has a cubic crystal structure, whereas Ti_5Si_3 has a hexagonal crystal structure. However, the annealing atmosphere during primary crystallization can affect the formation of the nanosized particles. This is most important to base elements that are prone to oxidation (e.g., Ti-based alloys). During the annealing of a $Ti_{38.5}Cu_{32}Co_{14}Al_{10}Zr_{5.5}$ alloy in vacuum [40], the amorphous alloy appears to be more resistant to crystallization. This implies that the transformation kinetics is reduced in a clean environment. The resultant partially crystallized microstructure consisted of 20-nm-sized particles with a body-centered structure and a lattice parameter very close to that of TiCo and TiNi intermetallic compounds (A2 type).

3.2.2. Nanocrystalline Nickel Alloys

A nanoscale mixed structure of crystalline and amorphous phases has also been found in a partially crystallized amorphous melt-spun $Ni_{58.5}Mo_{31.5}B_{10}$ alloy [41]. The primary crystallization of this amorphous alloy produces a microstructure of a 10–28-nm Ni(Mo) solid-solution phase embedded in an amorphous matrix. The maximum amount of Mo dissolved into Ni was found to be 20%. The remaining Mo and B segregate to the surrounding amorphous matrix, thereby increasing its crystallization temperature. Hence, the thermal stability of the nanocrystalline structure is increased because the intergranular amorphous layers prevent further grain growth.

3.2.3. Mechanical Properties

At present, detailed investigation of the mechanical properties of these nanoscale mixed structures has only been carried out on Al-based alloys. The mechanical properties of nanocrystalline Al alloys are very sensitive to the volume fraction of the α-Al phase and the solute contents, as shown in Figure 13. With increasing volume fraction V_f of the nanoscale α-Al particles, Young's modulus E and the hardness Hv increase, and the elongation decreases almost monotonically, while the tensile strength shows a maximum value of 1200 MPa for a volume fraction of 10% to 30% [22b]. The highest strength obtained is about 1200 MPa for an $Al_{88}Ni_{10}Y_2$ alloy. Similar results were obtained in the quaternary alloys with the highest strengths varying from 1460 to 1560 MPa, as shown in Table I.

The exceptionally high tensile strength is attributed to the presence of these nanoscale particles [27]. The nanoscale α-Al particles are too small to contain internal defects. The interface between the α-Al and amorphous phases has the following characteristics: (1) no faceted phases with stress concentration regions, (2) a highly dense atomic configuration, and (3) a relatively low interfacial energy between the amorphous (liquid-like) and α-Al phases. This interface structure enables a good transfer of applied load between the amorphous and α-Al phases, thereby suppressing the failure at the interface. Consequently, an

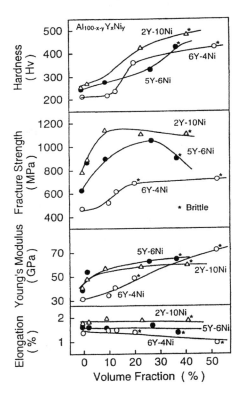

Fig. 13. Changes in σ_f, E, Hv, and ε_f as a function of the volume fraction of the α-Al phase for rapidly solidified $Al_{100-x-y}Y_xNi_y$ alloys. (Source: Reprinted with permission from [27]. © 1992 Japan Institute of Metals.)

Table I. Typical Alloy Systems and Highest Tensile Strengths of Nanocrystalline Al-Based Alloys Obtained by Rapid Solidification

Alloy	Structure	Preparation method	σ_f (MPa)
$Al_{88}Y_2Ni_8Mn_2$	Nano. α-Al + amorphous	Melt spinning	1470
$Al_{88}Ce_2Ni_9Fe_3$	Nano. α-Al + amorphous	Melt spinning	1560
$Al_{87}Ni_7Nd_3Cu_3$	Nano. α-Al + amorphous	Annealing of melt-spun ribbons	1460

(Source: Data from A. Inove et al., *Sci. Rep. Res. Inst. Tohoku Univ.*, 1996.)

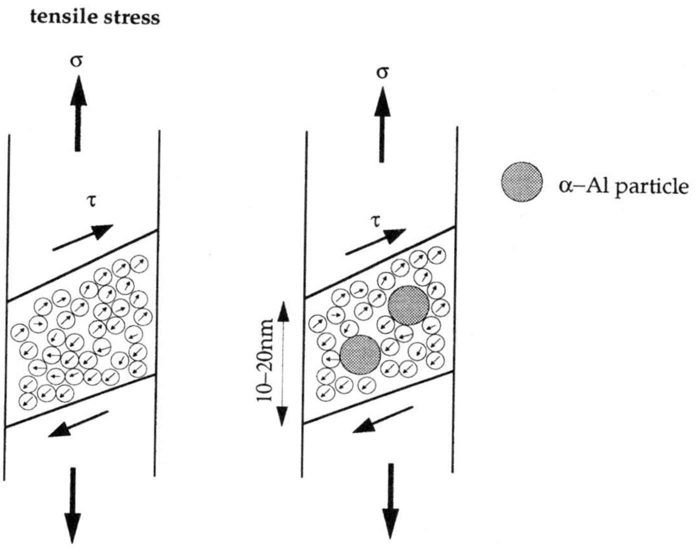

Fig. 14. Schematic diagram of the tensile deformation mode of an amorphous single-phase alloy and an amorphous phase containing nanoscale α-Al particles. (Source: Reprinted with permission from [22b]. © 1994 Japan Institute of Metals.)

increase in σ_f is achieved by capitalizing the high-strength α-Al particles with a perfect crystal structure. Furthermore, the presence of the nanoscale α-Al particles can influence the shear deformation of the amorphous matrix. As shown in Figure 14, it is known that an amorphous alloy is deformed along the maximum shear plane with a thickness of 10–20 nm, which is inclined by about 45° to the direction of the tensile load. Therefore, only when the particle size is comparable or smaller than the thickness of the shear deformation band can the particle act as an effective barrier against the subsequent shear deformation of the amorphous matrix.

Recently, Zhong et al. [33] have reported that the microhardness of the partially crystallized $Al_{86}Ni_{12}Y_2$ alloys obtained after primary crystallization is comparable to fully amorphous alloys of composition matching that of the residual amorphous matrix in the crystallizing alloys. This suggests that the contribution to the hardening is due to chemical solution hardening of the residual amorphous matrix resulting from solute enrichment. The same hardening mechanism is presumed to operate when nanophase composites are produced directly by quenching. Therefore, the improved strength in these nanocrystalline Al alloys is caused by a combination of particle strengthening and solution hardening mechanisms.

3.2.4. Nanocrystalline Soft Magnet

A new class of soft magnetic materials has been developed by exploring the primary crystallization of melt-spun amorphous Fe-based alloys with Fe content between 70 and 85 at%. Yoshizawa [42] found that the $Fe_{73.5}Si_{13.5}B_9Nb_3Cu_1$ (known as FINEMET) amorphous alloys transform from an amorphous structure to a mixed structure of α-Fe and residual amorphous phases on annealing at temperatures slightly above the onset of the primary crystallization ($\sim$520 °C), and the crystallized products exhibit good soft magnetic properties (i.e., high permeability, high magnetization, low core loss, and low coercivity). The α-Fe phase exists as 5–20-nm-sized particles. The crystalline volume fraction V_{cryst} ranges from about 50% to 80%, depending on the alloy composition and heat treatment. The amorphous layer thickness d can be estimated from the following simple geometric relationship between d, V_{cryst}, and the α-Fe crystal size D:

$$d \approx \frac{(1 - V_{cryst})D}{3} \tag{13}$$

Typically, this gives a thickness of $d \approx 1$–2 nm. The local chemical composition in the nanocrystalline Fe–Cu–Nb–Si–B alloy has been studied using FIM. Hono et al. [43] have reported that the Si partitioned into the α-Fe particles forming an Fe(Si) solid solution with a Si content of about 20 at% Si during primary crystallization. The Nb and B content segregate to the residual amorphous regions. This results in a positive magnetostriction (λ_s). The slow-diffusing Nb atoms lead to a sharp concentration profile, whereas the fast-diffusing B atoms give a flat concentration profile in the amorphous regions ahead of the growing α-Fe particles.

Once again, the primary crystallization of amorphous Fe–Cu–Nb–Si–B alloys involves the heterogeneous nucleation of the α-Fe phase. This is brought about by the addition Cu, which causes a chemical inhomogeneity of the amorphous matrix through cluster formation at the incipient stage of annealing. This is because Cu atoms have strong repulsive interatomic interactions with both Fe and Nb atoms, which provides a thermodynamic driving force for Cu clustering [35]. XAFS measurement on the formation of the nanocrystalline microstructure in $Fe_{73.5}Si_{13.5}B_9Cu_1Nb_3$ showed that the local structure around the Cu atoms in the alloy changes from an amorphous to a face-centered cubic structure prior to the precipitation of the α-Fe phase. The Cu-rich clusters can serve as nucleation sites and they trigger massive nucleation of α-Fe(Si) particles [44]. Subsequent growth of these particles involves the redistribution of elements. The Nb and B atoms are excluded from the crystallized region and they are enriched in the remaining amorphous phase, because they possess little solubility in the α-Fe(Si) phase. Thus, the additional Nb and the sharp concentration profile stabilize the amorphous phase against the formation of an intermetallic phase. Concurrently, grain growth of the α-Fe phase is suppressed. However, Naohara [45] has reported that quenched-in α-Fe(Si) nuclei can also be produced in Cu-free $Fe_{84-x}Si_6B_{10}Nb_x$ melt-spun alloys when the addition of Nb is in excess of 3 at%.

The addition of Ga has also been reported to assist the massive nucleation of the α-Fe phase in the primary crystallization of a melt-spun amorphous $Fe_{73}Si_{11}B_9Nb_3Ga_4$ alloy [46]. At present, the nucleation mechanism for an alloy containing Ga is still unclear. The addition of a refractory metal (X = Zr, Nb, Mo, V) to the alloy has been reported to influence the size of the α-Fe(Si) particles in partially crystallized $Fe_{73.5}Si_{13.5}B_9Cu_1Nb_2X_2$ alloys [47]. The average particle size and the volume fraction of α-Fe(Si) decrease in the order V > Mo > Nb > Zr for a given annealing condition. This is because of the increase in the thermal stability against crystallization with the addition of Zr. The addition of Mo and V, on the other hand, diminish the thermal stability of the amorphous phase.

Since then, other nanocrystalline Fe–M–B (M = Zr, Hf, Nb) alloys (NANOPERM) with a higher Fe content between 85 and 90 at% have been developed that exhibit superior

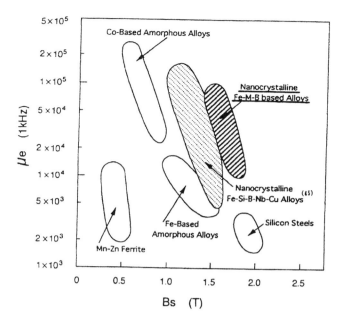

Fig. 15. Correlation between effective permeability (μ_e) at 1 kHz and saturation magnetization (B_s) for nanocrystalline Fe–M–B (M = Zr, Hf, Nb)-based alloys. The data on conventional soft magnetic alloys are also shown for comparison. (Reprinted with permission from [51]. © 1996 Institute for Materials Research Tohoku University.)

soft magnetic properties, as shown in Figure 15. In these alloys, the α-Fe phase is nearly pure Fe and the λ_s becomes negative. In Fe–Zr–B alloys, the nucleation of α-Fe particles is assisted by the formation of small medium-range ordering structure domains with no compositional fluctuation at temperatures below the crystallization temperature ($\sim$400 °C). The domain size increases with annealing time and these domains act as nucleation sites for primary α-Fe particles. At the nucleation and growth stage of α-Fe, it is clearly confirmed by FIM that the Zr and B atoms segregate to the amorphous matrix because they are not completely soluble in the α-Fe phase. Although Zr atoms are almost completely rejected into the amorphous matrix, some B atoms still remain in the particles. Eventually, this results in the enrichment of Zr and B in the remaining amorphous matrix. However, a sharp concentration gradient of Zr has been observed at the α-Fe/amorphous interface because Zr is the slowest diffusing species compared with other alloying elements. Consequently, a metastable local equilibrium is developed at the growing crystalline front and the growth kinetics of the α-Fe particles is predominantly controlled by the diffusion of Zr atoms. The maximum content of B near the interface appears to result from the enrichment of Zr at the interface because of the strongly attractive interaction between Zr and B atoms.

The small addition of Cu or Pd to Fe–M–B alloys has been found to reduce the size of the α-Fe particles [48]. For example, as shown by FIM, the Cu atoms in the $Fe_{89}Zr_7B_3Cu_1$ alloy form clusters but do not affect the redistribution of both Zr and B atoms. Hence, it is concluded that the addition of Cu to the Fe–Zr–B alloy plays a role similar to that in the Fe–Si–B–Nb–Cu alloy and the formation of Cu clusters enhances the massive nucleation of α-Fe particles. Varga et al. [49] have also studied the role of other nucleating additives (M = Cu, Ag, Au, Pd, Pt, Sb, Gb) in the formation of nanocrystalline structures and soft magnetic properties in $Fe_{86}Zr_7B_6M_1$. It was reported that Cu is the most effective nucleating agent. The addition of boron has been found to suppress the coarsening of the α-Fe phase because of the increased thermal stability of the residual amorphous phase and the suppression of the second stage of crystallization to form compound phases [50].

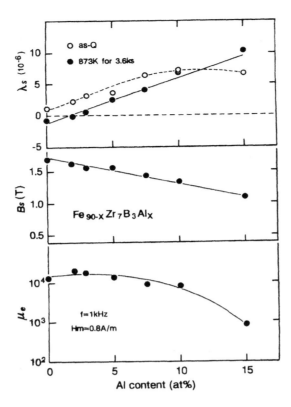

Fig. 16. Changes in λ_s, B_s, and μ_e as a function of Al content for the nanocrystalline $Fe_{90-x}Zr_7B_3Al_x$ annealed for 3.6 ks at 600 °C (873 K). (Reprinted with permission from [51]. © 1996 Institute for Materials Research Tohoku University.)

Recently, there has been some investigation into fabricating nanoscale α-Fe phases with a near-zero magnetostriction λ_s by forming Fe(Al) and Fe(Si) solid solutions in the Fe–Zr–B alloys using the small addition of Al or Si elements, respectively [51]. Figures 16 and 17 show the changes in the soft magnetic properties as a function of the Al and Si contents in the rapidly solidified Fe–Zr–B–Al and Fe–Zr–B–Si alloys, respectively. The advantages of quaternary $Fe_{88}Zr_7B_3Al_2$ and $Fe_{86}Zr_7B_3Si_4$ soft magnetic alloys consisting of a nanoscale α-Fe phase embedded in the amorphous matrix include the achievement of zero λ_s and improvement of permeability μ_e to 1.7×10^4 and saturation magnetization $B_s > 1.5$ T. In the partially crystallized $Fe_{86}Zr_7B_3Si_4$ alloy, the α-Fe phase contained approximately 96 at% Fe, 2 at% Zr, 1.5 at% Si, and less than 1 at% B, whereas the amorphous phase consisted of 7 at% Si, 17 at% Zr, and 2 at% B. The enrichment of Si in the residual amorphous phase is presumed to be caused by the strong interaction between Si and Zr atoms compared to that between the Si and Fe. The enthalpy of mixing between the Si and Zr is twice as high as that between Si and Fe [52]. When Zr is rejected from the α-Fe, Si would be attracted to the Zr-enriched amorphous phase although Si has high solubility in α-Fe.

3.2.5. Soft Magnetic Properties

The characteristics of good soft magnetic properties are high initial permeability μ_e, high saturation magnetization B_s, low coercive force H_c, low core loss, and near zero λ_s. This is closely associated with the nanoscale grain size of the α-Fe phase and the intergranular amorphous phase. Typical measured magnetic properties of this type of nanocrystalline Fe-based alloy are summarized in Table II. In the scale where the grain size is less than

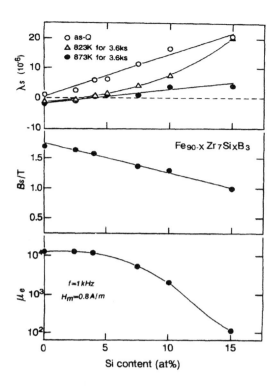

Fig. 17. Changes in λ_s, B_s, and μ_e as a function of Si content for the nanocrystalline $Fe_{90-x}Zr_7B_3Si_x$ annealed for 3.6 ks at 550 °C (823 K) and 600 °C (873 K). (Source: Reprinted with permission from [51]. © 1996 Institute for Materials Research Tohoku University.)

Table II. Magnetic Properties of Nanocrystalline Fe-Based Alloys and Other Conventional Soft Magnetic Alloys

Alloy	B_s (T)	μ_s at 1 kHz	Core loss[a] (W/kg)	$\lambda_s \times 10^6$	D_{fe}[b] (nm)
$Fe_{73.5}Si_{13.5}Nb_3B_9Cu_1$	1.24	100,000		2.1	12
$Fe_{84}Nb_{3.5}Zr_{3.5}B_8Cu_1$	1.53	100,000	0.06	0.3	8
$Fe_{84}Nb_7B_9$	1.59	50,000	0.1	0.2	9
$Fe_{90}Zr_7B_3$	1.70	29,000	0.17	−1.1	15
$Fe_{83}Nb_7B_9Cu_1$	1.52	57,000		1.1	8
Oriented Si-steel	1.80	2,400	0.73		

(Source: Data from A. Makino et al., *Sci. Rep. Res. Inst. Tohoku Univ.*, 1996.)

[a] At $f = 50$ Hz and $B_m = 1.4$ T.

[b] D_{fe} is the average size of the α-Fe particles.

the ferromagnetic exchange length (i.e., where the exchange interaction starts to dominate), H_c and the inverse initial permeability ($1/\mu_e$) are directly proportional to the average anisotropy $\langle K \rangle$ (i.e., they essentially determine the soft magnetic properties of the materials).

This is expressed in the following equation [54a]:

$$H_c \approx p_c \langle K \rangle / B_s \qquad (14)$$

where p_c is a dimensionless prefactor with a typical value of 0.1–0.2. Herzer [54b] has evaluated the average anisotropy $\langle K \rangle$ for nanocrystalline soft magnetic alloys on the basis of the random anisotropy model in which the randomly oriented grains are perfectly coupled through the exchange interaction. Accordingly, the $\langle K \rangle$ value for a three-dimensional sample can be written as

$$\langle K \rangle \approx \frac{K_1^4 D^6}{A^3} \tag{15}$$

where K_1 is the magnetocrystalline anisotropy constant of the grains, D is the grain size, and A is the exchange stiffness. The preceding expression shows that the $\langle K \rangle$ value is mainly dominated by D. It is believed that the existence of the residual amorphous phase in nanocrystalline Fe-based alloys can decrease the effective exchange stiffness between α-Fe particles, leading to a higher $\langle K \rangle$ value. This is because the residual amorphous phase can inhibit the exchange coupling between the α-Fe particles. However, the effective stiffness of the residual amorphous phase varies with the measurement temperature. It increases with decreasing measurement temperature because of the increasing magnetization of the amorphous phase at low temperature, leading to a decrease in the $\langle K \rangle$ value. Therefore, the soft magnetic properties of the nanocrystalline Fe–M–B alloys are also dominated by D and λ_s, together with the existence of the residual amorphous phase that affects the effective A value.

[57]Fe Mössbauer spectrometry of the nanocrystalline in $Fe_{86.5}Cu_1Zr_{6.5}B_6$ alloys [55] has revealed that the nanosized α-Fe particles are separated by a paramagnetic amorphous residual phase at the initial stage of primary crystallization. The chemical inhomogeneity in the structure caused by the clustering of Cu increases the Curie temperature. The relative content of the atoms inside these crystalline and amorphous zones is almost stable during the primary crystallization because the grain size does not change substantially. However, further crystallization at elevated temperatures reduces the fraction of the amorphous matrix and increases the portion of atoms with higher magnetic fields because of the propagation of ferromagnetic exchange interactions through the paramagnetic amorphous regions. This leads to high saturation magnetization B_s in the nanocrystalline Fe-based alloys.

The magnetostriction λ_s decreases significantly in nanocrystalline Fe–Cu–Nb–Si–B and Fe–M–B (M = Zr, Nb, Hf) alloys. The alloying elements in the α-Fe particles have a strong effect on the sign and value of λ_s, as shown by the addition of Al or Si to Fe–Zr–B alloys. It has been found that the magnetostriction of Co-doped α-Fe crystals in nanocrystalline $Fe_{57}Co_{21}Nb_7B_{15}$ is increased, leading to a higher coercive force. In the nanocrystalline microstructure, λ_s can be evaluated by the sum of the contributions from both the crystalline α-Fe and the residual amorphous phases using the following expression [56]:

$$\lambda_s \approx V_{cryst}\lambda_s^{fe} + (1 - V_{cryst})\lambda_s^{am} \tag{16}$$

where λ_s^{fe} and λ_s^{am} are the magnetostriction of the α-Fe and amorphous phases, respectively. A low magnetostriction is required to overcome the magnetoelastic anisotropy arising from internal mechanical stresses. Therefore, the small magnetostriction of the nanocrystalline state is closely related to the increase of initial permeability.

The high initial permeability (μ_e) of the nanoscale α-Fe phase is caused by the following four factors: (1) formation of the α-Fe phase with nearly zero magnetostriction (λ_s), (2) achievement of high magnetic homogeneity because of the small α-Fe particles in comparison to the magnetic domain walls, (3) small apparent magnetic anisotropy of the α-Fe phase resulting from ultrafine grain size, and (4) effective interaction of magnetic exchange coupling through a small amount of the thin intergranular residual amorphous layer.

Nanocrystalline Fe-based alloys exhibit low core loss. The classical eddy current loss W_c [57] is calculated by

$$W_c = \frac{(\pi t f B_m)^2}{6\rho D_m} \tag{17}$$

where t is the thickness of the sheet, f is the frequency, B_m is the maximum flux density, r is the electrical resistivity, and D_m is the density. However, the total eddy current loss W_{et} [57] is expressed as

$$W_{et} = W_a + W_c \tag{18}$$

where W_a is the anomalous eddy current loss, which varies with the frequency and the static hysteresis loss [57]. The anomaly factor η is calculated by

$$\eta = \frac{W_{et}}{W_c} \tag{19}$$

The anomaly factor η at 50 kHz and 1.0 T is evaluated to be 1.4 for the Fe–Zr–B nanocrystalline alloy with a direct-current (dc) remanence ratio of 0.44 and 5.7 for the amorphous alloy with a dc remanence ratio of 0.37 [57]. The small η comparable to that for the Co-based amorphous alloy appears to be a major reason for the low core losses of the nanocrystalline Fe–M–B (M = Zr, Hf, Nb) alloy. It is known that the η value is closely related to the magnetic domain structure of alloys, particularly the spacing of domains with 180° walls. At present, there is no report on the domain structure of the nanocrystalline Fe–M–B alloys. However, the curved domains with a width of about 100 μm separated by 180° walls have been observed in nanocrystalline $Fe_{73.5}Si_{13.5}B_9Nb_3Cu_1$ alloys with low core losses comparable to those of Co-based amorphous alloys with zero magnetostriction. Similar curve domains separated by 180° walls have been observed in nanocrystalline $Fe_{91}Zr_7B_2$ alloys using magnetic force microscopy (MFM). The domain wall thickness was estimated to be less than 2 μm from the MFM image. This result is consistent with the exchange correlation length (0.5 μm) evaluated from the measured magnetic properties. These results reveal that excellent soft magnetic properties are due to the averaging of the effects of the magnetocrystalline anisotropy over the order of 10^4 grains [58].

The nanocrystalline Fe-based alloys have other unique magnetic properties such as good piezomagnetic [59] and giant magnetoimpedance effects [60]. The piezomagnetic properties have been studied by the magnetic field dependence of the modulus of elasticity and magnetomechanical coupling measurement. The maximum value of the magnetomechanical coupling coefficient of the nanocrystalline $Fe_{73.5}Cu_1Nb_3Si_{13.5}B_9$ alloy was found to be 0.7 and the maximum elastic modulus changed from 50–60 to 170–180 GPa as a function of the applied magnetic field. This is believed to be due to the reduction of magnetostriction associated with the formation of a nanocrystalline structure. The giant magnetoimpedance effect (GMI) gives rise to large changes in the complex impedance upon the application of a dc magnetic field. The basic mechanism responsible for GMI is generally considered to be the skin depth, which is strongly dependent on the frequency of the exciting magnetic field, the transversal permeability, and the electrical resistivity [61]. The GMI ratio resulted from a combination of large permeability and high electrical resistivity, as found in a nanocrystalline microstructure [62]. The GMI effect has been found in many nanocrystalline $Fe_{73.5}Cu_1Nb_3Si_{13.5}B_9$ and $Fe_{86}Zr_7B_6Cu_1$ alloys [63, 64]. The GMI effect, together with a very high sensitivity at low fields, has opened up enormous potential applications in the field sensing and magnetic recording heads [65]. For nanocrystalline $Fe_{73.5}Cu_1Nb_3Si_{13.5}B_9$ alloys, a maximum magnetoimpedance ratio of -227% is obtained in the amorphous melt-spun ribbons after annealing at 550 °C for 3 h with alternating current (ac) at 300 kHz. The mechanism of the GMI effect in nanocrystalline materials is still a subject of further investigation. However, it has been found that the GMI is correlated with the high effective permeability associated with nanocrystalline structures.

These new quaternary soft magnetic alloys have tremendous potential applications including power transformers, data communication interface components, electromagnetic interference (EMI) prevention components, magnetic heads, sensors, magnetic shielding, and reactors. Furthermore, the soft magnetic materials are expected to be used in various kinds of magnetic parts of transformers, saturable reactors, choke cores, and so on [66].

One drawback of nanocrystalline alloys is their extreme brittleness. No winding or any kind of materials handling is possible on the final ribbons and the sample has to be encased in any thermal treatments. By replacing conventional annealing in an oven with Joule heating, where the ribbon is supplied with a current density on the order of some 10^7 A m^{-2} for times ranging between 10 and 100 s, it is possible to obtain high bending strains at fracture and higher initial permeability in Joule-heated $Fe_{73.5}Cu_1Nb_3Si_{13.5}B_9$ alloys [67, 68]

3.2.6. Nanocrystalline Permanent Magnet

A new class of nanocrystalline permanent magnetic $Fe_{89}Nd_7B_4$ alloys has been produced after heat treatment of the rapidly solidified amorphous structure at 800 °C for 60 s. The resultant microstructure consisted of three phases: 20–30-nm-sized α-Fe and 20-nm-sized tetragonal $Fe_{14}Nd_2B$ particles surrounded by the remaining amorphous phase with a thickness of 5 to 10 nm, as shown in Figure 18. The volume fractions of constituent phases in the Fe–Nd–B alloys are about 60% for the α-Fe phase, 20% for the remaining amorphous phase, and 20% for the $Fe_{14}Nd_2B$ phase. The Nd content is about 0.5% for the α-Fe phase and about 14 at% for the $Fe_{14}Nd_2B$ phase as measured using energy dispersive spectroscopy (EDX) [51]. The Nd content in the remaining amorphous phase is about twice that of the nominal Nd content, indicating that Nd is significantly enriched in the amorphous phase. The distribution of B in the nanocrystalline Fe–Nd–B alloy is similar to that in the Fe–Zr–B alloy because of a similar alloy composition. The enrichment of Nb and B elements near the interface causes the formation of a nanoscale mixed structure. It has been postulated that the formation of the $Fe_{14}Nd_2B$ phase is initiated by the enrichment of B in the preexisting Fe_3B phase that has been nucleated preferentially at the interface between the α-Fe and the surrounding amorphous matrix [51].

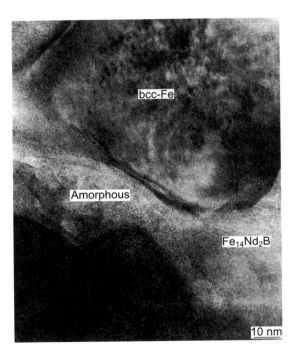

Fig. 18.   High-resolution transmission electron micrograph of the $Fe_{90}Nd_7B_3$ alloy annealed for 60 s at 800 °C showing a triplex microstructure consisting of 20–30-nm-sized α-Fe, 20-nm-sized $Fe_{14}Nd_2B$, and 5–10-nm-thick residual amorphous region. (Source: Reprinted with permission from [51]. © 1996 Institute for Materials Research Tohoku University.)

Table III. Comparison of the Hard Magnetic Properties for the Nanocrystalline Fe-rich Fe–Nd–B Magnet Containing an Intergranular Amorphous Phase with Those for Conventional Permanent Magnets

System	B_r (T)	H_c (kA/m)	$(BH)_{max}$ (kJ/m^3)
Ferrite magnet	0.4	312	30
Alnico magnet	0.9	112	42
SmCo$_5$ magnet	0.89	1360	151
Sm$_2$Co$_{17}$ magnet	1.14	800	239
Nd$_2$Fe$_{14}$B magnet	1.31	999	319
Nanocrystalline Fe$_{89}$Nd$_7$B$_4$	1.3	252	146
Nanocrystalline Fe$_{88}$Nb$_2$Pr$_5$B$_5$	1.23	270	110

(Source: Data from A. Inove et al., *Sci. Rep. Res. Inst. Tohoku Univ.*, 1996.)

3.2.7. Hard Magnetic Properties

Both the α-Fe and the amorphous phases exhibit soft magnetic properties, whereas the Fe$_{14}$Nd$_2$B exhibits hard magnetic properties. In this triplex nanostructure [69–71], the intergranular amorphous network phase has dual functions: (1) to provide an effective exchange magnetic coupling medium between the α-Fe and the α-Fe or tetragonal Fe$_{14}$Nd$_2$B phases, leading to an increase in remanence; and (2) to suppress the reversion of the magnetic domain walls in the central region of the soft magnetic α-Fe phase, leading to the achievement of a high coercive force H_c. The suppression may be the result of a combination of the inhomogeneity of the constituent elements at the crystalline/amorphous interface and the inhomogeneity of the ferromagnetic properties of the α-Fe and remaining amorphous phases. Consequently, one can regard the present Fe-rich Fe–Nd–B hard magnetic alloys as a multiple exchange-coupling-type magnet. Table III shows a comparison of the hard magnetic properties of the nanocrystalline Fe-rich Fe–Nd–B magnet containing an intergranular amorphous phase with those of conventional permanent magnets.

Recently, nanocrystalline Fe$_{88}$Nb$_2$Pr$_5$B$_5$ alloys produced from the crystallization of the melt-spun amorphous phase [72] have also been shown to exhibit superior hard magnetic properties. This is due to the fine nanoscale composite structure of the α-Fe and Fe$_{14}$Pr$_2$B phases with a grain size of 10–20 nm, which was achieved by the existence of the Nb- and Pr-enriched intergranular amorphous phase.

3.3. Nanoscale Mixed Structure of Quasicrystalline and Crystalline Phases

3.3.1. Nanocrystalline Aluminum Alloys

The icosahedral phase is formed in a number of rapidly solidified Al–TM (TM = Mn, Cr, V, Fe, Cu, Pd) alloys and it exhibits limited ductility at room temperature. The icosahedral phase has been reported [73] to comprise the Mackay icosahedral cluster containing 55 atoms, as shown in Figure 19, which are arranged through glue atoms to the three-dimensional quasiperiodical lattice. Consequently, by utilizing the large unit volume and a number of constituent atoms in the icosahedral structure, it is believed that nanoscale control of the icosahedral structure can improve the ductility and toughness at room temperature. This type of nanoscale mixed structure is predominantly based on nanosized icosahedral phases. This new type of microstructure was first reported in the melt-spun

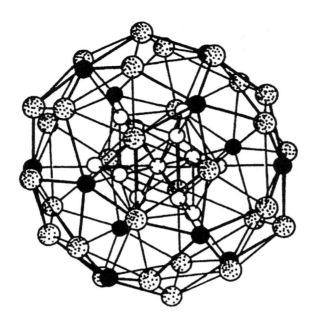

Fig. 19. Schematic diagram of the Mackay icosahedral cluster containing 55 atoms. (Source: Reprinted with permission from [73]. © 1995 American Physical Society.)

Al–Mn–Ln and Al–Cr–Ln (Ln = lanthanide metal) ternary alloys [74]. It consisted of nanoscale icosahedral (i.e., quasicrystalline) particles and an α-Al phase. Since then, other melt-spun Al–TM–Ln (TM = V, Cr, Mn, Fe, Mo, Ni) alloys have been found to have a similar nanoscale mixed structure. Recently, the application of high-pressure gas atomization to Al–Mn–TM and Al–Cr–TM (TM = Co, Ni) ternary alloys has caused the formation of a coexistent α-Al and quasicrystalline structure [75]. Figure 20 shows a typical nanoscale mixed microstructure of $Al_{92}Mn_6Ce_2$ that consists of 30–50-nm spherical icosahedral particles surrounded by a 10-nm layer of α-Al. The icosahedral particles appear to be distributed homogeneously and the surrounding Al phase has no high-angle grain boundary. The structural features of the homogeneous dispersion of the icosahedral particles and the absence of any high-angle grain boundaries are believed to result from the unique solidification mode. It is presumed [76–78] that the following solidification sequence occurred: liquid → primary icosahedral particles plus remaining liquid → primary icosahedral particles and α-Al phase. The primary precipitation of the icosahedral phase takes place as a result of the high homogeneous nucleation rate and low growth rates. Furthermore, the appearance of distinct reflection rings analogous to halo rings suggests that the nanoscale icosahedral particles have a slightly disordered structure. Further high-resolution transmission electron microscopy (TEM) investigation shows that icosahedral particles with a size of 10–30 nm have a disordered atomic configuration on a short-range scale less than 1 nm and an icosahedral atomic configuration on a long-range scale above about 3 nm.

3.3.2. Mechanical Properties

These mixed-phase alloys exhibit tensile fracture strengths s_f exceeding 1000 MPa combined with good ductility. This is believed to be the first evidence of the simultaneous achievement of high σ_f and good ductility in Al-based alloys containing more than 90 at% Al and the icosahedral phases as a main component having a volume fraction above 50%. The achievement of the high σ_f is independent of the kind of transition elements. The origin of the high tensile strength and good ductility in these alloys is attributed to the nonequilibrium short-range disorder and long-range icosahedral structure with the following characteristic features: (1) the existence of a natural affinity between the major (Al) and

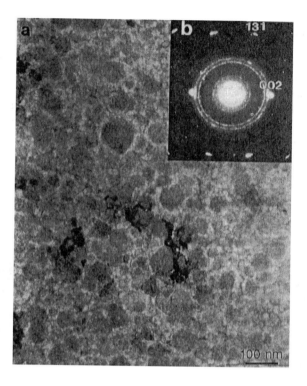

Fig. 20.   (a) Transmission electron micrograph and (b) selected-area diffraction pattern of a rapidly solidified $Al_{92}Mn_6Ce_2$ alloy. (Source: Reprinted with permission from [26]. © 1996 Institute for Materials Research Tohoku University.)

other minor elements, (2) the absence of a slip plane, (3) the existence of voids that enable the local movement of the constituent atoms, (4) an unfixed atomic configuration, leading to structural relaxation, and (5) the existence of Al–Al bonding pairs because of the Al-rich concentrations. These characteristic features are similar to those for metallic glasses with high tensile strength and good ductility. Furthermore, the existence of a ductile Al thin layer surrounding the icosahedral particles improves the ductility in these alloys because it provides an ease of sliding along the interface between the icosahedral and approximant crystalline phase.

The icosahedral-based structure produced by rapid solidification can be maintained up to 3.6 ks at 550 °C on annealing. The high thermal stability of this structure enables the production of bulk icosahedral-based alloys by extrusion of atomized icosahedral-based powders in the temperature range of 300 to 400 °C, well below the decomposition temperature of the icosahedral phase [75].

3.4. Nanoscale Mixed Crystalline and Crystalline Structure

3.4.1. Granular Nanocomposite

Direct production of nanocrystalline composite materials by melt spinning has been demonstrated successfully in the monotectic alloy systems exhibiting a liquid miscibility gap [79–81]. Melt spinning of near monotectic alloys leads to undercooling of the alloys followed by phase separation in the liquid phase, thereby producing a nanodispersed emulsion. Subsequently, solidification of the continuous liquid traps the liquid inclusions, producing a nanodispersed microstructure (also referred to as a granular nanocomposite). The nanodispersed solid phase in this case was formed by the heterogeneous nucleation during solidification of these trapped nanoscaled liquid inclusions. Therefore, these phases show an orientation relationship with the matrix. A typical microstructure of dispersion of

10–50-nm-sized Pb particles embedded in an α-Al matrix obtained by melt spinning of an Al–10 wt% Pb alloys is shown in Figure 21. The immiscible granular nanocomposite represents an interesting class of materials. Other examples of melt-spun granular nanocomposites have been fabricated from Al–Bi, Al–Pb, Cu–Pb, Zn–Pb, and Zn–Bi [79] in which the nanoparticle comprises the low-melting-point phase while the crystalline matrix comprises the high-melting-point phase. Table IV shows the size range of the nanoparticles in various systems as a function of the wheel surface velocity during melt spinning [79].

The matrix phase can be modified by selecting the appropriate type of alloying and this provides a great potential for alloy design. For example, melt spinning has been used to produce granular nanocomposites with Bi embedded in an amorphous Al–Fe–Si matrix [80].

Other granular nanocomposites that have attracted interest are based on Cu–Co [82, 83], Cu–Co–X (X = Fe, Ni, Mn) [84], Cu–Fe [85], Fe–Au, and Co–Au [86] alloys. These

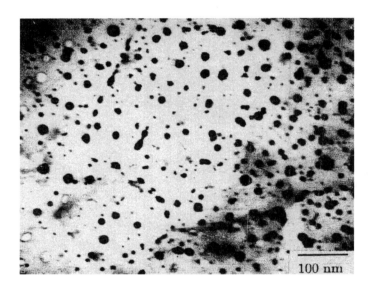

Fig. 21.   Transmission electron micrograph of a melt-spun Al–10 wt% Pb alloy showing nanodispersion of Pb particles in an Al matrix. (Source: Reprinted from [81], with permission of Elsevier Science.)

Table IV.   Size Range and Average Size of the Nanoparticle in Various Systems as a Function of the Wheel Surface Velocity during Melt Spinning

System	Wheel velocity (m/s)	Size range (nm)	Average size (nm)
Zn–10 wt% Pb	30	8–200	8
Zn–2 wt% Bi	15	15–75	25
Zn–10 wt% Bi	15	8–160	10
Al–10 wt% Pb	30	20–200	30
Al–2 wt% Pb	30	4–15	5
Al–8 wt% In	15	20–160	80
Cu–10 wt% Pb	15	100–400	100

(Source: Data from R. Goswami and K. Chattopadhyay, *Mater. Sci. Eng., A,* 1994.)

alloys exhibit giant magnetoresistance (GMR) when prepared by a combination of rapid solidification to form a supersaturated solid solution and subsequent heat treatment to cause decomposition into a granular nanocomposite structure. In the as-melt-spun state of the $Cu_{70}Co_{30}$ alloy, two types of face-centered cubic (fcc) Co-rich particles are present: large 100–300-nm-diameter particles, with a Cu-rich shell, formed by liquid phase separation; and smaller 15–40-nm-diameter disks formed via a monotectic reaction. These particles play little or no part in the magnetoresistance (MR) of these materials. On annealing, a very fine distribution of 5–7-nm-diameter disk-shaped precipitates, coherent with the matrix, is formed by a spinodal mechanism. It is these precipitates that are responsible for the large MR observed in the sample annealed at 450 °C.

However, a different microstructure is found in the as-melt-spun $Cu_{80}Fe_{20}$ alloy. Such a binary system has been reported to exhibit GMR in Cu–Fe multilayers. However, the melt-spun microstructure consisted of 0.1–0.5-mm droplet-shaped particles embedded in the Cu matrix. The droplet-like particles have an inner fine-scale nanostructure with particle size on the order of 20 nm. The occurrence of these nanostructured droplet particles can be explained as follows. As the liquid alloy of the composition $Cu_{80}Fe_{20}$ is rapidly quenched from $T = 1500$ °C down through the two-phase liquid $+ \gamma$-Fe region, the liquid phase decomposes to Fe-rich droplets within a Cu-rich liquid matrix. These Fe-rich droplets then solidify before the surrounding matrix. Because they are rich in iron, γ-Fe nucleates more easily within the droplets while rejecting excess Cu, thus generating the nanostructure within the solidified droplets, as shown in Figure 22. Solidification of the Cu-rich liquid matrix occurs subsequently, followed by the expected transformation of γ-Fe into α-Fe. The addition of boron to the Cu–Fe alloy suppresses the formation of α-Fe regions during the quench from the melt but not that of the droplet-type structures. During the decomposition of the undercooled liquid solution, boron atoms, which have a stronger affinity to Fe than to Cu, are redistributed preferentially in the Fe-rich zone. It appears that boron atoms stabilize the γ-Fe and suppress the α-Fe phase formation. Boron trapped in the metastable γ-Fe crystallites during the quench seems to retard their transformation into α-Fe.

Our present understanding of the GMR in layered structures is based on spin-dependent scattering at the interface between magnetic and nonmagnetic layers, as well as spin-dependent scattering in magnetic layers [87]. It is emphasized that high-density interfaces between magnetic and nonmagnetic materials give rise to GMR in the granular alloys. Song et al. [83] have reported that the melt-spun $Cu_{70}Co_{30}$ ribbon developed a maximum 4.2-K magnetoresistance of 22% following annealing for 1 h at 450 °C. The addition of a small amount of Ni to Co–Cu alloys is found to improve the GMR effect. Because Ni dissolves in both Co and Cu, the enhancement of the magnetoresistance ratio by the replacement of Ni may be due to the increase in the solubility limit of Co(Ni) in the Cu-rich matrix in the as-quenched state.

3.4.2. Equiaxed Nanocomposite

The production of equiaxed multiphase microstructures with an average grain size in the nanometer scale can be achieved by complete crystallization of rapidly solidified amorphous precursors. As referred to in a previous section, the primary crystallization of Al- and Fe-based alloys retains some residual amorphous phase. However, further heat treatment of the partially crystallized materials at elevated temperature provides sufficient thermodynamic driving force to complete the crystallization process by transforming the residual amorphous phase to a crystalline compound phase. Several workers have adopted this approach of complete crystallization of amorphous precursors to generate fully dense microstructures with an average grain size less than 50 nm in various Fe, Pd, and Ni alloys. In all cases, the amorphous precursors are transformed into a nanocrystalline equiaxed microstructure consisting of multiple crystalline phases either by primary crystallization or eutectic crystallization processes. Table V gives a summary of the minimum grain size achieved by complete crystallization of the amorphous alloys.

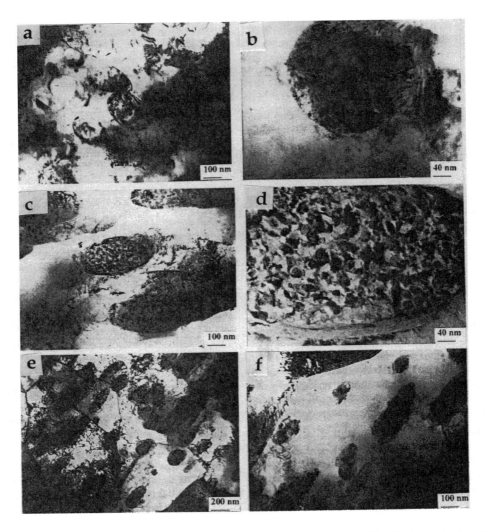

Fig. 22. Transmission electron micrographs of melt-spun $Cu_{80}Fe_{20}$ (a, b), $(Cu_{80}Fe_{20})_{99}B_1$ (c, d), and $(Cu_{80}Fe_{20})_{97}B_3$ (e, f). (Source: Reprinted with permission from [85]. © 1996 Trans Tech publications Ltd.)

Table V. Minimum Average Grain Size in Equiaxed Nanocomposites Produced by Complete Crystallization of Amorphous Precursors

System	Type	Crystalline phases	Minimum grain size d (nm)	$T_a/T_m{}^a$	Reference
$Ni_{80}P_{20}$	Eutectic	$Ni_3P + Ni(P)$	6–7	0.5	88
$Fe_{80}B_{20}$	Eutectic	$Fe_3B + Fe(B)$	8	0.46	89
$Fe_{40}Ni_{40}P_{14}B_6$	Eutectic	$(FeNi)_3 + FeNi(PB)$	9	0.55	90
$Fe_{78}B_{13}Si_9$	Primary	$Fe(Si) + Fe_3B$	21–22	0.51	91
$Fe_{60}Co_{30}Z_{10}$	Primary	$Fe(Co) + (FeCo)_2Zr$	15	0.51	92
$Pd_{78.1}Cu_{5.5}Si_{16.4}$	Primary	$Pd(Si) + (PdCu)_3Si$	19	0.62	93

(Source: Data from K. Lu, *Phys. Rev. B*, 1995.)

^{a} T_a = annealing temperature, T_m = melting point.

The thermodynamic aspects of the solid-state transformation from an amorphous phase to a nanocrystalline phase have been reviewed by Lu [88a–c]. This is based on the idea that the transformation involves the decomposition of the amorphous phase into nanometer crystallites and the interfaces. The interaction between the interface and the nanometer-sized crystallites is assumed to be negligible. The maximum fraction of the interface component can be calculated by equating the Gibbs free-energy change for the overall transformation to zero. If we let the atomic fraction of the interface be inversely proportional to the average grain size and assume the thickness of the interface to be independent of the grain size, the minimum grain size (d^*) can be deduced from the maximum interface fraction, which is related to the excess Gibbs free energies for the interface ΔG^i and the amorphous phase ΔG^a relative to the crystalline phase(s) according to the following expression [88c]:

$$d^* = \alpha \frac{\Delta G^i}{\Delta G^a} \tag{20}$$

where α is a constant. The ΔG^i is found to be dependent on the excess volume of the interface according to the quasiharmonic Debye approximation. The thermodynamic analysis has shown that the decrease in the excess volume of the interface can result in a significant refinement of the grain size in the crystallized products. The previous experimental data in Table V have shown that eutectic crystallization products consisted of smaller minimum grain size than the primary crystallization products. The reason is presumably due to the different crystallization mechanism. In eutectic crystallization, the amorphous alloy decomposes into a mixture of two equilibrium or metastable crystalline phases. The proportions of these phases give an overall eutectic composition. Such a transformation is controlled by the interface movement and no long-range diffusion ahead of the growing crystals. The two crystalline phases usually have a defined orientation relationship and the interface between these crystalline phases is either coherent or semicoherent such that the excess energy is small and the excess volume is low. However, primary crystallization, as described in the previous section, produces a different interface structure. The primary crystallization is a diffusion-controlled process and this leads to compositional pile-up ahead of the growing crystal front. Therefore, the crystallization of the residual amorphous phase involves heterogeneous nucleation and growth processes. The interface formed in primary crystallization is believed to be a high-energy state compared with other crystallization mechanisms. This implies that the interface has high excess volume. This is believed to be the reason for the larger grain size limits in primary crystallization than that in eutectic crystallization. Therefore, nanocrystalline alloys with small grain size can be achieved by crystallization from an amorphous phase when the interfacial excess volume is small.

A nanocrystalline $Fe_{73}B_{13}Si_9$ alloy has been found to exhibit enhanced oxidation resistance at a temperature range of 200 to 400 °C over its amorphous and coarse-grained crystalline counterparts with the same composition. The nanocrystalline microstructure consisted of an equiaxed mixture of α-Fe(Si) and Fe_2B phases with an average grain size of 30 nm. This microstructure is achieved by complete primary crystallization of the amorphous precursor. It has been proposed that the enhanced oxidation resistance can be attributed to the large fraction of interphase boundaries and the fast-diffusion character of the nanocrystalline materials. At elevated temperature, Si atoms in the α-Fe(Si) phase segregate to the interface and diffuse quickly to the surface of the sample along these interphase boundaries. Consequently, a large amount of Si atoms accumulate at the surface, where they oxidize to form a continuous SiO_2 film that prevents further oxidation [94].

3.5. Nanoscale Single-Phase Structure

So far, only nanoscale mixed structures with various phases produced by a combination of rapid solidification and thermal treatment have been reviewed. Currently, it is difficult to produce single-phase microstructures with average grain sizes less than 50 nm directly by rapid solidification processing. The most successful route to achieve a nanoscale single-phase structure is by polymorphous crystallization of an amorphous alloy. Polymorphous crystallization is similar to that found in eutectic crystallization. The process is controlled by the interface movement and there is no long-range diffusion ahead of the growing crystalline phase. Examples include NiZr$_2$ [95] and (Fe, Co)$_{33}$Zr$_{67}$ [96] alloys. The polymorphous crystallization of the amorphous (Co)$_{33}$Zr$_{67}$ alloy produces a single phase consisting of a tetragonal (Fe, Co)Zr$_2$ phase with an average grain size of 10.5 nm after annealing for 80 s at 478 °C (i.e., above the crystallization temperature). This type of microstructure provides the basis for the study of the grain growth mechanism [96] and Hall–Petch relationship [97] in the nanometer scale. The experimental data have shown that the grain growth kinetics for the nanocrystalline (Fe, Co)Zr$_2$ phase is as follows for the average grain size $\overline{D}$:

$$\overline{D} = c_{gg}(t - t_{cryst})^{1/3} \tag{21}$$

where c_{gg} is the constant dependent on certain physical parameters (i.e., mobility, grain boundary free energy) and t_{cryst} is the time for complete crystallization (i.e., until the entire volume is composed of very small grains in contact with one another). However, the microhardness measurement of the nanocrystalline microstructure has shown some controversial evidence of a negative Hall–Petch slope as the grain size approaches to the nanometer scale [97]. The reason for the softening in nanocrystalline alloys is still inconclusive.

4. CONCLUSIONS

Rapid solidification has been and continues to remain an important processing approach for materials. One of the main attractions is the flexibility that RSP offers for new approaches to material design and the fabrication of components with superior performance. Rapid solidification processing combined with controlled heat treatment is a powerful approach to generate novel nanocrystalline materials with unique high mechanical strength, excellent soft/hard magnetic properties, and enhanced oxidation resistance. It has become clear that processing conditions play a major role in the achievement of nanocrystalline metallic alloys. Furthermore, the selection of a suitable solute element is essential to cause the appearance of nanoscale mixed structures of fine particles embedded in an amorphous matrix produced by either direct rapid solidification or primary crystallization of the amorphous precursor. These effective solute elements have the following characteristic features: (1) high melting temperature, (2) large atomic size or large atomic size ratio among constituent elements, (3) large negative heat of mixing against the major element, and (4) nearly zero solubility limit against the major element.

Recent progress has yielded an improved comprehension of the nucleation and growth processes that enables a better understanding of the alloying effects on the formation of nanocrystalline microstructures. This provides new opportunities for the synthesis of unique microstructures in both structural and functional materials. It is, therefore, believed that new advanced materials exhibiting other novel properties can be fabricated by the modification of atomic configuration on a nanoscale. A variety of nanocrystalline microstructures have been reviewed and their exciting properties have been highlighted. As the microstructure reduces to the nanoscale, the physical properties deviate from those found in coarse-grained materials. This opens up questions on the validity of conventional theories to describe these physical properties. So far, on a laboratory scale, these nanocrystalline alloys have exhibited superior performance than their coarse-grained counterparts. To capitalize on their unique properties, it is essential to maintain the stability of the nanocrystalline microstructure during material processing and in service. The next challenge will be to focus on the processing of these materials in large quantity and economically without losing the nanocrystalline microstructure.

References

1. R. P. Andres, R. S. Averbach, W. L. Brown, L. E. Brus, W. A. Goddard III, A. Kaldor, S. G. Louie, M. Moscovits, P. S. Peercy, S. J. Riley, R. W. Siegel, F. Spaepen, and Y. Wang, *J. Mater. Res.* 4, 704 (1989).
2. V. Provenzano, N. P. Lonat, M. A. Imam, and K. Sadananda, *Nanostruct. Mater.* 1, 89 (1992).
3. R. Birringer, U. Herr, and H. Gleiter, *Suppl. Trans. Jpn. Inst. Met.* 27, 43 (1986).
4. Y. Yoshizawa and K. Yamauchi, *Mater. Trans., JIM* 31, 307 (1990).
5. A. L. Greer, in "Mechanical Properties and Deformation Behaviour of Materials Having Ultra-Fine Microstructures" (M. Nastasi, ed.), p. 53. Kluwer Academic, Norwell, MA, 1993.
6. H. Jones, in "Rapid Solidification of Metals and Alloys," Monograph No. 8. Institute of Metallurgists, London, 1982.
7. B. Cantor, in "Science and Technology of the Undercooled Melt" (P. R. Sahm, H. Jones, and C. M. Adam, eds.), p. 3. Nijhoff, Dordrecht, The Netherlands, 1986.
8. H. H. Liebermann, in "Amorphous Metallic Alloys" (F. E. Lubrosky, ed.), p. 26. Butterworths, London, 1983.
9. R. W. Cahn, in "Material, Science and Technology" (R. W. Cahn, P. Hassen, and E. J. Kramer, eds.), Vol. 9, p. 493. VCH Publishers, Weinheim, Germany, 1991.
10. C. Suryanarayana, in "Materials Science and Technology" (R. W. Cahn, P. Hassen, and E. J. Kramer, eds.), Vol. 15, p. 57. VCH Publishers, Weinheim, Germany, 1991.
11. S. Overshott, *Electron. Power* 25, 347 (1979).
12. T. R. Anthony and H. E. Cline, *J. Appl. Phys.* 50, 245 (1979).
13. I. T. H. Chang, B. Cantor, and A. G. Cullis, *J. Non-Cryst. Solids* 117–118, 263 (1990).
14. A. Lawley, in "Rapid Solidification Technology Source Book" (R. L. Ashbrook, eds.), p. 47. American Society of Metals, Metals Park, OH, 1983.
15. J. B. See and G. H. Johnston, *Powder Technol.* 21, 119 (1978).
16. O. S. Nichiporenko, *Sov. Powder Metall. Met. Ceram.* 15, 665 (1976).
17. H. Lubanska, *J. Met.* 22, 45 (1970).
18. M. Cohen, B. H. Kear, and R. Mehrabian, in "Rapid Solidification Processing: Principles and Technologies II" (R. Mehrabian, ed.), p. 1. Claitor's Publishing Division, Baton Rouge, LA, 1980.
19. O. Salas and C. G. Levi, *Int. J. Rapid Solid.* 4, 1 (1988).
20. J. N. Martin, R. D. Doherty, and B. Cantor, in "Stability of Microstructure in Metallic Systems," Cambridge University Press, p. 84, 1977.
21. A. L. Greer, *Mater. Sci. Eng., A* 133, 16 (1991).
22. (a) T. Matsumoto, *Mater. Sci. Eng., A* 179–180, 9 (1994). (b) A. Inoue, Y. Horio, Y. H. Kim, and T. Matsumoto, *Mater. Trans., JIM* 33, 669 (1992).
23. I. T. H. Chang, P. Svec, M. Gogebakan, and B. Cantor, *Mater. Sci. Forum* 225–227, 335 (1996).
24. H. Nagahama, K. Ohtera, K. Higashi, A. Inoue, and T. Matsumoto, *Mater. Philos. Mag. Lett.* 67, 225 (1993).
25. A. Inoue, H. Kimura, and K. Sasamori, in "Chemistry and Physics of Nanostructures and Related Non-Equilibrium Materials" (E. Ma, B. Frultz, R. Shull, J. Morral, and P. Nash, eds.), p. 201. TMS, Warrendale, PA, 1997.
26. A. Inoue, H. M. Kimura, K. Sasamori, and T. Matsumoto, *Sci. Rep. Res. Inst. Tohoku Univ., A* 42, 165 (1996).
27. Y. H. Kim, K. Hiraga, A. Inoue, T. Matsumoto, and H. H. Jo, *Mater. Trans., JIM* 35, 293 (1994).
28. A. Inoue, K. Nakazato, Y. Kawamura, A. P. Tsai, and T. Matsumoto, *Mater. Trans., JIM* 35, 95 (1994).
29. M. Matsuura, M. Sakurai, K. Suzuki, A. P. Tsai, and A. Inoue, *Mater. Sci. Eng., A* 226–228, 511 (1997).
30. A. Inoue, K. Ohtero, K. Kita, and T. Matsumoto, *Jpn. J. Appl. Phys.* 27, L2248 (1988).
31. K. Hono, Y. Zhang, A. Inoue, and T. Sakurai, *Mater. Trans., JIM* 36, 909 (1995).
32. B. Cantor, U. Köster, P. Duhaj, H. Matyja, and T. Kemeny, INCO-COPERNICUS (CIPACT 940155) Final Report, 1998.
33. A. L. Greer, Z. C. Zhong, X. Y. Jiang, K. L. Rutherford, and I. M. Hutchings, in "Chemistry and Physics of Nanostructures and Related Non-Equilibrium Materials" (E. Ma, B. Fultz, R. Sholl, J. Morral, and P. Nash, eds.), p. 5. TMS, Warrendale, PA, 1997.
34. M. Calin and U. Köster, *Mater. Sci. Forum* 269–271, 49 (1997).
35. A. R. Yavari and O. Drbohlav, *Mater. Sci. Forum* 225–227, 295 (1995).
36. X. Y. Jian, Z. C. Zhong, and A. L. Greer, *Mater. Sci. Eng., A* 226–228, 789 (1997).
37. S. G. Kim, A. Inoue, and T. Matsumoto, *Mater. Trans., JIM* 32, 875 (1991).
38. A. Inoue, N. Nishiyama, S. G. Kim, and T. Matsumoto, *Mater. Trans., JIM* 33, 360 (1992).
39. C. Seeger and P. L. Ryder, *Mater. Sci. Eng., A* 179–180, 641 (1994).
40. L. Battezzati, M. Baricco, P. Fortina, and W.-N. Myung, *Mater. Sci. Eng., A* 226–228, 503 (1997).
41. A. S. Aronin, G. E. Abrosimova, I. I. Zver'kova, Yu. V. Kir'janov, V. V. Molokanov, and M. I. Petrzhik, *Mater. Sci. Eng., A* 226–228, 536 (1997).
42. Y. Yoshizawa, S. Oguma, and K. Yamauchi, *J. Appl. Phys.* 64, 6044 (1988).
43. K. Hono, K. Hiraga, Q. Wang, A. Inoue, and T. Sakurai, *Acta Metall. Mater.* 40, 2137 (1992).
44. T. Naohara, *Metall. Mater. Trans. A* 27, 3424 (1996).
45. T. Naohara, *Acta Metall.* 46, 397 (1998).
46. T. Tomida, *Mater. Sci. Eng., A* 179–180, 521 (1994).

47. J. M. Borrego and A. Condo, *Mater. Sci. Eng., A* 226–228, 663 (1997).

48. A. Makino, T. Hatanai, A. Inoue, and T. Matsumoto, *Mater. Sci. Eng., A* 226–228, 594 (1997).

49. L. K. Varga, A. Lovas, L. Pogany, L. F. Kiss, J. Balogh, and T. Kemeny, *Mater. Sci. Eng., A* 226–228, 740 (1997).

50. K. Suzuki, A. Makino, A. P. Tsai, A. Inoue, and T. Matsumoto, *Mater. Sci. Eng., A* 179–180, 501 (1994).

51. A. Inoue, A. Takeuchi, A. Makino, and T. Matsumoto, *Sci. Rep. Res. Inst. Tohoku Univ., A* 42, 143 (1996).

52. R. Hultgen, P. D. Desai, D. T. Hawkins, M. Gleiser, and K. K. Kelley, eds., "Selected Values of Thermodynamic Properties of Binary Alloys." American Society of Metals, Metals Park, OH, 1973.

53. A. Makino, T. Hatanai, S. Yoshida, N. Hasegawa, A. Inoue, and T. Matsumoto, *Sci. Rep. Res. Inst. Tohoku Univ., A* 42, 121 (1996).

54. (a) G. Herzer, *IEEE Trans. Magn.* 26, 1397 (1990). (b) G. Herzer, *Mater. Sci. Eng., A* 113, 1 (1991).

55. M. Miglierini, Y. Labaye, N. Randriananatoandrom, and J.-M. Grenèche, *Mater. Sci. Eng., A* 226–228, 5589 (1997).

56. G. Herzer, *J. Magn. Magn. Mater.* 112, 258 (1992).

57. K. Suzuki, A. Makino, A. Inoue, and T. Matsumoto, *J. Appl. Phys.* 74, 3316 (1993).

58. K. Suzuki, D. Wexler, J. M. Cadogan, V. Sahajwalla, A. Inoue, and T. Matsumoto, *Mater. Sci. Eng., A* 226–228, 586 (1997).

59. Z. Kaczkowski, M. Muller, and P. Ruuskanen, *Mater. Sci. Eng., A* 226–228, 681 (1997).

60. M. Knobel, M. L. Sánchez, C. Gomez-Polo, A. Hernando, and P. Marín, *J. Appl. Phys.* 79, 1646 (1996).

61. L. V. Panina and K. Mohri, *Appl. Phys. Lett.* 65, 1189 (1994).

62. M. Knobel, M. L. Sánchez, C. Gomez-Polo, A. Hernando, and P. Marín, *J. Appl. Phys.* 79, 1646 (1996).

63. M. Knobel, J. Schoenmaker, J. P. Sinnecker, R. Sato Turtelli, R. Grössinger, W. Hofstetter, and H. Sassik, *Mater. Sci. Eng., A* 226–228, 536 (1997).

64. H. Q. Guo, C. Chen, M. Li. T. Y. Zhao, K. Z. Luan, B. G. Shen, Y. H. Liu, J. G. Zhao, L. M. Mei, and H. Kronmüller, *Mater. Sci. Eng., A* 226–228, 550 (1997).

65. F. L. A. Machado, C. S. Martins, and S. M. Rezende, *Phys. Rev. B* 51, 3926 (1995).

66. A. Makino, K. Suzuki, A. Inoue, and T. Matsumoto, *Mater. Trans., JIM* 32, 551 (1991).

67. P. Allia, M. Baricco, M. Knobel, P. Tiberto, and F. Vinai, *J. Magn. Mater.*, 133, 243 (1994).

68. K. Suzuki, D. Wexler, J. M. Cadogan, V. Sahjwalla, A. Inoue, and T. Matsumoto, *Mater. Sci. Eng., A* 226–228, 586 (1997).

69. A. Inoue, A. Takeuchi, A. Makino, and T. Matsumoto, *Mater. Trans., JIM* 36, 689 (1995).

70. A. Inoue, A. Takeuchi, A. Makino, and T. Matsumoto, *Mater. Trans., JIM* 36, 962 (1995).

71. A. Inoue, A. Takeuchi, A. Makino, and T. Matsumoto, *IEEE Trans. Magn.* MAG-31, 3626 (1995).

72. A. Kojima, F. Ogiwara, A. Makino, A. Inoue, and T. Matsumoto, *Mater. Sci. Eng., A* 226–228, 520 (1997).

73. Y. Elser and C. L. Henley, *Phys. Rev. Lett.* 55, 2883 (1985).

74. A. Inoue, M. Watanabe, H. M. Kimura, and T. Matsumoto, *Sci. Rep. Res. Inst. Tohoku Univ., A* 38, 138 (1993).

75. K. Kita, K. Saitoh, A. Inoue, and T. Matsumoto, *Mater. Sci. Eng., A* 226–228, 1004 (1997).

76. A. Inoue, M. Watanabe, H. M. Kimura, F. Takahashi, A. Nagata, and T. Matsumoto, *Mater. Trans., JIM* 33, 723 (1992).

77. M. Watanabe, A. Inoue, H. M. Kimura, T. Aiba, and T. Matsumoto, *Mater. Trans., JIM* 34, 162 (1993).

78. A. Inoue, M. Watanabe, H. M. Kimura, and T. Matsumoto, *Sci. Rep. Res. Inst. Tohoku Univ., A* 38, 138 (1993).

79. R. Goswami and K. Chattopadhyay, *Mater. Sci. Eng., A* 179–180, 198 (1994).

80. R. Goswami and K. Chattopadhyay, private communication.

81. R. Goswami and K. Chattopadhyay, *Mater. Sci. Eng., A* 226–228, 1012 (1997).

82. N. Kataoka, H. Endo, K. Fukamichi, and Y. Shimada, *Jpn. J. Appl. Phys.* 32, 1969 (1993).

83. X. Song, S. W. Mahon, B. J. Hickey, M. A. Howson, and R. F. Cochrane, *Mater. Sci. Forum* 225–227, 163 (1996).

84. N. Kataoka, I. J. Kim, H. Takeda, and K. Fukamichi, *Mater. Sci. Eng., A* 181–182, 888 (1994).

85. O. Drhohlav, W. J. Botta Filho, and A. R. Yavari, *Mater. Sci. Forum* 225–227, 359 (1996).

86. N. Kataoka, H. Takeda, I. J. Kim, and K. Fukamichi, *Sci. Rep. Res. Inst. Tohoku Univ., A* 39, 121 (1994).

87. D. M. Edwards, J. Mathon, and R. B. Muniz, *IEEE Trans. Magn.* 27, 3548 (1991).

88. (a) K. Lu, J. T. Wang, and D. Wei, *J. Appl. Phys.* 69, 522 (1991). (b) K. Lu, J. T. Wang, and D. Wei, *Scr. Metall. Mater.* 24, 2319 (1990). (c) K. Lu, *Phys. Rev. B* 51(1), 37 (1995).

89. A. L. Greer, *Acta Metall.* 30, 171 (1982).

90. D. G. Morris, *Acta Metall.* 29, 1213 (1981).

91. H. Y. Tong, J. T. Wang, B. Z. Ding, H. G. Jiang, and K. Lu, *J. Non-Cryst. Solids* 150, 444 (1992).

92. H. Q. Guo, T. Reininger, H. Kronmüller, M. Rapp, and V. Kh. Skumrev, *Phys. Status Solidi A* 127, 519 (1991).

93. P. G. Boswell and G. A. Chadwick, *Scr. Metall.* 70, 509 (1976).

94. H. Y. Tong, F. G. Shi, and E. J. Lavernia, *Scr. Metall. Mater.* 32, 511 (1995).

95. M. G. Scott, in "Amorphous Metallic Alloys" (F. E. Lubrosky, ed.), p. 144. Butterworths, London, 1983.

96. T. Spassov and U. Köster, *J. Mater. Sci.* 28, 2789 (1993).

97. X. D. Liu, M. Nagumo, and M. Umemoto, *Mater. Trans., JIM* 38, 1033 (1997).

Chapter 12

VAPOR PROCESSING OF NANOSTRUCTURED MATERIALS

K. L. Choy

Department of Materials, Imperial College, London, United Kingdom

Contents

Handbook of Nanostructured Materials and Nanotechnology, edited by H.S. Nalwa
Volume 1: Synthesis and Processing
Copyright © 2000 by Academic Press
All rights of reproduction in any form reserved.

ISBN 0-12-513761-3/$30.00

1. INTRODUCTION

Nanocrystalline materials are solid-state systems constituting crystals of sizes less than 100 nm in at least one dimension. In general, nanostructured or nanophase materials can be classified into four categories according to the shape of their structural constituents and to their chemical composition. These include (1) nanophase powder; (2) nanostructured film (including single-layer, multilayer, composite film, compositionally graded film, etc.); (3) monolithic nanostructured material; and (4) nanostructured composite. The nanocrystalline materials can be metals, ceramics, or composites containing crystalline, quasicrystalline, and/or amorphous phases.

Nanophase materials exhibit many exciting extraordinary properties, which are not found in conventional material. These include superplasticity, improved strength and hardness, reduced elastic modulus, higher electrical resistivity, and lower thermal conductivities. They also exhibit improved soft ferromagnetic properties and giant magnetoresistance effects. The significance of nanostructured materials is found in electrical, electronic, magnetic, superconductor, catalytic, structural ceramic, and functional applications. The quantum effects observed are the results of materials at the atomic and nanometer levels.

However, for many of these cases, it still needs to be ascertained whether the improved properties are due to new physical phenomena at small dimensions or to an extension of larger-scale systematics to small sizes [1].

The understanding of the extraordinary behavior of nanostructured materials requires detailed studies of the correlations between the processing, structure, and properties. These studies rely on the identification and development of appropriate (i) processing methods and (ii) suitable characterization methods and analytical tools for the nanocrystalline materials. This chapter focuses on the processing aspects of nanocrystalline materials and provides a brief review of the methods used and highlights the emergent technologies.

2. SELECTION CRITERIA OF THE APPROPRIATE PROCESSING TECHNIQUE

There are several criteria that one should consider when selecting the appropriate technique for the processing of nanostructured materials. The selected processing methods should be able to meet the following requirements:

1. Fabrication of nanocrystalline materials into useful sizes, shapes, and structures without loss of desirable nanometer-sized features
2. Production of large quantities of nanometer-scale materials cost effectively
3. Assurance of process reproducibility and ease of process control
4. Optimization of the process

For the processing of nanosized powders, the capability of the technique to control the particle agglomeration and particle size distribution is essential. The efficiency and the selectivity of the particle collection and the handling of ultrafine powders safely still need to be addressed.

Nonetheless, for the processing of nanocrystalline films, additional criteria such as the processing technique should not impair the properties of the substrate and the capability of coating the engineering components uniformly with respect to both size and shape, thus yielding improved end product quality.

The ability to produce nanocrystalline materials with well-controlled structure cost effectively will provide improved understanding of the process/structure/property relationships and pave the way for the successful exploitation of nanocrystalline materials commercially with improved properties.

3. WHY VAPOR PROCESSING TECHNIQUES?

Since Gleiter [2] first drew attention to the extraordinary properties of nanocrystalline materials fabricated using an inert-gas condensation method, many diverse methods for processing nanocrystalline materials have been reported. These include vapor processing routes, liquid phase/molten state methods, and wet chemical and solid-state routes. Figure 1 summarizes the different processing techniques available for the processing of nanostructured materials. There are atomic and morphological differences between the materials manufactured by the various techniques. Each method is particularly suited for particular nanocrystalline systems with specific shapes and volume.

Solid-state processing routes such as the mechanical milling-based methods involve mixing, grinding, calcination, and sintering. Although these methods involve relatively simple techniques, they are tedious and time consuming because of prolonged milling times and multiple cycles of processing, which are also prone to contamination from the milling media. However, these powders require further hot consolidation to form bulk samples. Furthermore, these powders can suffer from chemical and phase inhomogeneities. Extensive milling is expensive and limited to the processing of ultrafine powders. Nanostructured films and multilayer and functionally graded coatings cannot be produced using solid-state processing routes.

Wet chemical routes, such as the sol–gel, hydrothermal, and electrodeposition methods, require a high number of processing steps, including pretreatment, mixing, chemical reactions, filtration, purification, drying, and calcination during the fabrication of

Processing route	Processing methods	Nanocrystalline materials
Solid-state	• Mechanical Milling	powder
Liquid	• Sol–gel	powder/film
	• Sonochemistry	powder
	• Hydrothermal	powder
	• Electrodeposition	powder/film
	• Gas Atomization	powder
	• Laser Beam Melting	film
	• Melt Spinning	continuous ribbon
Vapor	• Chemical Vapor Deposition	powder/film
	• Physical Vapor Deposition	powder/film
	• Aerosol Processes	powder/film
	• Flame-Assisted Deposition	powder

Fig. 1. Classification of different processing techniques available for the processing of nanostructured materials.

ultrafine powder. These methods are tedious and can cause contamination. Waste treatment is difficult, especially when producing large quantities. Moreover, some wet chemical routes such as the hydrothermal method are limited to powder processing and cannot be used to produce nanostructured films and multilayer and functionally graded coatings.

This chapter aims to provide a brief overview of the vapor processing of nanostructured materials. There exist many distinctive advantages of using vapor processing techniques over other methods in the processing of nanostructured materials. It is obvious that vapor processing methods seem to be the only method that can provide highly pure materials with structural control at the atomic level or nanometer scale. Moreover, vapor processing routes can produce ultrafine powders, multilayer and functionally graded materials, and composite materials with well-controlled dimensions and unique structures at a lower processing temperature. The classification of vapor processing methods, including the recent discoveries of new vapor processing technologies, will be outlined in this chapter. The process principle, description of the processing technique, apparatus used, range of nanomaterials synthesized, and properties will be presented. Their advantages and limitations will be discussed, and their applications will be briefly reviewed.

4. CLASSIFICATION OF VAPOR PROCESSING TECHNIQUES

Vapor processing techniques can be classified into physical vapor deposition (PVD), chemical vapor deposition (CVD), aerosol-based processes, and flame-assisted deposition methods as shown in Figure 1. There are several variants of these techniques as summarized in Figure 2. For example, PVD can be subdivided further into evaporation, sputtering and ion plating processes based on the different ways of generating the gaseous species. Similarly, the classification of CVD processes can be based on different heating methods (e.g., thermally activated, photo, plasma, etc.) for the deposition reactions to occur or the type of precursor used. These have led to the development of different variants of CVD such as plasma-assisted CVD, laser-assisted CVD, metalorganic-assisted CVD, and so forth. Aerosol-based processing techniques can be subdivided into spray pyrolysis, electrostatic-assisted vapor deposition, and so on, based on the different aerosol generation methods used.

All of the preceding vapor processing methods have been used to produce films and coatings. There are numerous papers in the *Journal of the Electrochemical Society, Thin Solid Films, Chemical Vapor Deposition, Journal of Applied Physics*, and *Journal of Vacuum Science and Technology* that describe the process and applications of vapor processing techniques, giving further insight into recent developments. These techniques can be adapted to the manufacture of nanostructured materials in the form of either films or powders. Some of these techniques are better than others in providing precise control of the production of nanocrystalline materials and have the capability of scaling up for large-area or large-scale production. The advantages and disadvantages of these vapor processing techniques will be compared and discussed in the subsequent sections. Bulk nanocrystalline materials can be produced by *in situ* consolidation of the deposited powders.

5. PHYSICAL VAPOR DEPOSITION

5.1. Process Principles

The PVD process involves the creation of vapor phase species through (i) evaporation, (ii) sputtering, or (iii) ion plating. During transportation, the vapor phase species undergo collisions and ionization and subsequently condense onto a substrate where nucleation and

Variants of vapor processing techniques for the fabrication of nanophase materials

1. Physical Vapor Deposition
    (i) Evaporation
        (a) Inert-Gas Condensation
        (b) Electrical explosion wire
        (c) Laser ablation
        (d) Molecular beam epitaxy
    (ii) Sputtering

2. Chemical Vapor Deposition (CVD)
    (i) Thermally activated CVD
    (ii) Photo-assisted CVD
    (iii) Plasma-assisted CVD
    (iv) Metalorganic CVD
    (v) Atomic Layer Epitaxy Process

3. Aerosol-Based Processing Routes
    (i) Pyrosol
    (ii) Aerosol-assisted chemical vapor deposition
    (iii) Electrospraying-assisted deposition
        (a) Corona spray pyrolysis
        (b) Electrostatic spray pyrolysis/electrostatic spray deposition
        (c) Electrospray organometallic chemical vapor deposition
        (d) Gas-aerosol reactive electrostatic deposition
        (e) Electrostatic spray-assisted vapor deposition

4. Flame-Assisted Deposition
    (i) Counterflow diffusion flame synthesis
    (ii) Combustion flame-Chemical vapor condensation
    (iii) Sodium/halide flame deposition with in situ encapsulation process

Fig. 2.   Variants of the vapor processing techniques.

growth occur, leading to the formation of films. The process principles of PVD are summarized in Figure 3. For the formation of powders, the neutral and/or ionized vapor phase species will collide with the inert-gas molecules and undergo homogeneous gas phase nucleation to form powders that are eventually being removed and collected. Figure 4 shows the different methods of generating the vapor species, which give rise to a variety of PVD techniques such as evaporation, sputtering, and ion plating. The PVD processes take place in a vacuum. The vacuum environment plays an important role in the vapor flux and the deposition and growth of films. The three important aspects of the vacuum environment to thin-film deposition are: the pressure, expressed as the mean free path; the partial pressure ratio of reactive and sputtering gases in inert working gases; and the ratio of film vapor arrival to reactive gas impingement rate. Detailed descriptions of the PVD deposition mechanism are available in the literature [3–5].

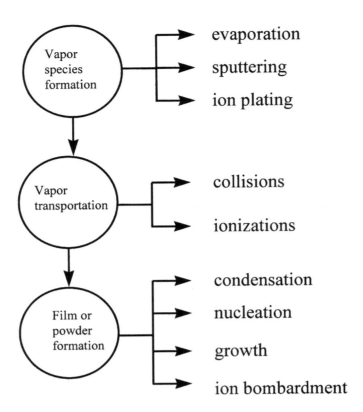

Fig. 3.   Process principles of PVD.

5.2. Advantages and Disadvantages

5.2.1. Advantages

The vacuum environment in the PVD process provides the ability to reduce gaseous/vapor contamination in the deposition system to a very low level. Therefore, ultrapure films or powders can be produced using PVD methods. During the fabrication of nanosized powders, the powders can be collected and compacted *in situ* into a bulk material with a high degree of cleanliness.

PVD is an atomistic deposition method that can provide good structural control by careful monitoring of the processing conditions. Moreover, the as-deposited materials are already nanocrystalline in nature and do not require any further milling to reduce the particle size or heat treatment to burn out the precursor complexes.

5.2.2. Disadvantages

In general, the disadvantages of PVD processes are as follows:

1. The deposition process needs to operate in the low-vapor-pressure range. Therefore, a vacuum system is required, which increases the complexity of the deposition equipment and the cost of production.
2. The synthesis of multicomponent materials is difficult, except for the laser ablation method. This is because different elements have different evaporation temperatures or sputtering rates. Many compound materials partially dissociate on thermal vaporization producing nonstoichiometric deposits. However, highly

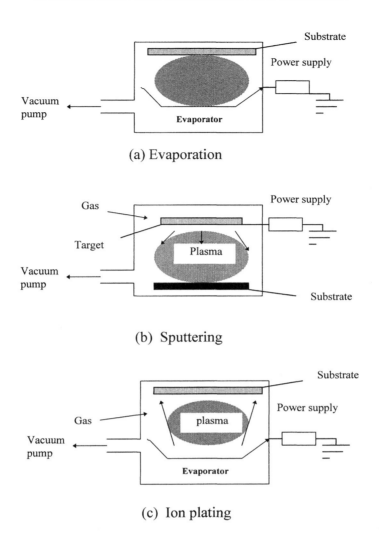

Fig. 4. Three basic PVD techniques: (a) evaporation, (b) sputtering, and (c) ion plating.

nonstoichiometric materials are beneficial for defect-related applications such as in sensors, fuel cells, ceramic membrane reactors, and oxidation catalysts.
3. PVD is a line-of-sight deposition process, which causes difficulty in producing nanocrystalline films on complex-shaped components, and has poor surface coverage.

5.3. Applications

The science of PVD can be traced back to the 1850s. During the 1950s, PVD techniques were widely used for the deposition of thin films for resistors and capacitors for telecommunications, microelectronic circuits, and optical coating applications. Today, PVD covers a wider range of commercial applications that include the deposition of various of metals, alloys, and compounds in the form of coatings and films for (i) optics (e.g., antireflection coatings), (ii) electronics (e.g., metal contacts), (iii) mechanics (e.g., hard coatings on tools), and (iv) protective coatings (e.g., corrosion, oxidation).

5.4. Evaporation

The basic evaporation process is shown in Figure 4a. The sources are generally made of refractory metal (e.g., W, Ta, and Mo) in the form of coils, rods, boats, or special-purpose designs. Material to be evaporated can be in the form of wires, rods, sheets, or powders placed in the evaporation sources. An intense heat is used to vaporize the source from a solid or molten state to a vapor state. The heat is provided by resistance, induction, arc, electron beam, or laser, which give rise to a variety of evaporation methods such as Joule heating evaporation, cathodic-arc deposition, electron beam evaporation, laser ablation, and so on. The vapor flux of the desired material condenses either onto the cooler substrate to form a solid film or onto a cool finger to form powders.

The process requirements that need to be considered are the compatibility of the evaporant and the power and the capacity availability. The evaporant compatibility seems to be the most difficult to achieve because many important evaporants (e.g., Al, Fe, Inconel, and Pt) dissolve all refractory metals to some extent. The pressure in the vacuum must be sufficiently low ($<10^{-4}$ mtorr), so that the mean free path (λ) of the vapor species is large; that is, evaporated atoms essentially travel in a straight line from the source to the substrate without colliding with the ambient gas molecules. This relationship can be written as

$$\lambda = \left(1/\sqrt{2}\sigma\right)(kT/P) \tag{1}$$

where σ is the collision cross section of the gas and T and P are the temperature and pressure of the gas, respectively.

The advantages of evaporation, in addition to those outlined in Sections 3 and 5.2.1, are as follows:

1. The material to be vaporized can be in any form and purity.
2. The residual gases and vapors in the vacuum environment are easily monitored.
3. The rate of vaporization is high.
4. The line-of-sight trajectories and point sources allow the use of deposition onto defined areas.
5. The cost of thermally vaporizing a given quantity of material is much less than that of sputtering the same amount of material.

The limitations of the evaporation method are the utilization of material may be poor. Many compound materials may partially dissociate on thermal vaporization, producing nonstoichiometric deposits.

Evaporation has been used by researchers to produce nanocrystalline films or powders. For example, Goodman and co-workers [6] deposited epitaxial MgO(100) films with thicknesses ranging from 2 to 100 monolayers by evaporating Mg onto Mo(100) at 300 K in 1.0×10^{-6} torr oxygen. Sasaki et al. [7] synthesized nanocrystalline Ni using the evaporation method. Small particles (10 nm in diameter) and aggregates were produced. The aggregation and coercivity were reduced with a lower evaporation temperature and a higher pressure in the evaporation chamber. They found that the evaporation rate was clearly dependent on the evaporation temperature; however, the particle size was almost independent of it.

5.4.1. Inert-Gas Condensation

Inert-gas condensation (IGC) involves the evaporation of materials using furnace or Joule heating sources into vaporized gaseous species, which are subsequently condensed onto a cold surface. Gleiter [10] adapted this technique to the fabrication of nanostructure materials. Figure 5 shows a schematic diagram of the IGC for the production of nanocrystalline powders. During the IGC process, the vaporized gaseous species lose their kinetic energy by colliding with the inert gas (e.g., He) molecules. The short collision mean free path

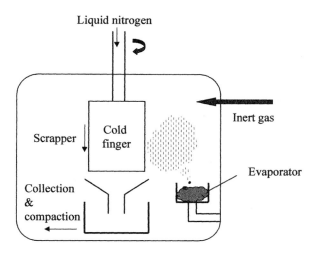

Fig. 5.   Schematic diagram of the inert-gas condensation apparatus for the production of nanocrystalline powders.

resulted in efficient cooling of the vapor species. Such cooling generates a high supersaturation of vapor locally, which leads to the homogeneous gas phase nucleation followed by cluster and particle growth through a coalescence mechanism [11]. The particles are transported via natural gas convection to a rotating cold finger, where they are collected via thermophoresis [12]. The particles are subsequently removed from the cold finger and assembled and compacted *in situ* into three-dimensional nanostructure compacts with an ultrahigh volume fraction of grain boundaries. The common processing conditions for the production of the smallest particle size while maintaining a high evaporation rate are a few hundred pascals of He and an evaporation temperature that corresponds to a vapor pressure of approximately 10 Pa for the evaporants [13]. Under such process conditions, the clusters formed normally have diameters within the range of 5 to 15 nm.

Guillou et al. [14] reported the preparation of nanocrystalline Ceria powders with narrow crystallite size distributions (3–3.5 nm diameter) by inert-gas condensation using thermal evaporation. No significant differences were observed between the powders collected in the different parts of the ultrahigh vacuum chamber. Particles develop cubic/octahedral shapes during annealing in the temperature range of 400 to 800 °C. The crystallites grow individually by a binary coalescence process and only very few grain boundaries were observed. The size of about 25% to 30% of all crystallites is not affected by sintering at 600 °C. Significant changes occur in the sample annealed at 800 °C when two populations of crystallites are formed.

Fougere et al. [15] developed a new consolidation device that was built to reduce the processing defect population inherent in samples and to minimize contamination. The Vickers microhardness of nanocrystalline Fe samples produced by a combination of inert-gas condensation and a new consolidation method is three to seven times higher than that of coarse-grained Fe.

The inert-gas condensation technique has also been used to produce nanocrystalline intermetallic compounds (Ni_3Al, NiAl, TiAl) with crystallite sizes in the range of 5 to 20 nm [16]. The as-prepared nanocrystalline Ni_3Al samples (24 at% Al) exhibited no superlattice reflection in X-ray diffraction (XRD) and transmission electron microscopy (TEM). Ordering occurred during annealing starting at 400 °C. As-prepared nanocrystalline NiAl samples (50 at% Al) were at least partially ordered. Vickers hardness measurements showed that nanocrystalline samples were substantially harder than polycrystalline, indicating that grain refinement caused strengthening. Grain growth could be inhibited using hot pressing at temperatures up to 650 °C under high vacuum as compared with annealing.

Provenzano and Holtz [17] at the Naval Research Laboratory used the inert-gas condensation method for the development of metal-based nanocomposites for high-temperature structural applications. The nanocomposite approach is based on a strengthening concept that involved the use of nearly immiscible constituents: a ductile matrix and a particulate reinforcing phase. Based on this approach, copper–niobium, silver–nickel, and copper–aluminum nanocomposites were produced, which displayed some degree of strength enhancement and high-temperature strength retention. However, the results clearly showed certain processing challenges associated with the issues of oxide contamination and consolidation of nanostructured metals and alloys by conventional processing. The highly reactive nature of nanocrystalline metals resulting from their high specific surface areas has led them to consider the potential of nanostructured metals and alloys for gas-reactive applications such as nanocrystalline palladium for hydrogen-sensing applications.

Daub and co-workers [18] used a quadruple mass spectrometer (QMS) to control the production of nanocrystalline metals by inert-gas condensation in a flow system. Nanocrystals (∼40 nm in size) were produced by the evaporation of zinc in a flow system. The metal atom concentration was monitored in relation to the argon carrier by a QMS.

The conventional evaporation process based on the Joule heating method is limited to low-melting-point or high-vapor-pressure materials, such as CaF_2 and MgO [19]. The deposition of refractory ceramic and high-temperature materials (e.g., Ti) may require the use of more powerful heating methods such as laser, electron beam, and so forth or other PVD methods like sputtering and ion plating. Furthermore, chemical reactions are likely to occur between most metal evaporants and the refractory metal crucibles, which changes the evaporation conditions. Nonhomogeneous temperature distributions are likely to occur in the molten metal using the Joule heating method, which may lead to unsatisfactory control of the evaporation. Moreover, it is difficult to control the stoichiometry of the deposited materials because the constituents of alloys tend to evaporate at different rates resulting from differences in vapor pressures, which produces films of variable composition.

Other PVD methods such as electrical explosion wire, laser ablation, molecular beam expitaxy, and sputtering methods have been employed to enable the deposition of high-temperature and more complex multicompoennt materials.

5.4.2. Electrical Explosion Wire

The electrical explosion wire method was first developed by Abrahams [20]. It was subsequently used by several researchers to fabricate metal, alloy, and ceramic powders. Kotov et al. [21, 22] employed the electrical explosion wire method for the synthesis of nano-sized ceramic powders in large quantities. The wire explosion is carried out at a discharge voltage of 15 kV and current in the range between 500 and 800 kA through the wire in a few microseconds at a chamber pressure of about 50 bar. The discharge can be repeated with a frequency of 1 Hz. The heating of Al, Ti, or Zr wire, for example, by a high-energy electric pulse to the evaporation point will disintegrate the wire into liquid spheres up to 50 μm in diameter that distribute and evaporate in the oxidizing atmosphere (Ar–O_2 mixture) and subsequently undergo combustion to form spherical oxide particles (<50 nm). Nanopowders can be produced at an average output rate of 1 kg/h with a power consumption of approximately 2 kWh/g. The specific surface of the powder is about 120 m^2/g and the powder normally contains a mixture of phases [22]. For example, rutile and anatase for TiO_2 and tetragonal and monoclinic phases for ZrO_2 can be produced. The mean particle size of the product decreases as the ratio of the electrical energy input over the energy necessary for evaporating the wire increases. Ivanov et al. [22] also reported the use of an improved pulse electromagnetic method for the compaction of nanosized powders by means of special concentrators and pressure pulses in the range of 1 to 5 GPa lasting 3 to 300 ms during the compaction. For example, the nanostructure of Al_2O_3 was preserved and a compact with a theoretical density of 62% to 83% was achieved under such compaction conditions.

5.4.3. Laser Ablation

During the laser ablation process, a focused laser beam irradiates on a target. The target becomes hot and initiates evaporation. Atoms and clusters are subsequently being ejected from the target and deposited onto a substrate, as shown in the schematic diagram in Figure 6. There are different types of lasers that can be employed for laser ablation, including the excimer laser, Nd:YAG laser, ruby laser, and CO_2 laser. The wavelength and pulse width of the laser are important. The wavelength of the ablation laser influences the absorption coefficient of the target material and the cross section of ambient gas excitation. The pulse width also plays an important role in the ablation mechanism [23]. The excimer laser is the most popular for laser ablation because of its short wavelength 193 nm (ArF) and small pulse width (6–12 ns). Various tentative models have been proposed to explain the mechanism of laser ablation. These models have been reviewed and summarized by Morimoto and Shimizu [23].

The distinctive advantages of laser ablation techniques over other deposition techniques are the capability of producing multicomponent materials with well-controlled stoichiometry. This is achieved by accurate transfer of the target composition to the coating. These features are especially important for the deposition of sophisticated material such as multi-component oxides used in superconducting, ferroelectric, and ferromagnetic applications.

The laser ablation technique has also been used for the synthesis and development of nanocrystalline films and powders. For example, Hu et al. [24] employed the laser ablation method for the deposition of nanocrystalline SnO_2 thin films using SnO_2 and Sn targets. Nanocrystalline SnO_2 films with a grain size between 4.0 and 5.2 nm were achieved by two methods: (1) crystallization of amorphous SnO_2 films at 400 °C and (2) oxidation of Sn films at 400 °C. The nanocrystalline SnO_2 films prepared with these methods exhibited much higher C_2H_5OH gas sensitivity in comparison with SnO_2 films prepared using SnO_2 and Sn targets at higher substrate temperatures with grain sizes of several tens of nanometers upon exposure to air containing 2000 ppm C_2H_5OH.

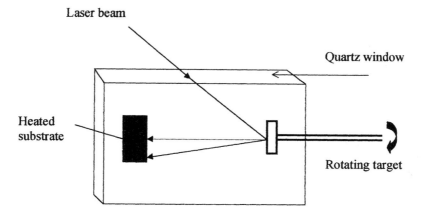

Fig. 6.   Schematic diagram of the laser ablation method.

Yamamoto and Mazumder [25] used the laser ablation method to synthesize nanocrystalline intermetallic NbAl$_3$ powders (with average particles about of 5.0 nm) by laser ablation. They investigated the production yield and ablation rate as a function of the He gas background pressure and laser pulse energy by measuring the weight of the produced powders and the weight loss of the target material before and after the experiment. They found that the ablation rate appeared to be relatively constant over the range of the He gas pressure. The production rate varied along with both He gas pressure and laser pulse energy. Optimum conditions for efficiently high production rates occurred at 1.0 torr of He gas pressure and 320 mJ of the laser pulse energy.

Laser ablation can also be performed in a reactive gas environment. For example, Goodwin et al. [26] used the reactive laser ablation of gallium metal in a N$_2$ atmosphere for the synthesis of GaN quantum dots. The GaN crystallites were about 2 nm in diameter, and followed a log-normal size distribution with a mean particle diameter of 12 nm. The GaN material exhibited combined bulk and quantum confined optical properties, which was consistent with a GaN particle size distribution that encompassed regions above and below the excitonic Bohr radius of GaN.

Nanocrystalline materials have also been synthesized using a hybrid PVD processes. For example, nanocrystalline TiC/amorphous carbon (a-C) composite films were synthesized at near room temperature using a hybrid process combining laser ablation of graphite and magnetron sputtering of titanium [27]. The deposited films consisted of 10-nm-sized TiC crystallites encapsulated in a sp^3-bonded a-C matrix. The composite films exhibited a hardness of about 32 GPa and a remarkable plasticity (40% in indentation deformation) at loads exceeding their elastic limit. They were also found to have a high scratch toughness in addition to a low (about 0.2) friction coefficient. Such properties make TiC/a-C composites potentially useful for surface wear and friction protection applications.

Although the laser ablation method is of great importance in the fabrication of multicomponent materials and it has also been used for the synthesis of nanostructured materials, its usage is limited to research and development at the laboratory level and small-scale production because of the high production cost involved. The possible availability of an improved pulsed laser at low cost in the future may lead to potential mass production for commercial applications.

5.4.4. Molecular Beam Epitaxy

Molecular beam epitaxy (MBE) is a refined form of vacuum evaporation. It was first developed for the controlled growth of III–V semiconductor epitaxial layers by Arthur [28] and Cho and Cheng [29]. This method involves neutral thermal energy molecular or atomic beams (Ga, Al, etc.) impinging on a hot crystalline substrate maintained in an ultrahigh vacuum environment that provides the required source of growth. Figure 7 shows a schematic diagram of an MBE system used for III–V growth. The evaporants are known as molecular beams when their mean free paths are greater than the chamber size (Knudsen regime, i.e., when the pressure $<10^{-4}$ torr). Epitaxy means that the growing layer derives its crystalline orientation from the underlying substrate [30]. The MBE process and growth mechanism have been reviewed in the literature [30, 31].

In the MBE method, the substrate temperature and beam intensities of each source are controlled separately for the epitaxy growth as compared to the conventional evaporation methods. The growth temperature is normally less than 66% of the melting point of the material and, thus, provides sufficient surface diffusion to allow layer-by-layer growth and well interfaces between layers, with minimal bulk interdiffusion [32]. The growth rate is well controlled and it is typically on the order of one molecular layer per second. The use of an ultrahigh vacuum system also facilitates *in situ* surface monitoring and characterization such as reflection high-energy electron diffraction, which can provide real-time growth

MBE Growth Chamber   **Analysis Chamber**

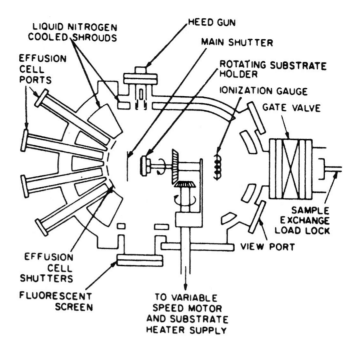

Fig. 7. Schematic diagram of an MBE system used for III–V growth. Reprinted with permission from A. Y. Cho and K. Y. Cheng, *Appl. Phys. Lett.* 38, 360 (1981). © 1981 American Institute of Physics.

monitoring and information on surface reconstruction and surface uniformity. The growth can be interrupted instantaneously using shutters over each molecular beam source. The preceding features provide the distinctive advantage of MBE for the precise controlled growth of heterostructures with layer thickness down to a single molecular layer, which is unrivaled by other growth methods. This has led to the development of nanostructure and electronic and optoelectronic devices with reduced dimension such as quantum well laser diodes and high-electron-mobility transistors [31].

MBE has been used by Bassani et al. [33] to deposit Si/CaF_2 multilayers that exhibit efficient visible luminescence at room temperature. The dimensions of the Si grains within the Si layers do not exceed 1.5 nm. The resulting optical pseudo-gap presents a large blue shift for decreasing Si layer thickness. The optical absorption results show evidence of quantum confinement of carriers in the low-dimensional Si structures in Si/CaF_2.

Notzel et al. [34] used MBE to fabricate low-dimensional nanostructures and quantum devices. They employed the concept of *in situ* formation of an array of nanometer-scale macrosteps or facets because of the natural evolution of nanometer-scale corrugations on the (311) surface. The growth using MBE was performed first by depositing a thick GaAs buffer layer until the surface is composed of (311) terraces 0.4 nm wide and two sets of $(3\ 3\ -1)$ and $(-3\ 1\ -3)$ facets of 1 nm height. This is followed by subsequent deposition of GaAs/AlAs as a double heterojunction or a multilayer with a GaAs layer less than 10 nm thick to form a structure consisting of well-ordered alternating thick and thin regions of GaAs and AlAs. The presence of strong anisotropy in the electronic properties indicates the formation of a GaAs quantum well wire superlattice. Moreover, high luminescence has been achieved in this type of structure.

MBE is a suitable method for the growth of compounds containing As. However, it has difficulty in the fabrication of structures containing phosphorus because of the allotropic nature of P, which makes the attainment of stable and reproducible flux from elemental P difficult. Such limitations have led to the introduction of gaseous group V sources to the MBE growth environment [35]. Recent developments have made use of gaseous feed stocks to replace conventional solid sources in MBE. This has improved control during the growth of III–V semiconductor heterostructures. Variants of MBE have been developed such as gas source MBE, which uses gaseous hydrides for the growth of group V elements; metalorganic MBE (MOMBE), which uses metal alkyls for the growth of group III elements; and chemical beam epitaxy (CBE), where gases are used for the growth of both group III and group V elements. MOMBE and CBE are new hybrid techniques of MOCVD and MBE that seem to be able to solve the limitations of the MBE method. However, such hybrid systems introduce greater complexity into the design of equipment and the growth kinetics. Tsang [36] has proposed the growth kinetics of MBE, MOCVD, and MOMBE.

The major drawback of MBE is that it involves the use of expensive and sophisticated equipment with high running cost. Moreover, production time could be lost because of the long period of time, required for the system to be pumped down to the required ultrahigh vacuum. There is limited potential for scale-up because of the large source-to-substrate distances and, hence, large reactor dimensions, which are required when deposition over large areas is attempted using liquid metal effusion ovens.

5.5. Sputtering

The sputter deposition process involves

1. Glow discharge to produce energetic particles (ions)
2. Momentum transfer from an indirect energetic projectile to a solid or molten target, resulting in the ejection of surface atoms or molecules to produce the sputtered species
3. Condensation of the sputtered species

A schematic diagram of the sputtering process is shown in Figure 4b.

These ejected atoms have much more energy and velocity than atoms that have been thermally evaporated. The number of atoms ejected per incident particle (ion), the so-called sputtering yield, S, is dependent on

1. Energy of the incident particle where the threshold is 5–25 eV for metals. S decreases with increasing energy because of increasing penetration depth.
2. Mass of incident particle, M. M increases with increasing S; more collisions will occur.
3. Heat of vaporization, H, of target materials. H decreases with increasing S.
4. Target temperature, T. S is insensitive to T except at very high temperatures ($>600\,°C$) during thermal evaporation.
5. Angle of incidence, θ. θ is the angle between the normal to the polycrystalline target surface and the beam direction. S increases approximately as $(\cos\theta)^{-1}$.
6. Target crystallinity. Atoms sputtered from a single-crystal target tend to be ejected predominantly along the preferred orientation of the close packing. S increases with densely packed crystallographic plane.

The process parameters include the applied target voltage and current, substrate-to-source separation distance, gas composition, pressure, flow rate, substrate conditioning, temperature, and bias.

Sputtering techniques include diode sputtering, radio frequency (rf) sputtering, and magnetron sputtering. Direct-current (dc) glow discharge sputtering can only be used for the sputtering of conductive targets. However, any target regardless of its conductivity can be sputtered using rf sputtering. Magnetron or triode sputtering is used to provide a higher current in the relatively inefficient ion source than that provided by glow discharges. In triode sputtering, a higher ionization of the plasma can be obtained by introducing electrons into the plasma from a thermionic emitter [37]. Ion beam sputtering (IBS) enables easy and independent control of the ion flux and ion energy during the deposition process. In addition, the angle of incidence of the ions to a surface can be varied.

Sputtering has a number of distinct advantages over evaporation. These include the capability of depositing high-melting-point materials such as ceramics and refractory metals, which are difficult to produce using evaporative methods. Furthermore, it can improve the purity of the deposited materials because no crucible for evaporation is used, which avoids undesirable chemical reactions and the formation of impurities. Such distinctive advantages, together with the simplicity, versatility, and flexibility of the process for modification and automation, has led to the popular use of sputtering for a wide range of commercial applications, including semiconductors, photovoltaic and optical devices, recording and automotive industries, sensors, decorative glasses, cutting tools, and so forth. Sputtering and its variants have also been used for the generation of nanostructured films and powders. Gouteff et al. [38] produced NdFeB thin films (29–1500 nm thick) using the sputtering method under a range of processing conditions to control the texture, coercivity, and magnetic saturation properties of the films. Magnetic force microscopy results showed that the all the films exhibited extensive interaction domain structures, similar to those observed in nanocrystalline bulk NdFeB. A nanostructured film of trilayer NdFeB/Fe/NdFeB with layer thicknesses of 18 nm/15 nm/18 nm was deposited using the sputtering method.

Losbichler et al. [39] deposited chemical compositionally graded Ti–B–N films onto austenite stainless steel and molybdenum sheets using a separated TiN_2/TiB_2 target during the unbalanced magnetron cosputtering process. Coatings consisting of face-centered cubic (fcc) and hexagonal close-packed (hcp) Ti–B–N phases with grain sizes ranging from 3 to 5 nm were deposited. The hardness of the Ti–B–N films was above 50 GPa and the hardness value was strongly influenced by varying the ion bombardment.

Wiedmann et al. [40] used dc magnetron sputtering to deposit TiB_2 coatings on steel and Si substrates. The coatings were subsequently vacuum annealed at 400 and 800 °C. The TiB_2 coatings consisted of a nanocrystalline structure with an average diameter of columnar grains between 20 and 50 nm, depending on the bias voltage. However, a nonstoichiometric B:Ti ratio was detectable and a high amount of Ar incorporation occurred.

He and co-workers [41] developed multifunctional ion beam-assisted deposition (IBAD) equipment with eight targets and four ion sources to synthesize diamond coatings and multilayers such as TiC/metal multilayers. The metal phases were Fe, Cr, Nb, Zr, Ni, Cu, and Al. The individual layers vary in thickness from 1 to 10 nm and the periods vary from 3 to 18 nm. The hardness and toughness of these laminated nanocrystallines depend on the properties on the modulation of the multilayers as well as on the characteristics of the metal phase.

Mei et al. [42] deposited nanocrystalline Pt films (grain size 8–9 nm) onto a liquid nitrogen-cooled Al_2O_3 substrate using rf diode sputtering. The diffusion coefficient was related to $\exp(E/kT)$ where E is the activation energy. By lowering the substrate temperature, the surface diffusion of the adsorbed species was limited and further crystallization and grain growth were prevented. Nanocrystalline S:H films (grain sizes 2–3 nm) were deposited onto a cooled substrate using rf magnetron sputtering in a hydrogen atmosphere [43].

Sattel et al. [44] formed nanocrystalline diamond by a hydrocarbon plasma beam deposition method. Diamond crystallites up to 40 nm in size were produced from a highly ionized plasma beam of acetylene for ion energies close to 100 eV per C atom and substrate temperatures above 450 °C. This work shows that diamond can be grown by physical vapor deposition from an ion-rich plasma, as well as by chemical vapor deposition from a radical-rich plasma. The formation mechanism is argued to be one of nucleation and growth rather than a stress-induced transformation from graphite.

Chou et al. [45] used a reactive sputter deposition method to produce nanocrystalline Al_2O_3 films with 10-nm average grain size. The sputter deposition was performed using an rf diode system and alumina target in an atmosphere of 4% oxygen in argon to avoid oxygen deficiency during the deposition. The substrate was maintained at a temperature below $60\,^\circ C$.

Nanocomposites. Lee et al. [46] used an rf cosputtering method to fabricate nanocomposite thin films consisting of Ag particles embedded in amorphous SiO_2 (Ag–SiO_2), nanocrystalline ZnO (Ag–ZnO), and amorphous Si (Ag–Si). Composite target sputtering materials were used in the deposition. Quantum-sized metal particles embedded in a nonmetallic matrix can be used to increase the local electromagnetic field within a nanocomposite film, leading to novel linear and nonlinear optical properties [47].

Powder. Performing the sputtering at high pressures can produce nanocrystalline particles. Ying [48] and Hann and Averback [49], for example, used a modified magnetron sputtering and inert-gas condensation method for the fabrication of nanocrystalline materials. The deposition equipment was similar to that shown in Figure 5, except that a dc/rf magnetron sputter source was used to replace the Joule heating source. Using this hybrid method and the controlled posttreatment, they produced nonstoichiometric CeO_{2-x}-based catalysts with approximately 5-nm grains, which exhibited superior catalytic activities on both SO_2 reduction by Co and CO oxidation by O_2 compared with conventional catalysts [50].

Sputtering is relatively simple and versatile, making it one of the most popular deposition methods. However, it involves the use of large and expensive targets, which limits the scope of its applications. Moreover, the sputtering method has low target material utilization in certain configurations. Typically, in conventional magnetron sputtering, only 25% to 30% of the total target material is expected to be sputtered [51]. However, the target utilization can be increased by flattening the magnetic field lines parallel to the target surface [52, 53]. Furthermore, compressive stresses may be generated in films during sputtering caused by the energetic species bombarding the substrate. In extreme cases, such stresses may cause the cracking and the delamination of the film from the substrate.

Although ion beam sputtering can be used for sputtering nonconductive targets, it requires neutralization of the ion beam to prevent charging at the target surface, which may decrease or completely stop the sputtering. The neutralization is normally performed by inserting a hot filament in the path of the ion beam. The hot filament needs frequent replacement as it will be sputtered away during operation and incorporated into the deposits, which could introduce undesirable impurities.

6. CHEMICAL VAPOR DEPOSITION

6.1. Process Principles

Chemical vapor deposition (CVD) involves thermal dissociation and/or chemical reaction of the gaseous reactants on or near the vicinity of a heated surface to form stable solid products. The deposition involves homogeneous and/or heterogeneous chemical reactions leading to the formation of powders or films. The chemical processes used in the CVD of thin films can be classified into the following types of reactions: thermal decomposition (pyrolysis), oxidation, reduction, hydrolysis, nitridation, disproportionation, photolysis, combined reactions, and so forth.

In general, the main steps occurring in the CVD process are summarized as follows using the deposition of a film as an example:

1. The active gaseous species is generated.
2. The gaseous species is transported into the reaction chamber.
3. The gaseous precursor undergoes gas phase reaction, forming an intermediate phase.

   a. At high temperature inside the reactor, homogeneous gas phase reactions occur whereby the intermediate species undergoes decomposition and/or chemical reaction, forming powders and volatile byproducts. The powders are collected on the substrate surface and may act as crystallization centers, and the byproducts are transported away from the deposition chamber.

   b. At temperatures below the dissociation of the intermediate phase, this intermediate species is diffused across the boundary layer (a thin layer close to the substrate surface).

4. The intermediate species is then absorbed onto the heated substrate, and heterogeneous reactions occur at the gas–solid (heated substrate) interface that produce the deposit and byproducts.
5. The deposit is diffused along the heated substrate surface, forming the crystallization center for subsequent growth of the film.
6. The gaseous byproducts are diffused away from the boundary layer.
7. The unreacted gaseous precursor and byproducts are transported away from the deposition chamber.

A schematic diagram of these steps is shown in Figure 8.

For the deposition of films and coatings, the process conditions are tailored to facilitate a heterogeneous reaction. A homogeneous gas phase reaction, on the other hand, is preferred for the production of powders. The main CVD process parameters include deposition temperature, pressure, input gas ratio, and flow rate. The deposition temperature is the dominant parameter. The influence of the temperature on the deposition rate can be

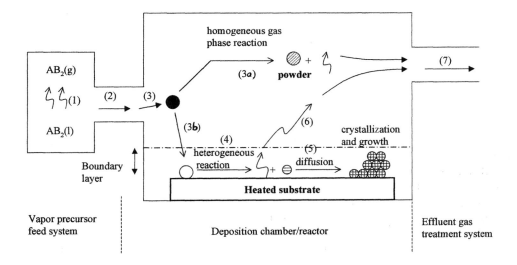

Fig. 8. Main steps occurring in the CVD process.

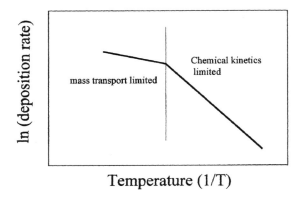

Fig. 9.   Arrhenius plot of the natural logarithm of the deposition rate versus the reciprocal temperature.

illustrated using an Arrhenius plot, as shown in Figure 9. The deposition rate increases with temperature. This is a result of the fact that high surface temperatures increase thermally activated reactions. This induces a higher growth rate and increased surface mobility of the atomic species and higher surface diffusion. The deposition increases rapidly in the chemical kinetics limited regions. At high deposition temperature, the deposition rate weakly depends on the temperature and depends more on the rate of diffusion of the active gaseous species through a boundary layer to the deposition surface. Therefore, it is mass transport limited. The deposition rate in the mass transport region could be increased by reducing the deposition pressure. This is because the supply of gaseous species to the substrate surface is enhanced at a reduced pressure.

CVD is a more complex method than PVD. It involves mass transport, heat transfer, and chemical reactions under nonequilibrium conditions. Modeling has been performed to relate deposition performance (e.g., deposition rate, uniformity, and film composition) to reactor geometry and process conditions (temperature, pressure, reactant concentrations, and gas flow rate). However, the process modeling or simulation such as described in [54, 55] is mostly based on a thermodynamic database, thermochemical equilibrium modeling, and transport phenomena modeling, which assumed the CVD system is in equilibrium, which is not the case, especially for the open CVD system where the deposition process is not in equilibrium. Therefore, the modeling results can only provide a guideline for the selection of appropriate process conditions, which need to be optimized experimentally by performing numerous test runs to reach suitable growth parameters, unlike the PVD process.

6.2. Advantages and Disadvantages

Although CVD is a complex system, it has distinctive advantages, such as the capability of producing highly dense and pure materials. Uniform films with good reproducibility and adhesion can be produced at reasonably high deposition rates. CVD is a non-line-of-sight process. Therefore, it can be used to coat complex-shaped components uniformly. Such distinctive features outweigh the PVD process. CVD process also has good throwing power and can deposit films with good conformal coverage. Ultrafine powders can also be produced using the CVD process.

However, the drawbacks of CVD method include the chemical and safety hazards caused by the toxic, corrosive, flammable, and/or explosive precursor gases. It is also relatively difficult to deposit multicomponent materials with well-controlled stoichiometry because different precursors have different vaporization rates.

6.3. Applications

The distinctive advantages of CVD outweigh its limitations. In general, CVD is a versatile deposition technique. It has become one of the main processing methods for the deposition of thin films and coatings for a wide range of applications. Some examples include the following [56]:

Semiconductors (Si, Ge, $Si_{1-x}Ge_x$, III–V, II–VI) used for microelectronics, opto-electronics, energy conversion devices

Dielectrics (SiO_2, Si_3N_4, AlN, etc.) used for microelectronics

Refractory ceramic materials (Al_2O_3, BN, $MoSi_2$, TiN, TiB_2, HfN, ZrO_2, etc.) used for hard coatings, protection against corrosion, oxidation, or as diffusion barriers

Metal films (Al, Au, Cu, Mo, Pt, W, TiN, silicides, etc.) used for microelectronics and for protective coatings

6.4. Chemical Vapor Deposition Systems

In general, CVD equipment consists of three main components, as shown in Figure 8: vapor precursor gas feed system, the reactor where the deposition occurs, and the effluent gas handling system to remove the byproduct and the unreacted precursor.

6.4.1. Chemical Precursor

The main classes of chemical compounds used in CVD reactions are inorganic compounds such as halides, hydrides, and halohydrides of metals and metalloids, as well as organometallic or metalorganic compounds.

6.4.2. Chemical Vapor Deposition Reactor

The CVD process can be performed in different types of reactors using a hot-wall or cold-wall system and at atmospheric or low pressure. The different heating methods give rise to a variety of CVD methods. The heat input can be in the form of thermal, light, plasma, and so on. This gives rise to a number of CVD techniques such as photo-assisted CVD, laser-assisted CVD, plasma-assisted CVD, and so forth.

6.4.3. Effluent Gas Handling System

The unreacted precursors and the corrosive byproduct such as HCl will be neutralized or trapped using a liquid nitrogen trap to prevent these gases from entering into the rotary or diffusion pump, which can cause damage to the pump. Inflammable gas such as hydrogen will be burnt off. The unreacted expensive precursors (e.g., BCl_3) may be collected at the outlet and recycled.

6.5. Thermally Activated Chemical Vapor Deposition

The thermally activated CVD process can be initiated using thermal energy such as resistance heating, infrared radiation, or rf heating. The process uses inorganic precursors such as halides. The thermally activated CVD can be subdivided further according to the pressure range where the deposition occurs. The deposition processes are known as atmospheric-pressure CVD (APCVD) or low-pressure CVD (LPCVD) if the deposition occurs at atmospheric or low pressure ($\sim$0.1–10 torr), respectively. The chemical reactions in both cases are basically the same. However, the mass transfer rate of the gaseous reactants becomes higher than the surface reaction rate during the LPCVD process. These two rates are of the same order of magnitude [57] in the APCVD process. Ultrahigh-Vacuum CVD (UHVCVD) ($<10^{-3}$ torr) has been developed [58] for the epitaxial growth of semi-conducting materials such as Si and SiGe alloys.

The use of thermally activated CVD for the fabrication of nanostructured materials has been demonstrated in the following examples. Stable electroluminescent devices were obtained from nanocrystalline Si thin films (15–30 nm) deposited on SiO_2 layers by LPCVD at temperatures between 580 and 610 °C [59]. The SiO_2 layers (5–20 nm) were thermally grown on Si using high-temperature oxidation. Frenklach and co-workers [60a] produced nanosized Si powders (100–400 Å in diameter) using pyrolysis of silane and disilane diluted in argon and hydrogen mixtures behind incident shock waves at temperatures ranging from 900 to 2000 K at pressures of 0.2 to 0.7 atm. The conversion of silane and disilane into silicon particles exhibited a pronounced maximum efficiency at about 1150 K, which was found to be influenced by the deposition pressure, the initial reactant concentration, and the addition of hydrogen. The produced particles were spherical, loosely agglomerated, and contained about 15% hydrogen on an atomic basis. The formation of silicon particles was monitored by the attenuation of laser beams of two different wavelengths, thereby determining the particle size, number density, and fractional yield.

6.6. Photo-assisted Chemical Vapor Deposition

Photoassisted chemical vapor deposition (PCVD) is a variant of the CVD process that relies on absorption of laser radiation in the substrate to raise its temperature and cause thermal decomposition/chemical reaction of the gaseous precursor to form the required products. Instead of having a thermally heated reactor, there is a provision of a window in the deposition chamber for optical access to the chamber via a suitably transmitting window for the excitation wavelength. The PCVD process can be performed at atmospheric or reduced pressure (e.g., 0.01–1 atm). There are a variety of photo sources for PCVD such as arc lamps, CO_2 lasers, Nd:YAG lasers, excimer lasers, and argon ion lasers.

The PCVD method has been developed for deposition at a lower temperature than in the thermally activated CVD process and for precise selected-area deposition such as laser writing of metal interconnects on an integrated circuit [60b]. The selected-area deposition can be achieved using laser scanning or projection imaging of a pulsed laser source for localized heating and to promote deposition reaction. The laser wavelength can be chosen to be highly absorbing in the substrate. The excitation energies are usually low (<5 eV), which is beneficial in avoiding film damage during deposition.

The selected-area deposition can be achieved using localized heating of the incident laser beam to promote thin-film deposition. This method may be too expensive for large-area deposition. However, its ability to provide greater control over film properties and localization of deposition is unrivaled by other CVD processes. A range of high-quality semiconductor film materials from silicon to III–V and II–VI compounds have been deposited. Currently, this method has also been used for the fabrication of nanocrystalline films and powders.

Nanocrystalline silicon powder was synthesized by the decomposition of SiH_4 gas using an excimer laser beam at a deposition pressure of 1–5 torr [61, 62]. The laser was operated at 193 nm, a repetition rate of 10–70 Hz, and a pulse duration of 24 ns. The powder was found to consist of nanocrystalline circular grains. The average size and size distribution of the grains were found to be strongly dependent on the pressure and flow rate of the gas mixture and on the repetition rate of the laser beam. The powder consists of spherical grains with an average diameter of 25 nm, and log-normal size distribution grains as small as 5 nm with a narrow size distribution could be obtained. The powder showed photoluminescence (PL) at wavelengths of 400–900 nm when excited at 488 and 330 nm.

Cao et al. [63] synthesized single-phase and stoichiometric TiN films using laser-assisted CVD by scanning linear deposition in a dynamic atmosphere on AISI 52100 bearing steel using $TiCl_4$, NH_3, C_2H_4, and H_2 as reactant gases induced by a CO_2 laser. The deposited films comprised about 2-μm equiaxed particles, each particle consisting of about 15-nm nanocrystalline grains. The average Knoop microhardness of the films was HK 1400, the highest being HK 1602.3; the wear resistance of the films was four times that of the substrates.

6.7. Plasma-Assisted Chemical Vapor Deposition

Conventional CVD and photoassisted CVD use thermal energy (heat) and light, respectively, as the activation sources for the chemical reactions, whereas plasma-assisted CVD (PACVD) uses electron energy (plasma) as the activation method. PACVD is also known as glow discharge chemical vapor deposition. This is because, by supplying electrical power at sufficiently high voltage to a gas at reduced pressure (<10 torr), breakdown of the gas occurred, generating a glow discharge plasma consisting of electrons, ions, and electronically excited species. The plasma will ionize and decompose the reactant gases at low temperature. Therefore, this process can be performed at a much lower deposition temperature than in thermally activated CVD ($\sim$500–1200 °C). However, plasma can only be generated at low pressure. Therefore, plasma-assisted CVD requires the use of a high-vacuum system and a more sophisticated reactor, which is often more expensive than the thermally activated CVD system.

The two commonly used PACVD modes are direct and remote. The direct PACVD reactors such as rf diode, microwave, and inductively coupled plasma involve gaseous precursors, inert carrier gas, and substrates being placed directly in the plasma source region. The remote PACVD methods generate a plasma away from the deposition zone. This can avoid film damage caused by energetic ions and electrons in the plasma.

The main advantage of PACVD over other thin-film deposition methods, including sputtering and evaporation, is that deposition can occur at relatively low temperatures on large areas. Ion bombardment can be substituted for deposition temperature to obtain the required film density. Such low-temperature deposition is important for applications that involve the use of temperature-sensitive substrates. The first commercial plasma reactor was developed in the 1970s for the deposition of Si_3N_4 passivation layers [64], replacing the silicon oxide/metal packaging technology and allowing the use of plastic packaging. With the availability of high-vapor pressure organometallic precursors at reasonable prices, the commercial applications of PACVD have been extended from semiconducting, dielectric, and metallic films to new applications including diamond deposition, diffusion barriers, optical filters, abrasion-resistant coatings on polymers, powder coatings, fiber coatings, and biomaterials [65]. Reviews on the PACVD methods are available in the literature [65–68].

The fabrication of diamond films cannot be realized using thermally activated CVD. The deposition of diamond films requires the use of a more intense energy form such as plasma to enable the growth of diamond films. For example, smooth (20–50-nm root mean square) diamond films with nanocrystalline structure were grown using the microwave plasma-assisted CVD method in Ar and fullerene (the carbon source) [69]. These films exhibited ultralow friction and wear properties. The friction coefficients of these films against Si_3N_4 balls were 0.04 and 0.12 in dry N_2 and air, respectively, comparable to that of natural diamond sliding against the same pin material, but were lower by factors of 5 to 10 than that afforded by rough diamond films grown in conventional $H_2 + CH_4$ plasmas. Nanocrystalline diamond films (average grain size 15 nm) were grown in an Ar–C_{60} microwave plasma [70].

Nanocrystalline diamond films (crystallites up to 40 nm in size) were grown by Sattel et al. [71] from a highly ionized plasma beam of acetylene for ion energies close to 100 eV per C atom and substrate temperatures above 450 °C. This work demonstrated that diamond would be grown by PVD from an ion-rich plasma as well as by CVD from a radical-rich plasma. The formation mechanism is argued to be one of nucleation and growth rather than a stress-induced transformation from graphite.

Plasma-assisted CVD has also been used for the fabrication of nondiamond materials. For example, Sugino et al. [72] deposited nanocrystalline boron nitride (BN) films consisting of hexagonal grains 3 nm in size using BCl_3 and N_2 as source gases. The electron emission characteristic of the Si tip array was much improved by coating with nanocrystalline BN film. The energy gap was estimated to be 6.0 eV from an ultraviolet–visible optical transmission measurement. The electrical resistivity was estimated to be 2×10^{11} and 1.3×10^2 Ω cm for undoped and sulfur-doped BN films, respectively. The tunneling barrier height was estimated to be 0.1 eV from the Fowler–Nordheim plot. *In situ* doped nanocrystalline BN, AlN, and GaN films were deposited by Werbowy and co-workers [73] using plasma-assisted CVD on silicon substrates. As a result, c-BN(n-type)/Si(p-type), as well as AlN(p-type)/Si(n-type) and GaN(p-type)/Si(n-type), heterojunction structures were fabricated.

Veprek et al. [74] deposited several micrometer-thick films of nanocrystalline Me(x)N/amorphous-Si_3N_4 (Me = Ti, W, V) materials using plasma-assisted CVD at a rate of 0.6–1 nm s^{-1} from the corresponding metal halides, hydrogen, nitrogen, and silane at deposition temperatures of less than or equal to 550 °C. A low chlorine content of less than or equal to 0.3 at% assured their stability against corrosion in air. The aim of the nanocrystalline/amorphous film concept was to avoid the formation and multiplication of dislocations in the nanocrystalline phase and to block crack propagation in a 0.3–0.5-nm thin amorphous layer. The hardness of these nanocrystalline films was greater than or equal to 50 GPa ($\sim$5000 kg mm^{-2}), the elastic modulus was greater than or equal to 500 GPa, and there was high stability against oxidation in air up to 800 °C. Moreover, they are thermodynamically more stable than diamond, c-BN, and C_3N_4 and can be prepared relatively easily.

Electron cyclotron wave resonance (ECWR) plasma sources produce highly dissociated radical fluxes at very low pressure ($\sim$10^{-4} torr), which enables deposition to occur at reasonable rates to minimize the gas phase nucleation of particles. For example, Scheib et al. [75] employed ECWR for plasma excitation during the deposition of hydrogenated nanocrystalline silicon (nc-Si:H) films. These films could be produced with high deposition rates up to 6.5 Å s^{-1} with pure SiH_4 as a process gas, in contrast to the conventional glow discharge technique where hydrogen dilution is needed for the formation of the crystalline phase, which leads to considerably lower deposition rates.

6.8. Metalorganic Chemical Vapor Deposition

CVD can also be classified according to the type of precursor used. The use of a metalorganic precursor, for example, has led to the development of metalorganic-assisted CVD (MOCVD). Metalorganic compounds contain metal atoms bonded to organic radicals. However, compounds having direct metal–carbon bonds are referred to as organometallics. This had led to the renaming of the process as organometallic CVD (OMCVD). For example, simple metal alkyls (methyl and ethyl derivatives) are the most common precursors for the growth of III–V compound semiconductors because they have high vapor pressures, the source temperatures are near room temperature, and the vapor precursors can be delivered using a carrier gas such as H_2. The deposition temperature is lower than hydrides or halohydrides, and it involves endothermic reactions; thus, cold-wall reactors with a single temperature zone can be used. However, metalorganic precursors tend to be very expensive compared to halides, hydrides, and halohydrides and they are not widely available commercially for some coating systems. Therefore, they often need to be synthesized specifically for certain applications. Furthermore, most metalorganics are volatile liquids and, thus, require accurate pressure control.

Metalorganic CVD (MOCVD) is a CVD process that uses metalorganic gas or liquid precursor systems, especially for the epitaxial growth of III–V as well as II–VI and IV–VI semiconducting materials with a specific arrangement of atoms in a regular array similar to that of the solid "substrate" upon which the film is deposited. Therefore, this CVD process is also referred to as organometallic vapor phase epitaxy (OMVPE) or metalorganic VPE (MOVPE). In addition, MOCVD has also been used to grow metallic films, dielectric films, and superconducting oxide thin films.

The MOCVD or OMCVD can be performed at atmospheric pressure (760 torr) and low pressure (20–200 torr). The thermal environment for the decomposition and/or deposition reaction of the precursors can be supplied using resistance heating or radio frequency. The common carrier gas and growth environment used during the deposition is hydrogen. A general discussion of the precursors employed in MOCVD has been compiled and summarized [76, 77]. The deposition of films using MOCVD has been reviewed [76, 78, 79]. There is also an international conference dedicated to metal organic vapour phase epitaxy, held biannually.

MOCVD has been used for the fabrication of nanostructured materials. Here are a few examples of such applications.

Gonsalves and co-workers [80] used the thermal decomposition of an amino precursor $[Ga_2(NMe_2)_6, Me = CH_3]$ in an ammonia atmosphere to produce nanostructured GaN powder consisting of approximately 50-nm-sized particles, which, in turn, were agglomerates of smaller particles with approximately 5-nm domain sizes. The photoluminescence (PL) emission spectrum of the GaN was found to be sensitive to the excitation wavelength exhibiting peaks at 378 and 317 nm. The PL excitation spectrum showed resonance in the 200–300-nm region. These PL results suggested the effect of quantum confinement in these GaN particles.

Gleiter [10] introduced the concept of inert-gas condensation (IGC) as a synthesis method of nonagglomerated nanoparticles by rapid condensation from the vapor phase using an evaporative source in a reduced-pressure environment as described in Section 5.4.1. The evaporant is suited to low-vapor-pressure and low-melting-point materials. Kear and co-workers [81, 82] have modified the IGC processing unit to accommodate the use of metalorganic precursors for the synthesis of nanophase materials. The modified process is known as chemical vapour condensation (CVC). The original evaporating source in IGC (Fig. 5) has been replaced by a hot-wall reactor in CVC to dissociate the metalorganic precursor/carrier gas to form a continuous stream of clusters or nanoparticles. This CVC process has expanded the choice of materials, which include SiC, Si_3N_4, Al_2O_3, TiO_2, and ZrO_2 and other refractory compounds [82] that are difficult to synthesize using the IGC method because of their high melting points and/or low vapor pressures. The success of the CVC process depends on the use of a low concentration of precursor in the carrier gas, rapid expansion of the gas stream through the uniformly heated tubular reactor, rapid quenching of the gas phase nucleated clusters or nanoparticles as they exit from the reactor tube, and a pressure in the reactor [81]. The synthesized powders are loosely agglomerated similar to the IGC process, which shows low-temperature sinterability as demonstrated in the fabrication of nanograined-matrix composites. For example, Al_2O_3 fiber weave infiltrated with nanoparticle ZrO_2 slurry was sintered at low temperature and achieved a fully dense composite at 1040 °C/16 ksi with no fiber-matrix reaction as compared to the use of conventional micrograined ZrO_2 powder, where sintering occurred at 1400 °C/16 ksi and exhibited a high degree of porosity, fiber damage, and the deleterious fiber–matrix interfacial reaction [83]. In addition, fiber–matrix interfaces in the nanograined-matrix composite seems to be effective in deflecting the cracks initiated by microhardness indentations.

6.9. Atomic Layer Epitaxy Process

Atomic layer epitaxy (ALE) is a variant of the CVD process that involves surface deposition for the controlled growth of epitaxial films and the fabrication of tailored molecular structures on the surfaces of solid substrates. "Monoatomic layer" can be grown in sequence by sequentially saturating surface reactions, which is a characteristic feature of ALE. Therefore, the desired coating thickness can be produced simply by counting the number of reaction sequences in the process. The surface reconstructions of the monolayer formed in the reaction sequence will influence the saturation mechanism and the saturation density obtained.

The ALE process can be performed at atmospheric pressure or with an inert gas as in CVD or in a vacuum system as in MBE. It can be considered as a special type of CVD or MBE process. The use of vacuum enables a variety of *in situ* surface analysis methods to be incorporated into the ALE equipment for the *in situ* analysis of the growth mechanism and the deposited surface structures. A review of the ALE process, as well as the reactants and reactor used, is available in [84].

ALE was initially developed for the growth of polycrystalline and amorphous thin films of ZnS and dielectric oxides for electroluminescent display devices [85]. The distinctive sequencing feature in ALE makes it an attractive method for the precise growth of crystalline compound layers, complex layered structures, superlattices, and layered alloys with precise interfaces. The ALE reaction sequences are normally performed in an "effective overdosing" condition to ensure complete saturation of the surface reaction to form the monoatomic layer. Furthermore, such effective overdosing also provides good conformal coverage that allows uniform coatings onto complex-shaped substrates. The sequencing in ALE also eliminates the gas phase reactions and enables a wider choice of reactants. The ALE process can also be scaled up for the deposition of high-quality thin films with excellent uniformity and reproducibility onto large-area substrates. The ALE method has been used to produce nanolaminate structures consisting of 3–20-nm-thick layers of two or three different oxide materials such as Ta_2O_5–HfO_2, Ta_2O_5–ZrO_5, Ta_2O_5–Al_2O_3, and so forth [86]. The leakage current of these nanolaminate structures was significantly reduced as compared to conventional dielectric films. The capability of ALE to produce layers at nanometer-level accuracy and the potential to scale up the process offer the potential application of ALE for the processing of thin-film electroluminescent displays.

7. AEROSOL-BASED PROCESSING ROUTES

7.1. Process Principles

Aerosol-based processing techniques involve the atomization of a liquid precursor into finely divided submicrometer liquid droplets (aerosol) that are distributed throughout a gas medium. The aerosol is subsequently delivered into a heated reaction zone, where the solvent is rapidly evaporated or combusted, and the intimately mixed chemical precursors are decomposed and/or undergo chemical reactions to yield the desired finished products.

7.2. Characteristics

The characteristics of aerosol-based processing techniques include the following:

1. A simple and inexpensive method unlike conventional powder processing routes that involve numerous steps such as pretreatment, mixing, chemical reaction, filtration, purification, drying, and calculation, which are tedious, time consuming, and require waste treatment. This method also has the potential to deposit quality films at low cost, outweighing conventional vacuum deposition methods (e.g., evaporation, sputtering, ion plating, plasma-assisted CVD, etc.), which are often too expensive for some industrial applications, such as thin-film coatings on large-area glass surfaces.

2. Synthesis of high-purity materials because of the absence of a liquid dispersion medium (possibly containing surfactants).
3. Good molecular mixing of chemical precursors, which enables (a) fabrication of multicomponent materials with well-controlled stoichiometry; and (b) small diffusion distances between reactants and intermediates, leading to rapid formation of the subsequent phases at relatively lower temperatures than in the solid-state sintering route.

Aerosol-based processing techniques can be used for the fabrication of ultrafine porous or nonporous homogeneous powders with narrow size distribution. The powders also tend to be unagglomerated.

7.3. Applications

The aerosol processing technique was first developed by Ebner in 1939 [87, 88] for the fabrication of advanced materials. However, the technology was not fully exploited until the late 1970s, leading to the development of a range of aerosol processing techniques based on different aerosol generation methods and/or heating methods. Currently, aerosol-based processing techniques are used for the fabrication of films and powders for electronic materials, high-performance ceramics, superconducting metal oxides, and catalysts. Thus far, the technology has been mainly applied to the fabrication of oxides and, in certain cases, sulfides and selenides as well.

There have been a few reviews on the aerosol processing of materials [89–91]. However, there is a lack of clear distinction and classification of the various aerosol processing techniques. Therefore, this section aims to provide a classification of these techniques. Furthermore, it covers the deposition mechanisms and suitability of these techniques for the processing of nanostructured materials.

7.4. Aerosol Generation Method

Aerosol processing methods involve the formation of aerosols. There are different types of aerosol generators as reviewed in [92, 93]. The aerosol generation methods will influence the size of the droplet and production rate, the particle size and distribution, and, hence, the nature and composition of the product. Each type has a different droplet formation mechanism, which produces different physical droplet properties. Pneumatic spray nozzles or humidifier units tend to generate submicrometer aerosols with large droplet size distribution. Therefore, in the processing of nanosized powders or nanostructured films, ultrasonic droplet generators and electrostatic atomization are preferred.

7.4.1. Ultrasonic Aerosol Generator

The ultrasonic aerosol generator is one of the common aerosol generators. It uses a piezo-electric transducer placed underneath a liquid precursor. When a high-frequency electric field is applied to the transducer, it vibrates and causes the formation of fine droplets. The properties of the aerosol will depend on the nature of the liquid precursor and the intensity and frequency of the ultrasonic beam. For example, this method generates water spray droplets with a mean diameter of 4.5 μm at 800 kHz; for butanol, it is 3.6 μm at the same excitation frequency [94]. A finer droplet size will enable complete vaporization and/or decomposition and facilitate a truly heterogeneous CVD reaction (Process III as described in Section 7.5), which is desirable for the deposition of dense film. The wavelength, λ, of the vibrations to the excitation frequency, f, can be described using Kevin's formula: $\lambda^3 = 2\pi\sigma/\rho f^2$, where ρ and σ are the density and surface tension, respectively. The diameter of the droplets can be determined using the equation $d = k[2\pi\sigma/\rho f^2]^{1/3}$ established by Lang [95], where k is a constant. The diameter of the droplets is a function of $\lambda(d = k'\lambda)$ and, hence, the ultrasonic frequency.

This method of aerosol generation produces droplets with a much narrower size distribution as compared to pneumatic spraying, as shown in Figure 10 [94]. Such narrow size distribution would lead to a better aerosol uniformity and, hence, coating quality.

This method of aerosol generation has been known for some time and has generated applications in different fields such as in medical inhalation and the production of calibrated powders. This type of droplet generator has also been used extensively for research and development (R&D) purposes in thin-film deposition and fabrication of fine powders. The generated aerosol is often conveyed by a carrier gas close to the substrate to be coated or into the reactor and is subsequently decomposed by pyrolysis to deposit thin films or powders. Another advantage of this method is that the gas flow rate is independent of the aerosol flow rate, which is not the case in pneumatic spraying.

This type of aerosol generator has been used extensively for R&D purposes. However, it is limited by the nature of the solution being sprayed. This is because the atomization process depends on setting a liquid film into motion, and the more viscous the liquid, and the more difficult it becomes to increase the energy input, making the atomization surface vibrate with greater amplitude. For large-scale coating or large-scale production of powders, multiple ultrasonic droplet generators may be required to form a large quantity of

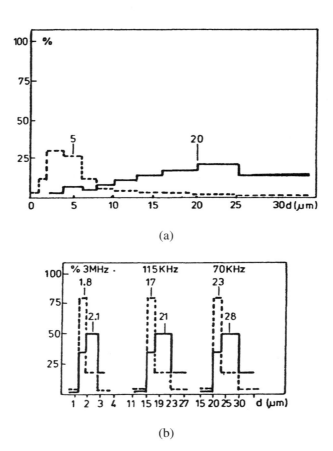

(a)

(b)

Fig. 10. Comparison of droplet distribution generated using (a) pneumatic spraying and (b) ultrasonic spraying: - - -, number distribution; —, volume distribution. Reprinted from *Thin Solid Films*, G. Blandenet et al., 77, 81 (© 1981), with permission from Elsevier Science.

fine droplets. However, a more detailed study and design of this droplet generator would be required as the existing one has a low production rate, which is not adequate for industrial applications.

7.4.2. Electrostatic Aerosol Generator and Electrospraying

Charged fine droplets can also be generated using the electrostatic atomization method, which involves the formation of a stable cone called a Taylor cone [96]. The cone is formed when the surface tension of the liquid precursor placed under an electric field balances with the electric force. The charged droplets are sprayed across an electric field and move toward the target object. This procedure increased the transfer efficiencies of sprayed droplets ranging from 60% to 90% and minimized the loss of chemical precursors to the surroundings.

The main advantages of electrostatic spraying are

1. High efficiency of material delivery to the substrate during deposition
2. Reduced consumption of liquid precursor
3. Reduced spray drift and, therefore, a safer process for both the operator and the surrounding environment

However, the main safety concerns with this technique are electrical shock caused by the high voltages required for electrostatic charging and electrical discharge arcs caused by improper equipment grounding.

Multiple electrostatic nozzles can be used for the generation of aerosol for large-area deposition of thin film. However, the challenge remains for the design of this type of aerosol generator that is capable of producing large aerosol output for the production of ultrafine powders.

7.5. Deposition Mechanism

The generated aerosol will undergo evaporation, decomposition, and/or chemical reactions to yield the desired finished product in a heated environment. The microstructure, composition and properties of the resulting deposited film depend on the deposition temperature, chemistry, and composition of the chemical precursors; the flow rate of the aerosol; the distance between the atomizer and the substrate; the atmosphere; and so forth. The deposition temperature is the most dominant process parameter. A variation of a few degrees in the deposition temperature will cause a change in the deposition mechanism and kinetics and, hence, the coating thickness and the quality of the finished product. From the reported literature [97–99], there are four possible mechanisms for the aerosol processing of films, depending on the deposition temperatures. These mechanisms are adapted here and an attempt is made to explain the various aerosol deposition processes, which are not well defined at the moment. The nature of the processes are explained and outlined as follows.

Process I. The aerosol precursor droplets are sprayed directly onto the heated substrate, followed by the removal of the solvent through evaporation and decomposition of the precursor to the finished product. Occasionally, the deposited film will be subjected to a further sintering step to achieve a dense crystalline film. In a way, this process has some similarities to the sol–gel process. The only difference is that, in aerosol processing, the precursor droplets are sprayed onto the substrate, whereas liquid precursor is being spin- or dip-coated onto the substrate in the sol-gel process. Both aerosol and sol–gel processing undergo subsequent evaporation, decomposition, and/or chemical reactions to form the required films. Dense thin films (<1 μm) with ultrafine crystalline structure can be produced using this deposition mechanism. To obtain a thick film, Process I has to be repeated several times in order to achieve the required thickness. Obviously, this is a time-consuming process for the deposition of thick film. Moreover, as the film thickness increases, the deposited film tends to be porous, and cracking or spalling of the film can occur because of repeated drying, decomposition, and/or sintering procedures. Therefore, this process mechanism may not be suitable for the deposition of a thick film with a nanocrystalline structure.

Process II. The solvent is evaporated close to the substrate surface, and the precursor precipitate is subsequently deposited onto the heated substrate and decomposed and/or undergoes chemical reactions to yield the desired materials.

Process III. The solvent is evaporated close to the substrate surface, and the precursor precipitate formed subsequently undergoes volatilization near the vicinity of the substrate and adsorption of the vapor onto the heated substrate surface, followed by the decomposition and/or chemical reactions to yield the desired materials. This mechanism is similar to the heterogeneous CVD deposition process, which tends to produce dense films with excellent coating adhesion.

Process IV. As the deposition/substrate temperature is very high, the decomposition and/or chemical reactions occur in the vapor phase, leading to homogeneous nucleation and, hence, formation of the stable particles, which are then deposited onto the heated substrates. The particles are then sintered on the heated substrates, leading to the formation of porous films, with poor adhesion.

Figure 11 depicts and summarizes these processes, with the effect of increasing temperature. Processes I, II, and III tend to produce films. The formation of films can involve either one or a combination of the preceding deposition mechanisms, which, in turn, will influence the microstructure of the films.

However, Process IV tends to produce particles in the gas phase and may lead to the deposition of porous and powdery films with poor adhesion. Therefore, Process IV needs to be avoided for film formation. However, Process IV can be adapted for the production of ultrafine powders. Uniform, spherical particles of known composition have been produced, reaction conditions can be controlled relatively easily, the size and composition can be predetermined (reproducibly obtained) for a given set of experimental conditions, and product

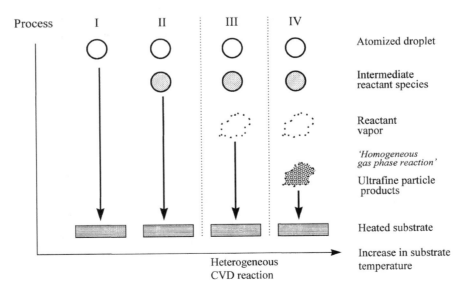

Fig. 11. Influence of the temperature on the aerosol deposition processes.

purity can generally be assured. In general, the optimization of the films or powders has been performed empirically by varying the process parameters. There is a lack of theoretical and experimental understanding of the aerosol dynamics, heat transfer, fluid flow, and kinetics of the deposition.

Various aerosol processing methods can be classified based on the preceding deposition mechanisms, as shown in Figure 11, as well as the heating method/type of reactor used to cause the thermal dissociation and/or chemical reactions. These include resistive heating, laser, flame, plasma, and so on. In addition, aerosol processes can also be classified according to the type of precursor used. For example, if metalorganic is used as the precursor, the process is known as aerosol-assisted metalorganic chemical vapor deposition (AAMOCVD). In general, nitrates, acetates, chlorides, and alkoxides are the common precursors, with alcohol (e.g., methanol, ethanol) or acetylacetone as solvents.

Most of the aerosol processing techniques are performed in an open atmosphere. However, some deposition occurs in a reduced pressure environment to increase the deposition rate by increasing the concentration of the active species near the vicinity of the decomposition zone and to facilitate the evaporation/decomposition of low volatile precursors. Furthermore, this provides a more suitable way of controlling the deposition atmosphere, especially for nonoxide deposition. The potential of these various aerosol-based vapor processing methods (e.g., spray pyrolysis, Pyrosol, aerosol-assisted chemical vapor deposition, electrostatic spray pyrolysis, electrostatic spray-assisted vapor deposition, aerosol-assisted flame synthesis) to produce nanostructured films and nanosized powders will be discussed.

7.6. Aerosol-Assisted Sol–Gel Thin-Film Deposition

This is an aerosol–gel process where an ultrasonically sprayed aerosol is directed toward a substrate at room temperature and atmospheric pressure such that the hydrolysis and polycondensation of alkoxide precursors (sol–gel polymerization) occur [100]. The deposited film will be subjected to a further annealing or sintering step to remove organic solvent and achieve dense crystalline film. The coating formation mechanism of this method occurs through Process I as described in Section 7.5. The premature sol–gel polymerization can be inhibited by controlling the vapor pressure of the solvent in the deposition reactor, which also prevents solvent evaporation during aerosol transport and maintains a low liquid precursor concentration during transport. Dense thin films (<1 μm) with ultrafine crystalline structure can be produced using this deposition method. This method is now being developed for deposition on large flat and curved surfaces. However, this process is not suitable for the deposition of thick film with nanocrystalline structure.

7.7. Spray Pyrolysis

Spray pyrolysis is also alternatively known as aerosol thermolysis, mist decomposition, evaporation decomposition of solutions, high-temperature aerosol decomposition, or aerosol high-temperature decomposition. The spray pyrolysis method occurs through Processes I–IV, depending on the processing temperature and the choice of chemical precursors.

7.7.1. Powder

Spray pyrolysis has been used extensively for the fabrication of ultrafine powders, which can be used as precursors for the formation of dense nanophase materials at a lower densification temperature or starting materials for thermal spraying.

7.7.1.1. Mechanism for the Formation of Powders

The formation of powders occurs through Process IV where the generated aerosol is delivered into a heated reactor where solvent evaporation, precursor precipitation, decomposition, and/or chemical reactions of the precursor occur. The desired powders can be collected by means of an electrostatic filter. The grain size is kept small by operating at temperatures that are sufficient for complete decomposition of the precursor but not high enough to cause excessive grain growth. The solvents used often contain liquid fuels such as alcohol, ethylene glycol, or solid fuel that is soluble in water, such as urea, that help the precursor to burst into smaller droplets during drying and facilitate the gas-to-particle conversion reaction of the powders resulting from combustion of the fuels.

One of the limitations of spray pyrolysis is the lack of morphological control over the particles, which could lead to the formation of hollow particles [90], especially when a concentration gradient is created during evaporation. Hollow particles can also be formed from precursors that melt after drying if the gases evolved by reaction cannot easily escape from the interior of the particle and, therefore, expand the particle volume. Gurav et al. [90] summarized that solid particle formation can be promoted by using high-solubility precursors, low evaporation rates, small droplet sizes, low solution concentration, long residence time, and precursors that form permeable salts. Thus, the choice of precursor with the desired physical properties is an important parameter for morphological control. For example, zirconyl hydroxychloride and zirconyl chloride resulted in the formation of solid ZrO_2 particles. Other salts formed hollow particles [101].

Gurav et al. [90] reviewed the literature on the process simulation and optimization of spray pyrolysis for particle formation, control of particle morphology, and reactor design. Most of the modeling involves droplet evaporation up to the onset of solute nucleation of large sprayed droplets (100–1000 μm) based on the basic transport momentum, heat, and mass correlation for droplets in a controlled atmosphere; heat and vapor transport and the effects associated with multiple particles are not taken into account. There is no active modeling performed after initial solute precipitation because of the complexity of describing the chemical decomposition and nucleation and growth [102]. Issues that still need to be addressed in the modeling are complex phase precipitation, diffusion, adsorption, and reaction processes that would require complex models [90].

7.7.1.2. Applications

Messing et al. [103] have reviewed the synthesis of ceramic powders using the spray pyrolysis method. Kodas [89] reviewed a wide range of powder materials. Selected examples of the application of spray pyrolysis for the synthesis of nanosized powders are summarized as follows.

Simple Oxide and Metallic Nanosized Powders. Dense, spherical palladium metal particles were produced by Plyum et al. [104] using spray pyrolysis at and above 900 °C in air and 800 °C in nitrogen, which are well below the melting point of palladium (1554 °C). At a lower temperature (500 °C), porous but not hollow aggregates of PdO (5–15 nm) were formed with surface areas of 30.2–32.8 m^2/g. Porous particles could be densified in the gas phase to form single-crystalline and fully dense powders. Solid particles of Pd formed at temperatures above 900 °C.

Mixed Oxide Nanosized Powders. Jayaram et al. [105] produced dense nanosized ZrO_2–Al_2O_3 from spray-pyrolyzed powders. Ultrafine amorphous powders of ZrO_2–Al_2O_3 were synthesized using spray pyrolysis of aqueous solutions of nitrates that had been subjected to high-pressure rapid-heating compaction at 700 °C where they underwent decomposition during heat treatment via a series of phase transformations. The use of high pressure results in a substantial decrease in the temperature required for each transformation. The resultant predominant phase is a tetragonal ZrO_2–Al_2O_3 solid solution, with a small amount of partitioned γ-alumina. Under optimum conditions, the crystallite size is 10–20 nm, with no evidence of porosity.

Alloyed Ultrafine Powders. Eroglu et al. [106] synthesized nanocrystalline 70 wt% Ni–30 wt% Fe alloy powders using spray pyrolysis from a solution of nickelcene $(C_2H_5)_2Ni$ and ferrocene $(C_2N_5)_2Fe$ dissolved in 2-methoxyethanol. The alloy powder consisted of ~10–80 nm particles. Thermodynamic analysis was used as a tool for gaining insight into the thermochemistry of the spray pyrolysis process.

Nanocomposite Powders. Carim et al. [107] produced a nanocrystalline $YBa_2Cu_3O_{7-x}$/Ag composite consisting of Ag particles (10–80 nm) dispersed uniformly within the 1-2-3 phase with nanometer grains (~10–80 nm). Chadda et al. [108] reported the synthesis of nanophase 1-2-4 superconducting powders by reacting 1-2-3/CuO composite powders (grain size 25–50 nm) at atmospheric pressure.

7.7.2. Films

Chamberlain and Skarman [109] used spray pyrolysis for the preparation of oxide and sulfide thin films for photovoltaic device applications. However, for electrical, optoelectrical, and functional applications, high-quality dense and oriented or epitaxed films with good conformal coverage are often required. However, it is difficult to obtain such films using the spray pyrolysis method. Although the vacuum techniques based on PVD and CVD can produce such coatings, they are often too expensive for commercial large-area and mass productions. Alternative methods such as Pyrosol CVD and aerosol-assisted chemical vapor deposition methods have been developed. These deposition methods occur through Process III, which facilitates truly heterogeneous CVD reactions for the deposition of dense film with controlled growth orientation and good conformal coverage.

7.8. Pyrosol

Pyrosol is an abbreviation of pyrolysis of an aerosol. This is a method patented in France and most other industrial countries [110]. This aerosol-based processing technique involves the use of ultrasonic spraying to generate an aerosol. The aerosol is conveyed by a carrier gas close to the substrate to be coated and is subsequently decomposed by pyrolysis to deposit thin films (metal oxides, sulfides, and metals). For the deposition of an oxide, air will be used as the carrier gas, whereas neutral gases, such as argon or nitrogen, will be used for nonoxide gases.

The main components used in the Pyrosol process are an ultrasonic atomizer, an aerosol spray nozzle, a pyrolysis reactor, and an exhaust gas system for the gas. Typical Pyrosol equipment is shown in Figure 12 [97]. The CVD Pyrosol process occurs through Process III to produce a dense film.

Thin films have been deposited onto glass, ceramic, or stainless steel using this method for optics, electronics, decoration, and solar energy conversion applications. The main application of this process is for metallic oxide films. Uniform and high-quality transparent conductive films of In_2O_3 and SnO_2 for electrical and optoelectrical applications have been produced using this simple and inexpensive method. In_2O_3-SnO_2 films exhibited resistivities of aproximately 2.2×10^{-4} Ω cm. The transmission of light ($\lambda = 0.4$–0.7 μm) for an approximately 6000-Å-thick In_2O_3 ($\varepsilon = 0.12$) film is 88% and 85% for SnO_2 ($\varepsilon = 0.2$). This method has been developed commercially for application as electrodes for display systems. Moreover, the process has also been used for the deposition of nonoxides such as sulfides (e.g., CdS) and noble metallic films such as Pt, Pd, Ru for use as catalysts for purifying engine exhaust fumes.

Strongly C-axis oriented ZnO thin films for microsensor devices and micromachined actuators were deposited using the Pyrosol CVD process [111]. The substrate temperature, deposition rate, and hygrometric degree of the carrier gas influence the crystalline orientations of the layers. The best texture ratio was obtained under a dry gas mixture N_2–O_2 at 495 °C and with a deposition rate of 35 A·min^{-1}.

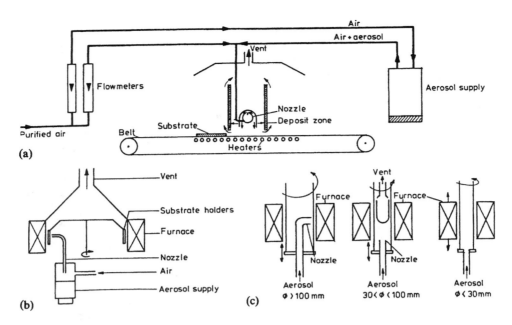

Fig. 12. Typical Pyrosol processing equipment: (a) conveyor furnace; (b) and (c) rotary furnaces. Reprinted from *Thin Solid Films*, G. Blandenet et al., 77, 81 (© 1981), with permission from Elsevier Science.

Multicomponent oxide films with oriented growth features have also been deposited using this technique. For example, [100]-oriented $Li_2B_4O_7$ thin films have been grown onto Si(111) substrates [112] for acoustic wave device applications because of their high electromechanical coupling constants and their low-temperature characteristics.

Organometallic precursors have also been used in the Pyrosol process, which alternatively can be called aerosol-assisted metalorganic chemical vapor deposition. For example, the epitaxial growth of substituted bismuth garnet thin films, BiDyGaIG, have also been deposited using gallium and iron acetylacetones $[Me(C_5H_7O_2)]$, bismuth triphenyl $[Bi(C_6H_5)_3]$, and $Dy(TMHD)_3$ dissolved in butanol to form a solution of 0.03 M. The deposition rate was 6 nm/min at a substrate temperature of 500–540 °C. The grain size is around 350 Å with a conventional annealing for 3 h at 650 °C [113, 114]. These films have potential applications for magnetooptic memory applications.

Films consisting of catalytic aggregates containing Pt or Pd nanoparticles (3–5 nm) dispersed onto SnO_2 grains (10–25 nm) for CO gas sensor applications can be produced using the Pyrosol method at 540 °C [115]. Figure 13 shows a high-resolution electron microscopy (HREM) image of the film. A distortion of the SnO_2 grain is observed because of the interaction between the Pt particle and the SnO_2 grain.

However, this technique has a coating thickness limitation (about 1 μm) that restricts the range of applications. Therefore, this technique cannot be used for the deposition of thick coatings for tribological applications.

7.9. Aerosol-Assisted Chemical Vapor Deposition

Aerosol-assisted chemical vapor deposition (AACVD) process involves the use of volatile precursors dissolved in liquids or solid particles in volatile precursors and suitable precursor decomposition temperatures that cause the deposition to occur through Process III. If a metalorganic precursor is used, the process is known as aerosol-assisted metalorganic chemical vapor deposition.

The AACVD process can be performed in an open atmosphere or a low-pressure reactor. It exhibits the following features:

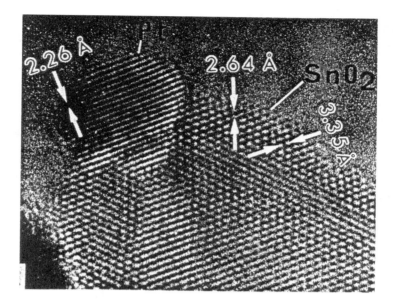

Fig. 13.   HREM of Pt (11%) SnO$_2$ films. Reprinted with permission from B. Gautheron et al., *J. Solid State Chem.* 102, 434 (1993). © 1993 Academic Press.

1. High deposition rates
2. Uniformity of thickness
3. Complete surface coverage
4. Precise compositional control
5. Deposition over three-dimensional devices, wires, or complex-shaped substrates

However, in some cases, a subsequent heat treatment is required to obtain the desired crystallinity, composition, or structure. This may be avoided by controlling the atmosphere in the vicinity of the substrate.

Films.   Examples of the fabrication of thin films using AACVD include InSe and In$_2$Se$_3$ [116], BaTiO$_3$ [117], and TiO$_2$ and Al$_2$O$_3$ [118, 119]. Gurav et al. [120] have synthesized fullerene (crystallite size was about 10 nm) and nanocomposites consisting of Rh-fullerene (20–50-nm-sized particles) using organometallic precursor solutions of [(1,5-cyclo-octadiane)RhCl]$_2$ and/or mixed fullerene extract (C80, C70) in toluene.

Powders.   Schulz et al. [121] produced film nanoparticles of CdTe (diameter 30–50 Å) by injecting a mixture of Cd(CH$_3$)$_2$, (n-C$_8$H$_{17}$)$_3$PTe, and (n-C$_8$H$_{17}$)$_3$P into (n-C$_8$H$_{17}$)$_3$PO at elevated temperatures. The nanoparticles were subsequently used for CdTe thin-film deposition by forming a colloidal suspension of CdTe nanoparticles in a suitable organic solvent and then spraying the colloid onto SnO$_2$-coated glass substrates at variable suspension temperatures.

7.10. Electrospraying-Assisted Deposition

Although Taylor [96] initiated studies of the behavior of liquid in an electric field in the 1960s, research on the use of electrostatic atomization/spraying for processing only began in the 1980s. Prior to the application of electrostatic spraying for the processing of materials, electrostatic spraying has been used successfully for paint spraying where electrostatic forces are used to move paint particles toward and adhere them on a target object. It also has been used in the agriculture industry where electrostatic forces are used to spray droplets toward vegetation.

To date, aerosol generated using the electrostatic aerosol generator as described in Section 7.4.2 has been explored by various research groups for the processing of films, powders, and composites. Different groups have adopted different technical names, for example, corona spray pyrolysis, electrostatic spray pyrolysis, electrostatic spray deposition, gas–aerosol reactive electrostatic deposition technique, electrospray organometallic chemical vapor deposition, and electrostatic spray-assisted vapor deposition. Each of the deposition processes is slightly different because of the nature of the chemical precursor used, the process conditions (e.g., deposition temperature, field strength, distance between the atomizer and substrate, etc.), and the design of the deposition equipment, which give rise to different deposition mechanisms (Processes I–IV as described in Section 7.5) and thereby lead to products of different quality. Some of the more successful techniques are proprietary and are the subjects of patents.

7.10.1. Corona Spray Pyrolysis

Early work on the use of electrospraying for processing has been investigated by Siefert [122, 123]. He reported the use of corona spray pyrolysis of metal halides in butyl acetate at 300–450 °C for the deposition of In_2O_3 and SnO_2 films (0.22–0.36 μm thick), which enhanced the low deposition efficiency of spray pyrolysis from a few percent to over 80%. The optical and electrical properties of the doped In_2O_3 and SnO_2 films were reported to be comparable with those reported in the literature. The paper showed the promise of corona spray pyrolysis as a new coating technique. However, since then, the author or the research group reported no further work. Therefore, whether there is any inherent limitation associated with this method remains unknown.

Unvala [124] employed the electrospraying of a metalorganic compound into a heated tube to generate ionized vapor that was transported into a CVD reactor where the decomposition and chemical reaction occur to deposit semiconducting materials for the manufacture of integrated circuits.

Rulison et al. [125] used electrospray atomization of high-concentration (similar to 400 g/L) hydrated yttrium nitrates dissolved in n-propyl alcohol solutions into fine aerosol. The aerosol was carried into an electric furnace where it resided for several seconds before being decomposed at 500 °C to form powders composed of dense, spheroidal, submicrometer, and nanocrystalline oxide particles.

7.10.2. Electrostatic Spray Pyrolysis/Electrostatic Spray Deposition

Schoonman et al. [126, 127] employed electrospraying for the deposition of $LiCoO_2$ and $LiMn_2O_4$ at a rate of 1–5 μm h^{-1}. The SEM micrographs revealed particulate deposition rather than atomistic deposition, which normally occurs in the CVD and PVD processes. The microstructure produced may be adequate for application as a cathode in rechargeable lithium. However, for the processing of nanophase materials, atomistic deposition is crucial; therefore, the chemistry of the precursors and the process conditions need to be tailored to facilitate deposition through Process III (i.e., heterogeneous CVD reaction).

The successes of electrospraying for aerosol processing of nanophase materials have been demonstrated in some recent studies. For example, Park and Burlitch [128] produced TiO_2 nanoparticles by electrospraying a Ti acetylacetonate precursor solution into a furnace reactor. Ultrafine particles with a mean particle size of 20 nm have been produced by controlling the concentration of the precursor and, hence, the hydrolysis and condensation reactions.

7.10.3. Electrospray Organometallic Chemical Vapor Deposition

Danek et al. [129] developed an electrospray organometallic chemical vapor deposition (ES-OMCVD) method for the fabrication of II–VI quantum dot composites. This is a

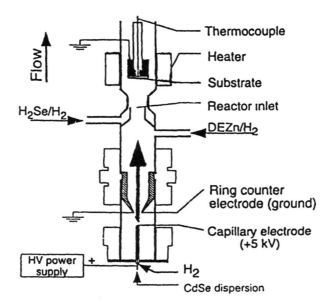

Fig. 14. Schematic diagram of the ES-OMCVD reactor. Reprinted with permission from M. Danek et al., *Appl. Phys. Lett.* 65, 2795 (1994). © 1994 American Institute of Physics.

method that combined electrospray and organometallic chemical vapor deposition (OM-CVD). Figure 14 shows a schematic diagram of the ES-OMCVD reactor. Nanocrystals of CdSe were transferred by electrospray into the OMCVD reactor and codeposited on a ZnSe matrix from the chemical reactions of hydrogen selenide and zinc at 150–250 °C. Highly monodispersed CdSe nanocrystals (∼0.5 nm) in a ZnSe matrix were produced. The absorption and emission properties of the composites could be finely adjusted in a broad special region by selecting the size of the nanocrystals.

7.10.4. Gas-Aerosol Reactive Electrostatic Deposition

Another experimental setup has been developed by Dobson et al. [130, 131] for the processing of nanocomposites, as shown in Figure 15. This method is known as gas–aerosol reactive electrostatic deposition and has been used for the fabrication of nanocomposites consisting of quantum-sized semiconductor compounds (groups II–VI and III–V) embedded in a wide variety of polymers. For example, cadmium nitrate in polyvinyl alcohol was atomized electrostatically and sprayed into a reaction chamber. Ions of Cd react with H_2S from the gas phase producing CdS nanoparticles (<5 nm) surrounded by the polymer molecules. The distribution of the crystallite size was less than 10%.

7.10.5. Electrostatic-Assisted Chemical Vapor Deposition

Choy and Bai [132a, b] have developed an electrostatic spray-assisted vapor deposition (ESAVD) method to deposit uniform films, which can offer significant reduction in production cost. This process involves spraying atomized precursor droplets across an electric field. By carefully tailoring the chemistry of the precursor and process conditions, the droplets will undergo chemical reaction in the vapor phase near the vicinity of the heated substrate (i.e., heterogeneous CVD reaction; Process III as shown in Figure 11) to deposit a stable solid film with excellent adhesion onto a substrate in a single production run. Figure 16 shows a schematic diagram of the deposition process. This method is capable of producing thin or thick strongly adherent coatings with well-controlled stoichiometry,

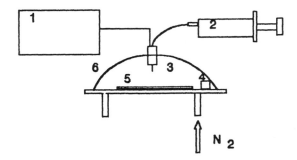

Fig. 15. Schematic diagram of the gas–aerosol reactive electrostatic deposition apparatus for nano-CdS/polymer films: 1. high-voltage supply; 2. syringe with a spraying solution; 3. capillary; 4. source of hydrogen sulfide; 5. substrate; 6. reaction chamber. Reprinted from *Thin Solid Films*, O. V. Salata et al., 251, 1 (© 1994), with permission from Elsevier Science.

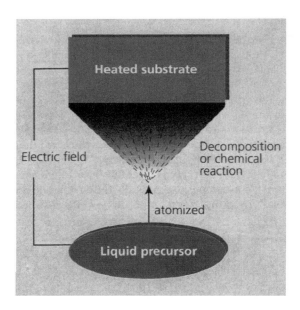

Fig. 16. Schematic diagram of the ESAVD deposition process. Reprinted with permission from K. L. Choy, *Materials World*, March 1998, 144 (© 1998 Materials World).

crystallinity, and texture. The method has the potential for molecular tailoring of multi-layer, composite, or gradient coatings. Other advantages of the process include its high deposition rate (e.g., 1–5 μm/min) and low deposition temperature (e.g., 200–650 °C). Simple oxides (e.g., Al_2O_3, SiO_2, SnO_2), multicomponent oxides [e.g., $La(Sr)MnO_3$, $BaTiO_3$, $PbTiO_3$, PZT], and doped oxides (e.g., Y_2O_3–ZrO_2, SnO_2–In_2O_3), as well as polymer films (e.g., polyvinylidene fluoride), have been deposited using sol-based precursors. This technique is now being exploited for the deposition of sulfide- and selenide-based nonoxide films. The technical and economical viability of this method for the manufacturing of improved solid oxide fuel cell components has been successfully demonstrated. Other spin-off commercial applications include reforming catalysts, ceramic membranes for selective gas separation, catalytic combustors, thermal barrier coatings, bioactive coatings, ferroelectric films for sensor, and memory devices.

This method can also be tailored to enable atomistic deposition for the fabrication of nanostructured materials. In principle, nanocrystalline powders, nanostructured thin films,

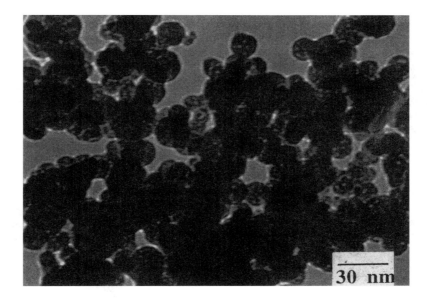

Fig. 17. TEM of homogeneous nanoscrystalline YSZ powders deposited at 500 °C. Reprinted with permission from K. L. Choy, *Materials World*, March 1998, 144 (© 1998 Materials World).

and nanocomposite coatings can be deposited using ESAVD-based methods. Figure 17 shows homogeneous nanocrystalline YSZ powders (mean particle size 20 μm) deposited at 500 °C. These nanocrystalline YSZ powders can reduce the sintering temperature of YSZ significantly to 900 °C as compared to commercial YSZ powders produced by Tosoh (Japan), which requires a sintering temperature of at least 1400 °C in order to achieve full density. Nanophase PZT films with an average grain size less than 100 nm have been deposited. The preliminary results indicate that the dielectric of this film is 320, which is comparable to that of PZT films deposited using CVD or PVD methods.

It has also been demonstrated that the ESAVD method is ideally suited to produce well-controlled compositions and nanometer-scaled textures of $CaO-P_2O_5-SiO_2$ and bioactive coatings on orthopedic alloys with complex designs and surface textures optimized for maximum osseomechanical integration. The flexibility of the process makes it feasible to tailor the composition, texture, and crystallinity of the gradient coating at the metal interface to produce maximum adherence and stability and optimal bioactivity.

8. FLAME-ASSISTED DEPOSITION

8.1. Process Principles

Flame-assisted deposition (also called flame synthesis or combustion flame synthesis) involves the combustion of liquid or gaseous precursors injected into diffused or premixed flames. Hydrogen diluted in argon is often used as the fuel for the flame because the combustion of hydrogen is a very fast process and it does not produce condensed species. Hydrocarbons could also be used as the primary fuel, but their combustion often leads to the formation of soot and can be quite slow. For the synthesis of oxide materials, liquid halide precursors such as $Al(CH_3)_3$ and $TiCl_4$ can be used together with an oxidizer stream consisting of oxygen diluted in argon for the synthesis of the oxides (e.g., Al_2O_3 and TiO_2). Halide precursors are often chosen because they have a relatively high vapor pressure at room temperature and they are relatively cheap.

The flame source and the combustion process provide a unique thermal environment and the energy necessary for vaporization, decomposition, and/or chemical reactions during the chemical processing of materials. Thus, this method enables the *in situ* chemical synthesis of powders (1–500 nm) or film in a single-step operation without postprocessing such as calcination. Moreover, the flame has a high temperature (typically 2000–3000 K) that allows the use of volatile and less volatile chemical precursors. This energy-efficient method offers rapid mixing of reactants on a molecular scale, thus reducing the processing time significantly. This method can avoid the time-consuming processing procedures and the consumption of large volumes of liquid precursor and the generation of a large quantity of byproducts encountered in conventional wet chemistry such as the sol–gel process.

The crystal structure, morphology, and particle size can be controlled by optimizing the processing parameters such as flame temperature and its distribution, precursor residence time in the flame, choice of precursor, and precursor concentration ratio. Introducing additives into the flames [133, 134] can also control the size, phase, and shape of the products.

This method can be distinguished from thermal spraying and its variants such as plasma spraying, which use solid precursors as starting material in the synthesis. Therefore, thermal spraying involves the use of a higher-energy thermal source (hydrogen fuel, plasma) as compared to flame-assisted deposition in order to melt the solid precursor to a molten or semimolten state before forming the required products.

8.2. Advantages and Disadvantages

The main advantage of using flame-assisted deposition (FAD) is that the desired products can be formed *in situ* without further heat treatment. For the fabrication of oxide materials, the process can be performed in an open atmosphere without the need for a sophisticated reactor or vacuum system as in the PVD or plasma-assisted CVD methods. Therefore, FAD-based methods are relatively low cost and have the potential to be scaled up for large-scale production, especially for the fabrication of powder.

The major drawback of FSD-based methods is the large temperature fluctuation during processing. This is due to the vast temperature gradient present in the flame. Therefore, these methods are less suitable for thin-film deposition and are mostly used commercially for the production of powders. However, the instability of the flame temperature causes poor control over the particle size, its distribution, and the microstructure and leads to inhomogeneity in the powders. This drawback may prevent the production of high-quality powders and limit their applications. This limitation may be overcome by the development of a more uniform flame such as reported by Katz et al. [135, 136] and Kear et al. [137, 138]. Katz et al. [135, 136] have developed a counterflow diffusion flame burner to produce a flame that is very flat and uniform in the horizontal plane. Kear et al. [137, 138] have produced a flat and uniform flame using a reduced-pressure flat flame burner. This type of burner may be more expensive because it involves the use of low pressure and a more sophisticated reactor design. However, it will be beneficial for the fabrication of ultrafine nonoxide materials, which are difficult to produce using conventional flame synthesis methods in an open atmosphere.

8.3. Applications

The advantages of using flame burners in the processing of materials were first identified by Cuer et al. [139, 140]. They demonstrated the feasibility of synthesizing Al_2O_3 and mixed oxides of SiO_2 and Al_2O_3 particles in an open flame burner. The benefit of using combustion flame synthesis for the production of oxide powders has long been recognized by the oxide powder industry. At present, tens of thousands of tons of oxide powders such as TiO_2 and SiO_2 are manufactured annually by the combustion of metal chloride precursors in hydrocarbon flames [141]. Ulrich [141], Kodas [142], Kriechbaum and Kleinschmit [143], and Pratsinis and Kodas [144] provided reviews on the flame synthesis of powders. Powders produced using this method have been used as starting materials for the fabrication of advanced engineering ceramics for both structural (e.g., rocket engines and combustors) and functional (e.g., piezoelectric devices, capacitors, thermistors, catalysts, and solar cells) applications [145].

8.4. Counterflow Diffusion Flame Synthesis

Chung and Katz [146] have developed a counterflow diffusion flame burner to produce a flame that is very flat and uniform in the horizontal plane. Figure 18 shows a schematic diagram of the counterflow diffusion flame burner. This burner consists of two vertically opposed tubes of rectangular cross section separated by a distance of 15 mm. The fuel flows upward from the lower tube and the oxidizer flows downward from the upper tube. A flame is generated in the region where the two opposed gas streams impinge. The distinctive advantage of this flame reactor design is that the flame is very flat and uniform in the horizontal plane. Thus, both temperature and concentration distributions are also uniform in the horizontal plane, and the gas flow along the stagnation streamline ($Y = 0$) can be considered as essentially one dimensional. This flatness also enables the accurate measurement and monitoring of temperature, vapor concentration, and light scattering intensity along the stagnation streamline. High-purity powders with high surface area and with the desired particle size, morphology, and crystalline structure can be fabricated by optimizing the processing parameters.

This type of flame reactor has been used extensively by Katz et al. [135, 136] to study the synthesis of mixed oxides. By simultaneous combustion of $TiCl_4$ and $SiCl_4$, they have

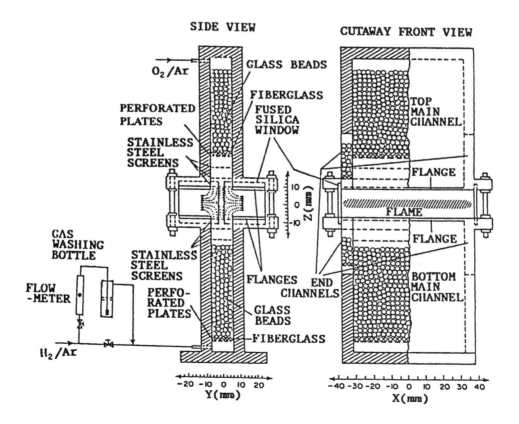

Fig. 18. Schematic diagram of the counterflow diffusion flame burner. Reprinted with permission from C. H. Hung and J. L. Katz, *J. Mater. Res.* 7, 1861 (1992). © 1992 Materials Research Society.

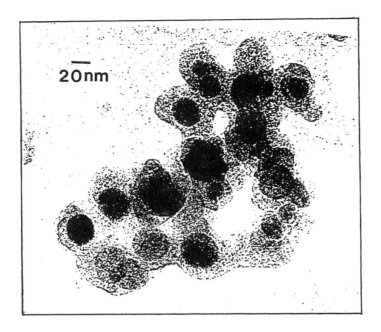

Fig. 19. Transmission electron micrograph of TiO$_2$ coated with SiO$_2$ produced using the counterflow diffusion flame burner. Reprinted with permission from C. H. Hung and J. L. Katz, *J. Mater. Res.* 7, 1861 (1992). © 1992 Materials Research Society.

obtained TiO$_2$ particles (30–38 nm) encapsulated by the SiO$_2$ coating layer (10–30 nm thick) in a single step. Figure 19 shows the TEM of TiO$_2$ coated with SiO$_2$ produced using the counterflow diffusion flame burner. Such a processing route of TiO$_2$ pigment is more economical and environmentally friendly than the conventional route, which involves combustion generation of TiO$_2$ powders followed by subsequent SiO$_2$ coating using silicate solutions. A model describing the particle growth processes for TiO$_2$, SiO$_2$, and TiO$_2$–SiO$_2$ has been proposed [135]. In addition, the *in situ* formation of nanostructured vanadium–phosphorous oxide (VPO) catalytic particles from VOCl$_3$ and PCl$_3$ liquid precursors in a hydrogen–oxygen flame using the counterflow diffusion flame burner has been demonstrated [136]. At low temperature (e.g., 1200 K), the powders collected are chainlike structures composed of particles 5–10 nm in diameter. However, the chainlike structures collapsed as the temperature increases (>1950 K), forming spherical particles (~30–50 nm in diameter) at high temperature.

Other researchers have also explored and developed flame synthesis methods for the fabrication of engineering ceramic materials. For example, Akhtar et al. [147] used flame reactors to synthesize nanostructured TiO$_2$ and Al-doped TiO$_2$ powders for use as pigments, pain pacifiers, catalysts, and engineering ceramics. Nanosized Si$_3$N$_4$ powders for advanced ceramics have been produced by Calcote et al. [148] using a mixture of SiH$_4$/NH$_3$/N$_2$ precursor in a premixed burner with a stabilized flame.

8.5. Combustion Flame Chemical Vapor Condensation

Kear et al. [137, 138] have synthesized nanopowders of silica, titania, and alumina at high production rates (up to 50 g/h) using a new, scalable combustion technique, which is a modification of the original chemical vapor condensation (CVC) process as described in Section 6.8, whereby the hot-wall reactor is replaced by a reduced-pressure flat flame reactor. The as-synthesized powders are loosely agglomerated and exhibit a narrow size distribution with a mean diameter between 10 and 20 nm.

8.6. Sodium/Halide Flame Deposition with *in situ* Encapsulation Process

Materials produced from nanopowders have many advantageous properties, including increased strength, high ductility, and improved fabricability. To maximize these benefits, the particulate must be 2–100 nm in size, unagglomerated, and uncontaminated. However, necking between particles tends to occur in particles that are produced using conventional aerosol synthesis routes, which will lead to the formation of hard agglomerates with poor densification characteristics. Such agglomeration, therefore, counteracts some of the advantages of nanostructured materials.

Axelbaum et al. [149] have developed a novel patented method for producing high-purity, unagglomerated nanosized powders based on the use of *in situ* encapsulation during the sodium/halide flame deposition process. This method can be used for the production of nonoxide nanosized powders such as Ti, TiB_2, and TiN, which are difficult to produce using conventional aerosol techniques, except using the expensive and sophisticated vacuum-based deposition techniques (such as inert-gas atomization).

In this novel approach [149, 150], a coflow burner has been developed to react sodium vapor with gaseous halides in a flame configuration similar to the hydrocarbon coflow flame. Sodium coflow flame is used in this method because the flame can be operated such that the products of combustion are all in condensed phase. In addition, unlike hydrocarbon flames, the sodium flame is hypergolic and is not stabilized by the premixed-flame mechanism.

For example, TiB_2 nanoparticles have been synthesized by the vapor phase reaction of sodium with a 1:2 mixture of $TiCl_4$ and BCl_3. Once the nucleation and growth of primary particles such as TiB_2 has proceeded to the desired size, the particles are then coated with a suitable material such as NaCl before they agglomerate. The flame is operated under conditions that lead to the condensation of the NaCl byproduct onto the particles. The NaCl coating acts to control the size and eliminate agglomeration of the particles in the flame because the core TiB_2 particles are encapsulated in NaCl. The particle encapsulation with NaCl also helps to protect the oxygen/moisture-sensitive core particles during postflame handling. It also helps to narrow the size distribution of the core particles. The NaCl coating can be efficiently removed from the core particles by washing with water/glycerine or by sublimation at 800 °C in a vacuum. TiB_2 particles with a mean diameter of less than 15 nm have been produced [151]. A thermodynamic analysis of the Ti/B/Cl/Na system indicates that near 100% yields of TiB_2 are possible with appropriate reactant concentrations, pressures, and temperatures. This method of encapsulation, with a removable coating material, is a general method of obtaining unagglomerated nanoparticles in a flame.

A range of nanosized nonoxide materials such as metals (Ti, TiN, TiB_2), ceramics, and composites (W/Ti, Al/AlN) have been fabricated using this novel method [152, 153]. For example, unagglomerated cubic or hexagonal nanoscale (~30 nm) W/Ti particles were fabricated in a sodium/halide flame using the novel *in situ* encapsulation process. When the particles were consolidated by hot pressing at 1000–1200 °C in Ar atmosphere, 96% dense consolidates with grain sizes of 30–40 nm were produced [152]. The sodium/halide flame and encapsulation process has the potential to be scaled up for large-scale synthesis of nonoxide nanosized powders.

9. CONCLUSIONS

This review has shown that PVD and CVD methods have the capability of producing nanophase materials. However, most of these vapor processing techniques involve the use of a vacuum system and sophisticated deposition chamber. Therefore, the drawbacks of these vapor processing techniques are the high production costs and the difficulty of fabricating nanophase materials cost effectively in large quantity. Consequently, high-quality nanostructured materials are costly and difficult to be scaled up for large-scale and mass production. Therefore, potential commercially high-payoff applications are likely to be focused and exploited by industry that use limited quantities of materials such as coatings, catalysts, and small engineering structures/devices (e.g., quantum dots, bearings). The recent development of novel and cost-effective vapor processing methods, especially those based on the aerosol and flame synthesis methods, offer cheaper alternatives to the conventional CVD and PVD techniques and may widen the scope of commercial applications of vapor processing of nanostructured materials.

References

1. R. D. Schull, *Nanostruct. Mater.* 2, 213 (1993).
2. H. Gleiter, in "Second Riso International Symposium on Metallurgy and Materials Science" (N. Hansen, A. Horsewell, T. Lefferes, and H. Lilholt, eds.), p. 15. Riso National Laboratory, Roskilde, Denmark, 1981.
3. D. A. Glocker and S. I. Shah, eds., "Handbook of Thin Film Process Technology." Institute of Physics Publishing, Bristol, UK, 1995.
4. J. L. Vosen and W. Kern, eds., "Thin Film Processes II." Academic Press, Boston, 1991.
5. L. I. Maissel and M. H. Francombe, eds., "An Introduction to Thin Films." Gordon and Breach, New York, 1973.
6. C. Xu and D. W. Goodman, in "Handbook of Nanophase Materials" (A. N. Goldstein, ed.), p. 250. Marcel Dekker, New York, 1997.
7. Y. Sasaki, K. Shiozawa, H. Tanimoto, Y. Iwamoto, E. Kita, and A. Tasaki, *Mater. Sci. Eng., A* 217, 344 (1996).
8. Y. A. Kotov and N. A. Yavorovski, *Fiz. Khim. Obrab. Mater.* 4, 24 (1978).
9. Y. A. Kotov, I. V. Beketov, A. M. Murzakaev, O. M. Samatov, R. Boehme, and G. Schumacher, *Mater. Sci. Forum* 255, 913 (1996).
10. H. Gleiter, in "Deformation of Polycrystals: Mechanism and Microstructures" (N. Hansen, A. Horsewell, T. Lefferes, and H. Lilholt, eds.), p. 15. Riso National Laboratory, Roskilde, Denmark, 1981.
11. C. G. Granqvist and R. A. Buhrman, *J. Appl. Phys.* 47, 2200 (1976).
12. R. W. Siegel, *Ann. Rev. Mater. Sci.* 21, 559 (1991).
13. R. W. Siegel and J. A. Eastman, *Mater. Res. Soc. Symp. Proc.* 132, 3 (1989).
14. N. Guillou, L. C. Nistor, H. Fuess, and H. Hahn, *Nanostruct. Mater.* 8, 545 (1997).
15. G. E. Fougere, J. R. Weertman, and R. W. Siegel, *Nanostruct. Mater.* 5, 127 (1995).
16. T. Haubold, R. Bohn, and R. Birringer, *Mater. Sci. Eng.* 153, 679 (1992).
17. V. Provenzano and R. L. Holtz, *Philos. Mag. B*, 76, 593 (1997).
18. O. Daub, W. Langel, C. Reiner, and L. Kienle, *Ber. Bunsen-Ges. Phys. Chem.* 101, 1753 (1997).
19. H. Hann and R. S. Averback, *J. Appl. Phys.* 67, 1113 (1990).
20. R. Abrams, "Production and Analysis of Radioactive Aerosols." University of Chicago Press, Chicago, 1946.
21. Y. A. Kotov and Y. Yavorovsky, *Fiz. Khim. Obrab. Mater.* 4, 24 (1978).
22. V. Ivanov, Y. A. Kotov, O. H. Samatov, R. Bohme, H. U. Karow, and G. Schumacher, *Nanostruct. Mater.* 6, 287 (1995).
23. A. Morimoto and T. Shimizu, in "Handbook of Thin Film Process Technology" (D. A. Glocker and S. I. Shah, eds.), A1.5. Institute of Physics Publishing, Bristol, UK, 1995.
24. W. S. Hu, Z. G. Liu, Z. C. Wu, and D. Feng, *Mater. Lett.* 28, 369 (1996).
25. T. Yamamoto and J. Mazumder, *Nanostruct. Mater.* 7, 305 (1996).
26. T. J. Goodwin, V. J. Leppert, S. H. Risbud, I. M. Kennedy, and H. W. H. Lee, *Appl. Phys. Lett.* 70, 3122 (1997).
27. A. A. Voevodin, S. V. Prasad, and J. S. Zabinski, *J. Appl. Phys.* 82, 855 (1997).
28. J. R. Arthur, *J. Appl. Phys.* 39, 4032 (1968).
29. A. Y. Cho and K. Y. Cheng, *Appl. Phys. Lett.* 38, 360 (1981).
30. S. A. Barnett and I. T. Ferguson, in "Handbook of Thin Film Process Technology" (D. A. Glocker and S. I. Shah, eds.), A2. Institute of Physics Publishing, Bristol, UK, 1995.
31. N. Inoe, Y. Kwamura, and Karimoto, in "Handbook of Nanophase Materials" (A. N. Goldstein, ed.), p. 83. Marcel Dekker, New York, 1997.
32. C. P. Flynn, *J. Phys. F: Met. Phys.* 18, L195 (1988).
33. F. Bassani, I. Mihalcescu, J. C. Vial, and F. A. d Avitaya, *Appl. Surf. Sci.* 117, 670 (1997).
34. R. Notzel, L. Daweritz, and Ploog, *J. Cryst. Growth* 127, 858 (1993).
35. M. B. Panish, *J. Electrochem. Soc.* 127, 2730 (1980).
36. W. T. Tsang, *Appl. Phys. Lett.* 45, 1234 (1984).

37. T. C. Tisone and P. D. Cruzan, *J. Vac. Sci. Technol.* 12, 677 (1975).
38. P. C. Gouteff, L. Folks, and R. Street, *J. Magn. Magn. Mater.* 177, 1241 (1998).
39. P. Losbichler, C. Mitterer, P. N. Gibson, W. Gissler, F. Hofer, and P. Wirbichler, *Surf. Coat. Technol.* 94–95, 297 (1997).
40. R. Wiedmann, H. Oettel, and M. Jerenz, *Surf. Coat. Technol.* 97, 313 (1997).
41. X. M. He, H. D. Li, C. H. Liu, and W. Z. Li, *J. Mater. Process. Technol.* 63, 902 (1997).
42. X. Mei, M. Tao, H. Tan, Y. Han, and W. Tao, *Mater. Res. Soc. Symp. Proc.* 286, 179 (1993).
43. S. Furukawa and T. Miyasato, *Phys. Rev. B* 38, 5726 (1988).
44. S. Sattel, J. Robertson, Z. Tass, M. Scheib, D. Wiescher, and H. Ehrhardt, *Diamond Relat. Mater.* 6, 255 (1997).
45. T. C. Chou, D. Adamson, J. Mardinly, and T. G. Nieh, *Thin Solid Films* 205, 131 (1991).
46. M. H. Lee, I. T. H. Chang, P. J. Dobson, and B. Cantor, *Mater. Sci. Eng., A* 179/180, 545 (1990).
47. D. E. Aspnes, *Am. J. Phys.* 50, 704 (1982).
48. J. Y. Ying, *J. Aerosol. Sci.* 24, 315 (1991).
49. H. Hann and R. S. Averback, *J. Appl. Phys.* 67, 1113 (1990).
50. A. Tschope, W. Liu, M. Flytzani-Stephanopoulos, and J. Y. Ying, *J. Catal.* 157, 42 (1995).
51. T. Van Vorous, *Solid-State Technol.* 19, 62 (1976).
52. T. Hata and Y. Kamide, *J. Vac. Sci. Technol., A* 5 2154 (1978).
53. B. W. Manlley, U.S. Patent 5,262,028, 1993.
54. M. Pons, C. Bernard, and R. Madar, *Surf. Coat. Technol.* 61, 274 (1993).
55. Y. J. P. Couderc, *J. Phys. IV* 3, 3 (1993).
56. L. Vescan, in "Handbook of Thin Film Process Technology" (D. A. Glocker and S. I. Shah, eds.), B1.4:28. Institute of Physics Publishing, Bristol, UK, 1995.
57. W. Kern and R. S. Rosler, *J. Vac. Sci. Technol.* 14, 1082 (1977).
58. T. S. Meyerson, *Appl. Phys. Lett.* 48, 797 (1986).
59. A. G. Nassiopoulou, V. Ioannou Sougleridis, P. Photopoulous, A. Travlos, V. Tsakiri, and D. Papadimitriou, *Phys. Status Solidi A* 165, 79 (1998).
60. (a) M. Frenklach, L. Ting, H. Wang, and M. J. Rabinowitz, *Isr. J. Chem.* 36, 293 (1996). (b) P. Burggraaf, *Semicond. Internat.* 116 (1988).
61. S. Berger, S. Schachter, and S. Tamir, *Nanostruct. Mater.* 8, 231 (1997).
62. S. Tamir and S. Berger, *Thin Solid Films* 276, 108, (1996).
63. L. X. Cao, Z. C. Feng, Y. Liang, W. L. Hou, B. C. Zhang, Y. Q. Wang, and L. Li, *Thin Solid Films* 257, 7 (1995).
64. R. S. Rosler, *Solid State Technol.* 67 (1991).
65. F. Jansen, in "Handbook of Thin Film Process Technology" (D. A. Glocker and S. I. Shah, eds.). Institute of Physics Publishing, Bristol, UK, 1995.
66. J. Hopwood, in "Handbook of Nanophase Materials" (A. N. Goldstein, ed.), p. 141. Marcel Dekker, New York, 1997.
67. G. Lucovsky and D. V. Tsu, in "Thin Film Processes II" (J. L. Vossen and W. Kern, eds.), p. 565. Academic Press, Boston, 1991.
68. R. Reif and W. Kern, in "Thin Film Processes II" (J. L. Vossen and W. Kern, eds.), p. 525. Academic Press, Boston, 1991.
69. A. Erdemir, C. Bindal, G. R. Fenske, C. Zuiker, A. R. Krauss, and D. M. Gruen, *Diamond Relat. Mater.* 5, 923 (1996).
70. A. Erdemir, M. Halter, G. R. Fenske, A. Krauss, D. M. Gruen, and S. M. Pimenov, *Surf. Coat. Technol.* 94–95, 537 (1997).
71. S. Sattel, J. Robertson, Z. Tass, M. Scheib, D. Wiescher, and H. Ehrhardt, *Diamond Relat. Mater.* 6, 255 (1997).
72. T. Sugino, K. Tanioka, S. Kawasaki, and J. Shirafuji, *Diamond Relat. Mater.* 7, 632 (1998).
73. A.Werbowy, J. Szmidt, A. Sokolowska, and A. Olszyna, *Diamond Relat. Mater.* 7, 397 (1998).
74. S. Veprek, M. Haussmann, S. Reiprich, L. Shizhi, and J. Dian, *Surf. Coat. Technol.* 87–88, 394 (1996).
75. M. Scheib, B. Schroder, and H. Oechsner, *J. Non-Cryst. Solids* 200, 895 (1996).
76. R. D. Dupuis, in "Handbook of Thin Film Process Technology" (D. A. Glocker and S. I. Shah, eds.), B1.1:1. Institute of Physics Publishing, Bristol, UK, 1995.
77. A. C. Jones, *J. Cryst. Growth* 129, 728 (1993).
78. J. P. Duchemin, *J. Vac. Sci. Technol.* 18, 753 (1981).
79. M. J. Ludowise, *J. Appl. Phys.* 58, R31 (1985).
80. K. E. Gonsalves, S. P. Rangarajan, G. Carlson, J. Kumar, K. Yang, M. Benaissa, and M. J. Yacaman, *Appl. Phys. Lett.* 71, 2175 (1997).
81. B. H. Kear and P. R. Strutt, *Nanostruct. Mater.* 6, 227 (1995).
82. Y. Chen, N. Glumac, B. H. Kear, and G. Skandan, *Nanostruct. Mater.* 9, 101 (1997).
83. S. Bose, Unpublished research, Pratt & Whitney, East Hartford, CT, 1994.
84. T. Suntola, in "Handbook of Thin Film Process Technology" (D. A. Glocker and S. I. Shah, eds.), B1:5.1. Institute of Physics Publishing, Bristol, UK, 1995.

85. T. Suntola and J. Antson, *Soc. Inform. Display Dig.* 108 (1980).

86. M. Ritala and M. Leskela, ESF Workshop on Thin Layers and Coatings from Aerosols, Grenoble, June 1997.

87. K. Ebner, U.S. Patent 2,155,119, 1939.

88. K. Ebner, West German Patent 877 196, 1953.

89. T. T. Kodas, *Angew. Chem. Int. Ed. Engl.* (1989).

90. A. Gurav, T. Kodas, T. Plyum, and Y. Xiong, *Aerosol Sci. Technol.* 19, 411 (1993).

91. S. E. Pratsinis and T. T. Kodas, in "Aerosol Measurement" (K. Willeke and P. Baron, eds.), Chap. 33. Van Nostrand–Reinhold, New York, 1992.

92. M. Kerker, *Adv. Colloid Interface Sci.* 5, 105 (1975).

93. O. G. Raabe, in "Fine Particle" (B. Y. H. Liu, ed.), pp. 60–110. Academic Press, 1975.

94. G. Blandenet, M. Court, and Y. Lagrade, *Thin Solid Films* 77, 81 (1981).

95. J. W. S. Rayleigh, "The Theory of Sound", Vol. 2, New York, 1945.

96. G. I. Taylor, *Proc. R. Soc. London Ser. A* 280, 383 (1964).

97. G. Blandenet, M. Court, and Y. Lagrade, *Thin Solid Films* 77, 81 (1981).

98. J. C. Viguie and J. Spitz, *J. Electrochem. Soc.* 122, 585 (1975).

99. I. Wiedmann, K. L. Choy, and B. Derby, *Br. Ceram. Proc.* 53, 133 (1994).

100. M. Langlet and C. Vautey, *J. Sol.–Gel Sci. Technol.* 8, 347 (1997).

101. S. C. Zhang, G. L. Messing, and M. Borden, *J. Am. Ceram. Soc.* 73, 61 (1990).

102. Y. Sano and R. B. Keey, *Chem. Eng. Sci.* 37, 881 (1982).

103. G. L. Messing, S. C. Zhang, and G. V. Jayanthi, *J. Am. Ceram. Soc.* 2707 (1993).

104. T. C. Plyum, S. W. Lyons, Q. H. Powell, A. Gurav, T. Kodas, and L. M. Wang, *Mater. Res. Bull.* 28, 369 (1993).

105. V. Jayaram, R. S. Mishra, B. Majumdar, C. Lesher, and A. Mukherjee, *Colloids Surf. A* 133, 25 (1998).

106. S. Eroglu, S. C. Zhang, and G. L. Messing, *Mater. Sci. Res.* 11, 2131 (1996).

107. A. H. Carim, P. Doherty, and T. T. Kodas, *Mater. Lett.* 8, 335 (1989).

108. S. Chadda, T. T. Kodas, T. L. Ward, A. Karim, D. Kroeger, and K. C. Ott, *J. Aerosol. Sci.* 22, 601 (1991).

109. R. R. Chamberlain and J. S. Skarman, *J. Electrochem. Soc.* 113, 86 (1966).

110. J. Spiz and J. C. Viguie, Fr. Patent 2,110,622, 1972; U.S. Patent 3,880,112, 1975; U.S. Patent 3,890,391, 1974; West German Patent 2,151,809, 1974; Jpn. Patent 83-845-7, 1971; Br. Patent 1,362,803, 1974.

111. M. Labeau, P. Rey, J. L. Deschanvres, J. C. Joubert, and G. Delabouglise, *Thin Solid Films* 213, 94 (1992).

112. V. Bornand, A. El Bouchikhi, Ph. Papet, and E. Philippot, *J. Phys. III* 7, 853 (1997).

113. J. L. Deschanvres and J. C. Joubert, *J. Magn. Mater.* 101, 224 (1991).

114. J. L. Deschanvres et al., *IEEE Trans. Mag.* 26, 187 (1990).

115. B. Gautheron, M. Labeau, G. Delabouglise, and U. Schmatz, *J. Solid State Chem.* 102, 434 (1993).

116. H. J. Gysling, A. A. Wernberg, and T. N. Blanton, *Chem. Mater.* 4, 900 (1992).

117. C. H. Lee and S. J. Park, *J. Mater. Sci. Mater. Electr.* 1, 219 (1990).

118. W. J. Desisto, Y. T. Qian, C. Hannigan, J. O. Edward, R. Kershaw, K. Dwight, and A. World, *Mater. Res. Bull.* 25, 183 (1990).

119. P. Wu, Y. M. Gao, R. Kershaw, K. Dwight, and A. World, *Mater. Res. Bull.* 25, 357 (1990).

120. A. S. Gurav, Z. Duan, L. Wang, M. J. Hampden-Smith, and T. T. Kodas, *Chem. Mater.* 4, 900 (1993).

121. D. L. Schulz, M. Pehnt, C. J. Curtis, and D. S. Ginley, *Mater. Sci. Forum* 225, 169 (1996).

122. W. Siefert, *Thin Solid Films* 121, 275 (1984).

123. W. Siefert, *Thin Solid Films* 120, 267 (1984).

124. B. A. Unvala, Br. Patent 2,192,901, 1988.

125. A. J. Rulison and R. C. Flagan, *J. Am. Ceram. Soc.* 77, 3244 (1994).

126. C. Chen, E. M. Kelder, P. J. J. M. van der Put, and J. Schoonman, *J. Mater. Chem.* 6, 765 (1996).

127. A. A. van Zomeren, E. M. Kelder, J. C. M. Marijnissen, and J. Schoonman, *J. Aerosol Sci.* 25, 1229 (1994).

128. D. G. Park and J. M. Burlitch, *Chem. Mater.* 4, 500 (1992).

129. M. Danek, K. F. Jensen, C. B. Murray, and M. G. Bawendi, *Appl. Phys. Lett.* 65, 2795 (1994).

130. P. J. Dobson, O. V. Salata, and P. J. Hull, Br. Patent 9,323,598.7, 1993.

131. O. V. Salata, P. J. Dobson, P. J. Hull, and J. L. Hutchison, *Thin Solid Films* 251, 1 (1994).

132. (a) K. L. Choy and W. Bai, Br. Patent 9,525,505.5. (b) K. L. Choy, *Materials World*, March 1998, p. 144.

133. B. S. Haynes, H. Jander, and H. G. G. Wagner, in "17th International Symposium on Combustion," Combustion Institute, Pittsburgh, PA, 1979, p. 1365.

134. M. K. Akhtar, S. E. Pratsinis, and S. V. R. Mastrangelo, *J. Am. Ceram. Soc.* 75, 3408 (1992).

135. C. H. Hung and J. L. Katz, *J. Mater. Res.* 7, 1861 (1992).

136. P. F. Miquel and J. L. Katz, *J. Mater. Res.* 9, 746 (1994).

137. N. G. Glumac, Y. J. Chen, G. Skandan, and B. Kear, *Mater. Lett.* 34, 148 (1998).

138. Y. J. Chen, N. G. Glumac, G. Skandan, and B. H. Kear, *ACS Symp. Ser.* 681, 158 (1998).

139. J. P. Cuer, J. Elston, and S. J. Teichner, *Bull. Soc. Chim. Fr.* 81 (1961).

140. J. P. Cuer, J. Elston, and S. J. Teichner, *Bull. Soc. Chim. Fr.* 89 (1961).

141. G. D. Ulrich, *Chem. Eng. News* 62, 22 (1984).

142. T. T. Kodas, *Angew. Chem. Int. Ed. Engl.* (1989).

143. G. W. Kriechbaum and P. Kleinschmit, *Adv. Mater.* 6, 330 (1989).
144. S. E. Pratsinis and T. T. Kodas, in "Aerosol Measurement" (K. Willeke and P. Baron, eds.), Chapt. 33. Van Nostrand–Reinhold, New York, 1992.
145. H. J. Sanders, *Chem. Eng. News* 62, 26 (1984).
146. S. L. Chung and J. L. Katz, *Combust. Flame* 61, 271 (1985).
147. M. K. Akhtar, S. E. Pratsinis, and S. V. R. Mastrangelo, *J. Mater. Res.* 9, 1241 (1994).
148. H. F. Calcote, W. Felder, D. G. Keil, and D. B. Olson, in "23rd International Symposium on Combustion," Combustion Institute, Pittsburgh, PA, 1990, p. 1739.
149. R. L. Axelbaum, L. J. Rosen, and D. P. DuFaux, U.S. Patent 5,498,446, 1996.
150. D. P. DuFaux and R. L. Axelbaum, *Combust. Flame* 100, 350 (1995).
151. R. L. Axelbaum, D. P. DuFaux, C. A. Fret, K. F. Kelton, S. A. Lawton, and L. J. Rosen, *J. Mater. Res.* 11, 948 (1996).
152. R. L. Axelbaum, J. I. Huertas, C. R. Lottes, S. Hariprasad, and S. M. L. Sastry, *Mater. Manufact. Proces.* 11, 1043 (1996).
153. R. L. Axelbaum, D. P. Dufaux, C. A. Frey, and S. M. L. Sastry, *Metall. Mater. Trans. B* 28, 1199 (1997).

Chapter 13

APPLICATIONS OF MICROMACHINING TO NANOTECHNOLOGY

Amit Lal

Department of Electrical and Computer Engineering, University of Wisconsin, Madison, Wisconsin, USA

Contents

Handbook of Nanostructured Materials and Nanotechnology, edited by H.S. Nalwa
Volume 1: Synthesis and Processing
Copyright © 2000 by Academic Press

ISBN 0-12-513761-3/$30.00

1. INTRODUCTION

The definition of micromachines changes from person to person. A definition requiring that at least one dimension of a micromachined device be in the micrometer range is generally accepted. Micromachining is a discipline that is even harder to define than micromachines. Micromachining is a rapidly evolving area of research, the scope of which seems to change constantly as new techniques to make micromachines are invented. Because of the ever-increasing set of tools and techniques available to micromachinists, it is difficult to summarize micromachining in a complete and all-encompassing manner. Hence, in this chapter, the focus will be on micromachining principle that seem to be common to most micromachine fabrication. In particular, focus will be on silicon bulk and surface micromachining techniques, which are commonly identified as being at the core of micromachining.

Micromachining has largely been developed in the hope of producing microelectromechanical systems (MEMS). MEMS are micromachines that can have mechanical and fluidic components integrated with electromagnetic fields and electronic circuits that have useful functions as sensors and actuators. Both academia and industry are exploring the possibilities of using micromachining to make useful devices. For example, micromachined devices such as pressure sensors and accelerometers have been available in the market for many years. However, a relatively unexplored area of research has been the use of micromachines to fabricate nanomachines, which could result in nanoelectromechanical systems (NEMS). This is what Richard Feynman had in mind three decades ago [1] when he made popular the notion of nanomechanical parts. Such devices can be defined as devices with at least one nanoscale dimension. Examples of such devices are micromachined probes for atomic force microscopy and tunneling microscopy. It seems certain that micromachines will play a significant role in the fabrication and operation of NEMS.

Figure 1 shows the size scale of devices fabricated by different fabrication techniques. With bulk micromachining, silicon wafers are usually etched through. Because the wafer thickness is typically 0.55 mm, minimum device sizes using these techniques are about a cubic millimeter. Using surface micromachining techniques of sacrificial layer etching, one can make devices that have dimensions of tens of micrometers and volumes of tens of cubic micrometers. To carry over these techniques, it seems certain that at least one limitation will be the lithography. Lithography constrains the smallest size features that can be reliably defined and, therefore, constrains the smallest mechanical component that could be fabricated. This limitation can be eliminated by using advanced lithography technologies such as X-ray lithography [2], electron beam lithography [3], and probe-based [scanning tunneling microscopy (STM), atomic force microscopy (AFM)] lithography [4].

Another limitation to NEMS could be the granularity of the available structural materials. For example, if polysilicon films with grain size of 100 nm are used as the structural material, devices smaller than 100 nm in dimension will have statistical existence. This potential problem can be eliminated by using either amorphous or crystalline materials.

Nanoscale devices tethered to substrates during fabrication will encounter tremendous forces during wet chemistry. The surface-to-volume ratio becomes much higher for nanoscale devices as compared to microscale devices. This large surface-to-volume ratio

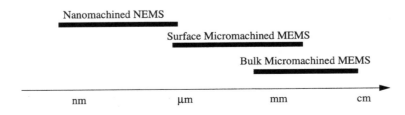

Fig. 1. Device sizes and associated fabrication technology.

would mean much larger viscous forces on the devices during any wet processing. This limitation could be eliminated by the use of purely dry processing techniques such as plasma etching and deposition and probe-based lithography.

Recently, several books covering micromachines and micromachining in great detail have been published [5–9]. The expert in the field of micromachining will, therefore, be best served by finding the details in these books. This chapter will summarize micromachining knowledge so that upon reading a novice is prepared to understand details in other references. Another goal of this chapter is to familiarize the reader with key micromachining concepts such that he or she can apply them to new materials and applications. To aid in this process, processing recipes are given in the appendices to help a person start micromachining.

It is also worth mentioning what this chapter does not cover. Associated with micromachines is a large field of mechanics and dynamics of micromechanical motion. Dynamics at small scales is in itself a subject and has been covered in other textbooks [6]. MEMS have many kinds of actuators and sensors that work on many physical and chemical principles and would require a chapter on just that subject. This chapter also does not cover issues in mechanical design of MEMS using computer-assisted design (CAD) tools, a rapidly emerging area of research. Finally, this chapter focuses on silicon micromachining, although techniques for micromachining other relevant materials such as III–V compounds, glasses, and plastics have been developed. There are also fabrication techniques using lasers [10], focused ion beams [6], electron discharges [6], and electron beams [3], which are not covered here. To find details pertaining to these topics, one can go to some of the textbooks, but, more important, to MEMS conference proceedings. One important international conference is the transducers' conference held every two years. A second conference is the Solid State Sensor and Actuator Workshop, also held every two years at Hilton Head Island, South Carolina. A third important conference is the IEEE MEMS conference held annually. A more recent MEMS meeting venue has been the SPIE conferences, also held annually.

2. BASIC PROCESSING TECHNOLOGY

Micromachining has been mostly developed in integrated-circuit fabrication laboratories. Most micromachining techniques utilize microfabrication procedures to fabricate microscale mechanical components, perhaps integrated with electronics. The basic tool kit of micromachining thus consists of microfabrication techniques such as lithography, thin-film deposition, chemical etching, and plasma deposition/etching.

Lithography is used to transfer a pattern onto a substrate using a master mask that contains the desired pattern [11]. The pattern can be transferred using electromagnetic waves (visible optics, X-rays, deep ultraviolet (UV), etc.) or electron beams. The mask material is typically a metal or polymer layer deposited on a glass plate. The masks are themselves patterned using a lithographic process that involves direct electron or photon exposure of photoresist in precision $x-y$ steppers [12].

Most lithography starts by spin coating the substrate with a photoresist. One alternative to spin coating is spraying the photoresist over the substrate, which is often used on surfaces that are not smooth. Bulk micromachining (to be discussed later) almost always produces three-dimensional features on which it is very hard to spin coat or even spray coat a photoresist. Another useful alternative in such cases is the use of electroplated negative resists that conformably coat arbitrarily shaped objects [13].

As shown in Figure 2, the photons or the electrons travel through the mask into an organic polymer called a photoresist. The exposed areas in the photoresist are chemically modified. In the case of a positive photoresist, exposed areas become soluble in developer solutions. In the case of a negative photoresist, exposure induces insolubility in developers.

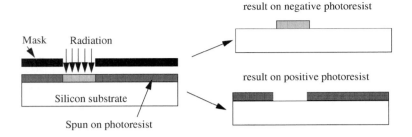

Fig. 2. Basic lithography process of pattern transfer onto negative and positive photoresist.

After exposure, the wafers are processed in a chemical called a developer, which dissolves the soluble sections of the photoresist pattern. The patterned photoresist is then baked at temperatures of approximately 120 °C for approximately 1 h to make it mechanically stronger and more chemically resistant. The exposed substrate areas are then etched by either wet or dry etchants. The mask material is also etched, but at a rate that is typically much lower than the substrate etch rate. The ratio of the substrate to the mask etch rate is called the etch-rate ratio. It is usually preferable to have very high etch rates, so that the mask layer thickness can be made much smaller than the substrate thickness to be etched.

As the substrate material is being etched vertically, the exposed substrate area at the mask edge can result in lateral etching (Fig. 3). This etching is called underetching and is usually undesirable to the extent that the substrate pattern is not an exact copy of the original lithography mask. Underetching is also dependent on the etch time and introduces statistical errors in device dimensions. Underetching also limits the line resolution or the smallest line one can define using lithography. However, underetching is also very useful because it forms the basis for surface micromachining described later in this chapter. There are certain etching techniques (such as dry etching) that can result in no undercutting at all. In general, etch profile engineering is of critical importance to micromachining.

3. MATERIALS

Although micromachining techniques are not limited by choice of materials, historically micromachining has developed around silicon and associated materials. The most fundamental material is crystalline silicon because it forms the substrate or the foundation for most micromachined devices.

3.1. Silicon Wafers

Whereas in integrated-circuit (IC) fabrication silicon is used mainly as a substrate, in micromachining it usually is the structural material of the mechanical device. Silicon wafers

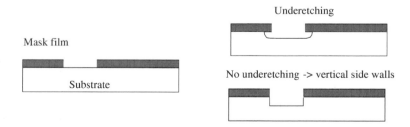

Fig. 3. Underetching or vertical side walls can be obtained upon etching.

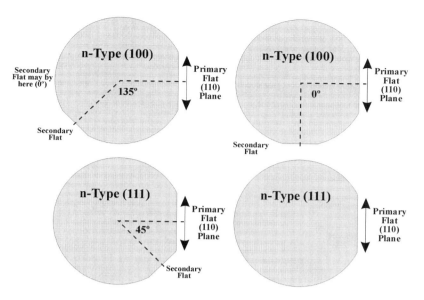

Fig. 4. Industry convention on wafer flats for ⟨100⟩- and ⟨111⟩-type wafers. Reprinted with permission from G. T. A. Kovacs, "Micromachined Transducers Sourcebook." (© 1998 McGraw-Hill).

are etched and bonded to other silicon devices to form composite silicon structures. In most applications, one uses standard wafers used by the IC industry because they tend to be the lowest in cost. Figure 4 shows the crystalline configurations in which silicon wafers are commercially available with the corresponding flat orientations. The standard thicknesses of silicon wafers increase with the wafer thickness. For example, 100-mm-diameter (4-in.) wafers are specified as 500–550 μm thick; 6-in. (150 mm) wafers are specified as 600–650 μm. The variations in the wafer thickness specifications can sometimes put limitations on the accuracy with which bulk-micromachined parts can be made.

Silicon atoms in a silicon crystal are arranged in the zinc-blende [14] crystal structure. Silicon crystals can be formed by simple repetition of the basic unit cell, which consists of atoms arranged in the zincblende structure. Each of the faces of the unit cell can be described by a vector such as ⟨100⟩, ⟨111⟩, ⟨110⟩, and so forth. One can buy silicon wafers whose surface consists of unit cells with the same face exposed everywhere. Commercial wafers come in three orientations of ⟨100⟩, ⟨110⟩ or ⟨111⟩. The ⟨110⟩ wafers have traditionally been used for the bipolar circuit process, while the ⟨100⟩ wafers are mostly used in metal oxide semiconductor (MOS) processes because of superior oxidation properties.

3.2. Thin Films

Thin films play the role of both a structural material and a fabrication aid. For example, in surface micromachines, micromechanical structures are formed from polysilicon thin films, whereas silicon oxide thin films are used as sacrificial layers and are removed in the final structure. There are at least three types of thin films that are central to micromachines. They are silicon dioxide, silicon nitride, and polysilicon thin films. Each type of film is utilized in different ways depending on the application. Furthermore, each type of film can be deposited in different ways. The choice of fabrication technique, in turn, determines the electrical, thermal, mechanical, chemical, and topographical properties of the thin films. Both the chemical structure and the morphology of thin films is affected by the growth process. In particular, the growth mechanism affects the stress in thin films, which is of great importance to MEMS structures. For a good discussion on the general properties of thin films, the reader is referred to [12, 15–17].

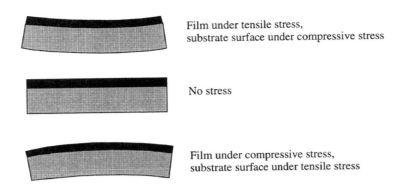

Film under tensile stress,
substrate surface under compressive stress

No stress

Film under compressive stress,
substrate surface under tensile stress

Fig. 5. A thin film can be in compressive or tensile stress, bending the wafer concavely or convexly.

Stress in thin films is a topic that is perhaps not well covered extensively in standard IC processing textbooks, but is of central importance in micromachined structures [18]. The thin-film stress (or strain) can be compressive, tensile, or neutral. A compressive stress in a thin film means that it is being held in compression by the substrate; that is, if the film is peeled away from the substrate, it will have dimensions greater than the substrate dimensions. Such a film, if deposited only on one side of the wafer, will bend the wafer, as shown in Figure 5. This compressive stress in the thin film induces a tensile stress in the substrate. Similarly, a thin film in tensile stress deposited only on one side of the wafer will bend the wafer, also as shown in Figure 5. Commercial stress-measuring tools measure the wafer curvature to estimate the thin-film stress. Typical stress levels in thin films are measured in hundreds of megapascals (MPa). It is noteworthy that stress levels in wafers can be large enough to fracture wafers. In MEMS, thin-film structures are often attached to the substrate at an anchor point. Any stress in the thin films results in bending, bowing, buckling, and other usually undesirable effects. However, built-in stress can also be used to self-assemble structures [19].

Stress in thin films occurs because of three factors:

1. *Interfacial stress.* If the atomic density of the deposited material is different from the substrate but the deposited material is chemically attached to the substrate, the molecules at the substrate interface are stressed. An example of this is thermally grown silicon dioxide in which oxygen diffuses into the silicon substrate and reacts with silicon atoms. The larger size of the SiO_2 molecules leads to compressive stress in the film and tensile stress in the substrate.

2. *Bulk thermal stress.* Most materials have nonzero thermal expansion coefficients. Hence, a film grown at high temperature wants to shrink when it is brought back to room temperature, resulting in tensile stress.

3. *Internal stress.* Films grown under conditions that result in atomic density variations of multiple species can have considerable internal stresses. Another source of internal stress is the presence of grain boundaries and interfacial stresses along them. All built-in stresses in thin films can vary with time and ambient temperature.

3.3. Important Thin-Film Materials

Silicon dioxide is perhaps the most widely used thin-film material in MEMS. It can serve as an electrical and thermal insulator, as a sacrificial layer, or as a stress compensator [20]. Some of the common ways of growing oxide thin films are discussed later. The stress level in most oxide films, with the exception of PECVD (Plasma-enhanced conformal vapor deposition) films, is compressive.

Thermal oxides are the easiest thin films to grow in a laboratory. Silicon substrate is thermally converted to silicon dioxide by reacting with dry oxygen or wet steam. Thermal oxidation is a well-characterized diffusion-controlled process covered in many texts. A disadvantage of thermal oxidation is the high temperatures (900 °C–1050 °C) needed to achieve films of the required thickness during a reasonable period of oxidation.

Another way to grow silicon dioxide is to deposit it in a reaction between silane and oxygen in a nitrogen-diluted plasma. This is called plasma enhanced conformal vapor deposition (PECVD). Although this process produces thin films at room temperatures, films can have submicrometer-sized pinholes that can adversely affect the micromachining process. Substrate roughness and cleanliness can affect film growth unpredictably. Furthermore, the films are generally not conformal; that is, the film growth rate is different on the corners or sides than on the plasma-electrode-facing surfaces. An advantage of PECVD silicon dioxide is the ability to change its stress from compressive to tensile by controlling the deposition substrate temperatures. Higher temperatures cause thermal expansion stresses, which are usually tensile. At low temperatures, the low surface mobility leads to disordered molecular arrangements and compressive stress.

Tetraethoxysilane (TEOS) oxide is an oxide formed by decomposing tetraethoxysilane over substrates at relatively high temperatures of 700 °C. The higher temperature results in higher surface mobility of the oxide molecules and, hence, more conformal films. TEOS films are dense and hard and are used in applications where the mechanical behavior of the thin film needs to be very stable.

Low-pressure conformal vapor deposition (LPCVD) oxides are deposited by decomposing silicon-containing gases in a low-pressure environment. The low pressure causes long mean-free paths and increased diffusivity of the reactive gases. This results in reaction-controlled, as opposed to diffusion-limited, film growth. Hence, uniform conformal film thickness over wafers is possible even at relatively low temperatures of 350 °C. Oxide formed in this way is also called low-temperature oxide (LTO). A great advantage of an LTO is that it can be deposited over thin films of low-melting-point aluminum, a commonly used metal in micromachining.

Silicon nitride thin films have unique material properties that have made their use in MEMS ubiquitous. Although in electronics it is mainly used as an insulating dielectric, in micromachining it is also used for its chemical inertness to common silicon etchants. It is used as a mask material against silicon etchants such as potassium hydroxide. It is deposited in PECVD or LPCVD reactors with reactions between silane or dichlorosilanes and ammonia with a nitrogen dilution. Usually, pressures in the range of 20 to 80 mtorr are used to deposit these films. Recipes for these depositions can be found in [17]. The PECVD process is carred out at room temperatures to 350 °C. These processing parameters result in stresses that range from several hundred megapascals tensile to compressive. As in the PECVD oxide films, PECVD nitride films also suffer from pinholes. These pinholes generally prohibit their use as a mask material for chemical etching of silicon. LPCVD films can be grown pinhole free and, hence, are used most often for masking silicon to chemical etchants. A unique feature of LPCVD nitride films is that their stress can be changed from tensile to compressive by controlling the gas flow rate ratio of the silicon-containing gas and ammonia. This affects the silicon content of the films, which is believed to affect the stress. Silicon-rich films (written as Si_xN_y) tend to produce lower-stress films, whereas more stoichiometric films (Si_3N_4) have large (as high as 1 GPa) tensile stresses. Typically, a flow ratio rate of silane to ammonia of 4:1 results in films with very low tensile stress (50–100 MPa) that can be used for micromachining applications [21]. This silicon nitride is commonly referred as a low-stress nitride in the MEMS literature.

Polysilicon or "poly" is a polycrystalline silicon thin film that is used by the IC industry as the gate material for complementary metal oxide semiconductor (CMOS) transistors and resistive material for on-chip resistors in integrated circuits. The fact that one can deposit polysilicon thin films with almost zero stress has made it an essential structural material for fabricating surface micromachines [22–24]. Although many deposition techniques (e.g.,

Table I. Important Materials in Micromachining

	Common deposition technique	Temperature (°C)	Stress condition	Comments on deposition and use
Silicon dioxide (SiO$_2$)	Thermal	900–1050	Compressive	Easy, conformal, high temperature
	PECVD	25–250	Tensile-compressive	Sensitive to temperature, pinholes
	LPCVD, LTO	350–700	Compressive	Conformal, can be deposited over Al
	TEOS	700	Compressive	Conformal, most stable
Silicon nitride (Si$_x$N$_y$)	PECVD	25–250	Tensile-compressive	Sensitive to temperature, pinholes
	LPCVD	700–850	Tensile, can be made zero by increasing silicon content	Conformal, chemically inert
Low-stress poly-silicon	LPCVD	600–650	Almost zero at 605 °C	Sensitive to deposition conditions
Metals: aluminum, tungsten	Evaporated, sputtered	25–300	Generally compressive	Tungsten used when high-temperature postprocessing is required
Photoresist-polyimide	Spin coat	200	Tensile	Very sensitive to aging and cure temperature

evaporation, sputtering, and PECVD) can be used to obtain a wide variety of silicon films with varying structural properties, the most relevant to MEMS is LPCVD poly deposition. LPCVD polysilicon films are deposited in an LPCVD reactor by pyrolyzing silane with nitrogen as a dilutant at temperatures ranging from 600 to 650 °C. At lower temperatures, the pyrolyzing rate is too low and, at higher temperatures, gas phase decomposition causes poorly structured films. Within the window of 600 to 650 °C, small temperature changes result in marked structural differences. At low temperatures, one obtains amorphous films, whereas, at higher temperatures, one obtains polycrystalline films. Furthermore, annealing of the amorphous silicon deposited at 600 °C can convert the silicon reproducibly into polysilicon films [12]. The deposition and annealing temperatures directly affect the grain size, which, in turn, controls the internal film stress. Gas phase doping using arsine or phosphine gases results in further modification of stress, and is often used as a control parameter to manipulate stress [25]. The sensitivity of the poly properties to deposition conditions has resulted in devices with widely varying mechanical properties from different micromachining laboratories. To ensure reproducible mechanical properties, one has to control the process parameters very diligently.

In addition to the previously mentioned thin films (see Table I), metal and organic thin films are also indispensable in micromachining. Most metals are deposited by evaporation or sputtering and result in compressive stresses. The choice of metal for interconnects is critical in cases where micromachining processes are integrated with electronics. If the micromachining process is performed after circuit fabrication, the metallization on the circuits has to be able to withstand polysilicon temperatures. Tungsten is one such metal that has been studied extensively for post-CMOS integration of polysilicon micromachines [26]. In addition to metals, organic films such as photoresist and polyimide are usually spun-on and tend to have shrinkage-related tensile stresses resulting from thermal postprocessing.

4. BULK SILICON MICROMACHINING

Micromachining techniques can be broadly categorized between bulk and surface etching techniques. Bulk micromachining involves etching of the silicon wafers, either isotropically or anisotropically, to form mechanical structures. Furthermore, bulk-etched structures formed from multiple silicon wafers can be bonded using several bonding techniques. The etching of silicon and other MEMS materials can be classified as shown in Table II. The etching of the silicon wafer can be done using wet etching or dry plasma–based etching, which can, in turn, etch materials isotropically or anisotropically. In what follows, the characteristics of wet and dry etching will be described.

4.1. Wet Silicon Micromachining

Bulk micromachining refers to selective bulk removal of the substrate. In the silicon micromachining literature, "bulk etching" is also synonymous with wet chemical etching of the substrate, even though there are dry ways of etching silicon. There are different families of silicon wet chemical etchants, which are either isotropic or anisotropic. Most of these etchants were discovered and recipes for their use originated in the 1950s in an effort to polish silicon wafers electrochemically [27–31].

In a typical silicon bulk etch, a masking layer that does not etch in the silicon etchant is first deposited over the silicon wafer. Although one would like to use photoresist as the masking material, it unfortunately is attacked and etched easily by most silicon etchants. Instead, photoresist-based lithography is used to define patterns in a suitable mask material such as silicon dioxide or silicon nitride (Fig. 6). Wet and dry etching methods for silicon dioxide and silicon nitrides, which are listed in Appendices 1 and 2, are used to etch the masking layers through the photoresist.

Most silicon etchants are used at an elevated temperature with a controlled concentration of the etchant in a solution with water as the main solvent. The etching temperature is typically kept below the etchant boiling point to avoid rapid loss by evaporation. The masked silicon wafer is placed in such a temperature-controlled etching chamber and the exposed silicon is put in contact with the electrolyte solution where it starts dissolving. The silicon undergoes a chemical reactions with the electrolyte in which charge transfer takes place. This charge transfer can be modeled using solid-state concepts of Fermi levels for different species in the silicon and the liquid electrolyte. The surface silicon atoms undergo chemical reactions that result in silicon removal from the solid to the liquid state.

The semiconductor–electrolyte interface is more complicated than the metal–electrolyte interface. The number of charge carriers (electrons and holes) at the silicon surface can actually be depleted or accumulated and, hence, reactions can be dependent on semiconductor doping and biasing conditions. Furthermore, discrete energy levels in the band gap caused by surface states can pin the Fermi level in the silicon band gap. Typical analysis of the etching process based on the energy–band gap formulation and charge transfer can be found in [32, 33]. Fortunately, most of the silicon etching can be understood by knowing the silicon chemistry with oxygen and fluorine. The important facts are that

Table II.   Classification of Silicon Etching Techniques Used in Micromachining

Etching	Wet	Dry
Isotropic	HNA	High-density (high-pressure) plasmas, vapor phase etching
Anisotropic	KOH, EDP, and other anisotropic etchants	Low-density plasmas, high-density plasmas with special chemistry

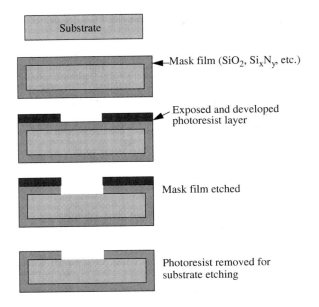

Fig. 6. Photoresist is easily attacked by many etchants that are used to etch a silicon substrate. A secondary masking layer that can withstand etchants is defined using a photoresist mask.

the Si–Si bond (free energy = 340 kJ/mol) is weaker than the Si–O bond (free energy = 452 kJ/mol), which is weaker than the Si–F bond (free energy = 565 kJ/mol) [34]. Depending on the concentration of fluorine, chlorine, and hydrogen ions in the liquid, different etching regimes can be established. Furthermore, oxidation of silicon by oxidizing agents (e.g., nitric acid and peroxide) can catalyze the further reaction of oxide removal by fluorine ions. The dynamics of the chemical reactions are also affected by the reaction rate and chemical species diffusivity toward the reaction front. Reaction-rate-controlled reactions result in high temperature sensitivity. Diffusion-rate-limited reactions are highly sensitive to stirring and mixing effects.

The wet etching techniques can be subdivided into isotropic and anisotropic etchants. The choice of which etchant to use is determined by the need for desired etching rates and geometry.

4.2. Isotropic Etching

Several chemical etchants etch silicon without regard to the crystalline orientation. An important family of etchants is the mixture of hydrofluoric acid, nitric acid, and acetic acid mixed with water [14]. This is called the HNA mixture. The nitric acid oxidizes the silicon surface, which is subsequently etched by the hydrofluoric acid with acetic acid used as a chemical buffer. Different etching rates and etched surfaces are obtained, depending on the concentrations of each chemical component used. Figure 7 shows the effect of chemical composition on etch behavior in a phase diagram representation. It can be seen that very high etch rates are possible in the polishing regime. One can dissolve an entire silicon wafer (550 μm thick) in HNA in approximately 20 min. However, these high etch rates are accompanied by high masking-layer etch rates. Even silicon nitride, which is usually chemically inert, is etched very fast and cannot be used reliably. The chemical reactions are also highly exothermic [35], resulting in higher reaction zone temperatures and, ultimately, extreme etch nonuniformity. Slower-etch-rate mixtures result in more uniform and controllable etching and are used regularly to etch silicon and polysilicon films in both micromachining and the integrated-circuit industry.

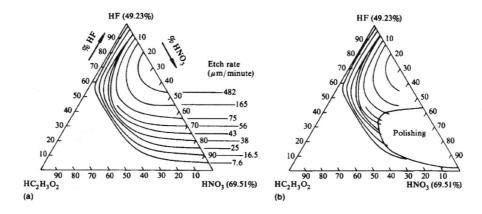

Fig. 7. Silicon etching in HNA mixtures. Reprinted with permission from W. R. Runyan and K. E. Bean, "Semiconductor Integrated Circuit Processing Technology." (© 1990 Addison-Wesley).

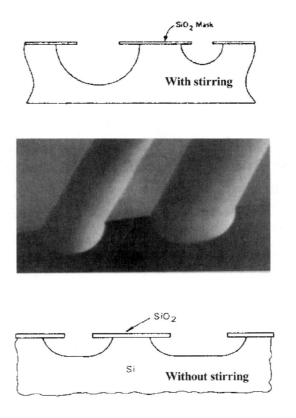

Fig. 8. Stirring effects during isotropic etching. Reprinted with permission from M. Madou, "Fundamentals of Microfabrication." (© 1997 CRC Press, Boca Raton, FL).

Placing the silicon wafer inside the etchant removes silicon equally in all directions, as shown in Figure 8. However, stirring causes a cross section as shown at the bottom of the figure. The stirring nourishes the depleted supply of reactant chemical species at the reaction front. Hence, a reaction-rate-limited etching as opposed to a diffusion-rate-limited etching occurs. One can also control the temperature to regulate the degree of reaction-rate-limited versus diffusion-rate-limited chemical reactions.

4.3. Anisotropic Etching

In the context of silicon bulk micromachining, anisotropic etching originates from the wide variation in silicon etching rates as a function of the exposed crystal planes. Figure 9 shows a diagram of etch rate versus crystal plane. Using the mask shown at the bottom of the figure, the different planes are exposed along an angular direction. The depth of etch at a particular angle indicates the etch rate of the crystal associated with the crystal angle. It is readily seen that the $\langle 111 \rangle$ direction has the lowest etching rate and $\langle 100 \rangle$ has the highest etch rate. Figure 10 shows the etch flowers for two etchants, EDP and KOH (to be discussed later), for $\langle 100 \rangle$ and $\langle 110 \rangle$ orientation wafers. The anisotropy between the $\langle 100 \rangle$ and $\langle 110 \rangle$ wafers and the $\langle 111 \rangle$ planes is readily observable and measurable. The origin of the anisotropy is still under investigation. One reason for the etch-rate difference is believed to be the surface atomic density variations as a function of the crystal plane direction. Band gap–based explanations for etch-rate anisotropy are also being investigated [33, 99].

To utilize anisotropic etching reproducibly, one has to align the wafer flat to the exposure mask reproducibly. Sometimes special jigs during contact lithography are used to align the mask to the wafer flat so that etching occurs with respect to a certain crystal direction. Most commonly, anisotropic etching is done on $\langle 100 \rangle$ or $\langle 110 \rangle$ wafers with $\langle 111 \rangle$ planes as etch stops. In the case of $\langle 100 \rangle$ wafers, the $\langle 111 \rangle$ crystal planes are oriented at a $54.74°$ angle to the wafer surface and produce a wall with that angle, as shown in Figure 11. Anisotropic etching on wafers with $\langle 110 \rangle$ orientation can produce vertical walls as shown in Figure 12.

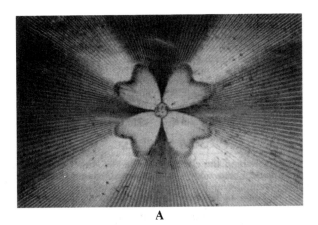

A

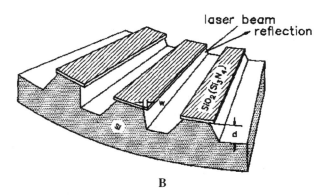

B

Fig. 9. Etch-rate anisotropy as a function of crystal plane exposure on a $\langle 100 \rangle$-oriented silicon wafer after etching in EDP. Reprinted with permission from H. Seidel, *J. Electrochem. Soc.* 177, 3612 (© 1990 Electrochemical Society).

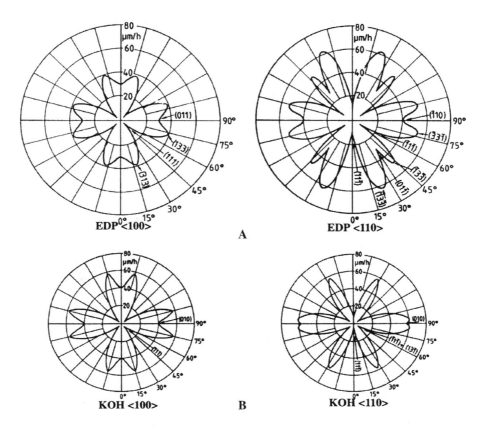

Fig. 10. Etch rate versus crystal orientation. Top: EDP (470 mL water, 1 L EDP, 176 g pyrocathechol at 95 °C). Bottom: KOH (50% solution at 78 °C). Left: ⟨100⟩ wafers. Right: ⟨110⟩ wafers. Reprinted with permission from H. Seidel, *J. Electrochem. Soc.* 177, 3612 (© 1990 Electrochemical Society).

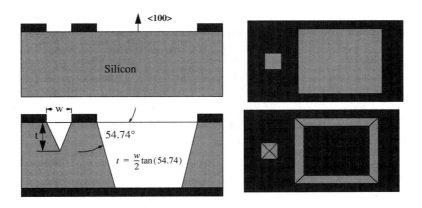

Fig. 11. Bottom shows the resulting self-terminated pyramid and through-wafer-etched cavity as a result of the mask in top with a ⟨100⟩ silicon wafer. Right side shows top view of wafer.

Two etch ratios are important in anisotropic etching. First is the anisotropy ratio, which is the ratio between the etch rate of the crystal plane to be etched (⟨100⟩ or ⟨110⟩) and the etch stop plane (⟨111⟩). The second is the ratio of the silicon etch rate to the mask-etch rate, called the mask-etch-rate ratio. As the etching rates for mask materials and substrate materials are Arrhenius functions of temperature with different activation energies, the

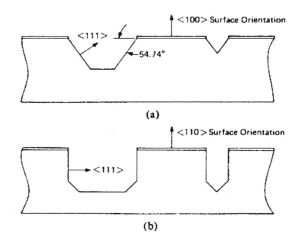

Fig. 12. Vertical walls can be obtained using ⟨110⟩ wafers. Reprinted with permission from K. Peterson, *Proc. IEEE* 70, 420 (© 1982 IEEE).

mask-etch-rate ratio and the anisotropy ratio are also functions of temperature. For a given desired etch depth and mask-etch-rate ratio, one can calculate the required mask thickness.

4.4. Masking Layers

Typical masking layers on silicon are thermally grown (dry or wet) oxide and low-stress silicon nitride thin films. Silicon dioxide films generally etch faster than silicon nitride films. This fact would suggest the use of silicon nitride films for masking material. However, silicon dioxide films have two distinct advantages as a masking material. First, oxides are more commonly available from ubiquitous thermal oxidation furnaces, whereas silicon nitride is deposited in the less common and hard to maintain LPCVD furnaces. Low-temperature plasma-deposited (PECVD) silicon dioxide or silicon nitride films are possible but usually contain pinholes. A second advantage of oxide films is the relative ease with which they can be dissolved away from the wafer after the anisotropic etch. Silicon dioxide films are rapidly (Appendix 1) etched in wet hydrofluoric acid. In contrast, silicon nitride thin films are harder to wet etch. Phosphoric acid at high temperatures (150 °C) is needed to etch it. Plasma etching of nitride (Appendix 2) is an alternative to phosphoric acid, but is generally nonuniform over a wafer. Evaporated, sputtered, or electroplated metal thin films can also be used as masking layers with selected anisotropic etchants (as described in the next section). Metal film deposition using evaporation and consequent etching using wet etches are very convenient and need low maintenance [17].

4.5. Selected Anisotropic Etchants

There are many anisotropic etchants that have been identified over the last 30 years. The factors affecting the etchant choice are the desired etch rate (determines the time of etch), masking material and its etch rate, temperature required, and chemical toxicity. Furthermore, the anisotropic etchants fall under two broad categories of organic and inorganic alkaline solutions.

4.5.1. Organic Etchants

Ethylene-diamene-pyrocatachol (EDP) is the most commonly used organic etchant. EDP is a highly toxic combination of carcinogenic, irritant, and pyrolytic chemicals, which micromachinists have used for decades under the protection of hoods, protective clothing, and masks. The EDP solutions also age rapidly and require frequent reformulation. It also discolors everything in contact with it (including Teflon). In spite of its toxicity, EDP is still widely used because of its somewhat guaranteed success in bulk etching. EDP-based solutions have the main advantage of possessing very low metal (Ag, Au, Cr, Cu, Ta) etch rates, permitting metals to be used as a mask material for anisotropic etching. EDP also has a very low silicon dioxide etch rate, allowing one to etch through a 500-μm-thick wafer with thin oxide coatings [7, 36]. For the case of etching $\langle 100 \rangle$ wafers at 110 °C in EDP, the silicon etches at an etch rate of approximately 0.85 μm/min, whereas the $\langle 111 \rangle$ planes, which are at a 54.74° angle, etch at roughly 0.02 μm/min. A typical mask layer of silicon dioxide etches at roughly 0.004 μm/h. The etch-rate ratio between the silicon and the oxide is approximately 12,000 and allows the capability to etch through a silicon wafer with 100-nm oxide film. The time duration for such an etch would be approximately 10 h.

Tetramethyl ammonium hydroxide (TMAH)/water organic etch solutions are more IC compatible because the solution does not decompose at temperatures below 130 °C. TMAH also has a low silicon dioxide etch rate, allowing one to use oxide masks. At low pH values [37], TMAH solutions etch aluminum at only 0.01 μm/min, making it possible to use aluminum as a masking material. The pH can be lowered by dissolving silicon in the bath. However, this increases the surface roughness and decreases the $\langle 100 \rangle$ etch rate. The aluminum selectivity is believed to originate from the formation of a thin layer of silicic acid on the aluminum, which prevents further aluminum oxidation. The disadvantage of TMAH is its low anisotropy ratio (12–50) as compared to that of KOH (500–1000). Hence, considerable underetching can occur with TMAH [38].

4.5.2. Inorganic Alkaline Etchants

The most common anisotropic etchant is the potassium hydroxide (KOH)/water solution used at 70–90 °C with KOH concentration in water varying from 20% to 60%. Concentration and temperature affect etch rates, etch-rate ratio, and surface roughness. Higher KOH concentrations result in lower etch rates and smoother surfaces. Higher KOH concentration means lower water ion concentrations, which are responsible for charge carrier transport from the wafer. A lower concentration of KOH results in higher etch rates and surface roughness as shown in Figure 13 [105, 106]. Recent studies have explored roughness versus crystalline plane orientation as well [39]. Etch rates in KOH at different concentrations and temperatures are shown in Figure 14. One can calculate the mask thickness required for a given depth of silicon etching using these data. For example, if one wishes to do a through-wafer etching with an oxide mask at 90 °C to achieve 2 μm/min, it can be shown that one requires 3 μm of oxide. This oxide thickness is nearly impossible to achieve in reasonable processing times using wet or dry oxidation. Silicon nitride films with extremely low etch rates are more commonly used with KOH etching solutions.

In addition to the microroughness, KOH etching can produce a bowing effect at the intersection of the $\langle 100 \rangle$ and the $\langle 111 \rangle$ planes (Fig. 15). This is believed to occur because of the presence of excess ionic reactive species near the edges where $\langle 111 \rangle$ planes do not consume the ions at all. Stirring the solution can reduce such nonuniformity. Stirring is also used to remove adhering reaction product gas bubbles from the silicon surface to produce smoother surfaces. The addition of isopropyl alcohol to the etch bath also results in smoother etch surfaces and a higher anisotropy ratio between the $\langle 100 \rangle$ and $\langle 111 \rangle$ etch rates [40]. Lower-density alcohol rises to the top of the etch bath and helps to maintain a uniform temperature distribution inside the etch bath. Metals can be added to the etch bath to increase the etch rates [7, pp. 68–69]. In addition to KOH, other alkaline etchants used for anisotropic etching are CsOH, NaOH, RbOH, and LiOH [6].

A safety concern of KOH and other alkaline solutions is that they can cause blindness if they come into direct contact with the eyes. Another disadvantage of alkaline etches is that the active ions (K^+, Na^+, Li^+) can diffuse into silicon as an impurity, wreaking havoc

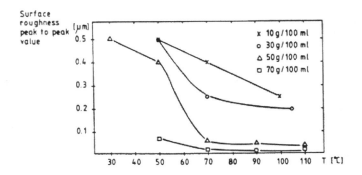

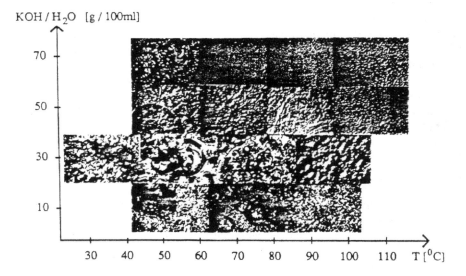

Fig. 13. Top graph shows measured surface roughness as a function of etch temperature and KOH concentration. Bottom shows the phase contrast microscope pictures of silicon etched at different temperatures and concentrations [105].

on any integrated-circuit process. Hence, these etchants are not considered IC compatible. Etchants such as TMAH are IC compatible and can be used in an IC foundry with little danger of contamination.

4.6. Corner Compensation

An often-desired structure in micromachining is an isolated rectangular mesa of silicon. However, the exposed convex intersection of two ⟨111⟩ planes etches rapidly because of the number of planes exposed. One cannot simply use a box as a mask and produce a mesa structure. To solve this problem, one has to design a mask using corner-compensating techniques [41]. Additional mask material at the corners is used to create additional silicon mass to be etched before reaching the desired corner. Designing the optimum mask presents a considerable challenge, and a number of masking schemes have been used. One such mask is shown in Figure 16 [42].

As in silicon etching, there are electrolytes that etch GaAs anisotropically. For example H_2O_2 is used to etch GaAs crystals anisotropically [43, 44].

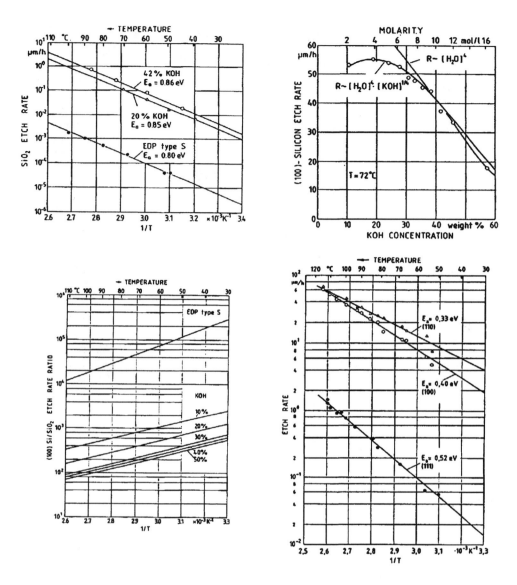

Fig. 14. Top left shows SiO_2 etching rates in EDP and KOH as a function of temperature. Top right shows $\langle 100 \rangle$ silicon etch rates at 72 °C for different KOH concentrations. Bottom left shows the anisotropy ratio as a function of etch temperature and etch formulation. These data can be used to calculate the activation energies. Bottom right shows the activation energies for EDP type S etching as a function of temperature and silicon orientation. Reprinted with permission from H. Seidel, The Mechanism of Anisotropic Silicon Etching and Its Relevance for Micromachining, presented at Transducers '87, pp. 120–125 (© 1987 IEEE).

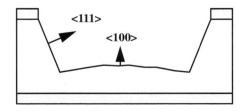

Fig. 15. Bowing effect in anisotropic etching. Higher reactant density results at corners of less reaction on $\langle 111 \rangle$ planes.

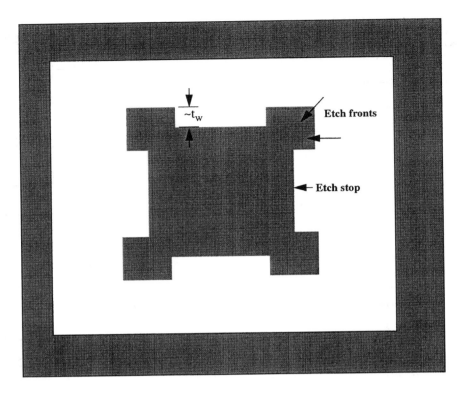

Fig. 16.   A possible corner-compensating mask used to produce sharp corners on silicon mesas.

4.7. Bulk-Micromachined Structures

Using anisotropic etching, one can generate prototypical structures that form the foundation of most micromechanical systems. One of the most common structures is the formation of a thin diaphragm or plate anchored by the silicon substrate. The membrane can be formed by many methods as shown in Figure 17. One method is to use a timed etch if the etch rate is known with confidence (Fig. 17a). However, etch rates vary considerably with fluctuating solution concentration and temperature. Hence, the membrane thickness reproducibility is no better than 5–10 μm. A second technique, as shown in Figure 17b, can be used to reproduce membrane thickness more effectively. A top window in the masking layer terminates into a pyramid early in etching. The mask opening is adjusted so that the pyramid height corresponds to the desired plate thickness. As the etch front from the bottom approaches the pyramid tip, a predetermined amount of light transmission through the etch front and the pyramid surface can be used to stop the etching. This technique can result in a membrane thickness reproducibility of 2–3 μm from wafer to wafer [45]. Another common way to obtain thin membranes is by using membrane materials that act as etch stops for the etch bath. As seen in Figure 17c, etching through the wafer can form silicon nitride or silicon dioxide membranes.

4.8. Doping Rate Effects on Etching Rate

Highly p^{++}-doped silicon can also act as an etch stop, resulting in p^{++}-doped silicon membranes as shown in Figure 17c. Highly doped p^{++}-type (p-type concentration $>10^{21}$ atoms/cm^3) areas of silicon behave as etch stops in the alkaline anisotropic etchants. The etch rate of doped p-type silicon as a function of doping concentration is shown in Figure 18. This agrees well with the electrochemical model of the anisotropic etching

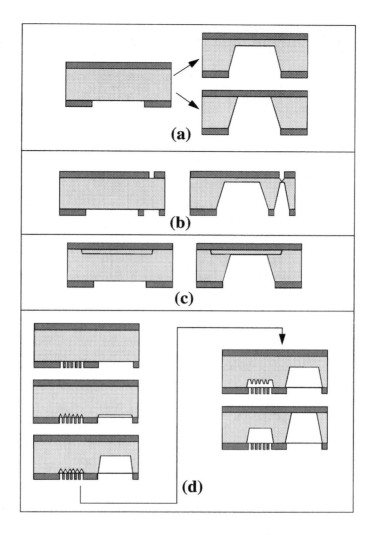

Fig. 17. Some techniques used to produce membranes and plates anchored by silicon substrates. (a) Timed etch produces silicon plates or etching to mask film produces thin-film membranes. (b) A self-terminating pyramid from top is used to stop etch when an optic signal is transmitted from the bottom. (c) Etch stops such as p+-doped silicon or electrochemically active layers can be used to stop etch. (d) The grill technique to obtain different-height cavities with one mask.

behavior, in which the access holes in the p++-doped silicon pin the Fermi level much lower than the redox potential in the etching bath, eliminating the charge exchange and the resulting chemical reaction [46]. Cantilevers and beams of p++-doped areas have been readily made using this process. The thickness of the doped region is usually limited to 2–3 μm because of diffusion-limited growth. As an exception, workers at Michigan [47] have developed an extended-time diffusion doping process that allows for doping to 15–20 μm. They have developed a dissolved wafer process utilizing deep p++ silicon as shown in Figure 19. The p+-doped silicon is at first bonded to a glass substrate using anodic bonding. Then the glass–silicon sandwich is etched in a silicon etchant that leaves the glass unetched. Usually, EDP is used, which leaves the glass substrate undamaged. The entire silicon wafer is dissolved leaving behind the p+-doped areas attached to the glass substrate. This process has been used to make a variety of sensors, including a tunneling pressure sensor [48].

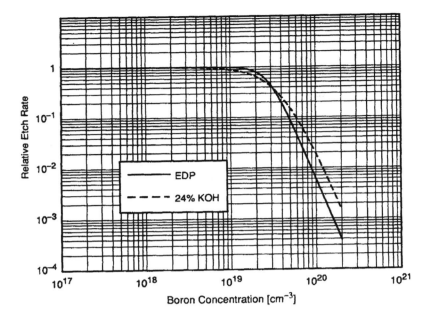

Fig. 18. Comparison of silicon etch rate in EDP and KOH solutions as a function of boron doping concentration. Reprinted with permission from C. M. Mastrangelo and W. C. Tang, "Semiconductor Sensor Technologies," in Semiconductor Sensors (© John Wiley & Sons, Inc.).

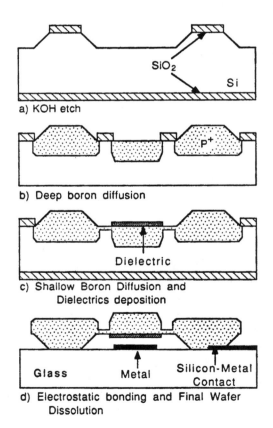

Fig. 19. Dissolved wafer process. Reprinted with permission from K. Suzuki et al., "A 1024-element High Performance Silicon Tactile Imager," IEEE IEDM, p. 674 (© 1988 IEEE).

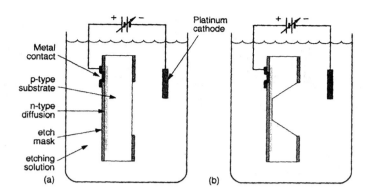

Fig. 20. Electrochemical etch bath setup, before and after etch. Reprinted with permission from C. M. Mastrangelo and W. C. Tang, "Semiconductor Sensor Technologies," in Semiconductor Sensors (© John Wiley & Sons, Inc.).

4.9. Electrochemical Etch Stops

By electrically biasing the silicon wafer to be etched, one can electrically control the population of charge carriers at the semiconductor–electrolyte interface. The ability to control the carriers can be used to modulate the electrochemical reaction (Fig. 20). Typically, a p–n junction is reverse biased. The voltage drop is largely across the p–n diode leaving the p-type silicon surface at open-circuit potential. The KOH etches the p-type silicon until it encounters the n-type silicon. The p–n junction is destroyed and the n-type silicon gets biased so as to eliminate any carriers on the surface stopping the etch completely [46]. An effective electrochemical reverse-biased diode stops the moving front near the diode depletion region. Such techniques can be used to control diaphragm thicknesses to within nanometers [49]. In addition to electrical biasing, one can bias the silicon by generating electrons and holes via photonic illumination. Hence, one has to take precautions of etching under dark conditions for reproducible membrane thicknesses.

4.10. Other Techniques

Often, it is desirable to etch cavities of two different heights simultaneously onto a silicon wafer. For example, one might want to etch a front that goes through the wafer while another etch front terminates at a desired height. One way to obtain this structure is simply to etch one cavity first and then redeposit and pattern the mask material for the second thickness. However, lithography steps are time consuming and expensive. It is also hard to spin-coat and expose photoresist on a wafer with deep features resulting from the first etching. Figure 17d shows the grill technique that can be used to achieve several different heights using *one* etching mask step. This method utilizes undercutting of the ⟨111⟩ planes, which occurs at a much lower etch rate than the primary etching planes. By using a mask grill that is undercut by two sides, one can slow down the etching front propagation as compared to one with no grill patterns. When the silicon is undercut completely, the sharp intersections with many exposed planes of silicon are etched at a very fast rate. The concave tips are etched away completely until a flat etch front is formed. By choosing different grill-opening-to-width ratios, one can obtain cavities of different heights [50].

In addition to membranes, one can form cantilevers made of silicon or etch stop materials by similar techniques. Figure 21 shows a cantilever of silicon nitride formed by exposing the silicon from the front side of the wafer [51]. In the case shown, the silicon nitride film was only 90 nm thick. By using ⟨110⟩ wafers, the cantilevers could be placed into contact with other surfaces and used to make ultrasensitive force measurements.

An illustrative example of an early (1979) bulk-micromachined accelerometer is shown in Figure 22 [52]. The p^+-doped areas on the front side are piezoresistors that convert the

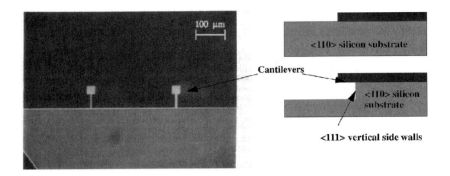

Fig. 21.   Very thin (∼900 Å thick) silicon nitride membranes fabricated using the process shown on right. Vertical side walls in ⟨110⟩ silicon wafers can be produced upon anisotropic etching. Reprinted with permission from S. Hoen et al., Fabrication of Ultrasensitive Force Detectors, presented at Solid-State Sensor and Actuator Workshop, Hilton Head, SC (© 1994 Transducers Research Foundation).

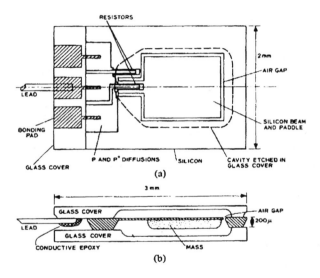

Fig. 22.   Top view and centerline cross section of bulk-micromachined accelerometer. Reprinted with permission from L. M. Roylance and J. B. Angell, *IEEE Trans. Electron Devices* ED-26, 1911 (© 1979 IEEE).

strain generated by an applied acceleration to the proof mask. Figure 23 shows the bulk-etched proof mask attached to the silicon frame. Because no corner compensation was used, the corners are etched inward as expected. The ability to attach a large mass to a very thin film strain gauge enables high sensitivities. However, the weak tether holding the mass can lead to the proof mass impacting the package under large accelerations.

5. DRY BULK SILICON CRYSTALLINE MICROMACHINING

Dry etching of silicon has several advantages over wet etching. First, dry etching eliminates the process of drying and cleaning wafers exposed to dangerous chemicals. The fluid turbulence associated with drying is particularly dangerous for etched silicon wafers with weakly tethered silicon structures. Second, dry etching tends to be a much more controllable process than wet etching, where most reaction rates are exponentially temperature dependent. By controlling the chemistry and the *density,* plasma etching can be more insensitive to temperature variations. Furthermore, vertical side walls with wet etching can

Fig. 23. Back side of the silicon-etched die with corners etched as expected without corner compensation. Reprinted with permission from L. M. Roylance and J. B. Angell, *IEEE Trans. Electron Devices* ED-26, 1911 (© 1979 IEEE).

be obtained with only a few wafer configurations ($\langle 110 \rangle$). Dry plasma etching has the potential of solving most of these problems. Dry plasma etching enables the possibility of vertical etch profiles while requiring no liquid contact. However, a marked disadvantage of dry etching is the complexity and high cost of equipment necessary to sustain well-controlled plasmas. Maintaining a uniform plasma density across large wafer areas is also a challenge and usually implies etching rate variations across the wafer.

5.1. Plasma Etching

The dry etching discussed in the context of MEMS can be classified into plasma and vapor phase etching. We will first discuss plasma, etching. In a typical etching plasma, the reactive gas is mixed with a dilutant gas and exposed to high-energy radio frequency (rf) electric and magnetic fields (Fig. 24). These fields ionize the gas molecules, creating electronics that further ionize the gas and result in a stable phase of ions and electrons. For plasma etching, the reactive ions have to diffuse toward the surface, diffuse on the surface, react with the surface atoms, and then diffuse away into the plasma. If the ions are very

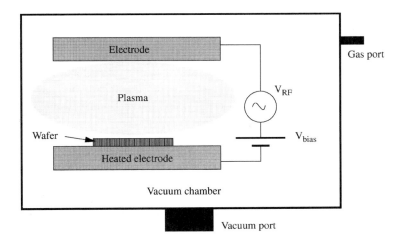

Fig. 24. Components of a typical plasma system.

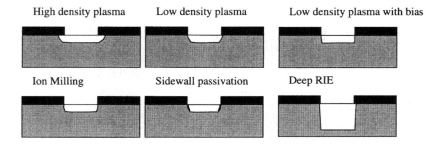

Fig. 25.   The gas pressure, bias, and chemistries produce different etch profiles in plasma etching.

energetic, pure physical etching can occur by ion–surface momentum transfer during impact. If the ions are not very energetic, chemical reactions of charged species take place, which are controlled by the plasma pressure, temperature, and rf power [16]. A few of the different kinds of etch profiles obtainable from plasma etching are shown in Figure 25 and described in the following sections.

5.1.1. Isotropic Etching

Isotropic plasma etching occurs when the plasma density or pressure is high. The high particle density results in an isotropic velocity distribution and isotropic etching. However, plasma etchants can never really be as isotropic as wet isotropic etchants. Because of the high reactivity of silicon with fluorine, silicon is most commonly etched in fluorine-based plasmas. Typically, SF_6 or CF_4 gases are mixed with oxygen in plasma. In addition to reacting with the silicon atoms directly, these gases form polymers that are etched by dilutant oxygen radicals. Furthermore, supply-limited reactions result in a loading effect where the more exposed areas of silicon are etched more slowly than the less exposed areas. The loading effect can be reduced by decreasing the reaction rate using a less reactive gas like chlorine, but at the cost of lower etch rates. Chlorine plasmas include gases such as $SiCl_4$, CCl_4, BCl_3, and Cl_2. To increase the etch rate with chlorine, one has to increase the ion bombardment energy by biasing the substrate, which leads to lower isotropy. The ions bombarding the surface have a higher net velocity perpendicular to the wafer.

5.1.2. Anisotropic Etching

One of the key advantages of plasma etching is the possibility of achieving vertical side walls. Momentum from the ions in a low-density plasma is transferred more favorably to the bottom surface of the etch front than to the side walls, because of longer mean-free paths. This phenomenon results in a very low side-wall etch rate and high bottom etch rates. Side-wall passivation can also occur because of organic polymerization on the side walls. Photoresist mask or trace organic gases in the plasma can provide the carbon supply for polymer formation.

The plasma etches described previously have low etch rates and low mask/substrate etch-rate ratios (Appendix 2). Maximum depths of 30–40 μm are obtainable using conventional masking films. To increase the etch rates and the etch-rate ratios, while maintaining vertical sidewalls, two new plasma etching techniques have been developed. These etchers are commercially available and they can etch deep trenches in silicon with very vertical side walls with etch rates as high as 1 μm/min. Both methods utilize very high plasma densities to boost the etching rates and side wall passivation to achieve vertical side walls. Collectively, both methods are referred to as deep reactive ion etching (deep-RIE) techniques [53, 54].

In the first method, high-density plasma is operated at cryogenic temperatures, resulting in not only silicon etching, but also the formation of a very thin passivating silicon dioxide film at the side walls. A 1-μm-thick silicon dioxide mask (Si/SiO$_2$ etch-rate ratio of 300:1) can be used to etch 300-μm-deep trenches with an aspect ratio of 15:1. This cryogenic chuck etch technology is being offered by Alcatel, Inc. Alternatively, STS Technologies, Inc., has developed a process in which very high density reactive species are created using inductively coupled plasmas. By using photoresist as a mask and source of organic molecules, side walls are passivated by polymerization. The typical etch-rate ratio of silicon to photoresist is 50:1 [53]. Hence, a 6-μm-thick photoresist can be used to etch 300 μm on the photoresist. Both the cryogenic chuck and the high-density plasma suffer from loading effects because of reaction-rate-limited etching. Although the deep-RIE methods promise through-wafer etching, they suffer from a large loading effect. Large etching areas etch more slowly than smaller etch areas. The masks have to be designed so that the etching areas do not vary much across the wafer.

Deep silicon etching has been used to make high-aspect-ratio structures. One such process flow is used to fabricate single-crystal silicon (SCS) micromachines [55, 56] (Fig. 26). First, silicon wafer is anisotropically etched in a deep RIE, resulting in pillars of silicon. PECVD oxide is then deposited conformably over the pillars. Using a plasma etch that is highly anisotropic, only the oxide on the top and bottom of the trenches is etched. The wafer is then thermally oxidized such that the oxidation fronts at the bottom of the trenches meet and form a sharp tip of single-crystal silicon. Sharp silicon tips are left behind after a hydrofluoric acid etch is used to dissolve the oxide. Because these tips can be formed in large high-density arrays, such devices have been proposed for high-density AFM and/or STM tips for scanning and data storage with nanometer precision [56].

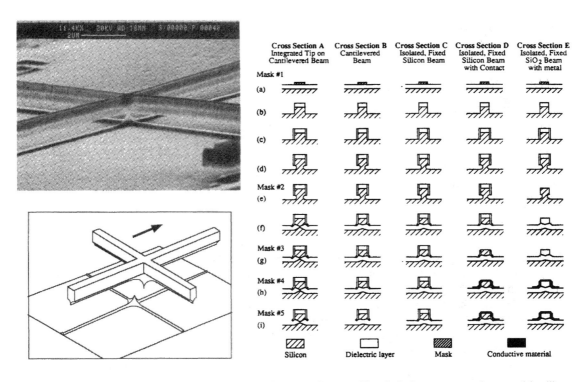

Fig. 26. Single-crystal silicon micromachining using deep plasma etching. Left shows a nanoprobe created in silicon. Right shows the process flow and cross sections. Reprinted with permission from J. J. Yao et al., *J. Microelectromech. Devices* 1, 14 (© 1992 IEEE).

5.2. Vapor Phase Etching of Silicon

In vapor etching, the etchant is vaporized and the molecules diffuse to the surface of the material to be etched. For example, hydrofluoric acid vapor has been used to underetch silicon dioxide. It has been used to clean surfaces inside LPCVD chambers before epitaxial growth of silicon films [57]. Furthermore, silicon etching using XeF_2 vapor has been known since the 1960s [58]. However, it was recently "rediscovered" in the micromachining context [59]. Unlike plasma etching, XeF_2 instantaneously decomposes on the silicon surface and the fluorine reacts with the silicon atoms. As in any pure chemical etching, XeF_2 silicon etch is isotropic. Most strikingly, XeF_2 is highly selective to silicon, not etching photoresist, metals, and silicon dioxide. Residual oxide can act as an etch stop and needs to be removed in a HF etch to obtain smooth XeF_2 etches. Polysilicon can be oxidized slightly and, hence, can be made not to etch in XeF_2 by simply letting it sit in air for 2 days.

A typical XeF_2 etch system consists of a pulsed (duty cycle of 50% at a frequency of one per minute) XeF_2 vapor exhausted into a chamber nominally held in a dry nitrogen environment. The pulsing reduces the very large loading effect observed in XeF_2 etching. Etch rates of 10 μm/min for small pieces versus 11 nm/min for 4-in. wafers have been observed. Better chamber designs capable of higher XeF_2 vapor pressures might reduce this effect.

The use of XeF_2 etching has been shown to be advantageous for creating microstructures by postprocessing on standard CMOS process wafers. Figure 27 shows a magnetic field sensor integrated in a standard CMOS process. Polysilicon underneath the patterned oxide and metal center plate was underetched using XeF_2 etching. Because XeF_2 is isotropic, it undercuts not only the desired platform, but also the polysilicon on the outer periphery of the sensor.

6. BONDING

The need for bonding silicon structures to other materials or silicon pieces came from packaging requirements. Glass–silicon bonding was developed to seal silicon micromachined devices with a glass cap. For example, an absolute pressure sensor requires a cavity with a well-defined and stable pressure sealed on one side of a micromachined plate. Although there are many kinds of bonding techniques, anodic, fusion, and adhesive bonding play a vital role in micromachined structures.

6.1. Anodic Bonding

Glass-to-silicon bonding originated from metal-to-glass (also called Emory) bonding. The glass wafer is placed on top of the silicon wafer in a vacuum environment to eliminate trapped air between the glass–silicon surface. A high electric field ($\sim 7 \times 10^6$ V/m) is placed across the glass–silicon sandwich at elevated temperatures (100–500 °C). The trapped ions, typically sodium and potassium in glass, migrate toward the interface under the influence of the electric field and the increased electronic mobility at high temperatures. Counteracting mirror charges develop in the silicon that form a strong electric field at the interface. When the silicon–glass sandwich is brought back to room temperature and the applied electric field removed, the ions are trapped at the interface. Hence, a permanent electric field holds the glass and the silicon together (Figure 28) [60]. The potential drop across the glass thickness starts out linear but becomes highly localized near the silicon–glass interface (Fig. 29). Because the glass and the silicon have different thermal expansions, the resulting interfacial thermal stresses can cause deformation of the bilayer structure at room

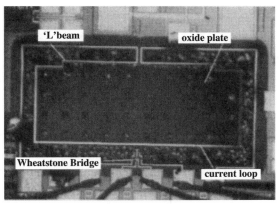

Top view photomicrograph of sensor. An oxide plate is suspended over an etched cavity in silicon substrate held by torsional support beams and L-shaped beams. On one side of the plate is a Wheatstone bridge and on the other a current loop comes onto on off from the plate. The dark rim around the structure is the XeF₂ etch front.

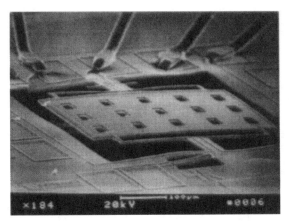

SEM of a smaller version of the resonant mechanical sensor. The oxide plate is seen to have a small stress gradient. The holes on the plate are for the purpose of reducing the etch time. This device was released using 50 one minute pulses of XeF₂ gas at room temperature. The etch pressure was ~ 2 torr.

Fig. 27.   Top shows the schematic diagram of the standard CMOS magnetic sensor fabricated using XeF₂ etching. Bottom shows the hanging oxide structure. Reprinted with permission from E. Hoffman et al., 3D Structures with Piezoresistive Sensors in Standard CMOS, presented at IEEE MEMS 1995, Amsterdam (© 1995 IEEE).

temperature. Hence, much early effort was spent in finding a glass with same net thermal expansion as silicon (2×10^{-6} ppm/°C) at the bonding temperature. One such glass is Corning 7740.

The electrostatic field at the interface in anodic bonding is high enough to bond the two materials permanently. Because the field strength is very high, the bonding process is not affected by the presence of surface irregularities and contamination. The glass

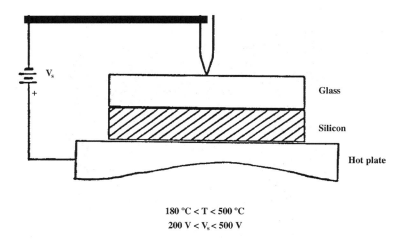

180 °C < T < 500 °C

200 V < V$_s$ < 500 V

Fig. 28. Schematic of the anodic bonding process. Reprinted with permission from W. H. Ko et al., in "Microsensors" (© 1990 IEEE).

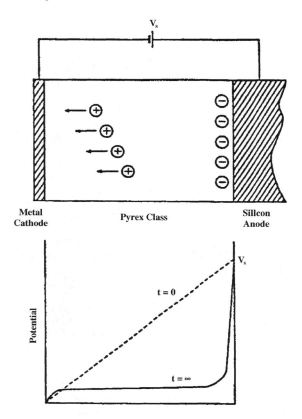

Fig. 29. Positive ions drift towards the cathode, while negative ions accumulate at the silicon–glass interface. The initial linear electric potential is changed to a cathodic distribution at the end of the bonding process. Reprinted with permission from W. H. Ko et al., in "Microsensors" (© 1990 IEEE).

deforms around any contaminant placed at the interface. This fact has been used to seal glass caps around metal interconnects, electrically connecting the inside of the cavity to the outside [60]. However, a disadvantage of anodic bonding is that the high electric field required to seal can damage any underlying electronics.

6.2. Low-Temperature Glass Bonding

In many applications, the high fields and temperatures required in anodic bonding are impractical. In this case, low-temperature glass bonding can be used. Many glasses that have low melting points can be deposited as thin films [61, 62] and then bonded to substrates when heated to the melting point. Thin films of phosphosilicate or borosilicate glass can be sputtered or spun on. The substrate to be bonded is placed under pressure and elevated temperatures on the glass film. Alternatively, relatively thick (greater than 25 μm) glass frit sheets can be purchased commercially. In general, the bond strengths obtained are not as high as those obtained with anodic or fusion bonding. Furthermore, the lack of bonding because of contaminant particles thicker than the glass film make thin-film bonding sometimes impractical.

6.3. Fusion Bonding

Fusion bonding is a more recent bonding technique in which surface-treated silicon surfaces are brought into contact with each other to form a medium-strength chemical bond. After high-temperature exposure at 1000 °C for 1 h, the intermediate chemical layers dissolve and a very strong Si–Si bond layer is formed [63]. Often, the resultant bond interface is indistinguishable from the surrounding crystal structure. Fusion bonding requires extreme cleanliness and wafer flatness to work reliably. Usually, one finds a sequence of steps and a location in a lab that works and rarely to changes anything that might spoil the "black art" nature of this procedure. Furthermore, the high temperatures required eliminate the possibility of bonding wafers with prefabricated circuits. Figure 30 shows an infrared image through a wafer sandwich of the fusion-bonded interface as a function of anneal temperature and duration. In the example, the time- and temperature-dependent nature of gas evolution at the interface is shown. At 600 °C, areas of bond failure appear, which disappear at higher temperatures. This is probably due to gas reactions with the substrate as is the case in gettering gases [64].

Fusion bonding has found a commercial application in silicon-on-insulator (SOI) wafer manufacturing. An oxidized wafer is fusion bonded to a blank wafer. Then one of the

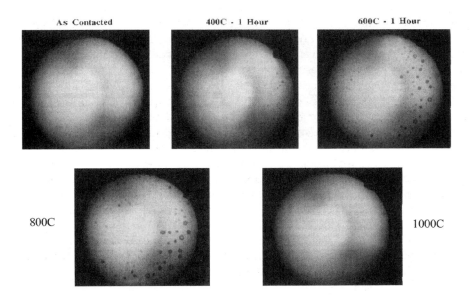

Fig. 30. Infrared image of intrinsic voids during fusion bonding as a function of anneal temperature. Reprinted with permission from M. A. Schmidt, Silicon Wafer Bonding for Micromechanical Devices, presented at Solid-State Sensor and Actuator Workshop, Hilton Head, SC (© 1994 Transducers Research Foundation).

wafers is chemically and mechanically polished to the desired silicon thickness. Such SOI wafers have found use as starting structures to fabricate silicon microstructures [65].

6.4. Eutectic Bonding

In eutectic bonding, one utilizes the positive free energy associated with the chemical reaction of a metal with silicon. When the metal and silicon are put together, the combination melts at a lower temperature than either metal or silicon alone. For example, gold reacts with silicon at low temperatures (363 °C) to form a AuSi eutectic [66]. Typically, gold is either evaporated, sputtered, or electroplated on the surface to be bonded to a silicon surface. When these areas are put in contact under vacuum and at the eutectic temperature, the metal diffuses into the silicon, forming the eutectic melt. When the interface is cooled, the melt solidifies, forming the bond layer. One has to prepare the silicon surface carefully by eliminating any surface contaminants and diffusion barriers such as native silicon dioxide.

6.5. Adhesive Bonding

Although technologically less exciting, glues and epoxies have been used to bond silicon parts together. A common problem with adhesives is that they are not easily photopatternable. Ultraviolet (UV) curable epoxies are an exception to this problem. Another issue is that one often needs to bond two parts with microscopic channels. Adhesives tend to flow and fill up these channels. One solution to this problem is to deliver highly viscous microdoses of adhesive through a silk-screening process [67–69]. Silk screening has been used to fabricate silicon needles by bonding silicon pieces with V grooves running along their length (Fig. 31).

7. SURFACE MICROMACHINING

As we have seen, bulk micromachining can be used to etch entire wafers. It is, in essence, a subtractive machining technique. In contrast, surface micromachining is an additive machining process. Figure 32 shows the basic procedure for fabricating a surface-micromachined cantilever. A sacrificial layer material is deposited first on a silicon wafer and coated with a passivation layer such as silicon nitride. Lithography is used to define areas where the sacrificial etch is removed selectively. A conformal thin film of the structural material is deposited over the entire wafer. The structural layer is patterned using lithography and chemical or plasma etch that is terminated on the sacrificial layer using a timed etch or selective etching chemistry. Then the entire wafer is etched in an etch that selectively etches the sacrificial layer but not the structural material. The etch-rate ratio of the sacrificial film to that of the structure has to be very high to maintain controllable structural layer thickness. Furthermore, the interfacial stresses between the sacrificial layer and the structural layer, as well as the internal stresses of the structural layer, have to be very low to avoid curling of the structural material after the release step. The structural material also has to be nearly defect free to reduce the surface roughness that typically results from the sacrificial layer etch. There are very few combinations of sacrificial/structural materials that match all these requirements.

The release etch time of a surface micromachine will be linearly dependent on the maximum dimension of the micromachined structure to be underetched. Hence, if one wants to etch very large area solid-plate structures, the etch times would be too long. The long etch times can also lead to a degraded structure, assuming the etch also chemically reacts with the structure. To reduce the etch time of large-area structures, etch holes (Fig. 32) are formed in the structure to give etchant access to the underlying sacrificial layer in a uniform way across the wafer. The etch holes are typically placed a distance d apart, where d

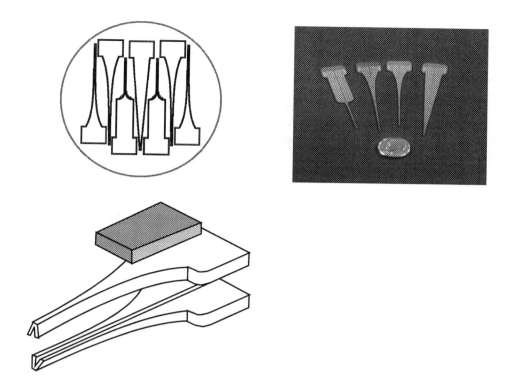

Fig. 31. Silicon ultrasonic horns. Top left shows a 4-in. wafer with different-shaped horns. Top right shows the etched horns. Bottom left shows the formation of a needle structure obtained by bonding two horns with V grooves. Reprinted with permission from A. Lal and R. M. White, Ultrasonically Driven Silicon Atomizer and Pump, presented at Solid-State Sensor and Actuator Workshop, Hilton Head, SC (© 1996 Transducers Research Foundation).

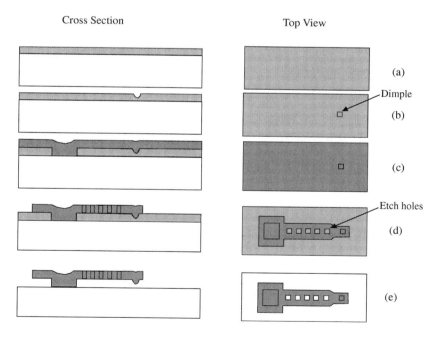

Fig. 32. The basic surface micromachining process used to fabricate surface-micromachined structures.

is twice the product of etch time and etch velocity. Typical etch rates of silicon dioxide can be found in the appendices.

Using aluminum as the structural material and silicon dioxide as the sacrificial material, Nathanson [70] fabricated surface micromachines as early as 1967. He was able to connect the resulting cantilevers to feedback amplifiers to create oscillators. The somewhat unpredictable and weak material properties of metals kept this technology in the textbooks for a long time. Work in the 1980s [71–73] led to the development of the LPCVD process that could deposit stress-free polysilicon films on silicon dioxide. Using polysilicon as the structural material, simple bridges and cantilevers could be fabricated. Once stress-free films of polysilicon were obtained, it took little time to realize a whole slew of surface-micromachined structures.

Polysilicon and phosphosilicate glass (PSG)–phosphorus-doped oxide are the most popular choices of structural and sacrificial materials. LPCVD PSG films are chosen for their much higher etch rates in hydrofluoric acid as compared to the undoped thermal or wet oxides. PSGs are patterned using lithography and a dry plasma etch to obtain vertical side walls. Polysilicon is deposited in an LPCVD tube with SiH_4 as the source gas and nitrogen and hydrogen as dilutants. The appropriate pressure (300–500 mtorr) and temperature (600–610 °C) result in reasonable deposition rates ($\sim$10 nm/min) [22]. This process results in slightly amorphous silicon film. A high-temperature anneal step is performed to recrystallize the amorphous silicon and reduce the stress at the same time. The phosphorus dopant in the polysilicon diffuses into the poly and further reduces the stress. The polysilicon is patterned and also etched in a plasma etch.

Various wet etches for the sacrificial etch have been tried and produce different final structures. Concentrated HF solutions give very fast etch rates, accompanied by high surface roughness. Dilute HF gives a slower etch rate but does not offer high selectivity between the polysilicon and the PSG. Because the HF PSG etch front moves laterally, very long etch times can result if one wants long structures. To solve this problem, one has to put etch holes in the polysilicon structure that effectively increase the etch front area and keep the etch time manageably small (Fig. 32).

A problem that plagued surface micromachines early was that of stiction between the released polysilicon structures and the substrate [74]. After the sacrificial layer has been etched away, the wafers are usually cleaned in water baths. During the wafer drying process, the water–air interface moves and eventually meets the structure. Because the polysilicon is hydrophobic, a surface tension force develops that pulls the released structure to the substrate. The resultant van der Waals or polymer residue–type bonding results in a stuck-on-substrate structure. One usually has to force the release using a probe tip. One technique to reduce stiction is by reducing the contact area between the released structure and the substrate. Dimples can be etched into the sacrificial layer that are reproduced in the structural layer (Fig. 32). These dimples act as contact pins that reduce the contact friction and adhesive forces.

Borrowing from the biology-lab technique of triple-point drying, a way to dry the postrelease wafers without stiction was developed [75]. The triple point refers to the simultaneous existence of solid, liquid, and gas phases. By placing the MEMS device in a triple-point material, one can eliminate the liquid–gas interface and the associated surface tension–driven stiction force. In the case of liquid CO_2 triple-point drying, the liquid water is replaced first by alcohol and then with liquid CO_2. Then the liquid CO_2 is driven to its triple point at high pressures (8 bar in a pressure chamber) at an elevated temperature (35 °C). Triple-point drying has virtually eliminated the problem of stiction. An alternative is sublimation drying. The structure is immersed in liquid CO_2 and cooled below the CO_2 freezing point. Then the dry ice is sublimated at a high temperature in the absence of H_2O, eliminating the liquid–air interface. Although release stiction problem is solved, in-use stiction due to friction induced changes still remains a major challenge.

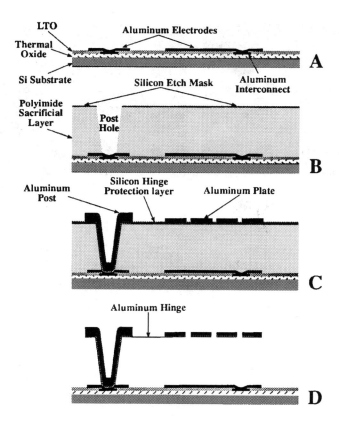

Fig. 33. An aluminum–polyimide surface micromachining process. Reprinted with permission from C. W. Storment et al., Dry Released Process of Aluminum Electrostatic Actuators, presented at the Solid-State Sensor and Actuator Workshop, Hilton Head, SC (© 1994 Transducers Research Foundation).

Other structural and sacrificial layer combinations can also be used to fabricate surface micromachines [76, 77]. One is the metal–polymer combination [78]. The metal structures are released by simply dissolving the organic layer. The metal layer can be evaporated, sputtered, or electroplated. The organic film can be photoresist or polyimide. The advantage of such materials is the low-temperature processing needed as opposed to the 600–700 °C temperatures needed with polysilicon deposition. An example of such a structure is shown in Figure 33. A recent example of a nanoscale surface-micromachined structure is the bridge of i-GaAs (undoped GaAs) formed by etching away an AlAs sacrificial layer [79, 80]. This bridge, shown in Figure 34, was used to measure the thermal conductance of the nanoscale-thick bridge.

7.1. Stress in Surface-Micromachined Structures

A class of surface-micromachined structures has evolved that measures the mechanical stress in the polysilicon films after release. An example of such a structure is the strain gauge shown in Figure 35. Upon release, any stress results in a net moment that bends the indicator beam. Both compressive stress and tensile stress can be measured using such structures. Another structure is the Guckel ring [81] shown in Figure 36. Any tensile stress in the polysilicon results in buckling of the center beam. Similarly, a simple beam anchored on both ends will buckle if compressive stress is present.

In addition to internal stresses in the structural material, stress gradients also play a crucial role in surface micromachines. For example, nonuniformity in doping or thickness variations at the anchors of surface micromachines can result in stress gradients at the anchor. These gradients cause the entire structure to bend out, as shown in Figure 36.

Fig. 34. Nanofabricated GaAs air bridge, suspended 1 μm over the substrate, was used to measure thermal conductance. Electron beam lithography was used to define 100-nm GaAs lines over intrinsic GaAs bridge. The overall dimension of the device is 1 mm^2. Reprinted with permission from T. S. Tighe et al., *Appl. Phys. Lett.* 70, 2687 (© 1997 American Institute of Physics).

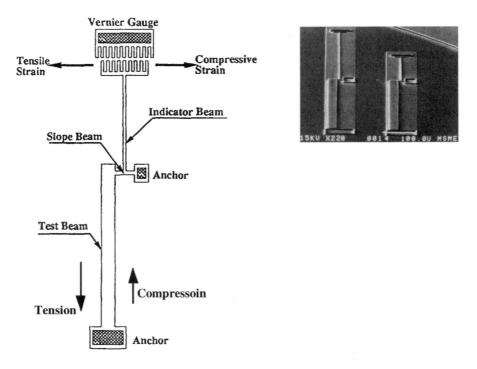

Fig. 35. A surface-micromachined passive strain gauge. Reprinted with permission from L. Lin et al., *J. Microelectromech. Systems* 6, 313 (© 1997 IEEE).

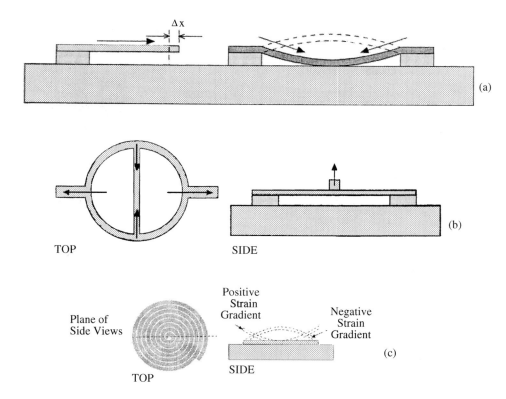

Fig. 36.   Micromechanical structures for testing internal stresses in thin films. (a) Compression in beam. (b) Tensile strain creates compression in center beam (Guckel rings). (c) Spiral for measuring strain gradient. Reprinted with permission from G. T. A. Kovacs, "Micromachined Transducers Sourcebook." (© 1998 McGraw-Hill).

7.2. Structures Formed from Polysilicon–Phosphosilicate Glass Combination

The basic sacrificial layer process has been used to fabricate many kinds of structures. Figure 37 shows the process flow of making an electrostatic motor [73, 82]. Figure 38 shows the resulting micromotor. Time-varying and phased electric fields are established across the rotor and the outer electrodes. The time-varying electrostatic forces resulting from the electric fields force the rotor to rotate.

Electrostatic forces have been used quite extensively in micromachined structures. The nonlinear nature of the electrostatic forces can be made linear using the comb structure shown in Figure 39 [83]. Electrostatic forces are generated by the applied electric fields between the combs. However, the force increase is linear because the capacitance increases linearly with electrode overlap increase. When the electric field is driven at the resonant frequency of the mechanical structure, large motions can be achieved.

Figure 39 also shows a thermal actuator that works similarly to a bimetallic actuator. Current is passed through the two arms to generate motion. By making one arm of the device thinner and, therefore, having a higher resistance, more heat and, hence, higher thermal expansion occurs in the thinner side. The differential expansion results in the entire structure's bending [84].

Figure 40, shows a pop-up structure that is made possible by a hinge, first created by Pister [85]. The second polysilicon hinge cover is patterned over the first polysilicon beam through etch holes. Such hinged structures have been used to transform surface-micromachined structures into three-dimensional structures. They have been used to

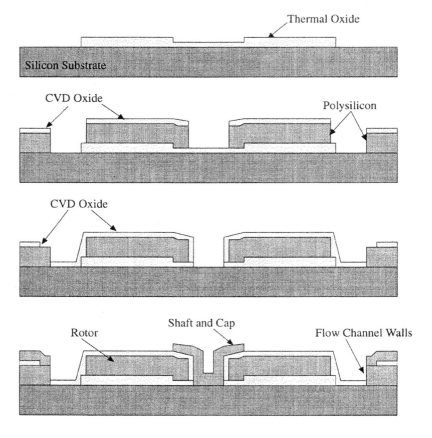

Fig. 37. A surface micromachining process for fabricating a rotor on a hub. Reprinted with permission from G. T. A. Kovacs, "Micromachined Transducers Sourcebook." (© 1998 McGraw-Hill).

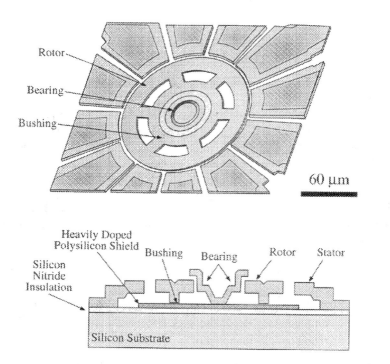

Fig. 38. The surface-micromachined micromotor with electrostatic actuation. Reprinted with permission from G. T. A. Kovacs, "Micromachined Transducers Sourcebook." (© 1998 McGraw-Hill).

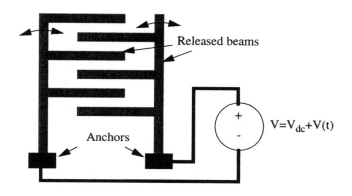

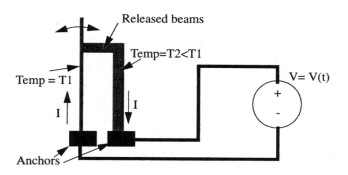

Fig. 39. Top shows a common electrostatic comb drive used to linearize the electrostatic force. Bottom shows a thermal actuator that utilizes thermal expansion.

fabricate on-chip hanging rf coils and capacitors [86]. Figure 41, shows structure in which most of the common polysilicon structures are integrated [66]. Electrostatic comb drives are configured as vibromotors to move the beam connected to the hinged structure. A metal coating on the polysilicon hinge reflects light at a controllable angle in this manner.

7.3. Sealed Structures

Often, it is necessary to make a sealed resonator structure for either low-pressure operation or isolation from the external environment. Resonant sensors generally require a vacuum atmosphere for high-quality factor operation. Many sealing techniques have been developed to accomplish the sealing of cavities. One way to make a sealed structure is shown in Figure 42. Etch holes are left in the polysilicon through which the PSG is etched away. An LPCVD nitride film is deposited that can seal the etch holes conformably, leaving the pressure of the deposition inside the sealed chamber [87, 88]. This sealing process has also been used to make surface-micromachined needles [107]. Instead of silicon nitride as the sealant, another option is to use a thermall oxide, as shown in Figure 42. This method results in much lower cavity pressure because the oxygen inside the cavity is consumed in a chemical reaction with the inside wall of the polysilicon [89].

7.4. Polysilicon Structure Direct Transfer

It would be desirable to transfer polysilicon microstructures to substrates that are more amenable to surface micromachining processes. For example, one might want to put a surface micromachined on an arbitrary-shaped steel bridge. Another application would be the

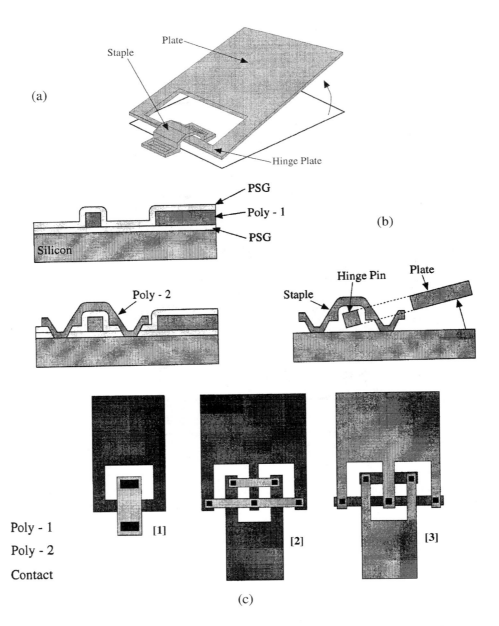

Fig. 40. The out-of-plane hinged structure. (a) The out-of-plane motion. (b) Process flow to fabricate a hinge. (c) Alternative hinge linkages. Reprinted with permission from G. T. A. Kovacs, "Micromachined Transducers Sourcebook." (© 1998 McGraw-Hill).

transfer of polysilicon structures to integrated circuits made on silicon or GaAs. A wafer level transfer of micromachines to another wafer would eliminate the tradeoffs of circuit quality versus micromachine quality in integrated microelectromechanical systems. One way to accomplish this is shown in Figure 43 [90]. A polysilicon cap with a stiffening rib is fabricated and attached to the substrate with thin tethers. Gold contact pads are defined on the cap edges as shown in the figure. The wafer was diced in dies and the poly cap was released in a HF etch and dried in a triple-point CO_2 etch. The dried dies were then put in contact with the second blank silicon die (with residual oxide removed). After applying pressure in a vacuum, the temperature was raised to the AuSi eutectic temperature of

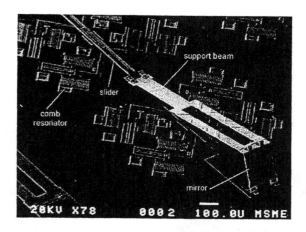

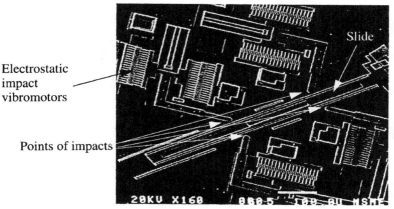

Fig. 41. A surface-micromachined optical mirror positioner actuated by vibromotors. Top shows the hinged mirror assembly. Bottom shows details of the vibromotor actuator. Reprinted with permission from M. J. Daneman et al., Linear Vibromotor-Actuated Micromachined Microreflector for Integrated Optics, presented at Solid-State Sensor and Actuator Workshop, Hilton Head, SC (© 1996 Transducers Research Foundation).

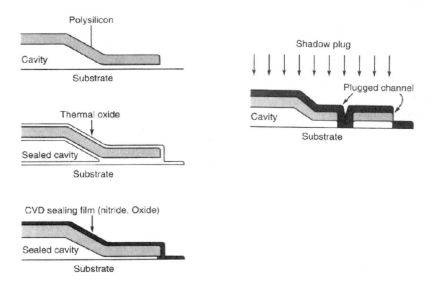

Fig. 42. Sealing techniques for surface microstructures. CVD film or thermal oxides are used to seal structures on the left. Metal is evaporated under vacuum to seal structure on the right. Reprinted with permission from C. M. Mastrangelo and W. C. Tang, "Semiconductor Sensor Technologies," in Semiconductor Sensors (© John Wiley & Sons, Inc.).

617

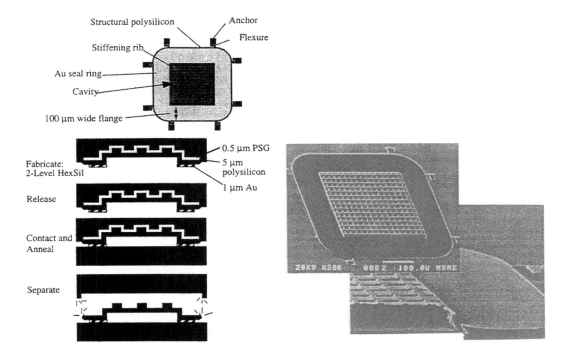

Fig. 43. Direct transfer of surface micromachined cap to silicon wafer via eutectic bonding. Left shows the process flow. Right shows SEM of the top view and section view of the cap. Reprinted with permission from M. B. Cohn et al., Wafer-to-Wafer Transfer of Microstructures for Vacuum Packaging, presented at Solid-State Sensor and Actuator Workshop, Hilton Head, SC (© 1996 Transducers Research Foundation).

370 °C. After eutectic bonding, the cap wafer was pulled by applying force normal to the dies and avoiding any shear. The weak tethers were broken and the cap was successfully left behind on the second die.

8. HYBRID BULK AND SURFACE MICROMACHINING

In this section, some illustrative examples of hybrid surface and bulk micromachining are presented. As the first example, Figure 44 shows a microgripper [91] that was used to pick up an individual bacterium. The process flow is also shown in the figure. The surface micromachines were defined first but not released. A protective nitride film was deposited over the entire wafer and the KOH etch cavity from the back and the front side were defined. The etch fronts from the top and the bottom meet, leaving the surface-micromachined structure intact.

An example of a hybrid deep-RIE etch and surface micromachining technique is illustrated by the Hexsil technology developed by Keller [92]. High-aspect-ratio structures are formed using deep-RIE in silicon wafers (Fig. 45). Sacrificial oxide is conformally deposited inside the grooves, followed by a polysilicon trench fill. If the trench width is very wide, open wells are left behind, which can be filled with electroplated nickel. Sacrificial etch in HF solutions results in the entire polysilicon structure being released. These polysilicon structures can be relatively thick (as thick as the silicon wafer) and can span over large areas. By controlling the polysilicon deposition process, proper stress can be designed into the structure such that it "pops out" upon release. Another claim of this technology is the reusability of the silicon mold.

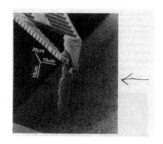

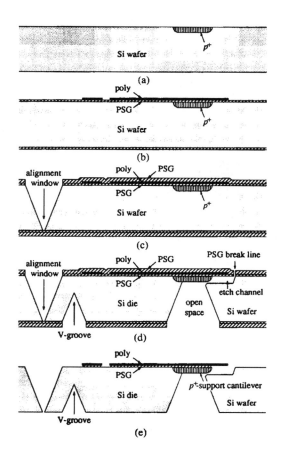

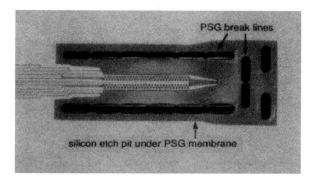

Fig. 44. Top left shows a microgripper holding a protozoan. Top right shows the process flow to fabricate the microgripper. Bottom shows the surface-micromachined comb drives and etch holes to undercut p^{++}-cantilever. Reprinted with permission from C. J. Kim et al., *J. Microelectromech. Systems* 1, 60 (© 1992 IEEE).

8.1. Porous Silicon Micromachines

Porous silicon is a term for uniformly etched silicon with pore diameters ranging from nanometer to micrometer dimensions and pore lengths that can be as long as millimeters. This combination of nanoscale and microscale makes porous silicon attractive for nanostructure fabrication [93]. Hence, extremely high aspect ratio devices are possible. Porous silicon can be fabricated by electrochemically etching silicon in a hydrofluoric acid

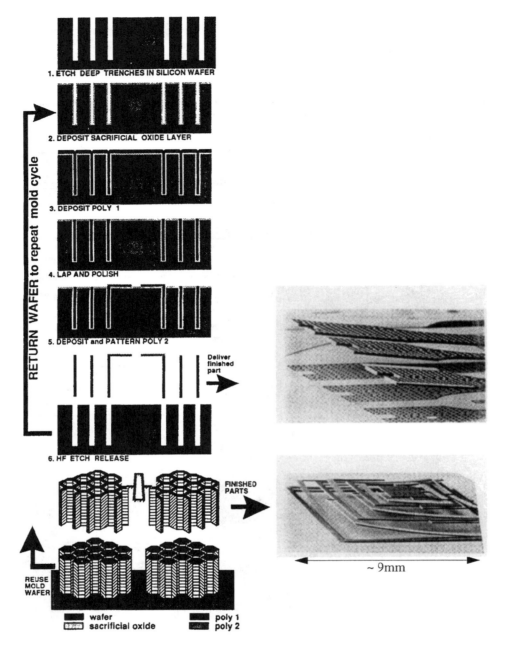

Fig. 45. Left shows process flow for fabricating three-dimensional polysilicon structures using the Hexsil process. Bottom shows a milliscale stage that self-assembles as a result of internal stress. Reprinted with permission from C. Keller and M. Ferrari, Milli-Scale Polysilicon Structures, presented at Solid-State Sensor and Actuator Workshop, Hilton Head, SC (© 1994 Transducers Research Foundation).

solution [93, 94]. The silicon oxidizes and then is etched by the HF acid. The pore size can be controlled by HF concentration or exposure to electron–hole-generating photonic sources. The subject of a pore formation mechanism is still unresolved. It is largely be-lieved that electric field concentration resulting from the very small radius of curvature at pore tips results in excessive etching and deep pore formation. The low wall etch rate has been attributed to either chemical passivation or diffusion-limited carrier concentra-tion [94].

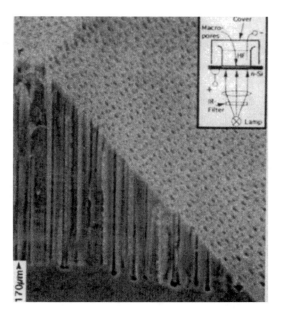

(A)

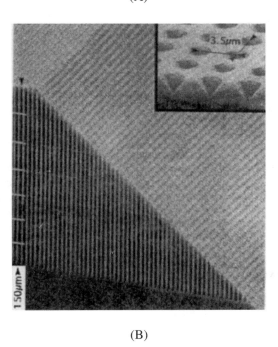

(B)

Fig. 46. Bulk porous silicon formation. (A) Random porous silicon formation. (B) Seeded porous silicon formation. Reprinted with permission from V. Lehmann, *J. Electrochem. Soc.* 140, 2836 (© 1993 Electrochemical Society).

Although pores on a planar surface produce randomly distributed pores, a surface with pore-initiating features can result in highly organized pores [95] (Fig. 46). Surface-micromachined structures using porous silicon have also been fabricated. Figure 47 shows a surface-micromachined polysilicon film sandwiched between silicon nitride films, with

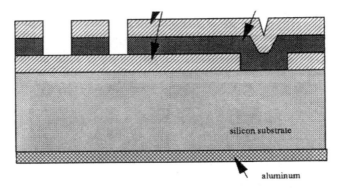

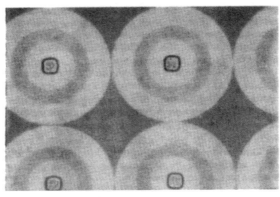

Fig. 47. Surface-micromachined porous silicon structures. Top shows cross section of polysilicon pattern to be anodized. Bottom shows the formation of a porous silicon plug formed during anodization. Reprinted with permission from R. C. Anderson et al., Laterally Grown Porous Polycrystalline Silicon: A New Material for Transducer Applications, presented at Transducers '91, San Francisco (© 1991 IEEE).

an anchor point to the silicon substrate. The silicon substrate is connected to an aluminum film on the back side of the wafer for electrical connection. The anodization circuit is completed by electrical contact to the aluminum film, through the silicon wafer and the poly. By varying the current condition from complete dissolution to porous silicon formation, Anderson et al. [96] were able to fabricate a porous silicon plug, forming a sealed chamber.

8.2. Electroplated Micromachines

Micromachining methods enable one to fabricate molds and, hence, provide a way to make micromechanical metallic parts by electroplating in micromachined molds. To fabricate thick micromachined metallic structures with a high aspect ratio, ways of creating deep molds have been invented. One way has been to spin-coat thick polymer coatings that can be exposed through their thickness without diffraction. To achieve this, one can use X-ray lithography to expose thick poly(methyl methacrylate) (PMMA) layers. X-rays are highly directional and do not diffract. The process of using X-rays to expose thick PMMA coatings and filling the resulting cavities with electroplated metal is called LIGA (X-ray Lithographie Galvanoformung Abformtechnik) [97]. The thick PMMA generally require very long exposures and special photoresist spinning apparatus. A LIGA-like process with electroplated aluminum to fabricate a gear is outlined in Figure 48. Figure 49 shows a complete microdynamometer made using LIGA technology [104]. Although the original behind LIGA was to make metal molds into which plastic parts could be embossed, most users use LIGA to make metal microelectromechanical parts.

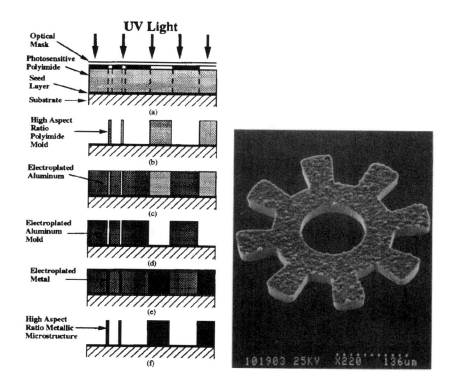

Fig. 48. Left shows the LIGA-like UV exposure process to make high-aspect-ratio metallic structures. Top shows a 45-μm-high aluminum gear made using electroplating. Reprinted with permission from A. B. Frazier and M. G. Allen, Uses of Electroplated Aluminum in Micromachining Applications, presented at Solid-State Sensor and Actuator Workshop, Hilton Head, SC (© 1994 Transducers Research Foundation).

Fig. 49. A microdynamometer fabricated using LIGA. It contains a 3-phase variable reluctance motor, idling gear, and magnetic break [104].

Appendix 1. Etch-rate data for wet etching (experiments done at UC Berkeley Microlab)

No.	Etchant: Concentration and conditions	Material to be etched	SCSi <100>	Poly n+	Poly undop	Wet ox	Dry ox	LTO undop	PSG unanl	PSG annld	Stoic nitride	Lows nitride	Al/ 2% Si	Sput tungs	Sput Ti	Sput Ti/W	OCG 820 PR	Olin HNT PR
1	Concentrated HF	Silicon oxides	–	0	–	23K 18K 23K	F	>14K	F	36K	140	52 30 52	42 0 42	<50	F	–	P,0	P,0
2	10:1 HF (room temperature)	Silicon oxides	–	7	0	230	230	340	15K	4700	11	3	2500 2500 12K	0	11K	<70	0	0
3	25:1 HF (room temperature)	Silicon oxides	–	0	0	97	95	150	W	1500	6	1	W	0	–	–	0	0
4	5:1 BHF (BHF = buffered hydrofluoric acid)	Silicon oxides	–	9	2	1000 900 1080	1000	1200	6800	4400 3500 4400	9	4 3 4	1400	<20 0.25 20	F	1000	0	0
5	Phosphoric acid (undiluted)	Silicon nitrides	–	7	–	0.7	0.8	<1	37	24 9 24	28 28 42	19 19 42	9800	–	–	–	550	390
6	Silicon etchant (126 HNO_3:60 H_2O:5 NH_4F) Room temperature	Silicon	1500	3100 1200 6000	1000	87	W	110	4000	1700	2	3	4000	130	3000	–	0	0
7	KOH (1 KOH:2 H_2O by weight) 80 °C	<100> Silicon	14K	>10K	F	77 41 77	–	94	W	380	0	0	F	0	–	–	F	F
8	Aluminum etchant – type A (16 H_3PO_4:1 HNO_3:1 Hac:2 H_2O) 50 °C	Aluminum	–	<10	<9	0	0	0	–	<19	0	2	6600 2600 6600	–	0	–	0	0

#	Etchant / Conditions	Target													
9	Titanium etchant (20H$_2$O:1 H$_2$O$_2$:1 HF), Room temperature	Titanium	–	12	–	120	W	W	2100	8	4	W	0 / 0 / <10	8800 / –	0	0
10	H$_2$O$_2$ (30%), Room temperature	Tungsten	–	0	0	0	0	0	0	0	0	<20	190 / 190 / 1000	0 / 60 / 60 / 150	<2	0
11	Pirahna (50H$_2$SO$_4$:1 H$_2$O$_2$), 120°C	Cleaning metals and organics	–	0	0	0	–	0	0	0	0	1800	– / 2400	–	F	F
12	Acetone, Room temperature	Photoresist	–	0	0	0	–	0	0	0	0	0	– / 0	–	>44K	>39K
13	HF Vapour, Room temperature & pressure	Silicon oxides	–	0	660	780	2100	1500	10	19	A	0 / A	A	–	P, 0	P, 0
14	XeF$_2$	Silicon	4600 / 2900 / 100K	1900 / 1100 / 2500	1800 / 1100 / 2300	0	0	0	0	120 / 120 / 180	2 / 0 / 2	0	800 / 440 / 1000	290 / 50 / 380	–	0

Note: All etch rates are given in angstroms. The top most etch rate was measured by Kirt Williams, the bottom two (if listed) were measured by others with slightly different conditions. Notation: "–" = test not performed, W = test not performed but known to work (etch rates > 100 Å/min), F = test not performed but known to be fast (etch rate > 10 kÅ/min), P = film peeling observed, A = Film was visibly attacked and roughened.

Appendix 2. Etch-rate data for plasma etching performed at the Microlab at UC Berkeley

No.	Dry etching recipe: Gas flows, plasma machine, power, pressure, and conditions	Material to be etched	Material														
			SC Si <100>	Poly n+	Poly undop	Wet ox	Dry ox	LTO undop	PSG unanl	PSG annld	Stoic nitride	Lows nitride	Al/ 2% Si	Sput tungs	Sput Ti	Sput Ti/W	OCG 820 PR	Olin HNT PR
1	SF$_6$ + He (175:50 sccm) LAM 480 plasma: 150 W, 375 mT gap = 1.35 cm, 13.56 MHz	Thin silicon nitrides	W	6400	7000 2000 7000	300 220 400	W	280	530	540	1300 830 2300	870	–	W	W 30 52	W 0 42	1500 1300 1500	1400
2	SF$_6$ + He (175:50 sccm) LAM 480 plasma: 250 W, 375 mT gap = 1.35 cm, 13.56 MHz	Thick silicon nitrides	W	8400	9200	800	W	770	1500	1200	2800 2100 4200	2100	–	W	W	W	3400 3100 3400	3100
3	CF$_4$ + CHF$_3$ + He (90:30:120 sccm) LAM 590 plasma: 450 W, 2.8 T gap = 0.38 cm, 13.56 MHz	Silicon oxides	W	1900 1400 1900	2100 1500 2100	4700 2400 4800	W	4500	7300 3000 7300	6200 2500 7200	1800	1900	–	W	W	W	2200	2000
4	CF$_4$ + CHF$_3$ + He (90:30:120 sccm) LAM 590 plasma: 850 W, 2.8 T gap = 0.38 cm, 13.56 MHz	Silicon oxides	W	2200 2200 2700	1700 1700 2100	6000 2500 7600	W	6400 6000 6400	7400 5500 7400	6700 5000 6700	4200 4000 6800	3800	–	W	W	W	2600 2600 6700	2900 2900 7200
5	Cl$_2$ + BCl$_3$ + CHCl$_3$ + N$_2$ (30:50:20:50 sccm) LAM 590: 850 W, 250 mT 60°C, 13.56 MHz	Aluminum	W	4500	W	680	670	750	W	740	930	860	6000 1900 6400	W	–	–	6300 3700 6300	6300 3300 6100
6	Cl$_2$ + He (180:400 sccm) LAM rainbow 4420: 275 W, 425 mT 40°C, gap = 0.80 cm, 13.45 MHz	Silicon	W 5000 5000	5700 3400 6300	3200 3200 3700	8 8 380	–	60	230	140	560	530	W	W	–	–	350 350 500	300
7	HBr + Cl$_2$ (70:70 sccm) LAM rainbow 4420: 200 W, 300 mT 40°C, gap = 0.80 cm, 13.45 MHz	Silicon	W	4500	W	680	670	750	W	740	930	860	6000 1900 6400	W	–	–	3000 2400 3000	2700
8	O$_2$ (50 sccm) Technics PEII-A: 400 W, 300 mT gap = 2.6 cm, 50 kHz sq. wave	Photoresist ashing	–	0	0	0	0	0	0	0	0	0	0	0	0	0	3400	3600

#	Process conditions	Film															
9	O_2 (50 sccm) Technics PEII-A: 50 W, 300 mT, gap = 2.6 cm, 50 kHz sq. wave	Descumming – organics removal	–	0	0	0	0	0	0	0	0	0	0	0	0	350	300
10	SF_6 + He (13:21 sccm) Technics PEII-A: 100 W, 300 mT, gap = 2.6 cm, 50 kHz sq. wave	Silicon nitrides	300 300 1000	730 730 800	670 670 760	310	350	370	610	480 230 480	820	620 550 800	–	W	W	690 690 830	630
11	CF_4 + CHF_3 + He (10:5:10 sccm) Technics PEII-A: 200 W, 300 mT, gap = 2.6 cm, 50 kHz sq. wave	Silicon nitrides	1100	1900	W	730	710	730	W	900	1300	1100	–	W	W	690	600
12	SF_6 (80 sccm) Tegal Inline 701: 200 W, 150 mT, 40°C, 13.56 MHz	Tungsten	W	5800	5400	1200 2000 2000	W	1200	1800	1500	2600	2300 1900 2300	–	2800 2800 4000	W	2400 2400 4000	2400
13	SF_6 (25 sccm) Tegal Inline 701: 125 W, 200 mT, 40°C, 13.56 MHz	Thin silicon nitrides	W	1700	2800	1100 1100 1600	W	1100	1400	1400	2800 2800 2800	2300	–	W	W	3400 2900 3400	3100
14	CF_4 + CHF_3 + He (45:15:60 sccm) Tegal Inline 701: 100 W, 300 mT, 40°C, 13.56 MHz	Thick silicon-rich nitrides	W	350	360	320	W	320	530	450	760	600	–	W	W	400	360

Note: All etch rates are given in angstroms. The top most etch rate was measured by Kirt Williams, the bottom two (if listed) were measured by others with slightly different conditions.
Notation: "–" = test not performed, W = test not performed but known to work (etch rates > 100 Å/min), F = test not performed but known to be fast (etch rate > 10 kÅ/min), P = film peeling observed, A = film was visibly attacked and roughened.

References

1. R. P. Feynman, *J. Microelectromech. Systems* 1, 60 (1992).
2. F. Cerrina, *Proc. IEEE* 84, 644 (1997).
3. S. Matsui, *Proc. IEEE* 84, 629 (1997).
4. J. Lydig, *Proc. IEEE* 84, 589 (1997).
5. G. T. A. Kovacs, "Micromachined Transducers Sourcebook." McGraw–Hill, New York, 1998.
6. M. Madou, "Fundamentals of Microfabrication." CRC Press, Boca Raton, FL, 1997.
7. P. Rai-Choudhary, "Handbook of Microlithography, Micromachining, and Microfabrication," Vol. 2. SPIE, Washington, DC, 1997.
8. I. Fujimasa, "Micromachines: A New Era in Mechanical Engineering." Oxford, University Press, New York, 1996.
9. S. M. Sze, "Semiconductor Devices: Physics and Technology." Wiley, New York, 1985.
10. D. Bauele, "Chemical Processing with Lasers." Springer-Verlag, New York, 1986.
11. W. R. Runyan and K. E. Bean, "Semiconductor Integrated Circuit Processing Technology." Addison–Wesley, Reading, MA, 1990.
12. S. Wolf and R. N. Tauber, "Silicon Processing for the VLSI Era," Vol. 1. Lattice Press, Sunset Beach, CA, 1986.
13. M. Heschel and S. Bouwstra, Conformal Coating by Photoresist of Sharp Corners of Anisotropically Etched Through-Holes in Silicon, presented at Transducers '97, Chicago, 1997.
14. K. E. Bean and W. R. Runyan, "Semiconductor Integrated Circuit Processing Technology." Addison–Wesley, Reading, MA, 1994.
15. S. M. Sze, "VLSI Technology." McGraw–Hill, New York, 1988.
16. J. L. Vossen, "Thin Film Processes." Academic Press, New York, 1978.
17. L. I. Maisel and R. Glang, "Handbook of Thin Film Technology." McGraw–Hill, New York, 1970.
18. R. S. Hijab and R. S. Muller, Residual Strain Effects on Large-Aspect Ratio Micro-Diaphragms, presented at IEEE MEMS Workshop, Salt Lake City, UT, 1989.
19. V. Aksyuk et al., Low Insertion Loss Packaged and Fiber-Connectorized Si Surface Micromachined Reflective Optical Switch, presented at Solid-State Sensor and Actuator Workshop, Hilton Head, SL, 1998.
20. S. S. Lee, R. P. Ried, and R. M. White, Piezoelectric Cantilever Microphone and Microspeaker, presented at Solid-State Sensor and Actuator Workshop, Hilton Head, SC, 1994.
21. W. G. Valkenberg et al., *J. Electrochem. Soc.* 132, 893 (1985).
22. L. S. Fan and R. S. Muller, As-Deposited Low-Strain LPCVD polysilicon, presented at International Workshop on Solid-State Sensors and Actuators, Hilton Head, SC, 1988.
23. H. Guckel et al., Processing Conditions for Polysilicon Films with Tensile Strain for Large Aspect Ratio Microstructures, presented at International Workshop on Solid-State Sensors and Actuators, Hilton Head, SL, 1988.
24. P. Krulevitch et al., Stress in Undoped LPCVD Polycrystalline Silicon, presented at Transducers '91, San Francisco, 1991.
25. T. Muraka and T. F. Retajczyk, *J. Appl. Phys.* 54, 2069 (1983).
26. W. Yun, in "Electrical Engineering and Computer Science." University of California Press, Berkeley, CA, 1992.
27. A. Reisman et al., *J. Electrochem. Soc.* 126, 1406 (1979).
28. H. Robbins and B. Schwartz, *J. Electrochem. Soc.* 107, 108 (1960).
29. A. M. Flynn et al., *J. Microelectromech. Devices* 1, 44 (1992).
30. E. Bassous, Type B EDP Etching, USA, 1975.
31. R. M. Finne and D. L. Klein, *J. Electrochem. Soc.* 114, 965 (1967).
32. H. Seidel, The Mechanism of Electrochemical and Anisotropic Silicon Etching and Its Applications, presented at Third Toyota Conference on Integrated Micro Motion Systems, Aichi, Japan, 1989.
33. H. Seidel, *J. Electrochem. Soc.* 177, 3612 (1990).
34. N. N. Greenwood and A. Earnshaw, "Chemistry of the Elements." Pergamon Press, Elmsford, NY, 1986.
35. B. Schwartz and H. Robbins, *J. Electrochem. Soc.* 108, 365 (1961).
36. K. Najafi, Silicon Micromachining: Key to Silicon Integrated Sensors, presented at Symposium on Sensor Science and Technology, Cleveland, OH, 1987.
37. O. Tabata, Anisotropy and Selectivity Control of TMAH, presented at Eleventh International Workshop on Micro Electromechanical Systems, Heidelberg, Germany, 1998.
38. U. Schnakenberg et al., *Sens. Actuators* (1990).
39. K. Sato et al., Characterization of Anisotropic Etching Properties of Single-Crystal Silicon: Surface Roughening as a Function of Crystallographic Orientation, presented at Eleventh International Workshop on Micro Electromechanical Systems, Heidelberg, Germany, 1998.
40. J. B. Price, Anisotropic etching of silicon with KOH–H_2O–Isopropyl Alcohol, presented at Semiconductor Silicon, Princeton, NJ, 1973.

41. S. C. Chang and D. B. Hicks, Street Corner Compensation, presented at IEEE Solid-State Sensor and Actuator Workshop, Hilton Head, SC, 1988.

42. H. Sandmaier et al., Corner Compensation Techniques in Anisotropic Etching of ⟨100⟩-Silicon Using Aqueous KOH, presented at Transducers '91, San Francisco, 1991.

43. R. E. Williams, "Modern GaAs Processing Methods." Artech House, Boston, 1990.

44. Z. L. Zhang et al., Submicron, Movable Gallium Arsenide Mechanical Structures and Actuators, presented at International Workshop on Microelectromechanical Systems, 1992.

45. A. Lal and R. M. White, Micro-Fabricated Acoustic and Ultrasonic Source-Receiver," presented at Transducers '93, Yokohama, Japan, 1993.

46. S. D. Collins, *J. Electrochem. Soc.* 144, 2242 (1997).

47. Y. B. Gianchandani and K. Najafi, *J. Microelectromech. Systems* 1, 77 (1992).

48. C. Yeh and K. Najafi, Bulk-Silicon Tunneling-Based Pressure Sensors, presented at Solid-State Sensor and Actuator Workshop, Hilton Head, SC, 1994.

49. B. Koek et al., *IEEE Trans. Electron Devices* 36, 663 (1989).

50. E. S. Kim, R. S. Muller and R. S. Hijab, *J. Microelectromech. Systems* 1, 95 (1992).

51. S. Hoen et al., Fabrication of Ultrasensitive Force Detectors, presented at Solid-State Sensor and Actuator Workshop, Hilton Head, SC, 1994.

52. L. M. Roylance and J. B. Angell, *IEEE Trans. Electron Devices* ED-26, 1911 (1979).

53. C. Linder, T. Tschan, and N. F. de Rooij, Deep Dry Etching Techniques as a New IC Compatible Tool for Silicon Micromachining, presented at Transducers '91, San Francisco, 1991.

54. C. Linder, T. Tschan, and N. F. de Rooij, *Sens. Mater.* 3, 311 (1992).

55. Z. L. Zhang and N. C. MacDonald, An RIE Process for Submicron, Silicon Electromechanical Structures, presented at Transducers '91, San Francisco, 1991.

56. J. J. Yao, J. C. Arney, and N. C. MacDonald, *J. Microelectromech. Devices* 1, 14 (1992).

57. A. E. T. Kuiper and E. G. C. Lathouwers, *J. Electrochem. Soc.* 139, 2594 (1992).

58. D. W. Oxtoby and N. H. Nachtrieb, "Principles of Chemistry." Saunders, Philadelphia, 1986.

59. E. Hoffman et al., 3D Structures with Piezoresistive Sensors in Standard CMOS, presented at IEEE MEMS 1995, Amsterdam, 1995.

60. W. H. Ko, J. H. Suminto, and G. H. Yeh, in "Microsensors" (R. S. e. a. Muller, ed.), IEEE, New York, 1990.

61. L. A. Field and R. S. Muller, *Sens. Actuators, A* 21–23, 935 (1990).

62. A. Hanneborg, *J. Micromech. Microeng.* 1, 139 (1991).

63. M. Shimbo et al., *J. Appl. Phys.* 60, 2987 (1986).

64. M. A. Schmidt, Silicon Wafer Bonding for Micromechanical Devices, presented at Solid-State Sensor and Actuator Workshop, Hilton Head, SC, 1994.

65. T. Nakamura, SOI Technologies for Sensors, presented at Transducers, Yokohama, Japan, 1993.

66. M. J. Daneman et al., Linear Vibromotor-Actuated Micromachined Microreflector for Integrated Optics, presented at Solid-State Sensor and Actuator Workshop, Hilton Head, SC, 1996.

67. A. Lal and R. M. White, Ultrasonically Driven Silicon Atomizer and Pump, presented at Solid-State Sensor and Actuator Workshop, Hilton Head, SC, 1996.

68. A. Lal and R. M. White, *Sens. Actuators, A* 54, 542 (1996).

69. A. Lal and R. M. White, Optimization of the Silicon/PZT Longitudinal Mode Resonant Transducer, presented at ASME World Conference, Dallas, TX, 1997.

70. H. C. Nathanson et al., *IEEE Trans. Electron Devices* ED-14 (1967).

71. R. T. Howe and R. S. Muller, *J. Electrochem. Soc.* 130, 1420 (1983).

72. H. Guckel and D. W. Burns, *IEEE IEDM* 176 (1986).

73. M. Mehregany et al., *IEEE Trans. Electron Devices* 35, 719 (1988).

74. D. J. Monk, D. S. Soane, and R. T. Howe, Sacrificial Layer SiO_2 Wet Etching for Micromachining Applications, presented at Transducers '91, San Francisco, 1991.

75. G. T. Mulhern, S. Soane, and R. T. Howe, Supercritical Carbon-Dioxide Drying from Microstructures, presented at Transducers, Yokohama, Japan, 1993.

76. L. Chen and N. C. MacDonald, Surface Micromachined Multiple Level Tungsten Microstructures, presented at Solid-State Sensor and Actuator Workshop, Hilton Head, SC, 1994.

77. C. W. Storment et al., Dry Released Process of Aluminum Electrostatic Actuators, presented at the Solid-State Sensor and Actuator Workshop, Hilton Head, SC, 1994.

78. A. B. Frazier and M. G. Allen, Uses of Electroplated Aluminum in Micromachining Applications, presented at Solid-State Sensor and Actuator Workshop, Hilton Head, SC, 1994.

79. T. S. Tighe, J. M. Worlock, and M. L. Roukes, *Appl. Phys. Lett.* 70, 2687 (1997).

80. K. Hjort, *J. Micromech. Microeng.* 6, 370 (1996).

81. H. Guckel, *Sens. Actuators, A* 28, 133 (1991).

82. L. S. Fan Y. C. Tai, and R. S. Muller, IC-Processed Electrostatic Micromotors, presented at IEEE International Electron Device Meeting, San Francisco, 1988.

83. W. C. Tang et al., *Sens. Actuators, A* 21–23, 328 (1990).

84. J. H. Comtois and V. M. Bright, Surface Micromachined Polysilicon Thermal Actuators Arrays and Applications, presented at Solid-State Sensor and Actuator Workshop, Hilton Head, SC, 1996.

85. K. S. J. Pister, *Sens. Actuators A* 33, 249 (1992).

86. L. Fan et al., Universal MEMS Platforms for Passive RF Components: Suspended Inductors and Variable Capacitors, presented at Eleventh International Workshop on Micro Electromechanical Systems, Heidelberg, Germany, 1998.

87. C. Mastrangelo and R. S. Muller, Vacuum-Sealed Silicon Micromachined Incandescent Light Source, presented at IEEE IEDM, 1989.

88. C. Liu and Y. C. Tai, Studies on the Sealing of Surface Micromachined Cavities Using Chemical Vapor Deposition Materials, presented at Solid-State Sensor and Actuator Workshop, Hilton Head, SC, 1994.

89. R. T. Howe, *J. Vac. Sci. Technol., B* 6, 1809 (1988).

90. M. B. Cohn et al., Wafer-to-Wafer Transfer of Microstructures for Vacuum Packaging, presented at Solid-State Sensor and Actuator Workshop, Hilton Head, SC, 1996.

91. C. J. Kim, A. P. Pisano, and R. S. Muller, *J. Microelectromech. Systems* 1, 60 (1992).

92. C. Keller and M. Ferrari, Milli-Scale Polysilicon Structures, presented at Solid-State Sensor and Actuator Workshop, Hilton Head, SC, 1994.

93. P. Steiner and W. Lang, *Thin Solid Films* 255, 52 (1995).

94. R. C. Anderson, in "Chemical Engineering." University of California Press, Berkeley, CA, 1991.

95. V. Lehmann, *J. Electrochem. Soc.* 140, 2836 (1993).

96. R. C. Anderson, R. S. Muller, and C. W. Tobias, Laterally Grown Porous Polycrystalline Silicon: A New Material for Transducer Applications, presented at Transducers '91, San Francisco, 1991.

97. W. Ehrfeld et al., Fabrication of Microstructures Using the LIGA Process, presented at IEEE Microrobots and Teleoperators Workshop, Hyannis, MA, 1987.

98. J. W. Judy, R. S. Muller, and H. H. Zappe, Magnetic Microactuation of Polysilicon Flexure Structures, presented at Solid-State Sensor and Actuator Workshop, Hilton Head, SC, 1994.

99. H. Seidel, The Mechanism of Anisotropic Silicon Etching and Its Relevance for Micromachining, presented at Transducers '87, pp. 120–125.

100. K. Suzuki, K. Najafi, and K. D. Wise, A-1024-element high performance silicon tactile imager, IEEE IEDM, 1988, p. 67.

101. K. Peterson, *Proc. IEEE* 70, 420 (1982).

102. C. M. Mastrangelo and W. C. Tang, Semiconductor sensor technologies, in "Semiconductor Sensors" (S. M. Sze, ed.).

103. L. Lim, A. P. Pisano, and R. T. Howe, *Microelectromech. Systems* 6, 313 (1997).

104. H. Guckel et al., Advances in Photoresist Based Processing Tools for 3-Dimensional Precision and Micromechanics, presented at Solid-State Sensor and Actuator Workshop, Hilton Head, SC, 1996.

105. L. Tenerz, Silicon Micromachining with applications in Sensors and Actuators, Ph.D. Thesis, 1989, Uppsala University. ISBN 91-554-2418-*X*.

106. E. D. Palik et al., *J. Appl. Phys.* 70, 3291 (1991).

107. K. S. Lebonitz, A. P. Pisano, and R. T. Hour, Permeable polysilicon etch-access windows for microshell fabrication, Transducers '95, Stockholm.

Index

HANDBOOK OF NANOSTRUCTURED
MATERIALS AND NANOTECHNOLOGY

Edited by H.S. Nalwa

Volume 1. SYNTHESIS AND PROCESSING

Volume 2. SPECTROSCOPY AND THEORY

Volume 3. ELECTRICAL PROPERTIES

Volume 4. OPTICAL PROPERTIES